Oxidative Stress in Heart Diseases

Sajal Chakraborti • Naranjan S Dhalla
Nirmal K Ganguly • Madhu Dikshit
Editors

Oxidative Stress in Heart Diseases

Springer

Editors
Sajal Chakraborti
Department of Biochemistry and Biophysics
University of Kalyani
Kalyani, West Bengal, India

Naranjan S Dhalla
Institute of Cardiovascular Sciences
St Boniface Hospital, Albrechtsen
Research Centre
Winnipeg, MB, Canada

Nirmal K Ganguly
Translational Health Science
and Technology (THSTI)
Department of Biotechnology
Faridabad, India

Madhu Dikshit
Translational Health Science
and Technology (THSTI)
Department of Biotechnology
Faridabad, India

ISBN 978-981-13-8272-7 ISBN 978-981-13-8273-4 (eBook)
https://doi.org/10.1007/978-981-13-8273-4

This Springer imprint is published by the registered company Springer Nature Singapore Pte Ltd.
The registered company address is: 152 Beach Road, #21-01/04 Gateway East, Singapore 189721, Singapore

Prof. Sugata Marjit

This book is dedicated to Prof. Sugata Marjit (born in December 4, 1959) for his outstanding leadership in promoting higher education and research in the state of West Bengal, India. He is the RBI Professor of Industrial Economics in the Centre for Studies in Social Sciences, Calcutta, and served as its Director. He also served the Calcutta University as its Vice Chancellor. He was the Chairman, Higher Education Council, the highest policy making body in higher education in the state of West Bengal, India, and showed his extraordinary commitment in developing and implementing different policies in modernizing higher education and research. He is the Recipient of several national and international awards and taught and researched at many reputed universities and institutions in the USA, Europe, and Asia as a Visiting Professor. Being inspired by his commitment and excellence, we dedicate this book to Prof. Sugata Marjit.

Preface

"Brooding over this immensity
I ask,
On this boundless land
Who rules over man's destiny?"

(From the Selected Poems of Mao Tse Tung)

About nine decades ago during the *Long March*, upon looking on to the setting sun from a mountain in Yenan, Mao's ingenious poetic stance swings momentarily from his revolutionary mind-set to the beauty and immensity of nature. Accordingly, we the scientists are astonished considering the complexity of physiological and biochemical mechanisms that interplay in our body, for instance, in the cardiovascular system, in response to different threats from environmental, behavioral, and occupational agents, drugs, genetic, and age-related disorders.

The quest for understanding the regulatory mechanisms of the functioning of the heart in health and disease in true scientific outlook started in 1883, when Sidney Ringer discovered that calcium has the potentiality to contract isolated rat hearts. This momentous observation initiated biochemical and physiological aspects of research on cardiovascular system in health and disease. The role of oxidants and redox imbalance in the progression of a variety of heart diseases gained tremendous impetus among biomedical scientists especially in the past two decades.

Reactive oxygen species (ROS) are primarily produced intracellularly, and their productions are inherent to normal physiology. Cells have evolved both enzymatic and nonenzymatic antioxidant defense mechanisms to scavenge excess ROS in order to maintain redox balance. A shift in redox homeostasis to an imbalance between ROS generation and endogenous antioxidant components results in oxidative stress, which has been implicated in the pathogenesis of many diseases including heart diseases. Data from different studies strongly support that dysregulation of redox signaling is integral to the pathogenesis of heart diseases such as ischemia, hypertension, atherosclerosis, diabetic cardiomyopathy, and heart failure.

This book contains 26 chapters which are divided into 4 sections. Dr. Gemma Figtree, Dr. Rajabrata Bhuyan, Dr. Carolina Panis, and Dr. Thomas Hansen elucidated different aspects of oxidant-induced heart diseases in general, while Dr. Cristina Vasselle, Dr. Veena Dhawan, Dr. Shantanu Sengupta, and Dr. Krishnapura Srinivasan exposed us with protective influence of novel therapeutics and antioxidants in mitigating oxidative stress-induced cardiovascular diseases. The impact of molecular biology and genetics on cardiovascular diseases is growing rapidly. A chapter, "Reactive Oxygen Species and Their Epigenetic Consequences in Heart Diseases," by Dr. Seema Bhargava provided important information in this area. The section on "atherosclerosis and ischemic heart disease" reflects novel insights provided by Dr. Perimur Bozaykut, Dr. Branislav Rovcanin, Dr. Fouad Zouein, Dr. Pasquale Pagliaro, Dr. Bodh Jugdutt, Dr. Jawahar Mehta, and Dr. Monika Bartekova. These authors in their chapters provided novel information pertaining to the risk factors for initiation and progression of atherosclerosis and pliable ways for its prevention.

The prevalence of diabetes has reached to a high level globally. Different characteristics of diabetes such as endothelial dysfunction and prothrombotic state are known to enhance the chance of plaque instability. Diabetes can cause myocardial damage, resulting in cardiovascular diseases and eventually heart failure. In view of the increasing importance of diabetes as a risk factor for vascular diseases, a section, titled "Diabetes-Induced Cardiovascular Dysfunction," has been incorporated in the book, where Dr. Belma Turan, Dr. Bhoomika Goyal Patel, Dr. Paras Mishra, Dr. Savita Bansal, Dr. Hobby Aggarwal, and Dr. Surekha Rani described the growing understanding of the interrelationship between diabetes and cardiovascular diseases.

Hypertension and heart failure are the most common diagnoses for cardiovascular patients. In this section, Dr. Pedro Gomes, Dr. Gerald Maarman, Dr. Sanjay Banerjee, and Dr. Lorenzo Calo provided contemporary description on the mechanisms of this important sub-specialty of cardiology.

This book will prove useful to those who wish to broaden their knowledge of cardiovascular research especially the role of oxidants and redox signaling in the pathophysiological aspects of different types of heart diseases. Thanks should go to all authors for their professional expertise, knowledge, and devoted scholarship. Thanks are also due to Dr. Madhurima Kahali and Dr. Uma Maheswari (Springer Nature, New Delhi) for their sincere cooperation and support during the preparation of this book.

Kalyani, India Sajal Chakraborti

Contents

Part I General Implications

1 Oxidative Stress and Cardiovascular Risk and Prevention in Children and Adolescents ... 3
Francesca Mastorci, Irene Traghella, Laura Sabatino, Alessandro Pingitore, Rudina Ndreu, and Cristina Vassalle

2 The Role of Redox Signalling in Cardiovascular Regeneration ... 19
Thomas Hansen, Soloman Saleh, Gemma A. Figtree, and Carmine Gentile

3 Oxidative Stress-Driven Cardiotoicity of Cancer Drugs ... 39
Thalita Basso Scandolara, Bruno Ricardo Pires, Rodrigo Kern, Vanessa Jacob Victorino, and Carolina Panis

4 Molecular Targets and Novel Therapeutics to Target Oxidative Stress in Cardiovascular Diseases ... 59
Veena Dhawan, Chetan Bakshi, and Riyaz Ahmad Rather

5 Dietary Antioxidants in Mitigating Oxidative Stress in Cardiovascular Diseases ... 83
Subhoshree Ghose, Swati Varshney, Rahul Chakraborty, and Shantanu Sengupta

6 Reactive Oxygen Species and Their Epigenetic Consequences in Heart Diseases ... 141
Seema Bhargava

7 Metabolic Signatures of Redox-Dependent Cardiovascular Diseases ... 159
Stephen T. Vernon, John F. O'Sullivan, and Gemma A. Figtree

8 Cardio Protective Influence of Dietary Spices Mediated Through Their Hypolipidemic and Antioxidant Potential 173
Krishnapura Srinivasan

9 Oxidative Stress and Modulation of Cardiac Kv1.5 Channel 191
Rajabrata Bhuyan and Sajal Chakraborti

Part II Atherosclerosis and Ischemic Heart Disease

10 Aging and Cardiovascular Diseases: The Role of Cellular Senescence 207
Perinur Bozaykut

11 Influence of Genetic Factor on Oxidative Stress Mediated Heart Damage 235
Branislav Rovcanin

12 Oxidative Stress in Cardiac Remodeling Post-Ischemia/Reperfusion: Friend or Foe? 253
Emna Abidi, Abdullah Kaplan, George W. Booz, and Fouad A. Zouein

13 Redox Aspects of Myocardial Ischemia/Reperfusion Injury and Cardioprotection 289
Pasquale Pagliaro, Saveria Femminò, and Claudia Penna

14 Role of Oxidative Stress in Myocardial Ischemia and Infarction 325
Bodh I. Jugdutt and Bernadine A. Jugdutt

15 Oxidant Stress in Atherosclerosis: Oxidatively Modified LDL and LOX-1 363
Ajoe John Kattoor and Jawahar L. Mehta

16 Role of Oxidative Stress and Cardiovascular Risk Factors in Ischemic Heart Disease 375
Monika Bartekova, Kristina Ferenczyova, Marek Jelemensky, and Naranjan S Dhalla

Part III Diabetes Induced Cardiovascular Dysfunction

17 Oxidative Stress and Labile Zinc in Heart Dysfunction Under Hyperglycemia 397
Belma Turan

18 Histone Deacetylases and Oxidative Stress: Role in Diabetic Cardiomyopathy 413
Bhoomika M. Patel

19 Regulating Inflammatory Cytokines in the Diabetic Heart 427
Santosh K. Yadav, Tyler N. Kambis, and Paras K. Mishra

20 Advanced Glycation End Products: A Potential Contributor of Oxidative Stress for Cardio-Vascular Problems in Diabetes 437
Savita Bansal, Pawan Kumar Kare, Ashok Kumar Tripathi, and Sri Venkata Madhu

21 Role of iNOS in Insulin Resistance and Endothelial Dysfunction 461
Hobby Aggarwal, Babu Nageswararao Kanuri, and Madhu Dikshit

22 Oxidative Stress Mechanisms in Type 2 Diabetes Induced Coronary Heart Disease 483
Keerthi Kupsal and Surekha Rani Hanumanth

Part IV Hypertension and Heart Failure

23 Role of Oxidative Stress in the Pathophysiology of Arterial Hypertension and Heart Failure 509
Teresa Sousa, Marta Reina-Couto, and Pedro Gomes

24 Pulmonary Hypertension and the Right Ventricle: The Roles of Mitochondrial Reactive Oxygen Species in Causing Further Right Ventricular Mitochondrial Changes 539
Gerald J. Maarman

25 Mitochondrial Dysfunction and Oxidative Stress: Focusing on Cardiac Hypertrophy and Heart Failure 551
Parmeshwar B. Katare, Hina L. Nizami, and Sanjay K. Banerjee

26 Oxidative Stress in Hypertension and Cardiovascular-Renal Remodeling: Focus on the Renin-Angiotensin-Aldosterone System 581
Giuseppe Maiolino, Verdiana Ravarotto, and Lorenzo A. Calò

About the Editors

Sajal Chakraborti is a Senior Professor of Biochemistry at the University of Kalyani, West Bengal, India. His research mainly focuses on the role of vasoactive agents in regulating vascular tone under oxidant and calcium signaling phenomena. He has been engaged in biochemistry teaching and research for over 40 years.

Naranjan S Dhalla is a Distinguished Professor at the University of Manitoba, Winnipeg, Canada. His expertise includes the subcellular and molecular basis of heart function in health and disease. He has been involved in multidisciplinary research and education to promote the scientific basis of cardiology for over 50 years.

Nirmal K Ganguly is a Distinguished Biotechnologist and former Director General of the Indian Council of Medical Research (ICMR) and is associated with Translational Health Science and Technology Institute (THSTI), Faridabad, India. He was presented with the *Padma Bhushan* Award (the 3rd highest civilian award in India) in the field of Medicine by the President of India in 2008.

Madhu Dikshit is a National Chair Fellow of the Translational Health Science and Technology Institute (THSTI), Faridabad, Haryana, India. She has contributed to the general area of biochemical pharmacology and has a special interest in cardiovascular pharmacology. She has been involved in research for over 40 years.

Part I

General Implications

1 Oxidative Stress and Cardiovascular Risk and Prevention in Children and Adolescents

Francesca Mastorci, Irene Traghella, Laura Sabatino, Alessandro Pingitore, Rudina Ndreu, and Cristina Vassalle

1.1 Introduction

It is now well established that non-communicable diseases, especially cardiovascular diseases (CVD) are major causes of mortality and morbidity both in industrialized and developing countries. Importance of CVD is given by a combination of factors that include obesity, physical inactivity, adoption of unhealthy dietary habits, and smoking. Fortunately, most cardiovascular (CV) clinical manifestations are preventable or at least can be delayed later in old age reducing events, morbidity, disability, and sanitary costs. Nonetheless, although it is never too late to improve unhealthy lifestyle habits, early initiation of prevention is more likely to be effective against onset and development of disease, delaying manifestations, and reducing adverse events [1]. In particular, as exposure to CV risk factors occurs from early ages, this strategy must be started and adjusted throughout the life of an individual. Thus, CVD prevention in at-risk subjects (primary prevention) as well as the prevention of the onset of risk factors in otherwise healthy subjects (primordial prevention), before intervention aimed to reduce the impact of a disease that has already occurred (secondary prevention), represent a social and sanitary priority (Fig. 1.1).

Fetal life, and neonatal physiology (which are not included in the present review) together with childhood and adolescence are all critical phases for the development of cardiometabolic risk (CMR) and later onset of atherosclerosis, hypertension and

F. Mastorci · L. Sabatino · A. Pingitore · R. Ndreu
Istituto di Fisiologia Clinica, CNR, Pisa, Italy

I. Traghella
Fondazione CNR-Regione Toscana G. Monasterio, Pisa, Italy

C. Vassalle (✉)
Istituto di Fisiologia Clinica, CNR, Pisa, Italy

Fondazione CNR-Regione Toscana G. Monasterio, Pisa, Italy
e-mail: cristina.vassalle@ftgm.it

S. Chakraborti et al. (eds.), *Oxidative Stress in Heart Diseases*,
https://doi.org/10.1007/978-981-13-8273-4_1

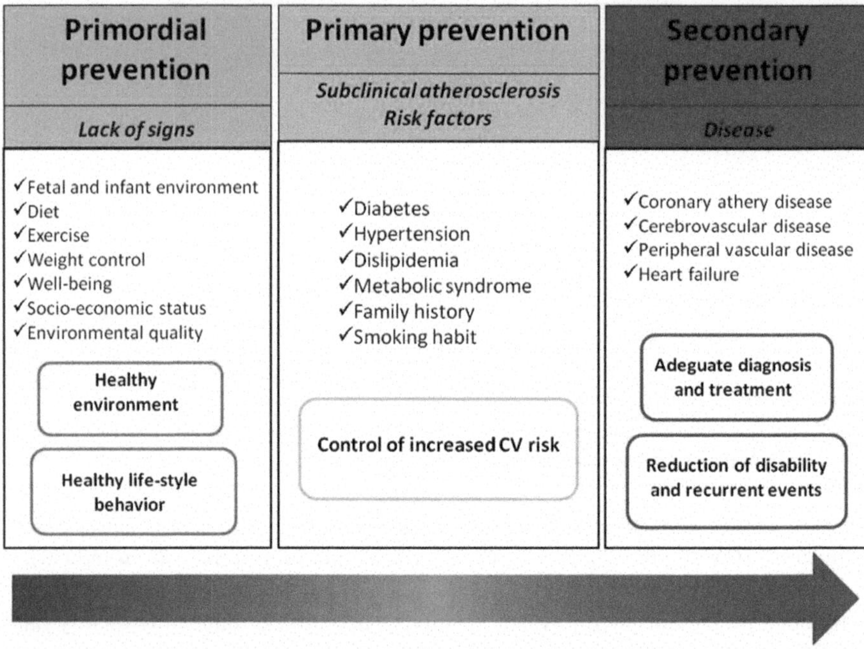

Fig. 1.1 Health Promotion and Cardiovascular Disease Prevention phases

diabetes [2]. Whether majority of children and adolescents are clearly free of manifest cardiometabolic disease, very few do not present any CMR factor, especially when considering sedentary lifestyle and unhealthy diet [2]. Thus, implementation of primordial prevention at this level is valuable, and for children and adolescents in whom risk factors emerge, the focus must be directed to promotion of healthful dietary habits and physical activity. Education to health from primary school may represent one pivotal approach to begin primordial prevention and should be included in public education along with the traditional education matter [3].

Since atherosclerosis begins with endothelial activation, oxidative stress is among the earliest series of adverse events, also closely related to futures of cardiometabolic risk in children and adolescents [4].

This review aims to provide an overview of all these aspects, focusing on the relationship between cardiometabolic parameters and oxidative stress.

1.2 Cardiovascular Primordial Prevention in Children and Adolescents

Childhood and adolescence are often considered the healthiest time of life. In most countries, especially developed ones, they are the points of lowest mortality across life course, placed between the early life mortality and cardiovascular disease

(CVD) in adulthood [5]. They are time windows in life where many health determinants are at their height, determining the so-called "ideal cardiovascular health (iCVH)". According to this new vision, starting from 2010, the American Heart Association defined a new concept of cardiovascular health, on the basis of four main behaviours (smoking status, body mass index-BMI, physical activity and diet) and three traditional cardiovascular risk factors (cholesterol, blood pressure and glucose).

Previous studies suggested that a high iCVH in adolescence is related to a better vascular health condition, whose main indicators for a primordial prevention are: a reduced vascular intima-media thickness and an augmented vascular elasticity [6]. Also, longitudinal studies have reported that high iCVH during adolescence is connected with a more positive cardiac structure and function, as well as a lower risk of hypertension and metabolic syndrome (MetS) in adulthood [7].

However, epidemiological, pathological, and risk factor data demonstrates that childhood and youth are susceptible stages of life since CVD may be deeply rooted in early life, and social determinants of health and many conventional risk factors start to accumulate in childhood and progress during the life course [8]. This is reflected in Barkers' 'developmental origin of adult disease' hypothesis, suggesting that adverse influences in early phases of development, including those in intrauterine life, can result in permanent changes in physiology and metabolism, resulting in increased disease risk in adulthood [9]. Accordingly, as the level of cardiovascular risk factors in children and young people is growing, with one-third of this population overweight or obese, the need to understand the potential future public health impact of the obesity epidemic in the currently younger critical.

The pathology of early stages of CVD in children and adolescents is a function of the same traditional risk factors that affect adults, but while CV prevention among adults is typically limited to primary or secondary prevention of CV outcomes, in children and adolescents primordial prevention is possible.

The Bogalusa Heart Study, collecting data and their lifetime consequences for 39 years, has recognised the role of risk factors in youth [10], proposing that the major etiologies of adult heart disease and atherosclerosis begin early in life, with evident anatomical changes by 5–8 years of age. Furthermore, one of the most important result in the Bogalusa study, in addition to the early identification of childhood risk factors, was their tracking into adulthood. A retrospective cohort study demonstrated, for example, that glucose homeostasis variables in childhood predicted adult diabetes correlating with CMR factors, and showing that the presence of multiple risk factors is associated with an even higher risk of atherosclerotic lesions [8].

Population-based data from prospective cohort studies, beginning in children and youth with a follow-up later in life, have shown that the modifiable risk factors associate with preclinical markers of CVD in adulthood, range from purely genetic to behavioural, psychosocial and environmental. This opens to the concept of primordial prevention of CVD as a valid tool either for CV epidemiologists or clinicians, in order to reduce the burden of CVD in the population before its clinical manifestations. So far, different policies have responded non adequately on how certain risk behaviours affect health outcomes. The more accredited strategies aiming to lower the CVD risk involve re-education, reorientation and motivation. To date, preventive strategies mainly concern nutrients intake and physical activity. The special Turku Coronary Risk Factor Intervention Project (STRIP), for example, has been able to provide important information on diet throughout childhood and its role in the primordial prevention of atherosclerosis, through dietary counselling for families, with clinical and public health implications [11].

On the contrary, psychological dimension is predominantly directed towards adolescents with mental disorders rather than the healthy ones. Instead, according to AHA 2020 Impact Goals, it is pivotal to examine all aspects, including psychosocial ones that have been identified as predictors of ideal cardiovascular health in adulthood [12].

In line with the American Heart Association's Social Determinants of Risk and Outcomes of Cardiovascular Disease Scientific Statement, common findings suggest that socioeconomic adversity and exposure to poor parenting practices during adolescence predict greater levels of cardiovascular risk factors in adulthood [13]. Also, the prospective Collaborative Perinatal Project, which examined psychosocial predictors of a favorable cardiovascular profile, showed that high childhood attention regulation, high cognitive ability, and a positive childhood environment were associated with a more favorable cardiovascular profile in adulthood [14].

Furthermore, socioeconomic and emotional factors, parental health behaviours, stressful events, and self-regulation data obtained from Cardiovascular Risk in Young Finns Study suggest a dose-response relationship between psychosocial factors in youth and CV health in adulthood [15]. In this perspective and on the basis of the clinical evidence that half of all CVD events, regardless of age, occur in subjects who are not classified at high risk, in accord with conventional risk factor profile [16], more recent epidemiological studies have focused on psychosocial factors potentially increasing CV risk [17]. According to this view, a foundational concept in adolescence is the "Positive Youth Development" that highlights the significance of psychosocial assets in understanding how life-long preservation of ideal cardiovascular health can actually occur. In this context, in children but mostly in adolescence it is necessary to promote social inclusion skills in school environment in order to reduce health risk behaviours [18]. In fact, socialization encourages physical activity and development of social support networks and prevents antisocial behaviours reducing unhealthy habits [19].

Current data show that small-scale interventions have begun to apply the concept of Positive Youth Development in CV health promotion among adolescents in school setting, although outcomes have not yet well characterized [1].

Given the central role of school in preventive approach, potentially effective interventions should include school-based programs, corroborating the idea that schools serve as "context of socialization" that influence students' developmental outcomes. Future studies should specifically aim in delivering school-based platforms adopting an "user-friendly language" in order to encourage children and adolescents to improve their health status and thus, evaluating the efficacy of interventions.

1.3 Oxidative Stress Biochemical Markers

Oxidative stress is defined as an imbalance between oxidants (reactive oxygen species – ROS) and antioxidants, leading to a disruption of redox signaling and control and molecular damage [20]. When the stress level exceeds defense capacity, a toxic effect may occur through the production of peroxides and free radicals; as major consequence nucleic acid, lipids, carbohydrates and proteins are damaged, cell health is compromised and generation of reactive species may lead to cell death by necrosis or apoptosis [21].

Nowadays is widely accepted a crucial role of oxidative stress in many disorders, like cardiovascular and neurodegenerative diseases, cancer, atherosclerosis and diabetes mellitus [22]. Biomarkers of oxidative stress are molecules modified by the interactions with ROS in the cellular microenvironment, or molecules of the antioxidant system that change in response to increased redox stress [23]. These biomarkers are indicators to assess the degree of oxidative stress and to evaluate antioxidant capacity.

Several markers of oxidative stress are available, but most are of limited value because of the instability of many reactive species. Furthermore, ROS are generally too reactive and have too short half-life to allow direct measurement in body fluids; for these reasons molecular products of ROS are generally considered more stable than ROS themselves, and are used to track ROS concentration, especially by measuring lipid peroxidation products and oxidized proteins [21]. Antioxidant biomarkers are included in the total antioxidant capacity (TAC) and divided into enzymatic biomarkers, such as catalase, glutathione peroxidase (GPX), superoxide dismutase (SOD), and non-enzymatic biomarkers like vitamins A, C, E, glutathione and uric acid [22].

Different oxidative stress biomarkers (e.g malondialdehyde-MDA, F2-isoprostanes, nitric oxide) have been found associated to obesity, BMI and adipokines and shown improvement in response to dietary or lifestyle modification in children and adolescents [24–26].

A recent longitudinal study (subjects aged 5–19), characterized by a mean 12.0 years follow-up, showed interesting correlations between childhood MetS severity and later measures of insulin resistance and uric acid, as oxidative stress biomarker, leading to a greater cardiometabolic risk [27].

Noninvasive tools to assess CMR could increase participation in screening and treatment programs and improve adherence to dietary and lifestyle intervention, especially in children and adolescent cohorts. Thus, the possibility of salivary collection and analysis of many oxidative stress biomarkers is valuable, especially in children and adolescents, because it is a simple, subject- and parent-friendly, and non-invasive specimen collection. In particular, it is very peculiar that for some oxidative stress biomarker a good correlation was found between salivary and serum concentrations tested in overweight/obese subjects or patients with MetS, suggesting that saliva sampling may be an useful surrogate for blood testing especially in CMR evaluation [28, 29].

Among the number of available oxidative stress biomarkers reflecting damage to lipids, proteins, carbohydrates, and nucleic acids, the most used are MDA, hydroperoxides, conjugated dienes, 4-hydroxynonenal, hydrocarbons, F2-isoprostanes, and oxidized-low density lipoproteins.

LIPIDS: are susceptible targets of oxidation because of their molecular structure [23]. Lipid peroxidation generates a variety of relatively stable products, mainly MDA, 4-hydroxy-2-nonenal, and isoprostanes [21].

MDA is generally quantified on plasma samples with a colorimetric assay named TBARS which uses thiobarbituric acid reacting with MDA, but this method has low specificity [23, 30]. Several commercial ELISA kits and high-performance liquid chromatography (HPLC) procedures are available for MDA detection [23].

Moreover, isoprostanes, generated from peroxidation of arachidonic acid, mostly present in cell membrane, can be measured in various samples using gas chromatography-mass spectrometry (GC/MS), liquid chromatography-mass spectrometry (LC/MS), enzyme-linked immunosorbent assays (ELISA) and radioimmunoassay (RIA) [23, 31].

PROTEINS: are major targets for ROS because of their great abundance in biological systems. Exposure of proteins to ROS may alter every level of protein structure; the major fate of oxidized proteins is catabolism by proteasomal and lysosomal pathways, but some functionally inactive proteins appear to be inadequately degraded, forming protein aggregates, and accumulating in separate compartments within cells or in the extracellular environment. The accumulation of such damaged material increases during the normal aging process, and may contribute to a range of human pathologies [21].

Oxidized-low density lipoproteins (OxLDL) play a central role in the pathogenesis of atherosclerosis and their accumulation inside the vascular wall also stimulates the production of proinflammatory cytokines including adhesion molecules such as intercellular adhesion molecule-1 (ICAM-1), vascular cell adhesion molecule-1 (VCAM-1) and endothelial selectin (E-selectin) [32]. Currently, there are different OxLDL ELISA kit available for research, using plasma as sample [33].

ANTIOXIDANT CAPACITY: total antioxidant capacity [TAC] can be estimated as a whole, or as its enzymatic (e.g CAT, GPx, SOD) or non-enzymatic (e.g. E, A, C, and GSH and uric acid) components in different biological samples, which can be easily estimated by colorimetric and ELISA assays [34].

1.4 Oxidative Stress and Cardiovascular Risk and Prevention in Children and Adolescents

As stated before, the four components of a healthy lifestyle most frequently focused in primordial prevention interventions include diet, physical activity, body weight maintaining (although clearly largely related to the two previous elements), and absence of smoking.

1.4.1 Smoking Habit

Although smoking habit is a minor problem in children, almost in developed countries, passive smoking may be more diffuse. As such, passive smoking has been associated with impaired serum lipids in children and adolescents exposed to second-hand smoke in the household, which presented significantly reduced high-density lipoprotein concentration [35]. More recent data showed that biomarkers of oxidative stress are elevated and inversely associated with impaired artery dilation in children exposed to passive smoking [36]. In another population of children and adolescents second-hand smoke exposure resulted associated with increased oxidative stress and decreased paraoxonase-1 [37]. Other data confirmed a passive smoking-related reduction of total antioxidant capacity, parallel to an increase of total peroxide levels in children (9–13 years), 61 of whom never been exposed to passive smoking, and 82 exposed to passive smoking at least 10 cigarettes per day for at least 1 year in their house [38]. Accordingly with smoking adverse effects, smoking habit in adolescents was found to be associated with increased oxidative damage [39].

Cigarette smoke–induced oxidative stress activates the endothelium through many mechanisms, including induction of adhesion molecule expression, macrophages and platelets [40]. Moreover, there is a reduction of NO cellular concentration, and increase of inflammatory and proatherogenic cytokines. Smoke contains a great number of different chemicals, including oxidants and prooxidants, capable of damage endothelial cell leading to apoptosis or necrosis [40]. Macrophages are activated by the expression of adhesion molecule receptors (eg, intracellular adhesion molecule, vascular cell adhesion molecule), adhesion and transendothelial migration are increased, as well as upload of oxidized lipids produced by smoke-increased oxidative stress, foam cells production, and lipid plaques [40]. Smoking induces an increase in smooth muscle cell proliferation and migration inducing intimal thickening [40]. Destruction of extracellular matrix is driven by increased expression of matrix metalloproteinases (MMPs) and reduced expression of tissue inhibitors of MMPs (TIMPs) [40].

1.4.2 Diet, Physical Activity, Body Size

Dietary habits, physical activity and weight of children and adolescents are closely related in a complex interactive pattern. In fact, some data suggested that active boys and girls frequently consumed breakfast cereals and fresh fruit, while sedentary girls eat more high fat foods and soft drinks, and that sedentary behaviour is inversely related to adherence to Mediterranean diet [41]. However, there are some determinants other than diet and physical activity which may affect weight and body size, including genetic factors, stress, sleep hours, and meal timing and number. In particular, psycho-emotional factors may be critical in children and adolescents, although generally neglected in preventive cardiology.

There are many data showing higher oxidative stress biomarker levels related to unhealthy dietary habit, physical inactivity, and weight gain in children and adolescents [42–44]. Clearly, energy intake must take into account age, gender, growth stage, body weight, and type and level of exercise of the specific child or adolescent, which cannot be considered as "small adults" since fundamental biological differences exist (e.g. lower sympatho-adrenal activity, immature hormonal system) [45]. Thus, results obtained in adults cannot be simply translated to children or adolescents.

Numerous clinical trials have showed the benefits of healthy eating patterns [34, 46]. Accordingly, children following a diet low in total fat (less than 30% of energy), saturated fat (8–10% of energy), and cholesterol (200–300 mg/d) presented reduced total cholesterol, LDL-C, and C-reactive protein [47]. In obese children (mean age, 9.18 ± 1.54 year), a dietary restriction-weight loss program (an equivalent to 70% of the needs for sex and stature age, subdivided in five meals per day, and including 50/60% carbohydrates, 25/30% lipids, and 10/15% proteins) improve oxidative stress, restoring levels of MDA and vitamin E to those shown by normal weight children [47].

Children and adolescents generally show an active participation in sports, however they rarely have a constant application in a sport, fact which might give the highest benefits. In any case, data available suggest evident positive effects of exercise training on oxidative status, increasing antioxidant capacity and reducing parameters belonging to the oxidant counterpart [48]. Clearly, these effects are influenced by different variables, such as training duration and load, type of sports etc. As an example, chronic endurance training (cycling and running) in adolescents (12–16 years) increased antioxidants (xanthine oxidase, glutathione and catalase activity) in comparison to controls, whereas this antioxidant upregulation correlates with an increase of endurance performance [43]. Accordingly, 1 year-training of running improved endurance performance, increasing anti-oxidation and lowering pro-oxidation parameters in adolescents (14–15 years) [49]. Moreover, physical activity has been shown as an effective tool in overweight children, where a 2-week training program consisting of 2–2.5 h/day (tennis, beach games, and gym-based exercises) induces significant reductions in lipids (except HDL cholesterol), and biomarkers of oxidative stress and endothelial activation (8-isoprostaglandin F2alpha, myeloperoxidase, soluble intracellular adhesion molecule, sE-selectin, C-reactive protein CRP, total matrix metalloproteinase-9, and cellular monocyte chemotactic protein-1 production) [50].

Taken into account that physical activity and healthy dietary habit have independent beneficial effects, the combination of exercise and dietary plans may enhance positive health-related changes. Accordingly, increased endothelial dysfunction-associated oxidative stress and inflammation were partially normalized by a 6-week diet and physical training intervention (aerobic physical activity for 60 min, 3 days a week for 6 consecutive weeks plus diet including 30% energy derived from fat, 15% from protein, and 55% from carbohydrate, with energy content based on the calorie requirement for height) in obese insulin-resistance adolescents (12–18 years), where changes in BMI, waist circumference, CRP, ox-LDL and MDA were inversely correlated with flow mediated dilation [51]. Moreover, short-term lifestyle changes (high-fiber, low-fat diet and 2–2.5 h daily exercise in a 2-week plan) in overweight children and adolescents (age 8–17) induced significant decrease in BMI, and markers of inflammation and oxidative stress (C reactive protein, sE-selectin, soluble inter-cell adhesion molecule 1, monocyte chemoattractant protein 1, matrix metalloproteinase 9, oxidative stress and lipid peroxidation, myeloperoxidase, 8-isoprostane) [50]. Recent data demonstrated that a 4-week exercise training, coupled with dietary restriction, improved BMI, lean body mass, fat body mass and fat mass ratio in male adolescents [52]. In addition, the activities of antioxidant enzymes (superoxide dismutase, and glutathione peroxidase) showed a significant increase [52].

It must be also considered that some adverse effects on oxidative stress, adipose tissue metabolism, and inflammation are influenced by pubertal hormonal changes. Children with growth hormone deficiency showed an unbalanced oxidant-antioxidant status with a reduced nitric oxide (NO) bioavailability and vascular reactivity [53]. Recent experimental data obtained in sedentary adolescent male rats, suggested that elevated testosterone may modulate oxidative stress status [54]. In this context, physical activity and dietary habit may beneficially influence oxidative stress, also modulating hormonal status during puberty [55–57]. Oxidative stress is associated with metabolic-related hormones in obese children and adolescents [58]. Specifically, a study evaluated pre- (11 years old, 28 normal-weight, 11 obese) and early- (11 years old, 25 normal-weight, 12 obese) pubertal Greek boys, for a wide panel of hormonal and oxidative and inflammatory panels (leptin, adiponectin, C reactive protein, IL-6, thiobarbitouric acid reactive substances, protein carbonyls, glutathione, oxidized glutathione, glutathione peroxidase, catalase, total antioxidant capacity) and results suggested that childhood obesity is associated with oxidative stress and inflammation [58]. Moreover, leptin and adiponectin predict negatively and positively anti-oxidation, respectively, while IL-6 predicts positively pro-oxidation and negatively and anti-oxidation [58]. Moreover, C reactive protein is increased and negatively associated with anti-oxidation in pre-pubertal obese boys [58]. Conversely, exercise and healthy dietary habit may improve oxidative stress in obese children and adolescents [25, 48].

1.5 Oxidative Stress Molecular and Genetics Biomarkers

Many metabolic processes produce free radicals, not necessarily dangerous for human body, since they often play a crucial role in defeating viral and bacterial infections. Oxidative stress occurs because of free radicals overload inside the cells and cell inability to dispose excessive oxidative products. In this case, oxidative stress induces damaging factors to produce lesions on DNA, promoting the loss of genetic information and genomic instability, which may lead to severe diseases such as cancer [59].

Since lesions to DNA are followed by cellular repair machinery activation, repair proteins are considered important biomarkers helping in the prediction of genotoxic stress level. An example is the phosphorylated H2AX (gamma-H2AX), one of the key molecules for the recognition and repair of DNA double strand breaks (DSBs). In fact, when DSB occurs within a nucleus, several H2AX molecules are rapidly phosphorylated on Ser-139, originating the so-called gamma-H2AX nuclear foci [60]. Therefore, gamma-H2AX focus assay is one of the principal methods to investigate early genome damage in cells exposed to various genotoxic agents, or in cells from people affected by cancer of other pathological conditions. Interestingly, this approach has been recently used to study the oxidative stress level and DNA damage in lymphocytes of adolescents affected by type 1 diabetes mellitus, in order to explore the possibility of employing the gamma-H2AX focus assay as molecular marker of future possible complications (*i.e.* cardiovascular disease or some type of cancer) in adulthood [61].

Oxidative stress is also demonstrated to be associated with the shortening of telomeres, specialized DNA/protein structures acting as final caps of chromosomes, preventing end-fusion and distinguishing the chromosome ends from DSBs [62]. Because of their low reduction potential, the G-rich telomere repeats are particularly prone to damage induced by reactive oxygen species, easily bringing to telomere shortening process. According to this idea, a recent study associated the leukocyte telomere length (LTL) to obesity at age 9, evaluating LTL in a cohort of young (age 4–5) Latino children in United States abusing of sugar-sweetened beverage consumption [63]. LTL has been demonstrated to be a good predictor of future obesity occurrence and, consequently, to development of cardiometabolic diseases [64].

Moreover, in a large case-control study on French obese children (average age, 11 years) a highly significant inverse association between obesity onset and mean LTL was demonstrated [65]. Mean LTL in the obeses is almost 25% less than controls, and LTL decreases with increasing age, weight, and height in both groups, with comparable effect sizes for all three variables in both cases and controls. Although the most evident difference between obese cohort and controls, it is reasonable to hypothesize that the association with LTL may reflect some other clinical characteristic of obesity, such as circulating lipid levels or inflammation. In fact, longitudinal studies in adults showed that LTL is positively associated with HDL-cholesterol levels [66]. Interestingly, the fact that obese children have an apparent biological age significantly greater than their effective age (lower LTL), suggests the urgency of intervention and support in order to reduce the highly probable risk of future disease at the earliest possible.

Moreover, several environmental factors leading to marked cellular oxidative stress are crucial in the determination of telomere attrition rate: air pollution [67] and second-hand smoke [68] are among the more significant contributors in DNA damage in relation to oxidative stress in young children. Telomere shortening in childhood is a potentially important biomarker in the evaluation of environmental impact, and damaging factors reasonably have stronger effects on children, when telomere attrition rate is at its maximum [67].

Among the different tests used to investigate early damage to genomic DNA, the single cell gel electrophoresis assay (Comet assay) is a simple and fast method for quantification of different types and levels of DNA lesions [69]. As an example, the comet assay has been used to evaluate DNA damaging caused by cigarette smoke on lymphocytes of passive smoking children. The increase in DNA fragmentation is shown by comet-looking nuclei, having the head of the comet made by unfragmented DNA and the tail showing the damaged DNA [68].

In summary, the characterization of molecular and genetic biomarkers for quantification of DNA damages in children and adolescents and the studies carried out to understand how the damage can be restored, may be of crucial significance in multiple disease prevention, including cancer and diabetes.

1.6 Conclusion

Considering the epidemic of obesity and cardiometabolic disease, prevention from early life is necessary. In particular, it is alarming the prevalence of obesity in childhood and adolescence, which has increased greatly during the last decades. Thus, improving the cardiovascular health of children and adolescence with a coordinated and comprehensive approach based on multifaceted and multidimensional perspective is a crucial public health priority and can yield incremental success in improving cardiovascular health assets into adulthood. In fact, in a society where population age is increasing, now more than ever, the focus must be directed to the reduction of cardiometabolic risk in future generations. Actually, children are at a crossroad, they can teach how improve lifestyle to their actual families of origin, as well as to their future ones, with a significant gain in terms of social, sanitary and economic aspects. In this radical change of mentality, school-based interventions can be effective in lifestyle improving and preventing the development of obesity in children. Schools can provide the education to practice healthy behaviours and achieve environment changes, giving tools for an active lifestyle and for a better food choices at home and outside of the scholastic context.

As oxidative stress is pivotal in the onset and development of cardiometabolic disease, also in children and adolescents, longitudinal trials will provide further clarification on the oxidative stress-related underlying mechanisms involved in CMR. Notably, the reduction of atherosclerosis progression and its clinical consequences in adulthood, as well as the mantainement of a balanced oxidative/antioxidant status, can be achieved efficaciously in childhood by a healthy diet adoption and active life, although it must be considered that effects of dietary and exercise interventions are closely dependent by children's maturity stage (hormonal status) and body weight (Fig. 1.2).

It is conceivable that oxidative stress biomarkers might serve in future as one early tool to predict risk of developing cardiometabolic and lifestyle-related diseases in childhood and adolescence and later in adulthood, as well as to formulate some mechanistic hypotheses. In this context, the availability of salivary measurements of several oxidative stress-related biomarkers can facilitate their evaluation as non-invasive sampling in large cohorts.

Taken together, available knowledge evidences the critical significance of primordial prevention in the context of cardiometabolic disease, but also to the importance of health campaigns at school levels, focused on the role of an healthy lifestyle factors, and the potential of evaluating oxidative stress status as cardiometabolic risk determinant (Fig. 1.3).

Fig. 1.2 Interrelationship between lifestyle, hormonal status, and oxidative stress in children and adolescents

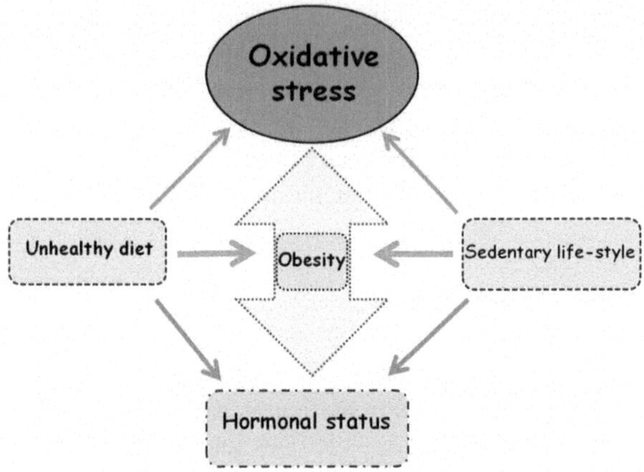

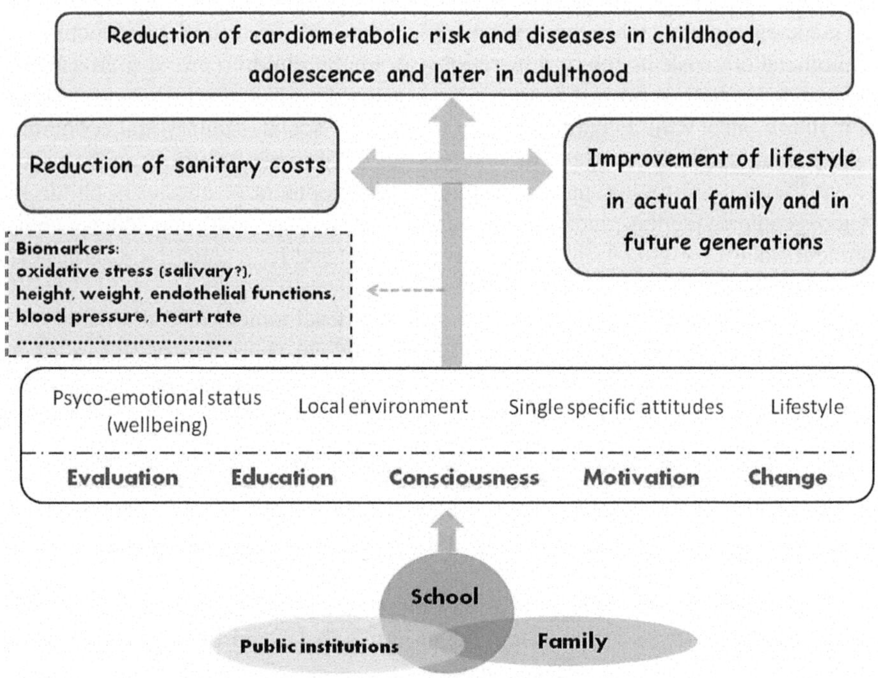

Fig. 1.3 Primordial prevention strategy in children and adolescents

References

1. Chung RJ, Touloumtzis C, Gooding H (2015) Staying young at heart: cardiovascular disease prevention in adolescents and young adults. Curr Treat Options Cardiovasc Med 17:61
2. Kelishadi R, Poursafa P (2014) A review on the genetic, environmental, and lifestyle aspects of the early-life origins of cardiovascular disease. Curr Probl Pediatr Adolesc Health Care 44:54–72
3. Weintraub WS, Daniels SR, Burke LE, Franklin BA, Goff DC Jr, Hayman LL, Lloyd-Jones D, Pandey DK, Sanchez EJ, Schram AP, Whitsel LP, American Heart Association Advocacy Coordinating Committee, Council on Cardiovascular Disease in the Young, Council on the Kidney in Cardiovascular Disease, Council on Epidemiology and Prevention, Council on Cardiovascular Nursing, Council on Arteriosclerosis, Thrombosis and Vascular Biology, Council on Clinical Cardiology, and Stroke Council (2011) Value of primordial and primary prevention for cardiovascular disease: a policy statement from the American Heart Association. Circulation 124:967–990
4. Incalza MA, D'Oria R, Natalicchio A, Perrini S, Laviola L, Giorgino F (2018) Oxidative stress and reactive oxygen species in endothelial dysfunction associated with cardiovascular and metabolic diseases. Vasc Pharmacol 100:1–19
5. Graham P (2004) The end of adolescence. Oxford University Press, New York
6. Pahkala K, Hietalampi H, Laitinen TT, Viikari JS, Rönnemaa T, Niinikoski H, Lagström H, Talvia S, Jula A, Heinonen OJ, Juonala M, Simell O, Raitakari OT (2013) Ideal cardiovascular health in adolescence: effect of lifestyle intervention and association with vascular intima-media thickness and elasticity (the Special Turku Coronary Risk Factor Intervention Project for Children [STRIP] study). Circulation 127:2088e96
7. Laitinen TT, Ruohonen S, Juonala M, Magnussen CG, Mikkilä V, Mikola H, Hutri-Kähönen N, Laitinen T, Tossavainen P, Jokinen E, Niinikoski H, Jula A, Viikari JS, Rönnemaa T, Raitakari O, Pahkala K (2017) Ideal cardiovascular health in childhood longitudinal associations with cardiac structure and function: the Special Turku Coronary Risk Factor Intervention Project (STRIP) and the Cardiovascular Risk in Young Finns Study (YFS). Int J Cardiol 230:304e9
8. Berenson GS, Srinivasan SR, Bao W, Newman WP 3rd, Tracy RE, Wattigney WA (1998) Association between multiple cardiovascular risk factors and atherosclerosis in children and young adults: the Bogalusa Heart Study. N Engl J Med 338:1650–1656. https://doi.org/10.1056/NEJM199806043382302
9. Barker DJP (1995) Fetal origins of coronary heart disease. BMJ 311:171–174
10. Berenson GS, Srinivasan SR (2010) Cardiovascular risk in young persons: secondary or primordial prevention? Ann Intern Med 153:202–203
11. Lapinleimu H, Viikari J, Jokinen E, Salo P, Routi T, Leino A, Rönnemaa T, Seppänen R, Välimäki I, Simell O (1995) Prospective randomized trial in 1,062 infants of diet low in saturated fat and cholesterol. Lancet 345:471–476
12. Berenson GS, Wattigney WA, Tracy RE, Newman WP 3rd, Srinivasan SR, Webber LS, Dalferes ER Jr, Strong JP (1992) Atherosclerosis of the aorta and coronary arteries and cardiovascular risk factors in persons aged 6 to 30 years and studied at necropsy (the Bogalusa Heart Study). Am J Cardiol 70:851–858
13. Taylor SE, Way BM, Seeman TE (2011) Early adversity and adult health outcomes. Dev Psychopathol 23:939–954. https://doi.org/10.1017/S0954579411000411
14. Appleton AA, Buka SL, Loucks EB, Rimm EB, Martin LT, Kubzansky LD (2013) A prospective study of positive early-life psychosocial factors and favorable cardiovascular risk in adulthood. Circulation 127:905–912

15. Pulkki-Råback L, Elovainio M, Hakulinen C, Lipsanen J, Hintsanen M, Jokela M, Kubzansky LD, Hintsa T, Serlachius A, Laitinen TT, Pahkala K, Mikkilä V, Nevalainen J, Hutri-Kähönen N, Juonala M, Viikari J, Raitakari OT, Keltikangas-Järvinen L (2015) Cumulative effect of psychosocial factors in youth on ideal cardiovascular health in adulthood: the Cardiovascular Risk in Young Finns Study. Circulation 131:245–253
16. Futterman LG, Lemberg L (1998) Fifty percent of patients with coronary artery disease do not have any of the conventional risk factors. Am J Crit Care 7:240–244
17. Pedersen SS, von Känel R, Tully PJ, Denollet J (2017) Psychosocial perspectives in cardiovascular disease. Eur J Prev Cardiol 24:108–115
18. Patton GC, Bond L, Carlin JB, Thomas L, Butler H, Glover S, Catalano R, Bowes G (2006) Promoting social inclusion in schools: a group-randomized trial of effects on student health risk behavior and well-being. Am J Public Health 96:1582–1587
19. O'Neil A, Scovelle AJ, Milner AJ, Kavanagh A (2018) Gender/sex as a social determinant of cardiovascular risk. Circulation 137:854–864
20. Sies H (2015) Oxidative stress: a concept in redox biology and medicine. Redox Biol 4:180–183
21. Dalle-Donne I, Rossi R, Colombo R, Giustarini D, Milzani A (2006) Biomarkers of oxidative damage in human disease. Clin Chem 52:601–623
22. Mastorci F, Vassalle C, Chatzianagnostou K, Marabotti C, Siddiqui K, Eba AO, Mhamed SAS, Bandopadhyay A, Nazzaro MS, Passera M, Pingitore A (2017) Undernutrition and overnutrition burden for diseases in developing countries: the role of oxidative stress biomarkers to assess disease risk and interventional strategies. Antioxidants 6:41
23. Ho E, Galougahi KK, Liu CC, Bhindi R, Figtree GA (2013) Biological markers of oxidative stress: applications to cardiovascular research and practice. Redox Biol 1:483–491
24. Codoñer-Franch P, Valls-Bellés V, Arilla-Codoñer A, Alonso-Iglesias E (2011) Oxidant mechanisms in childhood obesity: the link between inflammation and oxidative stress. Transl Res 158:369–384
25. Montero D, Walther G, Perez-Martin A, Roche E, Vinet A (2012) Endothelial dysfunction, inflammation, and oxidative stress in obese children and adolescents: markers and effect of lifestyle intervention. Obes Rev 13:441–455
26. McCrindle BW (2015) Cardiovascular consequences of childhood obesity. Can J Cardiol 31:124–130
27. Wang LX, Filipp SL, Urbina EM, Gurka MJ, DeBoer MD (2018) Longitudinal associations of metabolic syndrome severity between childhood and young adulthood: the Bogalusa Heart Study. Metab Syndr Relat Disord 16:208–214
28. De Giuseppe R, Cossellu G, Vigna L, Dicorato F, De Vita C, Venturelli G, Bamonti F, Maiavacca R, Farronato G (2015) Correlation between salivary and serum oxidized LDL levels: a pilot study on overweight/obese subjects. J Oral Pathol Med 44:884–887
29. Soukup M, Biesiada I, Henderson A, Idowu B, Rodeback D, Ridpath L, Bridges EG, Nazar AM, Bridges KG (2012) Salivary uric acid as a noninvasive biomarker of metabolic syndrome. Diabetol Metab Syndr 4:14
30. Meagher EA, FitzGerald GA (2000) Indices of lipid peroxidation in vivo: strengths and limitations. Free Radic Biol Med 28:1745–1750
31. Musiek ES, Yin H, Milne GL, Morrow JD (2005) Recent advances in the biochemistry and clinical relevance of the isoprostane pathway. Lipids 40:987–994
32. Tsimikas S (2006) Oxidized low-density lipoprotein biomarkers in atherosclerosis. Curr Atheroscler Rep 8:55–61
33. Mertens A, Holvoet P (2001) Oxidized LDL and HDL: antagonists in atherothrombosis. FASEB J 15:2073–2084
34. Pingitore A, Lima GP, Mastorci F, Quinones A, Iervasi G, Vassalle C (2015) Exercise and oxidative stress: potential effects of antioxidant dietary strategies in sports. Nutrition 31:916–922
35. Neufeld EJ, Mietus-Snyder M, Beiser AS, Baker AL, Newburger JW (1997) Passive cigarette smoking and reduced HDL cholesterol levels in children with high-risk lipid profiles. Circulation 96:1403–1407

36. Loffredo L, Zicari AM, Occasi F, Perri L, Carnevale R, Angelico F, Del Ben M, Martino F, Nocella C, De Castro G, Cammisotto V, Battaglia S, Duse M, Violi F (2018) Role of NADPH oxidase-2 and oxidative stress in children exposed to passive smoking. Thorax. pii: thoraxjnl-2017-211293
37. Kahraman FU, Torun E, Osmanoğlu NK, Oruçlu S, Özer ÖF (2017) Serum oxidative stress parameters and paraoxonase-1 in children and adolescents exposed to passive smoking. Pediatr Int 59(1):68–73
38. Kosecik M, Erel O, Sevinc E, Selek S (2005) Increased oxidative stress in children exposed to passive smoking. Int J Cardiol 100:61–64
39. Stephensen CB, Marquis GS, Douglas SD, Wilson CM (2005) Immune activation and oxidative damage in HIV-positive and HIV-negative adolescents. J Acquir Immune Defic Syndr 38:180–190
40. Messner B, Bernhard D (2014) Smoking and cardiovascular disease: mechanisms of endothelial dysfunction and early atherogenesis. Arterioscler Thromb Vasc Biol 34:509–515
41. Bibiloni Mdel M, Pich J, Córdova A, Pons A, Tur JA (2012) Association between sedentary behaviour and socioeconomic factors, diet and lifestyle among the Balearic Islands adolescents. BMC Public Health 12:718
42. Dennis BA, Ergul A, Gower BA, Allison JD, Davis CL (2013) Oxidative stress and cardiovascular risk in overweight children in an exercise intervention program. Child Obes 9:15–21
43. Llorente-Cantarero FJ, Gil-Campos M, Benitez-Sillero JD, Muñoz-Villanueva MC, Túnez I, Pérez-Navero JL (2012) Prepubertal children with suitable fitness and physical activity present reduced risk of oxidative stress. Free Radic Biol Med 53:415–420
44. Mager DR, Patterson C, So S, Rogenstein CD, Wykes LJ, Roberts EA (2010) Dietary and physical activity patterns in children with fatty liver. Eur J Clin Nutr 64:628–635
45. Liu M, Timmons BW (2016) The effect of acute exercise on neutrophil reactive oxygen species production and inflammatory markers in healthy prepubertal and adult males. Pediatr Exerc Sci 28:55–63
46. Van Horn L, Obarzanek E, Barton BA, Stevens VJ, Kwiterovich PO Jr, Lasser NL, Robson AM, Franklin FA Jr, Lauer RM, Kimm SY, Dorgan JF, Greenlick MR (2003) A summary of results of the Dietary Intervention Study in Children (DISC): lessons learned. Prog Cardiovasc Nurs 18:28–41
47. Mohn A, Catino M, Capanna R, Giannini C, Marcovecchio M, Chiarelli F (2005) Increased oxidative stress in prepubertal severely obese children: effect of a dietary restriction-weight loss program. J Clin Endocrinol Metab 90:2653–2658
48. Avloniti A, Chatzinikolaou A, Deli CK, Vlachopoulos D, Gracia-Marco L, Leontsini D, Draganidis D, Jamurtas AZ, Mastorakos G, Fatouros IG (2017) Exercise-induced oxidative stress responses in the pediatric population. Antioxidants (Basel) 6:6
49. Zalavras A, Fatouros IG, Deli CK, Draganidis D, Theodorou AA, Soulas D, Koutsioras Y, Koutedakis Y, Jamurtas AZ (2015) Age-related responses in circulating markers of redox status in healthy adolescents and adults during the course of a training macrocycle. Oxidative Med Cell Longev 2015:283921
50. Roberts CK, Chen AK, Barnard RJ (2007) Effect of a short-term diet and exercise intervention in youth on atherosclerotic risk factors. Atherosclerosis 191:98–106
51. Kelishadi R, Hashemi M, Mohammadifard N, Asgary S, Khavarian N (2008) Association of changes in oxidative and proinflammatory states with changes in vascular function after a lifestyle modification trial among obese children. Clin Chem 54:147–153
52. Li C, Feng F, Xiong X, Li R, Chen N (2017) Exercise coupled with dietary restriction reduces oxidative stress in male adolescents with obesity. J Sports Sci 35:663–668
53. Rothermel J, Reinehr T (2016) Metabolic alterations in paediatric GH deficiency. Best Pract Res Clin Endocrinol Metab 30:757–770
54. Sadowska-Krępa E, Kłapcińska B, Jagsz S, Nowara A, Szołtysek-Bołdys I, Chalimoniuk M, Langfort J, Chrapusta SJ (2017) High-dose testosterone enanthate supplementation boosts oxidative stress, but exerts little effect on the antioxidant barrier in sedentary adolescent male rat liver. Pharmacol Rep 69:673–678

55. Patel R, Shah G (2018) High-fat diet exposure from pre-pubertal age induces polycystic ovary syndrome (PCOS) in rats. Reproduction 155:141–151
56. Suzuki K, Takahashi M, Li CY, Lin SP, Tomari M, Shing CM, Fang SH (2015) The acute effects of green tea and carbohydrate coingestion on systemic inflammation and oxidative stress during sprint cycling. Appl Physiol Nutr Metab 40:997–1003
57. Paltoglou G, Fatouros IG, Valsamakis G, Schoina M, Avloniti A, Chatzinikolaou A, Kambas A, Draganidis D, Mantzou A, Papagianni M, Kanaka-Gantenbein C, Chrousos GP, Mastorakos G (2015) Antioxidation improves in puberty in normal weight and obese boys, in positive association with exercise-stimulated growth hormone secretion. Pediatr Res 78:158–164
58. Paltoglou G, Schoina M, Valsamakis G, Salakos N, Avloniti A, Chatzinikolaou A, Margeli A, Skevaki C, Papagianni M, Kanaka-Gantenbein C, Papassotiriou I, Chrousos GP, Fatouros IG, Mastorakos G (2017) Interrelations among the adipocytokines leptin and adiponectin, oxidative stress and aseptic inflammation markers in pre- and early-pubertal normal-weight and obese boys. Endocrine 55:925–933
59. Deng Y, Chang S (2007) Role of telomeres and telomerase in genomic instability, senescence and cancer. Lab Investig 87:1071–1076
60. Rogakou EP, Pilch DR, Orr AH, Ivanova VS, Bonner WM (1998) DNA double stranded breaks induce histone H2AX phosphorylation on serine 139. J Biol Chem 273:5858–5868
61. Gordon-Dseagu VLZ, Shelton N, Mindell JS (2013) Epidemiological evidence of a relationship between Type-1 diabetes mellitus and cancer: a review of the existing literature. Int J Cancer 132:501–508
62. Longhese MP (2008) DNA damage response at functional and dysfunctional telomeres. Genes Dev 22:125–140
63. Kjaer TW, Faurholt-Jepsen D, Mehta KM, Christensen VB, Epel E, Lin J, Blackburn E, Wojcicki JM (2018) Shorter preschool, leukocyte telomere length is associated with obesity at age 9 in Latino children. Clin Obes 8:88–94
64. Brouilette SW, Moore JS, AD MM, Thompson JR, Ford I, Shepherd J, Packard CJ, Samani NJ, West of Scotland Coronary Prevention Study Group (2007) Telomere length, risk of coronary heart disease, and statin treatment in the west of Scotland primary prevention study: a nested case-control study. Lancet 369:107–114
65. Buxton JL, Walters RG, Visvikis-Siest S, Meyre D, Froguel P, Blakemore JAIF (2011) Childhood obesity is associated with shorter leukocyte telomere length. Clin Endocrinol Metab 96:1500–1505
66. ChenW GJP, Kimura M, Brimacombe M, Cao X, Srinivasan SR, Berenson GS, Aviv A (2009) Leukocyte telomere length is associated with HDL cholesterol levels: the Bogalusa Heart Study. Atherosclerosis 205:620–625
67. Walton RT, Mudway IS, Dundas I, Marlin N, Koh LC, Aitlhadj L, Vulliamy T, Jamaludin JB, Wood HE, Barratt BM, Beevers S, Dajnak D, Sheikh A, Kelly FJ, Griffiths CJ, Grigg J (2016) Air pollution, ethnicity and telomere length in east London schoolchildren: an observational study. Environ Int 96:41–47
68. Zalata A, Yahia S, El-Bakary A, Elsheikha HM (2007) Increased DNA damage in children caused by passive smoking as assessed by comet assay and oxidative stress. Mutat Res 629:140–147
69. Tice RR, Agurell E, Anderson D, Burlison B, Hartmann A, Kobayashi H, Miyamae Y, Rojas E, Ryu J-C, Sasaki YF (2000) Single cell gel/comet assay: guidelines for in vitro and an in vivo genetic toxicological testing. Environ Mol Mutagen 35:206–222

The Role of Redox Signalling in Cardiovascular Regeneration

2

Thomas Hansen, Soloman Saleh, Gemma A. Figtree, and Carmine Gentile

2.1 Introduction: Heart Failure, Redox Signalling, and the Need for Regenerative Therapies

Heart disease is a major cause of morbidity and mortality globally. The major causes of heart failure, including ischaemic, hypertensive, alcoholic and valvular heart failure, are most prevalent in ageing Western society, affecting over 10% of adults aged over 65 in Australia [67].

Ischaemic heart disease (IHD) is not only the most common cause of all heart failure, but further is the most common cause of death worldwide [5]. There is a steadily increasing number of people living in the world with IHD- a consequence of both increasing obesity and diabetes, as well as increased survival of patients after myocardial infarction (MI) through improved early reperfusion strategies (thrombolysis and primary percutaneous coronary intervention). However, many of these individuals surviving heart attack are left with significant morbidity related to adverse left ventricular remodelling and left ventricular systolic dysfunction and associated chronic heart failure, occurring in approximately one in five patients [15].

Despite improving medical therapy, chronic heart failure carries a 5-year 50% mortality rate [67]. Existing therapies include beta-adrenergic receptor antagonists, angiotensin converting enzyme inhibitors, and aldosterone antagonists. These provide symptomatic relief and modest mortality benefits through the prevention of left ventricular remodelling; however, they do not replace the infarcted myocardium, and therefore are unable to restore cardiac function to pre-infarct baseline [65].

T. Hansen · G. A. Figtree (✉) · C. Gentile
The Kolling Institute, Royal North Shore Hospital, St Leonards, NSW, Australia

Sydney Medical School, The University of Sydney, Sydney, NSW, Australia
e-mail: gemma.figtree@sydney.edu.au

S. Saleh
School of Medical Sciences, Faculty of Medicine, University of New South Wales, Sydney, NSW, Australia

S. Chakraborti et al. (eds.), *Oxidative Stress in Heart Diseases*,
https://doi.org/10.1007/978-981-13-8273-4_2

Further advances have been made in device therapy, including left ventricular assist devices, resynchronization therapy, and automated implantable cardiac defibrillators [34]. Heart transplantation, perhaps the most definitive therapy for those with severe, end-stage dilated cardiomyopathy, is not a therapeutic approach with high enough scalability to accommodate the spiralling incidence of heart failure, due to the significant shortage of donors and complex nature of transplantation candidates [36]. Cardiac regeneration therapy though, offers a promising, relatively new approach to ameliorating cardiac function through either the direct replacement of cardiomyocytes, or by inducing their proliferation via modulation of endogenous signalling pathways.

Redox signalling is an important regulator of numerous cardiac signalling pathways involved in cardiac regeneration. Basal levels of reactive oxygen species (ROS) including superoxide (O_2^-) and hydrogen peroxide (H_2O_2) are important in physiological cell signalling. However, their imbalance can lead to oxidative stress, and often deleterious effects on the cardiomyocyte and its repair processes [26]. Despite this, clinical trials using antioxidant therapy in CVD have been largely disappointing, likely due to an indiscriminate, non-targeted approach, without regulating redox signalling in specific subcellular compartments.

Elevated levels of ROS are known to mediate myocardial ischaemia reperfusion (IR) injury and altered metabolic energetics and apoptotic cardiomyocyte cell pathways that drive heart failure [29]. It is also understood that many of the pathways that are altered in the transition from young to ageing heart are redox-sensitive [10]. However, ROS are also believed to be cardio-protective at basal physiological levels [8]. ROS are tightly correlated with cell oxygenation in specific cell types and compartments, with low levels favouring adult cardiomyocyte growth, and higher levels providing a nurturing environment for physiological stem cell differentiation. There is therefore a sound mechanistic basis for the use of cellular and molecular pharmacotherapies to modulate ROS and redox signalling to ameliorate heart failure and degenerative processes in the ageing heart. However, clinical stem cell trials have been limited by an often hostile (and redox-sensitive) myocardial niche, resulting in lack of cellular engraftment and proliferation. Through achieving a more complete understanding of the complexities of redox signalling in the cardiomyocyte microenvironment, we may uncover better targets its therapeutic potential for treating heart disease, and improving the efficiency of cellular and molecular approaches to cardiac regeneration.

2.2 Cellular Approaches to Cardiac Repair & Their Redox-Regulation

Significant research has been undertaken into the endogenous regenerative capacity of the human heart. Embryonic cardiomyocytes are capable of proliferation- but significant controversy has abounded regarding whether this capacity is retained throughout adulthood. Studies using carbon dating technologies have demonstrated that whilst cardiomyocyte turnover persists in adults, the rate of renewal is low

(~1–2%) and declines with age [3]. There is a fourfold increase in cardiomyocyte division in peri-infarct zones [66]- however, myocardial infarction causes the loss of approximately one billion cardiomyocytes in the adult heart, overwhelming this nominal gain in regenerative capacity [41]. Therefore, much focus has revolved around the potential of exogenously replacing cardiomyocytes with stem cells, and enhancing the proliferation and differentiation of these stem cells once engrafted- processes regulated by ROS. The following sections will summarise the different cell types that have been trialled for cardiac repair, and their redox-relevance.

2.2.1 Skeletal Myoblasts & Bone Marrow-Derived Mesenchymal Stem Cells:

Cellular therapy for cardiac regeneration first began with administration of skeletal myoblasts injected intramyocardially during coronary artery bypass grafting (CABG) surgery [51]. These cells initially yielded positive safety data and demonstrated excellent engraftment. However, they elicited concerns of being pro-arrhythmic, with downregulated N-cadherin and connexin-43 suggesting a lack of electrophysiological coupling [9, 63], later confirmed in the phase II MAGIC study [52], likely secondary to the delivery of multiple intramyocardial injections. A similar approach soon followed with bone marrow-derived cells and considerable excitement abounded after release of the seminal study demonstrating that injection of discrete haematopoietic progenitor cells regenerated 68% of infarcted mouse myocardium [56]. However, multiple studies have failed to reproduce these findings. The variable results produced from trials using bone marrow-derived cells has perhaps been attributed to impairment of these cells associated with patient-specific risk factors [16]. The randomised phase III CardiAMP trial is aiming to overcome this by trans-endocardial delivery of patient derived bone marrow cells that have been 'pre-screened' for certain criteria feasibly associated with an increased chance of therapeutic benefit (Table 2.1) [72].

A recent study demonstrated that treatment of bone-marrow derived mesenchymal stem cells (MSC) with bone morphogenetic protein-2 (BMP-2) stimulates their differentiation into functional cardiomyocytes [80]. BMP-2 is known to induce differentiation of other precursor cell types through ROS-mediated signalling pathways [44], highlighting the possibility that the same ROS-driven pathways occur and may be targeted in mesenchymal stem cells.

2.2.2 Cardiac Stem Cells

Cardiac stem cells (CSCs) were first identified and in the heart based upon expression of c-kit. c-kit-positive CSCs are multipotent, self-renewing, and capable of forming cardiomyocytes, smooth muscle cells, and endothelial cells [78].

CSCs can be propagated over long-term culture and maintained in an undifferentiated, self-renewing state [40]. Preclinical studies have demonstrated c-kit^{+}

Table 2.1 Outcomes of some recent and ongoing clinical trials using stem cell therapy to treat cardiovascular disease

Study	Cell type and dose	Delivery	Patient cohort	Primary outcome
CardiAMP [72]	Autologous, bone marrow-derived cells (pre-screened for specific pre-set criteria)	Percutaneous Intramyocardial	Refractory, chronic myocardial ischaemia.	Recruiting: Primary Outcome- change in time on treadmill using the Modified Bruce Protocol from baseline
ESCORT [53]	Embryonic stem cell derived CD15+ Isl-1+ progenitors	Fibrin hydrogel delivered directly onto epicardium	Ischaemic heart disease undergoing CABG and/or MVR	Met primary safety outcome measures
Dream-HF [74]	Allogeneic mesenchymal stem cells (STRO3+)	Transendocardial	Ischaemic and nonischaemic heart failure	Ongoing: Time to nonfatal, recurrent decompensated heart failure
SCIPIO [4]	cKit+ CSC $1*10^{6}$ cells	Intracoronary injection	Ischaemic cardiomyopathy (LVEF <40%)	Improvement in LVEF, reduction in infarct size compared to baseline
MSC-HF [49]	77.5 * 10^6 MSC cells	Intramyocardial injection	Severe IHD (LVEF <45%)	Improvement in LVESV and LVEF No difference in NYHA, 6-min walk test
MAGIC [52]	4–8*10^6 myoblasts or placebo	Intramyocardial injection	Severe IHD (LVEF <35%), MI, indication for coronary usrgery	No significant change in EF High dose had decreased LVESV Higher rate of arrhythmias
REPEAT [71]	Autologous bone marrow-derived cells	Intramyocardial infusion	Ischaemic cardiomyopathy (within 3 months of MI) LVEF < 45%	Mortality at 2 years-Recruiting
CONCERT-HF [73]	Mesenchymal stem cells (1.5 * 10^8) and/or Cardiac stem cells (5 * 10^6)	LV injection	Ischaemic cardiomyopathy (LVEF < 40%)	Primary outcome is change in LVEF/strain/other indices of cardiac function per MRI. Also, mortality, hospitalisation for HF. Recruiting

CSC-induced cardiomyocyte regeneration post-infarction via neo-proliferation and insulin-like growth factor 1 (IGF-1) secretion inhibiting apoptosis [20, 30]. Concern has been raised regarding c-kit$^+$ CSCs having predilection for differentiating into vascular cells rather than cardimyocytes, and becoming senescent in culture [51].

The SCIPIO trial showed that intracoronary infusion of autologous c-kit-positive CSCs improved left ventricular systolic function by 8.2% at 4 months post-infarct, and decreased 12-month infarct size [4] – it should be noted that this trial is the subject of a Lancet expression of concern regarding its data integrity [69].

Oxidative stress is detrimental to cardiac stem cells, and has been specifically shown to induce telomeric shortening and progenitor cell death [24].

2.2.3 ESCs and iPSCs

Whilst the aforementioned studies use cells derived from adult tissue sources, embryonic stem cells (ESCs) are cells taken from 4 to 6 day old embryos that are pluripotent (i.e., they have the capacity to develop into any cell type, given the appropriate stimulus or growth factors) [81]. Research using ESCs has been enormously promising, and already shown efficacy in several preclinical trials [42]. The primary advantage of these cells lie in their ability to flexibly direct the specific cellular differentiation pathway, allowing implantation at the time of preferred cardiac commitment. However, there are ethical and safety concerns surrounding the use of ESCs, specifically with the risk of teratoma formation from residual undifferentiated cells [9].

The ESCORT trial is an ongoing study using human ESCs to generate cardiovascular progenitors, characterised by dual expression of insulin gene enhancer protein ISL1 and stage-specific embryonic antigen 1 (SSEA1/FUT4). These cells are transplanted using a fibrin-scaffold delivered onto a patient's epicardium at time of coronary artery bypass grafting, using pericardium as a flap to provide trophic factors [53]. Preliminary results have determined that primary safety outcomes were met, although the study was not powered to demonstrate efficacy.

Induced pluripotent stem cells (iPSCs) have been reprogrammed from adult somatic cells to adopt the functional capacity of pluripotency. They have also demonstrated efficacy and safety in preliminary preclinical studies, and are not as ethically polarising [83]. Redox-status is known to play an important role in these cell types. Mice with diabetes-induced cardiomyopathy injected with human iPSCs demonstrated improved cardiac function mediated by an increase in antioxidant level and alteration of adverse cardiac remodelling pathways [82]. It would therefore be reasonable to infer that reducing oxidative burden in the specific milieu of iPSC-cardiomyocyte administration may improve proliferation and retention. However, it is likely that it is not only the level of ROS, but also their subcellular location that influences effects on cellular survival processes- that is to say that *compartmentalised and balanced* ROS production play an essential role in guiding cellular differentiation and proliferation. Further investigation into the exact mechanisms and subcellular localisation are necessary to manipulate the redox state and thereby facilitate engraftment of exogenously delivered iPSCs and other stem cells.

2.3 Redox-Related Changes in the Ageing and Failing Heart

Cardiomyocyte function and proliferative capacity progressively decline in the ageing heart. Gradual accumulation of direct oxidative damage and redox-sensitive post-translational modifications contribute to the phenotypic cardiovascular changes of both normal ageing and cardiac failure. This comes as no surprise, given that age is the principal risk factor for CVD [39]. This leads to: (i) reduced contractile and diastolic function; (ii) hypertrophy and adverse remodelling; and (iii) increased apoptosis in differentiated/progenitor populations [62]. The following section will detail some of the major changes that occur in the ageing and failing heart (summarised in Fig. 2.1), and how they impact on cardiac regeneration.

2.3.1 Hypoxia and Redox State

In the highly oxygenated milieu of the myocardium, shielding CSCs from oxidative stress poses a significant biological challenge. Cardiac stem cells must reside in specialised microdomains – *cardiac niches* – in which oxygen tensions are tightly regulated, and function to direct stem cell fate towards either quiescence or lineage commitment [60]. In the young mouse heart, a balance of proliferative 'lineage-committed' normoxic niches has been observed alongside quiescent hypoxic niches, which act as a reserve to replenish depleted populations [64]. However, the

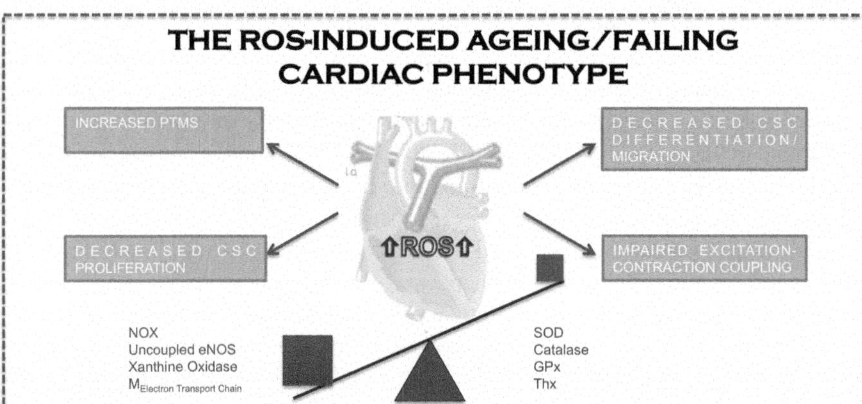

Fig. 2.1 Schematic representation of the changes involved in transition from healthy to ageing and/or failing heart. A primary increase in the ROS-generators that may occur in age and/or heart failure has a multitude of effects including an increase in post-translational modifications of key proteins including endothelial nitric oxide synthase (eNOS) and the sarcoplasmic-endoplasmic reticulum calcium ATPase (SERCA), and loss of proliferative and migratory potential of cardiac stem cells, and impaired cardiomyocyte excitation-contraction coupling. These collectively characterise the ageing/failing cardiac phenotype

proportion of niches maintained in a hypoxic state is now known to increase with age; these dormant niches force the remaining normoxic CSCs to become senescent through telomere erosion.

The control of stem cell lineage commitment is at least partially controlled by the 'metabolic switch' that determines either anaerobic glycolysis or oxidative phosphorylation [32]. B lymphoma Mo-MLV insertion region 1 (BMI1) is an important player in this metabolic switch [28]. BMI1 is a controller protein of cellular multipotency, mediating epigenetic modification of chromatin structure [76]. BMI1 expression prevents CSC commitment by down-regulating differentiation pathways, while H_2O_2-induced oxidative stress reverses this process [28]. This was mirrored in-vivo, where a heterozygous BMI1 knockout mouse model's ageing cardiac phenotype could be partially rescued by crossbreeding with transgenic G6DP$^+$ mice – known to reduce oxidative cardiac stress [55]. This phenotypic change was accompanied by a partial normalisation of DNA deactivation, suggesting the sensitivity of BMI1's commitment-repressive function to redox- and thus mitochondrial activity in the cell.

2.3.2 Neurohormonal Stimulation and Primary ROS Generators

The cardiovascular system has considerable neurohormonal control, which may become dysfunctional during ageing [10]. Angiotensin II (AngII) is well known for its role in the progression of cardiac failure, particularly in fibrosis and hypertrophy. In fact, inhibition of AngII is one of the few current pharmacotherapies shown to attenuate disease remodeling in heart failure and provide mortality benefit, at least partially through mediating a reduction in oxidative stress [6] Expression analysis of AngII and related proteins in mouse vasculature has revealed increasing AngII activity with ageing [85]. The study characterises AngII's communication with the redox system through the stimulation of Nox4 and subsequent generation of H_2O_2, as well as increasing O_2^- leading to eNOS uncoupling. This results in the activation of Transforming Growth Factor β (TGF-β), which triggers the transdifferentiation of cardiomyocytes to myofibroblasts leading to adverse remodelling [85].

Furthermore, a decrease in growth hormone (GH) activity with age contributes to cardiovascular disease in a redox-dependent manner [17]. Long-term replacement of GH in aged rats preserved diastolic function and reduced LV remodelling compared to controls. This appeared to be mediated in part by a reduction in both intracardiac and plasma AngII, and through a GH-mediated increase in insulin-like growth factor (IGF-1). On a broader clinical level, a below-median serum IGF-1 level doubled the risk of cardiac failure as compared to those above the median in the Framingham population [77]. Accordingly, mice with chronic hypopituitarism and thus low GH levels display reduced antioxidant capacity – namely superoxide dismutase and glutathione peroxidase [17]. This was accompanied by O_2^- and H_2O_2 increases in vascular and cardiac tissue, predominantly in the mitochondria.

2.3.3 Post-translational modifications

With the increased ROS of ageing and heart failure, there is an increasing propensity for post-translational modifications of key molecular proteins- one such modification is S-glutathionylation, characterised by the addition of a glutathione [19] adduct to proteins with reactive cysteine residues regulating cardiac ion conductivity, cardiac proliferative capacity and vascular function [7]. eNOS is a key membrane protein which is susceptible to glutathionylation and subsequent uncoupling in conditions of oxidative stress [12]. A recent elegant study using cardiac spheroids- a 3D co-culture model that biologically mimics the heart in vitro- demonstrated that doxorubicin likely exerts its cardiotoxic effects via eNOS uncoupling (secondary to ROS-mediated glutathionylation) [59]. This study also demonstrated that EC-derived nitric oxide protected against these ROS-driven changes, evidence that delivery of the correct supporting cells with contractile cardiac-committed cells would support their engraftment, by protecting against oxidative insults.

Beyond apoptosis and structural remodelling, the heart's contractile function decays with age on a cellular level [54]. Cardiac contractility is closely tied to redox state; unregulated and non-specific oxidative modifications compromise contractile function and contribute to cardiac failure, arrhythmias, and further oxidative dysfunction [38]. Cytosolic Ca^{2+} flux from the sarcoplasmic reticulum (SR) is a key determinant of contractility – conversely, diastole depends on the ability to clear cytosolic Ca^{2+} through re-sequestration in the SR or extracellular flux [61]. This is dictated in part by activity of the sarcoplasmic-endoplasmic reticulum calcium ATPase (SERCA), and the ryanodine receptor (RyR2) [38]. Both SERCA and RyR2 are subject to S-gluthathionylation in chronic cardiac disease [57]. The net effect is sarcoplasmic Ca^{2+} depletion and cytosolic overload. Reduced ability to clear cytosolic Ca^{2+} reduces diastolic filling, while depleted sarcomeric reserves compromise contractility.

2.4 Molecular Targets to Enhance Heart Regeneration

The emerging field of cardiac regenerative therapy involves the manipulation of molecular mechanisms to promote cellular renewal – often through a direct or indirect effect on oxidative signalling in cardiomyocytes and supporting cells. Here we discuss potential target pathways in cardiac regenerative therapy, with focus on points of regulatory control by the oxidative signalling system. While there is no singular redox-driven mechanism that determines stemness, several pathways have been characterised, with notable crosstalk and functional interdependence (Fig. 2.2).

Molecular therapies targeting cellular renewal offer a promising avenue towards addressing and reversing true disease progression beyond symptomatic relief. These approaches may stand alone, or serve as adjuncts for cell-based therapies. Here we discuss potential target pathways in cardiac regenerative therapy, with focus on points of regulatory control by the oxidative signalling system. Several redox-driven mechanisms directly relevant to stem cells and their 'stemness' are described, in addition to the capacity for future therapeutic targeting of these pathways.

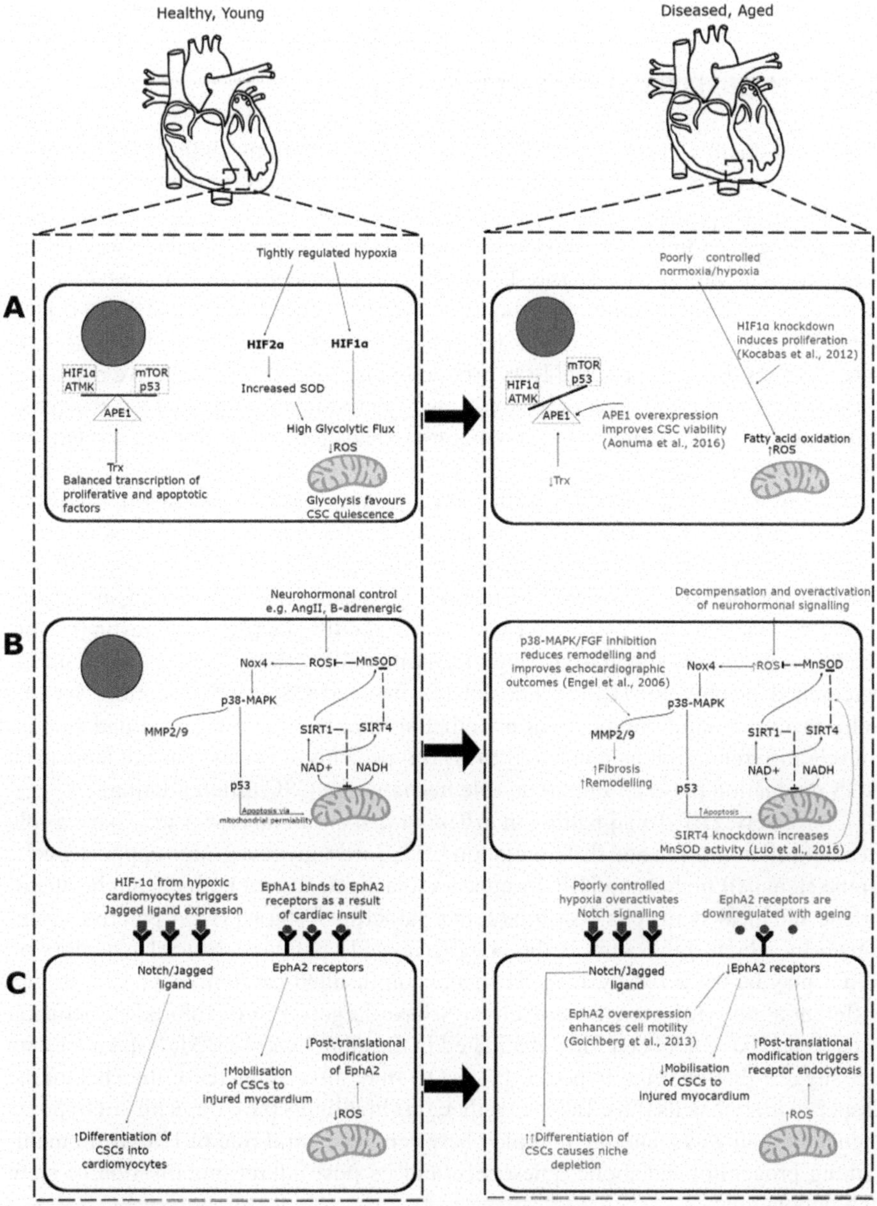

Fig. 2.2 Summary of molecular pathways influencing cardiac regeneration, and emerging therapeutic targets. (**a**) HIF and APE1 pathways control quiescence and apoptosis in CSCs; (**b**) SIRT and p38-MAPK pathways interact in a negative feedback loop on mitochondrial function, ROS, apoptosis, and remodelling in adult cardiomyocytes; (**c**) Eph and Notch1/Jagged pathways mediate migratory response of cardiac stem cells towards injured cardiomyocytes. Green inserts indicate promising therapies for improving regenerative capacity of the heart. *APE1* Apurinic/apyrimidinic endonuclease 1, *ATMK* Ataxia telangiectasia mutated kinase, *Eph* Ephrine, *HIF* Hypoxia Inducible Factor, *MMP* matrix metalloproteinase; *mTOR* Mechanistic target of rapamycin, *NAD⁺/NADH* Oxidised/reduced nicotinamide adenine dinucleotide, *NADHNox4* NADPH Oxidase 4, *ROS* Reactive oxygen species, *SIRT* Sirtuin, *SOD* Superoxide dismutase, *Trx* Thioredoxin

2.4.1 Hypoxia Inducible Factor 1-α

Hypoxia inducible factor-1α (HIF-1α) is an O_2-sensitive transcription factor responsive to tissue oxygenation and redox signalling; binding sites for molecular O_2 mark HIF-1α for ubiquitin proteasome degradation, and ROS have been shown to stabilise HIF-1α even under normoxic conditions [33]. HIF-1α upregulates key glycolytic enzymes and transporters – hexokinase, lactate dehydrogenase, and related proteins – whilst supressing mitochondrial activity; related isoform HIF-2α is activated similarly and induces redox defence mechanisms and antioxidant pathways [46]. As a result, mitochondrial ROS production is reduced, and oxidative resilience increased as the cell relies on glycolytic energy production. This exerts a protective effect against DNA damage and premature exhaustion and maintains a relatively quiescent state.

Accordingly, in the healthy adult mouse heart it has been shown that HIF-1α expressing CSCs had lower O_2 consumption as compared to cardiac fibroblasts, mirrored by increased viability in hypoxic or anoxic conditions [37]. Further, knockdown of HIF-1α in CSCs increased O_2 consumption while reducing glycolytic flux, and this was associated with a proliferative increase. Interestingly, this oxidative-metabolic signal is also significant in maintenance of the primitive undifferentiated phenotype; HIF-1α knockdown induced CSC differentiation, specifically favouring cardiomyocyte or endothelial lineages over fibroblast and smooth muscle differentiation as compared to wild type. This suggests that mitochondrial ROS production plays an important role in triggering CSC differentiation.

Conversely, HIF-1α signalling in *differentiated cardiomyocytes* may have a role in stem cell migration and differentiation [31]. Cardiomyocytes are supporting cells in the stem cell niche, and HIF-1α appears to activate/attract both distant haematopoetic and local stem cells post-infarct in a paracrine fashion [79]. A putative mechanism by which this occurs is the Notch1/Jagged1 pathway. Notch1 receptors on CSCs may be triggered by Jagged1 ligands on insulted cardiomyocytes to prompt differentiation. However, these pathways have largely been explored in neonatal models and have not been fully elucidated in the adult heart [25, 31, 80]. Other data yet suggest that extreme hypoxia (0.1% O_2) may directly increase the chemokine response in stem cells [68]. Taken together, this highlights the complexity of hypoxia signalling pathways, and the interplay between the central role of HIF-1α in maintaining progenitor metabolic quiescence and its post-infarct mobilisation of stem cell populations.

2.4.2 Apurinic/Apyrimidinic Endonuclease 1

CSC fate demonstrates a remarkable specificity to the nature, origin, and locale of a given oxidative stimulus [1]. Enzyme Apurinic/apyrimidinic Endonuclease (APE1) has a dual function of activating the DNA base excision repair pathway, and controlling an array of transcription factors implicated in both cardiac cellular survival and termination [35]. Notably, APE1 appears to be able to selectively induce transcription factors by acting as a molecular chaperone; in a stimulus-specific response, APE1 facilitates interaction between target factors and reducing agents such as thioredoxin or glutathione and subsequently maintains the factors reduced state. Targets of APE1 include maintainers of the primitive cardiac stem phenotype HIF-1α and Ataxia Telangiectasia Mutated Kinase, as well as the apoptosis-regulating factors p53 and the mTOR pathway of cell cycle control [11]. This suggests APE1 is of great interest as a therapeutic target for its role in controlling the balance between proliferative and apoptotic transcription.

In-vivo, the therapeutic impact of APE1 has been promising; murine APE1 overexpression has been shown to increase the viability of transplanted CSCs, while inhibiting apoptosis in surrounding untreated cardiomyocytes and reducing the host inflammatory response to the graft [1]. Overexpressed transplants showed significantly improved functional echo status over 4 weeks and reduced fibrotic change. This suggests that APE1 may have a role in both purely molecular therapies and as an adjuvant/preconditioning in cell-based approaches.

2.4.3 The Sirtuin Family

In the mitochondria, Nicotinamide Adenine Dinucleotide's (NAD^+) position in the electron transport chain makes the ratio of NAD^+ and NADH a functional indicator of oxidative phosphorylation and redox balance of the cell, and regulates several protective and regenerative mechanisms [48]. Signal transduction is largely through the post-translational and epigenetic processes deacetylation and mono-ADP-ribosylation, carried out by NAD^+-dependent enzymes. The Sirtuin enzyme family is known to catalyse both processes and have demonstrated a cardioprotective role – notably Sirtuin 1 (SIRT1) knockout induces in-vitro cardiomyopathy in a mouse model and conversely overexpression partially attenuates isoproterenol-induced cardiac hypertrophy [58]. Putative mechanisms are blocking pathological down-regulation of fatty acid oxidation, increased MnSOD activity and deacetylation of pro-apoptotic factor p53.

Conversely, SIRT4 has been shown to be a promising therapeutic target for inhibition. Notably, observed deacetylation activity is minimal by SIRT4, which primarily acts through inhibitory mono-ADP-ribosylation of glycolytic and fatty acid oxidation enzymes. Further, SIRT4 demonstrates a role in AngII-induced cardiac remodelling and hypertrophy, with overexpression exacerbating pathological phenotype and knockout attenuating AngII effect in in-vivo mice [43]. This may be explained by an effect on protective enzyme MnSOD – SIRT4 has been shown to inhibit the deacetylation-mediated increase in MnSOD activity by other members of the Sirtuin family.

2.4.4 Ephrine/EphA2

An important signalling pathway that guides stem cell commitment and migration is the family of Eph receptor tyrosine kinases and ephrin ligands [13], whereby ephrine A1 binds to EphA2, with resultant signal transduction promoting CSC mobilization at times of cardiac insult. In vivo and ex vivo studies have demonstrated that whilst levels of the EphA2 receptor protein remain constant as we age, elevated ROS lead to increased oxidative post-translational modifications of EphA2 in the ageing heart- this causes activation of Src-FK, EphA2 receptor endocytosis, and thus impaired trafficking of stem cells to sites of injury in the ageing heart [23]. Replacement of EphA2 in aged CSCs by over-expression studies corrected this pathway, restoring the young CSC phenotype, and capacity for CSC homeostasis and mobilisation, and is a promising molecular route to target to ameliorate or restore cardiac regenerative function.

2.4.5 p38-Mitogen Activated Kinase

The p38-Mitogen Activated Kinase (p38-MAPK) pathway is a redox-sensitive cascade upregulated in the context of accumulated oxidative stress as a means to remove damaged cells [84]. Specifically, Nox-4-derived O_2^- is known to target p38 in response to angiotensin II. The p38-MAPK then causes autophagy or apoptosis through induction of Bcl-2-associated X protein – a regulator protein directly triggering apoptosis through mitochondrial outer membrane permeabilization – via its activation of tumour suppressor p53 [2].

However prolonged activation of this pathway via mitochondrial overactivity, NOS uncoupling, ischaemia-reperfusion injury, or chronic disease may lead to adverse cardiac remodelling downstream of Nox-4. Following apoptotic cellular clearance, p38-MAPK induces matrix metalloproteinases MMP-2 and MMP-9 and thereby result in maladaptive remodelling and dilated cardiomyopathy [14, 27]. Experimentally, p38-MAPK inhibition has been observed to induce mitosis in infarcted rat cardiomyocytes *in vivo*, with a synergistic effect found with Fibroblast Growth Factor 1 inhibition [21]. Reduced pathological remodelling and fibrosis was also observed with p38-MAPK/FGF1 inhibition, alongside functional improvement as determined by echocardiographic LV fractional shortening at 1 and 14 days.

2.5 Future Directions & Emerging Therapies

2.5.1 Uncovering the "Secret-ome"

A recent study demonstrated that new proliferation of cardiomyocytes does occur in the adult heart, albeit at low levels- however, this cell division increases fourfold in peri-infarct zones, indicating that there are endogenous cardiac proliferative mechanisms that can be "switched on", given the right stimulation [66]. That stimulation may come in the acellular approach of the "secretome"- defined as the growth factors and chemo-attractant molecules produced by stem cell paracrine secretion [18].

Multiple studies have demonstrated that despite almost negligible levels of persistent cellular engraftment, there are *disproportionately increased and persistent* beneficial effects on cardiac function that can't reasonably be ascribed to engrafted cells [70]. This has led to a growing school of thought that it is the transient response to injected cells and its induced paracrine effects that are as important, if not more so than persistent cellular engraftment. It is therefore thought that in the future, identification of the exact cellular "secretome" inducing these paracrine effects will allow us to simply inject these signalling factors to achieve beneficial cardiac regenerative effects, thereby avoiding limitations of immune rejection, ethical concerns, teratoma formation, and issues of engraftment. Such an approach is currently the focus of the ongoing Ventrigel trial, which is testing the effects of endoventricular injections of a decellularised extracellular porcine matrix [75].

2.5.2 Tissue Engineering Approaches

The factors of paramount importance when considering cellular approaches to stem cell delivery include strategies to support the survival of delivered cells- i.e. promoting engraftment, giving co-supportive cells, and strategies to maximise electromechanical coupling. As alluded to earlier, clinical trials involving iPSCs have been successful in demonstrating improved cardiac function after injection into infarcted hearts [47]. However, a significant problem with iPSCs is their poor capacity for engraftment, with a significant loss in viability detected with time following injection attributed to the hostile microenvironment of infarcted myocardium lacking the necessary niche to support ongoing survival [80]. One strategy being developed to overcome this is the use of biomaterials to form synthetic scaffolds. These scaffolds, which are made of naturally derived materials including alginate, synthetic polymers and hydrogels, may help to recapitulate tissue structure, thereby either promoting cell retention once implanted or having therapeutic effects through the polymer's mechanical properties [45]. The biomaterials may also be used to fabricate 3D structures that are used as scaffolds for cardiac stem cells [22]. These 3D "bioprinted" material constructs assist with engraftment through acting as a physical scaffold to hold cells in place, and additionally provide co-supportive cells and optimise the electro-mechanical coupling with endogenous ventricular cardiomyocytes [50]. It has also been demonstrated that administration of these constructs are protected against oxidative stress by virtue of their three-dimensional structure, thereby improving engraftment and survival in the infarcted myocardium [24].

2.6 Conclusion

It is an exciting time in the world of cardiac regeneration therapy. Clinical trials using stem cells are all now supported by robust preclinical data, and have consistently demonstrated favourable safety profiles. Whilst the efficacy of clinical trials have yielded only marginally positive results, our new mechanistic insights in cardiac regenerative medicine is promising for the future.

From its historical beginnings 20 years ago in delivery of skeletal myoblasts, pioneering efforts in the cellular regeneration field have led us to an understanding that it is not only the cell type, but also cellular delivery strategies and the resulting engraftment that are of paramount importance. Nevertheless, further questions must be answered; what is the ideal cell type?; at what stage of cardiac-commitment should they be administered?; what delivery devices would allow optimal engraftment and functional integration to facilitate electromechanical coupling?; and how can we promote the endogenous self-regenerative properties of the heart with acellular secretome strategies?

At the core of all of these pathways is redox signalling. Of such pathophysiological significance in heart failure, the pre-clinical literature is strongly suggestive that strategies to promote cardiac regeneration, whether by cellular or molecular approaches, would benefit from altering redox state. It is of no doubt that as our understanding of reactive oxygen species builds, and we become cognisant of the redox balance that should be maintained in distinct subcellular compartments, so too will the effectiveness of antioxidant therapies to promote cardiomyocyte regeneration in the failing heart.

References

1. Aonuma T, Takehara N, Maruyama K, Kabara M, Matsuki M, Yamauchi A, Kawabe J, Hasebe N (2016) Apoptosis-resistant cardiac progenitor cells modified with apurinic/apyrimidinic endonuclease/redox factor 1 gene overexpression regulate cardiac repair after myocardial infarction. Stem Cells Transl Med 5:1067–1078
2. Arabacilar P, Marber M (2015) The case for inhibiting p38 mitogen-activated protein kinase in heart failure. Front Pharmacol 6:102
3. Bergmann O, Bhardwaj RD, Bernard S, Zdunek S, Barnabe-Heider F, Walsh S, Zupicich J, Alkass K, Buchholz BA, Druid H, Jovinge S, Frisen J (2009) Evidence for cardiomyocyte renewal in humans. Science 324:98–102
4. Bolli R, Chugh AR, D'Amario D, Loughran JH, Stoddard MF, Ikram S, Beache GM, Wagner SG, Leri A, Hosoda T, Sanada F, Elmore JB, Goichberg P, Cappetta D, Solankhi NK, Fahsah I, Rokosh DG, Slaughter MS, Kajstura J, Anversa P (2011) Cardiac stem cells in patients with ischaemic cardiomyopathy (SCIPIO): initial results of a randomised phase 1 trial. Lancet 378:1847–1857
5. Briceno N, Schuster A, Lumley M, Perera D (2016) Ischaemic cardiomyopathy: pathophysiology, assessment and the role of revascularisation. Heart 102:397–406
6. Brugts JJ, van Vark L, Akkerhuis M, Bertrand M, Fox K, Mourad JJ, Boersma E (2015) Impact of renin-angiotensin system inhibitors on mortality and major cardiovascular endpoints in hypertension: a number-needed-to-treat analysis. Int J Cardiol 181:425–429
7. Bubb KJ, Birgisdottir AB, Tang O, Hansen T, Figtree GA (2017) Redox modification of caveolar proteins in the cardiovascular system- role in cellular signalling and disease. Free Radic Biol Med 109:61–74
8. Cabrera-Fuentes HA, Aragones J, Bernhagen J, Boening A, Boisvert WA, Botker HE, Bulluck H, Cook S, Di Lisa F, Engel FB, Engelmann B, Ferrazzi F, Ferdinandy P, Fong A, Fleming I, Gnaiger E, Hernandez-Resendiz S, Kalkhoran SB, Kim MH, Lecour S, Liehn EA, Marber MS, Mayr M, Miura T, Ong SB, Peter K, Sedding D, Singh MK, Suleiman MS, Schnittler HJ,

Schulz R, Shim W, Tello D, Vogel CW, Walker M, Li QO, Yellon DM, Hausenloy DJ, Preissner KT (2016) From basic mechanisms to clinical applications in heart protection, new players in cardiovascular diseases and cardiac theranostics: meeting report from the third international symposium on "New frontiers in cardiovascular research". Basic Res Cardiol 111:69

9. Cambria E, Pasqualini FS, Wolint P, Gunter J, Steiger J, Bopp A, Hoerstrup SP, Emmert MY (2017) Translational cardiac stem cell therapy: advancing from first-generation to next-generation cell types. NPJ Regen Med 2:17
10. Cannata A, Camparini L, Sinagra G, Giacca M, Loffredo FS (2016) Pathways for salvage and protection of the heart under stress: novel routes for cardiac rejuvenation. Cardiovasc Res 111:142–153
11. Cesselli D, Aleksova A, Sponga S, Cervellin C, Di Loreto C, Tell G, Beltrami AP (2017) Cardiac cell senescence and redox signaling. Front Cardiovasc Med 4:38
12. Chen CA, Wang TY, Varadharaj S, Reyes LA, Hemann C, Talukder MA, Chen YR, Druhan LJ, Zweier JL (2010) S-glutathionylation uncouples eNOS and regulates its cellular and vascular function. Nature 468:1115–1118
13. Chen J, Song W, Amato K (2015) Eph receptor tyrosine kinases in cancer stem cells. Cytokine Growth Factor Rev 26:1–6
14. Cheng G, Wei L, Xiurong W, Xiangzhen L, Shiguang Z, Songbin F (2009) IL-17 stimulates migration of carotid artery vascular smooth muscle cells in an MMP-9 dependent manner via p38 MAPK and ERK1/2-dependent NF-kappaB and AP-1 activation. Cell Mol Neurobiol 29:1161–1168
15. Collins LR, Priest C, Caras I, Littman N, Kadyk L (2015) Proceedings: moving toward cell-based therapies for heart disease. Stem Cells Transl Med 4:863–867
16. Contreras A, Orozco AF, Resende M, Schutt RC, Traverse JH, Henry TD, Lai D, Cooke JP, Bolli R, Cohen ML, Moye L, Pepine CJ, Yang PC, Perin EC, Willerson JT, Taylor DA, Cardiovascular Cell Therapy Research N (2017) Identification of cardiovascular risk factors associated with bone marrow cell subsets in patients with STEMI: a biorepository evaluation from the CCTRN TIME and LateTIME clinical trials. Basic Res Cardiol 112:3
17. Csiszar A, Labinskyy N, Perez V, Recchia FA, Podlutsky A, Mukhopadhyay P, Losonczy G, Pacher P, Austad SN, Bartke A, Ungvari Z (2008) Endothelial function and vascular oxidative stress in long-lived GH/IGF-deficient Ames dwarf mice. Am J Physiol Heart Circ Physiol 295:H1882–H1894
18. Di Baldassarre A, Cimetta E, Bollini S, Gaggi G, Ghinassi B (2018) Human-induced pluripotent stem cell technology and cardiomyocyte generation: progress and clinical applications. Cells 7
19. Dumitrascu R, Kulcke C, Konigshoff M, Kouri F, Yang X, Morrell N, Ghofrani HA, Weissmann N, Reiter R, Seeger W, Grimminger F, Eickelberg O, Schermuly RT, Pullamsetti SS (2011) Terguride ameliorates monocrotaline-induced pulmonary hypertension in rats. Eur Respir J 37:1104–1118
20. Ellison GM, Torella D, Dellegrottaglie S, Perez-Martinez C, Perez de Prado A, Vicinanza C, Purushothaman S, Galuppo V, Iaconetti C, Waring CD, Smith A, Torella M, Cuellas Ramon C, Gonzalo-Orden JM, Agosti V, Indolfi C, Galinanes M, Fernandez-Vazquez F, Nadal-Ginard B (2011) Endogenous cardiac stem cell activation by insulin-like growth factor-1/hepatocyte growth factor intracoronary injection fosters survival and regeneration of the infarcted pig heart. J Am Coll Cardiol 58:977–986
21. Engel FB, Hsieh PC, Lee RT, Keating MT (2006) FGF1/p38 MAP kinase inhibitor therapy induces cardiomyocyte mitosis, reduces scarring, and rescues function after myocardial infarction. Proc Natl Acad Sci U S A 103:15546–15551
22. Gentile C (2016) Filling the gaps between the in vivo and in vitro microenvironment: engineering of spheroids for stem cell technology. Curr Stem Cell Res Ther 11:652–665
23. Goichberg P, Kannappan R, Cimini M, Bai Y, Sanada F, Sorrentino A, Signore S, Kajstura J, Rota M, Anversa P, Leri A (2013) Age-associated defects in EphA2 signaling impair the migration of human cardiac progenitor cells. Circulation 128:2211–2223

24. Goichberg P, Chang J, Liao R, Leri A (2014) Cardiac stem cells: biology and clinical applications. Antioxid Redox Signal 21:2002–2017
25. Guo J, Jie W, Shen Z, Li M, Lan Y, Kong Y, Guo S, Li T, Zheng S (2014) SCF increases cardiac stem cell migration through PI3K/AKT and MMP2/9 signaling. Int J Mol Med 34:112–118
26. Hafstad AD, Nabeebaccus AA, Shah AM (2013) Novel aspects of ROS signalling in heart failure. Basic Res Cardiol 108:359
27. Hao J, Du H, Li W, Liu F, Lu J, Yang X, Cui W (2016) Anthocyanins protected hearts against ischemic injury by reducing MMP-2 activity via Akt/P38 pathways. Am J Transl Res 8:1100–1107
28. Herrero D, Tome M, Canon S, Cruz FM, Carmona RM, Fuster E, Roche E, Bernad A (2018) Redox-dependent BMI1 activity drives in vivo adult cardiac progenitor cell differentiation. Cell Death Differ 25:807–820
29. Heusch G (2015) Mitochondria at the heart of cardiovascular protection: p66shc-friend or foe? Eur Heart J 36:469–471
30. Hong KU, Guo Y, Li QH, Cao P, Al-Maqtari T, Vajravelu BN, Du J, Book MJ, Zhu X, Nong Y, Bhatnagar A, Bolli R (2014) c-kit+ Cardiac stem cells alleviate post-myocardial infarction left ventricular dysfunction despite poor engraftment and negligible retention in the recipient heart. PLoS One 9:e96725
31. Hou J, Wang L, Long H, Wu H, Wu Q, Zhong T, Chen X, Zhou C, Guo T, Wang T (2017) Hypoxia preconditioning promotes cardiac stem cell survival and cardiogenic differentiation in vitro involving activation of the HIF-1alpha/apelin/APJ axis. Stem Cell Res Ther 8:215
32. Ito K, Suda T (2014) Metabolic requirements for the maintenance of self-renewing stem cells. Nat Rev Mol Cell Biol 15:243–256
33. Jung SN, Yang WK, Kim J, Kim HS, Kim EJ, Yun H, Park H, Kim SS, Choe W, Kang I, Ha J (2008) Reactive oxygen species stabilize hypoxia-inducible factor-1 alpha protein and stimulate transcriptional activity via AMP-activated protein kinase in DU145 human prostate cancer cells. Carcinogenesis 29:713–721
34. Katz JN, Waters SB, Hollis IB, Chang PP (2015) Advanced therapies for end-stage heart failure. Curr Cardiol Rev 11:63–72
35. Kelley MR, Georgiadis MM, Fishel ML (2012) APE1/Ref-1 role in redox signaling: translational applications of targeting the redox function of the DNA repair/redox protein APE1/Ref-1. Curr Mol Pharmacol 5:36–53
36. Kittleson MM (2018) Recent advances in heart transplantation. F1000Res:7
37. Kocabas F, Mahmoud AI, Sosic D, Porrello ER, Chen R, Garcia JA, DeBerardinis RJ, Sadek HA (2012) The hypoxic epicardial and subepicardial microenvironment. J Cardiovasc Transl Res 5:654–665
38. Kohler AC, Sag CM, Maier LS (2014) Reactive oxygen species and excitation-contraction coupling in the context of cardiac pathology. J Mol Cell Cardiol 73:92–102
39. Larson MG (1995) Assessment of cardiovascular risk factors in the elderly: the Framingham Heart Study. Stat Med 14:1745–1756
40. Leong YY, Ng WH, Ellison-Hughes GM, Tan JJ (2017) Cardiac stem cells for myocardial regeneration: they are not alone. Front Cardiovasc Med 4:47
41. Lin Z, Pu WT (2014) Strategies for cardiac regeneration and repair. Sci Transl Med 6:239rv231
42. Lundy SD, Gantz JA, Pagan CM, Filice D, Laflamme MA (2014) Pluripotent stem cell derived cardiomyocytes for cardiac repair. Curr Treat Options Cardiovasc Med 16:319
43. Luo YX, Tang X, An XZ, Xie XM, Chen XF, Zhao X, Hao DL, Chen HZ, Liu DP (2017) SIRT4 accelerates Ang II-induced pathological cardiac hypertrophy by inhibiting manganese superoxide dismutase activity. Eur Heart J 38:1389–1398
44. Mandal CC, Ganapathy S, Gorin Y, Mahadev K, Block K, Abboud HE, Harris SE, Ghosh-Choudhury G, Ghosh-Choudhury N (2011) Reactive oxygen species derived from Nox4 mediate BMP2 gene transcription and osteoblast differentiation. Biochem J 433:393–402
45. Mao AS, Mooney DJ (2015) Regenerative medicine: current therapies and future directions. Proc Natl Acad Sci U S A 112:14452–14459

46. Martin CM, Ferdous A, Gallardo T, Humphries C, Sadek H, Caprioli A, Garcia JA, Szweda LI, Garry MG, Garry DJ (2008) Hypoxia-inducible factor-2alpha transactivates Abcg2 and promotes cytoprotection in cardiac side population cells. Circ Res 102:1075–1081
47. Masumoto H, Ikuno T, Takeda M, Fukushima H, Marui A, Katayama S, Shimizu T, Ikeda T, Okano T, Sakata R, Yamashita JK (2014) Human iPS cell-engineered cardiac tissue sheets with cardiomyocytes and vascular cells for cardiac regeneration. Sci Rep 4:6716
48. Matasic DS, Brenner C, London B (2018) Emerging potential benefits of modulating NAD(+) metabolism in cardiovascular disease. Am J Physiol Heart Circ Physiol 314:H839–H852
49. Mathiasen AB, Qayyum AA, Jorgensen E, Helqvist S, Fischer-Nielsen A, Kofoed KF, Haack-Sorensen M, Ekblond A, Kastrup J (2015) Bone marrow-derived mesenchymal stromal cell treatment in patients with severe ischaemic heart failure: a randomized placebo-controlled trial (MSC-HF trial). Eur Heart J 36:1744–1753
50. Mawad D, Figtree G, Gentile C (2017) Current technologies based on the knowledge of the stem cells microenvironments. Adv Exp Med Biol 1041:245–262
51. Menasche P, Hagege AA, Scorsin M, Pouzet B, Desnos M, Duboc D, Schwartz K, Vilquin JT, Marolleau JP (2001) Myoblast transplantation for heart failure. Lancet 357:279–280
52. Menasche P, Alfieri O, Janssens S, McKenna W, Reichenspurner H, Trinquart L, Vilquin JT, Marolleau JP, Seymour B, Larghero J, Lake S, Chatellier G, Solomon S, Desnos M, Hagege AA (2008) The myoblast autologous grafting in ischemic cardiomyopathy (MAGIC) trial: first randomized placebo-controlled study of myoblast transplantation. Circulation 117:1189–1200
53. Menasche P, Vanneaux V, Hagege A, Bel A, Cholley B, Parouchev A, Cacciapuoti I, Al-Daccak R, Benhamouda N, Blons H, Agbulut O, Tosca L, Trouvin JH, Fabreguettes JR, Bellamy V, Charron D, Tartour E, Tachdjian G, Desnos M, Larghero J (2018) Transplantation of Human Embryonic Stem Cell-Derived Cardiovascular Progenitors for Severe Ischemic Left Ventricular Dysfunction. J Am Coll Cardiol 71:429–438
54. Munzel T, Gori T, Keaney JF Jr, Maack C, Daiber A (2015) Pathophysiological role of oxidative stress in systolic and diastolic heart failure and its therapeutic implications. Eur Heart J 36:2555–2564
55. Nobrega-Pereira S, Fernandez-Marcos PJ, Brioche T, Gomez-Cabrera MC, Salvador-Pascual A, Flores JM, Vina J, Serrano M (2016) G6PD protects from oxidative damage and improves healthspan in mice. Nat Commun 7:10894
56. Orlic D, Kajstura J, Chimenti S, Jakoniuk I, Anderson SM, Li B, Pickel J, McKay R, Nadal-Ginard B, Bodine DM, Leri A, Anversa P (2001) Bone marrow cells regenerate infarcted myocardium. Nature 410:701–705
57. Pastore A, Piemonte F (2013) Protein glutathionylation in cardiovascular diseases. Int J Mol Sci 14:20845–20876
58. Planavila A, Iglesias R, Giralt M, Villarroya F (2011) Sirt1 acts in association with PPARalpha to protect the heart from hypertrophy, metabolic dysregulation, and inflammation. Cardiovasc Res 90:276–284
59. Polonchuk L, Chabria M, Badi L, Hoflack JC, Figtree G, Davies MJ, Gentile C (2017) Cardiac spheroids as promising in vitro models to study the human heart microenvironment. Sci Rep 7:7005
60. Puente BN, Kimura W, Muralidhar SA, Moon J, Amatruda JF, Phelps KL, Grinsfelder D, Rothermel BA, Chen R, Garcia JA, Santos CX, Thet S, Mori E, Kinter MT, Rindler PM, Zacchigna S, Mukherjee S, Chen DJ, Mahmoud AI, Giacca M, Rabinovitch PS, Aroumougame A, Shah AM, Szweda LI, Sadek HA (2014) The oxygen-rich postnatal environment induces cardiomyocyte cell-cycle arrest through DNA damage response. Cell 157:565–579
61. Qin F, Siwik DA, Pimentel DR, Morgan RJ, Biolo A, Tu VH, Kang YJ, Cohen RA, Colucci WS (2014) Cytosolic H2O2 mediates hypertrophy, apoptosis, and decreased SERCA activity in mice with chronic hemodynamic overload. Am J Physiol Heart Circ Physiol 306:H1453–H1463
62. Redfield MM, Jacobsen SJ, Borlaug BA, Rodeheffer RJ, Kass DA (2005) Age- and gender-related ventricular-vascular stiffening: a community-based study. Circulation 112:2254–2262
63. Reinecke H, MacDonald GH, Hauschka SD, Murry CE (2000) Electromechanical coupling between skeletal and cardiac muscle. Implications for infarct repair. J Cell Biol 149:731–740

64. Sanada F, Kim J, Czarna A, Chan NY, Signore S, Ogorek B, Isobe K, Wybieralska E, Borghetti G, Pesapane A, Sorrentino A, Mangano E, Cappetta D, Mangiaracina C, Ricciardi M, Cimini M, Ifedigbo E, Perrella MA, Goichberg P, Choi AM, Kajstura J, Hosoda T, Rota M, Anversa P, Leri A (2014) c-Kit-positive cardiac stem cells nested in hypoxic niches are activated by stem cell factor reversing the aging myopathy. Circ Res 114:41–55
65. Sattler S, Fairchild P, Watt FM, Rosenthal N, Harding SE (2017) The adaptive immune response to cardiac injury-the true roadblock to effective regenerative therapies? NPJ Regen Med 2:19
66. Senyo SE, Steinhauser ML, Pizzimenti CL, Yang VK, Cai L, Wang M, Wu TD, Guerquin-Kern JL, Lechene CP, Lee RT (2013) Mammalian heart renewal by pre-existing cardiomyocytes. Nature 493:433–436
67. Stubbs SL, Crook JM, Morrison WA, Newcomb AE (2011) Toward clinical application of stem cells for cardiac regeneration. Heart Lung Circ 20:173–179
68. Tang YL, Zhu W, Cheng M, Chen L, Zhang J, Sun T, Kishore R, Phillips MI, Losordo DW, Qin G (2009) Hypoxic preconditioning enhances the benefit of cardiac progenitor cell therapy for treatment of myocardial infarction by inducing CXCR4 expression. Circ Res 104:1209–1216
69. The Lancet E (2014) Expression of concern: the SCIPIO trial. Lancet 383:1279
70. Tran C, Damaser MS (2015) Stem cells as drug delivery methods: application of stem cell secretome for regeneration. Adv Drug Deliv Rev 82–83:1–11
71. U.S. National Library of Medicine (2012) Compare the effects of single versus repeated intracoronary application of autologous bone marrow-derived mononuclear cells on mortality in patients with chronic post-infarction heart failure (REPEAT). Interventional (Clinical Trial)
72. U.S. National Library of Medicine (2018a) CardiAMP CMI randomized controlled pivotal trial in patients with chronic myocardial ischemia and refractory angina (CardiAMP CMI). Interventional 2018
73. U.S. National Library of Medicine (2018b) Combination of mesenchymal and C-kit+ cardiac stem cells as regenerative therapy for heart failure (CONCERT-HF). Interventional
74. U.S. National Library of Medicine (2018c) Efficacy and safety of allogeneic mesenchymal precursor cells (rexlemestrocel-L) for the treatment of heart failure. (DREAM HF-1). Interventional
75. U.S. National Library of Medicine (2018d) A study of VentriGel in post-MI patients
76. Valiente-Alandi I, Albo-Castellanos C, Herrero D, Arza E, Garcia-Gomez M, Segovia JC, Capecchi M, Bernad A (2015) Cardiac Bmi1(+) cells contribute to myocardial renewal in the murine adult heart. Stem Cell Res Ther 6:205
77. Vasan RS, Sullivan LM, D'Agostino RB, Roubenoff R, Harris T, Sawyer DB, Levy D, Wilson PW (2003) Serum insulin-like growth factor I and risk for heart failure in elderly individuals without a previous myocardial infarction: the Framingham Heart Study. Ann Intern Med 139:642–648
78. Vicinanza C, Aquila I, Scalise M, Cristiano F, Marino F, Cianflone E, Mancuso T, Marotta P, Sacco W, Lewis FC, Couch L, Shone V, Gritti G, Torella A, Smith AJ, Terracciano CM, Britti D, Veltri P, Indolfi C, Nadal-Ginard B, Ellison-Hughes GM, Torella D (2017) Adult cardiac stem cells are multipotent and robustly myogenic: c-kit expression is necessary but not sufficient for their identification. Cell Death Differ 24:2101–2116
79. Wang K, Ding R, Ha Y, Jia Y, Liao X, Wang S, Li R, Shen Z, Xiong H, Guo J, Jie W (2018a) Hypoxia-stressed cardiomyocytes promote early cardiac differentiation of cardiac stem cells through HIF-1alpha/Jagged1/Notch1 signaling. Acta Pharm Sin B 8:795–804
80. Wang YL, Zhang G, Wang HJ, Tan YZ, Wang XY (2018b) Preinduction with bone morphogenetic protein-2 enhances cardiomyogenic differentiation of c-kit(+) mesenchymal stem cells and repair of infarcted myocardium. Int J Cardiol 265:173–180
81. Yamanaka S, Li J, Kania G, Elliott S, Wersto RP, Van Eyk J, Wobus AM, Boheler KR (2008) Pluripotency of embryonic stem cells. Cell Tissue Res 331:5–22
82. Yan B, Singla DK (2013) Transplanted induced pluripotent stem cells mitigate oxidative stress and improve cardiac function through the Akt cell survival pathway in diabetic cardiomyopathy. Mol Pharm 10:3425–3432

83. Ye L, Chang YH, Xiong Q, Zhang P, Zhang L, Somasundaram P, Lepley M, Swingen C, Su L, Wendel JS, Guo J, Jang A, Rosenbush D, Greder L, Dutton JR, Zhang J, Kamp TJ, Kaufman DS, Ge Y, Zhang J (2014) Cardiac repair in a porcine model of acute myocardial infarction with human induced pluripotent stem cell-derived cardiovascular cells. Cell Stem Cell 15:750–761
84. Yokota T, Wang Y (2016) p38 MAP kinases in the heart. Gene 575:369–376
85. Yoon HE, Kim EN, Kim MY, Lim JH, Jang IA, Ban TH, Shin SJ, Park CW, Chang YS, Choi BS (2016) Age-associated changes in the vascular renin-angiotensin system in mice. Oxid Med Cell Longev 2016:6731093

Oxidative Stress-Driven Cardiotoicity of Cancer Drugs

3

Thalita Basso Scandolara, Bruno Ricardo Pires, Rodrigo Kern, Vanessa Jacob Victorino, and Carolina Panis

3.1 Introduction

Oxidative stress is one of the main events implicated in cell survival. Mitochondria are the main source of reactive species (ROS), and in pathologies in which its activity is enhanced, as cancer, the occurrence of redox imbalance is related with both cell survival and death [1, 2]. Mitochondria are also targeted by cytotoxic chemotherapy during patients' treatment, which may enhance the production of systemic oxidative stress.

The antineoplastic mechanism of action of most chemotherapeutic drugs is based on the generation of ROS [3–7]. Consequently, this process increases oxidative stress generation aiming to kill cancerous cells, but in contrast generates toxic mediators that also affect healthy cells by acting on diverse cellular molecules, such as DNA, lipid, and protein [6–10].

As consequence, cancer patients may experience various levels of cardiac toxicity and oxidative stress [7–11], which allow dividing cardiotoxicity into two main pathological events. The type I cardiotoxicity consists in irreversible and cumulative cardiac damage, and has been associated with the use of anthracyclines. The type II is associated with cell dysfunction, is not-cumulative and reversible, mainly related with trastuzumab-based treatment [12, 13]. Independently of the type of cardiotoxicity, oxidative stress has been implicated as one of the main cellular events related with cardiomyocyte damage [14–17]. In this context, the following topics present the main drugs associated with oxidative stress generation and cardiotoxicity development during cancer treatment.

T. B. Scandolara · B. R. Pires · R. Kern · V. J. Victorino · C. Panis (✉)
Laboratory of Tumor Biology, State University of West Paraná, Unioeste, Francisco Beltrão, Paraná, Brazil

S. Chakraborti et al. (eds.), *Oxidative Stress in Heart Diseases*,
https://doi.org/10.1007/978-981-13-8273-4_3

3.2 Physiopathology of Cardiac Damage in Cancer: Oxidative Stress at a Glance

Cancer is the second lead cause of death in industrialized countries, getting behind only from cardiac diseases. World Health Organization estimates that 17.7 million people die each year from cardiovascular diseases and 75% of deaths occur in low-income and middle-income countries [18]. Behind cardiovascular diseases, cancer was responsible for the death of 8.8 million people worldwide in 2015 [19].

In the last years, scientists have progressed in the studies whether tumor *per se* could be a risk factor that contribute to the development of cardiovascular diseases. Thus far, the mechanisms by which cancer and tumor microenvironment may be related to heart damage need to be better elucidated; still, it seems that the inflammatory status and production of reactive oxygen and nitrogen species (ROS/RNS) required for tumor progress may be a key component in this relationship.

The condition in which the same patient exhibit cardiac complications and cancer is usual, since both diseases is prevalent. Moreover, cancer patients can develop cardiac disease due to tumor *per se* or, most frequently discussed, to its therapy [20]. The importance of cardiotoxic side effects of anticancer therapies have increased and in order to better understand the implication of anticancer therapies in cardiac dysfunction and the mechanisms that are implicated, first it will be discussed whether tumor itself lead to cardiac alterations and damage, and how inflammation and oxidative stress status are influencing this process.

There is a lack of clinical data regarding cardiovascular diseases in patients diagnosed with cancer prior anticancer therapies [21]. Furthermore, clinical data regarding the involvement of oxidative stress on cancer- related cardiac dysfunction are scarce. To better understand the current knowledge regarding molecular mechanisms involved in cardiovascular complications in patients bearing tumor, it will be first discussed the state of the art of experimental models.

One of the most studied effect of cancer in individuals bearing advanced tumor is the development of cachexia. Experimental data shows that systemic impact of cancer is not only related to cachexia but to cardiac wasting as well, as loss of body mass reflects in organs and tissues. Significant cardiac atrophy is found in animals after tumor implantation concomitantly with an increased levels in pro-inflammatory cytokines as interleukine-6 (IL-6) and tumor necrosis factor-α (TNF-α) and decrease in cardiac function [22]. The structural and functional alterations in heart tissue caused by cancer itself are represented by a decrease in overall weight of the heart followed by a reduced left-ventricular ejection fraction and fraction shortening. A loss of left-ventricular end-diastolic diameter and reduction of stroke volume, cardiac output and fibrotic modelling are noticed in association to cachexia [23]. The combination of those events has a strong impact on patients' health, as it is responsible for an increase in mortality rate [23].

Beyond structural and functional impact of cancer-related cachexia on cardiac tissue, it also affects cardiomyocytes integrity. Studies have shown that the tumor growth declines contractile function in cardiomyocytes [24]. Moreover, cardiac insufficient performance in animal model of cancer- induced cachexia revealed an increased fibrosis between the myofibers and the altered composition of contractile proteins. Impaired mitochondrial integrity was found in heart muscles cells from animals bearing cancer- induced cachexia, which was characterized by irregular mitochondrial shape, size and dispersion, and disrupted mitochondrial structure [25]. Altogether, cancer status *per se* is capable of decrease heart function and structure. Here, several mechanisms seem to play a role in the development cardiac dysfunction in cachectic animals bearing tumor. Among them, one of the main pathways responsible for degrading muscles' proteins is the ubiquitin- proteasome via, which can be activated by oxidative stress. Ubiquitin- proteasome via mediates proteolysis contributing to cardiac muscle atrophy, which was shown in a study with cachectic animal inoculated with colon adenocarcinoma cells [26].

The literature shows clearly the development of cachexia as a consequence of advanced stages of cancer. Several factors may be involved in cancer- related cachexia and cardiac impairment. Pro-inflammatory cytokines such as IL-6, IL-1β and TNF-α are main contributors to heart failure [27]. In addition to inflammation, it is worth mentioning that redox regulation is also associated to cancer progression, where elevated levels of ROS are found in most types of tumors [28]. Therefore, combination of inflammatory status and oxidative stress revealed to be greater contributors to the development of cachexia. Oxidative stress is responsible for muscles loss both directly, by causing oxidative damage, or indirectly, by signaling degradation pathways as ubiquitin- proteasome system, as mentioned above [29]. ROS produced by cancer cells or tumor microenvironment act on cardiomyocytes lowering antioxidant defenses and causing cellular damage. Cardiac muscles cells of cancer-induced cachectic mice presented an increase in DNA oxidative damage characterized by up to 1.5 folder upper 8-hydroxy-2-deoxyguanosine. Here, the high levels of xanthine oxidase enzymatic pathway together with decreased activity of NADPH oxidase (NOX) and decreased superoxide dismutase (SOD) enzyme activity were main contributors as ROS source responsible for damaging cardiomyocytes [29].

The discussion about ROS role to cellular damage depends on its subcellular location, times, levels and appropriate duration. Therefore, it is well known that ROS can either contribute to cellular homeostasis or malignant transformation and death [30]. In cancer-induced cachexia, increased levels of ROS are able to react to cellular lipids yielding several metabolites that can regulate relevant cellular pathways, as proliferation, cell survival, invasion and metastasis to ensure cancer to advanced stages [31]. ROS attack to cellular lipids containing carbon-carbon double bounds is known as lipid peroxidation. Cell respond to lipid peroxidation depending on the circumstances above to promote cell survival or programmed death. Malondialdehyde (MDA) and hydroxynonenal (HNE) are low molecular weight aldehydes generated during lipid peroxidation process frequently used as oxidative stress markers. Between them, MDA is the most mutagenic product of lipid peroxidation, while HNE is the most cytotoxic [32]. Therefore, in cancer augmented levels of MDA and HNE can infer cellular damage while intermediate levels can mediate diverse cellular signaling pathways. It has been shown that cancer- related cachexia alters redox balance, as increased posttranslational modifications induced by ROS on proteins in cardiac muscles of cancer- induced cachectic rats, measured by MDA-protein adducts and HNE [33].

Experimental data shows that left side and right side of the heart may display distinct response to oxidative stress regarding damages to lipids, probably due to their distinguish metabolic activities. Higher levels of lipid hydroperoxides measured by chemiluminescence were detected on left side of rat's heart, while the right side of the heart showed a decreased lipid peroxide production after tumor implantation [34]. Therefore, it is important to state that heart's side may differ on adaptation response to cardiovascular diseases and response to oxidative stress.

The excessive reactive species may also promptly react with cellular protein resulting on the formation of carbonyl groups that can be implicated in the development of diseases, as cancer and cardiovascular diseases. Carbonylated protein content is commonly used as a marker of protein damage and they are increased in both right and left sides of heart from cancer- induced cachectic rat [34].

Taken together, available experimental data reveal the impact of ROS/ RNS in cardiac tissue of animals bearing tumor and how they are related to cardiomyocyte damage. In addition to increased ROS/RNS, an impaired antioxidant capacity caused by tumor itself can contribute to cardiac pathology. In heart muscles of cancer-induced cachectic rats it was found elevated levels of protein content of the antioxidant enzymes manganese-dependent superoxide dismutase (Mn-SOD) and catalase [33]. However, despite higher protein content of SOD, its activity is found to be 1.5- fold lower in heart of mice with cancer cachexia [29], which may result in an additional source of reactive species.

Clinical data regarding cancer- induced cardiac damage in patient prior treatment with anticancer therapies has been reported only in a few studies. The cardiovascular complications were evaluated in a research enrolling 555 patients diagnosed with cancer prior any treatment. Among them, breast cancer was prevalent within 146 patients (26.3%) followed by myeloproliferative neoplasia within 99 patients (17.8%). A subclinical functional impairment of cardiovascular system was configured in this study, as cancer patients displayed higher levels of circulating cardiovascular functional peptides, up to 100-fold upper reference limit and higher levels of cardiac morphological high- sensitive troponin T (hsTnT), a specific biomarker of cardiac injury. Furthermore, elevated cardiovascular peptides positively correlated with mortality independently of baseline characteristics of cancer patients [35].

There is a strong connection between the production of ROS and the inflammatory status. Tumor itself releases pro-inflammatory cytokines that can result into oxidative stress [27]. Inflammation is a hallmark of cancer and it mediates several processes in order to favor tumor progression and metastasis. Tumor itself release inflammatory cytokines which seems to be important for inducing a reactive oxygen species niche for favoring new mutations [31]. C- reactive protein (CRP) can predict cardiovascular mortality [36] and it is commonly used as a biomarker of acute inflammation, but it can be also associated to chronic inflammation [37].

The literature has been discussing the correlation of plasmatic CRP levels with future pathological cardiovascular events, as elevated levels of CRP over time may result in cardiovascular diseases such as atherosclerosis and chronic heart failure. Among the mechanisms by which CRP are associated to cardiovascular disease, it can be cited the induction of pro-inflammatory cytokines [37]. In fact, a followed-up study examining pro-inflammatory markers separately showed that CRP and the pro-inflammatory cytokine IL-6 were associated with all-cause and cancer-related mortality, with CRP also associated with cardiovascular mortality [36]. Moreover, IL-6 and CRP are both significantly increased in cancer patients with stage 4 disease compared to stage 1, 2 and 3. Besides, IL-6 and CRP correlated to circulating cardiovascular functional peptides in cancer patients prior to cardiotoxic therapy [35].

A highly significant increase in CRP levels in breast cancer patients (6.16 ± 1.17 IU/L) as compared to healthy controls (0.662 ± 0.189 IU/L) was evidenced in a study, indicating that even before chemotherapy, women bearing breast tumors are susceptible to cardiac complications [38]. Furthermore, augmented CRP can be observed in both early breast cancer stages (TNM I and II) as advanced breast cancer stages (TNM IIIc and IV) as compared to healthy women. Interestingly, besides increased CRP the data also showed that breast cancer patients disposal increased oxidative stress, which along with inflammation may be necessary for tumor to progress to advanced stages [39].

Besides CRP, creatine kinase-myocardial band (CKMB) is a well-known cardiac biomarker synthesized in carciomyocytes, which is widely used in the clinic routine due to its facility of detection, as it can be measured in biological fluid as serum and plasma. An increased in level of plasmatic CKMB was found in breast cancer patients as compared to healthy control individuals without cancer [38]. However, unlike CRP, augmented CKMB seems to be more evident in advanced breast cancer stages rather than early disease stage [39]. In fact, a study evaluating levels of serum creatine kinase (CK) and serum lactate dehydrogenase (LDH), biomarkers of cardiac injury, showed a non-significant enhancement of CK and LDH when comparing a group of breast cancer patients composed mainly by early stages and prior treatment to a healthy control group. In addition, despite the fact in this study the majority of breast cancer group was composed by early stages of disease, decreased activity of antioxidant capacity evaluated by catalase and glutathione reduced (GSH) in association with increased levels of nitric oxide (NO) and MDA were observed, indicating that those patients were under oxidative stress [40].

Altogether, the available literature indicates that biomarkers of cardiac damage in cancer patients prior treatment are elevated and results are more evident in cancer patients in advanced stages of disease. Data suggest that tumor *per se* features an inflammatory condition and oxidative stress that contributes to the increasing risk of developing cardiovascular diseases.

3.3 Chemotherapy-Induced Oxidative Stress and Cardiotoxicity

3.3.1 Doxorubicin

Therapy for cancer has progressed in recent years. However, severe treatment consequences have developed a new study front relying on comorbid illness of cancer survivors. Chemotherapy is a fundamental therapeutic approach for cancer treatment, although, many of the current chemotherapeutics have adverse side effects, including chronic cardiac disease and heart failure [41, 42]. Thus, understanding the effects of each drug on heart physiology would be important to avoid cardiotoxicity.

Doxorubicin (Dox) is considered one of the most effective antineoplastic drugs. It belongs to the family of Anthracyclines and since its approval by The Food and Drug Administration (FDA) in 1970s, it has been used to treat many hematological and solid tumors. Into the cancer cells, Dox forms complexes with DNA by intercalation between base pairs and inhibits DNA topoisomerase II activity, which ultimately led to blockage of DNA replication. But the main mechanism by which Dox induces cancer cell death is based on redox metabolism. Dox undergo to redox cycling (i.e. reduction and oxidation cycle between two forms of a molecule) catalyzed by the cytochrome P540 (CYP450) system, and the product of this reaction is the Dox Semiquinone radical. This radical is prone to cause oxidative damage in tumor cells by the release of iron from such cells [43]. Further, Dox-iron complex catalyzes oxygen and hydrogen peroxide into potent radicals, generating reactive oxygen species (ROS), which trigger important antitumoral responses in cancer cells [44, 45]. Thus, a balance between antitumoral and adverse effects of oxidative stress is necessary since the therapy should not be more harmful than the disease.

Despite Dox effectiveness, the major adverse effect observed in patients treated with Dox is severe cardiotoxicity (generally evidenced by ventricular dysfunction and clinical heart failure) provoked by increasing of oxidative stress in endomyocardial [46–48].

The cardiac tissue is highly susceptible to oxidation by Dox-induced ROS due to its intense oxidative phosphorylation activity and because the main antioxidant enzymes are present at lower levels in this than in other tissues [49, 50]. Dox is also able to induce a positive feedback loop of pro-inflammatory factors via TNF-α and IL-1β/NF-κB axis, which cooperates with ROS production [51]. In addition, ROS produced during Dox treatment stimulate lipid peroxidation, disruption of mitochondrial functions and damage the membranes of myocytes, which ultimately lead to cumulative and irreversible cardiotoxicity [44, 49].

Dox cardiotoxicity is cumulative, dose-related, and commonly has congestive heart failure and left ventricular dysfunction as main cardiac adverse events. Thus, some factors have to be considered to prevention of heart injury. The first described factors associated to risk for Dox-induced cardiotoxicity were advanced age and previous cardiovascular disease [52], however some studies reported that children are more vulnerable for its long-term effects [53–55]. Doses superior than 550 mg/m2 are described by increasing significantly the prevalence of cardiopathy [56, 57], however lower cumulative doses are reported by causing the same effects [58]. Interestingly, the schedule of administration is also a relevant factor, since continuous infusion than rapid intravenous infusion could reduce the cardiotoxicity. In combined chemotherapy, changing the sequence of administration can also reduce the cardiotoxicity, besides, complete cessation of alcohol and cocaine are also crucial for preservation of cardiac tissue [43].

Recent efforts have been taken to identify alternative methods able to reduce the adverse effect of Dox treatment. The liposomal delivery of Dox has been well-described as an important alternative to reduce cardiotoxicity [59, 60]. Another alternative are structural modifications to the Dox molecule (Epirubicin) or development of related drugs (Mitoxantron) to reduce their toxicity [61]. In addition, some compounds have been identified as cardioprotective agents and could be used since the beginning of treatment. In this context, anticoagulant agents, such as Dexrazoxane and Enoxaparin that were described by reducing the oxidative stress through decreasing oxidized iron levels, which make them promising for cardioprotection during Dox administration [53, 62–66]. Since Dox-induced cardiotoxicity is detected and the heart failure is instated, the standard treatment is angiotensin converting enzyme inhibitors and β-blockers. For severe cases, cardiac transplantation may be the last option [61]. Future research is needed to develop more effective compounds able to suppress selectively the redox mechanism on cardiac tissue. Therefore, methods for early detection and treatment of cardiotoxicity are crucial to reduce its occurrence.

Early identification of patients with risk of cardiotoxicity should be a starting point for cancer treatment, however this issue is considered just after detention of clinical myocardial damage. The most common noninvasive methods of monitoring cardiotoxicity are electrocardiography, Doppler echocardiography, rest and stress myocardial prefusion imaging, and left ventricular systolic function with radionuclide ventriculography [67, 68], although some authors consider them as low-sensitivity approaches [69]. An important advantage of these methods is that they can be conducted routinely together with therapeutics protocol to monitoring efficiently cardiopathies such as arrythmias, ischemic events and pericardial disease [61, 70].

Among the reported biomarkers, troponin I and T are useful in early detection of Dox cardiotoxicity. However, differential levels of troponins may not be identified in same cases [71–74]. Thus, although endomyocardial biopsy be invasive and high-costly, the sensitivity and specific found in this method still make it be the gold-standard. Hence, developing molecular biomarkers is essential for evaluation of early stages of cardiotoxicity, which would permit early interventions on treatment strategy.

In the route of personalized medicine, the risk of Dox-induced cardiotoxicity has to be calculated individually. Thus, data from each patient (e.g. age, previous cardiovascular disease, metabolic abnormalities, hypersensitivity to the drugs, previous radiation therapy, and genetic predisposition) and from each drug (e.g. cumulative dose, schedule, mechanism of delivery, route of administration, and combination) have to be considered before applying standardized protocols. Early detection and treatment of cardiotoxicity can significantly reduce the incidence of heart failure. In this context, Albini et al. [42] elegantly reviewed the importance of cardio-oncology for cancer treatment. In that manuscript, they propose a team flowchart to monitoring and avoiding cardiovascular complications during chemotherapy. These conducts would improve substantially the long-term survival and the life-quality of Dox-treated patients.

As Dox is widely used to treat solid and hematological tumors, the comprehension of the molecular mechanisms behind its action is fundamental to treat cancer patients. We currently know that the mechanism by which Dox induces myocardial injury is through the formation of free radicals, and that severe cardiotoxicity depends on the drugs' molecular site of action, their cumulative dose, and the cardiac condition of patients. In addition, we still have endomyocardial biopsy as the best way to monitor Dox-induced cardiotoxicity, even knowing that its invasive and high-costly nature limit its use. Thus, these findings should direct us to conduct strategic researches able to challenge and address these issues.

3.3.2 Trastuzumab

Trastuzumab (TRZ) is a monoclonal antibody (mAb) that targets against the extracellular domain of human epidermal growth factor receptor 2 (HER2), that belongs to the epidermal growth factor family of transmembrane receptors (ErbB family). This family consists of four transmembrane tyrosine kinase receptors: EGFR (HER1), ErbB2 (HER2), ErbB3 (HER3) and ErbB4 (HER4) [75, 76]. Although the other HER receptors are generally induce by ligands, the extracellular domain of HER2 can modify its conformation, dimerize and become activated even in the absence of a ligand, overexpression and/or mutation may induce this mechanism [77]. TRZ binds to the IV extracellular domain of HER2/ErbB2, preventing activation and inhibiting its signaling pathways, consequently disrupting differentiation, growth and survival of tumor cells [78]. Besides, cells treated with TRZ undergo arrest during G1 phase and might have a downregulation of HER2 [79].

The inclusion of TRZ therapy has become an invaluable tool since its approval and it is currently an universal and standard regiment in adjuvant treatment for HER2-overexpressive breast cancer, as well as in aggressive forms of HER2 positive gastric and gastroesophageal cancers [78, 80, 81]. However, TRZ is associated with an increased risk of cardiotoxicity, known as trastuzumab-induced cardiotoxicity (TIC) [82]. The mechanism that causes cardiotoxicity might be associated with the neuregulin-1/ErbB2 pathway, important for fetal heart development and known to be associated with signaling on adult heart, essential for survival of cardiomyocytes. ErbB2/ErbB4 promotes growth and cardioprotective signaling through activating phosphoinositide 3-kinase (PI3K), mitogen-activated protein kinase (MAPK) and focal adhesion kinase (FAK) survival pathways, reducing ROS production and inhibiting cardiomyocyte mitochondrial apoptosis [77, 83, 84]. When TRZ blocks HER2/ErbB2 survival pathway, it reduces tumor cell growth and survival but also severely harms cardiomyocytes by blocking cytoprotective mechanisms and enhancing cellular oxidative stress, inducing proapoptotic proteins such as Bcl-2 associated X protein (BAX) [83]. TIC appears to be dose/duration-independent and it is not associated with structural changes in cardiomyocytes, therefore, interrupting treatment reverse the symptoms in most cases [78, 85].

Some studies found that TRZ suppresses antioxidants enzymes and increases oxidative stress response by overproduction of ROS and higher expression of

TGF-β1 and IL-6, besides, erythrocyte death due to HER2 inhibition might be involved in this toxic environment that contribute to cardiac dysfunction [75, 86]. The patients under TRZ treatment are at risk for, symptomatic or not, left ventricular dysfunction, more specifically to left ventricular ejection fraction (LVEF) congestive heart failure (CHF) [75, 79, 82, 87]. However, cardiotoxicity is more common to occur in patients that already have risk for cardiac events, such as obesity, family history for cardiovascular disease, history of smoking and/or patients with clinical records for heart dysfunctions [85]. Prospective use of anthracyclines, another chemotherapy-drug and known for its expected and irreversible cardiotoxicity, *i.e* Doxurubicin (DOX), may lead to a higher risk of cardiac toxicity if TRZ treatment is added in the chemotherapeutic scheme [88]. The leading mechanism which DOX+TRZ mediates cardiac damage is increase of oxidative stress, and studies have proved that TRZ alone is not capable to cause cardiotoxicity, while DOX alone and combination therapy of DOX+TRZ lead to heart dysfunction, especially in LVEF [89].

Inhibition of HER2/ErbB2 by TRZ affects cells redox status and might trigger a pro-oxidant environment in the heart, corroborating to increasing of cell death by oxidative stress excessive damage, even though the systemic redox status is recovered [90]. It is suggested that myeloperoxidase, an enzyme secreted by leukocytes, which is pro-oxidant, is involved in the mechanism of ErbB2 inhibition by TRZ and could be a potential biomarker for oxidative stress and cardiotoxicity in breast cancer patients treated with this chemotherapy [91]. There is a new perspective that ErbB2 is not essential to maintain normal function of cardiomyocytes, but that it have a major role when these cells are under stress, independently of the stimulus, and so they might be important for protection against oxidative stress induced by chemotherapy-drugs as DOX and TRZ [92, 93].

Therefore, use of antioxidants have gained attention. Goyal and colaborators have found that DOX+TRZ therapy increased pro-apoptotic markers Bax and caspase 3, and treatment with N-Acetyl Cysteine Amide (NACA), an analog of N-Acetyl Cysteine (NAC), demonstrated that NACA is capable of decrease Bax and caspase 2 in mices treated with DOX+TRZ, offering a protective function [89]. A study *in vivo* with probucol, another antioxidant, attenuated cardiotoxicity induced by DOX+TRZ, while other found the same results but with melatonin (capable of scavenge oxygen-centered radicals and ROS [94]), mercaptoethylguanidine (a peroxynitrate scavenger) and 1400W (an iNOS inhibitor) [95].

3.3.3 Paclitaxel

Paclitaxel (PTX) is a taxane anticancer drug widely employed against solid tumors [96, 97]. Taxanes inhibits tumor proliferation through different pathways. Stabilization of microtubule [98], polymers by direct binding and effects propagated by redox-mediated reactions have been reported to be responsible for their anti-cancer activity [99].

PTX induces toxic effects on cancer cells in vitro through oxidative mechanisms, that is, the generation of oxidative stress by membrane-associated NADPH oxidase, giving rise to extracellular H_2O_2 [100, 101]. Moreover the evaluation of the total antioxidant capacity is interesting in PTX treatment because the depletion of intracellular antioxidants reduces the tumor resistance to (PTX) [102].

Patients under PTX treatment significantly increased plasma hydroperoxide levels when compared to breast cancer patients without chemotherapy [103]. Apart from, patients who undergo PTX treatment develop anemia immediately after chemotherapy infusion, suggesting oxidative stress as a probable causative mechanism [104]. Once PTX display an oxidative mechanism, it may be involved in the RBC hemolytic injury pathway [104, 105]

The mechanism whereby PTX affects mitochondria is not clear. Guigni et al. [106] hypothesize that chemotherapeutics such PTX may have cytotoxic/myotoxic effects through their ability to provoke mitochondrial dysfunction (reduced mitochondrial content and size, and increased expression and oxidation of peroxiredoxin) and oxidative stress. Another potential explanation for mitochondrial loss is the effect of PTX on microtubules, once microtubules serve as tracks to support kinesin-dependent mitochondrial movement and communication. Therefore, taxanes such as PTX may have detrimental effects duo to negative correlations of tubulin and detyrosinated tubulin to mitochondrial content and morphology.

Furthermore, Huang and collaborators [107] found evidences that indicates augmented expression of miR-4673 decreased cell viability and increased PTX-induced apoptosis. MiRNAs function as regulatory molecules in many physiological and pathological processes. In order to avoid the instauration of side effects due to oxidative stress, some authors suggests that PTX dosage must be limited [108].

In conclusion, undergoing PTX chemotherapy induce pro-oxidant status and consequently increased risk for several adverse effects.

3.3.4 5-Fluorouracil

5-Fluorouracil (5-FU) is in the antimetabolite and pyrimidine analog commonly used in the treatment of a variety of cancers, including those of the ovary, breast, gastrointestinal tract and head and neck [109, 110]. The 5-FU administered intravenously has a short half-life, however its active metabolites concentrate in cardiac and cancer cells, resulting in a prolonged exposure to the drug [111, 112]. A well-known first antitumor effect of 5-FU is blockage of DNA synthesis and cell proliferation through binding onto DNA and RNA [113]. Notwithstanding, potentiation of apoptosis is a secondary antitumor role of 5-FU.

5-FU induces apoptosis through enhanced generation of ROS [103]. Consequently, 5-FU induce oxidative stress in cardiomyocytes and endothelial cells [114]. Besides that, Xiao et al. [115] study suggested that 5-FU weakened the activity of antioxidant enzymes and the intracellular ROS content increased significantly.

Several reports about antineoplastic drugs report chemotherapy are accompanied by an oxidative balance perturbation [104, 116, 117]. Additionally, Rtibi et al. [118]

found that the 5-FU and capecitabine provoked drastic oxidative stress status in intestinal mucosa. These anticancer drugs caused depletion of the antioxidant enzymes activities such as superoxide dismutase (SOD), catalase (CAT) and glutathione peroxidase (GPx).

Furthermore, Bomfin et al. [119] suggests 5-FU causes oxidative stress and inflammation, mainly observed in the submandibular gland, which leads to periductal edema and cell death, resulting in alterations in the salivary flow rate and composition. Moreover, this study show that 5-FU modifies the saliva composition, decreasing lysozyme and SOD and increasing CAT. Differently of Campos et al. [119], which demonstrated that 5-FU injections caused a significant increase in CAT and peroxidase activities.

Still, Afrin et al. [120] demonstrated that the combination of 5-FU and Manuka honey (MH) enhanced the antiproliferative effects by inducing toxicity, up-regulating ROS and apoptotic markers. 5-FU combined with MH induced down-regulation of the transcription factor (NF-κB and Nrf2) and antioxidant enzyme activity (SOD, CAT, glutathione peroxidase and glutathione reductase), leading to more cell death by oxidative stress.

3.3.5 Etoposide

Etoposide (ETO) is a topoisomerase II inhibitor, commonly used alone or in combination with other anticancer agents to treat lung, ovarian, testicular, and various other cancers [121]. A topoactive drug, ETO inhibits the topoisomerase II–DNA cleavable complex, resulting in DNA damage, including double-stranded DNA breaks (DSBs), causing chromosomal aberration or apoptosis [122, 123].

In previous studies, it has been demonstrated that ETO increases the level of ROS in resting neutrophils [124]. Additionally, Shin et al. [125] demonstrated that ETO-induced cytotoxicity is executed through ROS generation. Moreover, Yadav et al. (2015) observed that both cellular and mitochondrial ROS were elevated on ETO treatments [126].

The reason for the strong cytotoxicity of ETO in hematopoietic precursors is the high activity of not only topoisomerase II, but also myeloperoxidase (MPO) [127]. MPO metabolizes ETO to highly toxic phenoxyl radicals. These radicals lead to a decrease in glutathione (GSH) and enhancement of oxidative stress [128]. Hence, Mahbub et al. [129] have demonstrated that polyphenols can synergistically enhance the action of the topoisomerase II inhibitors: doxorubicin and ETO in leukemia cells, for example.

ETO-induced ROS production may modulate mitochondrial function causing loss of mitochondrial membrane integrity and leakage of pro-apoptotic proteins cytochrome c, second mitochondria-derived activator of caspases, and AIF from mitochondria [130]. These pro-apoptotic proteins may initiate different apoptotic signaling causing demise of cancer cells. Furthermore, ETO induced a significant down-regulation of mRNA expression of the OGG1 repair gene and marked biochemical alterations characteristic of oxidative DNA stress, including enhanced lipid peroxidation and reduction in reduced glutathione [131].

3.3.6 Cyclophosphamide

Cyclophosphamide (CP) is an effective anti-cancer alkylating agent, while it also possesses a wide spectrum of cytotoxicity to normal cells [132]. CP metabolites such as phosphoramide mustard and acrolein can interact with DNA and induce the formation of DNA adducts that cause oxidative DNA damage [133, 134], protein, and lipid peroxidation [135, 136]. Regarding acrolein, one of CY metabolites, it initiates oxidative stress directly as well as be generated through lipid peroxidation, and thus contributes to a vicious circle [137]. The normal antioxidant system can be destroyed by active metabolites of CP, resulting in the accumulation of ROS [138]. Likewise, CP induces oxidative stress and cell death [139, 140].

Furthermore, studies have shown that CP mediates ROS generation [141] through multiple pathways including the activation of an intracellular signaling pathway important in regulating the cell cycle (e.g., PI3K/Akt/mTOR pathway), the NADPH complex (NADPH/NADP+), and the mitochondrial electron respiratory chain [142–144].

A decline in cellular level of glutathione has been considered to be indicative of oxidative stress mediated cellular damage produced by CP metabolites [145, 146]. Besides, Roy et al. [147] have also observed significant inhibition of superoxide dismutase, catalase and glutathione peroxidase activities in tumor bearing mice which might be due to the increase in the level of peroxides and reduction in the level of reduced glutathione [148].

3.4 Conclusions

Until now, virtually all chemotherapeutic drugs are capable to generate oxidative stress. Therefore, cancer patients will experience varying degrees of toxicity, and the cardiac is one of the most clinically relevant. Studies are helping to put together the players of oxidative-stress mediated toxicity, and will allow to developing strategies to minimize the undesirable results of chemotherapy without affect its antineoplastic effects.

References

1. Danial NN, Korsmeyer SJ (2004) Cell death: critical control points. Cell 116:205–219
2. Cory S, Adams JM (2002) The Bcl2 family: regulators of the cellular life-or-death switch. Nat Rev Cancer 2:647–656
3. Castaldo SA, Freitas JR, Conchinha NV, Madureira PA (2016) The tumorigenic roles of the cellular REDOX regulatory systems. Oxid Med Cell Longev:8413032
4. Okon IS, Zou M-H (2015) Mitochondrial ROS and cancer drug resistance: implications for therapy. Pharmacol Res 100:170–174
5. Kasapovic J, Pejic S, Todorovic A, Stojiljkovic V, Pajovic SB (2008) Antioxidant status and lipid peroxidation in the blood of breast cancer patients of different ages. Cell Biochem Funct 26:723–730

6. Deavall DG, Martin EA, Horner JM, Roberts R (2012) Drug-induced oxidative stress and toxicity. J Toxicol 2012
7. Gutteridge JMC, Halliwell B (2018) Mini-review: oxidative stress, redox stress or redox success? Biochem Biophys Res Commun 502:183–186. Academic Press
8. Varrichi G, Ameri P, Cadeddu C et al (2018) Antineoplastic drug-induced cardiotoxicity: a redox perspective. Front Physiol 9:167
9. Min K, Kwon O-S, Smuder AJ, Wiggs MP, Sollanek KJ, Christou DD et al (2015) Increased mitochondrial emission of reactive oxygen species and calpain activation are required for doxorubicin-induced cardiac and skeletal muscle myopathy. J Physiol 593:2017–2036
10. Albini A, Pennesi G, Donatelli F et al (2010) Cardiotoxicity of anticancer drugs: the need for cardio-oncology and cardio-oncological prevention. J Natl Cancer Inst 102(1):14–25
11. Schimmel KJ, Richel DJ, van den Brink RB, Guchelaar HJ (2004) Cardiotoxicity of cytotoxic drugs. Cancer Treat Rev 30(2):181–191
12. Florescu M, Cinteza M, Vinereanu D (2013) Chemotherapy-induced cardiotoxicity. Maedica (Buchar) 8(1):59–67
13. Mihalcea DJ, Florescu M, Vinereanu D (2017) Mechanisms and genetic susceptibility of chemotherapy-induced cardiotoxicity in patients with breast cancer. Am J Ther 24(1):e3–e11
14. Gorrini C, Harris IS, Mak TW (2013) Modulation of oxidative stress as an anticancer strategy. Nat Rev Drug Discov 12:931–947. Nature Publishing Group
15. Victorino VJ, Mencalha AL, Panis C (2015) Post-translational modifications disclose a dual role for redox stress in cardiovascular pathophysiology. Life Sci 129:42–47
16. Angsutararux P, Luanpitpong S, Issaragrisil S (2015) Chemotherapy-induced cardiotoxicity: overview of the roles of oxidative stress. Oxid Med Cell Longev 2015:795602
17. Rochette L, Guenancia C, Gudjoncik A, Hachet O, Zeller M, Cottin Y, Vergely C (2015) Anthracyclines/trastuzumab: new aspects of cardiotoxicity and molecular mechanisms. Trends Pharmacol Sci 36(6):326–348
18. WHO WHO. Cardiovascular Diseases [Internet]. (cited 2018, August 5). Available from: http://www.who.int/cardiovascular_diseases/en/
19. WHO WHO. Cancer [Internet]. (cited 2018, August 5]. Available from: http://www.who.int/cancer/en/
20. Yusuf SW, Razeghi P, Yeh ETH (2008) The diagnosis and management of cardiovascular disease in cancer patients. Curr Probl Cardiol 33:163–196
21. Murphy KT (2016) The pathogenesis and treatment of cardiac atrophy in cancer cachexia. Am J Physiol – Hear Circ Physiol [Internet] 310:H466–H477. Available from: http://ajpheart.physiology.org/lookup/doi/10.1152/ajpheart.00720.2015
22. Wysong A, Couch M, Shadfar S, Li L, Rodriguez JE, Asher S et al (2011) NF-κB inhibition protects against tumor-induced cardiac atrophy in vivo. Am J Pathol 178:1059–1068
23. Springer J, Tschirner A, Haghikia A, Von Haehling S, Lal H, Grzesiak A et al (2014) Prevention of liver cancer cachexia-induced cardiac wasting and heart failure. Eur Heart J 35:932–941
24. Xu H, Crawford D, Hutchinson KR, Youtz DJ, Lucchesi PA, Velten M et al (2011) Myocardial dysfunction in an animal model of cancer cachexia. Life Sci 88:406–410
25. Tian M, Nishijima Y, Asp ML, Stout MB, Reiser PJ, Belury MA (2010) Cardiac alterations in cancer-induced cachexia in mice. Int J Oncol 37:347–353
26. Tian M, Asp ML, Nishijima Y, Belury MA (2011) Evidence for cardiac atrophic remodeling in cancer-induced cachexia in mice. Int J Oncol 39:1321–1326
27. Belloum Y, Rannou-Bekono F, Favier FB (2017) Cancer-induced cardiac cachexia: pathogenesis and impact of physical activity (Review). Oncol Rep 37:2543–2552
28. Egea J, Fabregat I, Frapart YM, Ghezzi P, Görlach A, Kietzmann T et al (2017) European contribution to the study of ROS: a summary of the findings and prospects for the future from the COST action BM1203 (EU-ROS). Redox Biol 13:94–162
29. Hinch ECA, Sullivan-Gunn MJ, Vaughan VC, McGlynn MA, Lewandowski PA (2013) Disruption of pro-oxidant and antioxidant systems with elevated expression of the ubiqui-

tin proteosome system in the cachectic heart muscle of nude mice. J Cachexia Sarcopenia Muscle 4:287–293
30. Nathan C, Cunningham-Bussel A (2013) Beyond oxidative stress: an immunologist's guide to reactive oxygen species. Nat Rev Immunol 13:349–361
31. Mencalha A, Victorino VJ, Cecchini R, Panis C (2014) Mapping oxidative changes in breast cancer: understanding the basic to reach the clinics. Anticancer Res 34:1127–1140
32. Ayala A, Muñoz MF, Argüelles S (2014) Lipid peroxidation: production, metabolism, and signaling mechanisms of malondialdehyde and 4-hydroxy-2-nonenal. Oxid Med Cell Longev 2014
33. Marin-Corral J, Fontes CC, Pascual-Guardia S, Sanchez F, Olivan M, Argilés JM et al (2010) Redox balance and carbonylated proteins in limb and heart muscles of cachectic rats. Antioxid Redox Signal 12:365–380
34. Borges FH, Marinello PC, Cecchini AL, Blegniski FP, Guarnier FA, Cecchini R (2014) Oxidative and proteolytic profiles of the right and left heart in a model of cancer-induced cardiac cachexia. Pathophysiology 21:257–265
35. Pavo N, Raderer M, Hülsmann M, Neuhold S, Adlbrecht C, Strunk G et al (2015) Cardiovascular biomarkers in patients with cancer and their association with all-cause mortality. Heart 101:1874–1880
36. Singh-Manoux A, Shipley MJ, Bell JA, Canonico M, Elbaz A, Kivimaki M (2017) Association between inflammatory biomarkers and all-cause, cardiovascular and cancer-related mortality. CMAJ 189:E384–E390
37. Luan Y, Yao Y (2018) The clinical significance and potential role of c-reactive protein in chronic inflammatory and neurodegenerative diseases. Front Immunol 9:1–8
38. Panis C, Binato R, Correa S, Victorino VJ, Dias-Alves V, Herrera ACSA et al (2017) Short infusion of paclitaxel imbalances plasmatic lipid metabolism and correlates with cardiac markers of acute damage in patients with breast cancer. Cancer Chemother Pharmacol 80:469–478. Springer, Berlin/Heidelberg
39. Panis C, Victorino VJ, Herrera ACSA, Freitas LF, De Rossi T, Campos FC et al (2012) Differential oxidative status and immune characterization of the early and advanced stages of human breast cancer. Breast Cancer Res Treat 133:881–888
40. Amin KA, Mohamed BM, El-Wakil MAM, Ibrahem SO (2012) Impact of breast cancer and combination chemotherapy on oxidative stress, hepatic and cardiac markers. J Breast Cancer 15:306–312
41. Schultz PN, Beck ML, Stava C, Vassilopoulou-Sellin R (2003) Health profiles in 5836 long-term cancer survivors. Int J Cancer 104:488–495
42. Albini A, Pennesi G, Donatelli F, Cammarota R, De Flora S, Noonan DM (2010) Cardiotoxicity of anticancer drugs: the need for cardio-oncology and cardio-oncological prevention. J Natl Cancer Inst 102:14–25
43. Zhu H, Sarkar S, Scott L, Danelisen I, Trush MA, Jia Z et al (2016) Doxorubicin redox biology: redox cycling, topoisomerase inhibition, and oxidative stress. React Oxyg Species (Apex) 1:189–198
44. Minotti G, Menna P, Salvatorelli E, Cairo G, Gianni L (2004) Anthracyclines: molecular advances and pharmacologic developments in antitumor activity and cardiotoxicity. Pharmacol Rev 56:185–229
45. Pilco-Ferreto N, Calaf GM (2016) Influence of doxorubicin on apoptosis and oxidative stress in breast cancer cell lines. Int J Oncol 49:753–762
46. Singal PK, Iliskovic N (1998) Doxorubicin-induced cardiomyopathy. N Engl J Med 339:900–905
47. Bahadır A, Kurucu N, Kadıoğlu M, Yenilmez E (2014) The role of nitric oxide in doxorubicin-induced cardiotoxicity: experimental study. Turkish J Hematol 31:68–74
48. Paulides M, Kremers A, Stohr W, Bielack S, Jurgens H, Treuner J et al (2009) Prospective longitudinal evaluation of doxorubicin-induced cardiomyopathy in sarcoma patients: a report of the late effects surveillance system (LESS). Pediatr Blood Cancer 46:489–495

49. Zhou S, Palmeira CM, Wallace KB (2001) Doxorubicin-induced persistent oxidative stress to cardiac myocytes. Toxicol Lett 121:151–157
50. Sayed-Ahmed MM, Khattab MM, Gad MZ, Osman AMM (2001) Increased plasma endothelin-1 and cardiac nitric oxide during doxorubicin-induced cardiomyopathy. Pharmacol Toxicol 89:140–144
51. Guo RM, Xu WM, Lin JC, Mo LQ, Hua XX, Xi Chen P et al (2013) Activation of the p38 MAPK/NF-κB pathway contributes to doxorubicin-induced inflammation and cytotoxicity in H9c2 cardiac cells. Mol Med Rep 8:603–608
52. Von Hoff DD, Layard MW, Basa P, Davis HL Jr, Von Hoff AL, Rozencweig M et al (1979) Risk factors for doxorubicin-induced congestive heart failure. Ann Intern Med 91:710–717
53. Lipshultz SE, Rifai N, Dalton VM, Levy DE, Silverman LB, Lipsitz SR et al (2004) The effect of dexrazoxane on myocardial injury in doxorubicin-treated children with acute lymphoblastic leukemia. N Engl J Med 351:145–153
54. Lipshultz SE, Alvarez JA, Scully RE (2008) Anthracycline associated cardiotoxicity in survivors of childhood cancer. Heart 94:525–533
55. Trachtenberg BH, Landy DC, Franco VI, Henkel JM, Pearson EJ, Miller TL et al (2011) Anthracycline-associated cardiotoxicity in survivors of childhood cancer. Pediatr Cardiol 32:342–353
56. Myers CE (1988) Role of iron in anthracycline action. Organ Dir Toxicities Anticancer Drugs Dev Oncol:17–30
57. Jones LW, Haykowsky MJ, Swartz JJ, Douglas PS, Mackey JR (2007) Early breast cancer therapy and cardiovascular injury. J Am Coll Cardiol 50:1435–1441
58. Swain SM, Whaley FS, Ewer MS (2003) Congestive heart failure in patients treated with doxorubicin: a retrospective analysis of three trials. Cancer 97:2869–2879
59. Muggia FM (1997) Clinical efficacy and prospects for use of pegylated liposomal doxorubicin in the treatment of ovarian and breast cancers. Drugs 54:22–29
60. Iarussi D, Indolfi P, Casale F, Martino V, Di Tullio MT, Calabrò R (2005) Anthracycline-induced cardiotoxicity in children with cancer: strategies for prevention and management. Pediatr Drugs 7:67–76
61. Wu AH (2008) Cardiotoxic drugs: clinical monitoring and decision making. Heart 94:1503–1509
62. Seifert CF, Nesser ME, Thompson DF (1994) Dexrazoxane in the prevention of doxorubicin-induced cardiotoxicity. Ann Pharmacother 28:1063–1072
63. Swain BSM, Whaley FS, Gerber MC, Weisberg S, York M, Spicer D et al (1997) Cardioprotection with dexrazoxane for doxorubicin- containing therapy in advanced breast cancer. J Clin Oncol 15:1318–1332
64. Manduteanu I, Dragomir E, Voinea M, Capraru M, Simionescu M (2007) Enoxaparin reduces H2O2-induced activation of human endothelial cells by a mechanism involving cell adhesion molecules and nuclear transcription factors. Pharmacology 79:154–162
65. Young E (2008) The anti-inflammatory effects of heparin and related compounds. Thromb Res 122:743–752
66. Vejpongsa P, Yeh ETH (2014) Prevention of anthracycline-induced cardiotoxicity: challenges and opportunities. J Am Coll Cardiol 64:938–945
67. Haq MM, Legha SS, Choksi J, Hortobagyi GN, Benjamin RS, Ewer M et al (1985) Doxorubicin-induced congestive heart failure in adults. Cancer 56:1361–1365
68. Marchandise B, Schroeder E, Bosly A, Doyen C, Weynants P, Kremer R et al (1989) Early detection of doxorubicin cardiotoxicity: interest of Doppler echocardiographic analysis of left ventricular filling dynamics. Am Heart J 118:92–98
69. Dolci A, Dominici R, Cardinale D, Sandri MT, Panteghini M (2008) Biochemical markers for prediction of chemotherapy-induced cardiotoxicity systematic review of the literature and recommendations for use. Am J Clin Pathol 130:688–695

70. Yeh ETH, Tong AT, Lenihan DJ, Yusuf SW, Swafford J, Champion C et al (2004) Cardiovascular complications of cancer therapy: diagnosis, pathogenesis, and management. Circulation 109:3122–3131
71. Lipschultz SE, Rifai N, Sallan SE, Lipsitz SR, Dalton V, Sacks DB et al (1997) Predictive value of cardiac troponin T in pediatric patients at risk for myocardial injury. Circulation 96:2641–2648
72. Bryant J, Picot J, Baxter L, Levitt G, Sullivan I, Clegg A (2007) Use of cardiac markers to assess the toxic effects of anthracyclines given to children with cancer: a systematic review. Eur J Cancer 43:1959–1966
73. Reichlin T, Reichlin T, Hochholzer W, Hochholzer W, Bassetti S, Bassetti S et al (2009) Early Diagnosis of Myocardial Infarction with Sensitive Cardiac Troponin Assays. N Engl J Med 361:858–867
74. O'Brien PJ, Smith DEC, Knechtel TJ, MArchak MA, Pruimboom-Brees I, Brees DJ et al (2006) Cardiac troponin T is a sensitive, specific biomarker of cardiac injury in laboratory animals. Lab Anim Sci 40:153–171
75. Dirican A, Levent F, Alacacioglu A, Kucukzeybek Y, Varol U, Kocabas U et al (2014) Acute cardiotoxic effects of adjuvant trastuzumab treatment and its relation to oxidative stress. Angiology 65:944–949
76. Rochette L, Guenancia C, Gudjoncik A, Hachet O, Zeller M, Cottin Y et al (2015) Anthracyclines/trastuzumab: new aspects of cardiotoxicity and molecular mechanisms. Trends Pharmacol Sci 36:326–348
77. Hudis CA (2007) Trastuzumab — mechanism of action and use in clinical practice. N Engl J Med:39–51
78. Onitilo AA, Engel JM, Stankowski RV (2014) Cardiovascular toxicity associated with adjuvant trastuzumab therapy: prevalence, patient characteristics, and risk factors. Ther Adv Drug Saf 5:154–166
79. Gemmete JJ, Mukherji SK (2011) Trastuzumab (Herceptin). AJNR Am J Neuroradiol 32:1373–1374
80. Bonci A, Lupica CR, Morales M (2015) Trastuzumab interruption and treatment-induced carciotoxicity in early HER2-positive breast cancer. Breast Cancer Res Treat 149:489–495
81. Dokmanovic M, Wu WJ (2015) Monitoring trastuzumab resistance and cardiotoxicity: a tale of personalized medicine. 1st ed. Adv Clin Chem
82. Ayres LR, de Almeida Campos MS, de Oliveira Gozzo T, Martinez EZ, Ungari AQ, de Andrade JM et al (2015) Trastuzumab induced cardiotoxicity in HER2 positive breast cancer patients attended in a tertiary hospital. Int J Clin Pharm 37:365–372
83. Varga ZV, Ferdinandy P, Liaudet L, Pacher P (2015) Drug-induced mitochondrial dysfunction and cardiotoxicity. Am J Physiol – Hear Circ Physiol 309:H1453–H1467
84. Gorini S, De Angelis A, Berrino L, Malara N, Rosano G, Ferraro E (2018) Chemotherapeutic drugs and mitochondrial dysfunction: focus on doxorubicin, trastuzumab, and sunitinib. Oxid Med Cell Longev 2018:15
85. Sandoo A, Kitas GD, Carmichael AR (2015) Breast cancer therapy and cardiovascular risk: focus on trastuzumab. Vasc Health Risk Manag 11:223–228
86. Kabel AM, Elkhoely AA (2017) Targeting proinflammatory cytokines, oxidative stress, TGF-β1 and STAT-3 by rosuvastatin and ubiquinone to ameliorate trastuzumab cardiotoxicity. Biomed Pharmacother 93:17–26
87. Leung HW, Chan AL (2015) Trastuzumab-induced cardiotoxicity in elderly women with HER-2-positive breast cancer: a meta-analysis of real-world data. Expert Opin Drug Saf 14:1661–1671
88. Ürün Y, Utkan G, Yalcin B, Akbulut H, Onur H, Oztuna DG et al (2015) The role of cardiac biomarkers as predictors of trastuzumab cardiotoxicity in patients with breast cancer. Exp Oncol 37:53–57
89. Goyal V, Bews H, Cheung D, Premecz S, Mandal S, Shaikh B et al (2016) The cardioprotective role of N-acetyl cysteine amide in the prevention of doxorubicin and trastuzumab mediated cardiac dysfunction. Can J Cardiol

90. Lemos LGT, Victorino VJ, Herrera ACSA, Aranome AMF, Cecchini AL, Simão ANC et al (2015) Trastuzumab-based chemotherapy modulates systemic redox homeostasis in women with HER2-positive breast cancer. Int Immunopharmacol 27:8–14
91. Putt M, Hahn VS, Januzzi JL, Sawaya H, Sebag IA, Plana JC et al (2015) Longitudinal changes in multiple biomarkers are associated with cardiotoxicity in breast cancer patients treated with doxorubicin, taxanes, and trastuzumab. Clin Chem 61:1164–1172
92. Moilanen T, Jokimäki A, Tenhunen O, Koivunen JP (2018) Trastuzumab-induced cardiotoxicity and its risk factors in real-world setting of breast cancer patients. J Cancer Res Clin Oncol
93. Riccio G, Antonucci S, Coppola C, D'avino C, Piscopo G, Fiore D et al (2018) Ranolazine attenuates trastuzumab-induced heart dysfunction by modulating ROS production. Front Physiol 9:38
94. Reiter RJ, Mayo JC, Tan DX, Sainz RM, Alatorre-Jimenez M, Qin L (2016) Melatonin as an antioxidant: under promises but over delivers. J Pineal Res 61:253–278
95. Ozturk M, Ozler M, Kurt YG, Ozturk B, Uysal B, Ersoz N et al (2011) Efficacy of melatonin, mercaptoethylguanidine and 1400W in doxorubicin- and trastuzumab-induced cardiotoxicity. J Pineal Res 50:89–96
96. Victorino VJ, Panis C, Campos FC, Cayres RC, Colado-Simao AN, Oliveira SR et al (2013) Decreased oxidant profile and increased antioxidant capacity in naturally postmenopausal women. Age 35:1411–1421
97. Scripture CD, Figg WD, Sparreboom A (2005) Paclitaxel chemotherapy: from empiricism to a mechanism-based formulation strategy. Ther Clin Risk Manag 1:107–114
98. Jordan MA, Wilson L (2004) Microtubules as a target for anticancer drugs. Nat Rev Cancer 4:253–265
99. Victorino VJ, Pizzatti L, Michelletti P, Panis C (2014) Oxidative stress, redox signaling and cancer chemoresistance: putting together the pieces of the puzzle. Curr Med Chem 21:3211–3226
100. Hadzic T, Aykin-Burns N, Zhu Y, Coleman MC, Leick K, Jacobson GM et al (2010) Paclitaxel combined with inhibitors of glucose and hydroperoxide metabolism enhances breast cancer cell killing via H2O2-mediated oxidative stress. Free Radic Biol Med 48:1024–1033
101. Alexandre J, Hu Y, Lu W, Pelicano H, Huang P (2007) Novel action of paclitaxel against cancer cells: bystander effect mediated by reactive oxygen species. Cancer Res 67:3512–3517
102. Ramanathan B, Jan KY, Chen CH, Hour TC, Yu HJ, Pu YS (2005) Resistance to paclitaxel is proportional to cellular total antioxidant capacity. Cancer Res 65:8455–8460
103. Polk A, Vistisen K, Vaage-Nilsen M, Nielsen DL (2014) A systematic review of the pathophysiology of 5-fluorouracil-induced cardiotoxicity. BMC Pharmacol Toxicol 15:47
104. Panis C, Herrera AC, Victorino VJ, Campos FC, Freitas LF, De Rossi T et al (2012) Oxidative stress and hematological profiles of advanced breast cancer patients subjected to paclitaxel or doxorubicin chemotherapy. Breast Cancer Res Treat 133:89–97
105. Jezierska-Drutel A, Rosenzweig SA, Neumann CA (2013) Role of oxidative stress and the microenvironment in breast cancer development and progression. Adv Cancer Res 119:107–125
106. Guigni BA, Callahan DM, Tourville TW, Miller MS, Fiske B, Voigt T et al (2018) Skeletal muscle atrophy and dysfunction in breast cancer patients: role for chemotherapy-derived oxidant stress. Am J Physiol Cell Physiol
107. Huang HL, Shi YP, He HJ, Wang YH, Chen T, Yang LW et al (2017) MiR-4673 modulates paclitaxel-induced oxidative stress and loss of mitochondrial membrane potential by targeting 8-oxoguanine-DNA glycosylase-1. Cell Physiol Biochem 42:889–900
108. Granados-Principal S, Quiles JL, Ramirez-Tortosa CL, Sanchez-Rovira P, Ramirez-Tortosa MC (2010) New advances in molecular mechanisms and the prevention of adriamycin toxicity by antioxidant nutrients. Food Chem Toxicol 48:1425–1438
109. Goncalves A, Pierga JY, Ferrero JM, Mouret-Reynier MA, Bachelot T, Delva R et al (2015) UNICANCER-PEGASE 07 study: a randomized phase III trial evaluating postopera-

tive docetaxel-5FU regimen after neoadjuvant dose-intense chemotherapy for treatment of inflammatory breast cancer. Ann Oncol 26:1692–1697
110. Kim R, Hahn S, Shin J, Ock CY, Kim M, Keam B et al (2016) The effect of induction chemotherapy using docetaxel, cisplatin, and fluorouracil on survival in locally advanced head and neck squamous cell carcinoma: a meta-analysis. Cancer Res Treat 48:907–916
111. Miura K, Kinouchi M, Ishida K, Fujibuchi W, Naitoh T, Ogawa H et al (2010) 5-fu metabolism in cancer and orally-administrable 5-fu drugs. Cancers (Basel) 2:1717–1730
112. Lestuzzi C, Tartuferi L, Corona G (2011) Capecitabine (and 5 fluorouracil) cardiotoxicity. Metabolic considerations. Breast J 17:564–567
113. Wyatt MD, Wilson 3rd DM. Participation of DNA repair in the response to 5-fluorouracil. Cell Mol Life Sci 2009;66:788–799.
114. Lamberti M, Porto S, Zappavigna S, Addeo E, Marra M, Miraglia N et al (2014) A mechanistic study on the cardiotoxicity of 5-fluorouracil in vitro and clinical and occupational perspectives. Toxicol Lett 227:151–156
115. Xiao H, Xiong L, Song X, Jin P, Chen L, Chen X et al (2017) Angelica sinensis polysaccharides ameliorate stress-induced premature senescence of hematopoietic cell via protecting bone marrow stromal cells from oxidative injuries caused by 5-fluorouracil. Int J Mol Sci 18
116. Hess JA, Khasawneh MK (2015) Cancer metabolism and oxidative stress: insights into carcinogenesis and chemotherapy via the non-dihydrofolate reductase effects of methotrexate. BBA Clin 3:152–161
117. Sener MT, Sener E, Tok A, Polat B, Cinar I, Polat H et al (2012) Biochemical and histologic study of lethal cisplatin nephrotoxicity prevention by mirtazapine. Pharmacol Rep 64:594–602
118. Rtibi K, Selmi S, Grami D, Amri M, Sebai H, Marzouki L (2018) Contribution of oxidative stress in acute intestinal mucositis induced by 5 fluorouracil (5-FU) and its pro-drug capecitabine in rats. Toxicol Mech Methods 28:262–267
119. Bomfin LE, Braga CM, Oliveira TA, Martins CS, Foschetti DA, Santos AAQA et al (2017) 5-Fluorouracil induces inflammation and oxidative stress in the major salivary glands affecting salivary flow and saliva composition. Biochem Pharmacol 145:34–45
120. Afrin S, Giampieri F, Forbes-Hernandez TY, Gasparrini M, Amici A, Cianciosi D et al (2018) Manuka honey synergistically enhances the chemopreventive effect of 5-fluorouracil on human colon cancer cells by inducing oxidative stress and apoptosis, altering metabolic phenotypes and suppressing metastasis ability. Free Radic Biol Med 126:41–54
121. Bywater MJ, Pearson RB, McArthur GA, Hannan RD (2013) Dysregulation of the basal RNA polymerase transcription apparatus in cancer. Nat Rev Cancer 13:299–314
122. Pommier Y, Leo E, Zhang H, Marchand C (2010) DNA topoisomerases and their poisoning by anticancer and antibacterial drugs. Chem Biol 17:421–433
123. Saleh EM (2015) Inhibition of topoisomerase IIalpha sensitizes FaDu cells to ionizing radiation by diminishing DNA repair. Tumour Biol 36:8985–8992
124. Kapiszewska M, Cierniak A, Elas M, Lankoff A (2007) Lifespan of etoposide-treated human neutrophils is affected by antioxidant ability of quercetin. Toxicol Vitr 21:1020–1030
125. Shin HJ, Kwon HK, Lee JH, Anwar MA, Choi S (2016) Etoposide induced cytotoxicity mediated by ROS and ERK in human kidney proximal tubule cells. Sci Rep 6:34064
126. Yadav N, Kumar S, Marlowe T, Chaudhary AK, Kumar R, Wang J et al (2015) Oxidative phosphorylation-dependent regulation of cancer cell apoptosis in response to anticancer agents. Cell Death Dis 6:e1969
127. Haim N, Nemec J, Roman J, Sinha BK (1987) Peroxidase-catalyzed metabolism of etoposide (VP-16-213) and covalent binding of reactive intermediates to cellular macromolecules. Cancer Res 47:5835–5840
128. Kagan VE, Yalowich JC, Borisenko GG, Tyurina YY, Tyurin VA, Thampatty P et al (1999) Mechanism-based chemopreventive strategies against etoposide-induced acute myeloid leukemia: free radical/antioxidant approach. Mol Pharmacol 56:494–506
129. Mahbub AA, Le Maitre CL, Haywood-Small SL, Cross NA, Jordan-Mahy N (2015) Glutathione is key to the synergistic enhancement of doxorubicin and etoposide by polyphenols in leukaemia cell lines. Cell Death Dis 6:e2028

130. Yadav N, Chandra D (2014) Mitochondrial and postmitochondrial survival signaling in cancer. Mitochondrion 16:18–25
131. Attia SM, Ahmad SF, Harisa GI, Mansour AM, El Sayed el SM, Bakheet SA (2013) Wogonin attenuates etoposide-induced oxidative DNA damage and apoptosis via suppression of oxidative DNA stress and modulation of OGG1 expression. Food Chem Toxicol 59:724–730
132. Kim DJ, Kim EJ, Lee TY, Won JN, Sung MH, Poo H (2013) Combination of poly-gamma-glutamate and cyclophosphamide enhanced antitumor efficacy against tumor growth and metastasis in a murine melanoma model. J Microbiol Biotechnol 23:1339–1346
133. Sekeroglu V, Aydin B, Sekeroglu ZA (2011) Viscum album L. extract and quercetin reduce cyclophosphamide-induced cardiotoxicity, urotoxicity and genotoxicity in mice. Asian Pac J Cancer Prev 12:2925–2931
134. Wang LF, Gong X, Le GW, Shi YH (2008) Dietary nucleotides protect thymocyte DNA from damage induced by cyclophosphamide in mice. J Anim Physiol Anim Nutr 92:211–218
135. Esterbauer H, Schaur RJ, Zollner H (1991) Chemistry and biochemistry of 4-hydroxynonenal, malonaldehyde and related aldehydes. Free Radic Biol Med 11:81–128
136. Rezvanfar M, Sadrkhanlou R, Ahmadi A, Shojaei-Sadee H, Rezvanfar M, Mohammadirad A et al (2008) Protection of cyclophosphamide-induced toxicity in reproductive tract histology, sperm characteristics, and DNA damage by an herbal source; evidence for role of free-radical toxic stress. Hum Exp Toxicol 27:901–910
137. Shokrzadeh M, Ahmadi A, Naghshvar F, Chabra A, Jafarinejhad M (2014) Prophylactic efficacy of melatonin on cyclophosphamide-induced liver toxicity in mice. Biomed Res Int 2014:470425
138. Sengul E, Gelen V, Gedikli S, Ozkanlar S, Gur C, Celebi F et al (2017) The protective effect of quercetin on cyclophosphamide-Induced lung toxicity in rats. Biomed Pharmacother 92:303–307
139. Rehman MU, Tahir M, Ali F, Qamar W, Lateef A, Khan R et al (2012) Cyclophosphamide-induced nephrotoxicity, genotoxicity, and damage in kidney genomic DNA of Swiss albino mice: the protective effect of Ellagic acid. Mol Cell Biochem 365:119–127
140. Korkmaz A, Topal T, Oter S (2007) Pathophysiological aspects of cyclophosphamide and ifosfamide induced hemorrhagic cystitis; implication of reactive oxygen and nitrogen species as well as PARP activation. Cell Biol Toxicol 23:303–312
141. Chakraborty P, Roy SS, Basu A, Bhattacharya S (2016) Sensitization of cancer cells to cyclophosphamide therapy by an organoselenium compound through ROS-mediated apoptosis. Biomed Pharmacother 84:1992–1999
142. Chen XY, Xia HX, Guan HY, Li B, Zhang W (2016) Follicle loss and apoptosis in cyclophosphamide-treated mice: what's the matter? Int J Mol Sci 17
143. Silva A, Girio A, Cebola I, Santos CI, Antunes F, Barata JT (2011) Intracellular reactive oxygen species are essential for PI3K/Akt/mTOR-dependent IL-7-mediated viability of T-cell acute lymphoblastic leukemia cells. Leukemia 25:960–967
144. Park KR, Nam D, Yun HM, Lee SG, Jang HJ, Sethi G et al (2011) beta-Caryophyllene oxide inhibits growth and induces apoptosis through the suppression of PI3K/AKT/mTOR/S6K1 pathways and ROS-mediated MAPKs activation. Cancer Lett 312:178–188
145. Jonas CR, Puckett AB, Jones DP, Griffith DP, Szeszycki EE, Bergman GF et al (2000) Plasma antioxidant status after high-dose chemotherapy: a randomized trial of parenteral nutrition in bone marrow transplantation patients. Am J Clin Nutr 72:181–189
146. Patel JM, Block ER (1985) Cyclophosphamide-induced depression of the antioxidant defense mechanisms of the lung. Exp Lung Res 8:153–165
147. Roy SS, Chakraborty P, Biswas J, Bhattacharya S (2014) 2-[5-Selenocyanato-pentyl]-6-amino-benzo[de]isoquinoline-1,3-dione inhibits angiogenesis, induces p53 dependent mitochondrial apoptosis and enhances therapeutic efficacy of cyclophosphamide. Biochimie 105:137–148
148. Adams Jr. JD, Klaidman LK. Acrolein-induced oxygen radical formation. Free Radic Biol Med 1993;15:187–193.

Molecular Targets and Novel Therapeutics to Target Oxidative Stress in Cardiovascular Diseases

4

Veena Dhawan, Chetan Bakshi, and Riyaz Ahmad Rather

4.1 Introduction

Globally non-communicable diseases (NCDs) are by far the leading cause of death and disability [1, 2]. World Health Organization (WHO) states that cardiovascular diseases (CVDs) account for 17.7 million deaths globally, among NCDs [3]. CVD burden is significantly increasing worldwide due to an ageing population and spread of the Western diet and lifestyle.

CVDs are a group of pathologies that affect the cardiovascular system such as, atherosclerosis, heart failure (HF), myocardial infarction (MI) etc. They are referred to any disease affecting the blood vessels or heart, vascular diseases of the brain and kidney, and peripheral arterial disease. Although, significant understanding of biomolecular events prevailing the initiation and progression of CVDs has been made, much remains unknown. Moreover, it is now evident that cardiovascular function is affected by oxidative stress [4] in a way that it also provides ample sites for redox regulation, which includes synthesis and function of prostanoids [5], regulation of endothelial nitric oxide synthase (eNOS) activity and function [6], vascularization [7], control of thiol/disulfide state in aging-associated cardiovascular complications [8]. At present, it is known that autophagy is implicated in the development and propagation of various CVDs, such as cardiac hypertrophy, atrial fibrillation, atherosclerosis and HF [9–12]. Interestingly, the regulatory mechanisms of autophagy are primarily activated upon either reactive oxygen species (ROS) overproduction, or NO deficiency, or nutrient deprivation [13, 14].

V. Dhawan (✉) · C. Bakshi
Department of Experimental Medicine and Biotechnology, Postgraduate Institute of Medical Education and Research, Chandigarh, India

R. A. Rather
Department of Biotechnology, Wachemo University, Hossana, Ethiopia

S. Chakraborti et al. (eds.), *Oxidative Stress in Heart Diseases*,
https://doi.org/10.1007/978-981-13-8273-4_4

Oxidative stress can be defined as a state of imbalance between production of reactive oxygen and nitrogen species (RONS) and antioxidant defenses. In addition, it creates a state where oxidation process surpasses the antioxidant levels in the body. Alterations in the redox status of tissue, leads to formation of lipid peroxides, free radicals, and other harmful reactive oxygen species such as singlet oxygen and hypochlorite causing adverse effects to the cell, by damaging DNA, proteins and lipids, as well as induce the depletion of redox cofactors [e.g., tetrahydrobiopterin (BH4)] and low molecular-weight antioxidants [15]. Not only oxidative stress is cytotoxic, it also plays an important role in the modulation of intermediates that regulate essential functions of the cell membrane, which are crucial for survival. Furthermore, the oxidative stress can modulate intracellular redox status, thereby activating protein kinases, such as a series of receptor and non-receptor tyrosine kinases, protein kinase C (PKC), and the MAP kinase (MAPK) cascade resulting in varied cellular responses.

RONS are known to be produced by several endogenous and exogenous sources, and antioxidant defenses aid in neutralizing their toxic effects. At low concentration, antioxidants ameliorates the oxidation of macromolecules eg. lipids, DNA and proteins [16]. Superoxide dismutase (SOD), glutathione peroxidase, catalase, thioredoxin, and peroxiredoxin are enzymatic antioxidants [17], whereas non-enzymatic antioxidants include vitamin E, vitamin C, and glutathione [18]. In addition, uric acid and bilirubin are also known as antioxidants that are able to protect against CVDs [16]. Moreover, two important fat-soluble carotenoids, β-carotene and lycopene, can scavenge free radical to reduce fatty acid oxidation [19].

An active lifestyle along with moderate exercise not only prevents oxidative stress, but also provides protection against the onslaught of CVDs, type II diabetes, and metabolic syndrome [20]. Oxidative stress biomarkers may thus serve as a suitable diagnostic or therapeutic tool to explore novel therapies with antioxidant properties. Besides the endogenous antioxidant systems, resveratrol and other natural compounds, along with moderate exercise, can act as exogenous antioxidants positively affecting the damage incurred due to oxidative stress. However, future studies are warranted to test the efficacy of natural compounds and other novel interventions, both in young age and adulthood.

4.2 Oxidative Stress and CVDs

Oxidative stress ensues due to imbalance in the pro-oxidant/anti-oxidant status, causing accumulation of pro-oxidant species, thereby promoting oxidative damage to several biomolecules [21]. Oxidative stress not only induces cytotoxic effects but also influences many biological processes including inflammation and apoptosis. During these responses, nuclear factor-KB (NF-KB) and activator protein-1 (AP-1) genes (transcription factors) act as sensors of oxidative stress through regulation of their own oxidation and reduction. Thus oxidation and reduction cycling-induced chemical modification of transcription factors is called reduction-oxidation (redox) regulation.

Reactive nitrogen species (RNS) and reactive oxygen species (ROS) are referred to as reactive non-radical and radical by-products of nitrogen and oxygen, respectively [22, 23]. Reactive oxygen and nitrogen species are produced by all aerobic cells and play a crucial role in age-related disease processes including CVDs [24]. Additionally, RONS generation derives energy from organic molecules generated in various physiological processes such as immune defense, and signaling [25]. Several endogenous and exogenous sources generate RONS and antioxidant defenses aid in neutralizing their toxic effects. Endogenously RONS is produced by nicotinamide adenine dinucleotide phosphate (NADPH) oxidase, myeloperoxidase (MPO), lipoxygenase, and angiotensin II [26]. NADPH oxidase (Nox) is the predominant source of the superoxide anion (O_2^-) radical, during cellular respiration. SOD, an enzymatic antioxidant, dismutates most of the O_2^- into hydrogen peroxide (H_2O_2) [25]. Whereas, exogenous sources of RONS include alcohol, tobacco, heavy or transition metals, drugs (eg, gentamycin, cyclosporine and tacrolimus), cooking (eg, smoked meat, waste oil, and fat), air and water pollution, industrial solvents and radiation, that are metabolized into free radicals inside the body [27].

The adverse effects of oxidative stress to the cell are exerted by damaging macromolecules such as DNA, proteins, and lipids. Induction of DNA damage due to augmented oxidative stress occurs both directly by several mechanisms, including nucleotide base modification, single strand breaks, double strand breaks, and indirectly by inhibiting DNA-formamidopyrimidine (FAPY) glycosylase [28] respectively. Several mutagenic lesions such as 5-hydroxycytosine, cytosine glycol, glycol, 2-hydroxy adenine, 8-oxoadenine, and thymine, are the consequences of oxidative modification of DNA. However, the most mutagenic outcomes of oxidative stress induced DNA damage are 8-oxo-7,8-dihydro-guanine (8-oxoGua) and 8-oxo-7,8-dihydro-2′deoxyguanosine (8-oxodG) lesions, of which, 8-oxoGua is the most commonly recurring modification resulting in G-to-T transversion events [29]. Oxidative stress can also impair the process of translation and protein synthesis, thus alters cellular proteostasis [29]. Furthermore, protein folding can be impeded by oxidative stress, causing loss of protein function due to fragmentation [30]. Although, specialized cellular machinery recognizes and degrades these oxidized proteins; however, when excessive protein oxidation occurs, it leads to building up of toxic products causing cellular dysfunction [30]. Advanced glycation end products (AGEs) are generated in a process called glycoxidation, which occurs as a result of reaction between the arginine and lysine amino groups with the carbonyl groups of carbohydrates. Major AGEs generated are glucosepane, pentosidine, N∈-carboxymethyl-lysine, and hydroimidazolone [31]. The accumulation of AGEs is known to be pro-inflammatory and is associated with CVDs, diabetes and its complications, renal disease and thus may contribute to obesity [32–34]. Additionally, oxidation of lipids by ROS leads to the generation of aldehydes and hydroperoxides of lipids, which further contribute to cellular toxicity [35]. For instance, membrane permeability and fluidity are altered once plasma and organelle membrane lipids are oxidized [32]. Lipid peroxidation by hydroxyl and peroxyl radicals provide important targets eg. poly-unsaturated fatty acids (PUFAs), particularly linoleic and arachidonic acids. Oxidation of PUFAs leads to production of various reactive aldehydes, such as malondialdehyde (MDA), *trans*-4-hydroxy-2-nonenal (4-HNE), and isoprostanes (F2-IsoPs) [36].

The accumulation of above mentioned oxidized macromolecules intracellularly is implicated in the culmination of various NCDs, including cardiovascular diseases [26]. Moreover, several studies have documented that tolerance of the heart to oxidative stress declines with increasing age due to attenuated levels of enzymatic antioxidants (ie, SOD and glutathione peroxidase), hence influences the development of CVDs [37]. Inflammation is thought to play a pivotal role in both the initiation and development of several CVDs including atherosclerosis [38]. A high level of RONS due to continuous production attenuates endogenous antioxidant capacities in chronic inflammatory conditions. It has been observed that patients with chronic inflammation often represent elevated levels of oxidative stress biomarkers [39] and low blood levels of antioxidants [40]. This is generally due to an augmented demand generated under conditions of overwhelming RONS production by activated immune effector cells such as macrophages.

Numerous sources of RONS have been identified in the cardiomyocytes, such as the mitochondria [41], Nox (Nox2 and Nox4) [42], xanthine oxidase, uncoupled NO synthase (NOS), and monoamine oxidase-A [43]. In the myocardium, activation of Nox2 leads to augmented superoxide production resulting in disease progression [44], whereas, the Nox4 isoform is responsible for either a beneficial [45], or detrimental effect [46], depending on the model studied [47]. Mitochondria not only act as an amplifier to augment the burden of oxidative stress, but also play a crucial part in oxidative damage of the cardiovascular system by providing redox triggers. Mitochondrial RONS interacts with several other sources of oxidative stress, such as Nox, xanthine oxidase and an uncoupled NOS. Several investigations have implicated the role of $p66^{Shc}$ in ROS generation within mitochondria and its contribution in CVDs [48, 49]. Various studies have demonstrated that $p66^{Shc}$ is an essential regulator of the intracellular redox balance and levels of oxidative stress. Moreover, various investigations have elucidated crucial roles of PKC family members in programming aspects of HF pathogenesis.

4.3 Biomarkers of Oxidative Stress in CVDs

As defined by WHO, a biomarker is any substance, structure, or process that can either have an influence or predict the incidence of disease or its outcome [50]. A clinically useful biomarker should possess the specificity for a certain disease (bear a diagnostic or prognostic value), and should correlate well with the disease activity [36]. Oxidative stress biomarkers may serve as a guide in choosing the most efficient drugs/dose regimen for patients by providing crucial information about the effectiveness of a treatment. In addition, from a pathophysiological point of view if it is predominantly relevant, a biomarker may serve as a suitable therapeutic modality to identify novel treatments with antioxidant properties, [36].

4.4 Biomarkers of RONS-Induced Modifications in CVDs

1. **AGEs** levels serve as an independent risk factor for predicting CV mortality [51]. It has been observed that serum levels of AGEs show positive correlation with lipid profiles and atherosclerotic features in an age-dependent manner, suggesting that AGEs could serve as useful biomarkers in the context of atherosclerotic disease [52].
2. **Oxidized-LDL (ox-LDL)** levels are known to be associated with arterial stiffness and atherogenesis in the elderly [53]. Both the protein and the lipid portion of LDL undergo oxidative modifications resulting in the accumulation of cholesterol [54].
3. **MPO** is an enzyme released in inflammatory conditions by leukocytes and catalyzes the formation of several RONS. MPO levels can serve as independent predictors of endothelial dysfunction and CVD mortality. MPO levels are considered to be a major player in the development and destabilization of atherosclerotic plaque, due to its involvement in formation of ox-LDL [55]. Additionally, MPO leads to endothelial dysfunction by directly catalyzing the consumption of NO [56].

In addition to the above mentioned markers, phosphorylated vasodilator-stimulated phosphoprotein (P-VASP) and asymmetric dimethylarginine (ADMA) represent as markers of oxidative stress post-RONS-induced damage [36].

4. **ADMA** is an L-arginine analogue, which can uncouple nitric oxide synthase (NOS) isoenzymes by competing with L-arginine, resulting in generation of superoxide molecule instead of NO, thus exaggerating the burden of oxidative stress [57–59]. Hence, cellular ADMA levels *per se* can lead to the reduction of NO and ROS overproduction. Furthermore, a growing body of evidence from various studies suggests that ADMA levels influence the NO-ROS equilibrium in the developmental programming of CVD as well as cardiovascular outcome [59–63]. In particular, in a community-based study, "Invecchiare in Chianti Study" (InCHIANTI; aging in the Chianti area), higher ADMA levels were reported to independently predict all-cause and CV mortality in the randomly selected samples of 1155 elderly, in the age range of 65–102 years [64].
5. **P-VASP** is predominantly phosphorylated by cyclic guanosine monophosphate (cGMP)-dependent protein kinases. It is considered as the best-established biomarker for physiological cGMP signaling, and low P-VASP levels indicate pathological signaling [36]. In the "GENERATIONS" trial, P-VASP levels in human blood samples, were used to predict the efficacy of (or detect non-responders to) antiplatelet therapy [65, 66].

4.5 Oxidative Stress, Nutrient Signaling and Autophagy Cross-Talk in CVDs

A number of nutrient sensing signaling pathways including cyclic adenosine monophosphate (cAMP)-activated protein kinase (AMPK), mammalian target of rapamycin (mTOR), silent information regulator transcript (SIRT), peroxisome proliferator-activated receptors (PPARs) and PPAR coactivator-1α (PGC-1α), exist in the cardiovascular system. Among them, PGC-1α not only acts as a centre point for a nutrient-sensing cluster, it can also stimulate other nuclear receptors, such as estrogen-receptor related receptor (ERR) and nuclear respiratory factor (NRF) for regulating energy metabolism, in addition to PPARs [67]. Further, due to activation of AMPK by augmented NAD^+/NADH ratio, SIRT1 activation by increased mitochondrial AMP/adenosine triphosphate (ATP) ratio, or NO, enhanced activity of PGC-1α can promote mitochondria biogenesis [68, 69]. Interestingly, PGC-1α not only regulates mitochondrial biogenesis by interacting with nutrient sensing signals, but via the autophagy-lysosome machinery, it also leads to the degradation of mitochondria

A cellular catabolic process, autophagy results in the degradation of crucial organelles such as mitochondria, via their transportation to lysosomes [13]. PGC-1α can promote autophagy through regulation of transcription factor EB (TFEB). Furthermore, mTOR inhibits autophagy whereas AMPK by acting as the negative regulator of mTOR can promote autophagy [14]. Both mTOR and AMPK by acting antagonistically regulate unc-51-like kinase 1/2 (ULK1/2) activity via phosphorylation. AMPK can promote autophagy through SIRT1 activation which in turn deacetylates and activates several autophagy-related (Atg) proteins, such as Atg5, Atg7, and Atg8 in addition to activating ULK1/2 [70]. Importantly, ROS overproduction, NO deficiency, or nutrient deprivation primarily activate these regulatory pathways of autophagy [13, 14]. Since, in most mammalian cells, mitochondria are a major source of ROS, it is suggestive that close interrelationship exists among autophagy, mitochondrial ROS, NO production, and cellular apoptosis *versus* survival [71].

A number of studies implicate autophagy in the development and propagation of several cardiovascular disorders, such as HF, ischemia/reperfusion (I/R), atrial fibrillation, cardiac hypertrophy and atherosclerosis [9, 12, 72, 73]. For instance, Valentim et al. (2006) reported that inhibition of autophagy with 3-MA, augments viability of cells in I/R exposed cultured neonatal cardiomyocytes [74]. Likewise, in a mouse model of I/R, Beclin 1 knockdown enhanced cell death and also impaired autophagosome formation [75]. In addition, autophagy activation also promotes the initiation and progression of atherosclerotic plaques by promoting survival and de-differentiation of vascular smooth muscle cells (VSMCs) [76–78]. Furthermore, autophagic activity when in excess, can incite plaque rupture, thrombosis and promote acute clinical events [79]. As mentioned earlier, mitochondria are the major source of intracellular ROS [80]. Meanwhile, only autophagy can remove the damaged mitochondria; therefore, any condition related to heart that influences the removal of dysfunctional mitochondria signifies an important potential link between oxidative stress and CVDs. A study by Oka et al. (2012), demonstrated that DNA from impaired mitochondria, that escapes from lysosome-mediated degradation by autophagy leads to HF via a Toll-like receptor (TLR) 9-associated inflammatory response, suggesting existence of a mechanism involving potential crosstalk between ROS, autophagy and inflammation in CVDs [81]. Besides, in another study it was observed that the exposure to lipopolysaccharide (LPS) induced mitochondrial DNA damage, autophagy, ROS, LOX-1 expression and the NLRP3 inflammasome in human THP-1 macrophages. Ding et al. [82] reported that ROS inhibitors as well as autophagy inducers downregulate the expression of NLRP3 inflammasome, while, autophagy inhibition augments the expression of the NLPR3 inflammasome [82]. Recently, Hu and Zhang (2017) showed that high fat diet attenuates the ROS levels, cell death and intracellular abnormalities in Ca^{2+} signaling, whereas, it activates autophagy in a NF-KB/JNK-dependent fashion and improves cardiac function in TLR4-deficient mice as compared with wild type animals [83].

4.6 Oxidative Stress in CVDs and Therapeutics

A passable knowledge along with advanced technology has changed our concepts of the mechanisms on the role of oxidative stress in chronic diseases in the past so many years. Although, RONS and the ensuing oxidative stress have been studied in the milieu of damage to biologically vital targets like DNA, lipids, or proteins, many antioxidant-based clinical trials have failed to generate a positive outcome in the context of human disease, particularly atherosclerosis and CVDs. As a consequence, there has been an ever-increasing appreciation of how oxidative damage misbalances the target biomarkers and signaling molecules of vascular disease, involved in the pathophysiology of oxidative damage. Multifactorial etiology of CVDs makes it difficult to understand whether RONS intervention is apparent in all stages of the disease development, but therapeutic interventions have proved that naïve radicals greatly influence the CVD development. We are only now beginning to appreciate how these concepts facilitate the disease process.

The current regime of drugs used for vascular protection includes angiotensin-converting enzyme (ACE) inhibitors and statins. The latter drug, an accepted cholesterol inhibitor, does not directly act as an antioxidant, but indirectly hunt ROS by inhibition of 3-hydroxy-3-methyl-glutaryl-coenzyme A (HMG-CoA) reductase pathway [84, 85]. The effectiveness of statins in plummeting the frequency of cardiac clinical events and mortality is likely augmented by their impending antioxidant characteristics and their potential to increase endothelial NOS (eNOS) expression [86]. Statins though known for their lipid-lowering effects, also have the competence to improve the bioactivity of nitric oxide and stability of atherosclerotic plaques. Statins are important and play a pivotal role during chronic statin therapy for the coronary heart disease owing to their pleotropic effects. Apart from statins, an array of antioxidant vitamins like E, A and C, lycopene, quercetin, and β-carotene have been tested and validated for their preventive and therapeutic benefits in various CVD complications like I/R heart injury, ventricular remodelling, atherosclerosis, HF and MI [87]. Although, the association of oxidative stress in CVD origin has been noticeably expounded, the focal aspect that remains is the decoding of the specific mechanisms involved in CVD pathogenesis. Even though, several studies have been carried out to elucidate the role of oxidative stress-related mechanisms involved in CVDs, nevertheless additional work is still warranted to look for the promising targets in oxidative stress- related CVDs. The past strategies to suppress the oxidative tissue damage were formed on a drug-target basis wherein, free radicals were scavenged via chemical reactions. These drug entities have circumlocutory antioxidant effects, at the same time, potentiating endogenous cellular NO production. As mentioned earlier, statins apart from their lipid-lowering potential are competent in plummeting Nox-related ROS production. Additionally, they are shown to induce and activate the NOS [88]. However, the current strategies are solely technical and molecular-based, wherein a responsible gene is silenced or mutilated by molecular methods. Additionally, both statins and blockers of the angiotensin system decrease mitochondrial oxidative stress. Together, there is a general agreement that the useful impact of therapy with statin inhibitors is at least in part arbitrated by properties that are self-regulating and not dependent on the haemodynamic or cholesterol-lowering effects of these drugs.

The drug target-based strategies are receptive immediately eg, use of carvedilol (β-adrenergic blocker) antagonists of the AT1 receptor, ACE inhibitors and other antihypertensive agents has been advocated [89]. Use of conventional and newer antihypertensive drugs like allopurinol and oxypurinol, have been reported to reduce the incidence of stroke, MI, and other fatal CVDs [90]. PPAR homologue has also been found to reduce ROS production and decrease p22phox expression [91]. A ubiquinol-based compound MitoQ complexed with Triphenylphosphonium (TPP) has been used as mitochondria-targeted therapeutic antioxidant to limit the pathologic source of ROS production, thereby preventing oxidative DNA damage while reducing atherosclerotic plaques in ATM gene mice models [92]. Use of gp91phox (Nox2) homologue, inhibitors of vascular Nox via pharmacological or gene target methods also holds the promise to improve the endothelial function limiting hypertension. Genetic disruption of p47phox (involved in the formation of Nox) has been shown to inhibit the formation of atherosclerotic lesions in mice [93]. Tissue/organ-specific nano-formulated particle carrier for delivery of antioxidants can also serve as novel therapeutic modality as referred by Jain et al. (2015) [94]. Role of vitamins as antioxidants have been a prime choice in the present treatment regime, however, the patients at risk need to be screened for oxidative stress markers at an early stage. This is needed to obviate or lessen the further tissue damage created by free radicals in later stages of CVD. In patients with manifested CVD, they can be diagnosed by the dysregulated oxidant, antioxidant enzyme expression, and function and treated using pharmacological modulators or gene target therapeutics with novel strategies.

4.7 Oxidative Biomarkers and Antioxidant Therapy

The underlying phenomena of modulation of various redox enzymes leading to oxidative stress have elucidated various biomarkers for CVD. These include components of ox-LDL, oxidative phospholipids, and apoB containing lipoproteins. Lipoprotein-associated phospholipase A2 (Lp-PLA2) levels are modestly associated with high risk for CAD [95]. Since, MPO modified proteins and metalloproteinases are associated with collagen degradation, atherosclerotic lesions and non-fatal MI, serum MPO levels can serve as a suitable biomarker. Other markers of lipid peroxidation include F2α-isoprostanes, malondialdehyde, dienes, hydroperoxides, oxysterols and 7β-hydroxycholesterol. Although, thiobarbituric acid reactive substances (TBARS) are commonly used in the conventional assay, other spectrophotometric techniques like GC-MS, HPLC, and RIA are also used. F2α-isoprostanes have been used as gold standard for lipid peroxidation, a predictive marker of CAD using GC-MS technique. Likewise, 7β-hydroxycholesterol has a strong positive association with CVDs as shown by immunoassays [96]. Oxidative damage to proteins creates Amyloid precursor protein-mediated free radicals (AOPP) which are considered as critical biomarker endpoints of CVD. These AOPP have vast penetration power in tissue intima, hence are detected via ELISA immunofluorescence assay. Some studies have reported the formation of protein carbonyls which causes vascular tissue damage, however, it needs validation through extensive studies and subsequent evidence. The other important oxidation product is hydroxyeicosatetraenoic acid, rated an important marker in recent studies for CVD that can be detected in body fluids via spectrophotometric techniques [96]. Gene expression analysis of few components of Nox (gp91phox, p22phox, p47phox, p67phox and rac) in endothelial cells, has revealed an association with the ROS production, hence they represent a potential candidate as novel biomarkers for the oxidative stress-related vascular diseases. Use of fluorescent probes like dihydrochlorofluorescein diacetate (DCFH-DA), 4-amino-5-methylamino-2′, 7′-difluorofluorescein diacetate (DAF-FM DA) and 4,5-diaminofluorescein diacetate (DAF-2 DA) in flow cytometry have been used to detect ROS/RNS as indices of oxidative stress [97]. Aggregation of platelets with leukocytes generates activated platelets, thus reducing the general platelet count in Platelet-rich plasma-platelet concentrate. These aggregated platelets are the indices of mitochondrial mitophagy, a hallmark of redox imbalance-induced CVD [98]. Recently, a molecular marker of oxidative stress and CVD in monocytes and macrophages has been developed in the form of reversible protein-S-glutathionylation that can be measured by western blotting or ELISA tests using monoclonal anti-glutathione antibodies [99].

4.8 Trials on Antioxidant Vitamins

The role of Vitamin A, C, and E, carotenoids, lycopene and phytoconstituents as scavengers of free radical has been comprehensively studied as therapeutic modulators to attenuate CVD. Seven significant trials, which include Heart Outcomes Prevention Evaluation (HOPE) study (2545 subjects); α-tocopherol, β-carotene Cancer Prevention Study (ATBC; 27 271 males); Cambridge Heart Antioxidant Study (CHAOS; 2002 patients); Gruppo Italiano per lo Studiodella Sopravvivenza nell'Infarto Miocardico (GISSI)-Prevenzione trial (3658 patients); Medical Research Council/British Heart Foundation (MRC/BHF) Heart Protection Study (20536 adults); Primary Prevention Project (PPP; 4495 patients); and the Antioxidant Supplementation in Atherosclerosis Prevention (ASAP) Study (520 subjects), have been conducted, that positively reported that regular Vitamin E supplementation with a delayed release formulation of Vitamin C hampered the development of carotid atherosclerosis. Among these studies, the HOPE and HPS trials have not shown the significant benefits during atherosclerosis and plaque rupture due to the differences in their pattern of free radical scavenging [100]. Vitamins used as antioxidants by Lane et al. (2008), Stephen et al. (1996), Ashor et al. (2014) in their studies have found that vitamins alleviate arterial stiffness in adults, with lower incidences of peripheral arterial diseases and noticeable decline in non-fatal MI rate in the vascular disease patients [101–103]. The phytoconstituents of tomato juice along with lycopene can attenuate or reverse the oxidative stress-related parameters in CVD patients as reported by Gitenay et al. (2007) [104]. The synergistic effect of phytoconstituents also improved post-ischemic ventricular function, reduction in the size of myocardial infarct and cardiomyocyte apoptosis. The type, dosage of vitamin supplements and the duration also gain importance while contributing for their antioxidant effects irrespective of the traditional risk factors. In addition, the inability of vitamins to scavenge H_2O_2 or HOCl in hypertensive vascular damage, inhibit ROS production, inaccessibility to ROS radicals produced in intracellular compartments and organelles are the considerable factors to analyze the comprehensive effectiveness of antioxidant vitamin therapy in CVD.

Since, it is well-known that oxidative stress has a vital role in the pathophysiology of several CVDs, many studies were carried out to elucidate the therapeutic benefits of antioxidant therapy in various CVDs. However, apart from antioxidants, there are several candidates that may serve as potential candidate with therapeutic benefits against oxidative stress in CVDs.

4.9 Extremely Low-Frequency Pulsed Electromagnetic Field Exposure

For almost five decades, electric and electromagnetic fields (EMFs) in various forms have been employed to stimulate bone healing upon fractures as well as for the management of osteoporosis [105]. Recently, Ehnert et al. (2017) characterized the influence of ten defined (10 to 90.6 Hz) extremely low-frequency pulsed electromagnetic fields (ELF-PEMFs) on differentiation of human osteoblasts through ERK1/2-engagement, by producing harmless amounts of ROS, mainly $^{\bullet}O_2^-$ and H_2O_2 [106]. Similarly to p38 and JNK, an oxidative stress stimulus which is frequently generated by the mitochondrial respiratory chain directly affects ERK1/2 [107, 108]. Interestingly, in the above study it was observed that a single exposure to ELF-PEMF, triggered antioxidative defense mechanisms by increasing the mRNA and protein expression of CAT, SOD2, GSR and GPX3 as well as at enzyme activity levels. Furthermore, Raggi and colleagues (2008) showed that routine exposure of 27 min/day for 10 days to ELF-EMF, reduced the blood levels of oxidative stress biomarkers in healthy volunteers [109]. At present, a number of theories are trying to investigate the effect of EMFs, which include variations in ion flux and membrane potential, activity of voltage-sensitive enzymes, re-organization of the cytoskeleton, regulation of gene expression via EMF-responsive sequences, as well as alterations in the oxidative state of the cells [110–112]. Thus, ELF-PEMF therapy might prove to be an interesting adjunct to classical antioxidant therapy under oxidative stress conditions in CVDs.

4.10 γ-Glutamylcysteine Supplementation

Reduced glutathione (GSH) also referred to as the "master antioxidant", is a tripeptide (γ-L-glutamyl-L-cysteinyl glycine) that is produced in the cytoplasm of every cell at concentrations upto 10 mM [113]. Beside acting as a reducing agent and a reservoir of cysteine, it is implicated in several physiological processes including proliferation, apoptosis, thiol disulphide exchange and xenobiotic metabolism [114]. The intracellular homeostasis or concentration of GSH, is regulated by an active balance of synthesis, depletion, transport as well as oxidative stress status in some tissues [114, 115]. The *de novo* synthesis of cytosolic GSH takes place by two consecutive reactions catalyzed by ATP-dependent enzyme. In the first reaction, γ-glutamylcysteine (γ-GC) is produced from L-cysteine and L-glutamic acid through creation of an unusual γ-peptide bond by glutamate cysteine ligase (GCL). In the second reaction, GSH is generated as a result of addition of glycine to γ-GC by glutathione synthetase (GS) [116, 117]. Hence, an increase in susceptibility to oxidative stress results in manifestation of GSH deficiency, which is supposed to be a vital player in the initiation and progression of several chronic diseases [118]. In a recent human pilot study by Zarka and Bridge (2017), it has been shown that cellular GSH levels can transiently increase above homeostasis in human lymphocytes by oral supplementation of γ-GC [119]. Their finding that γ-GC can augment GSH levels in healthy human subjects suggests, that drugs or supplements capable of elevating levels of GSH could have therapeutic potential in treatment of numerous age-related chronic disorders including CVDs [120–122].

4.11 HSP72 Augmentation

During the occlusion of coronary arteries, myocytes due to ischemia undergo hypoxia and reperfusion results in further injury due to oxidative stress. The ischemic as well as oxidative stress leads to an irrevocable myocardial damage. Although, I/R injury can be curbed by pharmacological interventions, none of them have shown any significant efficiency in reduction of I/R injury in multicentre clinical trials [123–126]. A family of chaperone proteins known as heat-shock proteins (HSPs) ensure that the newly synthesized critical intracellular proteins are folded accurately and stabilized as well as also ensures the corrective refolding of proteins damaged due to oxidative and other cellular stresses. Both *in vivo* and *in vitro* studies have reported that the endogenously produced HSP72 demonstrates cardioprotective effects [127–131]. However, due to delayed induction and production of endogenous HSPs, the efficiency of approaches to augment the endogenous levels of HSP72 is reduced, such as in acute MI, where early intervention is necessary. A study by Tanimoto et al. (2017) in a rabbit model of I/R injury, demonstrated the cardioprotective effects of a single intravenous dose of HSP72 attached to a single-chain variable fragment (Fv) of monoclonal antibody 3E10 (3E10Fv) [132]. The Fv enables the rapid entry of HSP72 into cells with intact plasma membranes. It was observed that at the time of reperfusion, administration of single-dose of Fv-HSP72 fusion protein reduced one-half of myocardial apoptosis as well as improved the function of left ventricle in rabbits after myocardial I/R injury [132]. This study suggests that targeted biological agents (such as antibody) that assist in rapid transport of the therapeutic agent (such as HSP72) into cardiomyocytes might serve as a potential adjunct to I/R therapy in amelioration of ischemia and oxidative stress in CVDs.

4.12 Targeting Mitochondria to Curb ROS Generation

The activity of mitochondrial electron transport chain complexes is regulated via oxidants and electrophiles in an intricate manner. For instance, nitric oxide or its derivatives [133–135] and glutathione [136, 137] alter the activity of complex I (NADH ubiquinone oxidoreductase). Additionally, other complexes such as, complex II (succinate dehydrogenase) and complex V (ATP synthase), have been shown to be altered by RONS [138, 139]. Similarly, RONS can also modulate the catalytic activity of several proteins of mitochondrial matrix including $NADP^+$-isocitrate dehydrogenase [140], α-ketoglutarate dehydrogenase [141] and aconitase [142], along with the inter membrane space proteins such as creatine kinase [143] and cytochrome c [144]. In response to a greater demand in energy or stress, a limited amount of ROS is generated by mitochondria that act as signaling molecules to trigger an endogenous stress response inducing antioxidant enzymes such as catalase or superoxide dismutase, as well as other defence pathways against stress leading to detoxification of ROS [145]. On the contrary, intracellular damage is also endured due to high ROS levels [146].

Several approaches seem to have potential in apprehending the damage induced due to excessive mitochondrial oxidative stress. Firstly, mitochondria-targeted specific antioxidants such as MitoQ containing the covalently attached TPP cation to antioxidant ubiquinol, can reduce oxidative stress. MitoQ is a prospective therapeutic candidate for mitochondrial oxidative stress, due to its ability to decrease cardiac I/R injury [147]. An additional approach to eradicate detrimental ROS production is the elimination of impaired mitochondria. Recently, studies have primarily targeted the inhibition of Drp1 (dynamin-related protein 1), which is intricated in mitochondrial fission. Specifically, inhibiting mitochondrial fission by the inhibitor mdivi-1 protects the heart against I/R injury by reducing myocardial infarct size in mice undergoing I/R [148]. Another mechanism to eliminate dysfunctional or impaired mitochondria is upregulation of mitophagy by various molecular mechanisms, to prevent pathological ROS generation [149, 150]. Lastly, the proteasome machinery and the unfolded protein response represent other mechanisms that may have potential to maintain functional mitochondria.

4.13 Endoplasmic Reticulum Stress and the UPR Pathways in the Oxidative Stress-Induced Endothelial Dysfunction

The imbalance between ROS generation and antioxidants in endothelial cells can induce endothelial dysfunction (ED). ED is the initial pathogenic event of many CVDs as well as various metabolic diseases [151]. In addition to earlier mentioned sources of ROS, substantial evidence recognizes endoplasmic reticulum stress (ER stress) as another source of ROS [152]. Under normal conditions, the ER has limited antioxidant activity and is engaged in the tight regulation of folding and trafficking of secretory proteins [153]. A number of pathophysiological conditions could disrupt the ER proteostasis by inducing the accumulation of misfolded or unfolded proteins within the ER [154, 155]. This condition leads to ER stress resulting in the activation of unfolded protein response (UPR) pathways [156]. The UPR pathways aim to re-establish ER proteostasis through diverse consequences such as reducing ER protein load, potentiating the ER quality control, activating the ER-associated protein degradation machinery (ERAD), and, finally, activating autophagy [157]. Since, protein folding is coupled to ROS production, the increase in folding load during ER stress intensely promotes ROS generation and leads to exacerbated oxidative stress [158, 159]. The ER stress activates the UPR pathways by way of three transmembrane transducers: the activating transcription factor 6 (ATF6), the inositol-requiring kinase 1 (IRE1), and the pancreatic ER kinase (PERK) [160]. Since then, the correlations of ER stress and UPR to ED have been established in numerous studies in both animal and cellular models [161–163].

Two possible approaches can be utilized to neutralize oxidative stress-induced UPR. One is to directly manipulate the activity of specific UPR mediators. While other strategy entails the stimulation of auxiliary pathways potentiating the adaptive response against ER stress to relieve unfolding. A promising therapeutic approach to reduce ER stress is characterized by the augmentation of the folding capacity of ER chaperones or by means of chemical chaperones. Two such chemical chaperones are Sodium phenylbutyrate (PBA) and Tauroursodeoxycholate (TUDCA) that have been used for the treatment of primary biliary cirrhosis and urea-cycle disorders respectively, and are approved by the Food and Drug Administration (FDA) [164–167]. Interestingly, TUDCA and PBA have also shown cardioprotection effects and therapeutic application in certain CVDs such as I/R and atherosclerosis [168–170]. Another promising strategy to counteract ER stress-induced ED is the modulation of Bip/GRP78, PDI or Ero1 activity. In particular, a screening study carried out by Kudo et al. (2008), identified the compound BIX (Bip inducer X) as inducer of Bip/GRP78 expression via the ATF6 [171]. Moreover, intracerebral administration of BIX in ischemic mice reduced the infarction area suggesting its possible use also in an ischemic heart [171]. An additional possible approach is the modulation of Ero1 activity. In this regard, Blais et al. (2010) identified EN460 as a small Ero1α inhibitor, reporting that EN460 interacted specifically with the active form of Ero1α and prevented its reoxidation [172].

Alternative therapeutic strategy for mitigating ER stress is the modulation of individual UPR pathways such as PERK/eukaryotic initiation factor 2α (eIF2α) and IRE1/XBP1. With regard to the modulators of the PERK/eIF2α axis, salubrinal prevents the dephosphorylation of eIF2α through the inhibition of GADD34 and CReP (enzymes that direct the activity of the eIF2α protein phosphatase PP1) displayed significant protection from ER stress in several conditions [173–175] including MI [176, 177] and ox-LDL-mediated ED [178]. However, it has been observed that salubrinal could potentiate lipid-induced ER stress with cytotoxic outcome [179, 180] suggesting that salubrinal administration in CVDs needs further corroboration in clinical conditions. Additionally, or as another approach, to the inflection of PERK/eIF2α signaling, the inhibition of the IRE1/XBP1 pathway can also be adapted to impair UPR in ER stress-dependent diseases. IRE1/XBP1 signaling can be hindered by inhibiting either IRE1 kinase activity or IRE1 RNAse activity.

4.14 Imaging Free Radicals in CVDs by Immuno-Spin Trapping

At present, the methodology used to elucidate oxidative stress depend upon the detection of either the steady-state intermediates or end products of oxidative stress or quantification of alterations of an exogenous probing molecule, such as dihydroethidium (DHE), lucigenin, [(3-boronophenyl) methyl] triphenyl-phosphonium, monobromide (mitoB) [181, 182], mitochondria-targeted hydroethidine (MitoSOX), 10-acetyl-3,7-dihydroxyphenoxazine (AmplexRed) or various electron paramagnetic resonance (EPR) spin traps and probes [183–186]. Currently, majority of the techniques employed to identify biological free radicals in cells and tissues are not appropriate, either due to sensitivity limitations eg electron spin resonance (ESR), or manifest as artifacts that make the validity of the results uncertain (fluorescent probe-based analysis).

However, all of these difficulties were negated with the advancement of the immuno-spin trapping (IST) technique. The principal of IST method is based on the formation of DNA -and protein-5, 5-dimethyl-1-pyrroline N-oxide (DMPO) nitroxide radical adducts, owing to the reaction between DNA base-and amino acid-derived radicals with the spin trap DMPO, respectively. Due to their limited stability, these adduct decay to yield a very stable macromolecule-DMPO-nitrone product. The identification of this stable product can be achieved using anti-DMPO nitrone antibodies by nuclear magnetic resonance (NMR), mass spectrometry, or immunochemistry. Interestingly, all classes of macromolecule-derived free radicals generated in biological systems form DMPO-nitrone products, except for peroxyl radicals [187]. Moreover, the generation of macromolecular DMPO-nitrone adducts is not subjected to artifacts, which are recurrently detected with other techniques, since it is based on the specific reaction between free radical and spin trap DMPO.

Recently, Proniewski et al. (2018) quantified the oxidative modifications in cardiomyocytes and coronary endothelium in the heart of a murine model of HF (Tgαq∗44 mice) showing that IST signifies a unique technique for quantifying oxidative modifications [188]. The authors demonstrated that at the transition phase of HF in Tgαq∗44 mice, an increase in superoxide production leads to oxidative alterations both in cardiomyocytes and coronary endothelium, despite the compensatory activation of antioxidants [188]. Thus, IST can be considered as an effective, sensitive (a million times more than ESR), and easy technique to identify even low levels of free radicals generated, both *in vivo* and *in vitro*.

4.15 Conclusion

It is evident that oxidative stress is implicated in numerous age-related diseases including CVDs, neurodegenerative disorders, metabolic and inflammatory diseases. Recent studies have shown that the ER stress could be a new player in the promotion of the pro- or anti-oxidative pathways in CVDs by modulation of UPR pathways. Although, the UPR pathway restore ER proteostatis by promoting an adaptive response, the continuous activation of UPR leads to augmented oxidative stress and cell death. Nevertheless, it will be fascinating to study the clinical and biological effects of new therapeutic interventions in *in vivo* and *in vitro* models of ER stress-dependent ED and CVDs, considering their therapeutic potential. Additionally, interplay between inflammation, autophagy and oxidative stress, is implicated in CVDs, however their contribution to the disease progression may vary. Therefore, the preference of NAC or vitamins E/C to curb oxidative stress, and rapamycin or trehalose for inflection of autophagy may be considered as a choice of treatment, however the efficacy of such regimens warrant case by case evaluation. Along with moderate aerobic exercise, antioxidant therapy such as resveratrol and other phytochemicals may prevent against the adverse effects of oxidative stress. Considering the efficacy of the treatment, different types of RONS biomarkers identified in CVDs may provide important information about the most effective drugs/dose treatment regimens for patients. The efficacy of above mentioned novel molecules has been tested in models of various diseases, however, not much data is available from CVD models. Hence, further investigations are required in order to define the optimal targets for a particular clinical condition, to design novel drugs, and to prevent probable side effects as a result of oxidative stress-induced perturbations.

References

1. Reddy KS (2016) Global Burden of Disease Study 2015 provides GPS for global health 2030. Lancet 388:1448–1449
2. WHO (2013) Global action plan for the prevention and control of NCDs 2013/2020. World Health
3. WHO (2015) World health statistics 2015. World Health Statistics
4. Griendling KK, FitzGerald GA (2003) Oxidative stress and cardiovascular injury. Part II: animal and human studies. Circulation 108:2034–2040
5. Bachschmid M, Schildknecht S, Ullrich V (2005) Redox regulation of vascular prostanoid synthesis by the nitric oxide-superoxide system. Biochem Biophys Res Commun 338:536–542
6. Chen CA, Wang TY, Varadharaj S et al (2010) S-glutathionylation uncouples eNOS and regulates its cellular and vascular function. Nature 468:1115–1118
7. Roy S, Khanna S, Sen CK (2008) Redox regulation of the VEGF signaling path and tissue vascularization: hydrogen peroxide, the common link between physical exercise and cutaneous wound healing. Free Radic Biol Med 44:180–192
8. Altschmied J, Haendeler J (2009) Thioredoxin-1 and endothelial cell aging: role in cardiovascular diseases. Antioxid Redox Signal 11:1733–1740
9. Wang ZV, Ferdous A, Hill JA (2013) Cardiomyocyte autophagy: metabolic profit and loss. Heart Fail Rev 18:585–594
10. Ryter SW, Lee SJ, Smith A et al (2010) Autophagy in vascular disease. Proc Am Thorac Soc 7:40–47
11. Rifki OF, Hill JA (2012) Cardiac autophagy: good with the bad. J Cardiovasc Pharmacol 60:248–252
12. Lavandero S, Troncoso R, Rothermel BA et al (2013) Cardiovascular autophagy: concepts, controversies, and perspectives. Autophagy 9:1455–1466
13. Zhang J (2015) Teaching the basics of autophagy and mitophagy to redox biologists--mechanisms and experimental approaches. Redox Biol 4:242–259
14. Filomeni G, De Zio D, Cecconi F (2015) Oxidative stress and autophagy: the clash between damage and metabolic needs. Cell Death Differ 22:377–388
15. Elahi MM, Kong YX, Matata BM (2009) Oxidative stress as a mediator of cardiovascular disease. Oxid Med Cell Longev 2:259–269
16. Montezano AC, Touyz RM (2012) Molecular mechanisms of hypertension-reactive oxygen species and antioxidants: a basic science update for the clinician. Can J Cardiol 28:288–295
17. Gongora MC, Qin Z, Laude K (2006) Role of extracellular superoxide dismutase in hypertension. Hypertension 48:473–481
18. Traber MG, Stevens JF (2011) Vitamins C and E: beneficial effects from a mechanistic perspective. Free Radic Biol Med 51:1000–1013
19. Rimm EB, Stampfer MJ (2000) Antioxidants for vascular disease. Med Clin North Am 84:239–249
20. Baltaci SB, Mogulkoc R, Baltaci AK (2016) Resveratrol and exercise. Biomed Rep 5:525–530
21. D'Autréaux B, Toledano MB (2007) ROS as signaling molecules: mechanisms that generate specificity in ROS homeostasis. Nat Rev Mol Cell Biol 8:813–824

22. Powers SK, Ji LL, Kavazis AN et al (2011) Reactive oxygen species: impact on skeletal muscle. Compr Physiol 1:941–969
23. Adams L, Franco MC, Estevez AG (2015) Reactive nitrogen species in cellular signaling. Exp Biol Med (Maywood) 240:711–717
24. Venkataraman K, Khurana S, Tai TC (2013) Oxidative stress in aging--matters of the heart and mind. Int J Mol Sci 14:17897–17925
25. Genestra M (2007) Oxyl radicals, redox-sensitive signaling cascades and antioxidants. Cell Signal 19:1807–1819
26. Salisbury D, Bronas U (2015) Reactive oxygen and nitrogen species: impact on endothelial dysfunction. Nurs Res 64:53–66
27. Phaniendra A, Jestadi DB, Periyasamy L (2015) Free radicals: properties, sources, targets, and their implication in various diseases. Indian J Clin Biochem 30:11–26
28. Mühlenhoff U (2000) The FAPY-DNA glycosylase (Fpg) is required for survival of the cyanobacterium Synechococcus elongatus under high light irradiance. FEMS Microbiol Lett 187:127–132
29. Byrd AK, Zybailov BL, Maddukuri L et al (2016) Evidence that G-quadruplex DNA accumulates in the cytoplasm and participates in stress granule assembly in response to oxidative stress. J Biol Chem 291:18041–18057
30. Delanghe S, Delanghe JR, Speeckaert R et al (2017) Mechanisms and consequences of carbamoylation. Nat Rev Nephrol 13:580–593
31. Reynaert NL, Gopal P, Rutten EPA et al (2016) Advanced glycation end products and their receptor in age-related, non-communicable chronic inflammatory diseases; Overview of clinical evidence and potential contributions to disease. Int J Biochem Cell Biol 81(Pt B):403–418
32. Davis KE, Prasad C, Vijayagopal P et al (2016) Advanced glycation end products, inflammation, and chronic metabolic diseases: links in a chain? Crit Rev Food Sci Nutr 56:989–998
33. Mahajan N, Dhawan V (2013) Receptor for advanced glycation end products (RAGE) in vascular and inflammatory diseases. Int J Cardiol 168:1788–1194
34. Mahajan N, Malik N, Bahl A et al (2009) Receptor for advanced glycation end products (RAGE) and its inflammatory ligand EN-RAGE in non-diabetic subjects with pre-mature coronary artery disease. Atherosclerosis 207:597–602
35. Gaschler MM, Stockwell BR (2017) Lipid peroxidation in cell death. Biochem Biophys Res Commun 482:419–425
36. Frijhoff J, Winyard PG, Zarkovic N et al (2015) Clinical relevance of biomarkers of oxidative stress. Antioxid Redox Signal 23:1144–1170
37. Abete P, Napoli C, Santoro G et al (1999) Age-related decrease in cardiac tolerance to oxidative stress. J Mol Cell Cardiol 31:227–236
38. Libby P, Ridker PM, Maseri A (2002) Inflammation and atherosclerosis. Circulation 105:1135–1143
39. Fuchs D, Avanzas P, Arroyo-Espliguero R et al (2009) The role of neopterin in atherogenesis and cardiovascular risk assessment. Curr Med Chem 16:4644–4653
40. Murr C, Winklhofer-Roob BM, Schroecksnadel K et al (2009) Inverse association between serum concentrations of neopterin and antioxidants in patients with and without angiographic coronary artery disease. Atherosclerosis 202:543–549
41. Elas M, Bielanska J, Pustelny K et al (2008) Detection of mitochondrial dysfunction by EPR technique in mouse model of dilated cardiomyopathy. Free Radic Biol Med 45:321–328
42. Heymes C, Bendall JK, Ratajczak P et al (2003) Increased myocardial NADPH oxidase activity in human heart failure. J Am Coll Cardiol 41:2164–2171
43. Santos CX, Raza S, Shah AM (2016) Redox signaling in the cardiomyocyte: from physiology to failure. Int J Biochem Cell Biol 74:145–151
44. Sirokmány G, Donkó Á, Geiszt M (2016) Nox/Duox family of NADPH oxidases: lessons from knockout mouse models. Trends Pharmacol Sci 37:318–327

45. Zhang M, Brewer AC, Schröder K et al (2010) NADPH oxidase-4 mediates protection against chronic load-induced stress in mouse hearts by enhancing angiogenesis. Proc Natl Acad Sci U S A 107:18121–18126
46. Kuroda J, Ago T, Matsushima S et al (2010) NADPH oxidase 4 (Nox4) is a major source of oxidative stress in the failing heart. Proc Natl Acad Sci U S A 107:15565–15570
47. Burgoyne JR, Mongue-Din H, Eaton P et al (2012) Redox signaling in cardiac physiology and pathology. Circ Res 111:1091–1106
48. Martin-Padura I, de Nigris F, Migliaccio E et al (2008) p66Shc deletion confers vascular protection in advanced atherosclerosis in hypercholesterolemic apolipoprotein E knockout mice. Endothelium 15:276–287
49. Rota M, LeCapitaine N, Hosoda T et al (2006) Diabetes promotes cardiac stem cell aging and heart failure, which are prevented by deletion of the p66shc gene. Circ Res 99:42–52
50. WHO (2001) Biomarkers in risk assessment: validity and validation. Available from: http://www.who.int/iris/handle/10665/42363. Accessed 29 Nov 2017
51. Semba RD, Bandinelli S, Sun K et al (2009) Plasma carboxymethyl-lysine, an advanced glycation end product, and all-cause and cardiovascular disease mortality in older community-dwelling adults. J Am Geriatr Soc 57:1874–1880
52. Chang JB, Chu NF, Syu JT et al (2011) Advanced glycation end products (AGEs) in relation to atherosclerotic lipid profiles in middle-aged and elderly diabetic patients. Lipids Health Dis 10:228
53. Brinkley TE, Nicklas BJ, Kanaya AM et al (2009) Plasma oxidized low-density lipoprotein levels and arterial stiffness in older adults: the health, aging, and body composition study. Hypertension 53:846–852
54. Trpkovic A, Resanovic I, Stanimirovic J et al (2015) Oxidized low-density lipoprotein as a biomarker of cardiovascular diseases. Crit Rev Clin Lab Sci 52:70–85
55. Wang J, Tan GJ, Han LN et al (2017) Novel biomarkers for cardiovascular risk prediction. J Geriatr Cardiol 14:135–150
56. Tang WH, Katz R, Brennan ML et al (2009) Usefulness of myeloperoxidase levels in healthy elderly subjects to predict risk of developing heart failure. Am J Cardiol 103:1269–1274
57. Teerlink T, Luo Z, Palm F et al (2009) Cellular ADMA: regulation and action. Pharmacol Res 60:448–460
58. Tain YL, Huang LT (2011) Asymmetric dimethylarginine: clinical applications in pediatric medicine. J Formos Med Assoc 110:70–77
59. Tain YL, Hsu CN (2017) Toxic dimethylarginines: asymmetric dimethylarginine (ADMA) and symmetric dimethylarginine (SDMA). Toxins (Basel) 9:E92
60. Huang LT, Hsieh CS, Chang KA et al (2012) Roles of nitric oxide and asymmetric dimethylarginine in pregnancy and fetal programming. Int J Mol Sci 13:14606–14622
61. Böger RH, Maas R, Schulze F et al (2009) Asymmetric dimethylarginine (ADMA) as a prospective marker of cardiovascular disease and mortality–an update on patient populations with a wide range of cardiovascular risk. Pharmacol Res 60:481–487
62. Franceschelli S, Ferrone A, Pesce M et al (2013) Biological functional relevance of asymmetric dimethylarginine (ADMA) in cardiovascular disease. Int J Mol Sci 14:24412–24421
63. Alpoim PN, Sousa LP, Mota AP et al (2015) Asymmetric Dimethylarginine (ADMA) in cardiovascular and renal disease. Clin Chim Acta 440:36–39
64. Pizzarelli F, Maas R, Dattolo P et al (2013) Asymmetric dimethylarginine predicts survival in the elderly. Age (Dordr) 35:2465–2475
65. Geiger J, Teichmann L, Grossmann R et al (2005) Monitoring of clopidogrel action: comparison of methods. Clin Chem 51:957–965
66. Erlinge D, Gurbel PA, James S et al (2013) Prasugrel 5 mg in the very elderly attenuates platelet inhibition but maintains noninferiority to prasugrel 10 mg in nonelderly patients: the GENERATIONS trial, a pharmacodynamic and pharmacokinetic study in stable coronary artery disease patients. J Am Coll Cardiol 62:577–583
67. Finck BN, Kelly DP (2007) Peroxisome proliferator-activated receptor gamma coactivator-1 (PGC-1) regulatory cascade in cardiac physiology and disease. Circulation 115:2540–2548

68. Sugden MC, Caton PW, Holness MJ (2010) PPAR control: it's SIRTainly as easy as PGC. J Endocrinol 204:93–104
69. Lira VA, Brown DL, Lira AK et al (2010) Nitric oxide and AMPK cooperatively regulate PGC-1 in skeletal muscle cells. J Physiol 588:3551–3566
70. Dutta D, Calvani R, Bernabei R et al (2012) Contribution of impaired mitochondrial autophagy to cardiac aging: mechanisms and therapeutic opportunities. Circ Res 110:1125–1138
71. Scheitlin CG, Nair DM, Crestanello JA et al (2014) Fluid mechanical forces and endothelial mitochondria: a bioengineering perspective. Cell Mol Bioeng 7:483–496
72. Chiong M, Wang ZV, Pedrozo Z et al (2011) Cardiomyocyte death: mechanisms and translational implications. Cell Death Dis 2:e244
73. Nemchenko A, Chiong M, Turer A et al (2011) Autophagy as a therapeutic target in cardiovascular disease. J Mol Cell Cardiol 51:584–593
74. Valentim L, Laurence KM, Townsend PA et al (2006) Urocortin inhibits Beclin1-mediated autophagic cell death in cardiac myocytes exposed to ischaemia/reperfusion injury. J Mol Cell Cardiol 40:846–852
75. Ma X, Liu H, Foyil SR et al (2012) Impaired autophagosome clearance contributes to cardiomyocyte death in ischemia/reperfusion injury. Circulation 125:3170–3181
76. He C, Zhu H, Zhang W et al (2013) 7-Ketocholesterol induces autophagy in vascular smooth muscle cells through Nox4 and Atg4B. Am J Pathol 183:626–637
77. Martinet W, Schrijvers DM, Timmermans JP et al (2008) Interactions between cell death induced by statins and 7-ketocholesterol in rabbit aorta smooth muscle cells. Br J Pharmacol 154:1236–1246
78. Salabei JK, Cummins TD, Singh M et al (2013) PDGF-mediated autophagy regulates vascular smooth muscle cell phenotype and resistance to oxidative stress. Biochem J 451:375–388
79. Martinet W, De Meyer GR (2009) Autophagy in atherosclerosis: a cell survival and death phenomenon with therapeutic potential. Circ Res 104:304–317
80. Balaban RS, Nemoto S, Finkel T (2005) Mitochondria, oxidants, and aging. Cell 120:483–495
81. Oka T, Hikoso S, Yamaguchi O et al (2012) Mitochondrial DNA that escapes from autophagy causes inflammation and heart failure. Nature 485:251–255
82. Ding Z, Liu S, Wang X et al (2014) LOX-1, mtDNA damage, and NLRP3 inflammasome activation in macrophages: implications in atherogenesis. Cardiovasc Res 103:619–628
83. Hu N, Zhang Y (2017) TLR4 knockout attenuated high fat diet-induced cardiac dysfunction via NF-κB/JNK-dependent activation of autophagy. Biochim Biophys Acta 1863:2001–2011
84. Baigent C, Keech A, Kearney PM et al (2005) Efficacy and safety of cholesterol-lowering treatment: prospective meta-analysis of data from 90,056 participants in 14 randomised trials of statins. Lancet 366:1267–1278
85. Takemoto M, Node K, Nakagami H et al (2001) Statins as antioxidant therapy for preventing cardiac myocyte hypertrophy. J Clin Invest 108:1429–1437
86. Node K, Fujita M, Kitakaze M et al (2003) Short-term statin therapy improves cardiac function and symptoms in patients with idiopathic dilated cardiomyopathy. Circulation 108:839–843
87. Das S, Otani H, Maulik N et al (2005) Lycopene, tomatoes, and coronary heart disease. Free Radic Res 39:449–455
88. Brown DI, Griendling KK (2015) Regulation of signal transduction by reactive oxygen species in the cardiovascular system. Circ Res 116:531–549
89. Chu PY, Campbell MJ, Miller SG et al (2014) Anti-hypertensive drugs in children and adolescents. World J Cardiol 6:234–244
90. Hansson L, Lindholm LH, Ekbom T et al (1999) Randomised trial of old and new antihypertensive drugs in elderly patients: cardiovascular mortality and morbidity the Swedish Trial in Old Patients with Hypertension-2 study. Lancet 354:1751–1756
91. Inoue I, Goto S, Matsunaga T et al (2001) The ligands/activators for peroxisome proliferator-activated receptor alpha (PPARalpha) and PPARgamma increase Cu2+,Zn2+-superoxide dismutase and decrease p22phox message expressions in primary endothelial cells. Metabolism 50:3–11

92. Mercer JR, Yu E, Figg N et al (2012) The mitochondria-targeted antioxidant MitoQ decreases features of the metabolic syndrome in ATM+/-/ApoE-/- mice. Free Radic Biol Med 52:841–849
93. Barry-Lane PA, Patterson C, van der Merwe M et al (2001) p47phox is required for atherosclerotic lesion progression in ApoE(-/-) mice. J Clin Invest 108:1513–1522
94. Jain AK, Mehra NK, Swarnakar NK (2015) Role of antioxidants for the treatment of cardiovascular diseases: challenges and opportunities. Curr Pharm Des 21:4441–4455
95. Tsimikas S (2006) Oxidative biomarkers in the diagnosis and prognosis of cardiovascular disease. Am J Cardiol 98:9P–17P
96. Strobel NA, Fassett RG, Marsh SA et al (2011) Oxidative stress biomarkers as predictors of cardiovascular disease. Int J Cardiol 147:191–201
97. Zeng S, Zhou X, Ge L et al (2014) Monocyte subsets and monocyte-platelet aggregates in patients with unstable angina. J Thromb Thrombolysis 38:439–446
98. Kalyanaraman B, Darley-Usmar V, Davies KJ et al (2014) Measuring reactive oxygen and nitrogen species with fluorescent probes: challenges and limitations. Free Radic Biol Med 52:1–6
99. Ullevig SL, Kim HS, Short JD et al (2016) Protein S-glutathionylation mediates macrophage responses to metabolic cues from the extracellular environment. Antioxid Redox Signal 25:836–851
100. Lotufo PA (2003) After the hype, HOPE (and HPS): some lessons from the Women's Health Initiative trial, the Heart Outcomes Prevention Evaluation study and the Heart Prevention Study. Sao Paulo Med J 121:3–4
101. Lane JS, Magno CP, Lane KT et al (2008) Nutrition impacts the prevalence of peripheral arterial disease in the United States. J Vasc Surg 48:897–904
102. Stephens NG, Parsons A, Schofield PM et al (1996) Randomised controlled trial of vitamin E in patients with coronary disease: Cambridge Heart Antioxidant Study (CHAOS). Lancet 347:781–786
103. Ashor AW, Siervo M, Lara J et al (2014) Antioxidant vitamin supplementation reduces arterial stiffness in adults: a systematic review and meta-analysis of randomized controlled trials. J Nutr 144:1594–1602
104. Gitenay D, Lyan B, Rambeau M et al (2007) Comparison of lycopene and tomato effects on biomarkers of oxidative stress in vitamin E deficient rats. Eur J Nutr 46:468–475
105. Griffin XL, Warner F, Costa M (2008) The role of electromagnetic stimulation in the management of established non-union of long bone fractures: what is the evidence? Injury 39:419–429
106. Ehnert S, Fentz AK, Schreiner A et al (2017) Extremely low frequency pulsed electromagnetic fields cause antioxidative defense mechanisms in human osteoblasts via induction of •O(2)(-) and H(2)O(2). Sci Rep 7:14544
107. Cohen P (2009) Targeting protein kinases for the development of anti-inflammatory drugs. Curr Opin Cell Biol 21:317–324
108. Gehart H, Kumpf S, Ittner A et al (2010) MAPK signalling in cellular metabolism: stress or wellness? EMBO Rep 11:834–840
109. Raggi F, Vallesi G, Rufini S et al (2008) ELF magnetic therapy and oxidative balance. Electromagn Biol Med 27:325–339
110. Funk RH, Monsees T, Ozkucur N (2009) Electromagnetic effects – from cell biology to medicine. Prog Histochem Cytochem 43:177–264
111. Blank M, Goodman R (2009) Electromagnetic fields stress living cells. Pathophysiology 16:71–78
112. Cifra M, Fields JZ, Farhadi A (2011) Electromagnetic cellular interactions. Prog Biophys Mol Biol 105:223–246
113. Meister A, Anderson ME (1983) Glutathione. Annu Rev Biochem 52:711–760
114. Franco R, Schoneveld OJ, Pappa A et al (2007) The central role of glutathione in the pathophysiology of human diseases. Arch Physiol Biochem 113:234–258

115. Gould NS, Day BJ (2011) Targeting maladaptive glutathione responses in lung disease. Biochem Pharmacol 81:187–193
116. Anderson ME, Meister A (1983) Transport and direct utilization of gamma-glutamylcyst(e) ine for glutathione synthesis. Proc Natl Acad Sci U S A 80:707–711
117. Anderson ME (1998) Glutathione: an overview of biosynthesis and modulation. Chem Biol Interact 111–112:1-14.
118. Ballatori N, Krance SM, Notenboom S et al (2009) Glutathione dysregulation and the etiology and progression of human diseases. Biol Chem 390:191–214
119. Zarka MH, Bridge WJ (2017) Oral administration of γ-glutamylcysteine increases intracellular glutathione levels above homeostasis in a randomised human trial pilot study. Redox Biol 11:631–636
120. Cacciatore I, Cornacchia C, Pinnen F et al (2010) Prodrug approach for increasing cellular glutathione levels. Molecules 15:1242–1264
121. Pocernich CB, Butterfield DA (2012) Elevation of glutathione as a therapeutic strategy in Alzheimer disease. Biochim Biophys Acta 1822:625–630
122. Wu JH, Batist G (2013) Glutathione and glutathione analogues; therapeutic potentials. Biochim Biophys Acta 1830:3350–3353
123. Fokkema ML, Vlaar PJ, Vogelzang M et al (2009) Effect of high-dose intracoronary adenosine administration during primary percutaneous coronary intervention in acute myocardial infarction: a randomized controlled trial. Circ Cardiovasc Interv 2:323–329
124. Ibáñez B, Heusch G, Ovize M et al (2015) Evolving therapies for myocardial ischemia/reperfusion injury. J Am Coll Cardiol 65:1454–1471
125. Cung TT, Morel O, Cayla G et al (2015) Cyclosporine before PCI in patients with acute myocardial infarction. N Engl J Med 373:1021–1031
126. Heusch G, Gersh BJ (2017) The pathophysiology of acute myocardial infarction and strategies of protection beyond reperfusion: a continual challenge. Eur Heart J 38:774–784
127. Currie RW, Karmazyn M, Kloc M et al (1988) Heat-shock response is associated with enhanced postischemic ventricular recovery. Circ Res 63:543–549
128. Knowlton AA, Brecher P, Apstein CS (1991) Rapid expression of heat shock protein in the rabbit after brief cardiac ischemia. J Clin Invest 87:139–147
129. Donnelly TJ, Sievers RE, Vissern FL et al (1992) Heat shock protein induction in rat hearts. A role for improved myocardial salvage after ischemia and reperfusion? Circulation 85:769–778
130. Hutter MM, Sievers RE, Barbosa V et al (1994) Heat-shock protein induction in rat hearts. A direct correlation between the amount of heat-shock protein induced and the degree of myocardial protection. Circulation 89:355–360
131. Plumier JC, Ross BM, Currie RW et al (1995) Transgenic mice expressing the human heat shock protein 70 have improved post-ischemic myocardial recovery. J Clin Invest 95:1854–1860
132. Tanimoto T, Parseghian MH, Nakahara T et al (2017) Cardioprotective effects of HSP72 administration on ischemia-reperfusion injury. J Am Coll Cardiol 70:1479–1492
133. Dahm CC, Moore K, Murphy MP (2006) Persistent S-nitrosation of complex I and other mitochondrial membrane proteins by S-nitrosothiols but not nitric oxide or peroxynitrite: implications for the interaction of nitric oxide with mitochondria. J Biol Chem 281:10056–10065
134. Brown GC, Borutaite V (2004) Inhibition of mitochondrial respiratory complex I by nitric oxide, peroxynitrite and S-nitrosothiols. Biochim Biophys Acta 1658:44–49
135. Hill BG, Darley-Usmar VM (2008) S-nitrosation and thiol switching in the mitochondrion: a new paradigm for cardioprotection in ischaemic preconditioning. Biochem J 412:e11–e13
136. Kang PT, Zhang L, Chen CL et al (2012) Protein thiyl radical mediates S-glutathionylation of complex I. Free Radic Biol Med 53:962–973
137. Hurd TR, Requejo R, Filipovska A et al (2008) Complex I within oxidatively stressed bovine heart mitochondria is glutathionylated on Cys-531 and Cys-704 of the 75-kDa subunit: potential role of CYS residues in decreasing oxidative damage. J Biol Chem 283:24801–24815

138. Chen CL, Chen J, Rawale S et al (2008) Protein tyrosine nitration of the flavin subunit is associated with oxidative modification of mitochondrial complex II in the post-ischemic myocardium. J Biol Chem 283:27991–28003
139. Yarian CS, Rebrin I, Sohal RS (2005) Aconitase and ATP synthase are targets of malondialdehyde modification and undergo an age-related decrease in activity in mouse heart mitochondria. Biochem Biophys Res Commun 330:151–156
140. Benderdour M, Charron G, DeBlois D et al (2003) Cardiac mitochondrial NADP+-isocitrate dehydrogenase is inactivated through 4-hydroxynonenal adduct formation: an event that precedes hypertrophy development. J Biol Chem 278:45154–45159
141. McLain AL, Cormier PJ, Kinter M et al (2013) Glutathionylation of α-ketoglutarate dehydrogenase: the chemical nature and relative susceptibility of the cofactor lipoic acid to modification. Free Radic Biol Med 61:161–169
142. Bulteau AL, Lundberg KC, Ikeda-Saito M et al (2005) Reversible redox-dependent modulation of mitochondrial aconitase and proteolytic activity during in vivo cardiac ischemia/reperfusion. Proc Natl Acad Sci U S A 102:5987–5991
143. Stachowiak O, Dolder M, Wallimann T et al (1998) Mitochondrial creatine kinase is a prime target of peroxynitrite-induced modification and inactivation. J Biol Chem 273:16694–16699
144. Kim J, Rodriguez ME, Guo M et al (2008) Oxidative modification of cytochrome c by singlet oxygen. Free Radic Biol Med 44:1700–1711
145. Zarse K, Schmeisser S, Groth M et al (2012) Impaired insulin/IGF1 signaling extends life span by promoting mitochondrial L-proline catabolism to induce a transient ROS signal. Cell Metab 15:451–465
146. Yun J, Finkel T (2014) Mitohormesis. Cell Metab 19:757–766
147. Adlam VJ, Harrison JC, Porteous CM et al (2005) Targeting an antioxidant to mitochondria decreases cardiac ischemia-reperfusion injury. FASEB J 19:1088–1095
148. Ong SB, Subrayan S, Lim SY et al (2010) Inhibiting mitochondrial fission protects the heart against ischemia/reperfusion injury. Circulation 121:2012–2022
149. Kroemer G, Mariño G, Levine B (2010) Autophagy and the integrated stress response. Mol Cell 40:280–293
150. Andres AM, Hernandez G, Lee P et al (2014) Mitophagy is required for acute cardioprotection by simvastatin. Antioxid Redox Signal 21:1960–1973
151. Liao JK (2014) Linking endothelial dysfunction with endothelial cell activation. J Clin Invest 123:540–541
152. Cao SS, Kaufman RJ (2014) Endoplasmic reticulum stress and oxidative stress in cell fate decision and human disease. Antioxid Redox Signal 21:396–413
153. Ron D, Walter P (2014) Signal integration in the endoplasmic reticulum unfolded protein response. Nat Rev Mol Cell Biol 8:519–529
154. Han J, Kaufman RJ (2014) Physiological/pathological ramifications of transcription factors in the unfolded protein response. Genes Dev 31:1417–1438
155. Wang M, Kaufman RJ (2014) Protein misfolding in the endoplasmic reticulum as a conduit to human disease. Nature 529:326–335
156. Schröder M, Kaufman RJ (2014) ER stress and the unfolded protein response. Mutat Res 569:29–63
157. Walter P, Ron D (2014) The unfolded protein response: from stress pathway to homeostatic regulation. Science 334:1081–1086
158. Eletto D, Chevet E, Argon Y et al (2014) Redox controls UPR to control redox. J Cell Sci 127:3649–3658
159. Santos CX, Tanaka LY, Wosniak J et al (2014) Mechanisms and implications of reactive oxygen species generation during the unfolded protein response: roles of endoplasmic reticulum oxidoreductases, mitochondrial electron transport, and NADPH oxidase. Antioxid Redox Signal 11:2409–2427
160. Schröder M, Kaufman RJ (2014) The mammalian unfolded protein response. Annu Rev Biochem 74:739–789

161. Choi SK, Lim M, Byeon SH et al (2014) Inhibition of endoplasmic reticulum stress improves coronary artery function in the spontaneously hypertensive rats. Sci Rep 6:31925
162. Choi SK, Lim M, Yeon SI et al (2014) Inhibition of endoplasmic reticulum stress improves coronary artery function in type 2 diabetic mice. Exp Physiol 101:768–777
163. Kassan M, Galán M, Partyka M et al (2014) Endoplasmic reticulum stress is involved in cardiac damage and vascular endothelial dysfunction in hypertensive mice. Arterioscler Thromb Vasc Biol 32:1652–1661
164. Ji C, Kaplowitz N, Lau MY et al (2014) Liver-specific loss of glucose-regulated protein 78 perturbs the unfolded protein response and exacerbates a spectrum of liver diseases in mice. Hepatology 54:229–239
165. Ricobaraza A, Cuadrado-Tejedor M, Marco S et al (2014) Phenylbutyrate rescues dendritic spine loss associated with memory deficits in a mouse model of Alzheimer disease. Hippocampus 22:1040–1050
166. Ozcan U, Yilmaz E, Ozcan L et al (2014) Chemical chaperones reduce ER stress and restore glucose homeostasis in a mouse model of type 2 diabetes. Science 313:1137–1140
167. Remondelli P, Renna M (2014) The endoplasmic reticulum unfolded protein response in neurodegenerative disorders and its potential therapeutic significance. Front Mol Neurosci 10:187
168. Daosukho C, Chen Y, Noel T et al (2014) Phenylbutyrate, a histone deacetylase inhibitor, protects against Adriamycin-induced cardiac injury. Free Radic Biol Med 42:1818–1825
169. Erbay E, Babaev VR, Mayers JR et al (2014) Reducing endoplasmic reticulum stress through a macrophage lipid chaperone alleviates atherosclerosis. Nat Med 15:1383–1391
170. Cheng WP, Wang BW, Shyu KG (2014) Regulation of GADD153 induced by mechanical stress in cardiomyocytes. Eur J Clin Invest 39:960–971
171. Kudo T, Kanemoto S, Hara H et al (2014) A molecular chaperone inducer protects neurons from ER stress. Cell Death Differ 15:364–375
172. Blais JD, Chin KT, Zito E et al (2014) A small molecule inhibitor of endoplasmic reticulum oxidation 1 (ERO1) with selectively reversible thiol reactivity. J Biol Chem 285:20993–21003
173. Boyce M, Bryant KF, Jousse C et al (2014) A selective inhibitor of eIF2alpha dephosphorylation protects cells from ER stress. Science 307:935–939
174. Lee DY, Lee KS, Lee HJ et al (2014) Activation of PERK signaling attenuates Abeta-mediated ER stress. PLoS One 5:e10489
175. Sokka AL, Putkonen N, Mudo G et al (2014) Endoplasmic reticulum stress inhibition protects against excitotoxic neuronal injury in the rat brain. J Neurosci 27:901–908
176. Li RJ, He KL, Li X et al (2014) Salubrinal protects cardiomyocytes against apoptosis in a rat myocardial infarction model via suppressing the dephosphorylation of eukaryotic translation initiation factor 2α. Mol Med Rep 12:1043–1049
177. Liu Y, Wang J, Qi SY et al (2014) Reduced endoplasmic reticulum stress might alter the course of heart failure via caspase-12 and JNK pathways. Can J Cardiol 30:368–375
178. Hong D, Bai YP, Gao HC et al (2014) Ox-LDL induces endothelial cell apoptosis via the LOX-1-dependent endoplasmic reticulum stress pathway. Atherosclerosis 235:310–317
179. Cnop M, Ladriere L, Hekerman P et al (2014) Selective inhibition of eukaryotic translation initiation factor 2 alpha dephosphorylation potentiates fatty acid-induced endoplasmic reticulum stress and causes pancreatic beta-cell dysfunction and apoptosis. J Biol Chem 282:3989–3997
180. Ladrière L, Igoillo-Esteve M, Cunha DA et al (2014) Enhanced signaling downstream of ribonucleic Acid-activated protein kinase-like endoplasmic reticulum kinase potentiates lipotoxic endoplasmic reticulum stress in human islets. J Clin Endocrinol Metab 95:1442–1449
181. Chouchani ET, Pell VR, Gaude E et al (2014) Ischaemic accumulation of succinate controls reperfusion injury through mitochondrial ROS. Nature 515(7527):431–435
182. Logan A, Shabalina IG, Prime TA, Rogatti S, Kalinovich AV, Hartley RC, Budd RC, Cannon B, Murphy MP In vivo levels of mitochondrial hydrogen peroxide increase with age in mtDNA mutator mice. Aging Cell 13:765–768

183. Dikalov S, Griendling KK, Harrison DG (2007) Measurement of reactive oxygen species in cardiovascular studies. Hypertension 49:717–727
184. Villamena FA, Zweier JL (2004) Detection of reactive oxygen and nitrogen species by EPR spin trapping. Antioxid Redox Signal 6:619–629
185. Lee R, Margaritis M, Channon KM et al (2012) Evaluating oxidative stress in human cardiovascular disease: methodological aspects and considerations. Curr Med Chem 19:2504–2520
186. Zielonka J, Kalyanaraman B (2010) Hydroethidine- and MitoSOX-derived red fluorescence is not a reliable indicator of intracellular superoxide formation: another inconvenient truth. Free Radic Biol Med 48:983–1001
187. Dikalov SI, Mason RP (2001) Spin trapping of polyunsaturated fatty acid-derived peroxyl radicals: reassignment to alkoxyl radical adducts. Free Radic Biol Med 30:187–197
188. Proniewski B, Czarny J, Khomich TI et al (2018) Immuno-spin trapping-based detection of oxidative modifications in cardiomyocytes and coronary endothelium in the progression of heart failure in Tgαq*44 mice. Front Immunol 9:938

Dietary Antioxidants in Mitigating Oxidative Stress in Cardiovascular Diseases

5

Subhoshree Ghose, Swati Varshney, Rahul Chakraborty, and Shantanu Sengupta

5.1 Introduction

Cardiovascular disease (CVD) is a generalized term used to describe debilitating conditions of heart, blood and vasculature of the body including events like coronary artery disease, cerebrovascular disease, congestive heart failure, valvular heart disease, congenital heart disease, venous thromboembolism etc. [1]. Cardiovascular diseases are one of the leading causes of death in both developed and developing countries and accounts for approximately one third of deaths occurring all over the world [2]. In developing countries CVD occurs at an early age as compared to developed countries [3]. Both nutrition and environmental factors are linked to generation of reactive oxygen species (ROS) & reactive nitrogen species (RNS) [4] which are pivotal in the pathogenesis of cardiovascular diseases. Oxidative stress has been identified as one of the major factors for poor cardiovascular health [5] and altered ROS levels have been reportedly associated with atherosclerosis, stroke, congestive heart failure and other forms of cardiac diseases [6–8]. A balance between cellular levels of ROS and dietary intake of antioxidants is important for the maintenance of cardiovascular homeostasis [9].

A number of phytonutrients, vitamins and minerals, polyphenols, anthocyanins, flavonoids, saturated and unsaturated fat have shown beneficial implications in reducing cardiovascular risk [10]. Phytochemicals are known to have anti-inflammatory properties like inhibiting oxidation of LDL, vascular smooth muscle cell (VSMC) proliferation and platelet aggregation [11] Polyphenols show cardio

S. Ghose · S. Varshney · R. Chakraborty · S. Sengupta (✉)
Cardio-Respiratory Disease Biology, Genomics and Molecular Medicine Unit, CSIR-Institute of Genomics and Integrative Biology (CSIR-IGIB), New Delhi, India

Academy of Scientific &Innovative Research (AcSIR), Ghaziabad, India
e-mail: shantanus@igib.res.in

S. Chakraborti et al. (eds.), *Oxidative Stress in Heart Diseases*,
https://doi.org/10.1007/978-981-13-8273-4_5

protective activity by altering signaling pathways promoting anti-inflammatory mediators and limiting endothelial dysfunction [12]. Interestingly, they also alter DNA methylation status of key genes and regulate the expression of histone acetylases which are implicated in cardiac diseases [13, 14]. Apart from the traditional antioxidants, a few nonmetal trace elements like Selenium has shown to be a more potent antioxidant than vitamin E, C and β carotene, but is more toxic [15].

With this diverse range of antioxidants, it is important to understand the underlying molecular pathways affected by them. This is all the more important since several randomized intervention trials did not show beneficial effect of antioxidants [16]. For instance, low vitamin levels are associated with increased levels of homocysteine, a thiol amino acid implicated as an independent risk factor for CVD. However, supplementation of B vitamins had no beneficial outcomes in terms of reduction of CVD risk [17]. This could be due to the differences in absorption and transport, sample cohort selection, dose of antioxidants [18] which is likely to vary among individuals. The controversy regarding antioxidant supplementation thus warrants further- replication in large scale population trials.

Over the years, a number of drugs have been repurposed as antioxidants like NAC (N-acetyl-cysteine) and allopurinol and are used as therapeutic agents for CVDs [19]. NAC has been identified as an anti-thrombotic agent used in Type 2 diabetic patients [20]. On the other hand, allopurinol generally known as xanthine oxidase inhibitor is shown to improve left ventricular function and cardiac contractility [21]. Recent reports also claim that antioxidants confer better activities by modulating Th1/Th2 profile and it is observed that antioxidant trials sometimes fail because they are not supplemented taking into account their immune-oxidative properties [22]. Antioxidants, thus, employ several mechanisms to scavenge free radical stress inside the cell generated via cross talk of several signaling pathways. In this chapter, we broadly put forth the biochemical, pharmacokinetic and immune modulatory properties of dietary antioxidants and their downstream targets of action which have the potential to mitigate oxidative stress under abrogated cardiovascular conditions.

5.2 Dietary Antioxidants as Scavengers of Oxidative Stress: A Preventive Therapy

Mounting evidences suggest that oxidative stress characterized by accumulating levels of reactive oxygen species and reactive nitrogen species collectively called RONS is involved in the progression of cardio metabolic diseases [23, 24]. They are scavenged via antioxidants either enzymatically or non-enzymatically to mitigate oxidative stress in biological systems [25]. The enzymatic group of antioxidants present in the diet include catalase, superoxide dismutase and enzymes of glutathione thioredoxin system i.e. glutathione reductase, glutathione peroxidase and glutathione S transferase, which are involved in breakdown of hydrogen peroxide and hydroperoxides [26]. Superoxide dismutase utilize copper/zinc in the cytosol and manganese in mitochondria as cofactors for catalyzing the breakdown of superoxide anions into oxygen and hydrogen peroxide [27]. The non-enzymatic group on the other hand, comprises antioxidants which act on oxidative agents directly and can be acquired from dietary sources including vitamins (vitamin A, E, C), carotenoids, flavonoids, polyphenols, proline and many more [28]. The interplay of these antioxidants aid in survival against environmental stresses. In this section, we discuss the bioavailability, mechanism of action and therapeutic benefits of the natural antioxidants. The efficacy of the antioxidants can vary depending on several important factors both intrinsic (activation energy, rate constants, volatility, heat susceptibility, solubility, oxidation reduction potential) and extrinsic (radical chain reaction inhibitors, metal chelators, enzyme cofactors) properties [29].

5.2.1 Vitamins and Minerals

Vitamins and minerals owing to their antioxidant properties, play a major role in prevention of CVDs [30]. Antioxidant vitamins include vitamin C, vitamin E, folic acid, B group vitamins (B6 and B12), vitamin D, and coenzyme Q10. Vitamins are excellent free radical quenchers with the only difference stemming from their varying solubility in lipids and water. Vitamin C is water soluble and mostly abundant in body fluids [31] whereas vitamin A & E owing to their lipid soluble nature are mostly found in cell membranes and lipoproteins [32, 33]. Folic acid (Vitamin B9) also acts as a powerful anti-oxidant [34]. They are involved in lowering down the risk of several diseases like cancer, cardiovascular disease, diabetes and neurological diseases [35–38].

5.2.1.1 Vitamin C

Vitamin C (commonly known as ascorbic acid) is a hydrophilic molecule and is known to scavenge free radicals in aqueous environment within the cell [39]. It has a unique structure with two adjacent hydroxyl groups and a carbonyl, which makes it an excellent hydrogen or electron donor [40]. It therefore, acts as a co-factor of enzymatic reactions and as an anti-oxidant too. On oxidation, ascorbate is converted into ascorbate free radical (AFR), which does not undergo further oxidation but instead reduced back to ascorbate via NADH-dependent and independent mechanisms [41]. This increased oxidative condition leads to AFR accumulation, which results in reaction of two AFR molecules to produce one molecule of ascorbate and one molecule of dehydroascorbate (DHA). DHA can then be reduced back to ascorbate, or hydrolyzed to gulonic acid [42].

There is an inverse relationship between vitamin C and coronary artery disease and it has been reported that the concentration of ascorbate levels were lower in the aorta of atherosclerotic patients in comparison to healthy controls. Vitamin C inhibits LDL oxidation and it has been speculated that low vitamin C levels in the aorta could predispose the person to LDL oxidation leading to atherosclerosis [43]. The experimental models of atherosclerosis in guinea pigs, rabbits and rats are not well established. But since these animals cannot synthesize vitamin C and need supplementation to fulfill the requirements, it has been widely used to study ascorbic acid deficiency. Guinea pigs were fed with low vitamin C containing diet to induce vascular dysfunction. Low vitamin C diet along with high fat diet in guinea pigs is shown to be associated with increased severity of atherosclerosis [44, 45] and on supplementation resorption of lesion has been observed [46]. It has also been shown that with increased intake of vitamin C cholesterol deposition gradually decreased in scorbutic guinea pig model [47]. Various studies have also shown the development of aortic tissue abnormalities such as endothelial cell proliferation and intimal fibrotic plaque formation in chronic hypovitaminosis model of vitamin C [48]. On the other hand, anti- atherogenic effects of vitamin C supplementation using rabbit have been described using hypercholesterolemic animal models [49, 50]. The supplementation of vitamin C in rabbits is shown to reduce the lipid and cholesterol accumulation, intimal thickening, and lipid-laden foam cells and also led to lower severity of atherosclerosis in cholesterol fed diet [51]. Thus, low levels of vitamin C leads to vascular dysfunction in guinea pig based atherosclerotic models and supplementation of vitamin C inhibit cholesterol-induced atherosclerosis in rabbit models. In addition to these two-model systems, effect of vitamin C supplementation on rats and nonhuman primate has also been reported. In rats it is shown that vitamin C supplementation reduced cholesterol and phospholipid deposition [52]. The nonhuman primate done in monkeys suggested hypercholesterolemic effect of vitamin C deficiency [53]. The relationship between vitamin C status and total serum cholesterol level is complex. Low vitamin C levels seem to be associated to several cardiovascular risk factors, including high serum cholesterol, low HDL, hypertension, smoking, etc. [54–59]. However, it is not well understood whether these are instrumental in the development of atherosclerosis or is merely secondary markers for these risk factors. The epidemiological data on descriptive, dietary intake, blood based, case-control, prospective cohort studies suggest the possible benefit of vitamin C but the available data are controversial [60]. Various *in vivo* and *in vitro* studies have been carried out to investigate the possible modulation of cholesterol absorption, biosynthesis, catabolism, and excretion due to vitamin C. It has been shown that the hypercholesterolemic effect of chronic vitamin C deficiency is probably due to decreased catabolism of dietary cholesterol [61, 62]. Further, it is reported that the biosynthesis of cholesterol due to vitamin C deficiency is at best partially impaired due to inhibition of HMG-COA reductase, a rate limiting enzyme in cholesterol biosynthesis [63–67]. Vitamin C deficiency impairs the cholesterol catabolism to bile acid via cholesterol 7a-hydroxylase [68–75]. Interestingly, vitamin C does not have direct effect on cholesterol-7α-hydroxylase instead it decrease the microsomal cytochrome P-450 fraction specifically involved in cholesterol 7α-hydroxylation [69, 72, 76]. The high cholesterol levels in vitamin C deficiency could also be due to lipoprotein metabolism. Vitamin C deficiency has been shown to cause low plasma HDL cholesterol, low HDL to total cholesterol ratio [77–79], increased serum LDL levels mediated by decreased catabolic rate of LDL [79, 80] affect LDL receptor synthesis [81] and decrease activity of lipoprotein lipases with resultant hypertriglyceridemia [45, 82, 83].

5.2.1.2 Vitamin E

Vitamin E includes a group of lipophilic molecules mainly four tocopherols and four tocotrienols [84]. These molecules are lyophilic in nature due to the presence of long saturated and unsaturated phytyl chain in tocopherols and tocotrienols respectively [85, 86]. However, the antioxidant property of Vitamin E is contributed by the presence of a chromanol ring. Naturally occurring vitamin E s are α, β, γ, and δ defined by the position of methyl or proton groups attached to chromanol ring. α-tocopherol is the most abundant and biologically active form of vitamin E in plasma [87]. Due to its lipophilic nature it is mostly found in plasma membrane (mainly in Golgi membrane and lysosome) [88] and lipoproteins [89]. Vitamin E is known to exhibit various biological functions that are mediated by its both non-antioxidant (by modulating various signaling pathways) [90, 91], and anti-oxidant nature (by acting as free radical scavenger) [84]. Activities of various proteins involved in signal transduction such as protein kinase C and B, protein tyrosine kinases, 5-, 12-, and 15-lipoxygenases, cyclooxygenase-2, phospholipase A2, protein phosphatase 2A, protein tyrosine phosphatase, and diacylglycerol kinases are modulated by vitamin E [92, 93]. These proteins play a crucial role in smooth muscle cell proliferation [94, 95]. Vitamin E has an inhibitory effect on protein kinase C (PKC) mediated by the activation of phospho-serine/threonine phosphatase 2A (PP2A) which dephosphorylate PKC [92, 96]. The inhibition of PKC prevents smooth muscle cell proliferation [94, 95]. Vitamin E reduces oxidized LDL mediated apoptosis and potentially lower atherogenic effects via reduction of mitogen activated protein kinase (MAPK) [97]. It is also known to inhibit Protein Kinase B (PKB) and activate protein tyrosine phosphatase, both prevent cell proliferation and help in cell survival [93].

Another major role played by vitamin E is the enhancement of endothelial function by increasing the production of the prostanoids PGI1 and PGE2 in human aortic endothelial cells by increasing the release of arachidonic acid. [98]. The increased release of AA is due to increased phospholipase A(2) accompanied by the decrease in cyclo-oxygenase (COX) 1 and 2 activity.[98]. Vitamin E also increases nitric oxide dependent relaxation by increasing the phosphorylation of endothelial nitric oxide synthase (eNOS) at serine 1177, leading to increased nitric oxide (NO) levels [99] thus, preventing atherosclerosis by improving the endothelial dysfunction. Vitamin E also decreases vascular cell adhesion molecule-1 (VCAM-1) expression in hypercholesterolemic patients [100].

Vitamin E is also known for the regulation of inflammatory processes which is mediated via regulating expression of cell surface and cell adhesion proteins and inflammatory chemokines [91, 101–106]. Supplementation of vitamin E in humans has shown to decrease the levels of these cell adhesion molecules [100, 105, 107]. The pro-inflammatory cytokines, which are inhibited by Vitamin E includes TNF-α (tumor necrosis factor-α) [108] and IL-1β (interleukin-1 β) [90, 106]. Further, vitamin E plays a role in decreasing the expression of CD36 on the surface of macrophages thus reducing the uptake of oxidized LDL [109, 110]. Vitamin E also regulates the respiratory burst in macrophages by inhibiting the inducible NOS (iNOS) and NADPH oxidase [111, 112].

The most widely discussed role of vitamin E is its antioxidant functions to prevent the lipid peroxidation [113] and protecting polyunsaturated fatty acids present in the membrane. Vitamin E scavenges the lipid peroxyl radicals, further preventing the free radical generation and hence termination of the oxidative chain reaction [84]. Vitamin E oxidized in the process is reverted back by the action of Vitamin C and ubiquinols, preventing the accumulation of vitamin E radicals and peroxidation of lipids [114, 115]. Alpha tocopherol initiates the antioxidant cycle by transferring the H transfer to lipid peroxyl radicals. Due to the presence of polar phenol group of vitamin E, it does not diffuse in micelle but remain at the interface, which helps this molecule to be present at the interface of water membranes. This particular arrangement helps in reaction of vitamin E free radical with water-soluble antioxidants, which regenerate vitamin E [116, 117]

5.2.1.3 Vitamin A

Vitamin A consists of unsaturated lipophilic molecules that include retinol, retinal, retinoic acid and provitamins A carotenoids (mostly β-carotene) [118]. Retinol (alcoholic form) is the dietary form of the vitamin A [119]. Vitamin A is essentially obtained from the dietary sources such as liver, milk, leafy and root vegetables. All forms of vitamin A have a beta-ionone ring attached to a retinyl group. Vitamin A majorly exists as an ester (retinyl palmitate) in tissues, which is converted to an aldehyde (retinal), or as an acid (retinoic acid) form. Retinol can also be converted to retinal by oxidation, which can be in turn further oxidized to retinoic acid [119, 120]. Retinol and retinyl esters act as a precursor for the biologically active isoforms of retinoid, which includes all-trans (predominant forms), 11-cis, 13-cis, 9,13-di-cis, 9-cis, and 11,13-di-cis forms [120, 121]. Trans forms of retinoic acid influence the biological activities by activating certain members of the steroid hormone nuclear receptor family particularly RAR and RXR. The RARs and RXRs has subtypes α, β, and γ which binds to different forms of retinoids to regulate steroid hormone controlled mechanisms such as metabolism, inflammation, immune function, and development of sexual characteristics.

Vitamin A and its active metabolites also have both anti-oxidative and anti-proliferation properties such as prevention of angiogenesis, cellular growth and oxidative balance and thus are believed to be relevant for the atherosclerotic process and hence cardiovascular risk [121–124]. Several studies have highlighted the effect of trans retinoic acid on intimal hyperplasia following vascular injury using animal models [122]. All-trans retinoic acid have also been shown to inhibit proliferation, increase smooth muscle cell migration and decrease its differentiation at the site of vascular injury, reduce the carotid hyperplasia, endothelial cell mediated vasodilation, increased endothelial cell survival and reducing foam cell formation via RARα [119, 122, 125]. It is shown that since nuclear receptors RARs and RXRs are druggable targets, local stimulation of nuclear receptors in the vessel wall looks promising for potential treatment of atherosclerosis [126]. Retinoic acid also plays a role in inflammation by interfering innate and adaptive immune function by down-regulating B-cell and enhancing T-cell proliferation along with modulation of pro-inflammatory cytokines and C reactive proteins [120, 127, 128]. Retinoic acid through RAR receptors inhibit B cell apoptosis whereas via RXR receptors improves dendritic cell maturation and antigen-presentation in the presence of tumor necrosis factor (TNF) [120]. Retinoic acid also induces IL-1β expression and inhibits IL-1 receptor antagonist expression in activated monocytes. On the other hand in macrophages, it inhibits endotoxin, IFN-γ, IL-12, and TNF production [120, 128]. Retinoids also regulate the endothelial cell adhesion molecule (VCAM1) expression by suppressing TNF-stimulation that is an early stage inflammatory response in atherosclerosis [129]. CRP is an inflammatory biomarker produced at the site of vascular lesion is associated with CVD [130]. CRP production is mainly regulated by IL-6, whereas the IL-6 production is regulated by a transcription factor NF-IL6. Retinoic acid is known to antagonize NF-IL6 thus reducing CRP [120]. Vitamin A's also have antioxidative properties. It can bind to single oxygen species and free radicals forbidding their reaction with polyunsaturated fatty acids and thus reducing lipid peroxidation [119]. Vitamin A also subdues the activity of lipid peroxidation enzymes and prevents protein glycosylation in cell membranes [119, 123]. Vitamin A and CVD have been studied in various cohort studies and case control studies showing inverse relation of plasma retinol levels with CVD risk [131].

5.2.1.4 Vitamin D

Vitamin D (with biological active form 1α,25-dihydroxyvitamin D) is also a fat-soluble vitamin which functions primarily as a steroid hormone. The dietary forms of Vitamin D are vitamin D3 (also known as cholecalciferol) and vitamin D2 (ergocalciferol) obtained from animal and plant products respectively [132]. Vitamin D3 is mainly produced in skin on exposure to ultraviolet light [133]. The two step hydroxylation reactions by cytochrome P450 enzymes (CYP2R1 and CYP27B1) of vitamin D3 and D2 in liver and kidney produces biologically active form 1α,25-dihydroxyvitamin D [1,25(OH)2 D] [132]. The 25-hydroxyvitamin D [25(OH)D] is first produced which is the circulating biomarker of vitamin D status [134, 135]). Subsequent hydroxylation leads to formation of active form, 1α, 25-dihydroxyvitamin D [1,25(OH)2 D] or calcitriol [132, 136]. Vitamin D concentrations in arterial wall can be different from circulation as VSMCs, endothelial cells, macrophages, and dendritic cells can produce calcitriol via CYP27B1 activity [137, 138]. Calcitriol mediates its function by binding to Vitamin D Receptors (VDRs). VDRs belong to a nuclear receptor superfamily and genes with vitamin response elements in their promoter region response to VDRs. These genes are involved in controlling of various processes in CVD, such as stating from cell proliferation and differentiation, oxidative stress, matrix homeostasis, and cell adhesion etc. [139] VDRs have been found in all the major cardiovascular cell types, including VSMC, endothelial cells, cardiomyocytes, most immune cells, and platelets [140–143]. VDR binds calcitriol with high affinity and heterodimerizes with retinoid X receptor [144].

The hormonal derivative of vitamin D, calcitriol, not only influence calcium and phosphorus homeostasis but also influence various cardiovascular outcomes such as inhibition of VSMCs proliferation, preventing vascular calcification, down-regulation of proinflammatory cytokines and up-regulation of anti-inflammatory cytokines, controls parathyroid hormone secretion and regulate renin-angiotensin system negatively [145, 146]. Vitamin D through endocrine, paracrine, and autocrine regulate inflammatory processes [147]. Proinflammatory cytokines such as interleukin-1 (IL-1), IL-2, IL-6, IL-23, tumour necrosis factor-α, and interferon-γ are downregulated, and anti-inflammatory ones such as IL-4 and IL-10 are upregulated [148–150].

5.2.2 Minerals

Minerals play a very crucial role in maintaining cellular metabolism and proper functioning of cardiovascular system. While a number of metals like zinc, iron, copper, magnesium has shown associations with cardiovascular outcome, trace minerals like selenium has shown contentious effects in population trials. Zinc and selenium supplementation have shown to reduce lipid peroxidation, angiogenic and inflammatory markers in rat model [151]. Selenium, in combination with taurine, has also shown protective effect in rats against myocardial infarction where it curtails the expression of proinflammatory and proapoptotic factors and also restore contractility functions [152]. Selenium is known to regulate cardiomyocyte apoptosis through regulation of STAT3 transcription factor and mitochondrial energetics [153]. Selenoproteins also block contributing pathways of atherosclerosis majorly endothelial dysfunction, migration of monocytes, foam cell formation, vascular calcification and apoptosis [154]. Insufficiency of minerals like potassium and magnesium has been associated with predisposition of hypertension, a risk factor of myocardial infarction and stroke [155]. Meta-analysis also suggests a reciprocal relationship between dietary magnesium intake and risk of hypertension [156]. It is speculated that upon magnesium administration, N type calcium channels are inhibited which in turn, also inhibits the release of noradrenaline in spontaneously hypertensive rats [157]. Further, magnesium deficiency has been linked to insulin resistance in vitro [158, 159]. Furthermore, several studies have identified iron overload (also termed as hyperferritinemia) to be associated with inflammation and metabolic syndrome [160]. It is hypothesized that iron accumulation inside the arterial wall macrophages lead to the generation of ROS [161], and activation of NF-kβ with concomitant increase in TNF-α levels [162]. Interestingly, another study suggests that gender specific differences of cardiovascular risk could be explained by the reduced bioavailability of redox active Fe during menstrual period in women [163]. Elevated levels of copper have been shown to be a causative factor for ROS generation and subsequent oxidation of lipids, proteins, DNA & homocysteine [164]. In contrast, copper deficiency has been found to be mediate atherogenic processes, mechanisms of which have been attributed to increased protein glycation, alteration in levels of copper dependent antioxidant enzymes, deterioration of protein structure and functions maintaining structural integrity of heart and blood vessels [165].

5.2.3 Carotenes and Carotenoids

Carotenoids are fat soluble pigments synthesized by plants responsible for imparting protection to the plants from photo-oxidative damage and also impart color to fruits and vegetables [166]. They scavenge singlet molecular oxygen and peroxyl radicals [167]. β-carotenes are the most common form of carotenoids and have shown promising results in intervention trials. Other forms of carotenoids include lycopene, lutein, zeaxanthin which are not converted to pro vitamin A in humans [168]. On the contrary, β carotene is the major source of pro vitamin A in the diet which is found to be effective at reducing the rate of lipid peroxidation and hence could potentially reduce atherosclerotic risk [169]. Further, it improves immune function by protecting phagocytes from auto oxidative damage, enhancing T and B lymphocytes proliferative action, enhancing effector T cell function and macrophage and NK cell tumoricidal capacity [170]. A number of different biological reasons have been attributed to the atheroprotective effects of dietary carotenoids which involve lowering of serum cholesterol by inhibiting cholesterol biosynthetic enzyme, (3-hydroxy-3-methylglutaryl-CoA (HMG-CoA) reductase), increase in serum high density cholesterol (HDL), reduction of acyl-CoA-cholesterol acyl transferase (ACAT) activity which have been identified to be mediated via lycopene administration in animal models [171]. Interestingly, although carotene has been shown to be beneficial, consumption above dietary recommended doses has been shown to be detrimental and could lead to chronic diseases like cancer and cardiovascular disease [172]. Carotenoids are known to regulate oxidative stress via Nrf-2 pathway, thereby facilitating its translocation into the nucleus, and activating phase II enzymes and glutathione-S-transferases. In addition, they inhibit the production of proinflammatory cytokines (IL-8 or PGE-2) by interfering with NF-kβ pathway [173]. Alpha tocopherol and astaxanthin, (group of carotenoids synthesized by microalgae and phytoplanktons) showed improved plaque stability by decreasing macrophage infiltration and decreased apoptosis [174].

5.2.4 Polyphenols

Polyphenols are the most abundant form of antioxidant present in the diet with both pro and antioxidant properties and having contrasting physiological effects [175]. They are mostly present in two predominant forms, derivatives of benzoic acid and cinnamic acid [176]. The various kinds of polyphenols differ in their bioavailability which integrates different genetic and environmental variables [177]. Their mechanism of action go beyond controlling oxidative stress and involve inducing apoptosis and preventing tumor growth [178]. Polyphenols regulate the levels of oxidative stress biomarkers closely associated with cardiovascular diseases and are known to reduce the production of ROS and facilitate the production of nitric oxide (NO) in the aorta [179]. They are known to act as chelator of metal ions like Fe^{2+} and prevent the lipid oxidation caused by highly reactive hydroxyl radicals [180]. Polyphenols have been shown to reverse at least partially the damaging effects caused by high glucose and free fatty acids that tend to decrease eNOS phosphorylation and bioavailability of NO and increase endothelin-1 synthesis all of which are contributors of ROS production and endothelial dysfunction [181]. In a rat model fed with high cholesterol diet, reduced lipid levels were observed upon administration of polyphenols [182]. Dietary intervention with polyphenols in different animal model based studies has highlighted its anti-obesity and hypolipidemic properties [183]. In a myocardial ischemic injury model, polyphenols have shown athero protective benefits by inhibiting H_2O_2 induced oxidative stress through Akt/GSK-3β/caveolae pathway and also by preventing the activation of redox sensitive transcription factors NF-kβ/STAT-1 and subsequent PI-3K/Akt signaling pathway [184]. Also, polyphenols are also known to reduce cytosolic Ca^{2+} overload and improve myocardial contractility [185]

5.2.5 Flavonoids

Flavonoids are a group of natural phytochemicals belonging to the family of phenolic acids having antioxidant, antimicrobial, anti-allergenic and anti-inflammatory properties [186]. Flavonoid intake is associated with reduced LDL oxidation and endothelial injury thus reducing the risk of atherosclerosis [187]. It helps prevent blood clotting, lower blood cholesterol and improve insulin sensitivity [188, 189]. Quercetin, a flavonoid, is known to inhibit xanthine oxidase activity thereby reducing oxidative injury observed in ischemia-reperfusion [190]. Quercetin possesses iron chelating and iron stabilizing properties which again makes them useful as antioxidants [190]. In addition to free radical scavenging and metal ion chelating properties, flavonoids are known to act on several players of protein and lipid signaling pathways like phosphatidylinositol 3 kinase (PI3K), protein kinase B (Akt/PKB), tyrosine kinase, protein kinase C (PKC), and MAP kinase (MAPK) [191]. These exhibit their modulatory effects by altering phosphorylation status of their target genes and altering gene expression [192]. Flavonoids are known to bind to the ATP binding site for a number of proteins like mitochondrial ATPase, topoisomerases, calcium plasma membrane ATPase and protein kinase A and C [192]. They also prevent the activation of oxidative stress induced apoptosis by preventing the activation of JNK and activators of STAT pathway in endothelial cells [193]. Alternatively, they might regulate cytochrome c release during apoptosis by modulating mitochondria transition pore [192]. Flavonoids circumvent inflammation by modulating several pathways such as inhibition of arachidonic acid metabolism [194], reducing complement and platelet activation [195] and also by reducing the release of myeloperoxidases by neutrophils [196].

5.2.6 Anthocyanins

Anthocyanins are an important class of plant secondary metabolites known to protect plants against biotic and abiotic stress [197]. They are important for their cardioprotective activity as well as their protective effects against insulin resistance [198, 199]. Anthocyanin are broadly involved in three important processes of energy metabolism. They are substrates for complex I electron transfer system, mild uncouplers of oxidative phosphorylation and reduce cytochrome c release from mitochondria [199]. They are known to decrease of the activity of NADPH oxidase, reducing the production of superoxides [200]. The mRNA levels of several antioxidant enzymes like catalase, glutathione peroxidase, superoxide dismutase [201], heme oxygenase 1 [202], are also increased by anthocyanins. They have further been shown to inhibit palmitic acid induced ROS production in HUVEC cell line and also decreased p53 levels [179]. Anthocyanins have been shown to possess anti-inflammatory properties in ulcerative colitis patients where a 6 week long treatment with anthocyanin-rich bilberry extract lead to the reduction in the levels of TNF-α and MCP-1 [203]. Additionally, they are known to inhibit the expression of adhesion molecules (ICAM) involved in the recruitment of [204]. In an independent study conducted on hyperlipidemic ApoE knockout mice, anthocyanin extract has shown to improve lipid profile, inhibit vascular inflammation and monocyte migration by inhibiting expression of vascular cell adhesion molecule 1 (VCAM-1) [205]. Besides, anthocyanins inhibit the function of adipogenic transcription factors like peroxisome proliferator-activated receptor-γ (PPAR-γ), CCAAT/enhancer-binding proteins (C-EBPs), sterol regulatory element-binding protein-1c (SREBP-1c) and also reduce the expression of fatty acid binding protein, fatty acid synthase and leptin in 3T3-L1 pre adipocyte cell line, and is thus associated with reduced fat accumulation in adipocytes [206].

The cardioprotective abilities of these conventional antioxidants like vitamins and minerals, polyphenols, flavonoids and anthocyanin are attributed to their immune regulatory properties as well. However, there are a growing number of unconventional dietary antioxidants which mitigate the effects of excess ROS generation under different vascular abnormalities. According to a recent report, a **pecan rich diet** (containing mono and poly unsaturated fat, phenols, flavonoids, proanthocyanidins and essential minerals) was observed to reduce the risk of cardiometabolic diseases in a randomized controlled trial in healthy, middle aged and older adults [207]. This pecan rich diet also improved insulin sensitivity, inflammation and oxidative stress status [207]. Apart from this, a **Mediterranean diet** (rich in α-linolenic acid, ω 3 fatty acids, phytosterols, & folic acid) reportedly have the potential to reduce cardiovascular disease related mortalities [208]. This has also been identified as one of the major preventive therapy for hypertension and thrombosis [209]. As part of a flagship study in "**PREDIMED**" trial a modified form of Mediterranean diet comprising of extra virgin olive oil and almonds was found to reduce the levels of oxidized LDL and blood pressure [210]. A study based on

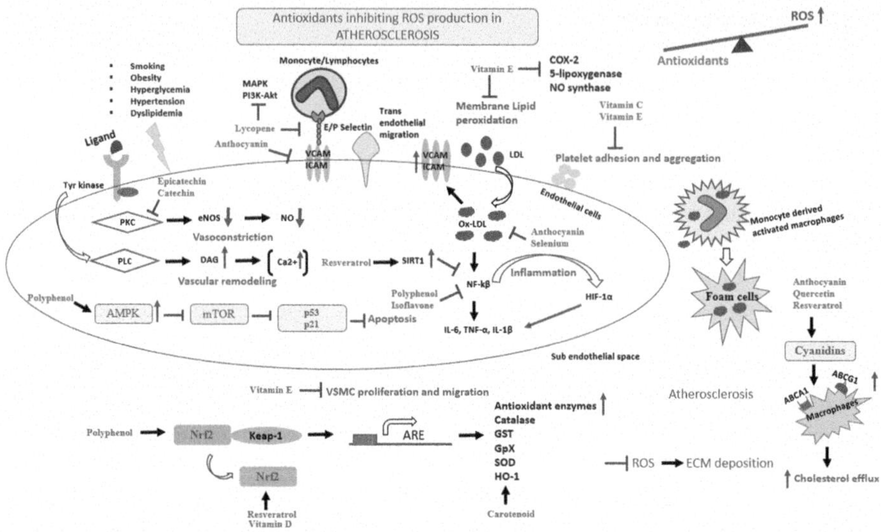

Fig. 5.1 The interplay of several signaling pathways contributing towards atherosclerosis and various antioxidants mitigating the levels of ROS through modulation of different signaling pathways

"PREDIMED" cohort showed that dietary intervention with Mediterranean diet is able to downregulate the expression of atherosclerotic and thrombotic genes playing important role in vascular foam cell formation, ventricular remodeling and inflammation [211]. They have shown inhibitory effects on inflammation related genes like cyclooxygenase 2 (COX-2), low density lipoprotein receptor related protein 1(LRP-1), monocyte chemo attractant protein (MCP-1) and thrombotic genes like tissue factor pathway inhibitor (TFPI) [212]. A number of preclinical and clinical studies have reported the antihypertensive, antithrombotic effects of the antioxidant **dietary flaxseed oil** (enriched in ω-3 fatty acid, α-linolenic acid and the antioxidant lignan secoisolariciresinol diglucoside) which exert beneficial effects in supplementation trials [213]. In streptozotocin-nicotinamide (STZ-NIC) induced diabetic rats, flaxseed oil is shown to upregulate the expression of antioxidant enzymes like catalase (CAT), SOD and GPx, and on the other hand downregulate the expression of heme oxygenase 1 and pro inflammatory genes (TNF-α, IL-6, MCP-1, IFN-γ, NF-κB) [214] (Fig. 5.1).

5.3 Molecular Mechanisms Employed by Antioxidants to Mitigate Oxidative Stress in Cardiovascular Diseases

Oxidative stress and inflammation are major pathological features of cardiovascular diseases [215]. In this section, the advantages of antioxidant supplementation for the prevention of different forms of CVD have been discussed.

5.3.1 Atherosclerosis

Dietary antioxidants inhibit LDL oxidation, cellular lipid peroxidation and control inflammation to mitigate the progression of atherosclerosis and associated vascular complications. Hyperlipidemic conditions, lead to redox imbalance and generation of excess reactive oxygen species leading to endothelial cell injury and atheroma formation [216]. Serum levels of homocysteine, cysteine, malondialdehyde have been identified as markers of oxidative stress which is presumably attenuated via exogenous antioxidant supplementation [217]. In the "**FLAVIOLA HEALTH STUDY**" it has been shown that cocoa flavonol supplementation for a month led to marked improvement in Framingham risk score and predicted a significant lowering of cardiovascular disease risk. In the trial conducted in healthy men and women which was also followed up and reported an improvement in endothelial function [218]. Sardine proteins and by products are known for their anti atherogenic, hypolipidemic, antioxidant properties in high fat diet induced hypercholesterolemic rat models [219]. They directly impact reverse cholesterol pathway by upregulating LCAT (lecithin cholesterol acyltransferase) activity. Moreover, they indirectly prevent lipoproteins from oxidation by increasing paraoxonase-1 (PON-1) activity [220]. Vitamin E has shown protective effects in LDLR knock out mice, by decreasing levels of MCP-1 and increasing levels of NO [221]. Isoflavones have been shown to reduce atherosclerosis progression by inhibiting estrogen induced tyrosine kinase activity, cytokine expression, smooth muscle cell proliferation and platelet aggregation without altering levels of plasma lipoproteins [222]. However, human trials have produced contradictory outcomes. Puerarin, a group of isoflavones have shown to exert atheroprotective effects via reducing the expression of monocyte adhesion molecules like VCAM-1, ICAM-1, MCP-1 & IL-8. Additionally, this antioxidant has also shown to activate ERK/KLF2 signaling axis which in turn regulates the expression of eNOS & thrombomodulin [223]. Citrus flavonoid supplementation has been shown to reverse phenotype associated with adiposity by up regulating Cpt-1α & Pgc-1α in the liver and increasing hepatic fatty acid oxidation in in LDLR KO mice fed with high cholesterol diet. They are also known to lower plasma lipids, activate insulin signaling [189]. Anti-atherogenic & hypolipidemic properties of quercetin flavonoid has been attributed to increased expression of scavenger receptor class B type 1 (SR-B1) and increased levels of LXRα and PPAR-γ. It significantly increased the expression of ABCA1/ABCG1 and SREBP-1c promoting cholesterol efflux. [224]. Similarly, another flavonoid apigenin, has been shown to increase the efflux of cholesterol from macrophages via upregulating the expression of ATP binding cassette transporter ABCA1 which results in lowered total cholesterol, cholesteryl ester and triglycerides in macrophages and also reduced levels of pro inflammatory cytokine expression [225]. In contrast, a study using a cocktail of dietary antioxidants (including vitamin C, vitamin E & β-carotene) in combination with genistein did not show regression of atherosclerotic plaque in old ApoE deficient mice fed with high chow diet [226].

5.3.2 Heart Failure

Several epidemiological studies have emphasized on the importance of nutritional modification to curtail the severity of heart failure patients. An inverse relationship has been noted between blood antioxidant status of β carotene and heart failure incidence in a population based study performed in Finland [227]. Similarly, prospective cohort studies also highlight that long chain plasma mono unsaturated fatty acids (MUFA) are positively related to HF incidence [228] whereas plasma polyunsaturated fatty acids (PUFA) are inversely associated with heart failure severity [229]. Although plenty of association studies have been reported about antioxidants in mitigating the risk of heart failure conditions but very limited studies have studied the potential mechanisms related to this. A recent review suggests that plant based diet (enriched in antioxidants, micronutrients and dietary fiber) improves myocardial contractility and cardiac function by increasing NO bioavailability, gut microbiome alteration and decreasing homocysteine levels [230]. Long term dietary supplementation with resveratrol, a well-known polyphenol (belonging to stilbene group) exhibit substantial improvement in parameters like left ventricular ejection fraction, LV end-systolic and end-diastolic volumes in cases of chronic heart failure [231]. Moreover, resveratrol has been proven to protect against cardiac hypertrophy and ventricular remodeling by inhibiting the activation of p38-MAPK/ERK signaling molecules, COX-2, iNOS activity, thereby reducing the production of ROS and reducing apoptosis [232]. Interestingly, it has been demonstrated that dietary ingestion of nitrate, which is a source of nitric oxide, improves muscle contractility problem in heart failure patients [233]. Baicalin, a flavone glycoside has been shown to inhibit cardiac fibrosis and extracellular collagen accumulation by activating AMP activated protein kinase (AMPK) signaling pathway which then inhibits mTOR pathway and mitigates cardiac hypertrophy and progression to heart failure [234]. A Quercetin derivative, has shown protective effects from cardiomyocyte injury by inhibiting expression of NADPH oxidase 4 (NOX4) and attenuating ROS production which also impacts cardiomyocyte apoptosis via inhibiting of MAPK signaling and p53 phosphorylation [235]. A recent study suggests that the flavonoid, isorhamnetin leads to attenuation of cardiac hypertrophy by blocking the activation of phosphatidylinositol-3-kinase (PI3K-Akt) pathway [236]. Sesamin, a well-known antioxidant rich in flavonoids, is known to restore cardiac hypertrophic phenotype via regulation of Sirtuin3/ROS pathway in a rat model that had undergone transverse aortic constriction (TAC) surgery [237] and shows anti-inflammatory actions via inhibition of MEK-ERK1/2 but does not act upon non canonical TGF-β signaling [237]. It has further been noted that sesamin improves cardiac function in a doxorubicin induced cardiotoxicity model via activation of Sirtuin-1 [238]. However, antioxidants rich in vitamins (vitamin E and vitamin C) and minerals fail to show cardioprotective benefits in coronary artery disease and heart failure patients which warrant further investigation about time and dosage of administration of these supplements [239].

5.3.3 Hypertensive Heart Disease

Hypertension is a chronic clinical condition which is associated with several cardiac diseases such as peripheral arterial diseases, myocardial infarction, heart failure, and stroke [240]. Oxidative stress plays a key role in progression of hypertension [240]. Endothelial nitric oxide synthase (eNOS) in general promotes the production of NO [241], but under vascular complications, eNOS becomes dysfunctional and produce superoxides rather than NO [241] Superoxide ions generated by ROS combines with NO and forms peroxynitrite (synthesized by eNOS). On the other hand, the peroxynitrite produces superoxides by destabilizing eNOS [242]. Tetrahydrobiopterin deficiency or its oxidation promotes uncoupling of eNOS and ROS production. In vascular endothelium, xanthine oxidase is one of the major sources of ROS which is mainly involved in the catalysis of last two steps of purine nucleotide biosynthesis. Angiotensin II, an activator of NADPH oxidase, also contributes to ROS production [243]. The pericarp extract of a tropical fruit "*Garcinia mangostana* Linn" (rich in polyphenols) is commonly used for therapeutic purposes and has been shown to protect against Nω-Nitro-l-arginine methyl ester (L-NAME) induced hypertension and cardiovascular remodeling. Though the underlying mechanism is not well understood but this GME extract potentially suppress p47phox NADPH oxidase subunit and concomitant iNOS expression resulting in increased bioavailability of NO [244]. In an experimental spontaneously hypertensive (SHR) wistar rat model, pomegranate extract (abundant in polyphenols) was able to restore morphological alterations induced in coronary arteries of female wistar rats by preventing coronary angiotensin converting enzyme (ACE) activity [245].

5.3.4 Cardiac Arrhythmia and Ventricular Remodeling

Cardiac arrhythmia is a clinical condition in which there is a disruption of normal heart rate or rhythm causing sudden cardiac death. The pathological processes underlying cardiac arrhythmia is still poorly understood. It promotes cardiac fibrosis and impaired gap junction function which results in reduced monocyte coupling. The mechanism behind ROS induced cardiac arrhythmia is not clearly understood, but abnormal splicing of cardiac sodium channels and activation of protein kinase C (PKC), c-Src tyrosine kinase, Ca^{2+}/CaM dependent kinase II [246] are thought to be responsible for ROS production. Inhibition of PKC can prevent cardiac sodium current reduction caused due to excess generation of mitochondrial ROS [247]. ROS can activate Ca^{2+}/CaM dependent kinase II and in pro oxidant conditions, if CaMKII is activated then two methionine residues are oxidized and sustained activation of CaMKII results in independent binding of Ca^{2+}/CaM [248]. Other probable mechanism through which CaMKII exerts arrhythmic effects are phosphorylation of RyR [249], shifting voltage dependence of Na^+ channel. [250]. **Allopurinol,** a popular dietary antioxidant has been found to be associated with lowered risk of ventricular arrhythmia and associated atrial fibrillation in a retrospective cohort study conducted in the US on elderly subjects [251]. It is speculated that due to its ability to inhibit the enzyme xanthine oxidase, it can mitigate oxidative stress and hence reduce the events of left ventricular dysfunction [251]. Interestingly, reports claim that vitamin C supplementation improves endothelial function by decreasing the levels of the enzyme creatinine kinase and also helps to control blood pressure [252, 253] Recent meta-analysis identifies the suitability of vitamin C to be used for the treatment of atrial fibrillation in several antioxidant trials performed worldwide [254]. Apart from that, **Epigallocatechin-3-gallate (EGCG)**, a particular group of polyphenols have shown anti arrhythmogenic effects in animal models of ischemia-reperfusion injury by directly modulating LA electrophysiological characteristics and calcium homeostasis. It is presumed that the ameliorative action exerted by this specific antioxidant is majorly by inhibition of a cGMP dependent protein kinase [255].

In summary, antioxidants employ various pathways to adapt to the oxidative stress generated by reactive oxygen and reactive nitrogen species in cases of CVDs (Table 5.1 and Fig. 5.2).

Table 5.1 Sources and biological functions of different forms of antioxidants

Type of the antioxidant	Dietary source	RDA (Recommended daily allowance)	Biochemical properties	Functions
Vitamin E	Olive, sunflower oil, Soybean, corn oil, wheat germ, almonds, hazelnuts etc.	15 mg/day	Fat soluble, include tocopherols and tocotrienols, peroxyl radical scavenger, alpha tocopherols constitute 80% of this vitamin group	Down regulates HMG-coA reductase, inactivation of protein kinase C, inhibition of platelet aggregation, protects lipids and prevents oxidation of PUFAs. (Sect. 5.2.1.2)
Vitamin C	Citrus fruits, brussels, sprouts, broccoli, raw bell peppers, strawberries etc.	60 -70 mg/day	Lipid soluble, maximum stability between pH 4-6,	Maintenance of collagen, synthesis of muscle carnitine, conversion of dopamine to norepinephrine. (Sect. 5.2.1.1)
Vitamin A	Cod liver oil, Turkey, chicken liver, capsicum, milk, bell pepper, spirulina, egg, apricot, tomatoes etc.	600-800 ug/day	Fat soluble,	Integrity of epithelial cells, reproduction, growth, immune system development (Sect. 5.2.1.3)

(continued)

Table 5.1 (continued)

Type of the antioxidant	Dietary source	RDA (Recommended daily allowance)	Biochemical properties	Functions
Alpha carotene	Yellow orange vegetables(carrots, pumpkin, sweet potatoes, winter squash), dark green vegetables (broccoli, spinach, parsley, avocado, green beans, green peas) etc.	15-16 mg/day	Pro vitamins, fat soluble	Ensure proper cell division by maintaining cell to cell communication, inhibitor of certain growth factors (N-myc) (Sect. 5.2.3)
Beta carotene	Crude palm oil, pumpkins, papayas, orange root vegetables, carrots, sweet potatoes, mangoes, cantaloupe etc.	2-7 mg/day	Distributed throughout the body, some form of it gets absorbed and circulates with lipoproteins	Facilitate communication between neighboring cells through forming pores within cell membranes and exchange of small molecules, possess anti carcinogenic properties (Sect. 5.2.3)
Flavonoids	Fruits, vegetables, nuts, nuts, seeds, spices, cranberries, apples, parsley, cocoa	150-200 mg/day	Secondary metabolites synthesized through phenylpropanoid pathway, water soluble & accumulate in cell vacuole, usually found in the form of glycosides and acyl glycosides	Involved in nodulation process in plants, protects plants against pathogens and herbivores, UV photoprotection (Sect. 5.2.5)
Polyphenols	Fruits and vegetables, green tea, black tea, red wine, coffee, chocolate, olives, soya, cereals, leguminous plants, grape seeds	1g/day	May acts as mutagens, pro oxidants, inhibitors of tyrosine kinase, Vary on their site of absorption inside human body	Inhibition of LDL oxidation, induce apoptosis and prevent tumor growth (Sect. 5.2.4)

(continued)

Table 5.1 (continued)

Type of the antioxidant	Dietary source	RDA (Recommended daily allowance)	Biochemical properties	Functions
Carotenoids	Carrots, corn, canaries, egg yolk, buttercups, bananas, alder, cottonwood, maple, alder, black cherry	6-10 mg/day	Fat soluble pigments present in plants and microorganisms, categorized in different classes based upon the structure and presence of conjugated double bond	Inhibition of oxidation of fats, supplementation improves immune function, cognitive function. (Sect. 5.2.3)
Anthocyanins	Blueberry, raspberry, black rice, black soybean, eggplant, Blood orange	12.5 mg/day	Water soluble phenolic flavonoid pigments derived from phenylalanine, common in nature as glycosides aglucons anthocyanidins	Anti-thrombotic, anti-inflammatory properties, inhibits platelet aggregation (Sect. 5.2.6)

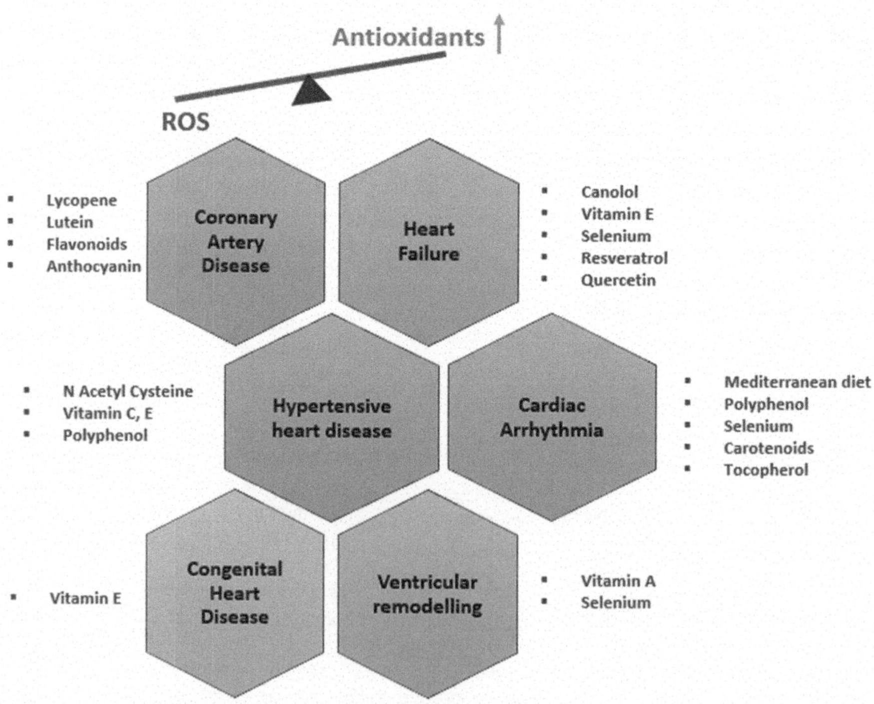

Fig. 5.2 Various antioxidants used in different forms of CVD

5.4 Population Trials and Experimental Model-Based Studies Emphasizing the Role of Dietary Antioxidants: Effect of Supplementation in Clinical Trials

There are many epidemiological studies, which suggest the role of antioxidants in the prevention of the risk of coronary artery diseases. These evidences from the basic research and epidemiological data provide clue for the protective role of antioxidants. Many potential cohort studies have shown inverse associations between dietary antioxidants and risk of heart diseases. But the results of many such clinical trials have been unpromising and unsuccessful to provide substantial evidences towards the protective effect of the antioxidants in cardiovascular disease. Thus, to ascertain whether the supplementation of antioxidant is beneficial or harmful is challenging. List of a few dietary antioxidants, their outcomes in different clinical trials conducted on various cardiovascular diseases is tabulated in Table 5.2.

5.5 Conclusions and Future Perspective

Precisely with the turn of the century, mortality rates due to acute cardiovascular events have grown enormously and antioxidant supplementation has been adopted as a strategy for prevention of cardiovascular diseases. A handful of dietary antioxidants have been identified to possess anti-inflammatory, anti-thrombotic, vasodilatory, anti-apoptotic properties which make them potential therapeutic candidates for the treatment of CVDs. Currently there are no accurate and reliable methods to determine the total antioxidant capacity (TAC) of any form of diet and identifying biomarkers reflective of antioxidant status of an individual becomes a challenging task. Systematic analysis of a number of clinical studies point towards the fact that Dietary total antioxidant capacity (DTAC) of any diet are highly correlated with several CVD risk factors like plasma triglyceride (TG), high density lipoproteins (HDL-C), low density lipoproteins (LDL-C) and more. But these observational trials need further replication in large cohorts to be considered for food fortification. Several trials on antioxidants did not really show cardioprotective effect since it is extremely difficult to identify an ideal cocktail of antioxidant which can mitigate the oxidative stress occurring in case of complex diseases. Interestingly, number of epigenetic changes happen at chromatin and micro RNA levels that are regulated by the repertoire of antioxidants present in our diet which ultimately manifest into pathophysiology of cardiac abnormalities. A few dietary antioxidants, their site of action and their downstream signaling mediators which are instrumental in maintaining the intracellular oxidative balance has been provided in Table 5.3. It however becomes important to understand the mechanistic role of dietary antioxidants in quenching the oxidative stress while deciphering the genetic and epigenetic

Table 5.2 Summary of various studies done with antioxidants and their Outcomes in heart diseases

Antioxidant types	Sub types of Antioxidants	Name of the study	Year	Heart disease	Result of the Study	PMID
Vitamins	Vitamin E	Cambridge Heart Antioxidant Study (CHAOS)	1996	CHD	Protective effect	[256]
	Vitamin E	Italian GISSI-Prevenzione trial (GISSI)	1999	Post-MI	No protective effect	[257]
	Vitamin E	Heart Outcomes Protection Study (HOPE)	2000	CVD	No beneficial effect	[258]
	Vitamin E	The Collaborative Group of the Primary Prevention Project (PPP)	2001	CVD	No beneficial effect	[259]
	Vitamin E	The Alpha-Tocopherol, Beta-carotene Cancer (ATBC) Cancer Prevention Study	1997	CVD	No effect of either beta carotene or Alpha tocopherol	[260]
	Vitamin E	Vitamin E consumption and the risk of coronary disease in women.	1993	CHD	No protective Effect	[261]
	Vitamin E	Cache County study	2007	CVD	Vitamin E use was unrelated to mortality	[262]
	Vitamin E	Secondary prevention with antioxidants of cardiovascular disease in end stage renal disease (SPACE)	2000	CVD	Vitamin E reduces CVD and MI	[263]
	Vitamin C	First National Health and Nutrition Examination Survey (NHANES I)	1992	CVD	Vitamin C intake lowers mortality risk due to CVD	[264]
	Vitamin C	Leukocyte ascorbic acid and its relationship to Coronary heart disease in man.	1980	CAD	Low levels of vitamin C leads to pathogenesis of CAD	[265]
	Vitamin C	Eastern Finland Study	1997	MI	Vitamin C deficiency leads to increased risk of CHD	[266]

	Vitamin C	Third National Health and Nutrition Examination Survey (NHANES III)	1999	Angina, MI, or stroke	Higher intakes of vitamin C may decrease risk of angina	[267]
	Vitamin C	Second National Health and Nutrition Examination Survey (NHANES II)	2001	CVD	Increased consumption may reduce the risk	[268]
	Vitamin C	Nurses' Health Study	2003	CHD	Vitamin C supplements lower risk for CHD	[269]
	Vitamin C	Iowa Women's Health Study42	2004	CVD and Diabetes	High vitamin C intake increase risk of CAD in postmenopausal women with diabetes	[270]
Minerals	Selenium	General population Eastern Finland	1982	CHD	Low serum selenium leads to higher risk of CHD	[271]
	Selenium	Men with high CVD risk	1983	AMI	Low serum selenium is not associated with higher risk of AMI	[272]
	Selenium	Rural men	1985	CHD	Low serum selenium is not associated with higher risk of AMI	[273]
	Selenium	Eastern Finland Heart Survey	1985	CHD	No association between serum selenium and the risk of death from CAD	[274]
	Selenium	First Tromsø Heart Study	1986	AMI	Low serum selenium is not does not predict the development of heart disease	[275]
	Selenium	General population	1987	CVD	Low serum Se is not clearly associated with risk of CVD death	[276]

(continued)

Table 5.2 (continued)

Antioxidant types	Sub types of Antioxidants	Name of the study	Year	Heart disease	Result of the Study	PMID
	Selenium	Second Tromsø Heart Study	1987	AMI	Low serum selenium is not associated with risk of myocardial infarction	[277]
	Selenium	Copenhagen Male Study	1992	CHD	Low serum selenium leads to increased risk of IHD	[278]
	Selenium	Physicians' Health Study	1995	AMI	No evidence for an association between increased plasma selenium and reduced risk of MI	[279]
	Calcium, magnesium, copper, ceruloplasmin, zinc, selenium, iron, ferritin, transferrin, alpha-tocopherol, retinol, folate, vitamin B12, malondialdehyde, orosomucoid and insulin	General elderly Population	1998	CVD	No association of Selenium with increased mortality	[280]
	Serum beta carotene, alpha tocopherol, selenium, and serum fatty acids in cholesterol esters	Men born in Uppsala in 1920–1924	2001	CVD	Partial association was established	[281]
	Selenium	Health Professionals Follow-Up Study	2003	CHD	No relation between selenium status and CHD	[282]

	Selenium	General population trial of Linxian	2004	CHD	Low serum selenium and high incidences of esophageal and gastric cardia cancer, heart disease, stroke, and total death	[283]
	Selenium	Etude du Vieillissement Arteriel (EVA)	2005	CVD	Trace element plays a role in health maintenance in aging individuals	[284]
	Selenium	The serum selenium concentration of patients with acute myocardial infarction	1986	AMI	Serum selenium concentration correlated negatively with minor injury of mitochondria during infarction	[285]
	Selenium	Serum selenium deficiency in myocardial infarction and congestive cardiomyopathy	1987	AMI	No significant difference	[286]
	Selenium	Relationship of serum selenium and antioxidants to plasma lipoproteins, platelet aggregability and prevalent ischemic heart disease in Eastern Finnish men.	1988	CHD	Low serum selenium is associated with IHD	[286]
	Selenium	Decreased selenium levels in acute myocardial infarction.	1989	AMI	Low levels of selenium were reported in infarction	
	Selenium	Decreased blood selenium and risk of myocardial infarction.	1990	AMI	Decreased blood selenium is a risk factor for CHD	[287]

(continued)

Table 5.2 (continued)

Antioxidant types	Sub types of Antioxidants	Name of the study	Year	Heart disease	Result of the Study	PMID
	Selenium	Selenium level in patients with acute myocardial infarct and in patients with severe angina pectoris without myocardial infarct	1995	AMI	Selenium concentration was significantly decreased MI	[288]
	Selenium	The EURAMIC Study	1997	AMI	Low levels of selenium are not an important risk factor of myocardial infarction	[289]
	Selenium	Etude du Viellissement Arterial (EVA) program	1997	AMI	Se and vitamin E levels were raised in cases of lipemia	[290]
	Selenium	Serum and urine selenium concentrations in patients with cardiovascular diseases and relationship to other nutritional indexes	1999	CHD	Urine Se concentrations of AMI patients was significantly low	[291]
	Selenium	Selenium levels and glutathione peroxidase activities in patients with acute myocardial infarction.	1999	AMI	MI exhibit lower plasma, erythrocyte and urinary Se	[292]
	Selenium	Increased plasma glutathione peroxidase activity in patients with acute myocardial infarction.	2001	AMI	No change in Se levels	[293]
	Selenium	Coronary Artery disease Patients Study (SETCAP)	2008	CAD	Sodium selenite supplementation helps increase antioxidant activity in CAD patients	[294]
	Selenium	Nutritional Prevention of Cancer Trial (1983-1996)	2006	CVD	No overall effect of selenium supplementation	[295]

	Selenium	Plasma levels of the antioxidant selenium and risk of myocardial infarction among U.S. physicians.	1995	MI	Increased plasma selenium levels reduce risk of MI	[279]
	Selenium	The Danish PRECISE (Prevention of Cancer by Intervention with Selenium) pilot study	2015	CVD	Supplementation showed no benefit over placebo	[296]
	Selenium	Effect of selenium supplementation after acute myocardial infarction.	1989	MI	Selenium decrease mortality rate	[297]
	Selenium	Selenium Treatment and Chagasic Cardiopathy (STCC): study	2014	Chagasic Cardiopathy	Selenium treatment reduces the progression of Chagas cardiopathy	[298]
	Selenium	Selenium supplementation for primary prevention of cardiovascular disease: proof of no effectiveness.	2014	CVD	No effectiveness	[299]
	Selenium	NHANES III	2010	CHD	Low selenium does not represent high-risk	[300]
	Selenium	Selenium supplementation does not improve vascular responsiveness in healthy North American men.	2009	CVD	No effect of selenium was observed	[301]
Carotenes and Carotenoids	Beta Carotene	Physicians Health Study (PHS)	1996	CVD	Neither benefit or harm	[302]

(continued)

Table 5.2 (continued)

Antioxidant types	Sub types of Antioxidants	Name of the study	Year	Heart disease	Result of the Study	PMID
	Carotenoids	Lipid Research Clinic Coronary Primary Prevention Trial and Follow-up (LRC-CPPT)	1994	CHD	Decreased risk of CHD	[303]
	Carotenoids	Massachusetts Health Care Panel Study	1995	CVD	Decreased risk of CVD	[304]
	Carotenoids	Nurses' Health Study	2003	CAD	Alpha- or beta-carotene intake reduce risk of CAD	[305]
	Carotenoids	Atherosclerosis Risk in Communities (ARIC) Study	1997	Atherosclerosis	Protective effect	[306]
Flavonoids	Flavonoids	Flavonoid intake and long term risk of coronary heart disease and cancer in the seven countries study	1995	CHD	Protective effect	[307]
	Flavonoids	Flavonoid intake and coronary mortality in Finland: a cohort study.	1996	CAD	Low intakes of flavonoids increase risks of coronary disease	[308]
	Flavonoids	Flavonoid intake and risk of chronic diseases	2002	IHD	Flavonoid intake reduces risk of disease	[309]
	Flavonoids	Dietary flavonoid intake and risk of cardiovascular disease in postmenopausal women	1999	CHD	Flavonoid intake was strongly associated with reduced risk of CHD death	[310]
	Catechins	Dietary catechins in relation to coronary heart disease death among postmenopausal women.	2001	CHD	Catechins typical of tea, was not associated with coronary heart disease death	[311]

	Flavonoids	Inverse association of tea and flavonoid intakes with incident myocardial infarction: the Rotterdam Study	2002	MI	An increased intake of tea and flavonoids provide primary prevention of IHD	[312]
	Flavonoids	Antioxidant flavonols and ischemic heart disease in a Welsh population of men: the Caerphilly Study.	1997	IHD	NA	[313]
	Flavonoids	Flavonoid intake and the risk of cardiovascular disease in women	2003	CVD	Flavonoid intake was not strongly associated with a reduced risk of CVD	[314]
	Flavonoids	Relation between intake of flavonoids and risk for coronary heart disease in male health professionals.	1996	CHD	No strong association	[315]
	Flavonoids	Dietary intakes of flavonols and flavones and coronary heart disease in US women.	2007	CHD	No strong association	[316]
	Flavonoids	Flavonoid intake and cardiovascular disease mortality: a prospective study in postmenopausal women	2007	CHD, CVD	Dietary intakes of flavanones, anthocyanidins, and certain foods rich in flavonoids were associated with reduced risk of death due to CHD, CVD, and all causes.	[317]

(continued)

Table 5.2 (continued)

Antioxidant types	Sub types of Antioxidants	Name of the study	Year	Heart disease	Result of the Study	PMID
	Flavanol-rich cocoa	Acute and chronic effects of flavanol-rich cocoa on vascular function in subjects with coronary artery disease: a randomized double-blind placebo-controlled study.	2006	CAD	Flavanol-rich cocoa does not modify vascular function	[318]
	Coffee and tea	Coffee and tea consumption and the prevalence of coronary heart disease in men and women: results from the Scottish Heart Health Study	1993	CHD	Positive relationship between coffee or tea consumption and coronary heart disease	[319]
	Flavonoids	Dietary flavonoids, antioxidant vitamins, and incidence of stroke: the Zutphen study	1996	Stroke	Flavonoids may protect against stroke	[320]
	Tea	Tea consumption and mortality after acute myocardial infarction.	2002	MI	Tea consumption is associated with lower mortality	[321]
	Coffee and tea	Coffee and tea intake and the risk of myocardial infarction	1999	MI	Only tea was associated with a lower risk.	[322]
	Soy isoflavones	Soy isoflavones lower serum total and LDL cholesterol in humans: a meta-analysis of 11randomized controlled trials.	2007	Lipid profile	Soy isoflavones significantly reduced serum total and LDL cholesterol but did not change HDL cholesterol	[323]
	Flavonoids	Cardiovascular effects of flavanol-rich chocolate in patients with heart failure.	2012	CHF	Flavanol-rich chocolate acutely improves vascular function in patients with CHF.	[324]

Polyphenols	Polyphenols	Cardiovascular actions of a standardized polyphenol concentrate on patients undergoing coronary bypass grafting: a randomized, double-blind, placebo-controlled study	2008	CVD	Polyphenols are beneficial	[325]
	Polyphenols	The regular consumption of a polyphenol-rich apple does not influence endothelial function: a randomized double-blind trial in hypercholesterolemic adults	2010	CVD	Polyphenol-rich apple does not improve vascular function	[326]
	Polyphenols	Total polyphenol excretion and blood pressure in subjects at high cardiovascular risk.	2011	CVD	Polyphenol intake, was negatively associated with BP and high cardiovascular risk	[327]
	Polyphenols	Effects of cranberry juice consumption on vascular function in patients with coronary artery disease.	2011	CAD	Acute benefits were found	[328]
	Polyphenols	Polyphenol-rich foods in the Mediterranean diet are associated with better cognitive function in elderly subjects at high cardiovascular risk	2012	CVD	Polyphenols provide better cognitive abilities in patients with CVD	[329]
	Polyphenols	Olive oil polyphenols decrease blood pressure and improve endothelial function in young women with mild hypertension.	2012	CVD	Polyphenol-rich olive oil can decrease BP and improve endothelial function	[330]

(continued)

Table 5.2 (continued)

Antioxidant types	Sub types of Antioxidants	Name of the study	Year	Heart disease	Result of the Study	PMID
	Polyphenols	Regular consumption of cocoa powder with milk increases HDL cholesterol and reduces oxidized LDL levels in subjects at high-risk of cardiovascular disease.	2012	CHD	Cocoa power with milk modulates the lipid profile	[331]
	Anthocyanin	Effects of anthocyanins on cardiovascular risk factors and inflammation in pre-hypertensive men: a double-blind randomized placebo-controlled crossover study.	2013	CVD	No other beneficial effects	[332]
	Polyphenols	Effects of red wine polyphenols and alcohol on glucose metabolism and the lipid profile: a randomized clinical trial.	2013	CVD	Beneficial effect of the non-alcoholic fraction of red wine (mainly polyphenols) on CVD	[333]
	Polyphenols	Diets naturally rich in polyphenols improve fasting and postprandial dyslipidemia and reduce oxidative stress: a randomized controlled trial.	2014	CVD	Polyphenols may positively affect cardiovascular disease risk.	[334]
	Polyphenols	Effects of total dietary polyphenols on plasma nitric oxide and blood pressure in a high cardiovascular risk cohort. The PREDIMED randomized trial.	2015	CVD	Polyphenols have protective effect	[335]

	Polyphenols	Intake of Total Polyphenols and Some Classes of Polyphenols Is Inversely Associated with Diabetes in Elderly People at High Cardiovascular Disease Risk.	2016	CVD	High intake of polyphenols has protective effect	[336]
	Pomegranate polyphenols	Effect of pomegranate extract on blood pressure and anthropometry in adults: a double-blind placebo-controlled randomized clinical trial.	2017	CVD	Might have protective effect	[337]
	Pomegranate polyphenols	Pomegranate supplementation improves cognitive and functional recovery following ischemicstroke: A randomized trial.	2018	Stroke	Provide protective effect	[338]
	Pomegranate polyphenols	Cardioprotective Effects of Pomegranate (Punicagranatum) Juice in Patients with Ischemic Heart Disease.	2017	MI	Provide protective effect	[339]
	Polyphenols	Inverse association between habitual polyphenol intake and incidence of cardiovascular events in the PREDIMED study.	2014	CVD	Greater intake of polyphenols was associated with decreased CVD risk	[340]
	Pomegranate polyphenols	Effects of pomegranate juice consumption on myocardial perfusion in patients with coronary heart disease	2005	MI	Provide protective effect	[341]

(continued)

Table 5.2 (continued)

Antioxidant types	Sub types of Antioxidants	Name of the study	Year	Heart disease	Result of the Study	PMID
	Polyphenols	Polyphenolic compounds from red grapes acutely improve endothelial function in patients with coronary heart disease.	2005	CHD	Has favorable effects of on the cardiovascular system.	[342]
	Polyphenols	Polyphenolic compounds from red grapes acutely improve endothelial function in patients with coronary heart disease.	2013	CVD	Has favorable effects of on the cardiovascular system	[343]
	Black Tea	Effect of black tea consumption on blood cholesterol: a meta-analysis of 15 randomized controlled trials.	2014	Lipid profile	Black tea might not have beneficial effects	[344]
Cocktail of Antioxidants	Vitamin C, carotene, and Vitamin E	Health Professionals Follow-Up Study (HPFS)	1993	CHD	Do not prove any casual effect	[345]
	Carotene, Vitamin C, and E	Longitudinal Population Study (Finland)	1994	CHD	Provide protection	[346]
	Vitamin E, C and Beta-carotene	MRC/BHF Heart Protection Study (HPS) trial	2002	CAD	No significant prevention	[347]
	Vitamin E or placebo, Beta-carotene or placebo, and Aspirin or placebo	Women's Health Study (WHS)	2004	CVD	No Beneficial effect was observed	[348], [349]
	Vitamin E, C, Folic acid and Vitamin B-12	The Women's Antioxidant Cardiovascular Disease Study (WACS)	2004	CVD	No Beneficial effect was observed	[350]
	Vitamin E, C And beta-carotene	Supplementation Vitamins Minerals and Antioxidant trial (SUVIMAX)	2004,2006	IHD	No Beneficial effect was observed	[351, 352]

	Vitamins E and A	Plasma vitamins E and A inversely correlated to mortality from ischemic heart disease in cross-cultural epidemiology.	1989	IHD	Beneficial effect was observed	[353]
	Vitamin E and C	Vitamin E and vitamin C supplement use and risk of all-cause and coronary heart disease mortality in older persons: The Established Populations for Epidemiologic Studies of the Elderly.	1996	CHD	Beneficial effect was observed	[354]
	Vitamins A, C, E, and Carotene.	Risk of angina pectoris and plasma concentrations of vitamins A, C, and E, and carotene.	1991	Angina Pectoris	Can be beneficial	[355]
	Vitamin A and Betacarotene	The Beta Carotene and Retinol Efficacy Trial (CARET)	1996	CVD	No benefit	[356]
	Beta-carotene, vitamin C, and vitamin E	Rotterdam Study	1999	MI	Protective effect of only beta carotene	[357]
	Tea and flavonoid	Rotterdam Study	2002	MI	Increased intake has protective effect	[312]
	Serum vitamins A, C, E, and B12, serum folate, red blood cell folate, serum carotenoids	National Health and Nutrition Examination Survey III	1988-1994	Angina pectoris	Serum carotenoid concentrations reduced risk for angina pectoris	[358]

(continued)

Table 5.2 (continued)

Antioxidant types	Sub types of Antioxidants	Name of the study	Year	Heart disease	Result of the Study	PMID
	Beta-carotene, vitamin C, and vitamin E	Rotterdam study	2001	Peripheral arterial disease	Vitamin C and E intake was significantly inversely associated with peripheral arterial disease but beta carotene	[359]
	Vitamin C, alpha-tocopherol, and provitamin A carotenoids	Atherosclerosis Risk in Communities Study (ARIC)	1995	Atherosclerosis	Dietary vitamin C and alpha-tocopherol may protect against atherosclerotic disease	[360]
	Vitamins A, E, and C	Iowa Women's Health Study	1996	CHD	Vitamin E but vitamins A and C was not associated with lower risks of dying from coronary disease.	[361]
	Vitamin E, C, and multivitamin supplements	Physicians Health Study	2002	MI,CVD	Vitamin E, C, or multivitamins are not associated with decreased risk of total CVD or CHD	[362]
	Vitamin E and vitamin C	Established Populations for Epidemiologic Studies of the Elderly	1996	CHD	Use of vitamins E and C lower risk of CHD mortality	[354]
	Retinol, total carotenoids, and alpha-, gamma-, and total tocopherol	Multiple Risk Factor Intervention Trial (MRFIT)	1998	MI or death from CHD	No significant protective effect	[363]
	Alpha-tocopherol, beta-carotene, vitamin C and selenium	Folate levels determine effect of antioxidant supplementation on micronuclei in subjects with cardiovascular risk	2004	CVD and MI	Folate deficiency may amplify the of disease	[364]

	Selenium and vitamin E	Effect of administration of selenium and vitamin E on heart failure and ventricular arrhythmias in patients with acute myocardial infarct	1999	Heart failure and ventricular arrhythmias	Difference in selenium levels was significant	[365]
	Beta-carotene, vitamins E and C, and multivitamins	Physicians' Health Study II	2000	CVD	NA	[366]
	Selenium and Coenzyme Q10	Supplementation with Selenium and Coenzyme Q10 Reduces Cardiovascular Mortality in Elderly with Low Selenium Status. A Secondary Analysis of a Randomized Clinical Trial	2016	CVD	High mortality in CAD patients with low selenium levels	[367]
	Coenzyme Q10, Zn, vitamin A, B-6, C, Selenium and E	Coenzyme Q10 and antioxidants in acute myocardial infarction.	1994	AMI	No significant changes were found	[368]
	Vitamin E, C, Selenium and beta carotene	Simvastatin and niacin, antioxidant Vitamins, or the combination for the prevention of coronary disease	2001	CVD	No significant changes were found	[369]
	α- and β-carotene and α-, δ- and γ-tocotrienol	Short term effects of palm-tocotrienol and palm-carotenes on vascular function and cardiovascular disease risk: A randomized controlled trial.	2016	CVD	No effects were observed on vascular function or CVD risk factors.	[370]

(continued)

Table 5.2 (continued)

Antioxidant types	Sub types of Antioxidants	Name of the study	Year	Heart disease	Result of the Study	PMID
	Omega 3 and Omega 3 plus Vitamin E	Various Effects of Omega 3 and Omega 3 Plus Vitamin E Supplementations on Serum Glucose Level and Insulin Resistance in Patients with Coronary Artery Disease.	2016	CAD	Co-administration of omega 3 and vitamin E can beneficially decrease serum insulin and insulin resistance in CAD patients.	[371]
	Vitamins C, E and beta-carotene	HAPIEE study	2016	CVD	Vitamin supplementation did not protective effects of antioxidant vitamin intake.	[372]
	Vitamins C and E	PREVEC Trial	2014	AMI	NA	[373]
	Vitamins C and E	MIVIT study	2009	AMI	Vitamins C and E reduces cardiac mortality.	[374] [375]
	Vitamins C and E	Physicians Health Study II	2008	CVD	Neither vitamin E nor vitamin C supplementation reduced the risk of major cardiovascular event	[376]
	n-3 polyunsaturated fatty acids	GISSI-Prevenzione trial.	2007	MI	No evidence was observed about the therapeutic benefit of n-3 PUFA consumption	[377]
	Ascorbic acid, vitamin E, and beta carotene	Women's Antioxidant Cardiovascular Study	2007	CVD	No overall effects of ascorbic acid, vitamin E, or beta carotene on cardiovascular events	[378]

	Folic acid and vitamin E	Folic acid and vitamin E supplementation effects on homocysteinemia, endothelial function and plasma antioxidant capacity in young myocardial-infarction patients.	2004	MI	No significant results were observed	[379]
	Omega-3 polyunsaturated fatty acids	Omega-3 polyunsaturated fatty acids and cardiovascular diseases	2003	CVD	Long-term administration of omega-3 PUFA significantly decreased the risk cardiovascular disease	[380]
	coenzyme Q10	Effect of coenzyme Q10 on risk of atherosclerosis in patients with recent myocardial infarction	2003	MI	CoQ10 may be beneficial in patients with recent MI in treatment	[381]
	Low-dose aspirin and Vitamin E	Collaborative Group of the Primary Prevention Project	2001	CVD	Vitamin E efficacy are not conclusive	[259]
	Alpha-tocopherol and beta-carotene	Randomized trial of alpha-tocopherol and beta-carotene supplements on incidence of major coronary events in men with previous myocardial infarction.	1997	MI	Risk of CHD increased in the on either beta-carotene or the combination of alpha-tocopherol and beta-carotene supplementation	[382]
	Eicosapentaenoic acid, docosahexaenoic acid, oleic acid, folic acid, and vitamins A, B-6, D, and E	Intake of fish oil, oleic acid, folic acid, and vitamins B-6 and E for 1-year decreases plasma C-reactive protein and reduces coronary heart disease risk factors in male patients in a cardiac rehabilitation program.	2007	CHD	Combination of dietary nutrients, reduced a variety of risk factors in MI patients	[383]

(continued)

Table 5.2 (continued)

Antioxidant types	Sub types of Antioxidants	Name of the study	Year	Heart disease	Result of the Study	PMID
	Atorvastatin, vitamin C, and vitamin E	St. Francis Heart Study	2005	CHD	Alpha-tocopherol, vitamin C, and low doses of atorvastatin has no effect on progression of coronary calcification.	[384]
	Vitamins C and E	Endothelial Assessment of Risk from Lipids in Youth (EARLY) Trial	2003	CAD	Vitamins C and E help in restoring endothelial function in hyperlipidemic children.	[385]
	Vitamin C and E	Antioxidant Supplementation in Atherosclerosis Prevention (ASAP) Study	2003	CAD	Vitamin E and vitamin C slows down atherosclerotic progression	[386]
	Vitamin E, vitamin C, vitamin A and fruits	Comparison of antioxidant efficacy of vitamin E, vitamin C, vitamin A and fruits in coronary heart disease: a controlled trial.	2001	CHD	All the antioxidant vitamins and fruits significantly oxidant load in CHD patients	[387]
	Vitamin C and grape-seed polyphenols	The combination of vitamin C and grape-seed polyphenols increases blood pressure: a randomized, double-blind, placebo-controlled trial	2005	Hypertension	Combinations of vitamin C and polyphenols should be taken cautiously	[388]

CHD Coronary Heart Disease, *CVD* Cardiovascular Disease, *CAD* Coronary Artery Disease, *MI* Myocardial Infarction, *AMI* Acute Myocardial Infarction, *IHD* Ischemic Heart Disease, *HF* Heart Failure, *CHF* Congestive Heart Failure

Table 5.3 Interaction of antioxidants with epigenetic modifiers

Antioxidant type	Target site of action	Possible effect	References
Sulforaphane	HDAC, NRF2/Keap1 pathway	Activation of Phase II detoxifying enzymes	[389]
Curcumin	HAT (p300/CBP), NF-kb, HDAC 1,3,8, DNMT1, miR-22	Induction of apoptosis, GATA4 down regulation	[390]
Epigallocatechin-3-gallate (EGCG)	p300 binding to IL-6, HAT, HDAC1	Suppression of inflammation, Reduced NF-kβ activation	[391]
Cocoa	DNMT & MTHFR	Global DNA methylation	[392]
Resveratrol	p300, SIRT1, HDAC11	Inhibition of MAPK, p53, activation of FOXO transcription factors	[393]
Selenium	SREBF1	Global DNA methylation indicated by LINE-1 methylation	[394]
Genistein	SIRT1	Tumor suppressor genes p21, p16 up regulation	[395]
Acetylsalicylic acid	ABCA1	Gene specific DNA methylation reduction	[395]
Quercetin	DNMT1, HDAC	Cell cycle inhibition, induction of apoptosis	[13]

changes that follow post supplementation. Well-designed human population studies are needed to assess the cardioprotective benefits of dietary antioxidants.

References

1. Stewart J, Manmathan G, Wilkinson P (2017) Primary prevention of cardiovascular disease: a review of contemporary guidance and literature. JRSM Cardiovasc Dis 6:2048004016687211
2. Butler D (2011) UN targets top killers. Nature 477:260–261
3. Yusuf S, Reddy S, Ounpuu S et al (2001) Global burden of cardiovascular diseases: part I: general considerations, the epidemiologic transition, risk factors, and impact of urbanization. Circulation 104:2746–2753
4. Liu G, Huang Y, Zhai L (2018) Impact of nutritional and environmental factors on inflammation, oxidative stress, and the microbiome. Biomed Res Int 2018:5606845
5. Khurana S, Piche M, Hollingsworth A et al (2013) Oxidative stress and cardiovascular health: therapeutic potential of polyphenols. Can J Physiol Pharmacol 91:198–212
6. Zhang M, Shah AM (2008) Reactive oxygen species in heart failure. In: Acute heart failure. Springer, London, pp 118–123
7. Shirley R, Ord EN, Work LM (2014) Oxidative stress and the use of antioxidants in stroke. Antioxidants (Basel) 3:472–501
8. Alfadda AA, Sallam RM (2012) Reactive oxygen species in health and disease. J Biomed Biotechnol 2012:936486
9. Poljsak B, Suput D, Milisav I (2013) Achieving the balance between ROS and antioxidants: when to use the synthetic antioxidants. Oxid Med Cell Longev 2013:956792
10. Sikand G, Kris-Etherton P, Boulos NM (2015) Impact of functional foods on prevention of cardiovascular disease and diabetes. Curr Cardiol Rep 17:39

11. Upadhyay S, Dixit M (2015) Role of polyphenols and other phytochemicals on molecular signaling. Oxid Med Cell Longev 2015:504253
12. Tangney CC, Rasmussen HE (2013) Polyphenols, inflammation, and cardiovascular disease. Curr Atheroscler Rep 15:324
13. Fang M, Chen D, Yang CS (2007) Dietary polyphenols may affect DNA methylation. J Nutr 137:223S–228S
14. Kiss AK (2017) Polyphenols and histone acetylation. In: Patel V, Preedy V (eds) Handbook of nutrition, diet, and epigenetics. Springer, Cham, pp 1–21
15. Baraboi VA, Shestakova EN (2004) Selenium: the biological role and antioxidant activity. Ukr Biokhim Zh (1999) 76:23–32
16. Ye Y, Li J, Yuan Z (2013) Effect of antioxidant vitamin supplementation on cardiovascular outcomes: a meta-analysis of randomized controlled trials. PLoS One 8:e56803
17. Ntaios G (2015) Homocysteine, B vitamins, and cardiovascular risk. In: Foods and dietary supplements in the prevention and treatment of disease in older adults. Elsevier, pp 309–318
18. Katsiki N, Manes C (2009) Is there a role for supplemented antioxidants in the prevention of atherosclerosis? Clin Nutr 28:3–9
19. Goszcz K, Deakin SJ, Duthie GG et al (2015) Antioxidants in cardiovascular therapy: panacea or false hope? Front Cardiovasc Med 2:29
20. Rushworth GF, Megson IL (2014) Existing and potential therapeutic uses for N-acetylcysteine: the need for conversion to intracellular glutathione for antioxidant benefits. Pharmacol Ther 141:150–159
21. Stull LB, Leppo MK, Szweda L et al (2004) Chronic treatment with allopurinol boosts survival and cardiac contractility in murine postischemic cardiomyopathy. Circ Res 95:1005–1011
22. Ajith Y, Dimri U, Dixit SK et al (2017) Immunomodulatory basis of antioxidant therapy and its future prospects: an appraisal. In: Inflammopharmacology
23. Brieger K, Schiavone S, Miller FJ Jr et al (2012) Reactive oxygen species: from health to disease. Swiss Med Wkly 142:w13659
24. Rochette L, Lorin J, Zeller M et al (2013) Nitric oxide synthase inhibition and oxidative stress in cardiovascular diseases: possible therapeutic targets? Pharmacol Ther 140:239–257
25. Kabel AM (2014) Free radicals and antioxidants: role of enzymes and nutrition. World J Nutr Health 2:35–38
26. Nordberg J, Arner ES (2001) Reactive oxygen species, antioxidants, and the mammalian thioredoxin system. Free Radic Biol Med 31:1287–1312
27. Bannister JV, Bannister WH, Rotilio G (1987) Aspects of the structure, function, and applications of superoxide dismutase. CRC Crit Rev Biochem 22:111–180
28. Ahmadinejad F, Geir Moller S, Hashemzadeh-Chaleshtori M et al (2017) Molecular mechanisms behind free radical scavengers function against oxidative stress. Antioxidants (Basel) 6. https://doi.org/10.3390/antiox6030051
29. Huang D, Ou B, Prior RL (2005) The chemistry behind antioxidant capacity assays. J Agric Food Chem 53:1841–1856
30. Adams AK, Wermuth EO, Mcbride PE (1999) Antioxidant vitamins and the prevention of coronary heart disease. Am Fam Physician 60:895–904
31. Abdullah M, Attia FN (2018) Vitamin C (ascorbic acid). In: StatPearls. StatPearls, Treasure Island
32. Kemnic TR, Coleman M (2018) Vitamin E deficiency. In: StatPearls. StatPearls, Treasure Island
33. Timoneda J, Rodriguez-Fernandez L, Zaragoza R et al (2018) Vitamin A deficiency and the lung. Nutrients 10. https://doi.org/10.3390/nu10091132
34. Ibrahim W, Tousson E, El-Masry T et al (2012) The effect of folic acid as an antioxidant on the hypothalamic monoamines in experimentally induced hypothyroid rat. Toxicol Ind Health 28:253–261
35. Qin X, Cui Y, Shen L et al (2013) Folic acid supplementation and cancer risk: a meta-analysis of randomized controlled trials. Int J Cancer 133:1033–1041

36. Li Y, Huang T, Zheng Y et al (2016) Folic acid supplementation and the risk of cardiovascular diseases: a meta-analysis of randomized controlled trials. J Am Heart Assoc 5. https://doi.org/10.1161/JAHA.116.003768
37. Song Y, Cook NR, Albert CM et al (2009) Effect of homocysteine-lowering treatment with folic Acid and B vitamins on risk of type 2 diabetes in women: a randomized, controlled trial. Diabetes 58:1921–1928
38. Reynolds EH (2014) The neurology of folic acid deficiency. Handb Clin Neurol 120:927–943
39. Nimse SB, Pal D (2015) Free radicals, natural antioxidants, and their reaction mechanisms. RSC Adv 5:27986–28006
40. Asard H, May J, Smirnoff N (2003) Vitamin C: its functions and biochemistry in animals and plants. Garland Science, New York
41. May JM, Qu ZC, Neel DR et al (2003) Recycling of vitamin C from its oxidized forms by human endothelial cells. Biochim Biophys Acta 1640:153–161
42. Aguirre R, May JM (2008) Inflammation in the vascular bed: importance of vitamin C. Pharmacol Ther 119:96–103
43. Jialal I, Fuller CJ (1995) Effect of vitamin E, vitamin C and beta-carotene on LDL oxidation and atherosclerosis. Can J Cardiol 11(Suppl G):97G–103G
44. Willis GC (1953) An experimental study of intimal ground substance in atherosclerosis. Can Med Assoc J 69:17–22
45. Fujinami T, Okado K, Senda K et al (1971) Experimental atherosclerosis with ascorbic acid deficiency. Jpn Circ J 35:1559–1565
46. Willis GC (1957) The reversibility of atherosclerosis. Can Med Assoc J 77:106–108
47. Ginter E, Bobek P, Babala J et al (1969) The effect of ascorbic acid on the lipid metabolism of guinea-pigs fed an atherogenic diet. Cor Vasa 11:65–73
48. Sulkin NM, Sulkin DF (1975) Tissue changes induced by marginal vitamin C deficiency. Ann N Y Acad Sci 258:317–328
49. Bocan TM, Mueller SB, Brown EQ et al (1992) Antiatherosclerotic effects of antioxidants are lesion-specific when evaluated in hypercholesterolemic New Zealand white rabbits. Exp Mol Pathol 57:70–83
50. Freyschuss A, Xiu R-J, Zhang J et al (1997) Vitamin C reduces cholesterol-induced microcirculatory changes in rabbits. Arterioscler Thromb Vasc Biol 17:1178–1184
51. Zaitsev VF, Myasnikov LA, Kasatkina LV et al (1964) The effect of ascorbic acid on experimental atherosclerosis. Cor Vasa 6:19–25
52. Nambisan B, Kurup PA (1974) The effect of massive doses of ascorbic acid and methionine on the levels of lipids and glycosaminoglycans in the aorta of weanling rats. Atherosclerosis 19:191–199
53. Mcintosh G, Richmond W, Himsworth R (1981) Vitamin C deficiency and hypercholesterolaemia in marmoset monkeys.
54. Gori AM, Sofi F, Corsi AM et al (2006) Predictors of vitamin B6 and folate concentrations in older persons: the InCHIANTI study. Clin Chem 52:1318–1324
55. Frei B, Forte TM, Ames BN et al (1991) Gas phase oxidants of cigarette smoke induce lipid peroxidation and changes in lipoprotein properties in human blood plasma. Protective effects of ascorbic acid. Biochem J 277. (Pt 1:133–138
56. Simon JA (1992) Vitamin C and cardiovascular disease: a review. J Am Coll Nutr 11:107–125
57. Trout DL (1991) Vitamin C and cardiovascular risk factors. Am J Clin Nutr 53:322S–325S
58. Bronte-Stewart B, Roberts B, Wells VM (1963) Serum cholesterol in vitamin C deficiency in man. Br J Nutr 17:61–68
59. Jacques PF, Hartz SC, Mcgandy RB et al (1987) Ascorbic acid, HDL, and total plasma cholesterol in the elderly. J Am Coll Nutr 6:169–174
60. Moser MA, Chun OK (2016) Vitamin C and heart health: a review based on findings from epidemiologic studies. Int J Mol Sci 17:1328
61. Ginter E (1972) Role of ascorbic acid in cholesterol metabolism (literature review). Vopr Pitan 31:25–32

62. Ginter E, Bobek P, Ovecka M (1968) Model of chronic hypovitaminosis C in guinea-pigs. Int Z Vitaminforsch 38:104–113
63. Rodwell VW, Nordstrom JL, Mitschelen JJ (1976) Regulation of HMG-CoA reductase. Adv Lipid Res 14:1–74
64. Bolker HI, Fishman S, Heard RD et al (1956) The incorporation of acetate-1-C14 into cholesterol and fatty acids by surviving tissues of normal and scorbutic guinea pigs. J Exp Med 103:199–205
65. Greene YJ, Harwood HJ Jr, Stacpoole PW (1985) Ascorbic acid regulation of 3-hydroxy-3-methylglutaryl coenzyme A reductase activity and cholesterol synthesis in guinea pig liver. Biochim Biophys Acta 834:134–138
66. Weight M, Kotzé J, De Klerk W et al (1974) The effect of dietary ascorbic acid on cholesterol biosynthesis in vitro. Int J Biochem 5:287–290
67. Holloway DE, Peterson FJ, Prigge WF et al (1981) Influence of dietary ascorbic acid upon enzymes of sterol biosynthesis in the guinea pig. Biochem Biophys Res Commun 102:1283–1289
68. Ginter E, Cerven J, Nemec R et al (1971) Lowered cholesterol catabolism in guinea pigs with chronic ascorbic acid deficiency. Am J Clin Nutr 24:1238–1245
69. Bjorkhem I, Kallner A (1976) Hepatic 7alpha-hydroxylation of cholesterol in ascorbate-deficient and ascorbate-supplemented guinea pigs. J Lipid Res 17:360–365
70. Harris WS, Kottke BA, Subbiah MT (1979) Bile acid metabolism in ascorbic acid-deficient guinea pigs. Am J Clin Nutr 32:1837–1841
71. Holloway DE, Rivers JM (1981) Influence of chronic ascorbic acid deficiency and excessive ascorbic acid intake on bile acid metabolism and bile composition in the guinea pig. J Nutr 111:412–424
72. Peterson FJ, Holloway DE, Duquette PH et al (1983) Dietary ascorbic acid and hepatic mixed function oxidase activity in the guinea pig. Biochem Pharmacol 32:91–96
73. Ginter E, Nemec R, Bobek P (1972) Stimulation of (26- 14 C)cholesterol oxidation by ascorbic acid in scorbutic guinea-pigs. Br J Nutr 28:205–211
74. Ginter E, Nemec R, Cerven J et al (1973) Quantification of lowered cholesterol oxidation in guinea pigs with latent vitamin C deficiency. Lipids 8:135–141
75. Hornig D, Weiser H (1976) Ascorbic acid and cholesterol: effect of graded oral intakes on cholesterol conversion to bile acids in guinea-pigs. Experientia 32:687–689
76. Kritchevsky D, Tepper SA, Story JA (1973) Influence of vitamin C on hydroxylation and side chain oxidation of cholesterol in vitro. Lipids 8:482–484
77. Horio F, Ozaki K, Oda H et al (1987) Effect of dietary ascorbic acid, cholesterol and PCB on cholesterol concentrations in serum and liver in a rat mutant unable to synthesize ascorbic acid. J Nutr 117:1036–1044
78. Kono K, Hayakawa M, Asai K et al (1988) Cholesterol metabolism in inherently scorbutic rats (ODS rats). J Nutr Sci Vitaminol (Tokyo) 34:35–45
79. Uchida K, Nomura Y, Takase H et al (1990) Effect of vitamin C depletion on serum cholesterol and lipoprotein levels in ODS (od/od) rats unable to synthesize ascorbic acid. J Nutr 120:1140–1147
80. Ginter E, Jurčovičcová M (1987) Chronic vitamin C deficiency lowers fractional catabolic rate of low-density lipoproteins in guinea pigs. Ann N Y Acad Sci 498:473–475
81. Aulinskas TH, Van Der Westhuyzen DR, Coetzee GA (1983) Ascorbate increases the number of low density lipoprotein receptors in cultured arterial smooth muscle cells. Atherosclerosis 47:159–171
82. Bobek P, Ginter E (1978) Serum triglycerides and post-heparin lipolytic activity in guinea-pigs with latent vitamin C deficiency. Experientia 34:1554–1555
83. Nambisan B, Kurup PA (1975) Ascorbic acid and glycosaminoglycan and lipid metabolism in guinea pigs fed normal and atherogenic diets. Atherosclerosis 22:447–461
84. Wang X, Quinn PJ (1999) Vitamin E and its function in membranes. Prog Lipid Res 38:309–336

85. Niki E, Kawakami A, Saito M et al (1985) Effect of phytyl side chain of vitamin E on its antioxidant activity. J Biol Chem 260:2191–2196
86. Atkinson J, Epand RF, Epand RM (2008) Tocopherols and tocotrienols in membranes: a critical review. Free Radic Biol Med 44:739–764
87. Engin KN (2009) Alpha-tocopherol: looking beyond an antioxidant. Mol Vis 15:855–860
88. Zhang Y, Turunen M, Appelkvist EL (1996) Restricted uptake of dietary coenzyme Q is in contrast to the unrestricted uptake of alpha-tocopherol into rat organs and cells. J Nutr 126:2089–2097
89. Bowry VW, Ingold KU, Stocker R (1992) Vitamin E in human low-density lipoprotein. When and how this antioxidant becomes a pro-oxidant. Biochem J 288(Pt 2):341–344
90. Devaraj S, Li D, Jialal I (1996) The effects of alpha tocopherol supplementation on monocyte function. Decreased lipid oxidation, interleukin 1 beta secretion, and monocyte adhesion to endothelium. J Clin Invest 98:756–763
91. Wu D, Koga T, Martin KR et al (1999) Effect of vitamin E on human aortic endothelial cell production of chemokines and adhesion to monocytes. Atherosclerosis 147:297–307
92. Mahoney CW, Azzi A (1988) Vitamin E inhibits protein kinase C activity. Biochem Biophys Res Commun 154:694–697
93. Zingg JM (2007) Modulation of signal transduction by vitamin E. Mol Aspects Med 28:481–506
94. Boscoboinik D, Szewczyk A, Hensey C et al (1991) Inhibition of cell proliferation by alpha-tocopherol. Role of protein kinase C. J Biol Chem 266:6188–6194
95. Tasinato A, Boscoboinik D, Bartoli GM et al (1995) d-alpha-tocopherol inhibition of vascular smooth muscle cell proliferation occurs at physiological concentrations, correlates with protein kinase C inhibition, and is independent of its antioxidant properties. Proc Natl Acad Sci U S A 92:12190–12194
96. Boudreau RT, Garduno R, Lin TJ (2002) Protein phosphatase 2A and protein kinase Calpha are physically associated and are involved in Pseudomonas aeruginosa-induced interleukin 6 production by mast cells. J Biol Chem 277:5322–5329
97. De Nigris F, Franconi F, Maida I et al (2000) Modulation by alpha- and gamma-tocopherol and oxidized low-density lipoprotein of apoptotic signaling in human coronary smooth muscle cells. Biochem Pharmacol 59:1477–1487
98. Wu D, Liu L, Meydani M et al (2004) Effect of vitamin E on prostacyclin (PGI2) and prostaglandin (PG) E2 production by human aorta endothelial cells: mechanism of action. Ann N Y Acad Sci 1031:425–427
99. Heller R, Hecker M, Stahmann N et al (2004) Alpha-tocopherol amplifies phosphorylation of endothelial nitric oxide synthase at serine 1177 and its short-chain derivative trolox stabilizes tetrahydrobiopterin. Free Radic Biol Med 37:620–631
100. Desideri G, Marinucci MC, Tomassoni G et al (2002) Vitamin E supplementation reduces plasma vascular cell adhesion molecule-1 and von Willebrand factor levels and increases nitric oxide concentrations in hypercholesterolemic patients. J Clin Endocrinol Metab 87:2940–2945
101. Yoshikawa T, Yoshida N, Manabe H et al (1998) alpha-Tocopherol protects against expression of adhesion molecules on neutrophils and endothelial cells. Biofactors 7:15–19
102. Faruqi R, De La Motte C, Dicorleto PE (1994) Alpha-tocopherol inhibits agonist-induced monocytic cell adhesion to cultured human endothelial cells. J Clin Invest 94:592–600
103. Yoshida N, Yoshikawa T, Manabe H et al (1999) Vitamin E protects against polymorphonuclear leukocyte-dependent adhesion to endothelial cells. J Leukoc Biol 65:757–763
104. Noguchi N, Hanyu R, Nonaka A et al (2003) Inhibition of THP-1 cell adhesion to endothelial cells by alpha-tocopherol and alpha-tocotrienol is dependent on intracellular concentration of the antioxidants. Free Radic Biol Med 34:1614–1620
105. Van Dam B, Van Hinsbergh VW, Stehouwer CD et al (2003) Vitamin E inhibits lipid peroxidation-induced adhesion molecule expression in endothelial cells and decreases soluble cell adhesion molecules in healthy subjects. Cardiovasc Res 57:563–571

106. Devaraj S, Jialal I (1999) Alpha-tocopherol decreases interleukin-1 beta release from activated human monocytes by inhibition of 5-lipoxygenase. Arterioscler Thromb Vasc Biol 19:1125–1133
107. Desideri G, Croce G, Marinucci MC et al (2002) Prolonged, low dose alpha-tocopherol therapy counteracts intercellular cell adhesion molecule-1 activation. Clin Chim Acta 320:5–9
108. Devaraj S, Jialal I (2005) Alpha-tocopherol decreases tumor necrosis factor-alpha mRNA and protein from activated human monocytes by inhibition of 5-lipoxygenase. Free Radic Biol Med 38:1212–1220
109. Munteanu A, Zingg JM, Azzi A (2004) Anti-atherosclerotic effects of vitamin E--myth or reality? J Cell Mol Med 8:59–76
110. Munteanu A, Taddei M, Tamburini I et al (2006) Antagonistic effects of oxidized low density lipoprotein and alpha-tocopherol on CD36 scavenger receptor expression in monocytes: involvement of protein kinase B and peroxisome proliferator-activated receptor-gamma. J Biol Chem 281:6489–6497
111. Khanduja KL, Avti PK, Kumar S et al (2005) Inhibitory effect of vitamin E on proinflammatory cytokines-and endotoxin-induced nitric oxide release in alveolar macrophages. Life Sci 76:2669–2680
112. Cachia O, Benna JE, Pedruzzi E et al (1998) alpha-tocopherol inhibits the respiratory burst in human monocytes. Attenuation of p47(phox) membrane translocation and phosphorylation. J Biol Chem 273:32801–32805
113. Yamauchi R (1997) Vitamin E: mechanism of its antioxidant activity. Food Sci Technol Int Tokyo 3:301–309
114. Niki E (1987) Interaction of ascorbate and alpha-tocopherol. Ann N Y Acad Sci 498:186–199
115. Brigelius-Flohe R (2009) Vitamin E: the shrew waiting to be tamed. Free Radic Biol Med 46:543–554
116. Laranjinha J, Cadenas E (1999) Redox cycles of caffeic acid, alpha-tocopherol, and ascorbate: implications for protection of low-density lipoproteins against oxidation. IUBMB Life 48:57–65
117. Niki E, Noguchi N, Tsuchihashi H et al (1995) Interaction among vitamin C, vitamin E, and beta-carotene. Am J Clin Nutr 62:1322S–1326S
118. Shils ME, Shike M (2006) Modern nutrition in health and disease. Lippincott Williams & Wilkins, Baltimore
119. Blomhoff R, Blomhoff HK (2006) Overview of retinoid metabolism and function. J Neurobiol 66:606–630
120. Rhee EJ, Nallamshetty S, Plutzky J (2012) Retinoid metabolism and its effects on the vasculature. Biochim Biophys Acta 1821:230–240
121. Solomons NW (2012) Vitamin A. In: Present knowledge in nutrition. Wiley-Blackwell, Ames, pp 149–184
122. Streb JW, Miano JM (2003) Retinoids: pleiotropic agents of therapy for vascular diseases? Curr Drug Targets Cardiovasc Haematol Disord 3:31–57
123. Rutkowski M, Grzegorczyk K (2012) Adverse effects of antioxidative vitamins. Int J Occup Med Environ Health 25:105–121
124. Raiten DJ, Namaste S, Brabin B et al (2011) Executive summary--biomarkers of nutrition for development: building a consensus. Am J Clin Nutr 94:633S–650S
125. Neuville P, Yan Z, Gidlof A et al (1999) Retinoic acid regulates arterial smooth muscle cell proliferation and phenotypic features in vivo and in vitro through an RARalpha-dependent signaling pathway. Arterioscler Thromb Vasc Biol 19:1430–1436
126. Kurakula K, Hamers AA, De Waard V et al (2013) Nuclear receptors in atherosclerosis: a superfamily with many 'Goodfellas'. Mol Cell Endocrinol 368:71–84
127. Ross AC (2007) Vitamin A supplementation and retinoic acid treatment in the regulation of antibody responses in vivo. Vitam Horm 75:197–222
128. Thurnham DI, Mccabe GP, Northrop-Clewes CA et al (2003) Effects of subclinical infection on plasma retinol concentrations and assessment of prevalence of vitamin A deficiency: meta-analysis. Lancet 362:2052–2058

129. Chadwick CC, Shaw LJ, Winneker RC (1998) TNF-alpha and 9-cis-retinoic acid synergistically induce ICAM-1 expression: evidence for interaction of retinoid receptors with NF-kappa B. Exp Cell Res 239:423–429
130. De Ferranti S, Rifai N (2002) C-reactive protein and cardiovascular disease: a review of risk prediction and interventions. Clin Chim Acta 317:1–15
131. Brazionis L, Walker KZ, Itsiopoulos C et al (2012) Plasma retinol: a novel marker for cardiovascular disease mortality in Australian adults. Nutr Metab Cardiovasc Dis 22:914–920
132. Kulda V (2012) Vitamin D metabolism. Vnitr Lek 58:400–404
133. Holick MF (2007) Vitamin D deficiency. N Engl J Med 357:266–281
134. Hollis BW (1996) Assessment of vitamin D nutritional and hormonal status: what to measure and how to do it. Calcif Tissue Int 58:4–5
135. Cashman KD, Van Den Heuvel EG, Schoemaker RJ et al (2017) 25-hydroxyvitamin D as a biomarker of vitamin D status and its modeling to inform strategies for prevention of vitamin D deficiency within the population. Adv Nutr 8:947–957
136. Holick MF, Schnoes HK, Deluca HF et al (1971) Isolation and identification of 1,25-dihydroxycholecalciferol. A metabolite of vitamin D active in intestine. Biochemistry 10:2799–2804
137. Razzaque MS (2011) The dualistic role of vitamin D in vascular calcifications. Kidney Int 79:708–714
138. Nagy L, Szanto A, Szatmari I et al (2012) Nuclear hormone receptors enable macrophages and dendritic cells to sense their lipid environment and shape their immune response. Physiol Rev 92:739–789
139. Bouillon R, Carmeliet G, Verlinden L et al (2008) Vitamin D and human health: lessons from vitamin D receptor null mice. Endocr Rev 29:726–776
140. Somjen D, Weisman Y, Kohen F et al (2005) 25-hydroxyvitamin D3-1alpha-hydroxylase is expressed in human vascular smooth muscle cells and is upregulated by parathyroid hormone and estrogenic compounds. Circulation 111:1666–1671
141. Merke J, Milde P, Lewicka S et al (1989) Identification and regulation of 1,25-dihydroxyvitamin D3 receptor activity and biosynthesis of 1,25-dihydroxyvitamin D3. Studies in cultured bovine aortic endothelial cells and human dermal capillaries. J Clin Invest 83:1903–1915
142. Guillot X, Semerano L, Saidenberg-Kermanac'h N et al (2010) Vitamin D and inflammation. Joint Bone Spine 77:552–557
143. Silvagno F, De Vivo E, Attanasio A et al (2010) Mitochondrial localization of vitamin D receptor in human platelets and differentiated megakaryocytes. PLoS One 5:e8670
144. Rosen CJ, Adams JS, Bikle DD et al (2012) The nonskeletal effects of vitamin D: an Endocrine Society scientific statement. Endocr Rev 33:456–492
145. Zittermann A, Schleithoff SS, Koerfer R (2005) Putting cardiovascular disease and vitamin D insufficiency into perspective. Br J Nutr 94:483–492
146. Shapses SA, Manson JE (2011) Vitamin D and prevention of cardiovascular disease and diabetes: why the evidence falls short. JAMA 305:2565–2566
147. Lahoute C, Herbin O, Mallat Z et al (2011) Adaptive immunity in atherosclerosis: mechanisms and future therapeutic targets. Nat Rev Cardiol 8:348–358
148. Cardus A, Panizo S, Encinas M et al (2009) 1,25-dihydroxyvitamin D3 regulates VEGF production through a vitamin D response element in the VEGF promoter. Atherosclerosis 204:85–89
149. Chen Y, Liu W, Sun T et al (2013) 1,25-Dihydroxyvitamin D promotes negative feedback regulation of TLR signaling via targeting microRNA-155-SOCS1 in macrophages. J Immunol 190:3687–3695
150. Helming L, Bose J, Ehrchen J et al (2005) 1alpha,25-Dihydroxyvitamin D3 is a potent suppressor of interferon gamma-mediated macrophage activation. Blood 106:4351–4358
151. Mousavi SN, Faghihi A, Motaghinejad M et al (2018) Zinc and selenium co-supplementation reduces some lipid peroxidation and angiogenesis markers in a rat model of NAFLD-fed high fat diet. Biol Trace Elem Res 181:288–295

152. Dallak M (2017) A synergistic protective effect of selenium and taurine against experimentally induced myocardial infarction in rats. Arch Physiol Biochem 123:344–355
153. Zhang C, Deng Y, Lei Y et al (2017) Effects of selenium on myocardial apoptosis by modifying the activity of mitochondrial STAT3 and regulating potassium channel expression. Exp Ther Med 14:2201–2205
154. Liu H, Xu H, Huang K (2017) Selenium in the prevention of atherosclerosis and its underlying mechanisms. Metallomics 9:21–37
155. Schutten JC, Joosten MM, De Borst MH et al (2018) Magnesium and blood pressure: a physiology-based approach. Adv Chronic Kidney Dis 25:244–250
156. Han H, Fang X, Wei X et al (2017) Dose-response relationship between dietary magnesium intake, serum magnesium concentration and risk of hypertension: a systematic review and meta-analysis of prospective cohort studies. Nutr J 16:26
157. Shimosawa T, Takano K, Ando K et al (2004) Magnesium inhibits norepinephrine release by blocking N-type calcium channels at peripheral sympathetic nerve endings. Hypertension 44:897–902
158. Ma J, Folsom AR, Melnick SL et al (1995) Associations of serum and dietary magnesium with cardiovascular disease, hypertension, diabetes, insulin, and carotid arterial wall thickness: the ARIC study. Atherosclerosis risk in communities study. J Clin Epidemiol 48:927–940
159. Gommers LM, Hoenderop JG, Bindels RJ et al (2016) Hypomagnesemia in type 2 diabetes: a vicious circle? Diabetes 65:3–13
160. Dongiovanni P, Fracanzani AL, Fargion S et al (2011) Iron in fatty liver and in the metabolic syndrome: a promising therapeutic target. J Hepatol 55:920–932
161. Mohammadifard N, Humphries KH, Gotay C et al (2017) Trace minerals intake: risks and benefits for cardiovascular health. Crit Rev Food Sci Nutr 59:1–13
162. She H, Xiong S, Lin M et al (2002) Iron activates NF-κB in Kupffer cells. Am J Physiol Gastrointest Liver Physiol 283:G719–G726
163. Sullivan JL (2003) Are menstruating women protected from heart disease because of, or in spite of, estrogen? Relevance to the iron hypothesis. Am Heart J 145:190–194
164. Galhardi CM, Diniz YS, Faine LA et al (2004) Toxicity of copper intake: lipid profile, oxidative stress and susceptibility to renal dysfunction. Food Chem Toxicol 42:2053–2060
165. Saari JT (2000) Copper deficiency and cardiovascular disease: role of peroxidation, glycation, and nitration. Can J Physiol Pharmacol 78:848–855
166. Riccioni G (2009) Carotenoids and cardiovascular disease. Curr Atheroscler Rep 11:434–439
167. Krinsky NI (2001) Carotenoids as antioxidants. Nutrition 17:815–817
168. Khoo H-E, Prasad KN, Kong K-W et al (2011) Carotenoids and their isomers: color pigments in fruits and vegetables. Molecules 16:1710–1738
169. Kellogg EW, Fridovich I (1975) Superoxide, hydrogen peroxide, and singlet oxygen in lipid peroxidation by a xanthine oxidase system. J Biol Chem 250:8812–8817
170. Bendich A (1989) Carotenoids and the immune response. J Nutr 119:112–115
171. Verghese M, Richardson JE, Boateng J, Shackelford LA, Howard C, Walker LT, Chawan CB (2008) Dietary lycopene has a protective effect on cardiovascular disease in New Zealand male rabbits. J Biol Sci 8:268–277
172. Paiva SA, Russell RM (1999) Beta-carotene and other carotenoids as antioxidants. J Am Coll Nutr 18:426–433
173. Kaulmann A, Bohn T (2014) Carotenoids, inflammation, and oxidative stress--implications of cellular signaling pathways and relation to chronic disease prevention. Nutr Res 34:907–929
174. Li W, Hellsten A, Jacobsson LS et al (2004) Alpha-tocopherol and astaxanthin decrease macrophage infiltration, apoptosis and vulnerability in atheroma of hyperlipidaemic rabbits. J Mol Cell Cardiol 37:969–978
175. Scalbert A, Johnson IT, Saltmarsh M (2005) Polyphenols: antioxidants and beyond. Am J Clin Nutr 81:215S–217S
176. Kurek-Gorecka A, Rzepecka-Stojko A, Gorecki M et al (2013) Structure and antioxidant activity of polyphenols derived from propolis. Molecules 19:78–101

177. Manach C, Scalbert A, Morand C et al (2004) Polyphenols: food sources and bioavailability. Am J Clin Nutr 79:727–747
178. Lambert JD, Hong J, Yang GY et al (2005) Inhibition of carcinogenesis by polyphenols: evidence from laboratory investigations. Am J Clin Nutr 81:284S–291S
179. Furuuchi R, Shimizu I, Yoshida Y et al (2018) Boysenberry polyphenol inhibits endothelial dysfunction and improves vascular health. PLoS One 13:e0202051
180. Hider RC, Liu ZD, Khodr HH (2001) Metal chelation of polyphenols. Methods Enzymol 335:190–203
181. Storniolo CE, Rosello-Catafau J, Pinto X et al (2014) Polyphenol fraction of extra virgin olive oil protects against endothelial dysfunction induced by high glucose and free fatty acids through modulation of nitric oxide and endothelin-1. Redox Biol 2:971–977
182. Hosoyamada Y, Yamada M (2017) Effects of dietary fish oil and apple polyphenol on the concentration serum lipids and excretion of fecal bile acids in rats. J Nutr Sci Vitaminol (Tokyo) 63:21–27
183. Lin JK, Lin-Shiau SY (2006) Mechanisms of hypolipidemic and anti-obesity effects of tea and tea polyphenols. Mol Nutr Food Res 50:211–217
184. Hsieh SR, Cheng WC, Su YM et al (2014) Molecular targets for anti-oxidative protection of green tea polyphenols against myocardial ischemic injury. Biomedicine (Taipei) 4:23
185. Liou YM, Hsieh SR, Wu TJ et al (2010) Green tea extract given before regional myocardial ischemia-reperfusion in rats improves myocardial contractility by attenuating calcium overload. Pflugers Arch 460:1003–1014
186. Pietta P-G (2000) Flavonoids as antioxidants. J Nat Prod 63:1035–1042
187. Yi L, Jin X, Chen C-Y et al (2011) Chemical structures of 4-oxo-flavonoids in relation to inhibition of oxidized low-density lipoprotein (LDL)-induced vascular endothelial dysfunction. Int J Mol Sci 12:5471
188. Nijveldt RJ, Van Nood E, Van Hoorn DE et al (2001) Flavonoids: a review of probable mechanisms of action and potential applications. Am J Clin Nutr 74:418–425
189. Burke AC, Sutherland BG, Telford DE et al (2018) Intervention with citrus flavonoids reverses obesity and improves metabolic syndrome and atherosclerosis in obese Ldlr(-/-) mice. J Lipid Res 59:1714–1728
190. Leopoldini M, Russo N, Chiodo S et al (2006) Iron chelation by the powerful antioxidant flavonoid quercetin. J Agric Food Chem 54:6343–6351
191. Spencer JP (2007) The interactions of flavonoids within neuronal signalling pathways. Genes Nutr 2:257–273
192. Williams RJ, Spencer JP, Rice-Evans C (2004) Flavonoids: antioxidants or signalling molecules? Free Radic Biol Med 36:838–849
193. Choi JS, Choi YJ, Shin SY et al (2008) Dietary flavonoids differentially reduce oxidized LDL-induced apoptosis in human endothelial cells: role of MAPK- and JAK/STAT-signaling. J Nutr 138:983–990
194. Ferrandiz ML, Alcaraz MJ (1991) Anti-inflammatory activity and inhibition of arachidonic acid metabolism by flavonoids. Agents Actions 32:283–288
195. Friesenecker B, Tsai AG, Intaglietta M (1995) Cellular basis of inflammation, edema and the activity of Daflon 500 mg. Int J Microcirc Clin Exp 15(Suppl 1):17–21
196. Ciz M, Denev P, Kratchanova M et al (2012) Flavonoids inhibit the respiratory burst of neutrophils in mammals. Oxid Med Cell Longev 2012:181295
197. Landi M, Tattini M, Gould KS (2015) Multiple functional roles of anthocyanins in plant-environment interactions. Environ Exp Bot 119:4–17
198. Belwal T, Nabavi SF, Nabavi SM et al (2017) Dietary anthocyanins and insulin resistance: when food becomes a medicine. Nutrients 9. https://doi.org/10.3390/nu9101111
199. Liobikas J, Skemiene K, Trumbeckaite S et al (2016) Anthocyanins in cardioprotection: a path through mitochondria. Pharmacol Res 113:808–815
200. Wallace TC (2011) Anthocyanins in cardiovascular disease. Adv Nutr 2:1–7
201. Han KH, Sekikawa M, Shimada K et al (2006) Anthocyanin-rich purple potato flake extract has antioxidant capacity and improves antioxidant potential in rats. Br J Nutr 96:1125–1133

202. Speciale A, Anwar S, Canali R et al (2013) Cyanidin-3-O-glucoside counters the response to TNF-alpha of endothelial cells by activating Nrf2 pathway. Mol Nutr Food Res 57:1979–1987
203. Lampe JW, Chang JL (2007) Interindividual differences in phytochemical metabolism and disposition. Semin Cancer Biol 17:347–353
204. Youdim KA, Mcdonald J, Kalt W et al (2002) Potential role of dietary flavonoids in reducing microvascular endothelium vulnerability to oxidative and inflammatory insults☆. J Nutr Biochem 13:282–288
205. Joo HK, Choi S, Lee YR et al (2018) Anthocyanin-rich extract from red chinese cabbage alleviates vascular inflammation in endothelial cells and apo E(-/-) mice. Int J Mol Sci 19. https://doi.org/10.3390/ijms19030816
206. Han MH, Kim HJ, Jeong JW et al (2018) Inhibition of adipocyte differentiation by anthocyanins isolated from the fruit of vitis coignetiae pulliat is associated with the activation of AMPK signaling pathway. Toxicol Res 34:13–21
207. Mckay DL, Eliasziw M, Chen CYO et al (2018) A pecan-rich diet improves cardiometabolic risk factors in overweight and obese adults: a randomized controlled trial. Nutrients 10. https://doi.org/10.3390/nu10030339
208. Widmer RJ, Flammer AJ, Lerman LO et al (2015) The Mediterranean diet, its components, and cardiovascular disease. Am J Med 128:229–238
209. Capurso C, Massaro M, Scoditti E et al (2014) Vascular effects of the Mediterranean diet part I: anti-hypertensive and anti-thrombotic effects. Vascul Pharmacol 63:118–126
210. Estruch R, Martínez-González MA, Corella D et al (2006) Effects of a Mediterranean-style diet on cardiovascular risk factors: a randomized trial. Ann Intern Med 145:1–11
211. Sleiman D, Al-Badri MR, Azar ST (2015) Effect of mediterranean diet in diabetes control and cardiovascular risk modification: a systematic review. Front Public Health 3:69
212. Llorente-Cortes V, Estruch R, Mena MP et al (2010) Effect of Mediterranean diet on the expression of pro-atherogenic genes in a population at high cardiovascular risk. Atherosclerosis 208:442–450
213. Parikh M, Netticadan T, Pierce GN (2018) Flaxseed: its bioactive components and their cardiovascular benefits. Am J Physiol Heart Circ Physiol 314:H146–H159
214. Jangale NM, Devarshi PP, Dubal AA et al (2013) Dietary flaxseed oil and fish oil modulates expression of antioxidant and inflammatory genes with alleviation of protein glycation status and inflammation in liver of streptozotocin-nicotinamide induced diabetic rats. Food Chem 141:187–195
215. Garcia N, Zazueta C, Aguilera-Aguirre L (2017) Oxidative Stress and Inflammation in Cardiovascular Disease. Oxid Med Cell Longev 2017:5853238
216. Hannan PA, Khan JA, Ullah I et al (2016) Synergistic combinatorial antihyperlipidemic study of selected natural antioxidants; modulatory effects on lipid profile and endogenous antioxidants. Lipids Health Dis 15:151
217. Liu H-H, Shih T-S, Huang H-R et al (2013) Plasma homocysteine is associated with increased oxidative stress and antioxidant enzyme activity in welders. Sci World J:2013, 370487
218. Sansone R, Rodriguez-Mateos A, Heuel J et al (2015) Cocoa flavanol intake improves endothelial function and Framingham Risk Score in healthy men and women: a randomised, controlled, double-masked trial: the Flaviola Health Study. Br J Nutr 114:1246–1255
219. Hamza-Reguig S, Louala S, Boualga A et al (2013) Effect of sardine protein on redox status in hypercholesterolemic rats. Nutr Food Sci 43:277–284
220. Affane F, Louala S, El Imane Harrat N et al (2018) Hypolipidemic, antioxidant and antiatherogenic property of sardine by-products proteins in high-fat diet induced obese rats. Life Sci 199:16–22
221. Cyrus T, Yao Y, Rokach J et al (2003) Vitamin E reduces progression of atherosclerosis in low-density lipoprotein receptor-deficient mice with established vascular lesions. Circulation 107:521–523
222. Adams MR, Golden DL, Anthony MS et al (2002) The inhibitory effect of soy protein isolate on atherosclerosis in mice does not require the presence of LDL receptors or alteration of plasma lipoproteins. J Nutr 132:43–49

223. Deng Y, Lei T, Li H et al (2018) ERK5/KLF2 activation is involved in the reducing effects of puerarin on monocyte adhesion to endothelial cells and atherosclerotic lesion in apolipoprotein E-deficient mice. Biochim Biophys Acta Mol Basis Dis 1864:2590–2599
224. Ren K, Jiang T, Zhao GJ (2018) Quercetin induces the selective uptake of HDL-cholesterol via promoting SR-BI expression and the activation of the PPARgamma/LXRalpha pathway. Food Funct 9:624–635
225. Ren K, Jiang T, Zhou HF et al (2018) Apigenin Retards Atherogenesis by Promoting ABCA1-Mediated Cholesterol Efflux and Suppressing Inflammation. Cell Physiol Biochem 47:2170–2184
226. Averill MM, Bennett BJ, Rattazzi M et al (2009) Neither antioxidants nor genistein inhibit the progression of established atherosclerotic lesions in older apoE deficient mice. Atherosclerosis 203:82–88
227. Karppi J, Kurl S, Mäkikallio TH et al (2013) Serum β-carotene concentrations and the risk of congestive heart failure in men: a population-based study. Int J Cardiol 168:1841–1846
228. Imamura F, Lemaitre RN, King IB et al (2013) Long-chain monounsaturated Fatty acids and incidence of congestive heart failure in 2 prospective cohorts. Circulation 127:1512–1521. 1521e1511-1518
229. Tavazzi L, Maggioni AP, Marchioli R et al (2008) Effect of n-3 polyunsaturated fatty acids in patients with chronic heart failure (the GISSI-HF trial): a randomised, double-blind, placebo-controlled trial. Lancet (London, UK) 372:1223–1230
230. Kerley CP (2018) Dietary patterns and components to prevent and treat heart failure: a comprehensive review of human studies. Nutr Res Rev 32:1–27
231. Ahmet I, Tae H-J, Lakatta EG et al (2016) Long-term low dose dietary resveratrol supplement reduces cardiovascular structural and functional deterioration in chronic heart failure in rats. Can J Physiol Pharmacol 95:268–274
232. Riba A, Deres L, Sumegi B et al (2017) Cardioprotective effect of resveratrol in a postinfarction heart failure model. Oxid Med Cell Longev 2017:6819281
233. Coggan AR, Peterson LR (2016) Dietary Nitrate and Skeletal Muscle Contractile Function in Heart Failure. Curr Heart Fail Rep 13:158–165
234. Xiao Y, Ye J, Zhou Y et al (2018) Baicalin inhibits pressure overload-induced cardiac fibrosis through regulating AMPK/TGF-beta/Smads signaling pathway. Arch Biochem Biophys 640:37–46
235. Wang Y, Zhong L, Liu X et al (2017) ZYZ-772 prevents cardiomyocyte injury by suppressing Nox4-derived ROS production and apoptosis. Molecules 22. https://doi.org/10.3390/molecules22020331
236. Gao L, Yao R, Liu Y et al (2017) Isorhamnetin protects against cardiac hypertrophy through blocking PI3K-AKT pathway. Mol Cell Biochem 429:167–177
237. Fan D, Yang Z, Liu F-Y et al (2017) Sesamin Protects Against Cardiac Remodeling Via Sirt3/ROS Pathway. Cell Physiol Biochem 44:2212–2227
238. Su S, Li Q, Xiong C et al (2014) Sesamin ameliorates doxorubicin-induced cardiotoxicity: involvement of Sirt1 and Mn-SOD pathway. Toxicol Lett 224:257–263
239. Chen QM, Alpert JS (2016) To supplement or not: a role for antioxidant vitamins in the management of heart failure? Am J Med 129:767–768
240. Brito R, Castillo G, Gonzalez J et al (2015) Oxidative stress in hypertension: mechanisms and therapeutic opportunities. Exp Clin Endocrinol Diabetes 123:325–335
241. Kawashima S (2004) The two faces of endothelial nitric oxide synthase in the pathophysiology of atherosclerosis. Endothelium 11:99–107
242. Laursen JB, Somers M, Kurz S et al (2001) Endothelial regulation of vasomotion in apoE-deficient mice: implications for interactions between peroxynitrite and tetrahydrobiopterin. Circulation 103:1282–1288
243. Taniyama Y, Griendling KK (2003) Reactive oxygen species in the vasculature: molecular and cellular mechanisms. Hypertension 42:1075–1081

244. Boonprom P, Boonla O, Chayaburakul K et al (2017) Garcinia mangostana pericarp extract protects against oxidative stress and cardiovascular remodeling via suppression of p47(phox) and iNOS in nitric oxide deficient rats. Ann Anat 212:27–36
245. Dos Santos RL, Dellacqua LO, Delgado NT et al (2016) Pomegranate peel extract attenuates oxidative stress by decreasing coronary angiotensin-converting enzyme (ACE) activity in hypertensive female rats. J Toxicol Environ Health A 79:998–1007
246. Sovari AA (2016) Cellular and Molecular Mechanisms of Arrhythmia by Oxidative Stress. Cardiol Res Pract 2016:9656078
247. Liu M, Liu H, Dudley SC Jr (2010) Reactive oxygen species originating from mitochondria regulate the cardiac sodium channel. Circ Res 107:967–974
248. Erickson JR, Joiner ML, Guan X et al (2008) A dynamic pathway for calcium-independent activation of CaMKII by methionine oxidation. Cell 133:462–474
249. Goff DC Jr, Lloyd-Jones DM, Bennett G et al (2014) 2013 ACC/AHA guideline on the assessment of cardiovascular risk: a report of the American College of Cardiology/American Heart Association Task Force on Practice Guidelines. Circulation 129:S49–S73
250. Aiba T, Hesketh GG, Liu T et al (2010) Na+ channel regulation by Ca2+/calmodulin and Ca2+/calmodulin-dependent protein kinase II in guinea-pig ventricular myocytes. Cardiovasc Res 85:454–463
251. Singh JA, Cleveland J (2017) Allopurinol and the risk of ventricular arrhythmias in the elderly: a study using US Medicare data. BMC Med 15:59
252. Wang ZJ, Hu WK, Liu YY et al (2014) The effect of intravenous vitamin C infusion on periprocedural myocardial injury for patients undergoing elective percutaneous coronary intervention. Can J Cardiol 30:96–101
253. Ashor AW, Lara J, Mathers JC et al (2014) Effect of vitamin C on endothelial function in health and disease: a systematic review and meta-analysis of randomised controlled trials. Atherosclerosis 235:9–20
254. Hemila H, Suonsyrja T (2017) Vitamin C for preventing atrial fibrillation in high risk patients: a systematic review and meta-analysis. BMC Cardiovasc Disord 17:49
255. Chang JH, Chang SL, Hong PD et al (2017) Epigallocatechin-3-gallate modulates arrhythmogenic activity and calcium homeostasis of left atrium. Int J Cardiol 236:174–180
256. Stephens NG, Parsons A, Schofield PM et al (1996) Randomised controlled trial of vitamin E in patients with coronary disease: Cambridge Heart Antioxidant Study (CHAOS). Lancet 347:781–786
257. Anonymous (1999) Dietary supplementation with n-3 polyunsaturated fatty acids and vitamin E after myocardial infarction: results of the GISSI-Prevenzione trial. Gruppo Italiano per lo Studio della Sopravvivenza nell'Infarto miocardico. Lancet 354:447–455
258. Yusuf S, Dagenais G, Pogue J et al (2000) Vitamin E supplementation and cardiovascular events in high-risk patients. N Engl J Med 342:154–160
259. De Gaetano G (2001) Low-dose aspirin and vitamin E in people at cardiovascular risk: a randomised trial in general practice. Collaborative Group of the Primary Prevention Project. Lancet 357:89–95
260. Anonymous (1994) The effect of vitamin E and beta carotene on the incidence of lung cancer and other cancers in male smokers. N Engl J Med 330:1029–1035
261. Stampfer MJ, Hennekens CH, Manson JE et al (1993) Vitamin E consumption and the risk of coronary disease in women. N Engl J Med 328:1444–1449
262. Bjelakovic G, Nikolova D, Gluud LL et al (2012) Antioxidant supplements for prevention of mortality in healthy participants and patients with various diseases. Cochrane Database Syst Rev 3:CD007176
263. Boaz M, Smetana S, Weinstein T et al (2000) Secondary prevention with antioxidants of cardiovascular disease in endstage renal disease (SPACE): randomised placebo-controlled trial. Lancet 356:1213–1218
264. Enstrom JE, Kanim LE, Klein MA (1992) Vitamin C intake and mortality among a sample of the United States population. Epidemiology 3:194–202

265. Ramirez J, Flowers NC (1980) Leukocyte ascorbic acid and its relationship to coronary artery disease in man. Am J Clin Nutr 33:2079–2087
266. Nyyssonen K, Parviainen MT, Salonen R et al (1997) Vitamin C deficiency and risk of myocardial infarction: prospective population study of men from eastern Finland. BMJ 314:634–638
267. Simon JA, Hudes ES (1999) Serum ascorbic acid and cardiovascular disease prevalence in U.S. adults: the Third National Health and Nutrition Examination Survey (NHANES III). Ann Epidemiol 9:358–365
268. Costa AL, Amato A, Loteta A et al (1979) Biological characteristics of various strains of Pseudomonas aeruginosa isolated from human pathological matter. G Batteriol Virol Immunol 71:55–61
269. Osganian SK, Stampfer MJ, Rimm E et al (2003) Vitamin C and risk of coronary heart disease in women. J Am Coll Cardiol 42:246–252
270. Lee DH, Folsom AR, Harnack L et al (2004) Does supplemental vitamin C increase cardiovascular disease risk in women with diabetes? Am J Clin Nutr 80:1194–1200
271. Salonen JT, Alfthan G, Huttunen JK et al (1982) Association between cardiovascular death and myocardial infarction and serum selenium in a matched-pair longitudinal study. Lancet 2:175–179
272. Miettinen TA, Alfthan G, Huttunen JK et al (1983) Serum selenium concentration related to myocardial infarction and fatty acid content of serum lipids. Br Med J (Clin Res Ed) 287:517–519
273. Virtamo J, Valkeila E, Alfthan G et al (1985) Serum selenium and the risk of coronary heart disease and stroke. Am J Epidemiol 122:276–282
274. Salonen JT, Salonen R, Penttila I et al (1985) Serum fatty acids, apolipoproteins, selenium and vitamin antioxidants and the risk of death from coronary artery disease. Am J Cardiol 56:226–231
275. Ringstad J, Thelle D (1986) Risk of myocardial infarction in relation to serum concentrations of selenium. Acta Pharmacol Toxicol (Copenh) 59(Suppl 7):336–339
276. Paine AJ (1978) Excited states of oxygen in biology: their possible involvement in cytochrome P450 linked oxidations as well as in the induction of the P450 system by many diverse compounds. Biochem Pharmacol 27:1805–1813
277. Ringstad J, Jacobsen BK, Thomassen Y et al (1987) The Tromso Heart Study: serum selenium and risk of myocardial infarction a nested case-control study. J Epidemiol Community Health 41:329–332
278. Suadicani P, Hein HO, Gyntelberg F (1992) Serum selenium concentration and risk of ischaemic heart disease in a prospective cohort study of 3000 males. Atherosclerosis 96:33–42
279. Salvini S, Hennekens CH, Morris JS et al (1995) Plasma levels of the antioxidant selenium and risk of myocardial infarction among U.S. physicians. Am J Cardiol 76:1218–1221
280. Marniemi J, Jarvisalo J, Toikka T et al (1998) Blood vitamins, mineral elements and inflammation markers as risk factors of vascular and non-vascular disease mortality in an elderly population. Int J Epidemiol 27:799–807
281. Kilander L, Berglund L, Boberg M et al (2001) Education, lifestyle factors and mortality from cardiovascular disease and cancer. A 25-year follow-up of Swedish 50-year-old men. Int J Epidemiol 30:1119–1126
282. Yoshizawa K, Ascherio A, Morris JS et al (2003) Prospective study of selenium levels in toenails and risk of coronary heart disease in men. Am J Epidemiol 158:852–860
283. Wei WQ, Abnet CC, Qiao YL et al (2004) Prospective study of serum selenium concentrations and esophageal and gastric cardia cancer, heart disease, stroke, and total death. Am J Clin Nutr 79:80–85
284. Akbaraly NT, Arnaud J, Hininger-Favier I et al (2005) Selenium and mortality in the elderly: results from the EVA study. Clin Chem 51:2117–2123
285. Oster O, Drexler M, Schenk J et al (1986) The serum selenium concentration of patients with acute myocardial infarction. Ann Clin Res 18:36–42

286. Auzepy P, Blondeau M, Richard C et al (1987) Serum selenium deficiency in myocardial infarction and congestive cardiomyopathy. Acta Cardiol 42:161–166
287. Beaglehole R, Jackson R, Watkinson J et al (1990) Decreased blood selenium and risk of myocardial infarction. Int J Epidemiol 19:918–922
288. Thiele R, Schuffenhauer M, Winnefeld K et al (1995) Selenium level in patients with acute myocardial infarct and in patients with severe angina pectoris without myocardial infarct. Med Klin (Munich) 90(Suppl 1):45–48
289. Kardinaal AF, Kok FJ, Kohlmeier L et al (1997) Association between toenail selenium and risk of acute myocardial infarction in European men. The EURAMIC Study. European Antioxidant Myocardial Infarction and Breast Cancer. Am J Epidemiol 145:373–379
290. Coudray C, Roussel AM, Mainard F et al (1997) Lipid peroxidation level and antioxidant micronutrient status in a pre-aging population; correlation with chronic disease prevalence in a French epidemiological study (Nantes, France). J Am Coll Nutr 16:584–591
291. Navarro-Alarcon M, De La Serrana HL-G, Perez-Valero V et al (1999) Serum and urine selenium concentrations in patients with cardiovascular diseases and relationship to other nutritional indexes. Ann Nutr Metab 43:30–36
292. Bor MV, Cevik C, Uslu I et al (1999) Selenium levels and glutathione peroxidase activities in patients with acute myocardial infarction. Acta Cardiol 54:271–276
293. Zachara BA, Ukleja-Adamowicz M, Nartowicz E et al (2001) Increased plasma glutathione peroxidase activity in patients with acute myocardial infarction. Med Sci Monit 7:415–420
294. Schnabel R, Lubos E, Messow CM et al (2008) Selenium supplementation improves antioxidant capacity in vitro and in vivo in patients with coronary artery disease The SElenium Therapy in Coronary Artery disease Patients (SETCAP) Study. Am Heart J 156:1201 e1201–1201 e1211
295. Stranges S, Marshall JR, Trevisan M et al (2006) Effects of selenium supplementation on cardiovascular disease incidence and mortality: secondary analyses in a randomized clinical trial. Am J Epidemiol 163:694–699
296. Cold F, Winther KH, Pastor-Barriuso R et al (2015) Randomised controlled trial of the effect of long-term selenium supplementation on plasma cholesterol in an elderly Danish population. Br J Nutr 114:1807–1818
297. Korpela H, Kumpulainen J, Jussila E et al (1989) Effect of selenium supplementation after acute myocardial infarction. Res Commun Chem Pathol Pharmacol 65:249–252
298. Alvarenga Americano Do Brasil PE, Pereira De Souza A, Hasslocher-Moreno AM et al (2014) Selenium Treatment and Chagasic Cardiopathy (STCC): study protocol for a double-blind randomized controlled trial. Trials 15:388
299. Brigo F, Storti M, Lochner P et al (2014) Selenium supplementation for primary prevention of cardiovascular disease: proof of no effectiveness. Nutr Metab Cardiovasc Dis 24:e2–e3
300. Eaton CB, Abdul Baki AR, Waring ME et al (2010) The association of low selenium and renal insufficiency with coronary heart disease and all-cause mortality: NHANES III follow-up study. Atherosclerosis 212:689–694
301. Hawkes WC, Laslett LJ (2009) Selenium supplementation does not improve vascular responsiveness in healthy North American men. Am J Physiol Heart Circ Physiol 296:H256–H262
302. Hennekens CH, Buring JE, Manson JE et al (1996) Lack of effect of long-term supplementation with beta carotene on the incidence of malignant neoplasms and cardiovascular disease. N Engl J Med 334:1145–1149
303. Morris DL, Kritchevsky SB, Davis CE (1994) Serum carotenoids and coronary heart disease. The Lipid Research Clinics Coronary Primary Prevention Trial and Follow-up Study. JAMA 272:1439–1441
304. Gaziano JM, Manson JE, Branch LG et al (1995) A prospective study of consumption of carotenoids in fruits and vegetables and decreased cardiovascular mortality in the elderly. Ann Epidemiol 5:255–260
305. Osganian SK, Stampfer MJ, Rimm E et al (2003) Dietary carotenoids and risk of coronary artery disease in women. Am J Clin Nutr 77:1390–1399

306. Iribarren C, Folsom AR, Jacobs DR Jr et al (1997) Association of serum vitamin levels, LDL susceptibility to oxidation, and autoantibodies against MDA-LDL with carotid atherosclerosis. A case-control study. The ARIC Study Investigators. Atherosclerosis Risk in Communities. Arterioscler Thromb Vasc Biol 17:1171–1177
307. Hertog MG, Kromhout D, Aravanis C et al (1995) Flavonoid intake and long-term risk of coronary heart disease and cancer in the seven countries study. Arch Intern Med 155:381–386
308. Knekt P, Jarvinen R, Reunanen A et al (1996) Flavonoid intake and coronary mortality in Finland: a cohort study. BMJ 312:478–481
309. Knekt P, Kumpulainen J, Jarvinen R et al (2002) Flavonoid intake and risk of chronic diseases. Am J Clin Nutr 76:560–568
310. Yochum L, Kushi LH, Meyer K et al (1999) Dietary flavonoid intake and risk of cardiovascular disease in postmenopausal women. Am J Epidemiol 149:943–949
311. Arts IC, Jacobs DR Jr, Harnack LJ et al (2001) Dietary catechins in relation to coronary heart disease death among postmenopausal women. Epidemiology 12:668–675
312. Geleijnse JM, Launer LJ, Van Der Kuip DA et al (2002) Inverse association of tea and flavonoid intakes with incident myocardial infarction: the Rotterdam Study. Am J Clin Nutr 75:880–886
313. Hertog MG, Sweetnam PM, Fehily AM et al (1997) Antioxidant flavonols and ischemic heart disease in a Welsh population of men: the Caerphilly Study. Am J Clin Nutr 65:1489–1494
314. Sesso HD, Gaziano JM, Liu S et al (2003) Flavonoid intake and the risk of cardiovascular disease in women. Am J Clin Nutr 77:1400–1408
315. Rimm EB, Katan MB, Ascherio A et al (1996) Relation between intake of flavonoids and risk for coronary heart disease in male health professionals. Ann Intern Med 125:384–389
316. Lin J, Rexrode KM, Hu F et al (2007) Dietary intakes of flavonols and flavones and coronary heart disease in US women. Am J Epidemiol 165:1305–1313
317. Mink PJ, Scrafford CG, Barraj LM et al (2007) Flavonoid intake and cardiovascular disease mortality: a prospective study in postmenopausal women. Am J Clin Nutr 85:895–909
318. Farouque HM, Leung M, Hope SA et al (2006) Acute and chronic effects of flavanol-rich cocoa on vascular function in subjects with coronary artery disease: a randomized double-blind placebo-controlled study. Clin Sci (Lond) 111:71–80
319. Brown CA, Bolton-Smith C, Woodward M et al (1993) Coffee and tea consumption and the prevalence of coronary heart disease in men and women: results from the Scottish Heart Health Study. J Epidemiol Community Health 47:171–175
320. Keli SO, Hertog MG, Feskens EJ et al (1996) Dietary flavonoids, antioxidant vitamins, and incidence of stroke: the Zutphen study. Arch Intern Med 156:637–642
321. Mukamal KJ, Maclure M, Muller JE et al (2002) Tea consumption and mortality after acute myocardial infarction. Circulation 105:2476–2481
322. Sesso HD, Gaziano JM, Buring JE et al (1999) Coffee and tea intake and the risk of myocardial infarction. Am J Epidemiol 149:162–167
323. Taku K, Umegaki K, Sato Y et al (2007) Soy isoflavones lower serum total and LDL cholesterol in humans: a meta-analysis of 11 randomized controlled trials. Am J Clin Nutr 85:1148–1156
324. Flammer AJ, Sudano I, Wolfrum M et al (2012) Cardiovascular effects of flavanol-rich chocolate in patients with heart failure. Eur Heart J 33:2172–2180
325. Simon JA, Hudes ES, Tice JA (2001) Relation of serum ascorbic acid to mortality among US adults. J Am Coll Nutr 20:255–263
326. Auclair S, Chironi G, Milenkovic D et al (2010) The regular consumption of a polyphenol-rich apple does not influence endothelial function: a randomised double-blind trial in hypercholesterolemic adults. Eur J Clin Nutr 64:1158–1165
327. Medina-Remon A, Zamora-Ros R, Rotches-Ribalta M et al (2011) Total polyphenol excretion and blood pressure in subjects at high cardiovascular risk. Nutr Metab Cardiovasc Dis 21:323–331
328. Dohadwala MM, Holbrook M, Hamburg NM et al (2011) Effects of cranberry juice consumption on vascular function in patients with coronary artery disease. Am J Clin Nutr 93:934–940

329. Valls-Pedret C, Lamuela-Raventos RM, Medina-Remon A et al (2012) Polyphenol-rich foods in the Mediterranean diet are associated with better cognitive function in elderly subjects at high cardiovascular risk. J Alzheimers Dis 29:773–782
330. Moreno-Luna R, Munoz-Hernandez R, Miranda ML et al (2012) Olive oil polyphenols decrease blood pressure and improve endothelial function in young women with mild hypertension. Am J Hypertens 25:1299–1304
331. Khan N, Monagas M, Andres-Lacueva C et al (2012) Regular consumption of cocoa powder with milk increases HDL cholesterol and reduces oxidized LDL levels in subjects at high-risk of cardiovascular disease. Nutr Metab Cardiovasc Dis 22:1046–1053
332. Hassellund SS, Flaa A, Kjeldsen SE et al (2013) Effects of anthocyanins on cardiovascular risk factors and inflammation in pre-hypertensive men: a double-blind randomized placebo-controlled crossover study. J Hum Hypertens 27:100–106
333. Chiva-Blanch G, Urpi-Sarda M, Ros E et al (2013) Effects of red wine polyphenols and alcohol on glucose metabolism and the lipid profile: a randomized clinical trial. Clin Nutr 32:200–206
334. Annuzzi G, Bozzetto L, Costabile G et al (2014) Diets naturally rich in polyphenols improve fasting and postprandial dyslipidemia and reduce oxidative stress: a randomized controlled trial. Am J Clin Nutr 99:463–471
335. Medina-Remon A, Tresserra-Rimbau A, Pons A et al (2015) Effects of total dietary polyphenols on plasma nitric oxide and blood pressure in a high cardiovascular risk cohort. The PREDIMED randomized trial. Nutr Metab Cardiovasc Dis 25:60–67
336. Tresserra-Rimbau A, Guasch-Ferre M, Salas-Salvado J et al (2016) Intake of total polyphenols and some classes of polyphenols is inversely associated with diabetes in elderly people at high cardiovascular disease risk. In: J Nutr, vol 146, pp 767–777
337. Stockton A, Farhat G, Mcdougall GJ et al (2017) Effect of pomegranate extract on blood pressure and anthropometry in adults: a double-blind placebo-controlled randomised clinical trial. J Nutr Sci 6:e39
338. Bellone JA, Murray JR, Jorge P et al (2018) Pomegranate supplementation improves cognitive and functional recovery following ischemic stroke: a randomized trial. Nutr Neurosci 13:1–6
339. Razani Z, Dastani M, Kazerani HR (2017) Cardioprotective Effects of Pomegranate (Punica granatum) Juice in Patients with Ischemic Heart Disease. Phytother Res 31:1731–1738
340. Tresserra-Rimbau A, Rimm EB, Medina-Remon A et al (2014) Inverse association between habitual polyphenol intake and incidence of cardiovascular events in the PREDIMED study. Nutr Metab Cardiovasc Dis 24:639–647
341. Sumner MD, Elliott-Eller M, Weidner G et al (2005) Effects of pomegranate juice consumption on myocardial perfusion in patients with coronary heart disease. Am J Cardiol 96:810–814
342. Lekakis J, Rallidis LS, Andreadou I et al (2005) Polyphenolic compounds from red grapes acutely improve endothelial function in patients with coronary heart disease. Eur J Cardiovasc Prev Rehabil 12:596–600
343. Li SH, Tian HB, Zhao HJ et al (2013) The acute effects of grape polyphenols supplementation on endothelial function in adults: meta-analyses of controlled trials. PLoS One 8:e69818
344. Wang D, Chen C, Wang Y et al (2014) Effect of black tea consumption on blood cholesterol: a meta-analysis of 15 randomized controlled trials. PLoS One 9:e107711
345. Rimm EB, Stampfer MJ, Ascherio A et al (1993) Vitamin E consumption and the risk of coronary heart disease in men. N Engl J Med 328:1450–1456
346. Knekt P, Reunanen A, Jarvinen R et al (1994) Antioxidant vitamin intake and coronary mortality in a longitudinal population study. Am J Epidemiol 139:1180–1189
347. Anonymous (2002) MRC/BHF Heart Protection Study of antioxidant vitamin supplementation in 20,536 high-risk individuals: a randomised placebo-controlled trial. Lancet 360:23–33
348. Lee IM, Cook NR, Manson JE et al (1999) Beta-carotene supplementation and incidence of cancer and cardiovascular disease: the Women's Health Study. J Natl Cancer Inst 91:2102–2106

349. Lee IM, Cook NR, Gaziano JM et al (2005) Vitamin E in the primary prevention of cardiovascular disease and cancer: the Women's Health Study: a randomized controlled trial. JAMA 294:56–65
350. Bassuk SS, Albert CM, Cook NR et al (2004) The Women's Antioxidant Cardiovascular Study: design and baseline characteristics of participants. J Womens Health (Larchmt) 13:99–117
351. Hercberg S, Galan P, Preziosi P et al (2004) The SU.VI.MAX Study: a randomized, placebo-controlled trial of the health effects of antioxidant vitamins and minerals. Arch Intern Med 164:2335–2342
352. Hercberg S (2006) The SU.VI.MAX study, a randomized, placebo-controlled trial on the effects of antioxidant vitamins and minerals on health. Ann Pharm Fr 64:397–401
353. Gey KF, Puska P (1989) Plasma vitamins E and A inversely correlated to mortality from ischemic heart disease in cross-cultural epidemiology. Ann N Y Acad Sci 570:268–282
354. Losonczy KG, Harris TB, Havlik RJ (1996) Vitamin E and vitamin C supplement use and risk of all-cause and coronary heart disease mortality in older persons: the Established Populations for Epidemiologic Studies of the Elderly. Am J Clin Nutr 64:190–196
355. Riemersma RA, Wood DA, Macintyre CC et al (1991) Risk of angina pectoris and plasma concentrations of vitamins A, C, and E and carotene. Lancet 337:1–5
356. Omenn GS, Goodman GE, Thornquist MD et al (1996) Effects of a combination of beta carotene and vitamin A on lung cancer and cardiovascular disease. N Engl J Med 334:1150–1155
357. Klipstein-Grobusch K, Geleijnse JM, Den Breeijen JH et al (1999) Dietary antioxidants and risk of myocardial infarction in the elderly: the Rotterdam Study. Am J Clin Nutr 69:261–266
358. Ford ES, Giles WH (2000) Serum vitamins, carotenoids, and angina pectoris: findings from the National Health and Nutrition Examination Survey III. Ann Epidemiol 10:106–116
359. Klipstein-Grobusch K, Den Breeijen JH, Grobbee DE et al (2001) Dietary antioxidants and peripheral arterial disease: the Rotterdam Study. Am J Epidemiol 154:145–149
360. Kritchevsky SB, Shimakawa T, Tell GS et al (1995) Dietary antioxidants and carotid artery wall thickness. The ARIC Study. Atherosclerosis Risk in Communities Study. Circulation 92:2142–2150
361. Kushi LH, Folsom AR, Prineas RJ et al (1996) Dietary antioxidant vitamins and death from coronary heart disease in postmenopausal women. N Engl J Med 334:1156–1162
362. Muntwyler J, Hennekens CH, Manson JE et al (2002) Vitamin supplement use in a low-risk population of US male physicians and subsequent cardiovascular mortality. Arch Intern Med 162:1472–1476
363. Evans RW, Shaten BJ, Day BW et al (1998) Prospective association between lipid soluble antioxidants and coronary heart disease in men. The Multiple Risk Factor Intervention Trial. Am J Epidemiol 147:180–186
364. Smolkova B, Dusinska M, Raslova K et al (2004) Folate levels determine effect of antioxidant supplementation on micronuclei in subjects with cardiovascular risk. Mutagenesis 19:469–476
365. Trankmann P, Thiele R, Winnefeld K et al (1999) Effect of administration of selenium and vitamin E on heart failure and ventricular arrhythmias in patients with acute myocardial infarct. Med Klin (Munich) 94(Suppl 3):78–80
366. Christen WG, Gaziano JM, Hennekens CH (2000) Design of Physicians' Health Study II--a randomized trial of beta-carotene, vitamins E and C, and multivitamins, in prevention of cancer, cardiovascular disease, and eye disease, and review of results of completed trials. Ann Epidemiol 10:125–134
367. Alehagen U, Alexander J, Aaseth J (2016) Supplementation with Selenium and Coenzyme Q10 Reduces Cardiovascular Mortality in Elderly with Low Selenium Status. A Secondary Analysis of a Randomised Clinical Trial. PLoS One 11:e0157541
368. Kuklinski B, Weissenbacher E, Fahnrich A (1994) Coenzyme Q10 and antioxidants in acute myocardial infarction. Mol Aspects Med 15. Suppl:s143–s147
369. Brown BG, Zhao XQ, Chait A et al (2001) Simvastatin and niacin, antioxidant vitamins, or the combination for the prevention of coronary disease. N Engl J Med 345:1583–1592

370. Stonehouse W, Brinkworth GD, Thompson CH et al (2016) Short term effects of palm-tocotrienol and palm-carotenes on vascular function and cardiovascular disease risk: a randomised controlled trial. Atherosclerosis 254:205–214
371. Saboori S, Djalali M, Yousefi Rad E et al (2016) Various Effects of Omega 3 and Omega 3 Plus Vitamin E Supplementations on Serum Glucose Level and Insulin Resistance in Patients with Coronary Artery Disease. Iran J Public Health 45:1465–1472
372. Stepaniak U, Micek A, Grosso G et al (2016) Antioxidant vitamin intake and mortality in three Central and Eastern European urban populations: the HAPIEE study. Eur J Nutr 55:547–560
373. Rodrigo R, Hasson D, Prieto JC et al (2014) The effectiveness of antioxidant vitamins C and E in reducing myocardial infarct size in patients subjected to percutaneous coronary angioplasty (PREVEC Trial): study protocol for a pilot randomized double-blind controlled trial. Trials 15:192
374. Jaxa-Chamiec T, Bednarz B, Herbaczynska-Cedro K et al (2009) Effects of vitamins C and E on the outcome after acute myocardial infarction in diabetics: a retrospective, hypothesis-generating analysis from the MIVIT study. Cardiology 112:219–223
375. Jaxa-Chamiec T, Bednarz B, Drozdowska D et al (2005) Antioxidant effects of combined vitamins C and E in acute myocardial infarction. The randomized, double-blind, placebo controlled, multicenter pilot Myocardial Infarction and VITamins (MIVIT) trial. Kardiol Pol 62:344–350
376. Sesso HD, Buring JE, Christen WG et al (2008) Vitamins E and C in the prevention of cardiovascular disease in men: the Physicians' Health Study II randomized controlled trial. JAMA 300:2123–2133
377. Marchioli R, Marfisi RM, Borrelli G et al (2007) Efficacy of n-3 polyunsaturated fatty acids according to clinical characteristics of patients with recent myocardial infarction: insights from the GISSI-Prevenzione trial. J Cardiovasc Med (Hagerstown) 8(Suppl 1):S34–S37
378. Cook NR, Albert CM, Gaziano JM et al (2007) A randomized factorial trial of vitamins C and E and beta carotene in the secondary prevention of cardiovascular events in women: results from the Women's Antioxidant Cardiovascular Study. Arch Intern Med 167:1610–1618
379. Assanelli D, Bonanome A, Pezzini A et al (2004) Folic acid and vitamin E supplementation effects on homocysteinemia, endothelial function and plasma antioxidant capacity in young myocardial-infarction patients. Pharmacol Res 49:79–84
380. Marchioli R (2003) Omega-3 polyunsaturated fatty acids and cardiovascular diseases. Minerva Cardioangiol 51:561–576
381. Singh RB, Neki NS, Kartikey K et al (2003) Effect of coenzyme Q10 on risk of atherosclerosis in patients with recent myocardial infarction. Mol Cell Biochem 246:75–82
382. Rapola JM, Virtamo J, Ripatti S et al (1997) Randomised trial of alpha-tocopherol and beta-carotene supplements on incidence of major coronary events in men with previous myocardial infarction. Lancet 349:1715–1720
383. Carrero JJ, Fonolla J, Marti JL et al (2007) Intake of fish oil, oleic acid, folic acid, and vitamins B-6 and E for 1 year decreases plasma C-reactive protein and reduces coronary heart disease risk factors in male patients in a cardiac rehabilitation program. J Nutr 137:384–390
384. Arad Y, Spadaro LA, Roth M et al (2005) Treatment of asymptomatic adults with elevated coronary calcium scores with atorvastatin, vitamin C, and vitamin E: the St. Francis Heart Study randomized clinical trial. J Am Coll Cardiol 46:166–172
385. Engler MM, Engler MB, Malloy MJ et al (2003) Antioxidant vitamins C and E improve endothelial function in children with hyperlipidemia: Endothelial Assessment of Risk from Lipids in Youth (EARLY) Trial. Circulation 108:1059–1063
386. Salonen RM, Nyyssonen K, Kaikkonen J et al (2003) Six-year effect of combined vitamin C and E supplementation on atherosclerotic progression: the Antioxidant Supplementation in Atherosclerosis Prevention (ASAP) Study. Circulation 107:947–953
387. Singhal S, Gupta R, Goyle A (2001) Comparison of antioxidant efficacy of vitamin E, vitamin C, vitamin A and fruits in coronary heart disease: a controlled trial. J Assoc Physicians India 49:327–331

388. Ward NC, Hodgson JM, Croft KD et al (2005) The combination of vitamin C and grape-seed polyphenols increases blood pressure: a randomized, double-blind, placebo-controlled trial. J Hypertens 23:427–434
389. Yuanfeng W, Gongnian X, Jianwei M et al (2015) Dietary sulforaphane inhibits histone deacetylase activity in B16 melanoma cells. J Funct Foods 18:182–189
390. Reuter S, Gupta SC, Park B et al (2011) Epigenetic changes induced by curcumin and other natural compounds. Genes Nutr 6:93–108
391. Li Y, Saldanha SN, Tollefsbol TO (2014) Impact of epigenetic dietary compounds on transgenerational prevention of human diseases. AAPS J 16:27–36
392. Crescenti A, Sola R, Valls RM et al (2013) Cocoa consumption alters the global DNA methylation of peripheral leukocytes in humans with cardiovascular disease risk factors: a randomized controlled trial. PLoS One 8:e65744
393. Gracia-Sancho J, Villarreal G Jr, Zhang Y et al (2010) Activation of SIRT1 by resveratrol induces KLF2 expression conferring an endothelial vasoprotective phenotype. Cardiovasc Res 85:514–519
394. Speckmann B, Schulz S, Hiller F et al (2017) Selenium increases hepatic DNA methylation and modulates one-carbon metabolism in the liver of mice. J Nutr Biochem 48:112–119
395. Vahid F, Zand H, Nosrat-Mirshekarlou E et al (2015) The role dietary of bioactive compounds on the regulation of histone acetylases and deacetylases: a review. Gene 562:8–15

Reactive Oxygen Species and Their Epigenetic Consequences in Heart Diseases

6

Seema Bhargava

6.1 Introduction: Reactive Oxygen Species

Highly reactive molecules with an unpaired electron are known as free radicals. By virtue of their unpaired electrons, these molecules are very unstable and survive for a very small fraction of a second (approximately 10^{-9} to 10^{-12} seconds), colliding with other molecules and either donating or accepting an electron within this time frame. Depending on the molecular structure and stability of the other molecule, they either create another free radical or are quenched by it. Thus, a chain of events occurs whereby the unpaired electron is transmitted from one molecule to another till it gets paired [1].

In biological systems, the most damaging free radicals are the oxygen radicals – also termed as the reactive oxygen species [ROS]. Amongst these, the most commonly occurring ones are the following:

(a) Superoxide depicted as O_2^-
(b) Hydroxyl depicted as $\cdot OH$
(c) Perhydroxyl depicted as $\cdot O_2H$

Of these, a and c are less reactive than b, but because they have a longer half-life, they are able to not only react longer with target molecules, but also they may react with molecules far from their site of production.

The most common example of molecular alteration by a free radical is the formation of lipid peroxides by peroxidation of 1,4-diene structure of polyunsaturated fatty acids in the cell membrane and plasma lipids. The ensuing biological damage due to lipid peroxides leads to atherosclerosis. Similarly, other molecules (e.g. bases of the DNA) and structures (e.g. vascular endothelium) can also be subject to damage with alternate pathological outcomes.

S. Bhargava (✉)
Department of Biochemistry, GRIPMER, Sir Ganga Ram Hospital, New Delhi, India

S. Chakraborti et al. (eds.), *Oxidative Stress in Heart Diseases*,
https://doi.org/10.1007/978-981-13-8273-4_6

6.2 Where Do These Free Radicals Come From?

They are derived from our body contents under varying circumstances as detailed below:

- Routinely, we are exposed to ultraviolet rays and X-rays from the atmosphere and also while undergoing some investigative procedures. These rays can lyse water resulting in the formation of hydroxyl radicals. Metal ions (Cu^{+}, Co^{2+}, Ni^{2+} and Fe^{3+}) can react with oxygen and hydrogen peroxide to generate hydroxyl radicals.
- Nitric oxide is often known as the endothelium derived relaxation factor and is important in cell signalling. In itself, it is also a free radical and also serves as a source of hydroxyl radicals from peroxynitrite which is the product of interaction between nitric oxide and superoxide.
- Macrophages are the main defenders of the respiratory paths so that any injury here leads to their activation. This leads to an increased utilization of glucose by the pentose phosphate pathway yielding more NADPH. This is then oxidised in a process termed 'respiratory burst' with the release of superoxide. Primarily, this cytotoxic agent is meant to combat micro-organisms, but it can also damage the normal host cell [1, 2].
- When an infection (even a mild one) occurs, free-radicals are produced to combat the pathogens but they also increase damage to the lipids considerably as evidenced by increase in the circulating lipid peroxides.
- The process of intramitochondrial oxidation of reduced flavin coenzymes involves the microsomal electron transport system wherein the flavin semiquinone radicals form transient oxygen radicals as intermediates. Though these are only transient intermediates and not the final products in the reaction, their unpredictable nature allows some of them to leak. This accounts for the daily production of about 1.5 mol of reactive oxygen species in the human body.

6.3 Antioxidant Systems

Though the ROS are harmful to many molecules and structures in our body, it would be pertinent to mention here that in physiological amounts, reactive oxygen species have a very important role as signalling molecules in the regulation of homeostasis in several physiological processes [3, 4].

Following the corollary of physics – Newton's third law of motion which states that every action has an equal and opposite reaction, our body functions in such a way that it has a countering mechanism for every type of insult. Thus, the body has an inbuilt antioxidant system to combat the excess of free radicals. These are in the form of enzymes (viz superoxide dismutase, catalase, glutathione peroxidase), vitamins (viz. A, C, E) and other molecules (e.g. albumin, thiols, uric acid).

Certain micronutrients like zinc, selenium, copper, iron and manganese can also contribute to the antioxidant system by virtue of being co-factors for these enzymes. In addition, the metal ions which are reactive in themselves, are normally bound to their carrier proteins rendering them inactive – thus minimizing the spontaneous production of free radicals [1].

It is pertinent to note here that some of the antioxidants also behave as pro-oxidants

In the healthy state, the constant dynamic relationship between the free radicals and the antioxidants remains in equilibrium as shown in Fig. 6.1. When this equilibrium is disrupted, it results in pathology.

The mechanisms of action of ROS are varied. These include direct oxidative injury to endothelium of blood vessels/target tissue, lipids, proteins, and DNA. The interaction of ROS with DNA is a part of signal transduction and, when physiological, is termed as epigenetics. Any imbalance in the dynamic equilibrium between the ROS and the antioxidant system, will be reflected in this process as a deviation from the normal gene expression and will lead to several pathological situations.

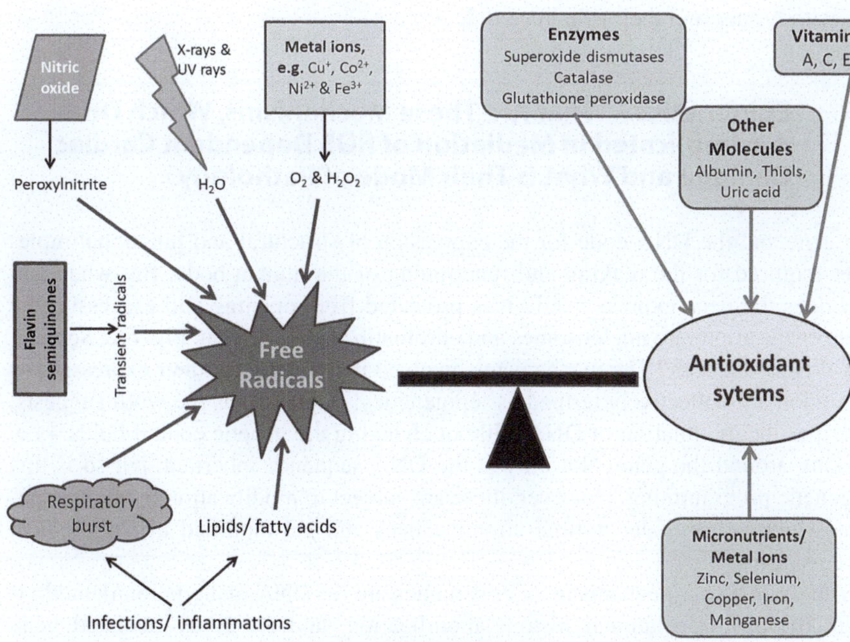

Fig. 6.1 Reactive oxygen species and the antioxidant system in the healthy state
In the healthy state, free radicals (whether endogenous or exogenous) and antioxidant system of the human body are in dynamic equilibrium with each other allowing for the required functions of the latter without pathological consequences

6.4 Why Is the Heart More Susceptible to Oxidative Damage?

To cater to the constant requirement of energy by the tirelessly contracting and relaxing cardiac muscles, it is known that the cardiomyocyte harbors a high density of mitochondria which are the powerhouse of cells as oxidative phosphorylation occurs therein. The mitochondria and the NOXs (NADPH oxidases) are also cellular sources of ROS. There are seven such enzymes (NOX 1-5 and DUOX ½) of which NOXs 1, 2, 4 and 5 are expressed in the cardiomyocyte. In addition, there are several other significant sources of ROS in the cardiac tissue – nitric oxide synthase (NOS), xanthine oxidoreductases, cyclo- and lipo-oxygenases, members of the cytochrome P450 system and some peroxisomal oxidases. Consequently, enzymes of the antioxidant systems, like superoxide dismutase (SOD) and catalase, are also highly expressed in cardiac tissue. With so many players in the maintenance of physiological ROS levels in the myocardium, the smallest imbalance between oxidants and antioxidants will be immediately reflected in the functioning of the cardiomyocytes, making the heart the first target of such an imbalance [3, 4].

In this chapter we propose to outline the epigenetic mechanisms of ROS interaction with tissues and their role in cardiac disease.

6.5 Epigenetics – What Are These Mechanisms, Which Ones Are Implicated in Mediation of ROS Dependent Cardiac Damage and What Is Their Mode of Pathology

The gene and the DNA code for the expression of structural and functional molecules required for the making and functioning of the human body. But when and how does the gene express itself? It is protected from opening and expressing by their organisation into nucleosomes and chromatin as described in Fig. 6.2. So then, how do they express? The mechanisms involved in allowing for their expression or repression are collectively termed as 'epigenetics'. The term as the word suggests, describes the modulation of DNA without changing the genetic code it holds – i.e. working around the gene. Not only is the DNA sequence inherited, but so is this epigenetic programming. However, these are subject to modification by many reactions which occur under normal circumstances and get modified in pathological circumstances.

The three broad mechanisms of epigenetics are (a) DNA methylation/demethylation, (b) post-translational histone modification, and (c) RNA-mediated gene silencing. These processes are intricately interwoven into transcription of genes in the form of initiating, propagating and silencing reactions. When we talk about DNA methylation and histone modification, it would be pertinent here to understand the spatial and functional relationship of these structures [5].

Each nucleus of each cell in our body contains 6 feet of DNA! So how is this organized within a 6μm organelle of the cell? Obviously, it would have to be folded upon itself in a manner that would maintain the structure of the folds. Thus, the

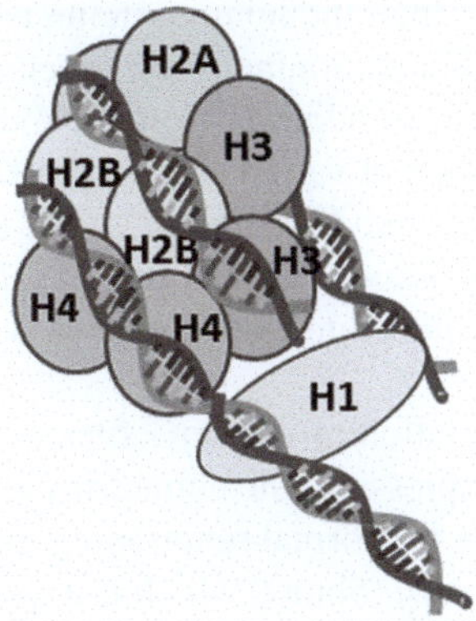

Fig. 6.2 Schematic representation of a nucleosome
Typically, 8 histone proteins co-ordinate to form one spool for DNA strands to wind around. And to stabilize the spool there is a ninth histone protein. Histones are designated as H1, H2, H3 and H4 with H2 having 2 types – H2A and H2B. A pair of each of these positively charged H2A, H2B, H3 and H4 together form the core of histones around which the negatively charged DNA winds. H1 associates with this globule to stabilise it. Thus, each nucleosome has 9 histone proteins – 8 in the core and the ninth at the periphery, as depicted in this figure

double-stranded double helical DNA is spooled several times over a congregation of proteins- the histone proteins (Fig. 6.2). Several of these are longitudinally aligned to form chromosomes. Each such organized spool appears as a clump on the chromosome (like beads on a string) and are termed nucleosomes. The chromosomes are laterally aligned to each other. This allows for the 6-foot long DNA strands to be accommodated in the 6μm nucleus of the cell – in less than 1/50000 of its length! Truly, nature is amazing and organized! [6]

For the DNA to undergo replication and for expression of genes, the histone protein complex has to be dissociated so that the DNA is released and may replicate. Under normal circumstances, this involves the epigenetic modification of this histone-DNA complex. It is evident, therefore, that epigenetic modifications are required for gene expression. Thus, any molecule/ion (e.g. free radicals or ROS) that alters these mechanisms will result in an altered gene expression.

Going back to the types of epigenetic mechanisms, lets talk about **DNA methylation** first as this was the first epigenetic mechanism discovered. Methylation of the DNA and its associated histones is the key to epigenetic modifications which direct the functioning of the genome by modifying the chromatin structure. The methylation of DNA typically occurs at the 5-cytosine-phosphate-guanine-3′ site (CpG) to form 5-methyl cytosine (5mC). It is processed by the DNA methyl transferases 1, 3A and 3B (DNMT1/3A/3B), all of which use S-adenosyl methionine (SAM) as the methyl donor. Once SAM donates its methyl group, it becomes S-adenosyl homocysteine (SAH) which is a potent feedback inhibitor of these SAM-dependent DNMTs. For these methylations to continue, SAM needs to be regenerated. This happens in 3 steps – first the SAH is hydrolysed to yield adenosine and homocysteine; then the homocysteine is methylated in a folate and B_{12} dependent reaction to yield methionine; finally methionine is converted to SAM in an ATP-dependent reaction. In addition to these methyl transferases, there are others which enable methylation at varied sites of the DNA/RNA, i.e. the peptidyl side chains of glutamate/glutamine/histidine or the N-termini or C-termini. Gene repression is most commonly associated with the presence of 5mC in gene regulatory regions. Methylations at other sites are mostly associated with gene expression [7].

As important as the methylation of the CpG sites of our genome is for stability of genome inheritance and expression, the countering reaction – i.e. DNA demethylation – is equally imperative for normal functioning. It starts with the oxidation of 5mC to 5-hydroxymethylcytosine (5hmC) which is a spontaneous process mediated by a group of oxygenases called the ten eleven transferase (TET) proteins. These also further catalyse the 5hmC to 5-formylcytosine and then 5-carboxylcytosine (5fc and 5cc). Thymine DNA glycosylase (TDG) may then excise the bases from these molecules and subject them to base excise repair (BER) resulting in DNA demethylation [8, 9] (Fig. 6.3).

Thus, like most processes in the body, the DNA methylation-demethylation is also a dynamic process dependent on the activity of the DNMTs and the TETs. This would avoid DNA hypermethylation and consequent gene silencing. It is evident, therefore, that interference in the dynamics of DNMT/TET would alter gene expression/repression [10].

Histone modifications are post-translational and can occur in several ways, most common among them being acetylation/deacetylation, methylation/demethylation and ubiquitinylation/sumoylation. Methylation usually occurs on the lysine and arginine residues on H3 and H4. Lysine methylations are dependent on a group of enzymes called histone lysine methyl trnasferases (HKMTs). Each of the enzymes in this group are specific to the site of the lysine on the histone protein, mostly targeting the lysines in the N-terminal tails. They exhibit the SET domain which is the enzyme activity site and contains the aromatic determinant (Y or F) which determines whether it is a monomethylating or trimethylating enzyme. The arginine methyl transferases form a family of protein methyl transferases referred to as PRMTs. The histone demethylases mostly target lysine.

Histone acetylation/deacetylation too occurs on the lysine residues on the N-terminal tail, leading to an alteration in the binding of the histone protein to the DNA strand and thereby altering the expression thereof. Acetylation is directed by the enzymes histone acetyl transferases (HATs) and involves mainly H3 and H4. The histone deacetylases (HDACs) direct the deacetylation of all 4 core histones – H2A, H2B, H3 and H4. Typically, deacetylation restores the positive charge on the lysine and, therefore, stabilises the chromatin thereby enhancing gene repression. Conversely, acetylation destabilises the chromatin and favours transcription [11].

Histones also undergo ubiquitylation and sumoylation. Whereas the methylation/demethylation and the acetylation/deacetylation reactions result in small alterations

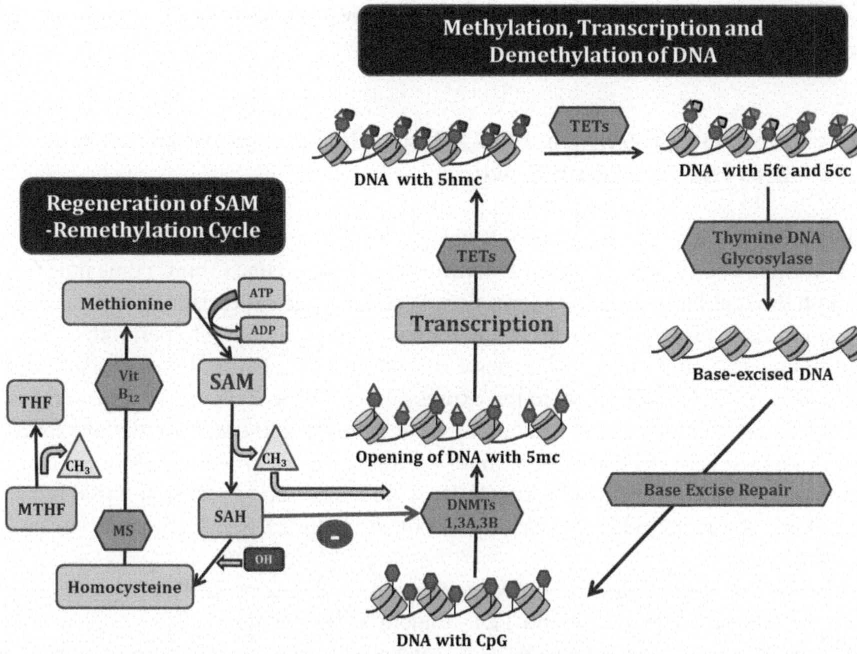

Fig. 6.3 Schematic representation of DNA Methylation/Demethylation Reactions
Methylation of DNA by DNMTs opens it, allowing transcription, which is followed by demethylation by TETs, base excision by thymine DNA glycosylase and then base excision repair, making the DNA available for another cycle. The methyl groups for methylation by DNMTs are obtained from SAM which is then regenerated through homocysteine in the remethylation cycle. Its demethylated form – SAH – subjects DNMTs to feedback inhibition.
THF-tetrahydrofolate; MTHF-methyl tetrahydrofolate; MS-methionine synthase; SAM-S adenosyl methionine; SAH-S adenosyl homocysteine; ATP-adenosine triphosphate; ADP-adenine diphosphate; CH_3-methyl group; CpG-cytosine-p-guanine; DNMTs-DNA methyl transferases; TETs-ten-eleven transferases; 5mc-5 methyl cytosine; 5hmc-5 hydroxymethyl cytosine; 5fc-5 formyl cytosine; 5cc-5 carboxycytosine

in the amino acid side chains, ubiquitylation involves the covalent attachment of the 76-amino acid ubiquitin to the histone protein, mostly H2A, H2B and H3. The end-result is variable – some leading to gene silencing and others to transcription initiation. Sumoylation is a similar covalent histone modification that occurs on all core histone proteins. Both these are mediated by 3 enzymes E1, E2 and E3 which are the activating, conjugating and ligating enzymes, respectively [12].

The numerous histone modifications account for a controlled genetic expression. The "Histone Code Hypothesis" suggests that epigenetic modification of histones is at the helm of the structure and transcription of chromatin. This is further accentuated by the 'cross-talk' amongst themselves, many of the enzymes being present in multiple distinct complexes. In addition, the controlled genetic expression-repression is further fine-tuned by the interaction between DNA methylations and histone modifications. Further, there are transcription factors associated with regulation of gene expression which are themselves subject to alterations due to epigenetic modification of their genes [13].

6.6 How Do ROS Impact DNA Methylations/Demethylations and Histone Modifications?

ROS have opposing effects on the normal epigenetic mechanisms. The presence of •OH promotes oxidation of 5mC to 5-hydroxymethyl cytosine (5hmC) thus interfering with the regulation of gene expression. Whether gene expression is enhanced or repressed depends on the site of methylation/demethylation/hypermethylation. Also, these ROS are known to decrease the activity of DNMTs by decreasing the production of SAM, and promote the expression of TETs. Regeneration of SAM is dependent on the methylation of homocysteine to methionine with methylene tetrahydrofolate (MTHF) as the methyl donor. In the presence of increased ROS, folate is diverted to the folate shuttle for endothelial nitric oxide synthesis, making it unavailable as a methyl donor with consequent reduced SAM regeneration and hypomethylation.

All histone methylases are SAM-dependent and consequently subject to the same fate as DNA methylases and TETs under oxidative stress.

On the other hand, demethylases of histone are ferrous (Fe^{2+}) and αketoglutarate (αkg) dependent dioxygenases. In presence of oxidative stress, the regeneration of ferric (Fe^{3+}) from Fe^{2+} is inhibited, impairing the functioning of these enzymes by decreasing TETs (through the decreased action of hypoxia inducible factor [HIF] prolyl oxidases) and increasing the expression of DNMTs (Fig. 6.4).

Yet again, $\cdot O_2H$ also cause single-strand and double-strand DNA breaks rendering them available for epigenetic modification. Another mechanism involved is the oxidation of the guanosine residues to form 8-oxo-2′-deoxyguanosine (8-oxodG). When the formation of 8-oxodG is limited (i.e. oxidative stress is minimal), the DNA glycosylase can remove these residues, the gap being filled by base excision repair (BER). However, when the oxidative stress persists with the excessive formation of 8-oxodG, hypomethylation of the adjacent cytosines occurs [10, 11, 14, 15].

In addition to the above-mentioned effects of ROS of methylation of DNA and histones, the latter are modulated by several other mechanisms leading to either open chromatin or closed chromatin further resulting in overexpression or repression of gene transcriptions, respectively.

6.7 Congenital Heart Disease

Chromatin regulation, exemplified above, is the key to the expression/repression modulation of all genes including those related to cardiac development in utero as well as during functioning/malfunctioning in postnatal life.

The heart has several types of tissues and systems – (a) the myocardium made up of myocytes which are inherently contractile even without nervous stimulation, (b)

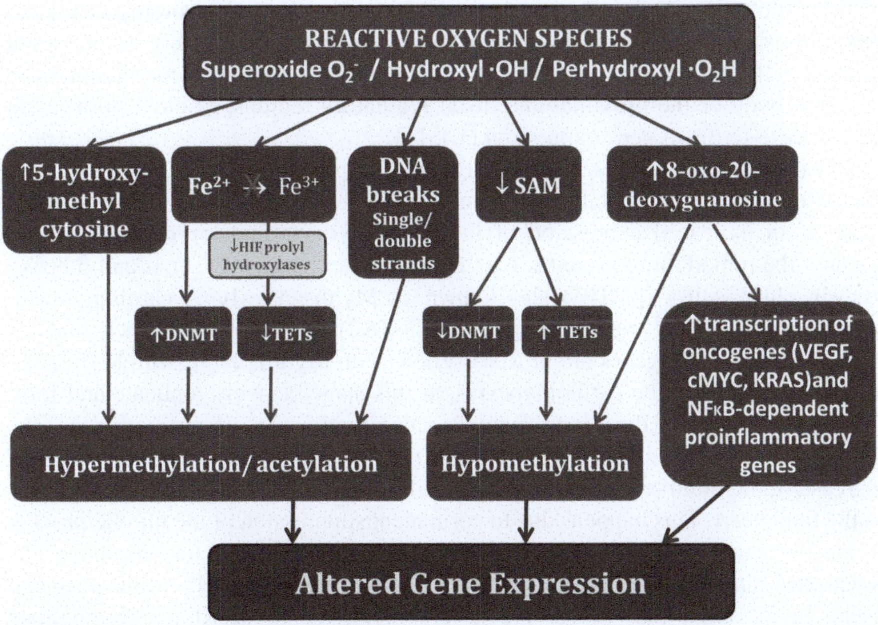

Fig. 6.4 Epigenetic mechanisms of ROS
The major epigenetic effects of reactive oxygen species is altered gene expression through either hypermethylation or hypomethylation. The mechanisms for these are several: (a) by increased formation of 5-hydroxymethyl cytosine, (b) by decreased generation of ferric from ferrous leading to increased expression of DNMTs and decreased activity of TETs, (c) breaks in single as well as double stranded DNA allowing for increased methylation, and (d) decreased regeneration of SAM leading to decreased DNMT activity and increased expression of TETs

the conduction system, (c) the valvular system, (d) the endocardium made of endothelial cells which line the myocardium (including the valves) from the inside and provide a continuum with the blood vessels, and (e) the pericardium, a thin fluid-containing double layered sac engulfing the heart, protecting it while allowing for the normal expansion and contraction of diastole and systole. It follows therefore that for each component of each type of cell there are coding genes (over 11,000) which are each individually, precisely and co-operatively regulated, resulting in the development of an organ that is physiologically and hemodynamically efficient to pump blood for 70–90 years (with the increase in longevity)! And, considering that on each gene there are multiple sites which are subject to epigenetic modification, imagine the myriad of pathological possibilities that can result from the smallest impact on these regulations, starting from the embryological stage to adulthood to ageing!

During development (and disease), the epigenetic modification of DNAs and histones is highly dynamic, though the exact mechanisms through which TETs and 5mc act are still elusive. Gilsbach et al as well as Greco et al conducted, separately, very comprehensive experiments to demonstrate the intricacies of these epigenetic methylations and hydroxymethylations on the development of cardiomyocytes and their component proteins at different stages – in the embryo, during normal adult life and in the presence of disease. They elucidated that epigenetic modifications in the gene body or the proximal and distal regulatory regions, resulted in different outcomes in terms of gene expression [10, 14].

The cardiomyocytes develop from progenitors early in the embryo and have limited cell division in postnatal life. Normally, in the early embryo – first 6–8 weeks – there is an increased expression of the cardiomyocyte and morphology related genes – the remodelling proteins, e.g. myosin light chain (MYL1), α and β heavy myosin chain genes (βMHC) also known as Myh6 and Myh7, cardiomyocyte-specific sarcoplasmic reticulum gene (SERCA2). The miRNA which are co-expressed with these genes are miRNAs 208a (with Myh6), 208b (with Myh7) and 499 (with Myh7b). The cardiomyocytes, at this stage, actively replicate and form the basic structure of the heart. Thereafter, cardiomyocytes do not replicate – the growth in the heart is due to the increased size of each cardiomyocyte rather than an increase in the number of cells. Thus, the adult heart has as many cardiomyocytes as the fetal heart. This happens due to permanent suppression of the mitotic process of the cardiomyocyte just before birth through the formation of the terminally differentiated binucleated cardiomyocytes which comprise about 70% of the postnatal heart. In the latter part of cardiac development, the myocardial trabeculations develop in the cardiac jelly of the embryo. One of the major players in all stages of cardiac development is the Brahma-related gene 1 (BRG1) which acts in a time- and tissue-specific manner. It regulates cardiac gene expression, tissue growth and differentiation through promotion of cardiomyocyte proliferation by maintaining Bmp10 and suppressing p57, the former being a key factor for myocardial proliferation while the latter is a cyclin-dependent kinase inhibitor that prevents cellular proliferation. In addition, BRG1 also regulates the trabeculation by modulating the expression of Adamts 1, a secreted matrix metalloproteinase (MMP) which uses versican (a component of cardiac jelly required for development) as substrate [16–19].

To some extent in embryonic life and more so in postnatal life, the cardiomyocyte dramatically increases in size to meet the physiological requirements of the fast growing organism. However, it also responds similarly to injury or pathology. For example, if there is a chronic ventricular overload, the cardiomyocyte grows beyond its normal size to compensate for the extra work it has to do, initially compensating for the pathology but later leading to cardiomyopathy.

Whereas in the early stages of development, the expression of genes coding for the myocardial development (especially myosin and actin filaments) predominate, in the later stages, expression of the proteins involved in the electrical connectivity responsible for the cardiac contractile functions and energy generation (especially those for fatty acid metabolism) predominate [20].

Broadly speaking, the genes that are preferentially expressed in the different stages of cardiac development are listed below:

A. Early cardiac development MYL1 (myosin light chain 1), Myh3 (myosin heavy chain 3), Myh6, Myh7, MYBPC1 (myosin binding protein C), SLN (sarcolipin) and ACTG2 (actin gamma) – which code for sarcoplasmic reticulum, actin and essential myosin contractile proteins.
B. Late cardiac development FABP4 (fatty acid binding protein 4), FABP2, NRAP (nebulin-related anchoring protein), APOA2 (apolipoprotein A2) and ACE2 (angiotensin converting enzyme 2) – which encode cardiac fatty acid metabolism and structural remodelling.

It would follow, therefore, that epigenetic effects of oxidative stress during early stages of cardiac development could result in structural defects of the heart, e.g. atrial and ventricular septal defects, foramen ovale, anomalies of the bicuspid and tricuspid valves, developmental absence of right or left heart, coarctation of aorta, cardiac dilatation, etc. Similarly, epigenetic modulations due to oxidative stress during latter weeks of cardiac development could lead to defects in the conduction system (arrhythmias), contractility of the heart and its energy-generating metabolism. Either of these types of congenital epigenetic consequences of oxidative stress could also lay the foundation for several types of cardiac diseases that appear in later life, including coronary artery anomalies which may lead to coronary insufficiency, or valvular defects which lead to improper blood flow and oxygenation of cardiac as well as peripheral tissues, or conduction defects which lead to arrhythmias or heart failure.

6.8 Other Cardiac Diseases

Cardiac pathology can be of three major types: (a) alterations in the normal electrical pathways of the heart that allow for normal synchronous contraction and relaxation of the different compartments of the heart – this would lead to arrhythmias and fibrillations, (b) inability to maintain a normal stroke volume which could be a result of altered physiology of the myocardium as well as due to alterations in the electrical pathways, and (c) decrease or absence of coronary arterial blood flow resulting in ischemia of the myocardium [16].

6.8.1 Arrythmias

The normal electrical conduction in the cardiomyocytes depends on the structure and function of the atria and ventricles, so that changes in structural as well as functional remodeling can lead to arrhythmias. Typicallly, the Ca^{2+} and K^{+} ion channels of the cardiomyocyte as well as the intercellular channels modulate the flow of ions across the cell membrane and maintain normal individual and composite electrical function of the cardiomyocytes. Each channel is a complex of proteins and hence its function is governed by appropriate expression of each of these proteins. The sequence of events during the generation and propagation of an electrical pulse in the heart starts with the L-Type Ca^{2+} channel (LTCC) mediated Ca^{2+} influx into the cardiomyocyte. This induces release of Ca^{2+} from the sarcoplasmic reticulum (SR) via the ryanodine receptor 2 (RyR2). An SR- Ca^{2+}-adenosine triphosphate (SERCA2a) maintains homeostasis by enabling re-uptake of Ca^{2+} into the SR. This SERCA2a is under the inhibitory control of an SR associated protein, phospholamban (PLB). The initial event of increased influx of Ca^{2+} into the cytosol corresponds to systole whereas the homeostatic event of reabsorption of Ca^{2+} into the SR corresponds to diastole. Another channel involved in the latter process is the Na^{+}- Ca^{2+} exchanger (NCX). SR Ca^{2+} storage is depleted by Ca^{2+}/calmodulin kinase II (CaMKII)-dependent phosphorylation of RyR2, leading to increased cytosolic Ca^{2+} in the cardiomyocyte, and hence it plays a role in arrhythmias. Hence, it is evident that expression of each of these proteins or complex of proteins regulates the events of cardiac electrical activity [21].

Several studies have demonstrated that ROS inhibit LTTC current and SERCA2 activity, and increase activity of NCX and RYR, thus implicating ROS in epigenetic and direct remodelling of cardiac electrical activity. Hudasek et al showed an increased expression of the pore-forming α-1 subunit of LTTC by ROS leading to changes in the ionic influx and outflux in the cardiomyocytes. In the RYR, ROS modulate the cysteine residues which alter its interaction with triadin, a transmembrane protein responsible for the RYR's Ca2+ sensitivity [22–25].

The whole cascade of events that are initiated by oxidative stress include structural as well as electrical remodelling of the heart which culminates in arrhythmias. Many of these are mediated through altered expression of several miRNAs, which are small non-coding RNAs involved in RNA silencing and post-translational gene expression. For example, miRNAs 21, 29, 30, 133 affect structural genes leading to increased expression of TGFβ (transforming growth factor β), ERK (extracellular signal related kinase), p38 (a protein kinase), JNK (Janus N-terminal kinase) and CTGF (connective tissue growth factor), and decreased expression of DUSP8 (dual specificity protein phosphatase). These result in the increased activity of the MAPKs (mitogen associated protein kinases) due to decreased dephosphorylation (low DUSP8) and along with increased TGFβ, p38 and JNK, there is cardiac hypertrophy and increased collagen leading to fibrosis and structural remodelling. On the other hand, miRNAs 1, 19, 21, 26, 130, 133, 145, 328, 499 closely regulate the ion channels and their functioning through altered expression of LTTC, SERCA, NCX, RYR2, CaMK II – all leading to altered ionic movement across the sarcolemma of the cardiomyocyte and consequent electrical remodeling. All these miRNAs are altered under conditions of oxidative stress [21].

Here, we would, once again, like to direct the reader's attention to the fact that even in physiological concentrations ROS function similarly, but through our antioxidant mechanisms we are able to mitigate their effects, unless their concentration becomes pathological.

6.8.2 Hypertrophic Cardiomyopathy (HCM)

The number of cardiomyocytes in the heart does not usually alter much after birth due to the permanent suppression of mitotic processes of the cardiomyocyte just before birth. However, as listed above, the mitosis, meiosis and differentiation of the cardiomyocyte is induced by genes whose expression is altered by oxidative stress. There is indirect evidence that ROS modulate expression of the histone demethylases, especially LSD1 (lysine-specific demethylase 1) which promotes myogenic proliferation and differentiation, leading to cardiac hypertrophy. At the same time, LSD1 activation is accompanied by a burst of H_2O_2 and ROS which results in 8oxodG formation with consequent induction of the BER enzymes promoting repair. Thus, opposing mechanisms are induced by ROS which might lead to self-containment of the pathology, exemplifying the double-edged sword-like nature of ROS [3].

For increased hemodynamic requirements during exercise and pregnancy, there is an increase in the size of the heart, mainly through the increased size of the cardiomyocyte to enable it to meet the increased pumping demands. This is reversible and exemplifies the body's adaptation to altered requirements. A similar phase of reversible adaption occurs in various chronic conditions – hypertension, diabetes, ageing, obesity, myocardial infarction. However, these adaptive changes become pathological due to the chronic nature of these conditions, so that the myocardium loses its plasticity and ability to adapt to further systemic demands.

As mentioned above, ROS are associated with cardiac hypertrophy through increased expression of histone demethylases; however, when these events are chronic, they become pathological and are accompanied by an inflammatory component whereby NOX are overexpressed leading to further production of ROS – and a vicious pathological cycle is initiated. These ROS are also known to act through the altered expression of some miRNAs. Cardiospecific miRNAs 1 and 2 are overexpressed in the hearts that already show cardiomyopathy, whereas the miRNAs 9 and 448 are downregulated in the dystrophic heart even before cardiomyopathy is evident. The miRNA 448 has been shown to downregulate Ncf1 gene which codes for the neutrophil cytosolic factor which is itself a component protein of the NADPH oxidase enzyme system. Hence, a downregulation of miRNA 448 will be associated with an upregulation of Ncf1 and NADPH oxidase which are an integral part of mitochondrial functioning, leading to further ROS generation. Whereas in the leucocytes, ROS production aids in cytotoxic activity against invading pathological microorganisms, in the cardiomyocyte they give rise to a cascade of events resulting in cardiac dystrophy as described above [26].

6.8.3 Myocardial Infarction

A decreased blood flow to any part of the myocardium will first lead to insufficiency and then infarction in that area. Apart from disease states like obesity, hypertension and diabetes, old age is also a risk factor for myocardial infarction due to coronary vessel narrowing or occlusion.

Myocardial infarction has three components – inflammation, disturbed blood flow and arterial remodelling, which usually occur in this sequence. Excess ROS formation contributes to all three components through various mechanisms including direct injury to the vessel wall and lipid peroxidation, and hence, it is impossible to separate the direct effects from the epigenetic events. The epigenetic mechanisms of this vascular remodelling due to ROS include increased expression of MMP2, transforming growth factor (TGF) β, vascular cell adhesion molecule-1 (VCAM1), fibronectin (FN1), P-selectin (SELP) and E-selectin (SELE). VCAM1 and the selectins promote adhesion of the monocytes especially at the branch points of the vessels where they then attract adhesion of more leucocytes. There is an increased expression of NOX in these cells as well as the vascular endothelium, the fibroblasts and smooth muscle cells of the myocardium, which furthers inflammation through increased production of ROS, especially O_2•. Damage to the vascular wall is initiated by oxidation of the LDL (low density lipoprotein) particles by ROS, which are then engulfed by macrophages to form foam cells. These migrate into the subintimal layer – the tunica media. The altered vascular endothelial cells, foam cells and cardiomyocytes exhibit increased expression of MMPs which lead to degradation of the extracellular matrix in the vessel wall resulting in rupture of the atherothrombotic area of the vessel wall. In addition, the increased expression of TGFβ results in proliferation of the injured tunica media cells leading to increased thickness of the vessel wall and, therefore, decreased lumen and altered flow mechanics therein. Thus, ROS contribute to the initiation and propagation of atherosclerosis and myocardial infarction through dual pathways. Paradoxically, it has been demonstrated that surgical reperfusion of the blocked coronary arteries, by primary percutaneous coronary intervention, may be followed by further damage to the vascular endothelium which may result in arrhythmias. Whether or not ROS are involved in this process has not yet been demonstrated [27].

6.9 Therapeutic Implications

Since ROS can function pathologically directly on molecules as well as through epigenetic mechanisms, therapy would also have to address both aspects. Hence, it would involve administration of anti-oxidants as well as modification of the epigenetic effects.

With emerging technologies, the identification of epigenetic mechanisms and specific involvement of miRNAs is becoming easier. Since each player in each mechanism of epigenetics has a specific function, the future holds a lot of promise in terms of identification of diseases in the very early stages as miRNAS are found in the body fluids much before the disease is manifest. It would also allow for development of disease-specific therapies by targeting these molecules and events. In fact, one may go a step further and indicate evolution of patient specific therapies through cardiomyocytes developed from the patient's own embryonic stem cells. However, it would be pertinent to bear in mind that all the events and co-regulations mentioned above have been demonstrated in animal models and maybe in embryonic stem cells; very few have been demonstrated in the adult or embryonic human heart. Hence, more experimental evidence in humans is required which could become feasible due to the presence of miRNAs in the body fluids which could be an ethically acceptable medium for diagnosis [28].

As for administration of antioxidants is concerned, it has to be carefully monitored. ROS also act physiologically to modulate gene expression and are themselves modulated by inherent antioxidant mechanisms. However, when they are in excess of their physiological concentrations, they promote pathological expression/repression of genes, and that is when antioxidant administration may benefit the patient. Therefore, laboratory estimation of ROS, antioxidants and maybe miRNA to assess the pathological/physiological level of ROS needs to be performed, at least, for patients likely to be under oxidative stress. Further experiments are also warranted to demonstrate the efficacy of any proposed management.

Though several antioxidant molecules are commercially available, their efficacy is sometimes limited by their pharmacodynamics, which could even differ from preparation to preparation and from patient to patient. Some examples are – vitamin C, vitamin E, CoQ10, reserveterol, β-carotene, lycopene, quercetin, etc. These have shown variable results in patients of coronary artery disease or those at risk for the same. However, to improve their delivery to the target tissue, some scientists have suggested a different mode of formulation, e.g. nanoparticles and liposomes [29].

Table 6.1 Cardiological consequences of epigenetic modulation of genes by ROS

Stage of oxidative stress (ROS)	Target genes for epigenetic modulation by ROS	Associated Cardiac disease
Early stage of embryonal cardiac development	MYL1, Myh3, Myh6, Myh7, SLN, MYBPC1, ACTG2, BRG1, Adamts1	Atrial and ventricular septal defects, foramen ovale, anomalies of the bicuspid and tricuspid valves, developmental absence of right or left heart, coarctation of aorta, cardiac dilatation
Late stage of embryonal cardiac development	SERCA2, FABP4, FABP2, NRAP, Apo A2, ACE2	Arrythmias, coronary artery anomalies, valvular defects
After birth/ adult heart	LTTC, SERCA2A, CAMK II, RYR2, NCX, TGFβ, JNK	Arrythmias (atrial /ventricular)
	LSD1, NOX, Ncf1 miRNAS 1,2,9,448	Hypertrophic cardiomyopathies, cardiac dystrophy, congestive cardiac failure
	MMP2, TGFβ, VCAM1, FN1, SELP, SELE,	Coronary artery remodelling and myocardial infarction

6.10 Summary

The cardiac malformations/diseases consequent to epigenetic modulations due to ROS depend on the stage of development when oxidative stress occurs (for congenital cardiac diseases) and the inherent antioxidant mechanisms, i.e. the balance between oxidants and anti-oxidants in the affected human system. Amongst congenital malformations, structural defects result from oxidative stress in the early developmental stages, whereas and functional deficits result from oxidative stress in the later stages of cardiac development. In the adult heart, arrhythmias, cardiomyopathies, congestive heart failure and coronary insufficiency can all result from oxidative stress through epigenetic mechanisms. The genes involved in each of these situations are detailed in Table 6.1.

Establishment of therapeutic management is yet in its infancy, but could be promising.

References

1. Bender D (2015) Free radicals and antioxidant nutrients. In: Rodwell VW, Kennelly PJ, Bender DA, Weil PA, Botham KM (eds) Harper's illustrated biochemistry, 13th edn. McGraw Hill, New York
2. Iles KE, Forman HJ (2002) Macrophage signalling and respiratory burst. Immunol Res 26(1–3):95–105
3. Kietzmann T, Petri A, Shvetsova A, Gerhold JM, Gorlach A (2017) The epigenetic landscape related to reactive oxygen species formation in the cardiovascular system. Br J Pharmacol 174:1533–1554
4. Petry A, Weitnauer M, Gorlach A (2010) Receptor activation of NADPH oxidases. Antioxid Redox Signal 13:467–487

5. Gao S, Li C, Chen L, Zhou X (2017) Actions and mechanisms of reactive oxygen species and oxidative system in semen. Mol Cell Toxicol 13:143–154
6. Annunziato A (2008) DNA packaging: nucleosomes and chromatin. Nat Educ 1(1):26
7. Le DD, Fujimori DG (2012) Protein and nucleic acid methylating enzymes: mechanisms and regulation. Curr Opin Chem Biol 16:507–515
8. Branco MR, Ficz G, Reik W (2011) Uncovering the role of 5-hydroxymethylcytosine in the epigenome. Nat Rev Genet 13:7–13
9. Rasmussen KD, Helin K (2016) Role of TET enzymes in DNA methylation, development, and cancer. Genes Dev 30:733–750
10. Greco CM, Kunderfranco P, Rubino M, Larcher V, Carullo P, Anselmo A et al (2016) DNA hydroxymethylation controls cardiomyocyte gene expression in development and hypertrophy. Nat Commun 7:12418
11. Niu Y, DesMarais TL, Tong Z, Yao Y, Costa M (2015) Oxidative stress alters global histone modification and DNA methylation. Free radical biology & medicine 82:22–28. https://doi.org/10.1016/j.freeradbiomed.2015.01.028
12. Bannister AJ, Kouzarides T (2011) Regulation of chromatin by histone modifications. Cell Research 21(3):381–395. https://doi.org/10.1038/cr.2011.22
13. Jenuwein T, Allis CD (2001) Translating the histone code. Science 293(5532):1074–1080
14. Kietzmann T, Petry A, Shvetsova A, Gerhold JM, Gorlach A (2017) The epigenetic landscape related to reactive oxygen species formation in cardiovascular system. Br J Pharmacol 174:1533–1554
15. Balasubramanian S, Hurley LH, Neidle S (2011) Targeting Gquadruplexes in gene promoters: a novel anticancer strategy? Nat Rev Drug Discov 10:261–275
16. Gilsbach R, Preissl S, Grüning BA, Schnick T, Burger L, Benes V, Wurch A, Bonish U, Gunther S, Fleishmann BK, Schubeler D, Hein L (2014) Dynamic DNA methylation orchestrates cardiomyocyte development, maturation and disease. Nat Commun 5:5288. https://doi.org/10.1038/ncomms6288
17. Thornburg K, Jonker S, O'Tierney P, Chattergoon N, Louey S, Faber J, Giraud G (2011) Regulation of the cardiomyocyte population in the developing heart. Prog Biophys Mol Biol 106(1):289–299. https://doi.org/10.1016/j.pbiomolbio.2010.11.010
18. Han P, Hang CT, Yang J, Chang CP (2011) Chromatin remodelling in cardiac development and physiology. Circ Res 183(3):378–396. https://doi.org/10.1161/CIRCRESAHA.110.224287
19. Van Rooij E, Quiat D, Johnson BA, Sutherland LB, Qi X, Richardson KA, Kelm RJ Jr, Olson EN (2009) A family of microRNAs encoded by myosin genes governs myosin expression and muscle performance. Dev Cell 17(5):662–673. https://doi.org/10.1016/j.devcel.2009.10.013
20. Iruretagoyena JI, Davis W, Bird C, Olsen J, Radue R, Teo Broman A, Kendziorski C, BonDurant SS, Golos T, Bird I, Shah D (2014) Metabolic gene profile in early human fetal heart development. Mol Hum Reprod 20(7):690–700. https://doi.org/10.1093/molehr/gau026
21. Lee S, Choi E, Cha MJ, Hwang KC (2014) Looking into a conceptual framework of ROS-miRNA-atrial fibrillation. Int J Mol Sci 15:21754–21776. https://doi.org/10.3390/ijms151221754
22. Hudasek K, Brown ST, Fearon IM (2004) H2O2 regulates recombinant Ca2+ channel alpha1C subunits but does not mediate their sensitivity to acute hypoxia. Biochem Biophys Res Commun 135–141(22):318
23. Morris TE, Sulakhe PV (1997) Sarcoplasmic reticulum Ca2+-pump dysfunction in rat cardiomyocytes briefly exposed to hydroxyl radicals. Free Radic Biol Med 22:37–47
24. Goldhaber JI (1996) Free radicals enhance Na+/Ca2+ exchange in ventricular myocytes. Am J Physiol 271:H823–H833
25. Zima AV, Blatter LA (2006) Redox regulation of cardiac calcium channels and transporters. Cardiovasc Res *71(2)*:310–*321*. https://doi.org/10.1016/j.cardiores.2006.02.019
26. Kyrychenko S, Kyrychenko V, Ma B, Ikeda Y, Sadoshima J, Shirokova N (2015) Pivotal role of miR 448 in the development of ROS-induced cardiomyopathy. Circ Res 108:324–334. https://doi.org/10.1093/cvr/cvv238

27. He F, Zuo L (2015) Redox roles of reactive oxygen species in cardiovascular diseases. Int J Mol Sci 16:27770–27780. https://doi.org/10.3390/ijms161126059
28. Chaturvedi P, Tyagi SC (2014) Epigenetic mechanisms underlying cardiac degeneration and regeneration. Int J Cardiol 173(1):1–11. https://doi.org/10.1016/j.ijcard.2014.02.008
29. Jain AK, Mehra NK, Swarnakar NK (2015) Role of antioxidants for the treatment of cardiovascular diseases: challenges and oppurtunities. Curr Pharm Des 21(30):4441–4455

Metabolic Signatures of Redox-Dependent Cardiovascular Diseases

7

Stephen T. Vernon, John F. O'Sullivan, and Gemma A. Figtree

7.1 Introduction

Dysregulated redox signalling plays a central role in the development and progression of cardiovascular disease (CVD). To date only a small number of biomarkers that reflect cellular redox status have been identified and these markers have not been utilised in the clinical setting. Metabolomics is an unbiased approach that allows the identification and quantification of small molecules within a biological fluid. Advances in metabolomic platform technologies as well as bioinformatic approaches may allow for the rapid identification and utilisation of novel biomarkers that accurately reflect intracellular redox status relevant to the development of CVD. In addition to utility in both clinical diagnosis and monitoring disease status, the identification of such markers may lead to greater understanding of the biological processes and pathways involved in CVD development. Specific metabolic signatures may also better reflect sub-types and stages of CVD that are currently considered and treated as a single entity, this may improve precision in both risk prediction and disease treatment.

S. T. Vernon · G. A. Figtree (✉)
Cardiothoracic and Vascular Health, Kolling Institute and Department of Cardiology, Royal North Shore Hospital, Northern Sydney Local Health District, Sydney, NSW, Australia

Sydney Medical School, Faculty of Medicine and Health, The University of Sydney, Sydney, NSW, Australia
e-mail: gemma.figtree@sydney.edu.au

J. F. O'Sullivan
Sydney Medical School, Faculty of Medicine and Health, The University of Sydney, Sydney, NSW, Australia

Charles Perkins Centre, The University of Sydney, Sydney, NSW, Australia

Heart Research Institute, Sydney, NSW, Australia

S. Chakraborti et al. (eds.), *Oxidative Stress in Heart Diseases*,
https://doi.org/10.1007/978-981-13-8273-4_7

7.2 What Is Metabolomics?

Omics studies allow incorporation and analysis of large amounts of data that represent the entirety of a particular biological parameter within a biological fluid or tissue. Currently, available omics technologies allow the global study of: genetic mutations and polymorphisms (genomics), gene expression (transcriptomics), proteins (proteomics), small molecules (metabolomics), immunophenotyping, and lipids (lipidomics) in biological fluids or tissues [1]. Metabolomics is the systematic identification and quantification of small molecules (up to 1.5 kDa) in biological fluids that reflects the state of the organism at a particular point in time [2]. The metabolome can be determined at the cellular, tissue, organ, or whole organism scale depending on what sample is obtained for assessment.

Metabolites in a biological system are diverse and dynamic, reflecting the array of metabolic activity at a point in time that is affected by various environmental and genetic factors and the interaction between the two [3]. High-throughput technologies, such as ultra-performance liquid chromatography mass spectrometry (UPLC-MS), allow the quantification of thousands of circulating metabolites across multiple pathways. This approach is promising as it has the ability to capture the complexity of metabolic networks and is not limited to a single enzymatic reaction or pathway [4, 5]. Oxidative stress reflects a loss of homeostasis between ROS production and cellular antioxidants. There are currently only a small number of circulating metabolites that have been shown to reflect intracellular redox state (see below). It is feasible that recent advances in metabolomic platform technology and bioinformatics approaches may allow the identification of more specific circulating metabolic markers and signature patterns that reflect specific intracellular abnormalities relevant to redox signalling.

7.3 Technology for Measuring and Analysing Metabolites

Currently available analytical platforms used to identify and quantify metabolites in metabolomic analysis include: nuclear magnetic resonance (NMR) spectroscopy, direct infusion mass spectrometry, ultra-performance liquid chromatography mass spectrometry (UPLC-MS), and gas chromatography mass spectrometry. These NMR and mass spectrometry methods rely on different fundamental physical phenomena to resolve and quantify molecules. NMR and UPLC-MS are the two predominant platforms used in metabolomic studies. NMR spectroscopy is a quantitative, non-destructive technique. NMR allows a single sample to be analysed by multiple experiments/assays and thus provide detailed information on solution-state molecular structures [4].

Unselective NMR experiments can be used to profile metabolites in order to establish whether several metabolites are linked to the outcomes of interest. Alternatively, standard NMR sequences can be used to record resonances for a given nuclei (commonly ^{1}H) for further chemometric analysis. Specific NMR pulse sequences can also be used to select subsets of metabolites. Metabolite molecular dynamics (such as ligand–protein binding) can also be assessed through the interpretation of NMR spin relaxation and molecular diffusion properties. UPLC-MS combines physical separation obtainable by liquid chromatography with mass analysis by mass spectrometry. A pressurized aqueous liquid biofluid sample is run through a column filled with solid adsorbent material. Each molecule type interacts differently with the adsorbent material causing different flow rates for different metabolites, this can then be detected upon elution [4]. The measurements obtained utilising these techniques requires bioinformatic approaches to analyse the data. There are a number of software packages available to perform such analyses. New data science approaches utilising network analytics and machine learning techniques allow for multi-omics data that incorporate multiple individual omics platforms to be utilised in biomarker discovery as well as approaches to uncovering novel biological pathways [1].

7.4 Metabolic Signatures

Metabolic signatures identified through metabolomic analysis may be relatively simple involving a small number of metabolites or may be complex and may include permutations of hundreds or even thousands of metabolites. These diverse metabolic signatures have a vast array of potential utility including: early disease detection and diagnosis; disease activity and treatment monitoring; and in identifying new biological pathways and potential treatment targets. Systems biology provides a platform to try to unpack the underlying relationships, interconnected networks and mechanisms contained within the complex signatures.

Using advanced unbiased bioinformatic approaches including the incorporation of network analysis and machine learning techniques there is now the potential to integrate metabolomic as well as proteomic, lipidomic and genomic data to unravel novel biological pathways and therapeutic targets [1]. Mendelian randomization studies, that utilise specific germline genetic variants (single nucleotide polymorphisms) that have been shown to be associated with a risk factor or biomarker (e.g. a metabolite), are able to test for causal associations and therefore predict the likely effectiveness of targeting specific pathways with new therapeutics [6]. Such approaches should also be able to help direct future candidate based research towards pathways with causal effects (Fig. 7.1).

7.5 Insights from Application of LC-MS Measures of Metabolites in Cellular Models

LC-MS methods have become valuable tools for the measurement of intracellular metabolites. Redox-related metabolites such as nicotinamide co-factors (NAD, NADP, NADH, and NADPH) are readily assessable using these techniques. Glutathione exists in both reduced (GSH) and oxidized (GSSG) states and reflect redox state and cellular health [7]. The above polar metabolites are favourably

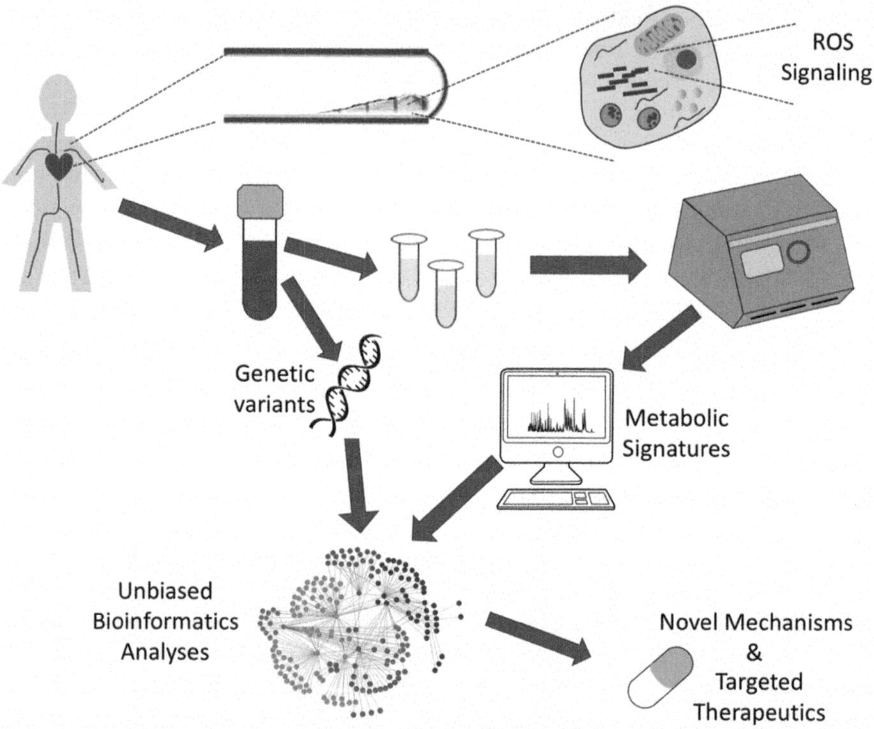

Fig. 7.1 Unbiased approaches to identifying metabolic signatures of redox dependent heart disease. ROS signalling occurs at the intracellular and extracellular level. To date few circulating biomarkers have been identified that accurately reflect intracellular redox status. Metabolomic assessment of plasma utilising NMR or MS platforms can be analysed using bioinformatic techniques including network analysis and machine learning to identify metabolic signatures of cardiovascular disease. Other biological parameters such as genetic variants (single nucleotide polymorphisms) can be incorporated with metabolomic data in an unbiased manner using advanced bioinformatic approaches to identify novel biological pathways, disease mechanisms and therapeutic targets. Systems biology approaches may also be applied to cell and tissue specific samples to identify and further characterise biological pathways

retained using a hydrophilic column with a bonded amide phase. This stationary phase is also suitable for measuring related energetic compounds such as AMP, ADP, and ATP. Advances in metabolite extraction directly from the culture plate for adherent cells, or using a cell suspension for non-adherent cells, has allowed a reliable and robust method of stably retrieving these fragile metabolites. Central carbon metabolism, e.g. measurement of the entire set of TCA cycle intermediates, gives a reliable measure of mitochondrial respiration. Other metabolites, such as ADMA – the major endogenous inhibitor of NOS – are important for a comprehensive assessment of the redox state. Finally, recent advances in stable extraction of cellular organelles such as mitochondria enhance the fidelity of redox signal in a cell.

7.6 Circulating Metabolic Markers Associated with Cardiovascular Disease

A number of metabolic markers of cardiovascular disease have been identified to date. One study utilizing untargeted metabolomics techniques in a general population cohort identified four metabolites that predicted cardiovascular events independently of traditional cardiovascular risk factors: lysophosphatidylcholine 18:1, lysophosphatidylcholine 18:2, monoglyceride 18:2 and sphingomyelin 28:1 [8]. Another study utilising a targeted metabolomic approach in patients with hypertrophic cardiomyopathy who were undergoing an alcohol septal ablation procedure that intentionally causes an area of myocardial infarction identified a four-metabolite signature of acute myocardial infarction, consisting of: aconitic acid, hypoxanthine, trimethylamine N-oxide, and threonine [9]. A recent metanalysis of studies utilising metabolomics to identify metabolic signatures of incident cardiovascular disease identified: acylcarnitine, dicarboxylacylcarnitine, trimethylamine N-oxide (TMAO), amino acids such as phenylalanine, glutamate, and several lipid classes, as being associated with cardiovascular disease [10]. Many of these metabolic markers and their associated pathways are now the focus of candidate based research trying to elucidate the underlying biological processes and potential novel therapeutic targets.

7.7 Gut Microbiome and Cardiovascular Disease

The human microbiome consists of more than 100 trillion microorganisms belonging to 300–500 different species living within each human [11]. It is now increasingly recognised that gut microflora act as a 'filter' for one of the most frequent environmental exposures, food. There is marked variation in the way food is digested within the gastrointestinal tract as a result of the distinct microbial flora within individuals' gastrointestinal tracts. This produces a variety of gut microbe-derived metabolites some of which have been demonstrated to be biologically active in the host. For example, TMAO has been identified using metabolomic approaches to be a strong predictor of coronary artery disease [12]. Circulating TMAO levels increase 4–8 h after ingestion of phosphatidylcholine and/or L-carnitine which are found in abundance in red meat and animal studies have shown both a role of the microbiota affecting the production of and circulating levels of TMAO but also an effect on atherosclerosis progression [12]. Consistent with this, vegetarians have a distinct microbiota and produce less TMAO compared to omnivores [13]. This is consistent with findings that dietary exposure effects the microbiota composition and function. Microbial transplantation studies also confirm the involvement of gut microbe-dependent trimethylamine (TMA)/TMAO in the development of atherosclerosis [14]. It is likely that the TMA/TMAO pathways are one of many microbe-dependent pathways involved in cardiovascular disease pathogenesis. Systems biology methodologies that combine immunomic, proteomic, metabolomic and genomic data should allow for a more integrated understanding of the complex relationship between microbes, diet/environmental factors and the host. For example, Elliott et al. utilised systems biology approaches to identify a metabolic reaction network of human adiposity/obesity. This included the identification of nine gut microbial co-metabolites associated with body mass index reflecting five different host-gut transformation microbial pathways [15].

7.8 Urine Metabolite Signatures

7.8.1 Diabetic Renal Disease

The prevalence of type two diabetes mellitus is increasing throughout the developed world. Diabetic renal disease, one of the most common complications of diabetes mellitus, is increasing in prevalence as a consequence [16], mediated through both vascular dysfunction, as well as direct effects on the glomerulus and nephron. Diabetic renal disease is an important step in deteriorating cardiovascular health of the individual, feeding back to contribute to myocardial fibrosis and hypertrophy and heart failure, as well as vascular dysfunction and atherosclerosis. The identification of diabetic renal disease has traditionally relied on monitoring serum creatine and urine microalbumin levels. However, there may be other metabolic markers of early diabetic renal disease detectable in plasma or urine, which, may allow additional targeted strategies for prevention. Metabolomic studies utilising plasma from diabetes mellitus and ESRD individuals have identified accumulation of metabolites secondary to alterations in branched-chain and aromatic amino acid metabolism [17, 18]. As the kidney concentrates and excretes many metabolites, urine metabolomic assessment may allow unique metabolic signatures to be identified. It is likely that there are urinary metabolic signatures that are unique to various mechanisms of renal dysfunction including diabetic nephropathy. One study utilising urinary metabolomics in healthy controls and diabetics with and without diabetic renal disease identified 13 metabolites that are differentially expressed in diabetic renal disease. The 13 differential metabolites identified were: 3-hydroxy isovalerate, Aconitic acid, Citric acid, 2-ethyl 3-OH propionate, Glycolic acid, Homovanillic acid, 3-hydroxy isobutyrate, 2-methyl acetoacetate, 3-methyl adipic acid, 3-methyl crotonyl glycine, 3-hydroxy propionate, Tiglylglycine, and Uracil. Twelve of the 13 differentially expressed metabolites have been linked to mitochondrial metabolism with 10 exclusively produced by mitochondria, suggesting that renal mitochondrial activity is suppressed in diabetic renal disease [19].

7.8.2 Obesity

The prevalence of obesity is also increasing across the globe [20]. As previously mentioned, there are complex metabolite associations with adiposity involving an extensive set of biochemical pathways and physiological processes, including gut microbial–human co-metabolism. One study, once again utilising metabolomic assessment of urine, identified 29 metabolites significantly associated with body mass index independently of potential confounders. These metabolites included: urinary glycoproteins and *N*-acetyl neuraminate (related to renal function), trimethylamine, dimethylamine, 4-cresyl sulfate, phenylacetylglutamine and 2-hydroxyisobutyrate (gut microbial co-metabolites), succinate and citrate (tricarboxylic acid cycle intermediates), ketoleucine and the ketoleucine/leucine ratio (linked to skeletal muscle mitochondria and branched-chain amino acid metabolism), ethanolamine (linked to skeletal muscle turnover), and 3-methylhistidine (linked to skeletal muscle turnover and meat intake) [15].

7.9 Central Role of Redox in Cardiovascular Disease

Dysregulated production and degradation of reactive oxygen species (ROS), including Superoxide anion (O_2^-), hydrogen peroxide (H_2O_2) and hydroxyl radicals ($\cdot OH$), have been implicated in multiple cardiovascular disease processes including atherosclerosis, hypertension, myocardial hypertrophy, heart failure and ischaemia-reperfusion injury [7, 21]. Reactive nitrogen species (RNS), such as nitric oxide (NO) and peroxynitrite ($ONOO^-$) are also involved in redox signalling in both physiological and pathological processes.

Under physiological conditions the main source of ROS production is within mitochondria, primarily via the electron transport chain (ETC). ROS production by the ETC depends upon a number of physiological and pathological factors including: the metabolic state of the mitochondria and degree of tissue oxygenation. Other sources of ROS relevant to the cardiovascular system include: nicotinamide adenine dinucleotide phosphate (NADPH) oxidases (Noxs), Xanthine oxidase, cytochrome p450 enzymes, and uncoupled NO synthases (NOS) [22].

Tightly regulated production of ROS alters intracellular molecules and provides signals that may alter cellular phenotype both acutely and chronically [23]. Physiological processes within the cardiovascular system regulated by ROS include: nitric oxide (NO) regulation of endothelium-dependent microvascular and epicardial vasodilatation, NO inhibition of platelet aggregation and adhesion; S-Glutathionylation (formation of a disulphide bridge between reactive cysteine residue and the tripeptide glutathione) mediated redox regulation of a number of key cellular proteins including endothelial nitric oxide synthase (eNOS), ryanodine receptor, SERCA and Na^+/K^+ pump, resulting in changes to intracellular Na^+ and Ca^{2+} handling effecting function such as myocyte contractility.

In addition to their physiological role, mitochondrial produced ROS are also implicated in the pathophysiology of a variety of cardiovascular and other diseases. Redox signalling is implicated in many pathological pathways relevant to the cardiovascular system including: vascular smooth muscle proliferation, atherosclerosis, angiogenesis, cardiac hypertrophy, fibrosis, and remodelling via reversible and irreversible modification of proteins and DNA [23].

7.10 Compartmentalisation of Redox Signalling Pathways

Redox signalling refers to the processes by which ROS and RNS induce site-specific and reversible modifications to proteins that influence function through steric effects [21]. There are two main cellular compartments where ROS signalling takes place: the mitochondria, and the caveolae. Redox signalling is influenced by: the site of ROS production, the precise ROS species, local ROS concentration, and cell/compartment specific antioxidants [22]. Major ROS sources in the heart and other tissues include the mitochondrial electron transport chain (ETC), other mitochondrial and metabolic enzymes, nicotinamide adenine dinucleotide phosphate (NADPH) oxidases (Noxs), and uncoupled NO synthases (NOS). The caveolae are specialized membrane invaginations that allow the interaction of ROS with redox-regulated proteins. Caveolae are ~50–100 nm in diameter and contain signature proteins called caveolins which act as scaffolds providing spatial as well as temporal regulation for multiple signalling pathways including protein/protein and redox-protein interaction [24, 25]. Caveolae are found in multiple cell types within the cardiovascular system including: endothelial cells, smooth muscle cells, fibroblasts and macrophages [26] They are dynamic, forming and receding in response to varied stimuli such as ischaemia [27]. Of the >245 proteins known to localise in the caveolae, including ion channels, pumps, and receptors, >80% are thought to have susceptible cysteines that are redox modified [24, 28]. Thus, the potential impact on cellular function of dysregulated redox signalling in the caveolae is profound, and measurements in the circulation that may reflect this compartment would have immense potential as biologically relevant biomarkers in cardiovascular disease. This has been a challenge using standard approaches, but may benefit from advances in both metabolomic technologies and bioinformatic analytical approaches of metabolite signatures.

7.11 Which Metabolite Profiles/Signatures Change Under Redox Stress?

Lipids, proteins, carbohydrates, and DNA are all susceptible to oxidative modification under redox stress [7]. Some of these modifications have functional effects, for example inhibition of enzymatic activity, whilst others appear to be functionally silent. There are currently no good circulating biomarkers that closely represent sub-cellular dysregulated redox signalling in the tissue, arterial wall or myocardium. Improved prognostic ability, and potentially early disease detection and risk stratification may be achieved through a more reflective biomarker of the dysregulated signalling microdomain. Whilst there is no currently identified circulating redox biomarker reflecting the intricacies of sub-cellular redox dysregulation in cardiovascular disease, there are some markers that are either directly or indirectly related to redox state that have been associated with cardiovascular diseases. Metabolomics platforms allow for the measurement of metabolites that are related to or result from lipid and protein oxidation. The enzymes involved in these pathways may also be measured using traditional techniques including enzyme activity assays [29].

7.11.1 Lipid Oxidation

F2-isoprostanes (prostaglandin-like substances) are produced independently of COX enzymes by ROS induced peroxidation of arachidonic acid [30]. F2-isoprostanes are detectable in all biological fluids and may be identified using metabolomic approaches. They are elevated in disease states characterised be elevated ROS [31] including: in association with cigarette smoking, hypercholesterolaemia and diabetes mellitus [30]. F2-isoprostanes have been shown to affect platelet aggregation, vasoconstriction and platelet aggregation, and levels are increased in atherosclerotic lesions [32, 33]. They have also been shown to be elevated in myocardial reperfusion and to correlate with left ventricular dilatation and heart failure severity [34, 35].

Malondialdehyde (MDA) (generated via peroxidation of polyunsaturated fatty acids) are proposed to play a key role in atherogenesis. MDA impairs macrophage action by inducing lysine-lysine cross links in apolipoprotein B fractions of OxLDL [36]. MDA levels can be measured using metabolomic platforms including high performance liquid chromatography (HPLC) however they are more commonly assessed using a commercial ELISA assays that actually measures Thiobarbituric acid-reactive substances (TBARS). ELISA assays for TBARS have a strong correlation with direct HPLC of MDA. Serum levels of TBARS predicts major adverse cardiovascular events, including cardiovascular mortality and myocardial infarction, independent of traditional cardiovascular risk factors and inflammatory markers [37].

4-hydroxynonenal (4-HNE) is also produced via lipid peroxidation and is assessable by metabolomic platforms. 4-HNE results from the reaction of OH^- with lipids. 4-HNE is highly reactive with proteins and 4-HNE produced in the vascular wall exerts a paracrine effect on perivascular fat resulting in activation of peroxisome proliferator-activated receptor-γ signalling in the perivascular fat. This in turn leads to adipokine and adiponectin (antioxidants) release in the perivascular fat which subsequently has a paracrine effect back on the vascular wall reducing NADPHoxidase activity, thereby improving eNOS coupling [38]. This feedback cascade demonstrates the complexities of redox signalling and redox regulation within and between tissues.

7.11.2 Protein Oxidation

Free circulating 3-nitrotyroseine (3-NO-Tyr) is a predictor of cardiovascular risk independent of traditional risk factors. Statin therapy has also been demonstrated to alter circulating 3-NO-Tyr levels [39]. Carbonyl compounds formed by oxidation of a few amino acid side chains via the addition of aldehydes may be used as markers of severe protein oxidation/damage. Carbonyl compounds accumulate during: aging, ischemia/reperfusion [40], diabetes, and obesity [41]. S-glutathionylation, the formation of a mixed disulphide bond between glutathione and the cysteine residue on proteins is both stable and reversible. S-glutathionylation of various proteins has an array of functional effects including regulation of eNOS, and the Na^+-K^+ pump, as well as non-functional effects. However, access to these key proteins, and measurement of their modification in the circulation in a manner that reflects their maladaptive redox modifications in situ, is challenging [7]. One promising approach involves measuring S-glutathionylation of the Na^+-K^+ pump in circulating erythrocytes that has been demonstrated to mirror S-glutathionylation of the Na^+-K^+ pump in myocardial cells in patients with heart failure [7], although this still relies on more candidate approaches than achieved with broad metabolomic platforms.

7.11.3 Myeloperoxidase

Myeloperoxidase (MPO) is an abundant heme enzyme located within an array of inflammatory cells including activated neutrophils, macrophages and monocytes. MPO contributes to the production of a variety of ROS by catalysing the conversion hydrogen peroxide (H_2O_2) to hydroxyl radicals ($\cdot OH$), ($ONOO^-$), hypochlorous acid (HOCL) and nitric dioxide (NO_2). MPO. MPO levels are an independent predictor of cardiovascular events in patients presenting to the Emergency Department with chest pain [42]. MPO levels also predict the development of coronary artery disease in "healthy" individuals [43] and increases in MPO levels have been associated with accelerated atheroma progression in diabetes mellitus patients [44]. MPO is an enzyme and is therefore not directly assessable using metabolomic techniques however given it's known role in the production of ROS and association with cardiovascular disease it is likely that MPO levels contribute to metabolic signatures of CVD and this relationship is a focus of ongoing research.

7.12 Therapeutic Relevance of Redox Signalling in Cardiovascular Disease

Multiple clinical trials testing the utility of anti-oxidant intake have failed to demonstrate a reduction in cardiovascular events or death [45]. This may reflect an inability of dietary anti-oxidants to effectively and efficiently target specific redox pathways involved in cardiovascular disease. Many of the evidence-based cardiovascular drugs in clinical practice, including: statins, beta-blockers, and ACE inhibitors, are known to effect redox-dependent signalling pathways which may contribute to their demonstrated utility in preventing cardiovascular events and mortality [24, 46]. A greater understanding of redox signalling pathways and markers that reflect this may allow novel treatments designed to specifically target redox dysregulation in cardiovascular disease. In addition, specific metabolic signatures identified through metabolomic assessment may be transferrable to the clinical setting and be incorporated in personalised medicine approaches.

7.13 Conclusion

Advances in metabolomics technologies, and bioinformatics capabilities will assist in identification and precise measurement of both candidate and unsuspected metabolites in the circulation that reflect dysregulated redox signalling and may be of relevance to clinical practice and our efforts to improve cardiovascular health.

References

1. Vernon ST, Hansen T, Kott KA, Yang JY, O'Sullivan JF, Figtree GA (2019) Utilizing state-of-the-art "omics" technology and bioinformatics to identify new biological mechanisms and biomarkers for coronary artery disease. Microcirculation 26:e12488
2. Nicholson JK, Lindon JC, Holmes E (1999) 'Metabonomics': understanding the metabolic responses of living systems to pathophysiological stimuli via multivariate statistical analysis of biological NMR spectroscopic data. Xenobiotica 29(11):1181–1189
3. Nedic Erjavec G, Konjevod M, Perkovic M, Svob D, Tudor L, Barbas C et al (2018) Short overview on metabolomic approach and redox changes in psychiatric disorders. Redox Biol 14:178–186
4. Dona AC, Coffey S, Figtree G (2016) Translational and emerging clinical applications of metabolomics in cardiovascular disease diagnosis and treatment. Eur J Prev Cardiol 23(15):1578–1589
5. Zamboni N, Saghatelian A, Patti GJ (2015) Defining the metabolome: size, flux, and regulation. Mol Cell 58(4):699–706
6. Burgess S, Harshfield E (2016) Mendelian randomization to assess causal effects of blood lipids on coronary heart disease: lessons from the past and applications to the future. Curr Opin Endocrinol Diabetes Obes 23(2):124–130
7. Karimi Galougahi K, Antoniades C, Nicholls SJ, Channon KM, Figtree GA (2015) Redox biomarkers in cardiovascular medicine. Eur Heart J 36(25):1576–1582
8. Ganna A, Salihovic S, Sundstrom J, Broeckling CD, Hedman AK, Magnusson PK et al (2014) Large-scale metabolomic profiling identifies novel biomarkers for incident coronary heart disease. PLoS Genet 10(12):e1004801
9. Lewis GD, Wei R, Liu E, Yang E, Shi X, Martinovic M et al (2008) Metabolite profiling of blood from individuals undergoing planned myocardial infarction reveals early markers of myocardial injury. J Clin Invest 118(10):3503–3512
10. Ruiz-Canela M, Hruby A, Clish CB, Liang L, Martínez-González MA, Hu FB (2017) Comprehensive metabolomic profiling and incident cardiovascular disease: a systematic review. J Am Heart Assoc 6(10):e005705
11. Guarner F, Malagelada J-R (2003) Gut flora in health and disease. Lancet 361(9356):512–519
12. Wang Z, Klipfell E, Bennett BJ, Koeth R, Levison BS, DuGar B et al (2011) Gut flora metabolism of phosphatidylcholine promotes cardiovascular disease. Nature 472(7341):57–63
13. Tang WHW, Wang Z, Levison BS, Koeth RA, Britt EB, Fu X et al (2013) Intestinal microbial metabolism of phosphatidylcholine and cardiovascular risk. N Engl J Med 368(17):1575–1584
14. Wilson Tang WH, Hazen SL (2017) The gut microbiome and its role in cardiovascular diseases. Circulation 135(11):1008–1010
15. Elliott P, Posma JM, Chan Q, Garcia-Perez I, Wijeyesekera A, Bictash M et al (2015) Urinary metabolic signatures of human adiposity. Sci Transl Med 7(285):285ra62

16. Rosolowsky ET, Skupien J, Smiles AM, Niewczas M, Roshan B, Stanton R et al (2011) Risk for ESRD in type 1 diabetes remains high despite renoprotection. J Am Soc Nephrol 22(3):545–553
17. Wang TJ, Larson MG, Vasan RS, Cheng S, Rhee EP, McCabe E et al (2011) Metabolite profiles and the risk of developing diabetes. Nat Med 17:448–453
18. Newgard CB, An J, Bain JR, Muehlbauer MJ, Stevens RD, Lien LF et al (2009) A branched-chain amino acid-related metabolic signature that differentiates obese and lean humans and contributes to insulin resistance. Cell Metab 9(4):311–326
19. Sharma K, Karl B, Mathew AV, Gangoiti JA, Wassel CL, Saito R et al (2013) Metabolomics reveals signature of mitochondrial dysfunction in diabetic kidney disease. J Am Soc Nephrol 24(11):1901–1912
20. The GBDOC, Ng M, Fleming T, Robinson M, Thomson B, Graetz N et al (2014) Global, regional and national prevalence of overweight and obesity in children and adults 1980–2013: a systematic analysis. Lancet (London, England) 384(9945):766–781
21. Madamanchi NR, Runge MS (2013) Redox signaling in cardiovascular health and disease. Free Radic Biol Med 61:473–501
22. Burgoyne JR, Mongue-Din H, Eaton P, Shah AM (2012) Redox signaling in cardiac physiology and pathology. Circ Res 111(8):1091–1106
23. Finkel T (1999) Signal transduction by reactive oxygen species in non-phagocytic cells. J Leukoc Biol 65(3):337–340
24. Bubb KJ, Birgisdottir AB, Tang O, Hansen T, Figtree GA (2017) Redox modification of caveolar proteins in the cardiovascular system- role in cellular signalling and disease. Free Radic Biol Med 109:61–74
25. Williams TM, Lisanti MP (2004) The caveolin proteins. Genome Biol 5(3):214
26. Patel HH, Insel PA (2008) Lipid rafts and caveolae and their role in compartmentation of redox signaling. Antioxid Redox Signal 11(6):1357–1372
27. Tsutsumi YM, Horikawa YT, Jennings MM, Kidd MW, Niesman IR, Yokoyama U et al (2008) Cardiac-specific overexpression of caveolin-3 induces endogenous cardiac protection by mimicking ischemic preconditioning. Circulation 118(19):1979–1988
28. Wypijewski KJ, Tinti M, Chen W, Lamont D, Ashford MLJ, Calaghan SC et al (2015) Identification of caveolar resident proteins in ventricular myocytes using a quantitative proteomic approach: dynamic changes in caveolar composition following adrenoceptor activation. Mol Cell Proteomics 14(3):596–608
29. Pulli B, Ali M, Forghani R, Schob S, Hsieh KLC, Wojtkiewicz G et al (2013) Measuring myeloperoxidase activity in biological samples. PLoS One 8(7):e67976
30. Morrow JD, Awad JA, Boss HJ, Blair IA, Roberts LJ (1992) Non-cyclooxygenase-derived prostanoids (F2-isoprostanes) are formed in situ on phospholipids. Proc Natl Acad Sci U S A 89(22):10721–10725
31. Griffiths HR, Moller L, Bartosz G, Bast A, Bertoni-Freddari C, Collins A et al (2002) Biomarkers. Mol Aspects Med 23(1–3):101–208
32. Kromer BM, Tippins JR (1996) Coronary artery constriction by the isoprostane 8-epi prostaglandin F2 alpha. Br J Pharmacol 119(6):1276–1280
33. Patrono C, FitzGerald GA (1997) Isoprostanes: potential markers of oxidant stress in atherothrombotic disease. Arterioscler Thromb Vasc Biol 17(11):2309–2315
34. Delanty N, Reilly MP, Pratico D, Lawson JA, McCarthy JF, Wood AE et al (1997) 8-epi PGF2 alpha generation during coronary reperfusion. A potential quantitative marker of oxidant stress in vivo. Circulation 95(11):2492–2499
35. Mallat Z, Philip I, Lebret M, Chatel D, Maclouf J, Tedgui A (1998) Elevated levels of 8-iso-prostaglandin F2alpha in pericardial fluid of patients with heart failure: a potential role for in vivo oxidant stress in ventricular dilatation and progression to heart failure. Circulation 97(16):1536–1539
36. Halliwell B (2000) Lipid peroxidation, antioxidants and cardiovascular disease: how should we move forward? Cardiovasc Res 47(3):410–418

37. Walter MF, Jacob RF, Jeffers B, Ghadanfar MM, Preston GM, Buch J et al (2004) Serum levels of thiobarbituric acid reactive substances predict cardiovascular events in patients with stable coronary artery disease: a longitudinal analysis of the PREVENT study. J Am Coll Cardiol 44(10):1996–2002
38. Antonopoulos AS, Margaritis M, Coutinho P, Shirodaria C, Psarros C, Herdman L et al (2015) Adiponectin as a link between type 2 diabetes and vascular NADPH oxidase activity in the human arterial wall: the regulatory role of perivascular adipose tissue. Diabetes 64(6):2207–2219
39. Shishehbor MH, Aviles RJ, Brennan M et al (2003) Association of nitrotyrosine levels with cardiovascular disease and modulation by statin therapy. JAMA 289(13):1675–1680
40. Dalle-Donne I, Rossi R, Colombo R, Giustarini D, Milzani A (2006) Biomarkers of oxidative damage in human disease. Clin Chem 52(4):601–623
41. Bollineni RC, Fedorova M, Bluher M, Hoffmann R (2014) Carbonylated plasma proteins as potential biomarkers of obesity induced type 2 diabetes mellitus. J Proteome Res 13(11):5081–5093
42. Brennan ML, Penn MS, Van Lente F, Nambi V, Shishehbor MH, Aviles RJ et al (2003) Prognostic value of myeloperoxidase in patients with chest pain. N Engl J Med 349(17):1595–1604
43. Meuwese MC, Stroes ES, Hazen SL, van Miert JN, Kuivenhoven JA, Schaub RG et al (2007) Serum myeloperoxidase levels are associated with the future risk of coronary artery disease in apparently healthy individuals: the EPIC-Norfolk Prospective Population Study. J Am Coll Cardiol 50(2):159–165
44. Kataoka Y, Shao M, Wolski K, Uno K, Puri R, Tuzcu EM et al (2014) Myeloperoxidase levels predict accelerated progression of coronary atherosclerosis in diabetic patients: insights from intravascular ultrasound. Atherosclerosis 232(2):377–383
45. Jialal I, Devaraj S (2003) Antioxidants and atherosclerosis: don't throw out the baby with the bath water. Circulation 107(7):926–928
46. Shah AM, Channon KM (2004) Free radicals and redox signalling in cardiovascular disease. Heart 90(5):486–487

Cardio Protective Influence of Dietary Spices Mediated Through Their Hypolipidemic and Antioxidant Potential

8

Krishnapura Srinivasan

8.1 Introduction

Spices are common ingredients of our food, being used to render our food more appealing by imparting typical flavour, colour and pungency. Spices are also understood to offer many medicinal properties [1] and a few spices are hence particularly employed in the indigenous systems of medicine and folk remedies. The chemical constituents of spices responsible for their typical quality attributes such as pungency, aroma, flavor, etc. – termed as spice active principles – are coincidentally responsible for the health beneficial physiological effects also (Bioactive compounds) in many instances [1]. The health beneficial effects of dietary spices include digestive stimulant action, lipid-lowering (particularly cholesterol-lowering) effects, cardio protective property, gallstone preventive influence, anti-inflammatory potential, and cancer preventive effects.

The role of high concentrations of cholesterol in circulating blood on the progression of atherosclerosis and coronary heart diseases is well understood. In this context, while many dietary spices have been evaluated for a cholesterol lowering effect in different experimental situations involving animal models and clinical trials, garlic, onion, fenugreek seeds, turmeric and red pepper are found to be significantly effective as hypocholesterolemic spices under conditions of hypercholesterolemia/hyperlipidemia [2]. Garlic, fenugreek seeds, and onion are found effective in hyperlipidemic patients. The hypolipidemic efficacy of turmeric (and curcumin), red pepper (and capsaicin) and of onion and garlic has been exhaustively reviewed by several authors. While the above spices exert their cardio protective effect primarily through hypocholesterolemic potential, the anti-thrombotic influence, antioxidant effect particularly in cardiac tissue and suppression of low-density lipoprotein (LDL) oxidation, anti-obesity/thermogenic influence, and

K. Srinivasan (✉)
Department of Biochemistry, CSIR - Central Food Technological Research Institute, Mysore, India

S. Chakraborti et al. (eds.), *Oxidative Stress in Heart Diseases*,
https://doi.org/10.1007/978-981-13-8273-4_8

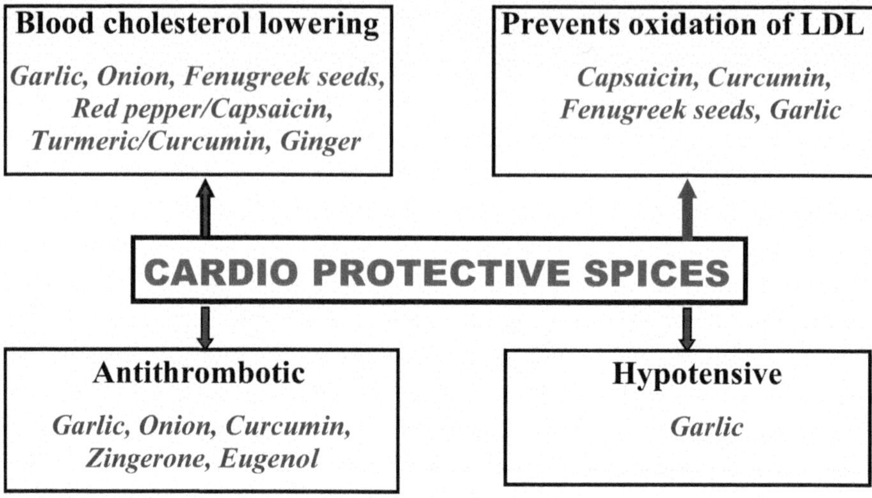

Fig. 8.1 Anti-atherogenic and Cardio protective effect of spices

anti-hypertensive influence exerted by specific spices also contribute to their cardio protective attribute [3]. A substantial body of evidence suggests that oxidative stress in myocardium contribute to cardiovascular disease. More than the cholesterol carried by it, the oxidative modification of LDL under conditions of oxidative stress is thought to play a significant role in the development of atherosclerotic process. The antioxidant property of a few specific spices is of particular interest in the context of the impact of oxidative modification of LDL in the development of atherosclerosis. Garlic, onion, curcumin of turmeric, cuminaldehyde of cumin, eugenol of cloves, and zingerone of ginger are understood to inhibit platelet aggregation (Fig. 8.1).

8.2 Moderation of Hypercholesterolemia/Hyperlipidemia by Dietary Spices

8.2.1 Garlic (*Allium sativum*)

Garlic with its characteristic flavour when used in foods, is also valued for its diverse medicinal properties including favourable physiological influences on CVD and cancer [4–6]. Dietary garlic has been unequivocally associated with a benereduction in blood cholesterol (particularly LDL-associated cholesterol) and also triglyceride levels [7]. More than 30 clinical studies have established that daily consumption of one clove of garlic (equivalent to 600–900 mg of dehydrated garlic powder) will have a consistent 10–12% cholesterol-lowering effect [8]. These clinical studies with garlic powder and its extracts include randomized, placebo-controlled, and double-blind trials, have demonstrated that garlic powder or its preparations also produce significant reduction in blood pressure in addition to lowering total cholesterol and triglycerides. The beneficial effect of garlic orally fed for 6 months on blood lipids has been demonstrated in patients with coronary heart disease [9]. The essential oil of garlic is shown to have distinct hypolipidemic effect in patients with coronary heart disease. In moderately hypercholesterolemic subjects, dietary supplementation with aged garlic extract is reported to exert a greater effect on blood lipid profile and blood pressure as compared to fresh garlic [10]. Simultaneous consumption of garlic with fish oil had a higher benefit on serum LDL-cholesterol and triglyceride concentrations as well as on the ratios of total cholesterol to HDL-cholesterol and LDL-cholesterol to HDL cholesterol [11] (Fig. 8.2).

Animal studies employing rats and dogs have also indicated that garlic or its ingredients suppress cholesterol synthesis in the liver, lower serum cholesterol by particularly reducing LDL-associated cholesterol, and lower serum triglycerides [12–15]. Aqueous garlic extracts diminish hepatic cholesterol biosynthesis through inhibiting the activity of hydroxymethyl glutaryl-CoA reductase [16]. The hydrophilic organosulfur compounds of garlic are also shown to have inhibitory effects on triglyceride and fatty acid synthesis in cultured rat hepatocytes [17]. These compounds impair triglyceride biosynthesis partly due to decreased *de novo* fatty acid synthesis by inhibiting the activity of fatty acid synthetase.

Garlic also exhibits anti-thrombotic and blood pressure lowering properties, both of which also contribute to cardiovascular protection independent of its hypocholesterolemic property. The anti-atherosclerotic property of garlic is mainly attributed to the sulfur compound allicin produced upon crushing of the garlic. Garlic extracts and components – ajoene (4,5,9-trithiododeca-1,6,11-triene-9-oxide),

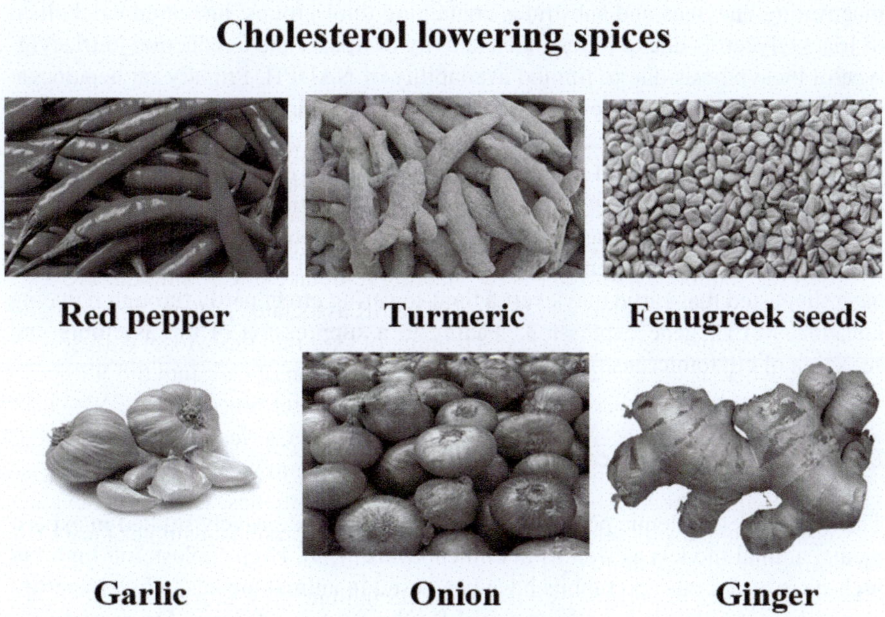

Fig. 8.2 Blood cholesterol lowering spices

2-vinyl-4H-1,3-dithio and diallyl trisulfide have shown potent antithrombotic effect by inhibiting human platelet aggregation [18, 19]. Aged garlic extract has been endowed with anti-clotting, as well as reductions in blood pressure. The anti-platelet aggregation, the anti-platelet adhesion and the anti-proliferation properties of aged garlic extracts are believed to contribute to the cardiovascular protection more than their hypolipidemic influence [20].

Garlic is shown to reduce blood clotting by inhibiting platelet aggregation better than aspirin [21], thus preventive to heart attacks. It facilitates the regression of fatty plaque deposits in blood vessels and reverses arterial blockages caused by atherosclerosis. Dietary supplementation with allicin (9 $mg.kg^{-1}$) reduced the atherosclerotic plaque by >50% in apolipoprotein E-deficient and LDL receptor knockout mice [22]. Allicin was also shown to beneficially affect atherosclerosis by lipoprotein modification and inhibition of LDL uptake and degradation by macrophages. Water-soluble protein of garlic (16% of diet) and garlic oil (100 $mg.kg^{-1}.day^{-1}$) are reported to have significantly lowered hyperlipidemia in rats [23], primarily due to a decrease in hepatic cholesterogenesis. Garlic protein has also been observed to be a hypolipidemic agent in alcohol-induced hyperlipidemia in rats, by facilitating increased cholesterol degradation to bile acids and neutral sterols and mobilization of triacylglycerols [24].

Several studies have elucidated the mechanism(s) of cholesterol-lowering effect of garlic. The hypolipidemic effect of garlic is ruled out to be mediated through the thyroid as indicated by a study on cholesterol and garlic oil-fed Sprague Dawley rats [25]. Based on a study involving rats maintained on high sucrose and alcohol diets [26], the mechanism of hypolipidemic effect of garlic oil involves diallyl disulfide inactivating enzymes and substrates containing thiol groups, increased catabolism of triacylglycerols due to upregulated activity of lipase, and a reduction in triacylglycerol biosynthesis due to limited availability of NADPH. Primary rat hepatocyte cells pretreated with garlic extracts showed decreased incorporation of [^{14}C]-acetate into cholesterol and fatty acids [17]. This suggested that the hypocholesterolemic effect of garlic is mediated through decreased hepatic cholesterogenesis, whereas the triacylglycerol-lowering effect is due to an inhibition of fatty acid synthesis. Studies on the effect of garlic on the expression of the microsomal triglyceride transfer protein (MTTP) gene evaluated *in vitro* in cell lines and *in vivo* in rats [27] have suggested that garlic may exert a lipid-lowering effect partly through reducing intestinal MTTP gene expression, leading to a suppression of the assembly and secretion of chylomicrons from intestine.

8.2.2 Onion (*Allium cepa*)

The hypocholesterolemic potential of onion has been extensively studied in experimental animal models as well as in clinical trials [4, 5]. High cholesterol/ high fat/ high sucrose-fed rats and rabbits have been used in animal models, while hyperlipidemic human subjects have been tested for the efficacy of onion. Decreased concentration of cholesterol in blood and liver has been evidenced as a result of intake of onion or its essential oil in several studies involving high sucrose-/high fat-/ethanol-fed rats. Similarly, reduction in blood cholesterol concentration has been observed in rabbits maintained on high cholesterol diet upon feeding onion or its essential oil. Onion has been evidenced to reduce blood cholesterol in a number of human trials involving normal subjects as well as in hyperlipidemic patients. Onion juice reversed the elevated serum cholesterol levels and of plasma fibrinogen and also decreased fibrinolytic activity in rabbits fed a high-cholesterol diet for 6 months [28, 29]. Onion juice (equivalent of 25 g of onion.kg^{-1}.day^{-1}) prevented the increase in serum cholesterol in high-cholesterol diet-fed rabbits.

The influence of onion in reversing the deleterious effect of a high-cholesterol diet in the erythrocytes of rabbit has been reported [30, 31]. Dietary onion completely suppressed the shape change and aggregation of erythrocytes. Dietary onion not only showed hypocholesterolemic effect; whole onion and its various fractions exerted hypocholesterolemic effect in erythrocyte membranes in rats, and this was accompanied by changes in the erythrocyte membrane enzymes [32]. The lipid-lowering potential of S-methyl cysteine sulfoxide (SMCS) isolated from onion (200 mg.kg^{-1} for 45 days) has been documented in Sprague-Dawley rats maintained on a 1% cholesterol diet [33]. The study indicated that SMCS reduced endogenous lipogenesis and increased the catabolism of lipids. Administration of aqueous onion extract significantly reduced serum, liver, and aorta triglycerides in sucrose-fed rabbits [34]. These effects of onion have been attributed to the inherent sulfur compounds, which oxidize free thiol compounds as well as those bound in proteins. The hypolipidemic effect of dietary onion has also been demonstrated in streptozotocin diabetic rats [35]. These authors have also reported amelioration of diabetic renal lesions by dietary onion attributable to the hypocholesterolemic and antioxidant influences [36].

8.2.3 Fenugreek Seeds (*Trigonella foenum-graecum*)

Fenugreek seeds are commonly used as a spice for seasoning. Both fenugreek seeds and leaves have been exhaustively studied for their influence on body cholesterol levels in several human trials and in different experimental animal models [37]. Fenugreek seeds are shown to exhibit hypocholesterolemic property in hyperlipidemic rats induced with either a high fat diet [38] or a high cholesterol diet [39, 40]. Defatted fenugreek seed was found to be effective in diabetic hypercholesterolemia in dogs [41] as well as in humans [40]. Varying levels of fenugreek seeds have been tested by researchers and a 50% dietary level produced as much as 42% decrease in serum cholesterol in normal and 58% decrease in hydrogenated fat-fed rats [38]. The efficacy of fenugreek seeds fed at 15%, 30%, and 60% dietary levels has revealed a significant prevention of the rise in serum cholesterol levels in high cholesterol-fed rats [39]. Serum LDL-associated cholesterol was particularly reduced along with hepatic cholesterol; and this was accompanied with an increase in the fecal excretion of bile acids and neutral sterols. Thus, fenugreek seeds bring about the hypocholesterolemic effect via depletion of cholesterol stores in the liver through stimulating the conversion of hepatic cholesterol to bile salts, and facilitating the increased excretion of fecal bile acids and neutral sterols.

The hypocholesterolemic property of fenugreek seeds is exhibited exclusively by the fiber and saponin components. The efficacy of the defatted fraction (fiber) of fenugreek seeds has been reported in diabetic hypercholesterolemia in dogs [41]. While the fiber component is predominantly responsible for the hypocholesterolemic/ hypolipidemic action, a fenugreek subfraction rich in steroid saponins and diosgenin has been observed to exhibit hypocholesterolemic effect in diabetic dogs. Fenugreek seeds and leaves (given for 3 weeks) produced a significant reduction in serum cholesterol level in diabetic subjects in addition to a beneficial effect on blood glucose and serum insulin responses. The hypocholesterolemic and hypotriglyceridemic effect of fenugreek seeds has been demonstrated in both insulin-dependent and noninsulin-dependent diabetic subjects as well as in diabetic rats [42–44]. Dietary fenugreek seeds not only improved insulin sensitivity in rats maintained on a high-fat/high-sucrose (HFS) diet, but also distinctly reduced triglyceride, cholesterol, and phospholipid levels in the liver [45].

Beneficial influence of fenugreek leaves (10 g.1500 g^{-1}) on serum cholesterol has been evidenced in rabbits, as revealed by an increase in HDL-associated cholesterol [46]. Dietary supplementation of fenugreek leaves showed a lipid-lowering effect in diabetes-induced hyperlipidemia in streptozotocin-administered rats [47]. Consumption of the sprouted fenugreek seeds (18 g.day^{-1} for 1 month) resulted in a significant reduction in total and LDL-associated cholesterol in human subjects [48].

8.2.4 Red Pepper (*Capsicum annuum*)/Capsaicin

Red pepper is valued for its characteristic pungency attributable to its bioactive compound capsaicin. The diverse health beneficial physiological influences of red pepper and capsaicin have been recently reviewed [49]. The modulatory influence of red pepper or its pungent principle capsaicin on lipid metabolism is very well documented by investigators. Dietary red pepper (included up to 5%) or capsaicin (included up to 0.015%) has been consistently shown to lower serum and liver cholesterol in high fat-fed rats. The hypolipidemic influence of dietary capsaicin (0.15, 1.5, and 15 mg% in the diet for 1 week) has also been established in sucrose-induced hypertriglyceridemic rats [50]. The hypocholesterolemic efficacy of red pepper (5%) or capsaicin (15 mg%) has also been documented in animals fed atherogenic high (1%) cholesterol diet [51]. Hepatic cholesterol was lower in red pepper or capsaicin fed rats, which was accompanied by an enhanced fecal excretion of free cholesterol and of bile acids. The antihypercholesterolemic influence of capsaicin also resulted in reversing the changes in membrane lipid profile in the erythrocytes [52].

In rabbits maintained on a 0.5% cholesterol diet, capsaicin lowered the plasma cholesterol, triglycerides, and total cholesterol/HDL-cholesterol ratio compared to cholesterol fed controls [53]. Turkeys maintained on a capsaicin containing feed (2–3 mg.kg^{-1} for 9 days) along with 0.5% cholesterol displayed significantly lowered serum cholesterol [54]. The beneficial effect of capsicum oleoresin (75 mg. $kg^{-1}.day^{-1}$) has been reported in hypercholesterolemic gerbils [55]. Capsaicin oleoresin not only reduced serum cholesterol and triglyceride levels, but also prevented the accumulation of cholesterol and triglycerides in the liver tissue and aorta. This was accompanied by a higher fecal excretion of cholesterol and triglycerides. Capsaicinoids were found to be beneficial in reducing plasma cholesterol and decreasing aortic plaque in high-cholesterol fed situation in hamsters [56]. Dietary capsaicinoids increased the fecal excretion of total acidic sterols possibly mediated by up-regulation of cholesterol-7α-hydroxylase and decreasing cholesterol absorption. Decreased cholesterol absorption as well as increased fecal excretion of cholesterol and bile acids are the possible mechanisms of this cholesterol lowering action of capsaicin. A decrease in plasma LDL-cholesterol is facilitated by an upregulated expression of LDL receptors in the liver [57]. Dietary capsaicin is understood to stimulate the conversion of cholesterol to bile acids through upregulating the activity of cholesterol-7α-hydroxylase in the liver [58].

8.2.5 Turmeric (*Curcuma longa*)/Curcumin

Turmeric (*Curcuma longa*) is a popular spice used in foods contributing an attractive yellow colour. Turmeric is also valued for its numerous medicinal properties in the indigenous system of medicine and folk remedies in India. The hypocholesterolemic action of turmeric (0.004% in the diet) is reported in experimental animals with a significant reduction in both serum and liver cholesterol [59, 60]. This effect has been observed in conditions of both adequate protein and low-protein diets which also contained 10% hydrogenated fat, and also in hypercholesterolemia-induced rats by feeding cholesterol-enriched diet.

The anti-hypercholesterolemic efficacy of dietary curcumin has been evidenced in rats maintained on an atherogenic high-cholesterol diet. Curcumin (0.1–0.5%) present in the hypercholesterolemic diet lowered serum and liver cholesterol concentrations and also appeared to normalize the serum α- and β-lipoprotein levels furing cholesterol feeding [61]. These effects were observed concomitant with an increased fecal excretion of bile acids and cholesterol. Dietary curcumin also countered the cholesterol enrichment in membranes of erythrocytes [52]. These beneficial effects of dietary curcumin were also seen in rats rendered hypertriglyceridemic by feeding high sucrose [50]. The hypocholesterolemic property of curcumin in high-cholesterol diet fed rats is believed to be resulting from a suppressing effect on cholesterol absorption, stimulated degradation cholesterol to bile acids and the subsequent elimination [62]. Curcumin in the diet (0.5% w/w) decreased serum LDL-associated cholesterol and increased serum HDL-associated cholesterol, thus reducing the atherogenic indices. Curcumin (0.05% in diet) exhibited a hypolipidemic effect in high-fat diet fed hamsters by increasing the activity of plasma paraoxonase enzyme, ratio of HDL-cholesterol to total cholesterol and ratio of apo A-I to apo B, and activity of hepatic fatty acid oxidation machinary with simultaneous inhibition of fatty acid and cholesterol biosynthesis in the liver [63].

Hypocholesterolemic and hypotriglyceridemic action of dietary curcumin (0.5%) has also been demonstrated in diabetic situation in streptozotocin administered rats [64]. Tetrahydrocurcumin (THC) (80 $mg.kg^{-1}$ for 45 days), one of the active metabolites of curcumin significantly increased plasma insulin titres and reduced blood glucose in streptozotocin-nicotinamide-induced diabetic rats [65]. The hypoglycemic and hypolipidemic effects of THC were even higher than that of curcumin. The oxidation of low-density lipoprotein is understood to play an important role in the development of atherosclerosis, more than the cholesterol associated with it. Ethanol-aqueous extract of turmeric decreased the susceptibility of LDL to oxidative modification in atherosclerotic rabbits [66], suggesting that turmeric could be useful in the management of CVD.

8.2.6 Ginger (*Zingiber officinale*)

Ginger rhizomes impart characteristic pungency and piquant flavour and hence are extensively consumed in foods, beverages, carbonated drinks, liquors and as a preserve in sugar syrup [67]. Several reports claim the potential of ginger in suppressing body's cholesterol and lipid accumulation. The lipid-lowering property of ginger contributes to an effective weight management and hence in lowering the risk of CVD. Dietary 0.5% ginger oleoresin is reported to reduce serum and liver cholesterol levels in rats maintained on high (1%) cholesterol diet for 20 days, and this was accompanied by a higher excretion of fecal cholesterol [68]. Beneficial hypocholesterolemic effect has also been reported in rats fed 10% ginger along with a high 1% cholesterol diet for 24 days [69]. The hypocholesterolemic effect of ginger oleoresin is inferred to be mediated through an interference with cholesterol absorption.

The anti-hypercholesterolemic influence of aqueous ginger extract (100, 200, and 400 $mg.kg^{-1}$) has been observed in terms of a decrease in serum total cholesterol, LDL-cholesterol, and triglycerides in hypercholesterolemic rats [70]. Orally administered alcoholic extract of ginger (200 $mg.kg^{-1}$ for 20 days) lowered serum cholesterol and triglycerides while HDL-cholesterol was particularly increased in streptozotocin-induced diabetic rats, thus evidencing the potential of ginger in moderating diabetic dyslipidemia [71]. Ginger is understood to inhibit HMG-CoA reductase and activate LDL-receptors in diabetic rats [72]. Aqueous ginger extract (500 $mg.kg^{-1}$ orally administered for 4 weeks) lowered serum cholesterol, platelet thromboxane-B_2 and prostaglandin-E_2 production in rats [26], thus suggesting the antithrombotic and anti-inflammatory property of ginger. 6-Gingerol, the key phytochemical of ginger is believed to target cholesterol homeostasis and fatty acid oxidation with hypocholesterolemic and anti-obesogenic consequences. Gingerol is also understood to prevent High Fat Diet-induced hyperlipidemia in rats by modulating the expression of appropriate enzymes involved in cholesterol homeostasis [73].

High blood cholesterol levels being a risk factor for the etiology of CVD, the cholesterol lowering property of ginger rhizomes suggests its cardio protective potential. Thc hypotensive property, vasodilator and cardio-suppressant and stimulant effects of ginger rhizome extract have been reported in isolated endothelium-intact rat aorta [74]. The cardio protective potential of ginger extract has also been shown in myocardial infarction induced with isoproterenol in Wistar rats [75]. Pretreatment with ginger extract (400 $mg.kg^{-1}$ for 4 weeks) significantly decreased cardiac markers of infarction such as troponin protein, and activities of creatine kinase-MB, lactate dehydrogenase, aspartate and alanine aminotransferases.

8.2.7 Other Spices Evaluated for Influence on Lipid Homeostasis

The anti-hypertensive potential of an aqueous extract of cumin (*Cuminum cyminum*) seed (200 $mg.kg^{-1}.day^{-1}$) and its role in arterial endothelial nitric oxide synthase expression, inflammation, and oxidative stress has been reported in renal hypertensive rats [76]. Cumin reduced the systolic blood pressure in hypertensive rats, and improved plasma nitric oxide. This was accompanied by the up-regulation of the expression of iNOS, Bcl-2, TRX1, and TRXR1; and down-regulation of the expression of Bax, TNF-α, and IL-6. Cumin seeds thus augment endothelial functions and ameliorate inflammatory and oxidative stress in hypertensive rats. Paraoxanase-1 plays a protective role against the oxidative modification of plasma lipoproteins and hydrolyzes lipid peroxides in atherosclerotic lesions. Cumin extract is reported to significantly decrease the level of oxidized LDL while upregulating the activity of paraoxonase in serum [77].

Daily administration of ether extract mango ginger (*Curcuma amada*) corresponding to 100 g of this rhizome spice for 21 days produced a significant reduction in serum cholesterol in hypercholesterolemic rabbits [78]. The hypotriglyceridemic property of mango ginger has also been reported in induced hypertriglyceridemic rats [79]. The anti-hypercholesterolemic influence of dietary 10% mango ginger along with a high 1% cholesterol containing diet for 5 weeks decreased serum total and LDL-cholesterols and increased the HDL-cholesterol, and also lowered hepatic cholesterol in experimental rats [80], which was accompanied by an increased output of bile acids.

Hypocholesterolemic efficacy of asafetida has been demonstrated in rats fed asafetida powder (1.5%) along with high (1%) cholesterol diet [81]. Liver cholesterol level was also significantly lowered by dietary asafetida. Intestinal absorption of dietary cholesterol in rats maintained on 2% dietary asafetida along with a high-cholesterol diet is reported to be significantly lower; this was accompanied with a higher fecal excretion of cholesterol and bile acids [82].

8.3 Potentiation of Hypolipidemic Influence by Spices

Dietary garlic has been shown to potentiate the hypolipidemic influence of tender cluster beans (*Cyamopsis tetragonoloba*) in high cholesterol-fed rats [83]. The anti-hypercholesterolemic effect of dietary cluster beans was accompanied by a significant decrease in the concentrations of cholesterol and triglycerides in the liver. This lipid-lowering effect in serum and the hepatic tissue was higher in the case of rats provided a dietary intervention with a combination of cluster beans and garlic. This higher beneficial influence was envisaged because of the difference in the mechanisms exerted by garlic and the soluble fibre-rich cluster beans. Similarly, the beneficial effects of dietary soluble fiber-rich tender cluster beans in weight control and adverse changes in body lipid profile were potentiated by capsaicin of red chilli co-administered to high-fat fed rats [84].

8.4 Synergistic Hypolipidemic and Antioxidant Effects of Dietary Fibre-Rich Fenugreek (*Trigonella foenum-graecum*) Seeds and Garlic in Conditions of Hypercholesterolemia and Hyperlipidemia

Translational studies on experimental animals as well as clinical investigations indicate that hypercholesterolemia, especially of LDL-associated cholesterol, lowered HDL-associated cholesterol, hypertension, thrombotic tendency and oxidative stress in the myocardial tissue are the major risk factors for CVD [85]. Plant foods which provide abundant amounts of dietary fiber and antioxidant phytochemicals are advocated for the prevention of CVD. While the hypocholesterolemic potential of fenugreek seeds are well understood in recent decades, the mechanism of action of *Allium* spices – garlic or onion in exerting the hypocholesterolemic and antioxidant action is different from that of the former and is mediated through their sulphur compounds [2]. Hence, there is a possibility of an additive effect when these spices (fenugreek seeds and Allium spice) are consumed in combination.

Among the dietary interventions made with 10% fenugreek seed or 2% garlic or their combination along with a high-cholesterol diet in Wistar rats, the hypocholesterolemic effect particularly of LDL-cholesterol produced was more by the combination of the two spices [86]. The reversal of the elevated cholesterol: phospholipid ratio and atherogenicity index by these dietary interventions was also higher in the case of fenugreek+garlic. The elevated cholesterol content in the heart tissue was also beneficially modulated by dietary fenugreek and garlic, with a higher benefit from the combination. High cholesterol diet induced increase in lipid peroxides was countered by these two dietary spices individually and in combination, which was accompanied by restoration of the antioxidant molecule vitamin E in the myocardial tissue, the effect being highest with fenugreek + garlic, as compared to the two individual spices [86]. This study indicated that fenugreek seeds and garlic will exert higher cardio protective influence under hypercholesterolemic situation when consumed together.

The hypolipidemic and antioxidant influences of dietary (10%) fenugreek seeds and dietary (2%) garlic, individually and in combination for 8 weeks have also been evaluated in high-fat diet (HFD) fed Wistar rats [87]. These dietary interventions produced a significant reversal of increased concentration of triglycerides and LDL-cholesterol in serum caused by HFD. HFD-induced increase in atherogenicity index and the increase in triglycerides and cholesterol: phospholipid ratio in the heart tissue were reversed by dietary fenugreek+garlic. In addition, the elevated lipid peroxides in circulation and cardiac tissue of HFD-fed rats was countered by dietary fenugreek+garlic [87]. Dietary fenugreek+garlic increased the activities of antioxidant enzymes in blood and heart, and so also of antioxidant molecules in HFD-fed situation. Thus, these two spices may have a higher cardio protective influence when consumed together.

LDL oxidation being a key factor in the arteriosclerotic process, the potential of dietary fenugreek seeds and garlic to impede LDL oxidation has been examined in high-cholesterol diet fed rats [88]. Iron-induced oxidation of LDL *in vivo* in ferrous sulfate administered animals was considerably lowered by dietary pretreatment with fenugreek seeds and garlic included along with a high-cholesterol diet. Copper-induced oxidation of isolated LDL in *vitro* was also significantly lesser in fenugreek or fenugreek+garlic fed rats. These spices significantly lowered lipid peroxides generated in plasma and heart in ferrous iron (II)-administered rats. Thus, these two dietary spices are protective to LDL oxidation under normal situation as well as in hypercholesterolemic situation; the protective effect of the combination of fenugreek and garlic being greater than that of these individual spices. The protective effect of dietary fenugreek and garlic on LDL oxidation is also suggestive of their cardio protective potential.

8.5 Cardio Protective Effect of Fenugreek Seeds and Garlic in Experimental Myocardial Infarction Through Hypolipidemic Influence and Alleviation of Oxidative Stress

Garlic bulbs and garlic oil have been shown to exert cardio protective influence in experimentally induced myocardial infarction in rats [89, 90]. Garlic oil produced a marked reversal of all the metabolic changes observed during myocardial infarction induced by isoproterenol [89]. Oxidative stress in the myocardium as a result of increased production of free radicals and decreased levels of antioxidants damages the myocardial cells leading to CVD [91]. Garlic oil exerted its effects by modulating lipid peroxidation and enhancing antioxidant enzyme systems. Significant countering of the increased serum free-iron content, decreased plasma iron-binding capacity, and ceruloplasmin, increased lipid peroxides, associated with decreased activities of antioxidant enzymes in the heart with isoproterenol-induced myocardial necrosis was evident in garlic oil treatment. The protective role of garlic oil on isoproterenol-induced myocardial infarction has been further evaluated in rats [90]. Oral treatment of garlic oil for 60 days (75 $mg.kg^{-1}$) significantly countered the depletion of endogenous antioxidants and increase in marker enzymes (Aspartate and alanine aminotransferases, LDH and CPK) in the serum. The study demonstrated that the cardio protective effects of garlic oil in isoproterenol-induced oxidative damage are mediated through inhibition of lipid peroxidation and augmentation of the endogenous antioxidant machinery.

The cardio protective influence of dietary fenugreek seeds and garlic has been evaluated both individually and as a combination in rats with isoproterenol-induced myocardial infarction [92]. Dietary interventions with 10% fenugreek seeds or 2% freeze-dried garlic powder, or their combination ameliorated the disturbed activities of cardiac marker enzymes in serum and heart and pathological changes in the myocardial tissue. This beneficial effect was found to be higher with the combination of fenugreek seeds and garlic invariably conforming to an additive effect.

Since oxidant stress plays a major role in the etiology of myocardial complications, another animal study with experimentally induced myocardial infarction in Wistar rats has examined the beneficial effect of dietary fenugreek seeds and garlic on the oxidative stress in blood and heart tissue under this condition [93]. Increased concentration of circulatory troponin, disturbed activities of ATPases in the cardiac tissue, increased serum iron, decreased ceruloplasmin, elevated lipid peroxides, depleted antioxidant molecules, and altered activities of antioxidant enzymes in the heart in isoproterenol-induced myocardial infarction were countered by dietary fenugreek seeds, garlic, and fenugreek+garlic. The cardio protective effect was generally higher in the case of dietary intervention with the combination of fenugreek seeds and garlic.

8.6 Amelioration of Cardiac Damage by Dietary Fenugreek Seeds and Onion in Streptozotocin-Induced Diabetic Rats

Dietary intervention with fenugreek seeds are shown to considerably decrease lipid peroxidation and counter the alteration in circulatory antioxidant molecules in streptozotocin-induced diabetic rats [94] and to the restore tissue antioxidant molecules in diabetic condition [95]. Hyperglycemia is a metabolic abnormality that increases the cardiovascular complication in diabetic patients by increased oxidative stress. It has been recently shown that dietary fenugreek seeds and onion significantly alleviate the oxidative stress and lipid abnormalities in the cardiac tissue of streptozotocin-induced diabetic rats [96]. Dietary interventions with fenugreek or onion considerably lowered oxidative stress, the combination producing a higher effect. In addition to hypocholesterolemic effect, dietary fenugreek, onion, or fenugreek+onion countered the elevated cholesterol and triglycerides in the heart tissue under diabetic condition, the beneficial effect being higher with the combination of the two spices, invariably amounting to an additive effect. The cardio protective effect thus observed was also corroborated by the apparent restoration of histopathological abnormalities in the heart tissue.

The mechanism of cardio protective influence of dietary fenugreek seeds and onion in hyperglycemia-mediated cardiac damage has been further investigated in streptozotocin-induced diabetic rats [97]. The observed cardio protective effect of these two spices was mediated through their potential to block the renin-angiotensin system (RAS), which is presumably a consequence of reduced activation of Angiotensin-converting enzyme (ACE) and Angiotensin Type 1 receptor (AT_1) in the heart tissue. Fenugreek+onion produced an additive effect on the protein and mRNA expressions of ACE and AT_1. Dietary fenugreek seeds, onion, and their combination were found to ameliorate the upregulated expression of type IV collagen, fibronectin, Bax, 4-hydroxynonenal, iNOS and metabolites of nitric oxide. There was also a restoration of the disturbed ratio of polyunsaturated fatty acids to saturated fatty acids and activities of cardiac marker enzymes in blood and in the cardiovascular system. This cardio protective effect was higher with the combination of fenugreek and onion, being additive (iNOS expression) or even synergistic (cardiac Bax and type IV collagen expression and circulatory marker enzymes. Thus, the combination of fenugreek seeds and onion offer a higher beneficial influence in ameliorating the risk of cardiac damage accompanying diabetes mellitus.

References

1. Srinivasan K (2005) Role of spices beyond food flavouring: nutraceuticals with multiple health effects. Food Rev Int 21:167–188
2. Srinivasan K (2013) Dietary spices as beneficial modulators of lipid profile in conditions of metabolic disorders and diseases. Food Funct 4:503–521
3. Srinivasan K (2008) Reason to season: spices as functional food adjuncts with multiple health effects (A technical review). Indian Food Ind 27(5):36–47
4. Fenwick GR, Hanley AB (1985) The genus allium – part-3. CRC Crit Rev Food Sci Nutr 23:1–73
5. Carson JF (1987) Chemistry and biological properties of onion and garlic. Food Rev Int 3:71–103
6. Jain MK, Apitz-Castro R (1994) Garlic: a matter for heart. In: Charalambous G (ed) Spices, herbs and edible fungi. Elsevier Science, Amsterdam, pp 309–327
7. Kleijnen J, Knipschild P, Terriet G (1989) Garlic, onions and cardiovascular risk factors. Br J Clin Pharmacol 28:535–544
8. Gore JM, Dalen JE (1994) Cardiovascular disease. JAMA 271:1660–1661
9. Bordia A (1981) Effect of garlic on blood lipids in patients with coronary heart disease. Am J Clin Nutr 34:2100–2103
10. Steiner M, Khan AH, Holbert D, Lin RIS (1996) A double-blinded crossover study in moderately hypercholesterolemic men that compared the effect of aged garlic extracts and placebo administration on blood lipids. Am J Clin Nutr 64:866–870
11. Adler AJ, Holub BJ (1997) Effect of garlic and fish oil supplementation on serum lipid and lipoprotein concentration in hypercholesterolemic men. Am J Clin Nutr 65:445–450
12. Chi MS (1982) Effects of garlic products on lipid metabolism in cholesterol-fed rats. Proc Soc Exp Biol Med 171:174–178
13. Das SN, Pramanik AK, Mitra SK, Mukherjee BN (1982) Effect of garlic pearls on blood cholesterol level in normal dogs. Indian Vet J 59:937–938
14. Kamanna VS, Chandrasekhara N (1984) Hypocholesterolemic activity of different fractions of garlic. Indian J Med Res 9:580–583
15. Bakhsh R, Chugtai MID (1985) Comparative study of onion and garlic on serum cholesterol, liver cholesterol, prothrombin time and fecal sterols excretion in male albino rats. J Chem Soc Pak 7:285–288
16. Gebhardt R (1991) Inhibition of cholesterol biosynthesis by water-soluble garlic extract in primary cultures of rat hepatocytes. Arzneim-Forsch/Drug Res 41:800–804
17. Liu L, Yeh YY (2001) Water-soluble organosulfur compounds of garlic inhibit fatty acid and triglyceride syntheses in cultured rat hepatocytes. Lipids 36:395–400

18. Apitz-Castro R, Cabrera S, Cruz MR et al (1983) Effects of garlic extract and of three pure components isolated from it on human platelet aggregation, arachidonate metabolism, release reaction and platelet ultrastructure. Thromb Res 32:155–169
19. Block E, Ahmad S, Jain MK et al (1984) Ajoene: a potent antithrombotic agent from garlic. J Am Chem Soc 106:8295–8296
20. Lin RI (1994) In: Goldber I (ed) Phytochemicals and antioxidants in functional foods. Chapman & Hall, London, pp 393–449
21. Sharma CP, Nirmala NV (1985) Effects of garlic extract and of three pure components isolated from it on human platelet aggregation, arachidonate metabolism, release reaction and platelet ultrastructure – comments. Thromb Res 37:489–490
22. Gonen A, Harats D, Rabinkov A et al (2005) The antiatherogenic effect of allicin: possible mode of action. Pathobiology 72:325–334
23. Mathew BC, Daniel RS, Augusti KT (1996) Hypolipidemic effect of garlic protein substituted for casein in diet of rats compared to those of garlic oil. Indian J Exp Biol 34:337–340
24. Rajasree CR, Rajamohan T, Augusti KT (1999) Biochemical effects of garlic protein on lipid metabolism in alcohol fed rats. Indian J Exp Biol 37:243–247
25. Chaudhuri BN, Mukherjee SK, Mongia SS, Chakravarty SK (1984) Hypolipidemic effect of garlic and thyroid function. Biomed Biochim Acta 43:1045–1047
26. Adoga I (1987) The mechanism of the hypolipidemic effect of garlic oil extract in rats fed on high sucrose and alcohol diets. Biochem Biophys Res Commun 142:1046–1052
27. Lin MC, Wang EJ, Lee C et al (2002) Garlic inhibits microsomal triglyceride transfer protein gene expression in human liver and intestinal cell lines and in rat intestine. J Nutr 132:1165–1168
28. Sharma KK, Choudhury NK, Sharma AL (1975a) Studies on hypocholesterolemic activity of onion – effect on serum cholesterol in rabbits maintained on high cholesterol diet. Indian J Nutr Diet 12:388–391
29. Sharma KK, Choudhury NK, Sharma AL (1975b) Long term effect of onion on experimentally induced hypercholesterolemia and consequently decreased fibrinolytic activity in rabbits. Indian J Med Res 63:1629–1634
30. Vatsala TM, Singh M (1981) Relationship between plasma cholesterol level and erythrocytes shape in rabbits on atherogenic diet and onion extracts. Curr Sci 50:211–213
31. Vatsala TM, Singh M (1982) Effects of onion in induced atherogenesis in rabbits: II. Reduction of lipid levels in the eye. Curr Sci 51:230–232
32. Ahluwalia P, Mohindroo A (1989) Effect of oral ingestion of different fractions of *Allium cepa* on the blood and erythrocyte membrane lipids and certain membrane-bound enzymes in rats. J Nutr Sci Vitaminol 35:155–161
33. Kumari K, Augusti KT (2007) Lipid lowering effect of S-methyl cysteine sulfoxide from Allium cepa Linn in high cholesterol diet fed rats. J Ethnopharmacol 109:367–371
34. Sebastian KL, Zacharias NT, Philip B, Augusti KT (1979) The hypolipidemic effect of onion (Allium cepa) in sucrose fed rabbits. Indian J Physiol Pharmacol 23:27–30
35. Babu PS, Srinivasan K (1997) Influence of dietary capsaicin and onion on the metabolic abnormalities associated with diabetes mellitus. Mol Cell Biochem 175:49–57
36. Babu PS, Srinivasan K (1999) Renal lesions in streptozotocin induced diabetic rats maintained on onion and capsaicin-containing diets. J Nutr Biochem 10:477–483
37. Srinivasan K (2006) Fenugreek (*Trigonella foenum-graecum*): a review of health beneficial physiological effects. Food Rev Int 22:203–224
38. Singhal PC, Gupta RK, Joshi LD (1982) Hypocholesterolemic effect of *T. foenum-graecum*. Curr Sci 51:136–137
39. Sharma RD (1984) Hypocholesterolemic activity of fenugreek (*T. foenum-graecum*): an experimental study in rats. Nutr Rep Int 30:221–231
40. Sharma RD (1986) An evaluation of hypochoesterolemic factor of fenugreek seeds (*T. foenum-graecum*) in rats. Nutr Rep Int 33:669–677
41. Valette G, Sauvaire Y, Baccon JC, Ribes G (1984) Hypocholesterolemic effect of fenugreek seeds in dogs. Atherosclerosis 50:105–111

42. Sharma RD, Raghuram TC, Sudhakar Rao N (1990) Effect of fenugreek seeds on blood glucose and serum lipids in type I diabetes. Eur J Clin Nutr 44:301–306
43. Sharma RD, Raghuram TC (1990) Hypoglycemic effect of fenugreek seeds in non-insulin dependent diabetic subjects. Nutr Res 10:731–739
44. Murthy RR, Murthy PS, Prabhu K (1990) Effects on blood glucose and serum insulin levels in alloxan-induced diabetic rats by a fraction of *T. foenum-graecum*. Biomedicine 10:25–29
45. Muraki E, Chiba H, Tsunoda N, Kasono K (2011) Fenugreek improves diet-induced metabolic disorders in rats. Horm Metab Res 43:950–955
46. Chaturvedi V, Pant MC (1987) Effect of feeding Trigonella foenum-graecum leaves on serum cholesterol, triglycerides and high density lipoprotein cholesterol in the normal rabbits. Curr Sci 56:600–601
47. Annida B, Stanely-Mainzen P (2004) Supplementation of fenugreek leaves lower lipid profile in streptozotocin induced diabetic rats. J Med Food 7:153–156
48. Sowmya P, Rajyalakshmi P (1999) Hypocholesterolemic effect of germinated fenugreek seeds in human subjects. Plant Foods Hum Nutr 53:359–365
49. Srinivasan K (2016) Biological activities of red pepper (*Capsicum annuum*) and its pungent alkaloid – capsaicin: a review. Crit Rev Food Sci Nutr 56:1488–1500
50. Srinivasan MR, Satyanarayana MN (1988) Influence of capsaicin, eugenol, curcumin and ferulic acid on sucrose induced hypertriglyceridemia in rats. Nutr Rep Int 38:571–581
51. Sambaiah K, Satyanarayana MN (1980) Hypocholesterolemic effect of red pepper and capsaicin. Indian J Exp Biol 18:898–899
52. Kempaiah RK, Srinivasan K (2002) Integrity of erythrocytes of hypercholesterolemic rats during spices treatment. Mol Cell Biochem 236:155–161
53. Negulesco JA, Young RM, Ki P (1983) Capsaicin lowers plasma cholesterol and triglycerides of lagomorphs. Artery 12:301–311
54. Ki P, Negulesco JA, Murnane M (1982) Decreased total serum myocardial and aortic cholesterol levels following capsaicin treatment. IRCS Med Sci 10:446–447
55. Gupta RS, Dixit VP, Dobhal MP (2002) Hypocholesterolemic effect of the oleoresin of *Capsicum annum* L. in gerbils (Meriones hurrianae Jerdon). Phytother Res 16:273–275
56. Liang YT, Tian XY, Chen JN et al (2013) Capsaicinoids lower plasma cholesterol and improve endothelial function in hamsters. Eur J Nutr 52:379–388
57. Negulesco JA, Noel SA, Newman HA et al (1987) Effect of pure capsaicinoids (capsaicin and dihydrocapsaicin) on plasma lipids and lipoprotein concentrations of turkey poults. Atherosclerosis 64:85–90
58. Srinivasan K, Sambaiah K (1991) Effect of spices on cholesterol-7a-hydroxylase activity and on serum and hepatic cholesterol levels in the rat. Int J Vitam Nutr Res 61:364–369
59. Bhuvaneswaran C, Kapur OP, Sriramachari S et al (1963) Rat growth studies on vanaspathi coloured with turmeric extract using an adequate diet. Food Sci 12:182–184
60. Bhuvaneswaran C, Sriramachari S, Jayaraj AP et al (1963) Rat growth studies on vanaspathi coloured with turmeric extract using a low protein diet. Food Sci 12:185–187
61. Rao DS, Chandrasekhara N, Satyanarayana MN, Srinivasan M (1970) Effect of curcumin on serum and liver cholesterol levels in the rat. J Nutr 100:1307–1316
62. Arafa HM (2005) Curcumin attenuates diet-induced hypercholesterolemia in rats. Med Sci Monit 11:228–234
63. Jang EM, Choi MS, Jung UJ et al (2008) Beneficial effects of curcumin on hyperlipidemia and insulin resistance in high-fat-fed hamsters. Metab Clin Exp 57:1576–1583
64. Babu PS, Srinivasan K (1997) Hypolipidemic action of curcumin, coloring principle of turmeric (Curcuma longa) in streptozotocin induced diabetic rats. Mol Cell Biochem 166:169–175
65. Pari L, Murugan P (2007) Antihyperlipidemic effect of curcumin and tetrahydrocurcumin in experimental type 2 diabetic rats. Ren Fail 29:881–889
66. Ramirez-Tortosa MC, Mesa MD, Aguilera MC et al (1999) Oral administration of a turmeric extract inhibits LDL oxidation and has hypocholesterolemic effects in rabbits with experimental atherosclerosis. Atherosclerosis 147:371–378

67. Govindarajan VS (1982) Ginger—chemistry, technology & quality evaluation: parts I & II. Crit Rev Food Sci Nutr 17:189–258
68. Gujral S, Bhumra H, Swaroop M (1978) Effect of ginger (*Zingiber officinale*) oleoresin on serum and hepatic cholesterol levels in cholesterol fed rats. Nutr Rep Int 17:183–189
69. Giri J, Devi TKS, Meerarani S (1984) Effect of ginger on serum cholesterol levels. Indian J Nutr Diet 21:433–436
70. El-Rokh SM, Yassin NA, El-Shenawy SM, Ibrahim BM (2010) Antihypercholesterolaemic effect of ginger rhizome (*Zingiber officinale*) in rats. Inflammopharmacology 18:309–315
71. Bhandari U, Kanojia R, Pillai KK (2005) Effect of ethanolic extract of Zingiber officinale on dyslipidaemia in diabetic rats. J Ethnopharmacol 97:227–230
72. Akash MS, Rehman K, Tariq M, Chen S (2015) *Zingiber officinale* and type 2 diabetes mellitus: evidence from experimental studies. Crit Rev Eukaryot Gene Expr 25:91–112
73. Naidu PB, Uddandrao VV, Naik RR et al (2016) Ameliorative potential of gingerol: promising modulation of inflammatory factors and lipid marker enzymes expressions in HFD induced obesity in rats. Mol Cell Endocrinol 419:139–147
74. Ghayur MN, Gilani AH, Afridi MB, Houghton PJ (2005) Cardiovascular effects of ginger aqueous extract and its phenolic constituents are mediated through multiple pathways. Vasc Pharmacol 43:234–241
75. Amran AZ, Jantan I, Dianita R, Buang F (2015) Protective effects of the standardized extract of *Zingiber officinale* on myocardium against isoproterenol-induced biochemical and histopathological alterations in rats. Pharm Biol 53:1795–1802
76. Kalaivani P, Saranya RB, Ramakrishnan G et al (2013) *Cuminum cyminum*, a dietary spice, attenuates hypertension via endothelial nitric oxide synthase and NO pathway in renovascular hypertensive rats. Clin Exp Hypertension 35:534–542
77. Samani KG, Farrokhi E (2014) Effects of cumin extract on oxldl, paraoxanase 1 activity, FBS, total cholesterol, triglycerides, HDL-C, LDL-C, apo A1, and apo B in in the patients with hypercholesterolemia. Int J Health Sci 8:39–43
78. Pachauri S, Mukherjee SK (1970) Effect of *Curcuma longa* and *Curcuma amada* on the cholesterol levels in experimental hypercholesterolemia of rabbits. J Res Indian Med 5:27–31
79. Srinivasan MR, Chandrasekhara N (1992) Effect of mango ginger (*Curcuma amada*) on lipid status in normal and hypertriglyceridemic rats. J Food Sci Technol 29:130–132
80. Srinivasan MR, Chandrasekhara N, Srinivasan K (2008) Cholesterol lowering activity of mango ginger (Curcuma amada Roxb.) in induced hypercholesterolemic rats. Eur Food Res Technol 227:1159–1163
81. Kamanna VS, Chandrasekhara N (1982) Effect of garlic (*Allium sativum*) on serum lipoproteins and lipoprotein cholesterol levels in albino rats rendered hypercholesterolemic by feeding cholesterol. Lipids 17:483–488
82. Srinivasan MR, Srinivasan K (1995) Hypocholesterolemic efficacy of garlic smelling flower – *Adenocalymma alliaceum* in experimental rats. Indian J Exp Biol 33:64–65
83. Pande S, Srinivasan K (2012a) Potentiation of the hypolipidemic influence of dietary tender cluster bean (*Cyamopsis tetragonoloba*) by garlic in cholesterol-fed rats. Food Chem 133:798–805
84. Pande S, Srinivasan K (2012b) Potentiation of hypolipidemic and weight reducing influence of dietary tender cluster bean (*Cyamopsis tetragonoloba*) in combination with capsaicin in high-fat fed rats. J Agric Food Chem 60:8155–8162
85. Amani R, Sharifi N (2012) Cardiovascular disease risk factors. In: Gaze D (ed) The cardiovascular system—physiology, diagnostics and clinical implications. InTech, Rijeka, pp 279–310
86. Mukthamba P, Srinivasan K (2015a) Beneficial hypolipidemic influence of a combination of dietary fenugreek (*Trigonella foenum-graecum*) seeds and garlic (*Allium sativum*) in induced hypercholesterolemic rats. Eur Food Res Tech 240:1049–1058
87. Mukthamba P, Srinivasan K (2016a) Hypolipidemic and antioxidant influence of dietary fenugreek (*Trigonella foenum-graecum*) seeds and garlic (*Allium sativum*) in high-fat fed rats. Food Biosci 14:1–9

88. Mukthamba P, Srinivasan K (2016b) Protective effect of dietary fenugreek (*Trigonella foenum-graecum*) seeds and garlic (*Allium sativum*) on iron and copper induced oxidation of low-density lipoprotein in rats. J Basic Clin Physiol Pharmacol 27:39–47
89. Saravanan G, Prakash J (2004) Effect of garlic (*Allium sativum*) on lipid peroxidation in experimental myocardial infarction in rats. J Ethnopharmacol 94:155–158
90. Senthilkumar GP, Moses MF, Sengottuvelu S et al (2010) Cardio protective activity of garlic (*Allium sativum*) in isoproterenol induced rat myocardial necrosis: A biochemical and histoarchitectural evaluation. Int J Pharm Sci Nanotechnol 2:779–784
91. Das DK, Maulik N (1995) Protection against free radical injury in the heart and cardiac performance. In: Sen CK, Packer L, Hanninen O (eds) Exercise and Oxygen Toxicity. Elsevier Science, Amsterdam, p 129
92. Mukthamba P, Srinivasan K (2015b) Beneficial hypolipidemic influence of dietary fenugreek (*Trigonella foenum-graecum*) seeds and garlic (*Allium sativum*) in experimental myocardial infarction. Food Funct 6:3117–3125
93. Mukthamba P, Srinivasan K (2017) Beneficial influence of dietary fenugreek (*Trigonella foenum-graecum*) seeds and garlic (*Allium sativum*) on oxidative stress in experimental myocardial infarction. Food Sci Hum Wellness 6:77–87
94. Ravikumar P, Anuradha CV (1999) Effect of fenugreek seeds on blood lipid peroxidation and antioxidants in diabetic rats. Phytother Res 13:197–201
95. Anuradha CV, Ravikumar P (2001) Restoration of tissue antioxidants by fenu-greek seeds (*Trigonella foenum-graecum*) in alloxan-diabetic rats. Indian J Physiol Pharmacol 45:408–420
96. Pradeep SR, Srinivasan K (2017) Abrogation of oxidative stress by a combination of dietary fenugreek (*Trigonella foenum-graecum*) seeds and onion (*Allium cepa*) in streptozotocin-induced diabetic rats. Appl Physiol Nutr Metab 42:816–828
97. Pradeep SR, Srinivasan K (2018) Alleviation of cardiac damage by dietary fenugreek (*Trigonella foenum-graecum*) seeds is potentiated by onion (*Allium cepa*) in experimental diabetic rats via blocking renin-angiotensin system. Cardiovasc Toxicol 18:221–231

Oxidative Stress and Modulation of Cardiac Kv1.5 Channel

9

Rajabrata Bhuyan and Sajal Chakraborti

9.1 Introduction

Oxidative stress and its mechanisms play major role in human physiology including ageing and many pathogenesis such as cancer, heart diseases, atherosclerosis and other chronic diseases. The oxidative stress can be defined as condition when the oxidant metabolites or oxygen radicals exert toxic effects due to their increased production or altered cellular metabolisms [1–3]. Among the major heart problems, the sudden cardiac death (SCD) and the cardiac arrhythmias are always considered as severe health issue caused due to stress. The reactive oxygen species (ROS) is known to be associated with the development and progression of myocardial dysfunction and subsequent cardiovascular tissue injury [4–7]. However, the molecular targets of these ROS have always remained obscure. In last decades, the importance of in cardiac dysfunction under varieties of pathophysiological conditions have been explored in depth. The above studies showed that the significance of cell redox state is necessarily required for balancing the level of ROS and nitric oxide (NO) in cell [2, 8–16]. Furthermore, understanding the reduction-oxidation imbalance that mechanistically contribute to several cardiac disorders such as hypertension, atherosclerosis, ischemic heart disease, cardiomyopathies, cardiac hypertrophy and congestive heart failure have to be broaden in future.

R. Bhuyan (✉)
BIF Centre, Department of Biochemistry and Biophysics, University of Kalyani,
Kalyani, West Bengal, India

S. Chakraborti
Department of Biochemistry and Biophysics, University of Kalyani,
Kalyani, West Bengal, India
e-mail: Sajchak9@klyUniv.ac.in

S. Chakraborti et al. (eds.), *Oxidative Stress in Heart Diseases*,
https://doi.org/10.1007/978-981-13-8273-4_9

9.2 Role of Ion channels in cardiac stress

The balancing of ROS and nitric oxide (NO) level has considerable impact on ion channel function in cardiac cells [17]. Ion channels are the multimeric molecular machines, located in plasma membrane that have multiple roles such as ion conduction, membrane potential regulation, cell communication and signalling [18–21]. In cardiac tissue, numbers of ion channels including Na^+, K^+, Ca^{2+} and TRPV channels are expressed maintaining and regulating the regular physiology of heart [22–24]. Amongst them, most of these channels are already declared as targets and off-targets for various therapeutic interventions [25–28]. ROS are known to induce posttranslational modification in ion channels through nitrosylation, and/or nitration at some specific amino acids that can directly/indirectly modulate the channel function by affecting many signaling pathways. However, the modulation of these ion channels due to the oxidative stress rapidly emerging, which enables the discovery of novel regulatory and pathological circuits for the treatment of several diseases. Interestingly more reports are available on ion channels involving in cardiac oxidative stress-related diseases [29–33].

Membrane potential in cardiac cells is maintained by the above said ion channels with large conductance of Na^+, K^+, Ca^{2+} ions. The intracellular Ca^{2+} level plays important role in balancing the system and any changes in its level can cause mechanical mechanical dysfunction of the ion channels [17, 34, 35]. During the state of depolarization, the sequential calcium influx by the voltage-gated Ca^{2+} channels is responsible to generate the essential electromechanical coupling for regular contraction of cardiac muscle [36]. Subsequently, the K^+ channels (mostly the voltage-gated) repolarise back the cell resting potential state by the outward current flow before a new excitation can occur. The most prominent voltage-gated K^+ channel (Kv) that play the role during early and plateau phase repolarization are the ultra rapid delayed rectifier channel (IK_{ur}) [37–40]. Attentions have been drawn in recent years to these ion channels participating in maintaining the membrane potential duration in heart as targets of oxidative stress for modulating their gating properties. Moreover, their over-expression and malfunctioning have been implicated in numerous health issues such as restricted blood flow, cell injury, long QT syndrome and heart failure including atrial fibrillation (AF) [41–43].

9.3 Kv1.5 in Heart

Kv1.5 is the fifth member of *Shaker*-type voltage gated potassium channel family, encoded by KCNA5 gene in human. It's known to express in a wide range of tissues including heart, aorta, pancreatic b-cells, skeletal muscle, and brain [44–47]. In human heart cells, membrane repolarization is initiated by Kv1.5 and followed by other Kv channels. The Kv1.5 channels mainly participate in repolarizing the ultra-rapid delayed-rectifier current, and any alternation of their function can lead to cardiac arrhythmia such as atrial flutter and AF along with sudden cardiac death [37, 38, 48]. In contrast to other IK_{ur} channels, Kv1.5 has been detected in human atrial cells, but not in ventricles [36, 40]. By this reason, the Kv1.5 is regarded as a promising drug target for antiarrhythmic treatment. Usually, the prolongation of action potential duration by blocking the Kv1.5 channel can restore the normal cardiac rhythm, which is considered as the basic therapeutic strategy of AF [49]. Studies revealed that sulfenic acid modification at a conserved cysteine residue of Kv1.5 (C581) under prolonged oxidative stress can induce arrhythmia [50]. There are also several reports on mutations on Kv1.5 that can alter channel function leading to AF. In this chapter, we will discuss about some mutations and modulation of Kv1.5 channel function.

9.4 Structure of Kv1.5

The Kv1.5 channel shares more than 70% of identity with the other members of *Shaker*-type ion channels [51–53]. It contains an extended N-terminal region located to be intracellular, which is a bit larger in comparison to other popular members of this family (Fig. 9.1a). Like any other member of *Shaker*-related family, the Kv1.5 is comprised of homotetrameric segments arranged in a square like conformation. Each monomer contains six transmembrane helices (known as S1–S6), cytoplasmic unit and extracellular loops necessary for channel methylation. The first four helices (S1–S4) that sense the membrane potential difference are known as voltage sensing domains (VSDs), and the last two transmembrane segments (S5 & S6) forms the ion conduction pore. These segments are arranged in such a way that the VSDs remain at the periphery of the channel surrounded by the pore at centre (Fig. 9.1b, c, d, e). The S4 helix contains positively charged amino acids i. e. five arginines and one lysine in its every third position, known as gating charge residues [54, 55].

During membrane polarization, the S4 helix senses the voltage and all the gating charge residues start to move up in the positive potential along with a large conformational change in VSDs. These structural changes create a driving force that is transferred to the pore for channel activation via S4-S5 linker [56–60]. This S4-S5 linker is a α-helix formed by amphipathic amino acids remains parallel to the membrane [61]. Unlike most of the other members of Kv1 family, the Kv1.5 contains two amino acids alterations in this linker; a glutamine is replaced by a lysine at position 425, and at 428 a glutamine takes the place of lysine. Due to these alterations, interaction between S4-S5 linker and S6 helix is lost and less driving force is required for channel activation. Again, the Lys425 of Kv1.5 forms a firm salt-bridge interaction with Glu433 of S5, holding the pore forming helices. The above interaction sustains the previous concept of channel open state conformation, which suggests that the S4–S5 linker in activated condition must be extended to the N-terminal regions of S5. These features are believed to make the Kv1.5 a distinguished ion channel of IK_{ur} category, where very slow or almost no inactivation takes place [53].

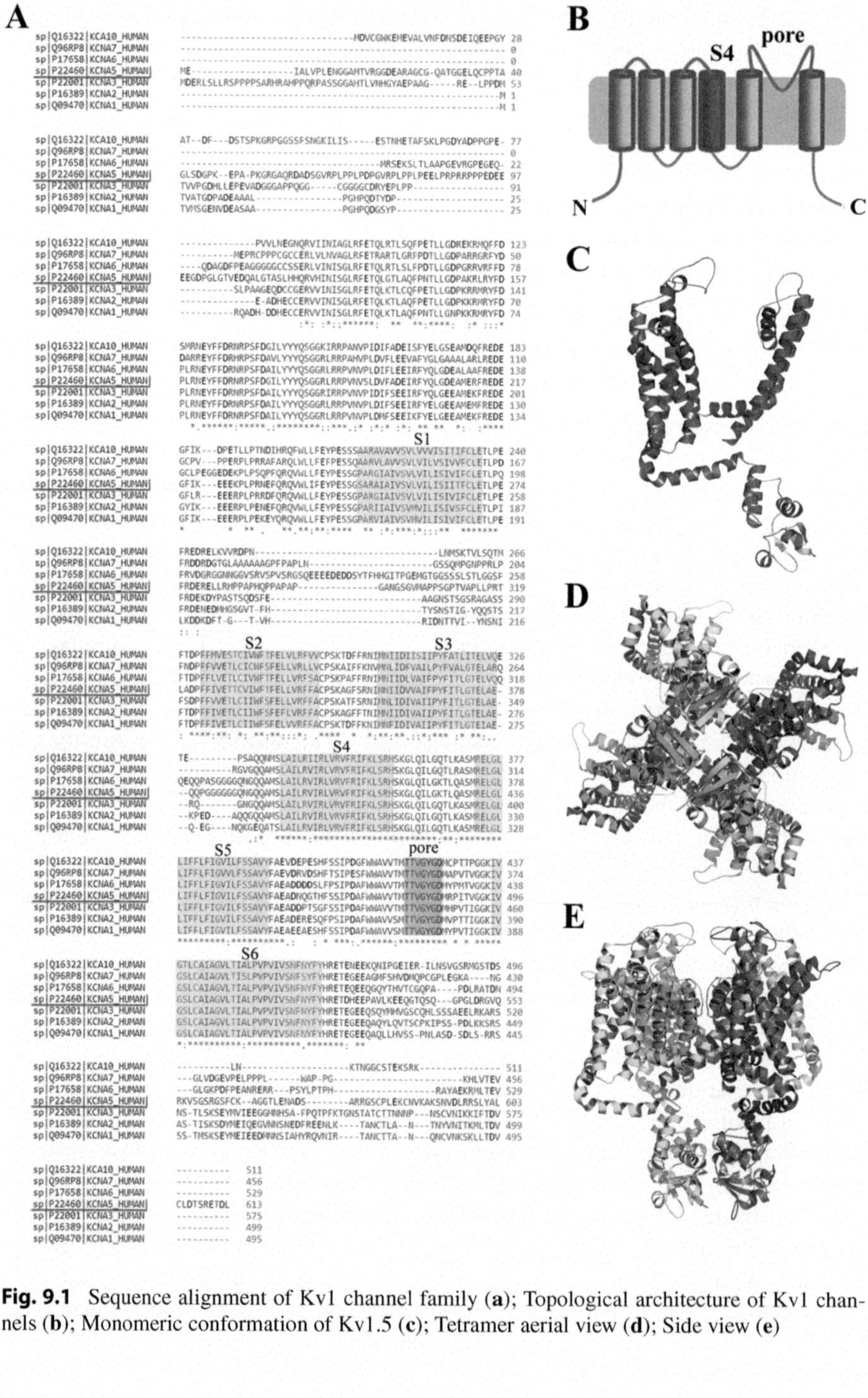

Fig. 9.1 Sequence alignment of Kv1 channel family (**a**); Topological architecture of Kv1 channels (**b**); Monomeric conformation of Kv1.5 (**c**); Tetramer aerial view (**d**); Side view (**e**)

9.5 Mutations in Kv1.5

Several mutations are reported in Kv1.5, associated with AF or heart failure. Among them, the E375X is the stress-provoked most crucial one resulting loss of channel activity [62]. From the rest, E33V and P91L are identified in patients with cardiac attack [63]. Christophersen et al. have studied six genetic variations in KCNA5 among AF patients and pointed three mutations Y155C, D469E and P488S, that induce channel malfunction [64]. Another study reported three novel mutations T527 M, A576V and E610K, that had pivotal role in familial AF for consistent Kv1.5 loss-of-function [65]. Here, we performed a comparative analysis of these point mutations that are associated with AF or heart failure.

There are several programs and web servers available to study the effects or functional changes of a protein after the mutation occurs. PolyPhen-2 (http://genetics.bwh.harvard.edu/pph2/) is one of the leading server that characterize the functional changes in mutated region using a structure-homology-based method. It takes the protein ∗.fasta sequence as input and calculates position - specific independent counts (PSIC) scores for the variant sites on the basis of sequence substitution site and profile analysis of homologous sequences. Based on the PSIC score difference, it predicts whether a mutation is "BENIGN" or "POSSIBLY DAMAGING" or "PROBABLY DAMAGING" using threshold of Naive Bayes probabilistic score [66]. Similarly, the PROVEAN (Protein Variation Effect Analyzer) server (http://provean.jcvi.org) is known for predicting the functional effects of mutation from protein sequence by filtering sequence variants. Mutation are regarded as "deleterious" that gives the PROVEAN score equal or below the threshold of −2.5, and the above are considered as "neutral" [67]. On the basis of structural features, the DUET server uses two distinct algorithms (mCSM and SDM) to predict the protein functional effects on the basis of change in protein stability ($\Delta\Delta G$) [68]. The mCSM applies a graph based prediction model of signatures to represent the protein 3D model [69]; whereas, the SDM uses a statistical potential energy function [70]. For DUET server, the 3D model of Kv1.5 from our previous study was taken [53].

The mutational analysis by PolyPhen-2 and PROVEAN server predicted the E33V, P91L and A576V mutations to be neutral or to induce no such functional alterations in protein functions; whereas, the E610K mutation was predicted to be "PROBABLY DAMAGING" by PolyPhen-2 server only. We tested the most crucial E375X with four different amino acids with different properties. Here, the prediction servers noted them all as damaging or destabilizing the protein conformation (Table 9.1). From the rests, the D469E and T527 M were found with comparatively less damaging by their scores. However, the SDM server predicted them to positively increase the protein stability. The P488S was suggested to induce reduced stability by all the sequence and structure based methods (Table 9.1).

Table 9.1 Mutation analysis of Kv1.5 channel detected in AF patients

Mutations	By sequence properties		By structure properties	
	PolyPhen-2 with PSIC score	PROVEAN with score	mCSM (ΔΔG in Kcal/mol)	SDM (ΔΔG in Kcal/mol)
E33V	0.019 (BENIGN)	−0.401 (Neutral)	–	–
P91L	0.00 (BENIGN)	0.841 (Neutral)	–	–
E375D	0.096 (BENIGN)	−1.739 (Neutral)	−1.39 (Destabilizing)	−1.37 (Reduced stability)
E375K	0.635 (POSSIBLY DAMAGING)	−3.113 (Deleterious)	−0.095 (Destabilizing)	−0.86 (Reduced stability)
E375R	0.486 (POSSIBLY DAMAGING)	−4.054 (Deleterious)	−0.003 (Destabilizing)	−0.56 (Reduced stability)
E375A	0.766 (POSSIBLY DAMAGING)	−4.309 (Deleterious)	−0.735 (Destabilizing)	−0.02 (Reduced stability)
E375I	0.486 (POSSIBLY DAMAGING)	−5.412 (Deleterious)	−0.064 (Destabilizing)	−0.57 (Reduced stability)
D469E	0.842(POSSIBLY DAMAGING)	−3.176 (Deleterious)	−0.748 (Destabilizing)	1.25 (Increased stability)
P488S	1.000 (PROBABLY DAMAGING)	−7.230 (Deleterious)	−2.116 (Destabilizing)	−0.4 (Reduced stability)
T527M	0.954 (POSSIBLY DAMAGING)	−4.964 (Deleterious)	−0.251 (Destabilizing)	0 (Increased stability)
A576V	(BENIGN) 0.01	0.542 (Neutral)	–	–
E610K	0.982 (PROBABLY DAMAGING)	−1.034 (Neutral)	–	–

9.6 Modulation of cardiac Kv1.5 channel

Among the other heart problems, the AF is regarded as the most severe type of cardiac arrhythmias, which is caused by irregular excitation of heart pacemaker cells, and coupled with sudden heart failure, stroke and ageing related disorders [41–43]. The "atrial stabilization" or prolongation of action potential duration is a probable solution of AF [49]. In this regard, the modulation of Kv1.5 channel by small compounds has been one of the prime interest for researchers in developing antiarrhythmic drugs.

During recent years, several attempts have been made to develop the selective Kv1.5 blockers that have the potential in AF treatment [36, 39, 40, 71–73]. Initially, two research groups started working for the selective blockade of Kv1.5 by using mathematical models of action potentials in human atrial cells [74–76]. They suggested that the reduction of IK_{ur} and selective inhibition of IK_{ur} can be effective in producing prolonged or stable action potential in healthy individuals. These above studies apparently resulted in finding several antiarrhythmic drugs (Table 9.2).

Among these, the Benzocaine [77], Bupivacaine [78], Candesartan [79], Eprosartan [79], Irbesartan [80][, Losartan [81], Clotrimazole [82], Loratadine [83], Rupatadine [84], Terfenadine [85], Papaverine [86] and Erythromycin [87] are

Table 9.2 List of known antiarrhythmic drugs against Kv1.5

Drug	Kv1.5 (IC50, μM)
Amiodarone	22.9
Bepridil	6.6
Clofilium	0.14
Diltiazem	29.2
Dronedarone	1.0
Flecainide	101
SSR149744C	2.7
Propafenone	4.4
Verapamil	3.2
Benzocaine	901
Bupivacaine	7.4
Candesartan	0.1
Eprosartan	1
Irbesartan	0.1
Losartan	1
Clotrimazole	1.99
Loratadine	0.8
Rupatadine	2.4
Terfenadine	1.1
Papaverine	43.4
Erythromycin	26

known as class III drugs that inhibit multiple cardiac K^+ along with the other IK_{ur}. Apart from these known drugs, there are many antiarrhythmic agents available that show stronger inhibition on IK_{ur} channel modulation, but induces several adverse effects like long QT syndrome [88, 89]. Furthermore to develop safe and more selective antiarrhythmics leads, the derivatives of anthranilic acid amides [90, 91], biphenyls & bisaryls [92, 93], indanes [94], pyridazinones & phosphine oxides [95], and tetrahydroindolone semicarbazones [96] have been considered. Bhuyan and Seal have also proposed some drug-like compounds for Kv1.5 modulation through the computational studies [53]. On the basis of Quantitative Structure Activity Relationship (QSAR) and Pharmacophore based survey, the above structural classes of compounds commonly possess three hydrophobic moieties in a triangular arrangement, which is necessary to occlude the Kv1.5 channel pore [97]. In absence of crystal structure, computational docking and electrophysiology on mutants have been performed to identify the ligand binding residues of Kv1.5. The above studies revealed that the Kv1.5 blockers are possibly interact at the channel pore by many hydrophobic residues (T479, T480, I502, V505, I508, A509, and L510, V512, P513, and V516), close to the PVP region of S6 helix and the selectivity filter [53, 78, 98–101]. Recent molecular modeling, virtual screening, and dynamics study on Kv1.5 modulation illustrated that the residues T479, T480 commonly participate in electrostatic interaction, whereas the others contribute impressive Van der walls or non-polar interaction. Overall, the whole hydrophobic patch is responsible for Kv1.5 inhibition [53].

9.7 Conclusion

Ion channels and transporters are well known to be associated with oxidative stress in human health and disease. These proteins are required for basic cellular physiology and many of them are already declared as important drug targets in several pathologies. AF is one of such cardiac arrhythmia that is rapidly growing to be an expanding epidemic issue world-wide. More studies on ion channels in heart associated with oxidative stress should be helpful in detecting and developing new therapeutic strategies for many cardiac diseases.

References

1. McCord JM, Fridovich I (1969) Superoxide dismutase. An enzymic function for erythrocuprein (hemocuprein). J Biol Chem 244:6049–6055. https://doi.org/10.1016/0003-2697(69)90079-7
2. Ceconi C, Boraso A, Cargnoni A, Ferrari R (2003) Oxidative stress in cardiovascular disease: myth or fact? Arch Biochem Biophys 420:217–221
3. Chandrasekaran A, Idelchik M del PS, Melendez JA (2017) Redox control of senescence and age-related disease. Redox Biol 11:91–102
4. Kukreja RC, Hess ML (1992) The oxygen free radical system: From equations through membrane-protein interactions to cardiovascular injury and protection. Cardiovasc Res 26:641–655. https://doi.org/10.1093/cvr/26.7.641
5. Singal PK, Khaper N, Palace V, Kumar D (1998) The role of oxidative stress in the genesis of heart disease. Cardiovasc Res 40:426–432. https://doi.org/10.1016/S0008-6363(98)00244-2
6. Kukreja RC, Hess ML (1994) Free Radicals, Cardiovascular Dysfunction and Protective Strategies. R. G. Landes Co., Austin
7. Tomaselli GF, Barth AS (2010) Sudden cardio arrest: Oxidative stress irritates the heart. Nat Med 16:648–649
8. Stuehr DJ, Kwon NS, Nathan CF (1990) FAD and GSH participate in macrophage synthesis of nitric oxide. Biochem Biophys Res Commun 168:558–565. https://doi.org/10.1016/0006-291X(90)92357-6
9. Marletta MA (1989) Nitric oxide: biosynthesis and biological significance. Trends Biochem Sci 14:488–492
10. Ferrari R, Guardigli G, Mele D et al (2004) Oxidative stress during myocardial ischaemia and heart failure. Curr Pharm Des 10:1699–1711. https://doi.org/10.2174/1381612043384718
11. Nishiyama Y, Ikeda H, Haramaki N et al (1998) Oxidative stress is related to exercise intolerance in patients with heart failure. Am Heart J 135:115–120. https://doi.org/10.1016/S0002-8703(98)70351-5
12. Keith M, Geranmayegan A, Sole MJ et al (1998) Increased oxidative stress in patients with congestive heart failure. J Am Coll Cardiol 31:1352–1356
13. Mallat Z, Philip I, Lebret M et al (1998) Elevated levels of 8-iso-prostaglandin F(2α) in pericardial fluid of patients with heart failure: a potential role for in vivo oxidant stress in ventricular dilatation and progression to heart failure. Circulation 97:1536–1539. https://doi.org/10.1161/01.CIR.97.16.1536
14. Tappel AL (1980) Vitamin E and selenium protection from in vivo lipid peroxidation*. Ann N Y Acad Sci 355:18–31. https://doi.org/10.1111/j.1749-6632.1980.tb21324.x
15. Weitz ZW, Birnbaum AJ, Skosey JL et al (1991) High breath pentane concentrations during acute myocardial infarction. Lancet 337:933–935. https://doi.org/10.1016/0140-6736(91)91569-G

16. Sobotka PA, Brottman MD, Weitz Z et al (1993) Elevated breath pentane in heart failure reduced by free radical scavenger. Free Radic Biol Med 14:643–647. https://doi.org/10.1016/0891-5849(93)90145-K
17. Choudhary G, Dudley SC (2002) Heart failure, oxidative stress, and ion channel modulation. Congest Heart Fail 8:148–155. https://doi.org/10.1111/j.1527-5299.2002.00716.x
18. Hille B (2001) Ion channels of excitable membranes. Sinauer Associates. Sunderland, MA 1375
19. Hille B (1992) Ionic channels of excitable membranes, Ed 2. Sinauer Association Inc., Sunderland, Massachusetts
20. Purves D, Augustine GJ, Fitzpatrick D et al (2004) Neuroscience, Ed 3. Sinauer Associates. Inc., USA
21. Hille B, Catterall WA (2012) Electrical excitability and ion channels. In: Basic neurochemistry, pp 63–80
22. Roden DM, Balser JR, George AL Jr, Anderson ME (2002) Cardiac ion channels. Annu Rev Physiol 64:431–475. https://doi.org/10.1161/circep.108.789081
23. Grant AO (2009) Cardiac ion channels. Circ Arrhythm Electrophysiol 2:185–194
24. Priest BT, McDermott JS (2015) Cardiac ion channels. Channels (Austin) 9:352–359. https://doi.org/10.1080/19336950.2015.1076597
25. Ackerman MJ, Clapham DE (1997) Ion channels–basic science and clinical disease. N Engl J Med 336:1575–1586. https://doi.org/10.1056/NEJM199705293362207
26. Belardinelli L, Antzelevitch C, Fraser H (2004) Inhibition of late (sustained/persistent) sodium current: a potential drug target to reduce intracellular sodium-dependent calcium overload and its detrimental effects on cardiomyocyte function. Eur Hear J Suppl 6:I3–I7. https://doi.org/10.1016/S1520-765X(04)80002-6
27. Hondeghem LM, Katzung BG (1984) Antiarrhythmic agents: the modulated receptor mechanism of action of sodium and calcium channel-blocking drugs. Annu Rev Pharmacol Toxicol 24:387–423. https://doi.org/10.1146/annurev.pa.24.040184.002131
28. Carmeliet E, Mubagwa K (1998) Antiarrhythmic drugs and cardiac ion channels: mechanisms of action. Prog Biophys Mol Biol 70:1–72. https://doi.org/10.1016/S0079-6107(98)00002-9. [pii]
29. Annunziato L, Pannaccione A, Cataldi M et al (2002) Modulation of ion channels by reactive oxygen and nitrogen species: a pathophysiological role in brain aging? Neurobiol Aging 23:819–834. https://doi.org/10.1016/S0197-4580(02)00069-6
30. Liu Y, Gutterman DD (2002) Oxidative stress and potassium channel function. Clin Exp Pharmacol Physiol 29:305–311
31. Takahashi K, Kakimoto Y, Toda K, Naruse K (2013) Mechanobiology in cardiac physiology and diseases. J Cell Mol Med 17:225–232
32. Fedele F, Mancone M, Chilian WM et al (2013) Role of genetic polymorphisms of ion channels in the pathophysiology of coronary microvascular dysfunction and ischemic heart disease. Basic Res Cardiol 108:387. https://doi.org/10.1007/s00395-013-0387-4
33. Ramírez A, Vázquez-Sánchez AY, Carrión-Robalino N, Camacho J (2016) Ion channels and oxidative stress as a potential link for the diagnosis or treatment of liver diseases. Oxidative Med Cell Longev 2016:1–17. https://doi.org/10.1155/2016/3928714
34. Rourke BO, D a K, Tomaselli GF et al (1999) Mechanisms of altered excitation-contraction coupling in experimental studies. Heart 84:562–570. https://doi.org/10.1161/01.RES.84.5.562
35. Pogwizd SM, Qi M, Yuan W et al (1999) Upregulation of Na+/Ca2+ exchanger expression and function in an arrhythmogenic rabbit model of heart failure. Circ Res 85:1009–1019. https://doi.org/10.1161/01.RES.85.11.1009
36. Amos GJ, Wettwer E, Metzger F et al (1996) Differences between outward currents of human atrial and subepicardial ventricular myocytes. J Physiol 491:31–50. https://doi.org/10.1113/jphysiol.1996.sp021194

37. Fedida D, Wible B, Wang Z et al (1993) Identity of a novel delayed rectifier current from human heart with a cloned K+ channel current. Circ Res 73:210–216. https://doi.org/10.1161/01.RES.73.1.210
38. Feng J, Wible B, Li G-R et al (1997) Antisense oligodeoxynucleotides directed against Kv1.5 mRNA specifically inhibit ultrarapid delayed rectifier K+ current in cultured adult human atrial myocytes. Circ Res 80:572–579. https://doi.org/10.1161/01.RES.80.4.572
39. Wang Z, Fermini B, Nattel S (1993) Sustained depolarization-induced outward current in human atrial myocytes. Evidence for a novel delayed rectifier K+ current similar to Kv1.5 cloned channel currents. Circ Res 73:1061–1076. https://doi.org/10.1161/01.RES.73.6.1061
40. Li GR, Feng J, Yue L et al (1996) Evidence for two components of delayed rectifier K+ current in human ventricular myocytes. Circ Res 78:689–696
41. Alpert JS, Petersen P, Godtfredsen J (1988) Atrial fibrillation: natural history, complications, and management. Annu Rev Med 39:41–52. https://doi.org/10.1146/annurev.me.39.020188.000353
42. Chugh SS, Blackshear JL, Shen WK et al (2001) Epidemiology and natural history of atrial fibrillation: clinical implications. J Am Coll Cardiol 37:371–378. https://doi.org/10.1016/S0735-1097(00)01107-4
43. Peters NS, Schilling RJ, Kanagaratnam P, Markides V (2002) Atrial fibrillation: strategies to control, combat, and cure. Lancet 359:593–603
44. MacDonald PE, Wheeler MB (2003) Voltage-dependent K + channels in pancreatic beta cells: role, regulation and potential as therapeutic targets. Diabetologia 46:1046–1062. https://doi.org/10.1007/s00125-003-1159-8
45. Roberds SL, Tamkun MM (1991) Cloning and tissue-specific expression of five voltage-gated potassium channel cDNAs expressed in rat heart. Proc Natl Acad Sci 88:1798–1802
46. Roe MW, Worley JF, Mittal AA et al (1996) Expression and function of pancreatic beta-cell delayed rectifier K+ channels. Role in stimulus-secretion coupling. J Biol Chem 271:32241–32246. https://doi.org/10.1074/jbc.271.50.32241
47. Su J, Yu H, Lenka N et al (2001) The expression and regulation of depolarization-activated K+ channels in the insulin-secreting cell line INS-1. Pflugers Arch Eur J Physiol 442:49–56. https://doi.org/10.1007/s004240000508
48. Snyders DJ, Tamkun MM, Bennett PB (1993) A rapidly activating and slowly inactivating potassium channel cloned from human heart. Functional analysis after stable mammalian cell culture expression. J Gen Physiol 101:513–543. https://doi.org/10.1085/jgp.101.4.513
49. Gerlach U, Brendel J, Lang H-J et al (2001) Synthesis and activity of novel and selective IKs-channel blockers. J Med Chem 44:3831–3837. https://doi.org/10.1021/JM0109255
50. Svoboda LK, Reddie KG, Zhang L et al (2012) Redox-sensitive sulfenic acid modification regulates surface expression of the cardiovascular voltage-gated potassium channel Kv1.5. Circ Res 111:842–853. https://doi.org/10.1161/CIRCRESAHA.111.263525
51. Bhuyan R, Seal A (2015) Conformational dynamics of shaker-type Kv1.1 ion channel in open, closed, and two mutated states. J Membr Biol 248:241–255. https://doi.org/10.1007/s00232-014-9764-7
52. Bhuyan R, Seal A (2015) Molecular dynamics of Kv1.3 ion channel and structural basis of its inhibition by scorpion toxin-OSK1 derivatives. Biophys Chem 203–204:1–11. https://doi.org/10.1016/j.bpc.2015.04.004
53. Bhuyan R, Seal A (2017) Dynamics and modulation studies of human voltage gated Kv1.5 channel. J Biomol Struct Dyn 35:380–398. https://doi.org/10.1080/07391102.2016.1144528
54. Ashcroft FM (2000) Ion channels and disease. Academic, London
55. Durell SR, Hao Y, Guy HR (1998) Structural models of the transmembrane region of voltage-gated and other K+channels in open, closed, and inactivated conformations. J Struct Biol 121:263–284. https://doi.org/10.1006/jsbi.1998.3962
56. Bezanilla F, Perozo E, Stefani E (1994) Gating of shaker K+ channels: II. The components of gating currents and a model of channel activation. Biophys J 66:1011–1021. https://doi.org/10.1016/S0006-3495(94)80882-3

57. Zagotta WN, Hoshi T, Aldrich RW (1994) Shaker potassium channel gating. III: evaluation of kinetic models for activation. J Gen Physiol 103:321–362. https://doi.org/10.1085/jgp.103.2.321
58. Smith-Maxwell CJ, Ledwell JL, Aldrich RW (1998) Role of the S4 in cooperativity of voltage-dependent potassium channel activation. J Gen Physiol 111:399–420. https://doi.org/10.1085/jgp.111.3.399
59. Jensen MO, Borhani DW, Lindorff-Larsen K et al (2010) Principles of conduction and hydrophobic gating in K+ channels. Proc Natl Acad Sci 107:5833–5838. https://doi.org/10.1073/pnas.0911691107
60. Jensen M, Jogini V, Borhani DW et al (2012) Mechanism of voltage gating in potassium channels. Science 336:229–233. https://doi.org/10.1126/science.1216533
61. Long SB, Tao X, Campbell EB, MacKinnon R (2007) Atomic structure of a voltage-dependent K+channel in a lipid membrane-like environment. Nature 450:376–382. https://doi.org/10.1038/nature06265
62. Olson TM, Alekseev AE, Liu XK et al (2006) Kv1.5 channelopathy due to KCNA5 loss-of-function mutation causes human atrial fibrillation. Hum Mol Genet 15:2185–2191. https://doi.org/10.1093/hmg/ddl143
63. Nielsen NH, Winkel BG, Kanters JK et al (2007) Mutations in the Kv1.5 channel gene KCNA5 in cardiac arrest patients. Biochem Biophys Res Commun 354:776–782. https://doi.org/10.1016/j.bbrc.2007.01.048
64. Christophersen IE, Olesen MS, Liang B et al (2013) Genetic variation in KCNA5: impact on the atrial-specific potassium current IKur in patients with lone atrial fibrillation. Eur Heart J 34:1517–1525. https://doi.org/10.1093/eurheartj/ehs442
65. Yang Y, Li J, Lin X et al (2009) Novel KCNA5 loss-of-function mutations responsible for atrial fibrillation. J Hum Genet 54:277–283. https://doi.org/10.1038/jhg.2009.26
66. Adzhubei I, Jordan DM, Sunyaev SR (2013) Predicting functional effect of human missense mutations using PolyPhen-2. Curr Protoc Hum Genet 76:7.20.1–7.20.41. https://doi.org/10.1002/0471142905.hg0720s76
67. Choi Y, Chan AP (2015) PROVEAN web server: a tool to predict the functional effect of amino acid substitutions and indels. Bioinformatics 31:2745–2747. https://doi.org/10.1093/bioinformatics/btv195
68. Pires DEV, Ascher DB, Blundell TL (2014) DUET: a server for predicting effects of mutations on protein stability using an integrated computational approach. Nucleic Acids Res 42:W314–W319. https://doi.org/10.1093/nar/gku411
69. Pires DE, Ascher DB, Blundell TL (2014) mCSM: predicting the effects of mutations in proteins using graph-based signatures. Bioinformatics 30:335–342. https://doi.org/10.1093/bioinformatics/btt691
70. Pandurangan AP, Ochoa-Montaño B, Ascher DB, Blundell TL (2017) SDM: a server for predicting effects of mutations on protein stability. Nucleic Acids Res 45:W229–W235. https://doi.org/10.1093/nar/gkx439
71. Feng J, Xu D, Wang Z, Nattel S (1998) Ultrarapid delayed rectifier current inactivation in human atrial myocytes: properties and consequences. Am J Phys 275:H1717–H1725
72. Nattel S, Yue L, Wang Z (1999) Cardiac ultrarapid delayed rectifiers: a novel potassium current family of functional similarity and molecular diversity. Cell Physiol Biochem 9:217–226. https://doi.org/10.1159/000016318
73. Schaffer P, Pelzmann B, Bernhart E et al (1998) Estimation of outward currents in isolated human atrial myocytes using inactivation time course analysis. Pflugers Arch Eur J Physiol 436:457–468. https://doi.org/10.1007/s004240050657
74. Nygren A, Leon LJ, Giles WR (2001) Simulations of the human atrial action potential. Philos Trans R Soc A Math Phys Eng Sci 359:1111–1125
75. Nygren A, Fiset C, Firek L et al (1998) Mathematical model of an adult human atrial cell: the role of K+ currents in repolarization. Circ Res 82:63–81. https://doi.org/10.1161/01.RES.82.1.63

76. Courtemanche M, Ramirez RJ, Nattel S (1999) Ionic targets for drug therapy and atrial fibrillation-induced electrical remodeling: insights from a mathematical model. Cardiovasc Res 42:477–489. https://doi.org/10.1016/S0008-6363(99)00034-6
77. Delpón E, Caballero R, Valenzuela C et al (1999) Benzocaine enhances and inhibits the K+current through a human cardiac cloned channel (Kv1.5). Cardiovasc Res 42:510–520. https://doi.org/10.1016/S0008-6363(99)00043-7
78. Franqueza L, Longobardo M, Vicente J et al (1997) Molecular determinants of stereoselective bupivacaine block of hKv1.5 channels. Circ Res 81:1053–1064. https://doi.org/10.1161/01.RES.81.6.1053
79. Caballero R, Delpón E, Valenzuela C et al (2001) Direct effects of candesartan and eprosartan on human cloned potassium channels involved in cardiac repolarization. Mol Pharmacol 59:825–836. https://doi.org/10.1124/mol.59.4.825
80. Moreno I, Caballero R, González T et al (2003) Effects of irbesartan on cloned potassium channels involved in human cardiac repolarization. J Pharmacol Exp Ther 304:862–873. https://doi.org/10.1124/jpet.102.042325
81. Caballero R, Delpón E, Valenzuela C et al (2000) Losartan and its metabolite E3174 modify cardiac delayed rectifier K+currents. Circulation 101:1199–1205. https://doi.org/10.1161/01.CIR.101.10.1199
82. Iftinca M, Waldron GJ, Triggle CR, Cole WC (2001) State-dependent block of rabbit vascular smooth muscle delayed rectifier and Kv1.5 channels by inhibitors of cytochrome P450-dependent enzymes. J Pharmacol Exp Ther 298:718–728
83. Lacerda AE, Roy ML, Lewis EW, Rampe D (1997) Interactions of the nonsedating antihistamine loratadine with a Kv1.5-type potassium channel cloned from human heart. Mol Pharmacol 52:314–322
84. Caballero R, Valenzuela C, Longobardo M et al (1999) Effects of rupatadine, a new dual antagonist of histamine and platelet-activating factor receptors, on human cardiac Kv1.5 channels. Br J Pharmacol 128:1071–1081. https://doi.org/10.1038/sj.bjp.0702890
85. Delpón E, Valenzuela C, Tamargo J (1999) Blockade of cardiac potassium and other channels by antihistamines. Drug Saf 21(Suppl 1):11–8–7
86. Choe H, Lee Y-K, Lee Y-T et al (2003) Papaverine blocks hKv1.5 channel current and human atrial ultrarapid delayed rectifier K^+ currents. J Pharmacol Exp Ther 304:706–712. https://doi.org/10.1124/jpet.102.042770
87. Rampe D, Murawsky MK (1997) Blockade of the human cardiac K+ channel Kv1.5 by the antibiotic erythromycin. Naunyn Schmiedeberg's Arch Pharmacol 355:743–750. https://doi.org/10.1007/PL00005008
88. Camm AJ, Savelieva I (2004) Advances in antiarrhythmic drug treatment of atrial fibrillation: where do we stand now? Heart Rhythm 1:244–246
89. Choudhury A, Lip GY (2004) Antiarrhythmic drugs in atrial fibrillation: an overview of new agents, their mechanisms of action and potential clinical utility. Expert Opin Investig Drugs 13:841–855. https://doi.org/10.1517/13543784.13.7.841
90. Decher N, Pirard B, Bundis F et al (2004) Molecular basis for Kv1.5 channel block. Conservation of drug binding sites among voltage-gated K+channels. J Biol Chem 279:394–400. https://doi.org/10.1074/jbc.M307411200
91. Peukert S, Brendel J, Pirard B et al (2004) Pharmacophore-based search, synthesis, and biological evaluation of anthranilic amides as novel blockers of the Kv1.5 channel. Bioorg Med Chem Lett 14:2823–2827. https://doi.org/10.1016/j.bmcl.2004.03.057
92. Bachmann A, Gutcher I, Kopp K et al (2001) Characterization of a novel Kv1.5 channel blocker in Xenopus oocytes, CHO cells, human and rat cardiomyocytes. Naunyn Schmiedeberg's Arch Pharmacol 364:472–478. https://doi.org/10.1007/s002100100474
93. Decher N, Kumar P, Gonzalez T et al (2006) Binding site of a novel Kv1.5 blocker: a "foot in the door" against atrial fibrillation. Mol Pharmacol 70:1204–1211. https://doi.org/10.1124/mol.106.026203

94. Peukert S, Brendel J, Pirard B et al (2003) Identification, synthesis, and activity of novel blockers of the voltage-gated potassium channel Kv1.5. J Med Chem 46:486–498. https://doi.org/10.1021/jm0210461
95. Brendel J, Peukert S (2003) Blockers of the Kv1.5 channel for the treatment of atrial arrhythmias. Curr Med Chem Cardiovasc Hematol Agents 1:273–287. https://doi.org/10.2174/1568016033477441
96. Wu S, Fluxe A, Janusz JM et al (2006) Discovery and synthesis of tetrahydroindolone derived semicarbazones as selective Kv1.5 blockers. Bioorg Med Chem Lett 16:5859–5863. https://doi.org/10.1016/j.bmcl.2006.08.057
97. Li Y, Starrett JE, Meanwell NA et al (1997) The discovery of novel openers of Ca2+-dependent large-conductance potassium channels: pharmacophore search and physiological evaluation of flavonoids. Bioorg Med Chem Lett 7:759–762. https://doi.org/10.1016/S0960-894X(97)00076-0
98. Caballero R, Moreno I, González T et al (2002) Putative binding sites for benzocaine on a human cardiac cloned channel (Kv1.5). Cardiovasc Res 56:104–117. https://doi.org/10.1016/S0008-6363(02)00509-6
99. Snyders J, Knoth KM, Roberds SL, Tamkun MM (1992) Time-, voltage-, and state-dependent block by quinidine of a cloned human cardiac potassium channel. Mol Pharmacol 41:322–330
100. Snyders DJ, Ycola SW (1995) Determinants of antiarrhythmic drug action: electrostatic and hydrophobic components of block of the human cardiac hKv1.5 channel. Circ Res 77:575–583. https://doi.org/10.1161/01.RES.77.3.575
101. Yeola SW, Rich TC, Uebele VN et al (1996) Molecular analysis of a binding site for quinidine in a human cardiac delayed rectifier K^+ channel. Circ Res 78:1105–1114

Part II

Atherosclerosis and Ischemic Heart Disease

Aging and Cardiovascular Diseases: The Role of Cellular Senescence

10

Perinur Bozaykut

Abbreviations

AAA	Abdominal aortic aneurysm
CVD	Cardiovascular disease
DDR	DNA damage response
DNA	Deoxyribonucleic acid
EC	Endothelial cells
ER	Endoplasmic reticulum
ETC	Electron transport chain
HFpEF	Heart failure with a preserved ejection fraction
IL	Interleukin
Keap1	Kelch-like ECH associated protein 1
LV	Left ventricular
MCP	Monocyte chemoattractant protein 1
mTOR	Mammalian target of rapamycin
NAD	Nicotinamide adenine dinucleotide
NFκB	Nuclear factor kappa B
NO	Nitric oxide
NOS	Nitric oxide synthase
NOX	NADPH oxidases
Nrf2	Nuclear factor erythroid 2–related factor 2:
oxLDL	Oxidized low-density lipoproteins:
PGC-1α	PPAR-γ coactivator 1 alpha
ROS	Reactive oxygen species
SAHF	Senescence-associated heterochromatin foci
SAMP8	Senescence accelerated mice prone 8

P. Bozaykut (✉)
Department of Molecular Biology and Genetics, Acibadem Mehmet Ali Aydinlar University, Istanbul, Turkey

S. Chakraborti et al. (eds.), *Oxidative Stress in Heart Diseases*,
https://doi.org/10.1007/978-981-13-8273-4_10

SASP	Senescence-associated secretory phenotype:
SA-β-gal	Senescence-associated beta-galactosidase
SIPS	Stress-induced premature senescence
SIRT	Sirtuin
SMC	Smooth muscle cells
SOD	Superoxide dismutase
TGFβ	Transforming growth factor β
TNFα	Tumor necrosis factor
VSMC	Vascular smooth muscle cell

10.1 Introduction

As life-expectancy increases, healthy aging is becoming an important problem [1]. Chronological age is the significant risk factor for the cardiovascular diseases (CVDs) and CVDs remain to be the major health problem in the elderly people [2]. CVDs such as heart failure, diabetes, and atherosclerosis are also termed as age-associated diseases [3]. The vascular system deteriorates within aging process shows the powerful link between age and vascular disorders [4]. Aging is also associated with oxidative stress, endoplasmic reticulum (ER) stress, inflammation, apoptosis and mitochondrial dysfunction, all of which are associated with CVD pathogenesis [5]. Among these aging etiologies, low-grade chronic inflammation and elevated oxidative stress are suggested to be most significant mechanisms to contribute disease pathology [6].

Aging process includes several complex mechanisms that are interconnected. Among these mechanisms, cellular senescence recently gains interest having a pivotal role in the progression of cardiovascular pathology [7, 8]. Over a half century ago, cellular senescence was originally described by Hayflick et al. in human lung fibroblasts [9]. They showed the limited replication capacity of human primary fibroblast cells which then, enter to permanent cell cycle arrest. In accordance with cell cycle arrest, alterations of gene expression changes lead to secretion of pro-inflammatory proteins which is called as the senescence-associated secretory phenotype (SASP) [10]. SASP is especially essential in CVDs for leading to chronic inflammation and therefore, tissue remodeling [4]. Increasing evidence also have demonstrated that cellular senescence is an essential step in the development of vascular aging [11]. It has been shown that the vasculature and the myocardium undergo changes with aging and cellular senescence termed as "vascular senescence". Vascular senescence leads to cardiac morbidity and mortality that eventually promote atherosclerosis [12], systolic cardiac dysfunction [13] and systemic metabolic dysfunction [14].

With aging, large arteries go into progressive alterations in the mechanical properties such as endothelial dysfunction [15]. Vascular endothelial cells (ECs) are identified as an essential part of the vascular wall and therefore, are important for the maintenance of cardiovascular homeostasis [16]. Impairments in endothelial function significantly contribute to arterial inflammation, lesion formation, deterioration in vasodilation, and loss of compliance. Endothelial dysfunction is also accepted as the biomarker for future cardiac pathology since it is an early event in the development of atherosclerotic plaque [17]. Well-known clinical risk factors for CVDs, such as high blood pressure and oxidative stress lead to endothelial dysfunction [18]. Therefore, identification of mechanisms that underlies EC senescence and vascular aging may develop next-generation therapy strategies for cardiovascular diseases. Aging process is associated with changes in the vasculature at the cellular and molecular levels and microvascular changes may be seen through the onset of vascular remodeling. Specifically, increased oxidative stress and low-grade chronic inflammation are possible underlying mechanisms that may trigger physiological and early vascular aging [19, 20].

This review addresses the cellular senescence in aging-related diseases exploring in-depth role of oxidative stress with the focus on vascular system. It is reviewed the mechanisms underlying vascular senescence and the context of senescent cells in the vessels and usefulness of clearance of these senescent cells as next generation therapies for cardiovascular diseases. It is discussed the importance of fully understanding of aging mechanisms could emerge new strategies against vascular pathologies.

10.2 Cellular Senescence

Cellular senescence is described as an irreversible process in which cells enter permanent cell cycle arrest and typically, followed by a severe damage [9, 10]. Cellular senescence of cells is suggested to play a protective role in cancer process since cell cycle arrest occurs as a response to DNA damage could avoid tumor progression [21]. Therefore, in cancer process, cellular senescence functions as a vital safeguard against hyperproliferative states of the cells. On the other hand, it is known that cellular senescence can be also detected in other cases than cancer, like aging and aging-associated disorders. Therefore, whether the senescence would be beneficial or detrimental, depends on the status and the age of the organism.

As described by Hayflick and Moorehead, cultured cells have a finite lifespan [9] and it was also shown that critical loss of telomeres causes cellular senescence when somatic cells undergo many cell divisions [22]. The discovery of senescence during serial passaging of human diploid fibroblasts started the speculation that senescence is a natural process during aging. In addition to fibroblast cells, cellular senescence has been observed in various cell types including epithelial cells, endothelial cells, lymphocytes and chondrocytes and more on [23]. The cells enter a state of cell cycle arrest following serial passaging of cells, termed as "replicative senescence" or "Hayflick limit", has been characterized by the telomere shortening [4]. Many studies have also reported the increase of senescent cells with age and aging pathologies and have also shown the relationship of senescence to other biological processes including cardiovascular pathologies [24].

However, the cellular senescence could be induced by other non-telomeric signals which include various types of stressors, mainly, oxidative stress [25]. When there is more stress than the physiological stress range, a different type of cellular senescence can also occur which is referred as "stress-induced premature senescence" (SIPS). SIPS also includes the other candidate drivers of senescence including oncogene signaling such as Ras, disrupted chromatin, DNA damage, intense mitogenic signals, oncogene activation, metabolic stress, and stress owing to cell culture conditions [26]. It is worth noting that telomere damage occurs at SISP as well, however, only to some extent [27].

Cellular senescence is also defined by another phenomenon which is termed as SASP [28]. This phenomenon is characterized by the secretion of many biologically active proteins that affect both the senescent cell itself and adjacent cells. The SASP includes proinflammatory cytokines that are particularly crucial in driving aging and the pathogenesis of age-associated diseases. The SASP develops consequence of extensive chromatin remodeling during senescence and which in turn suppresses the nuclear lamina protein, lamin B1 transcription [29]. Inhibition of lamin B1 leads to increase in inflammatory cytokines and chemokines, alterations in the production of growth factors, proteases secretion, and as a result, leads to reactive oxygen species (ROS) generation [29]. Activation of inflammation by chronic SASP causes disruption in normal tissue structure and function [28], and further leads to death of cells around them, tissue remodeling, and attraction of immune elements. Innate immunity activation leads to the removal of component cells which is increases during aging, potentially contributing to senescent cell accumulation in old age [30]. The fact that activation of inflammation by SASP underlies many age-associated pathologies and even may drive cancer.

Recent studies reported the effect of cellular senescence by its growth arrest phenotype and SASP factors as an important contributor in the pathology of age-associated disorders. In addition, it is also claimed that age-associated disease development by positive feedback mechanisms is maintained mainly by SASP factors. The overview of oxidative stress related mechanisms and its effect on the cardiovascular diseases by the cellular senescence induction will be discussed in the following sections [31].

10.2.1 Signaling Pathways of Cellular Senescence

Deterioration of various organ functions during cellular senescence is accompanied by several molecular mechanisms [7]. Although the molecular mechanism of the cellular senescence is not fully clarified yet, recent studies underlined several signaling pathways. Lessons from what happens during senescence show the important role of p53 pathway [32]. The p53 protein which is known to protect the genome by the inhibition of tumorigenesis, also have pivotal roles in apoptosis, cell cycle control and DNA repair [33]. Additionally, it is well known that p53 is effective through many other biological processes including, autophagy, antioxidant defenses and angiogenesis [34, 35]. However, studies have shown that in aging process, p53 pathway leads to increase in the levels of ROS and some other stresses as well, such as

DNA damage and oncogenic stress [32] and therefore, is one of the well-studied mechanism related to cellular senescence. As previously discussed, during replicative senescence incomplete replication of telomeres leads to shortening of telomeres as a result of cell cycle arrest. Accordingly, when telomere shortening is extensive, chromosomal stability and DNA replication is disturbed leading to DNA damage. At this point, DNA damage induces cellular senescence via p53/p16 signaling pathway is suggested to play the main roles during replicative senescence [34].

However, stress induced SIPS is triggered independent of cell cycle arrest and telomere shortening which critically erodes telomeres. Signals from various stresses including oxidative stress, UV, oncogenic stress, metabolic stress (Fig. 10.1) results in DNA damage leading to senescence is also regulated by the p53 or p16 pathways. Among stress factors, especially ROS is known to have crucial role for the induction of senescence in vascular cells since ROS can induce senescence either dependent or independent of telomere shortening by driving DNA lesions [36]. Other less

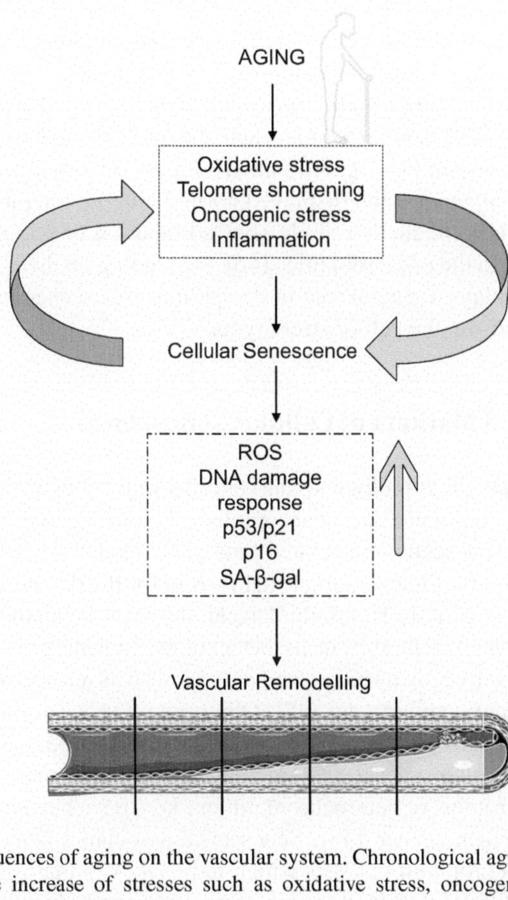

Fig. 10.1 Consequences of aging on the vascular system. Chronological aging induces cellular senescence via the increase of stresses such as oxidative stress, oncogenic stress, elevated telomere attrition and inflammation. Cellular senescence is associated with an increase of ROS, p53/p21, p16 and SA-β-gal that results in pathological changes in the vascular system

characterized stressors for vascular cells is the components of SASP itself which function via transforming growth factor β (TGFβ) signaling. As mentioned before, SASP takes crucial role during the pathology of the disorders since it affects nearby cells in the vessels [31]. The decision to activate p53 or p16 pathway depends either on the cell type or the stress type. p16 pathway is mainly activated almost in all cells during senescence progress [37]. On the other hand, when there is DNA damage and telomere dysfunction, p53 pathway is the preferred pathway. However, general cellular stresses such as mitogenic stress lead to activation of p16 signaling pathway [38, 39]. This type of senescence is typically linked to vascular aging since vessel wall-resident cells such as ECs, smooth muscle cells (SMCs), fibroblasts might not replicate as much as replicative senescence. On the other hand, in some cases such as when endothelial cell repopulation is occurred following angioplasty, both replicative and nonreplicative senescence can be observed [40, 41].

The role of p53 signaling in cardiovascular/heart diseases has been shown with several studies. Increased level of p53 protein in old vessels, failing hearts and in the visceral fat of obese patients has been reported. The role of cellular stress triggered by p53 pathway has also been elicited in the pathology of aging and age-associated diseases such as atherosclerosis, diabetes, obesity and heart failure [13, 14, 35, 42]. On the other hand, some other studies have shown its beneficial effect during aging. A study with Trp53/Cdkn2a transgenic mice showed resistance to carcinogenesis with an increased lifespan [43]. Additionally, "Super p53" mice was also shown to be resistant to carcinogenesis and displayed normal glucose tolerance on a standard diet [44, 45]. In addition, another study showed deletion of p53 or p21 increased cellular senescence in the progeroid mice [46]. Depending on these studies, p53/p21 signaling pathway plays a crucial role in the cellular senescence process, however, the response depends on the cell or stress type.

10.2.2 Biological Markers of Cellular Senescence

Although, there is no direct biological marker reflecting cellular senescence, identification of cellular senescence can be accomplished by using combination of several markers. Of them, senescence-associated beta-galactosidase (SA-β-gal) activity is the most widely used biological marker that is used for the detection cellular senescence [47]. SA-β-gal activity is mainly detected by immunohistochemical method in cell culture and less preferably in tissue sections. In senescent cells, lysosomal beta-galactosidase activity can be detected only in pH 6 as a result of marked expansion of the lysosomal compartment [47]. Other common used markers depend on the signaling status of the cell such as increased levels of p53, $p16^{Ink4a}$, p21, p38 mitogen-activated protein kinase. In some cases, epigenetic markers such as high mobility group A proteins or heterochromatin markers have been used as biological markers of cellular senescence [48]. Senescence-associated heterochromatin foci (SAHF) are heterochromatin markers which have DNA domains stained by 4′,6′-diamidino-2-phenylindole (DAPI) and have enriched histone methylations (H3K9me). Additionally, the modified histone, γH2AX contributes to DNA repair and stabilization and can be used as a proof of DNA damage linked by telomere-induced foci [24, 48]. In addition, the presence of senescence can be supported by the appearance of the cells which display flat morphology and vacuolated cells with enlarged nucleoli, by the microscopy. The senescent cells also exhibit stable growth arrest although they are metabolically active [37].

SASP factors including interleukin-6 (IL-6), growth factors, proteases and other pro-inflammatory factors can be detected as well in the presence of cellular senescence [49]. SASP factors can be easily detected form the media of cell culture by using ELISA method or expression analysis of the factors by RNA and protein expressions of the tissues. The levels of ROS are yet, another biological marker of senescent cells. The detection of these markers could be performed in days to weeks depending on the development process of cellular senescence [36].

10.3 Oxidative Stress and Cellular Senescence

The effect of ROS which is explained by the "Free radical theory of ageing" has been speculated for a long time as one of the main contributor of aging pathology in mammals. Many studies on various species have reported the effects of oxidative stress suggesting ROS has a central role on age-associated diseases and possibly on lifespan. Elevated oxidative stress levels, which is higher than normal cellular levels, damage macromolecules and promote cellular senescence. Extensive ROS production results in the oxidative modification of biomolecules including proteins, DNA and lipids which in turn leads to cellular and vascular dysfunction [50]. The consequences of ROS includes other mechanisms such as apoptosis or autophagy, however, the fate of oxidative stress depends on the duration and level of the oxidative stress [51].

Oxidative stress caused by factors such as exposure to oxygen, hydrogen peroxide or tert-butylhydroperoxide can induce SIPS in cells [52, 53]. Other stressors such as oncogenes (H-RasV12) can also result in oxidative stress induction [54] and lead to oncogene induced SIPS by DNA damage response (DDR) [55]. The cellular senescence induced by oxidative stress has two main pathways as mitochondrial and non-mitochondrial pathways. These two pathways are likely to merge at some point by several molecular factors such as p53, pRB, p16 and p21 [56, 57]. p53/p21 pathway has been suggested to act as the main molecular player when the stress is related to DDR [55], however, p16 pathway also has been shown to be activated during DDR damage (Fig. 10.1). Several studies also showed that cellular senescence that is induced by ROS also involves a positive feedback pathway that results in the amplification of senescence factors. In this case, when senescence is induced by ROS, SASP factors would lead to even more increased oxidative stress and, thus, to increased senescence [58], and in turn, generated ROS would lead to more mitochondrial mutations and ROS that finally would result in ultimate senescent phenotype [56, 57].

Although some mechanisms have been proposed to link oxidative stress and senescence, the exact mechanisms haven't fully cleared yet. A study conducted with a senescent mice model [59], reported that the nuclear factor erythroid 2–related

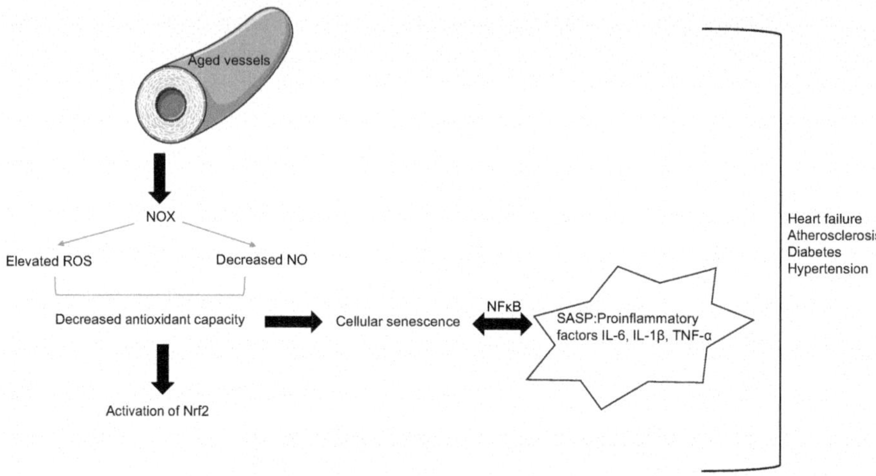

Fig. 10.2 Development of cardiovascular diseases by the effect of oxidative stress. Aged vessels are characterized by elevated reactive oxygen species (ROS) levels and decreased nitric oxide (NO) levels modulated by nitric oxide synthase NADPH oxidases (NOX). Decreased antioxidant capacity induces cellular senescence that promotes senescence-associated secretory phenotype (SASP) factors that lead to pathogenesis of heart failure, atherosclerosis, diabetes and hypertension

factor 2 (Nrf2) pathway could be the responsible molecular factor in the generation of oxidative stress in senescent animals. Nrf2 is a transcription factor which binds to cytosolic Kelch-like ECH associated protein 1 (Keap1) in its inactive form in normal cell conditions and activated under oxidative stress conditions. Activated Nrf2 translocates into the nucleus where it binds antioxidant response element and induces the gene expression of antioxidant and phase II enzymes. Studies have shown that the effect of Nrf2 signaling on the inhibition of oxidative stress in adverse cardiac remodeling via its antioxidant role and Nrf2 silencing has also shown to lead to the activation of proinflammatory genes such as IL-1β and tumor necrosis factor (TNFα). Therefore, any impairment in Nrf2 signaling is suggested to lead to an increase in the cardiac disease severity by the activation of inflammation factors (Fig. 10.2) [60].

10.4 The Mitochondrial Pathway of Oxidative Stress and Cellular Senescence

Mitochondria has the central role on the production of ROS by the escape of electrons through electron transport chain (ETC), therefore, mitochondrial dysfunction has been associated to senescence and related disease pathologies [61]. ROS produced by mitochondria results in the mutations of mitochondrial DNA leading to mitochondrial function defects. As a result, especially in aging process, mitochondrial dysfunction results in biological function decline and leads to age-associated heart disorders [62].

Recent findings reported the fact that oxidative stress can result in DDR by various mechanisms such as telomeric and non-telomeric mechanisms. Oxidative stress can also induce ROS production by the positive feedback mechanism [63] and the molecular players that is known to responsible for the positive feedback mechanism includes p53-dependent signaling pathway such as p21, GADD45A, p38 and TGFβ [57]. On the other, as it is previously discussed, oxidative stress can be generated also by non-mitochondrial mechanisms which leads to cellular senescence in collaboration with the mitochondrial ROS. Recently, it was reported that the presence of mitochondrial ROS is required for the cellular senescence generation which is shown by typical markers of senescence such as SA-β-gal and SASP factors [51]. In addition, uncoupling of mitochondria suggested to lead to proteasomal degradation and autophagy however, no common markers of senescent phenotype were detected [51]. The study also suggested mTOR could be the responsible pathway to drive DDR for the induction of cellular senescence. Recent findings also suggested the role of mitochondria on senescent phenotype not only by ROS production but also other factors including mitochondrial dynamics, altered redox state and metabolism and impaired ETC [64]. It is also worth noting that mitochondrial dysfunction has a regulative role on SASP as well which induces growth arrest as the senescent phenotype [57].

It has been reported that a proper model to study age-associated cardiovascular diseases is the accelerated senescent model of mice named as "senescence accelerated mice prone 8" (SAMP8). A study of SAMP8 showed increased mitochondrial dysfunction and cardiac ROS in the mitochondria [61]. Various studies have also reported impairment of mitochondria is related to apoptosis [65] and it was shown that hearts of SAMP8 mice is affected by apoptosis significantly by caspase3 pathway during aging process [66]. The reports of senescent mice model also confirm that aging-associated diseases and typically cardiac remodeling are related ROS, mitochondrial dysfunction, and apoptosis.

Sirtuins (SIRTs) which are also known as Nicotinamide adenine dinucleotide (NAD+)-dependent histone/protein deacetylases are controlled by NAD+ physiological levels in the cell. It has been reported that NAD+ and SIRT activity is decreased by aging, and this decrease is responsible for mitochondrial dysfunction. Reduced NAD+ levels result in ROS production either via PGC-1α (PPAR-γ coactivator 1 alpha) dependent and independent pathways [67]. On the other hand, AMPK increases SIRT1 activity via affecting NAD+/NADH ratio [68]. Another study reported elevated NAD+ levels inhibit senescent phenotype in muscle, neural and other adult stem cells [69]. Overall, these reports demonstrate the importance of NAD+ and SIRTs as molecular players in mitochondrial dysfunction and oxidative stress production for healthspan. All information above suggest that mitochondrial ROS is crucial in process of cellular senescence and related aging pathologies.

10.5 Oxidative Stress, Senescence and Cardiovascular Pathologies

Previous sections explain common processes of ROS and oxidative stress levels, their mechanisms and responses against various conditions that result in dysfunction of the cellular and tissue physiology. Many studies have suggested the role of cellular senescence in the development of age-associated pathologies [21, 46] that has been linked with oxidative stress, mitochondrial dysfunction and telomere shortening [70] all of which are considered as the markers of aging [71]. Changes in the level of ROS has been suggested to be important in the association of oxidative stress with aging and aging-associated heart and cardiovascular diseases [72]. In the following sections, the effects of oxidative stress and cellular senescence on aging-associated pathologies including heart failure, atherosclerosis, hypertension and diabetes will be described (Fig. 10.2).

Processes of oxidative stress are identified in ageing vessels [73, 74] including elevated vascular ROS levels and decreased nitric oxide levels leading to the formation of injurious peroxynitrite in aorta of aged rodents [50]. Oxidative stress typically promotes several molecular events of vascular aging such as vascular dysfunction, fibrosis and calcification, altered calcium homeostasis, activation of redox-sensitive or pro-inflammatory factors, and activation of cellular senescence and autophagy in ECs and VSMCs. Increased ROS levels could be reversed by superoxide dismutase (SOD) mimetics, such as tempol, leading to decrease in endothelial impairment in old rodents suggest the essential role of oxidative stress in age-associated endothelial dysfunction [75]. The alterations in cellular anti-oxidant systems by aging such as the decrease of SOD as an antioxidant enzyme, also take crucial role. Nrf2, which is previously described as the master transcription factor regulating anti-oxidant genes, is also downregulated by the reduced anti-oxidant capacity leading to extensive dysfunction of the cells [76]. Finally, these processes are followed by low-grade chronic inflammation via NFκB molecular pathway in aged vessels (Fig. 10.2) [63].

Telomere shortening and dysfunction which have crucial role during replicative senescence also related with CVDs [77]. Particularly, telomere dysfunction of senescent cells has been associated with chronic ROS production. Relatively, it is suggested that the senescence phenotype SASP leads to degenerative and proliferative activities in the cells and their component cells which is also important in the pathology of cardiovascular disorders [78]. It has been shown that senescent human umbilical venous endothelial cells (HUVECs) contributed to endothelial dysfunction that resulted in atherosclerosis development [79]. Senescence of HUVECs is further proposed to elevate the expression of pro-inflammatory cytokines that could lead to a progressive development in the pathogenesis of CVDs. It has been also demonstrated that pro-inflammatory molecules including TNF-α and IL-6 levels are elevated during aging process [79].

p66Shc, which is an adaptor protein, has been suggested to control oxidative stress and be involved in CVD pathogenesis [80]. It was shown in fibroblast cells that p66Shc controls various cellular fates including apoptosis and senescence [81].

On the other hand, several studies have shown that p66Shc silencing decreased the levels of ROS under stress conditions. In addition, knockout mice model of p66Shc was shown to extend lifespan. Elevated levels of oxidative stress and decreased NO were shown to lead to vessel impairment [82] and elevated expression of p66Shc was shown in coronary artery disease patients [83]. Another study in knockdown mice model of p66Shc showed that myocardial injury was decreased, and resistance to endothelial dysfunction was increased due to low oxidative stress [49]. Therefore, it is suggested that the regulators of cellular senescence such as p66Shc are essential to develop therapeutic interventions for CVDs.

EC senescence which can be induced by several factors has important role on cellular homeostasis and relatively, on vascular aging and diseases. Although the exact mechanism of EC senescence is not known, it is suggested that ROS levels are crucial for the generation of senescent ECs and vascular aging [84]. It is well known that senescence-induced vascular aging is crucial in disorders such as atherosclerosis, diabetes, and hypertension [3]. p53 which is the target of SIRT1 is also regulated by oxidative stress levels and, therefore, antiaging mechanisms are controlled by ROS at some level [85]. Particularly, the p53-p21-Rb molecular pathway is suggested to have an important role for the generation of senescent cells in various cellular stimuli. When activated, p53 protein induces the expression of p21 that leads to cell cycle arrest and activation of cell cycle repressor retinoblastoma (Rb) [86]. A very recent study also explained the relation of EC senescence and vascular aging with excess ROS levels and decreased SIRT1 levels by the activation of p53-p21-Rb pathway [87]. Another recent study on SAMP8 as a vascular aging model showed the increased levels of ROS and inflammation in perivascular adipose tissue which resulted in the vascular dysfunction [88]. These studies suggested that association of increased oxidative stress levels and senescence have big impact on the vascular aging and dysfunction through the senescence of vascular cells and, therefore, on the development of CVDs.

10.6 Heart Failure

Among age-associated diseases, heart failure has a high prevalence in old people [89] and, therefore, well-established therapies for severe heart failure is urgent. Age-associated heart failure is also observed without known risk factors, such as hypertension, obesity, diabetes, or atherosclerotic pathologies [3, 90]. %50 of heart failure patients develop the disease without systolic dysfunction and this type is named as "heart failure with a preserved ejection fraction" (HFpEF). HFpEF is the common type of heart failure in aging populations and a major clinical problem since its mechanism is still not fully known. HFpEF pathology is suggested to be related with cardiac endothelial cell remodeling [91] and endothelial inflammation [92]. In addition, several studies have suggested the pathological influence of senescent phenotype in the development of heart failure. A recent study showed the important role of EC senescence in SAMP8 model (46) and another recent study showed the increase of oxidative stress biomarkers in the hearts of senescent mice model [93]. Therefore, the physiological aging and correlatively cellular senescence process are suggested to elevate the risk of heart failure.

EC senescence has been proposed to have a critical role in the failing heart although the mechanism has not been fully explained yet. It was shown that p53 level in cardiac aging is elevated in mouse model of left ventricular (LV) pressure overload. In addition, LV pressure overload leads to capillary rarefaction, tissue hypoxia, and cardiac dysfunction [35]. Another recent study by LV pressure overload model also showed the inflammation and remodeling in vascular ECs induced via p53 molecular pathway [13]. Increased p53 expression induces inflammation and exacerbates the intercellular adhesion molecule (ICAM)−1 expression leading to cardiac dysfunction in endothelial cells. It has been suggested that sympathetic nervous system is activated in heart failure [94] and the sympathetic nervous system/ROS axis leads to elevated p53 expression in LV pressure overload [13]. Additionally, it has been suggested that accumulation of p53 in ECs lead to deterioration of cardiac function, resulting in angiogenesis and failing heart [95]. These studies suggested that senescent ECs could potentially become therapeutic target for cardiac dysfunction of failing heart.

In addition to elevated p53 and p21 expression, telomere shortening as another characteristic of cellular senescence plays a critical role in heart pathologies [96]. Correlatively, decreased levels of telomerase activity was reported in ECs of people with coronary heart disease [97] and in circulating leukocytes of chronic heart failure [98]. It is also suggested that telomere shortening is related to cardiovascular diseases independent of known vascular risk factors [99, 100].

As half of the HFpEF patients are diagnosed with a preserved ejection fraction, the rest develops the disease with other well-known risk factors such as obesity, hypertension, diabetes, and aging. Coronary microvascular inflammation is another most established risk factor for the development of HFpEF [91], and a recent study showed that senescent ECs have an influence on HFpEF. A study performed by SAMP8 showed that when accelerated senescent mice fed with high-fat diet, cardiac cellular senescence and inflammation are significantly elevated along with HFpEF physiological alterations [101]. It is not surprising that cardiac cellular senescence leads to vascular dysfunction and inflammation and accordingly, to the pathology of HFpEF. Thus, inhibition of senescent EC generation is suggested as a possible therapeutic intervention for the treatment of HFpEF.

10.7 Structural Changes of Arteries with Aging and Atherosclerosis

Although aging is still not considered as a pathological condition, aging-related arterial remodeling is suggested to be one of the pathological determinant of cardiovascular disease. During cardiovascular pathology, the arterial walls are damaged by the increased oxidative stress levels that leads to the generation of oxidized low-density lipoproteins (oxLDL). The initial and pivotal step of atherosclerosis progress is the infiltration of oxLDL to the subendothelial space of the arterial wall

[102]. Meanwhile, monocytes are attracted, which in turn, transform into lipid-loaded foam cell macrophages. More monocytes are attracted with the help of pro-inflammatory factors that leads to the accumulation of more inflammatory cells and results in the formation of lesions and plaques. The plaque stability is determined by the content of the plaques and aged arteries have thicker intima/media by two- to three-fold than young arteries [2, 3].

Aging modifies SMCs structure to a more synthetic phenotype that contributes to the development of atherosclerosis. Both intima and media of the artery get thickened through aging is accompanied by increased collagen and decreased elastin generation which lead to impairment in the integrity of the arteries [103, 104]. In addition, calcification is yet another phenotype of aging arteries and, therefore, plaques become more severe by aging. A study of aged rabbits showed that high fat diet resulted in more developed plaques when compared to the young rabbits [105]. All these characteristics of vascular dysfunction by old age increase the risk for atherosclerosis and recently, cellular senescence has been proposed to have the pivotal role in the pathology.

Growing evidence have showed the presence of senescent cells in the vessel content and vascular senescence is linked by disorders such as atherosclerosis, intimal hyperplasia, hypertensive arteries, aneurysms and diabetic arteries [4]. Although the mechanisms of cellular senescence in the vascular dysfunction and development of atherogenesis are not fully understood, it is known that oxidative stress is one of the major contributor. A study suggested that increased oxidative stress levels are caused by the elevated TNF-α levels which also modulates the inflammation process by NFκB activation. NFκB is a redox sensitive transcription factor which regulates many inflammation processes in the arteries and the main controller of SASP [106, 107]. Moreover, in a recent study that compared the young and elderly people, NFκB expression was found to be significantly increased in elderly people [108]. In addition, a protein named as Klotho inhibits cellular senescence and prolongs lifespan of mice and its suppression results in the development of the atherogenesis [109, 110].

It is a well known fact that cellular senescence increases over time in the presence or absence of atherosclerotic process [79, 111]. Particularly, advanced plaques show senescent cell phenotype SASP and the presence of common senescence markers such as SA-β-Gal, p16Ink4a, p53, and p21 expression [112]. Several studies have also suggested the emergence of cellular senescence mechanisms in the pathology of atherosclerosis among other cellular fates [113]. Both ECs and SMCs were reported in patients with abdominal aortic aneurysm (AAA) [111]. In addition, both ECs and SMCs were also reported to be induced by well-known stressors such ROS and angiotensin II [77, 114]. The presence of cellular senescence in ECs has been shown in atherosclerotic human coronary arteries [115] and thoracic aorta [116] by the SA-β-gal activity. Senescent ECs are associated to be induced by aortic flow impairment in the atherosclerotic mouse model and the molecular pathway responsible for senescence is suggested to be p53 signaling [117]. Typically, ECs are suggested to take a critical role for several vascular functions such as angiogenesis and coagulation and ECs show loss of function by aging process. Most importantly, decreased nitric oxide synthase (NOS) activity results in reduced nitric oxide (NO) generation leading to deterioration in vasodilation and cardiovascular pathology. Additionally, the impairments in ECs function result in oxidative stress and inflammation phenotypes and, therefore, it is suggested that cellular senescence contributes these properties [105]. During replicative senescence, progressive generation of senescent ECs is typically important since it initiates SASP phenotype resulting in tissue remodeling and elevates pro-inflammatory cytokines [28]. Senescence of ECs are, therefore, essential to have causal role in chronic inflammation and tissue remodeling. It was also demonstrated that aging-related loss of function in ECs is linked to telomere shortening [115]. In aortic aneurysm samples, elevated oxidative stress and telomere attrition were also shown in ECs as well [111]. Since replicative senescence is proven to be a common characteristic of aging, increased ROS levels and decreased NO in ECs cells contribute to the occurrence of vascular senescence [105].

In atherosclerosis patients, cellular senescence of VSMCs was also shown in atherosclerotic lesions of patients with coronary artery disease, AAA, and peripheral artery disease [115]. Another study demonstrated the SA-β Gal activity, p16, and p21 expression and IL-6 as SASP phenotype in SMCs of carotid artery plaques [118]. Additionally, angiotensin II treatment resulted in senescent vascular SMC generation in ApoE −/− mice suggests the relation of well-known stressors with senescence [119]. As articulated above, the switch of SMCs to a more synthetic form is partly explained by the impairment of TGF-β pathway. In stress conditions, SMCs was shown to have increased inducible NOS, ICAM-1 and angiotensinogen in aging process [105, 120]. Smooth muscle 22α protein is suggested to be marker for senescent SMCs and a recent study demonstrated that it contributes to senescent cell generation by the inhibition of p53 degradation [121]. Moreover, it was previously demonstrated that VSMCs from the aneurysms display oxidative DNA damage [111]. As expected, senescence of VSMCs has common senescent phenotype such as increased of pro-inflammatory cytokines, growth factors, and matrix metalloproteases all of which contribute to the vascular dysfunction.

Some of atherosclerotic plaques are also characterized by telomere shortening which is a hallmark of senescent phenotype [12]. VSMCs that are shown to be senescent by elevated p16 and p21 and SA-β-gal activity are reported to have telomere shortening in the atherosclerotic plaques. Shorter telomeres are suggested to be result of oxidative stress induced by DNA damage and leading senescence in VSMCs [122]. Since cellular senescence functions to initiate atherosclerosis process, it is not surprising that the decreased levels of telomeric repeat-binding factor-2 (Trf2) in VSMCs contributes to the plaque development in ApoE −/− mice. However, increased plaque growth by the knockout of senescence associated genes such as p53, p21, or p19Arf suggests the anti-atherosclerotic role of senescence [105, 123]. Relatively, it was demonstrated that decreased p16Ink4a and p14Arf expressions in human and mouse studies elevated the atherosclerosis development [86]. On the other hand, a very recent study of Childs et al. showed the deleterious effect of senescent cells in the progression of atherosclerosis by using both transgenic and pharmacological models [124]. In the study, the clearance of senescent cells in Ldlr−/− mice demonstrated the pivotal role of senescence during atherogenesis. It was suggested that even early stages of the plaques contain senescent cells and accumulate in the subendothelial space. Accumulation of senescent macrophages in the early atheroma leads to elevated inflammation factors and, therefore, contributes to the development of the atherogenesis. On the other hand, clearance of senescent foam cell macrophages was shown to reduce plaque formation. It was also suggested that in the severe plaques, senescent phenotype results in the plaque instability not only by inflammation factors and chemokines but also contributes to the plaque development by the increased expression of matrix metalloproteases and clearance of the senescent cells resulted in the regression of plaque growth and remodeling [124].

In addition to ECs and SMCs, immune cells show senescence phenotypic properties as well. A report conducted in old people who have the higher risk of heart diseases were shown to be characterized by the telomere attrition [125]. Another study showed elevated oxidative stress and inflammation factors in the monocytes of atherosclerotic patients [126]. A recent study also established the important role of cellular senescence for the pathogenesis of atherosclerosis in the macrophages by the deletion of senescent cells [124]. Additionally, it was previously shown that senescent macrophages are driven by p16 signaling pathway and cellular senescence has the central role in the occurrence of senescent phenotypes in the macrophages [127]. Collectively, studies have suggested the central role of cellular senescence in the atherosclerosis and therefore, senolytic agents as promising interventions for combating the disease.

10.8 Hypertension

Another well-known risk factor for cardiovascular diseases is hypertension, therefore, the association of vascular senescence and hypertension have gained interest in the recent years. The reports have suggested that typically increased blood pressure over time has the most significant effect on the development of hypertension within aging [128]. Other characteristics of hypertension such as vascular dysfunction, inflammation, extracellular matrix deposition are also common features of aging as well [129]. The presence of premature vascular aging was also shown in the young hypertensive individuals and it is suggested that hypertension fastens the vascular aging process similar to accelerated aging syndrome, progeria [130]. Therefore, it is proposed that understanding of the relationship of vascular deterioration with hypertension during aging is crucial.

The presence of senescent cells were demonstrated in vessels of the patients with hypertension. The study suggested the role of p53/p21 signaling in the telomere attrition by the uncapping of telomeres [131]. There is also growing evidence of animal studies that support the role of cellular senescence in hypertension. In a transgenic mouse model of aging produced by the defect of nucleotide excision repair genes, senescent ECs and VSMCs formation were shown to be increased along with elevated hypertension and vascular dysfunction [132]. Another report showed that hypertension resulted in the inhibition of cell cycle and activation of aortic p16 signaling both in the transgenic rat model and in humans [133]. Furthermore, the inhibition of NOS resulted in higher p16 levels in the arteries and in the development of hypertension suggesting the relation of cellular senescence with hypertension [134]. NO which acts as vasodilator contributes to modulation of blood pressure and accordingly has anti-hypertensive features. Several studies have shown that NO donor decreases the levels of senescent ECs and activates telomerase activity in aging process [135]. In normal conditions, endothelium releases NO that modulates vasodilation, however, in case of aging related hypertension, other molecules such as endothelin-1, angiotensin II and superoxide anions are secreted as well. These substances are known to lead to defects in the vasodilation exacerbating vascular remodeling through aging [136].

The common role of oxidative stress and inflammation has been explained in various cardiovascular pathology. Several inflammation factors including C-reactive protein, IL-6, TNF-α, and IL-1β are shown to be increased in hypertension patients [137]. On the other hand, since ROS affects vascular physiology, it is proposed to have essential effects on hypertension during aging by promoting cellular senescence of vascular cells [73, 74]. Clinical reports have revealed that vascular O_2^- generation supports hypertension development by the increased blood pressure [138]. In addition, other human studies have suggested the increased oxidative damage in plasma and urine samples [139] and increased levels of O_2^- and H_2O_2 in VSMCs in hypertension [140]. Various studies have demonstrated the role of ROS generated by NADPH oxidases (NOX) in vascular dysfunction through aging process (Fig. 10.2). Particularly, NOX1 and NOX2 were found to be elevated in old and hypertensive rat vessels and vascular dysfunction was reversed by NOX inhibition [141]. Therefore, among other oxidases that also generate ROS, NOXs are suggested to play key role for the hypertension related cardiovascular pathology [142]. These findings suggest that cellular senescence is driven by various factors takes a critical role in the development of hypertension and vascular remodeling.

10.9 Diabetes

Similar to other cardiovascular diseases, chronic inflammation is also related to obesity and suggested as a critical contributor to insulin resistance [143]. Clinical studies have demonstrated elevated SASP factors including IL-6, IL-8 and MCP-1 in obese patients [144, 145] and IL-6, IL-1 as the major markers of diabetes [146]. Another study demonstrated elevated SASP factors both in blood and in the vessels of diabetic patients [147]. The SASP factor MCP-1 is suggested to lead to macrophage infiltration related insulin resistance and responsible mechanism of this process is proposed to be autophagy [148].

A recent study also demonstrated that cellular senescence induced by p53 molecular pathway drives inflammation in diabetes [149]. In addition, animal studies have shown the presence of senescent vascular cells in the diabetes models and it is speculated that accumulated senescent cells in obesity could be the key drivers of diabetes [150]. In diabetic rat and mice, the presence of cellular senescence was also demonstrated in the aorta, suggesting that hyperglycemia may lead to senescent EC accumulation [151]. The mechanisms responsible for the vascular senescence are suggested to be increased oxidative stress, decreased NO levels and SIRT1 activation.

Dysfunction of adipose tissue is also proposed as a crucial contributor to cellular senescence linked diabetes. Accumulated senescent preadipose cells were confirmed both in young and old obese patients suggests the key role of senescence in adipogenic function [152]. SASP which is another senescent phenotype also results in the impairment of adipogenesis process and, therefore, results in insulin response [153]. It was also demonstrated that senescent cells lead to lipodystrophy in accelerated aging mouse model, however, consequences of senescent cell removal in diabetes have not been fully explained yet. In addition, adipose dysfunction could lead to fat deposition in various tissues such as liver and heart resulting in the pathogenesis of other diseases including atherosclerosis [154].

Senescent pancreatic β-cells are also suggested to play critical role in the development of type 2 diabetes. It was shown that high fat diet resulted in cellular senescence that induced β-cell function loss in mice [155]. Another study of mice demonstrated the association of type 2 diabetes with cell cycle inhibition which is a common phenotype of cellular senescence [156]. Furthermore, deletion of p53-dependent apoptosis in mice led to senescent cell accumulation and dysfunctional β-cells formation resulting in accelerated development of diabetes [157]. Although more evidence is required to clarify the contribution of cellular senescence on diabetes, several studied already provided information about the increased senescent cells in damaged tissues in diabetes process [155, 158]. It is suggested that accumulated senescent cells during diabetes represent a critical risk factor for the inflammatory state and, therefore, for cardiovascular diseases.

10.10 Future Directions for Therapy of Cardiovascular Diseases

Emerging data have identified the critical role of vascular senescence during aging for the development of cardiovascular diseases. Particularly, studies that have indicated the clearance of senescent cells extends both healthspan and lifespan suggests senolytic agents as a promising candidate for the therapy of age-associated cardiovascular pathology. The recent study by Childs et al. is breakthrough in the senolytic research for the treatment of atherosclerosis. In the study, the presence of senescent cells was shown to be detrimental at all stages of atherosclerosis and selective removal of these cells reversed the plaque formation. Furthermore, various studies have also supported the effect of selective clearance of senescent cells in the vascular homeostasis suggesting that senolytic agents hold a therapeutic paradigm [4].

The term senolytic is first described by Zhu et al. for the compounds that selectively kill senescent cells [159]. It was proposed that accumulation of senescent cells could be resulted from resistance to apoptosis [160] therefore, researchers have focused on the compounds that selectively target apoptosis of senescent cells. Some of the well-known pharmacological agents are already repurposed for senescent clearance. Of them, dasatinib is an approved drug for use in patients with chronic myelogenous leukemia that inhibits tyrosine kinase signaling [161]. The effect of dasantinib was shown in the clearance of senescent cell in preadipocytes cell culture. In addition, quercetin which is a bioflavonoid antioxidant was reported to kill senescent ECs in a selective way [162]. Morever, combination therapy of dasantinib and quercetin was also effective in the clearance of senescent cells in tissues of mice and increased healthspan via reduced cardiac damage in aged mice [159]. Inhibitor of anti-apoptotic proteins (ABT263) was identified as another potential senolytics which cleared senescent hematopoietic stem cells and reversed aging phenotype in mice [163]. However, the molecular pathways underlying the apoptosis of senescent cells has not been cleared yet and further studies are needed to determine the side effects of the therapy.

SASP is the senescent phenotype which transforms cells into a pro-inflammatory status and many reports have identified the damage of SASP to the tissues in aging-associated diseases. Since SASP secretes inflammatory molecules and contributes to the disease pathogenesis, it is suggested to be potential therapeutic target for the treatment of selective cellular senescence. SASP leads to an increase in the metabolic activity of SMCs and inhibitor of glycolysis was reported to clear senescent SMCs. However, the specificity of that treatment and its translation to humans is questioned to be safe and nontoxic [4, 164].

Another potential approach against cellular senescence related cardiovascular diseases would be targeting oxidative stress. Antiaging therapy against inhibition of cellular senescence such as sirtuins was previously studied and it was shown that increase in SIRT1 expression reduced the cellular senescence in SMCs and ECS resulting in lifespan extension [165, 166]. Another antioxidant that was suggested as a potential approach is NAD fueling that regulates metabolic pathways as well [167]. NAD therapy was suggested to lead to SMC translocation in addition to reducing accumulation of senescent SMC and ECs. Especially, migration of SMC in straight-line manner was indicated to be beneficial to suppress vascular damage [168]. Finally, it was stated that statins are also effective in the suppression of cellular senescence and DDR pathways in atherosclerosis [169].

Regarding the critical role of oxidative stress and inflammation in the senescence related cardiovascular diseases, it is important to clarify the underlying mechanisms. As mentioned in the text previously, the transgenic model SAMP8 has the accelerated senescence phenotype. Aged SAMP8 mice display elevated oxidative stress and ER stress, inflammation, vascular dysfunction features, therefore, it is suggested to be convenient model to investigate the vascular homeostasis in aging. Recently, exciting data have been obtained from the SAMP8 studies, though, more work is required to determine potential therapies for age-associated cardiovascular diseases [5, 60, 170].

10.11 Conclusion

Suggested pathological role of vascular senescence linked to oxidative stress in cardiovascular disorders such as heart failure, diabetes and atherosclerosis is explained throughout this chapter. Particularly, senescence of ECs and VSMCs cells have been shown to be important in the disease pathology however, immune cell senescence and pancreatic β-cells senescence received considerable attention as well. Recently, various studies have also focused on the selective clearance of senescent cells to reverse aging associated cardiac damage and the findings are promising and exciting. Especially, removal of cardiac senescent cells could be an emerging approach since vascular cells are pivotal in the maintenance of vascular homeostasis [105].

Oxidative stress has been suggested to be critical in the aging associated pathologies for a long time, however, recently, its potential effect on cellular senescence to contribute disease development has gained more interest. Various stresses including ROS generation and DDR were shown to have ultimate role on the cellular homeostasis leading to detrimental effects on the cardiac tissues. Therefore, antioxidant molecules such as quercetin have been repurposed to reverse the damage of cellular senescence in the recent years. In addition to antioxidants, various pharmacological agents have been recently studied as senolytic agents, though more evidence is required for the side effects of the senolytics before translation into the clinical studies [171]. On the other hand, approaches to suppress cellular senescence, rather than clearing senescent cells, are suggested to be another potential therapy. There has been no clinical trial for the therapy of senescent cell clearance in cardiovascular diseases, however, cell culture and animal studies have indicated its pivotal potential to hold a next generation therapy for vascular pathology.

Acknowledgment This work was supported by The Scientific and Technological Council of Turkey (TUBITAK) 2219 Grant Program.

References

1. Forman DE, Alexander K, Brindis RG, Curtis AB, Maurer M, Rich MW, Sperling L, Wenger NK (2016) Improved cardiovascular disease outcomes in older adults. F1000Res 5(F1000 Faculty Rev):112
2. Lakatta EG (2003) Arterial and cardiac aging: major shareholders in cardiovascular disease enterprises: part III: cellular and molecular clues to heart and arterial aging. Circulation 107:490–497
3. Lakatta EG (1993) Cardiovascular regulatory mechanisms in advanced age. Physiol Rev 73(2):413–467
4. Hao Y, Pickering GP (2016) Cellular senescence and vascular disease: novel routes to better understanding and therapy. Can J Cardiol 32:612–623
5. Forman K, Vara E, Garcia C, Kireev R, Cuesta S, Escames G, Tresguerres JA (2011) Effect of a combined treatment with growth hormone and melatonin in the cardiological aging on male SAMP8 mice. J Gerontol A Biol Sci Med Sci 66:823–834

6. Cevenini E, Caruso C, Candore G, Capri M, Nuzzo D, Duro G, Rizzo C, Colonna-Romano G, Lio D, Di Carlo D, Palmas MG, Scurti M, Pini E, Franceschi C, Vasto S (2010) Age-related inflammation: the contribution of different organs, tissues and systems. How to face it for therapeutic approaches. Curr Pharm Des 16:609–618
7. Lopez-Otin C, Blasco MA, Partridge L, Serrano M, Kroemer G (2013) The hallmarks of aging. Cell 153(6):1194–1217
8. de Cabo R, Carmona-Gutierrez D, Bernier M, Hall MN, Madeo F (2014) The search for antiaging interventions: from elixirs to fasting regimens. Cell 157(7):1515–1526
9. Hayflick L, Moorhead PS (1961) The serial cultivation of human diploid cell strains. Exp Cell Res 25:585–621
10. Storer M, Mas A, Robert-Moreno A et al (2013) Senescence is a developmental mechanism that contributes to embryonic growth and patterning. Cell 155:1119–1130
11. Childs BG, Durik M, Baker DJ, Deursen JMV (2015) Cellular senescence in aging and age-related disease: from mechanisms to therapy. Nat Med 21(12):1424–1435
12. Wang JC, Bennett M (2012) Aging and atherosclerosis: mechanisms, functional consequences, and potential therapeutics for cellular senescence. Circ Res 111(2):245–259
13. Yoshida Y, Shimizu I, Katsuumi G, Jiao S, Suda M, Hayashi Y et al (2015) p53-induced inflammation exacerbates cardiac dysfunction during pressure overload. J Mol Cell Cardiol 85:183–198
14. Yokoyama M, Okada S, Nakagomi A, Moriya J, Shimizu I, Nojima A et al (2014) Inhibition of endothelial p53 improves metabolic abnormalities related to dietary obesity. Cell Rep 7(5):1691–1703
15. North BJ, Sinclair DA (2012) The intersection between aging and cardiovascular disease. Circ Res 110(8):1097–1108
16. Drummond GR, Sobey CG (2014) Endothelial NADPH oxidases: which NOX to target in vascular disease? Trends Endocrinol Metab 25(9):452–463
17. Kovacic JC, Moreno P, Nabel EG, Hachinski V, Fuster V (2011) Cellular senescence, vascular disease, and aging: part 2 of a 2-part review: clinical vascular disease in the elderly. Circulation 123:1900–1910
18. Corti R, Fuster V, Badimon JJ (2003) Pathogenetic concepts of acute coronary syndromes. J Am Coll Cardiol 41:7S–14S
19. Chen B, Lu Y, Chen Y, Cheng J (2015) The role of Nrf2 in oxidative stress-induced endothelial injuries. J Endocrinol 225:R83–R99
20. Reiter RJ, Tan DX, Galano A (2014) Melatonin: exceeding expectations. Physiology (Bethesda) 29:325–333
21. Campisi J (2013) Aging, cellular senescence, and cancer. Annu Rev Physiol 75:685–705
22. Bodnar AG, Ouellette M, Frolkis M et al (1998) Extension of life-span by introduction of telomerase into normal human cells. Science 279:349–352
23. Tan FC, Hutchison ER, Eitan E et al (2014) Are there roles for brain cell senescence in aging and neurodegenerative disorders? Biogerontology 15:643–660
24. van Deursen JM (2014) The role of senescent cells in ageing. Nature 509:439–446
25. Dimri GP (2005) What has senescence got to do with cancer? Cancer Cell 7:505–512
26. Correia-Melo C, Hewitt G, Passos JF (2014) Telomeres, oxidative stress and inflammatory factors: partners in cellular senescence? Longev Healthspan 3(1):1
27. Hewitt G, Jurk D, Marques FD et al (2012) Telomeres are favoured targets of a persistent DNA damage response in ageing and stress-induced senescence. Nat Commun 3:708
28. Tchkonia T, Zhu Y, van Deursen J, Campisi J, Kirkland JL (2013) Cellular senescence and the senescent secretory phenotype: therapeutic opportunities. J Clin Invest 123:966–972
29. Barascu A, Le Chalony C, Pennarun G et al (2012) Oxidative stress induces an ATM-independent senescence pathway through p38 MAPK-mediated lamin B1 accumulation. EMBO J 31:1080–1094
30. Sebastian C, Lloberas J, Celada A (2009) Molecular and cellular aspects of macrophage aging. In: Fulop T (ed) Handbook on immunosenescence. Springer, Dordrecht, pp 919–945

31. Acosta JC, Banito A, Wuestefeld T et al (2013) A complex secretory program orchestrated by the inflammasome controls paracrine senescence. Nat Cell Biol 15:978–990
32. Stewart SA, Weinberg RA (2006) Telomeres: cancer to human aging. Annu Rev Cell Dev Biol 22:531–557
33. Lane DP (1992) Cancer. p53, guardian of the genome. Nature 358(6381):15–16
34. Green DR, Chipuk JE (2006) p53 and metabolism: inside the TIGAR. Cell 126(1):30–32
35. Sano M, Minamino T, Toko H, Miyauchi H, Orimo M, Qin Y et al (2007) p53- induced inhibition of Hif-1 causes cardiac dysfunction during pressure overload. Nature 446(7134):444–448
36. Chen Q, Fischer A, Reagan JD, Yan LJ, Ames BN (1995) Oxidative DNA damage and senescence of human diploid fibroblast cells. Proc Natl Acad Sci U S A 92:4337–4341
37. Jeyapalan JC, Sedivy JM (2008) Cellular senescence and organismal aging. Mech Ageing Dev 129:467–474
38. Beausejour CM, Krtolica A, Galimi F, Narita M, Lowe SW, Yaswen P et al (2003) Reversal of human cellular senescence: roles of the p53 and p16 pathways. EMBO J 22(16):4212–4222
39. Campisi J (2005) Senescent cells, tumor suppression, and organismal aging: good citizens, bad neighbors. Cell 120(4):513–522
40. Voghel G, Thorin-Trescases N, Farhat N et al (2007) Cellular senescence in endothelial cells from atherosclerotic patients is accelerated by oxidative stress associated with cardiovascular risk factors. Mech Ageing Dev 128:662–671
41. Voghel G, Thorin-Trescases N, Farhat N et al (2008) Chronic treatment with N-acetyl-cystein delays cellular senescence in endothelial cells isolated from a subgroup of atherosclerotic patients. Mech Ageing Dev 129:261–270
42. Varela I, Cadinanos J, Pendas AM, Gutierrez-Fernandez A, Folgueras AR, Sanchez LM et al (2005) Accelerated ageing in mice deficient in Zmpste24 protease is linked to p53 signalling activation. Nature 437(7058):564–568
43. Matheu A, Maraver A, Klatt P, Flores I, Garcia-Cao I, Borras C et al (2007) Delayed ageing through damage protection by the Arf/p53 pathway. Nature 448(7151):375–379
44. Garcia-Cao I, Garcia-Cao M, Martin-Caballero J, Criado LM, Klatt P, Flores JM et al (2002) "Super p53" mice exhibit enhanced DNA damage response, are tumor resistant and age normally. EMBO J 21(22):6225–6235
45. Tomas-Loba A, Flores I, Fernandez-Marcos PJ, Cayuela ML, Maraver A, Tejera A et al (2008) Telomerase reverse transcriptase delays aging in cancer-resistant mice. Cell 135(4):609–622
46. Baker DJ, Weaver RL, van Deursen JM (2013) p21 both attenuates and drives senescence and aging in BubR1 progeroid mice. Cell Rep 3(4):1164–1174
47. Dimri GP, Lee X, Basile G et al (1995) A biomarker that identifies senescent human cells in culture and in aging skin in vivo. Proc Natl Acad Sci U S A 92:9363–9367
48. Salama R, Sadaie M, Hoare M, Narita M (2014) Cellular senescence and its effector programs. Genes Dev 28(2):99–114
49. Carpi A, Menabo R, Kaludercic N et al (2009) The cardioprotective effects elicited by p66(Shc) ablation demonstrate the crucial role of mitochondrial ROS formation in ischemia/reperfusion injury. Biochim Biophys Acta 1787:774–780
50. Fan Q, Chen L, Cheng S, Li F, Lau WB, Wang Le F et al (2014) Aging aggravates nitrate-mediated ROS/RNS changes. Oxidative Med Cell Longev 376515:2014
51. Correia-Melo C, Marques FD, Anderson R et al (2016) Mitochondria are required for proageing features of the senescent phenotype. EMBO J 35:724–742
52. Chen Q, Fischer A, Reagan JD et al (1995) Oxidative DNA damage and senescence of human diploid fibroblast cells. Proc Natl Acad Sci U S A 92:4337–4341
53. Parrinello S, Samper E, Krtolica A et al (2003) Oxygen sensitivity severely limits the replicative lifespan of murine fibroblasts. Nat Cell Biol 5:741–747
54. Colavitti R, Finkel T (2005) Reactive oxygen species as mediators of cellular senescence. IUBMB Life 57:277–281
55. d'Adda di Fagagna F, Reaper PM, Clay-Farrace L et al (2003) A DNA damage checkpoint response in telomere-initiated senescence. Nature 426:194–198

56. Passos JF, Nelson G, Wang C et al (2010) Feedback between p21 and reactive oxygen production is necessary for cell senescence. Mol Syst Biol 6:347
57. Wiley CD, Velarde MC, Lecot P et al (2016) Mitochondrial dysfunction induces senescence with a distinct secretory phenotype. Cell Metab 23:303–314
58. Kuilman T, Peeper DS (2009) Senescence-messaging secretome: SMS-ing cellular stress. Nat Rev Cancer 9:81–94
59. Bayram B, Ozcelik B, Grimm S, Roeder T, Schrader C, Ernst IM, Wagner AE, Grune T, Frank J, Rimbach G (2012) A diet rich in olive oil phenolics reduces oxidative stress in the heart of SAMP8 mice by induction of Nrf2-dependentgene expression. Rejuvenation Res 15:71–81
60. Karuppagounder V, Arumugam S, Thandavarayan RA, Sreedhar R, Giridharan VV, Watanabe K (2016) Molecular targets of quercetin with anti-inflammatory properties in atopic dermatitis. Drug Discov Today 21:632–639
61. Rodriguez MI, Carretero M, Escames G, Lopez LC, Maldonado MD, Tan DX, Reiter RJ, Acuna-Castroviejo D (2007) Chronic melatonin treatment prevents age-dependent cardiac mitochondrial dysfunction in senescence-accelerated mice. Free Radic Res 41:15–24
62. Miquel J (1998) An update on the oxygen stress-mitochondrial mutation theory of aging: genetic and evolutionary implications. Exp Gerontol 33:113–126
63. Spencer NF, Poynter ME, Im SY, Daynes RA (1997) Constitutive activation of NF-kappa B in an animal model of aging. Int Immunol 9(10):1581–1588
64. Ziegler DV, Wiley CD, Velarde MC (2015) Mitochondrial effectors of cellular senescence: beyond the free radical theory of aging. Aging Cell 14:1–7
65. Nakahara H, Kanno T, Inai Y, Utsumi K, Hiramatsu M, Mori A, Packer L (1998) Mitochondrial dysfunction in the senescence accelerated mouse (SAM). Free Radic Biol Med 24:85–92
66. Sreedhar R, Giridharan VV, Arumugam S, Karuppagounder V, Palaniyandi SS, Krishnamurthy P, Quevedo J, Watanabe K, Konishi T, Thandavarayan RA (2016) Role of MAPK-mediated endoplasmic reticulum stress signaling in the heart during aging in senescence-accelerated prone mice. Biofactors 42(4):368–375
67. Gomes AP, Price NL, Ling AJ et al (2013) Declining NAD(+) induces a pseudohypoxic state disrupting nuclear-mitochondrial communication during aging. Cell 155:1624–1638
68. Ruderman NB, Xu XJ, Nelson L et al (2010) AMPK and SIRT1: a long-standing partnership? Am J Physiol Endocrinol Metab 298:E751–E760
69. Zhang H, Ryu D, Wu Y et al (2016) NAD+ repletion improves mitochondrial and stem cell function and enhances life span in mice. Science 352(6292):1436–1443
70. Passos JF, Saretzki G, Ahmed S et al (2007) Mitochondrial dysfunction accounts for the stochastic heterogeneity in telomere-dependent senescence. PLoS Biol 5:e110
71. Blackburn EH, Epel ES, Lin J (2015) Human telomere biology: a contributory and interactive factor in aging, disease risks, and protection. Science 350:1193–1198
72. Rodriguez MI, Escames G, Lopez LC, Garcia JA, Ortiz F, Lopez A, Acuna-Castroviejo D (2007) Melatonin administration prevents cardiac and diaphragmatic and diaphragmatic mitochondrial oxidative damage in senescence-accelerated mice. J Endocrinol 194:637–643
73. Harman D (1956) Aging: a theory based on free radical and radiation chemistry. J Gerontol 11:298–300
74. Harman D (2006) Free radical theory of aging: an update: increasing the functional lifespan. Ann N Y Acad Sci 1067:10–21
75. Tatchum-Talom R, Martin DS (2004) Tempol improves vascular function in the mesenteric vascular bed of senescent rats. Can J Physiol Pharmacol 82(3):200–207
76. Ungvari Z, Bailey-Downs L, Sosnowska D, Gautam T, Koncz P, Losonczy G (2011) Vascular oxidative stress in aging: a homeostatic failure due to dysregulation of NRF2-mediated antioxidant response. AmJ Physiol Heart Circ Physiol 301:H363–H372
77. Zhao W, Zheng XL, Peng DQ, Zhao SP (2015) Myocyte enhancer factor 2A regulates hydrogen peroxide-induced senescence of vascular smooth muscle cells via microRNA-143. J Cell Physiol 230(9):2202–2211

78. Fyhrquist F, Saijonmaa O, Strandberg T (2013) The roles of senescence and telomere shortening in cardiovascular disease. Nat Rev Cardiol 10:274–283
79. Minamino T, Miyauchi H, Yoshida T et al (2004) The role of vascular cell senescence in atherosclerosis: antisenescence as a novel therapeutic strategy for vascular aging. Curr Vasc Pharmacol 2:141–148
80. Migliaccio E, Giorgio M, Mele S et al (1999) The p66shc adaptor protein controls oxidative stress response and life span in mammals. Nature 402:309–313
81. Suski JM, Karkucinska-Wieckowska A, Lebiedzinska M et al (2011) p66Shc aging protein in control of fibroblasts cell fate. Int J Mol Sci 12:5373–5389
82. Cosentino F, Francia P, Camici GG et al (2008) Final common molecular pathways of aging and cardiovascular disease: role of the p66Shc protein. Arterioscler Thromb Vasc Biol 28:622–628
83. Franzeck FC, Hof D, Spescha RD et al (2012) Expression of the aging gene p66Shc is increased in peripheral blood monocytes of patients with acute coronary syndrome but not with stable coronary artery disease. Atherosclerosis 220:282–286
84. Rodríguez-Mañas L, El-Assar M, Vallejo S, López-Dóriga P, Solís J, Petidier R, Montes M, Nevado J, Castro M, Gómez-Guerrero C (2009) Endothelial dysfunction in aged humans is related with oxidative stress and vascular inflammation. Aging Cell 8(3):226–238
85. Mavrogonatou DE, Sklirou A, Kletsas D (2015) Oxidative stress inhibits the proliferation, induces premature senescence and promotes a catabolic phenotype in human nucleus pulposus intervertebral disc cells. Eur Cell Mater 30:89
86. Muñozespín D, Serrano M (2014) Cellular senescence: from physiology to pathology. Nat Rev Mol Cell Biol 15(7):482–496
87. Ke Y, Li D, Zhao M et al (2018) Gut flora-dependent metabolite Trimethylamine-N-oxide accelerates endothelial cell senescence and vascular aging through oxidative stress. Free Radic Biol Med 116:88–100
88. Agabiti-Rosei C et al (2017) Effect of long-term treatment with melatonin on vascular markers of oxidative stress/inflammation and on the anticontractile activity of perivascular fat in aging mice. Hypertens Res 40:41–50
89. Borlaug BA (2014) The pathophysiology of heart failure with preserved ejection fraction. Nat Rev Cardiol 11(9):507–515
90. Shih H, Lee B, Lee RJ, Boyle AJ (2011) The aging heart and post-infarction left ventricular remodeling. J Am Coll Cardiol 57(1):9–17
91. Mohammed SF, Redfield MM (2015) Response to letters regarding article, "coronary microvascular rarefaction and myocardial fibrosis in heart failure with preserved ejection fraction". Circulation 132(16):e206
92. Paulus WJ, Tschope C (2013) A novel paradigm for heart failure with preserved ejection fraction: comorbidities drive myocardial dysfunction and remodeling through coronary microvascular endothelial inflammation. J Am Coll Cardiol 62(4):263–271
93. Rebrin I, Zicker S, Wedekind KJ, Paetau-Robinson I, Packer L, Sohal RS (2005) Effect of antioxidant-enriched diets on glutathione redox status in tissue homogenates and mitochondria of the senescence-accelerated mouse. Free Radic Biol Med 39:549–557
94. Cohn JN, Levine TB, Olivari MT, Garberg V, Lura D, Francis GS et al (1984) Plasma norepinephrine as a guide to prognosis in patients with chronic congestive heart failure. N Engl J Med 311(13):819–823
95. Gogiraju R, Xu X, Bochenek ML, Steinbrecher JH, Lehnart SE, Wenzel P et al (2015) Endothelial p53 deletion improves angiogenesis and prevents cardiac fibrosis and heart failure induced by pressure overload in mice. J Am Heart Assoc 4(2):e001770
96. Morgan RG, Ives SJ, Lesniewski LA, Cawthon RM, Andtbacka RH, Noyes RD et al (2013) Age-related telomere uncapping is associated with cellular senescence and inflammation independent of telomere shortening in human arteries. Am J Physiol Heart Circ Physiol 305(2):H251–H258
97. Satoh M, Ishikawa Y, Takahashi Y, Itoh T, Minami Y, Nakamura M (2008) Association between oxidative DNA damage and telomere shortening in circulating endothelial progenitor cells

obtained from metabolic syndrome patients with coronary artery disease. Atherosclerosis 198(2):347–353
98. van der Harst P, van der Steege G, de Boer RA, Voors AA, Hall AS, Mulder MJ et al (2007) Telomere length of circulating leukocytes is decreased in patients with chronic heart failure. J Am Coll Cardiol 49(13):1459–1464
99. Haycock PC, Heydon EE, Kaptoge S, Butterworth AS, Thompson A, Willeit P (2014) Leucocyte telomere length and risk of cardiovascular disease: systematic review and meta-analysis. BMJ 349(3):g4227
100. Salpea KD, Humphries SE (2010) Telomere length in atherosclerosis and diabetes. Atherosclerosis 209(1):35–38
101. Gevaert AB, Shakeri H, Leloup AJ, van Hove CE, de Meyer GRY, Vrints CJ et al (2017) Endothelial senescence contributes to heart failure with preserved ejection fraction in an aging mouse model. Circ Heart Fail 10(6):e003806
102. Navab M, Berliner JA, Watson AD et al (1996) The Yin and Yang of oxidation in the development of the fatty streak. A review based on the 1994 George Lyman Duff Memorial Lecture. Arterioscler Thromb Vasc Biol 16:831–842
103. Spina M, Garbisa S, Hinnie J, Hunter JC, Serafini-Fracassini A (1983) Age-related changes in composition and mechanical properties of the tunica media of the upper thoracic human aorta. Arteriosclerosis 3(1):64–76
104. Gerstenblith G, Frederiksen J, Yin FC, Fortuin NJ, Lakatta EG, Weisfeldt ML (1977) Echocardiographic assessment of a normal adult aging population. Circulation 56(2):273–278
105. Katsuumi G, Shimizu I, Yoshida Y, Minamino T (2018) Vascular senescence in cardiovascular and metabolic diseases. Front Cardiovasc Med 5:18
106. Chien Y, Scuoppo C, Wang X et al (2011) Control of the senescence-associated secretory phenotype by NF-kappaB promotes senescence and enhances chemosensitivity. Genes Dev 25:2125–2136
107. Freund A, Patil CK, Campisi J (2011) p38MAPK is a novel DNA damage response-independent regulator of the senescence-associated secretory phenotype. EMBO J 30:1536–1548
108. Csiszar A, Wang M, Lakatta EG et al (2008) Inflammation and endothelial dysfunction during aging: role of NF-kappaB. J Appl Physiol (1985) 105:1333–1341
109. Eren M, Boe AE, Murphy SB et al (2014) PAI-1-regulated extracellular proteolysis governs senescence and survival in Klotho mice. Proc Natl Acad Sci U S A 111:7090–7095
110. Kuro-o M (2008) Klotho as a regulator of oxidative stress and senescence. Biol Chem 389:233–241
111. Cafueri G, Parodi F, Pistorio A, Bertolotto M, Ventura F, Gambini C et al (2012) Endothelial and smooth muscle cells from abdominal aortic aneurysm have increased oxidative stress and telomere attrition. PLoS One 7(4):e35312
112. Gorenne I, Kavurma M, Scott S, Bennett M (2006) Vascular smooth muscle cell senescence in atherosclerosis. Cardiovasc Res 72:9–17
113. Wang J, Uryga AK, Reinhold J et al (2015) Vascular smooth muscle cell senescence promotes atherosclerosis and features of plaque vulnerability. Circulation 132:1909–1919
114. Herbert KE, Mistry Y, Hastings R, Poolman T, Niklason L, Williams B (2008) Angiotensin II-mediated oxidative DNA damage accelerates cellular senescence in cultured human vascular smooth muscle cells via telomere dependent and independent pathways. Circ Res 102(2):201–208
115. Minamino T, Miyauchi H, Yoshida T et al (2002) Endothelial cell senescence in human atherosclerosis: role of telomere in endothelial dysfunction. Circulation 105:1541–1544
116. Lin JR, Shen WL, Yan C, Gao PJ (2015) Downregulation of dynamin-related protein 1 contributes to impaired autophagic flux and angiogenic function in senescent endothelial cells. Arterioscler Thromb Vasc Biol 35(6):1413–1422
117. Warboys CM, de Luca A, Amini N, Luong L, Duckles H, Hsiao S et al (2014) Disturbed flow promotes endothelial senescence via a p53-dependent pathway. Arterioscler Thromb Vasc Biol 34(5):985–995

118. Gardner SE, Humphry M, Bennett MR, Clarke MCH (2016) Senescent vascular smooth muscle cells drive inflammation through an interleukin-1α-dependent senescence-associated secretory phenotype. Atherosclerosis 244:e5
119. Kunieda T, Minamino T, Nishi J et al (2006) Angiotensin II induces premature senescence of vascular smooth muscle cells and accelerates the development of atherosclerosis via a p21-dependent pathway. Circulation 114:953–960
120. Faggiotto A, Ross R, Harker L (1984) Studies of hypercholesterolemia in the nonhuman primate. I. Changes that lead to fatty streak formation. Arteriosclerosis 4(4):323–340
121. Miao SB, Xie XL, Yin YJ, Zhao LL, Zhang F, Shu YN et al (2017) Accumulation of smooth muscle 22α protein accelerates senescence of vascular smooth muscle cells via stabilization of p53 in vitro and in vivo. Arterioscler Thromb Vasc Biol 37(10):1849–1859. https://doi.org/10.1161/ATVBAHA.117.309378
122. Matthews C, Gorenne I, Scott S, Figg N, Kirkpatrick P, Ritchie A et al (2006) Vascular smooth muscle cells undergo telomere-based senescence in human atherosclerosis: effects of telomerase and oxidative stress. Circ Res 99(2):156–164
123. Mercer J, Figg N, Stoneman V, Braganza D, Bennett MR (2005) Endogenous p53 protects vascular smooth muscle cells from apoptosis and reduces atherosclerosis in ApoE knockout mice. Circ Res 96:667–674
124. Childs BG, Baker DJ, Wijshake T, Conover CA, Campisi J, van Deursen JM (2016) Senescent intimal foam cells are deleterious at all stages of atherosclerosis. Science 354(6311):472–477
125. Cawthon RM, Smith KR, O'Brien E, Sivatchenko A, Kerber RA (2003) Association between telomere length in blood and mortality in people aged 60 years or older. Lancet 361(9355):393–395
126. Calvert PA, Liew TV, Gorenne I, Clarke M, Costopoulos C, Obaid DR et al (2011) Leukocyte telomere length is associated with high-risk plaques on virtual histology intravascular ultrasound and increased proinflammatory activity. Arterioscler Thromb Vasc Biol 31(9):2157–2164
127. Cudejko C, Wouters K, Fuentes L, Hannou SA, Paquet C, Bantubungi K et al (2011) p16INK4a deficiency promotes IL-4-induced polarization and inhibits proinflammatory signaling in macrophages. Blood 118(9):2556–2566
128. Wang M, Monticone RE, Lakatta EG (2010) Arterial aging: a journey into subclinical arterial disease. Curr Opin Nephrol Hypertens 19(2):201–207
129. Barja G (2013) Updating the mitochondrial free radical theory of aging: an integrated view, key aspects, and confounding concepts. Antioxid Redox Signal 19(12):1420–1445
130. Kotsis V, Stabouli S, Karafillis I, Nilsson P (2011) Early vascular aging and the role of central blood pressure. J Hypertens 29(10):1847–1853
131. Morgan RG, Ives SJ, Walker AE, Cawthon RM, Andtbacka RH, Noyes D et al (2014) Role of arterial telomere dysfunction in hypertension: relative contributions of telomere shortening and telomere uncapping. J Hypertens 32(6):1293–1299
132. Durik M, Kavousi M, van der Pluijm I, Isaacs A, Cheng C, Verdonk K et al (2012) Nucleotide excision DNA repair is associated with age-related vascular dysfunction. Circulation 126(4):468–478. https://doi.org/10.1161/CIRCULATIONAHA.112.104380
133. Westhoff JH, Hilgers KF, Steinbach MP, Hartner A, Klanke B, Amann K et al (2008) Hypertension induces somatic cellular senescence in rats and humans by induction of cell cycle inhibitor p16INK4a. Hypertension 52(1):123–129
134. Boe AE, Eren M, Murphy SB, Kamide CE, Ichimura A, Terry D et al (2013) Plasminogen activator inhibitor-1 antagonist TM5441 attenuates Nω-nitro-L-arginine methyl ester-induced hypertension and vascular senescence. Circulation 128(21):2318–2324
135. Vasa M, Breitschopf K, Zeiher AM, Dimmeler S (2000) Nitric oxide activates telomerase and delays endothelial cell senescence. Circ Res 87(7):540–542
136. Rubio-Ruiz ME, Perez-Torres I, Soto ME, Pastelin G, Guarner-Lans V (2014) Aging in blood vessels. Medicinal agents FOR systemic arterial hypertension in the elderly. Ageing Res Rev 18:132–147

137. Singh T, Newman AB (2011) Inflammatory markers in population studies of aging. Ageing Res Rev 10:319–329
138. Nabha L, Garbern JC, Buller CL, Charpie JR (2005) Vascular oxidative stress precedes high blood pressure in spontaneously hypertensive rats. Clin Exp Hypertens 27:71–82
139. Ward NC, Hodgson JM, Puddey IB, Mori TA, Beilin LJ, Croft KD (2004) Oxidative stress in human hypertension: association with antihypertensive treatment, gender, nutrition, and lifestyle. Free Radic Biol Med 36:226–232
140. Touyz RM, Schiffrin EL (2001) Increased generation of superoxide by angiotensin II in smooth muscle cells from resistance arteries of hypertensive patients: role of phospholipase D-dependent NAD(P)H oxidase-sensitive pathways. J Hypertens 19:1245–1254
141. Wind S, Beuerlein K, Armitage ME, Taye A, Kumar AH, Janowitz D et al (2010) Oxidative stress and endothelial dysfunction in aortas of aged spontaneously hypertensive rats by NOX1/2 is reversed by NADPH oxidase inhibition. Hypertension 56(3):490–497
142. Montezano AC, Burger D, Ceravolo GS, Yusuf H, Montero M, Touyz RM (2011) Novel Nox homologues in the vasculature: focusing on Nox4 and Nox5. Clin Sci (Lond) 120(4):131–141
143. Esser N, Legrand-Poels S, Piette J, Scheen AJ, Paquot N (2014) Inflammation as a link between obesity, metabolic syndrome and type 2 diabetes. Diabetes Res Clin Pract 105:141–150
144. Kim CS, Park HS, Kawada T et al (2006) Circulating levels of MCP-1 and IL-8 are elevated in human obese subjects and associated with obesity-related parameters. Int J Obes (Lond) 30:1347–1355
145. Utsal L, Tillmann V, Zilmer M et al (2012) Elevated serum IL-6, IL-8, MCP-1, CRP, and IFN-g levels in 10- to 11-year-old boys with increased BMI. Horm Res Paediatr 78:31–39
146. Pradhan AD, Manson JE, Rifai N, Buring JE, Ridker PM (2001) C-reactive protein, interleukin 6, and risk of developing type 2 diabetes mellitus. JAMA 286:327–334
147. Schneider DJ, Sobel BE (2012) PAI-1 and diabetes: a journey from the bench to the bedside. Diabetes Care 35:1961–1967
148. Xu H, Barnes GT, Yang Q et al (2003) Chronic inflammation in fat plays a crucial role in the development of obesity-related insulin resistance. J Clin Invest 112:1821–1830
149. Stern DM, Yan SD, Yan SF, Schmidt AM (2002) Receptor for advanced glycation endproducts (RAGE) and the complications of diabetes. Ageing Res Rev 1:1–15
150. Palmer AK et al (2015) Cellular senescence in type 2 diabetes: a therapeutic opportunity. Diabetes 64:2289–2298
151. Chen J, Brodsky SV, Goligorsky DM et al (2002) Glycated collagen I induces premature senescence-like phenotypic changes in endothelial cells. Circ Res 90:1290–1298
152. Tchkonia T, Morbeck DE, Von Zglinicki T et al (2010) Fat tissue, aging, and cellular senescence. Aging Cell 9:667–684
153. Hamm JK, el Jack AK, Pilch PF, Farmer SR (1999) Role of PPAR gamma in regulating adipocyte differentiation and insulin-responsive glucose uptake. Ann N Y Acad Sci 892:134–145
154. Tchkonia T, Corkey BE, Kirkland JL (2006) Current views of the fat cell as an endocrine cell: lipotoxicity. In: Bray GA, Ryan DH (eds) Overweight and the metabolic syndrome: from bench to bedside (endocrine updates). Springer, New York, pp 105–118
155. Sone H, Kagawa Y (2005) Pancreatic beta cell senescence contributes to the pathogenesis of type 2 diabetes in high-fat diet-induced diabetic mice. Diabetologia 48:58–67
156. Uchida T, Nakamura T, Hashimoto N et al (2005) Deletion of Cdkn1b ameliorates hyperglycemia by maintaining compensatory hyperinsulinemia in diabetic mice. Nat Med 11:175–182
157. Tavana O, Puebla-Osorio N, Sang M, Zhu C (2010) Absence of p53-dependent apoptosis combined with nonhomologous end-joining deficiency leads to a severe diabetic phenotype in mice. Diabetes 59:135–142
158. Verzola D, Gandolfo MT, Gaetani G et al (2008) Accelerated senescence in the kidneys of patients with type 2 diabetic nephropathy. Am J Physiol Renal Physiol 295:F1563–F1573
159. Zhu Y, Tchkonia T, Pirtskhalava T et al (2015) The Achilles' heel of senescent cells: from transcriptome to senolytic drugs. Aging Cell 14:644–658
160. Wang E (1995) Senescent human fibroblasts resist programmed cell death, and failure to suppress bcl2 is involved. Cancer Res 55:2284–2292

161. Montero JC, Scoanc S, Ocana A, Pandiella A (2011) Inhibition of SRC family kinases and receptor tyrosine kinases by dasatinib: possible combinations in solid tumors. Clin Cancer Res 17:5546–5552
162. Bruning A (2013) Inhibition of mTOR signaling by quercetin in cancer treatment and prevention. Anti Cancer Agents Med Chem 13:1025–1031
163. Chang J, Wang Y, Shao L, Laberge RM, Demaria M, Campisi J et al (2016) Clearance of senescent cells by ABT263 rejuvenates aged hematopoietic stem cells in mice. Nat Med 22(1):78–83
164. Raez LE, Papadopoulos K, Ricart AD et al (2013) A phase I dose-escalation trial of 2-deoxy-D-glucose alone or combined with docetaxel in patients with advanced solid tumors. Cancer Chemother Pharmacol 71:523–530
165. van der Veer E, Ho C, O'Neil C et al (2007) Extension of human cell lifespan by nicotinamide phosphoribosyltransferase. J Biol Chem 282:10841–10845
166. Borradaile NM, Pickering JG (2009) Nicotinamide phosphoribosyltransferase imparts human endothelial cells with extended replicative lifespan and enhanced angiogenic capacity in a high glucose environment. Aging Cell 8:100–112
167. Canto C, Auwerx J (2012) Targeting sirtuin 1 to improve metabolism: all you need is NAD(ş)? Pharmacol Rev 64:166–187
168. Yin H, van der Veer E, Frontini MJ et al (2012) Intrinsic directionality of migrating vascular smooth muscle cells is regulated by NAD(ş)biosynthesis. J Cell Sci 125:5770–5780
169. Mahmoudi M, Gorenne I, Mercer J et al (2008) Statins use a novel Nijmegen breakage syndrome-1-dependent pathway to accelerate DNA repair in vascular smooth muscle cells. Circ Res 103:717–725
170. Dong W, Quo W, Wang F, Li C, Xie Y, Zheng X, Shi H (2015) Electroacupuncture upregulates SIRT1-Dependent PGC-1alpha expression in SAMP8 mice. Med Sci Monit 21:3356–3362
171. Pole A, Dimri M, Dimri G (2016) Oxidative stress, cellular senescence and aging. AIMS Mol Sci 3(3):300–324

Influence of Genetic Factor on Oxidative Stress Mediated Heart Damage

11

Branislav Rovcanin

11.1 Introduction

Oxidative stress develops as a result of the imbalance between enhanced production of free radicals (prooxidants) and decreased efficiency of antioxidative mechanisms. The initiation and progression of heart diseases are greatly dependent upon oxidative stress, since the cardiovascular system represents one of the targets for prooxidative injury. Coronary blood vessels and cardiomyocytes can be significantly damaged in oxidative stress conditions, leading to the further aggravation and disease onset [1]. The tissue damage caused by oxidative stress can result as a direct effect of reactive oxygen species (ROS) on biomolecules and may also achieve delicate effects on the modulation of cellular metabolic, signaling, transport and survival processes [2, 3]. The susceptibility to oxidative stress mediated heart damage is greatly determined by genetic factors which often represent functional gene polymorphisms of enzymes that control the extent of redox metabolism.

We are on the edge of comprehending the welfares of Precision Medicine in cancer treatment through the application of genetic data to guide the selection of specific therapy [4]. Modern genotyping technology is continuously providing an insight into underlying genetic mechanisms of cardiovascular diseases. The ability of clinical trials that evaluate the genetic risk factors for cardiovascular diseases show a proof-of-principle for the novel treatment benefits and they are currently above the use of genetics in the prevention of cardiovascular diseases [5]. Clinically severe cardiovascular phenotypes usually develop as a result of a single gene disorder and their inheritance pattern is monogenic (Mendelian forms of hypertension) while the majority of diseases includes the polygenic contribution. The development of common heart diseases represents the multifactorial model and it encompasses an interaction between various genetic and environmental factors. It has been

B. Rovcanin (✉)
Center for Endocrine Surgery, Clinical Center of Serbia, Faculty of Medicine,
University of Belgrade, Belgrade, Serbia

S. Chakraborti et al. (eds.), *Oxidative Stress in Heart Diseases*,
https://doi.org/10.1007/978-981-13-8273-4_11

demonstrated that there is a considerable inter individual variability in the relative contribution of genetic and environmental factors for disease development and symptom expression [6]. Using the studies which rely on animal models and its translation to human pathological entities enabled the identification of candidate genes responsible for susceptibility to various heart diseases. A large-scale genome-wide association studies (GWAS) are used in identification of genes accountable for increased heart disease risk and they are placed into quality control criteria for genotyping and phenotyping [7]. Enhancements in genome sciences provided the development of DNA microarray methodology which owns the high-throughput power in evaluation of gene sequence and their expression. By cDNA microarray chips it is possible to evaluate the simultaneous expression of a vast number of genes and to acquire the information which genes are down-regulated or up-regulated in various physiological and pathological conditions. This technology is especially useful in examination of global events in genomic expression during the heart disease onset and progression both *in vivo* and *in vitro* [8]. Moreover, epigenetic mechanisms which include regulation by microRNAs and chemical modifications of DNA and histones modulate the gene expression without a presence of DNA mutation. These mechanisms and DNA re-arrangements, such as copy-number variants are also recognized as contributing factors to the genetic basis of oxidative stress mediated heart damage [9].

The aim of this paper was to present the most important knowledge about the genetics of oxidative stress that influences the predisposition and the extent of cardiac damage in different pathological entities. Next to the genetics of the most common heart disease, the heritable basis of oxidative stress in atrial fibrillation and congestive heart failure were also presented (Fig. 11.1).

11.2 Gene Polymorphisms Associated with Common Heart Diseases

Some gene is said to be polymorphic if it has more than one allele in certain population. These polymorphisms can have a major impact on the function of certain protein in terms of their structure or activity, or they can be without any visible or measurable phenotype effect. In the following text, the most important polymorphisms of redox metabolism genes are shown according to the weight of their pathogenicity and clinical effects.

11.2.1 Glutathione Redox Cycle

It is well known that glutathione redox cycle represents the vital component of the cellular antioxidative protection system against oxidative injury. Glutathione peroxidases (GPx) 1–6 represent the cellular enzymatic defense against hydrogen peroxide free radicals, and its sufficient activity is critical for the preservation of redox homeostasis [10]. The GPx-1 codes the isoform, which is the most abundant GPx

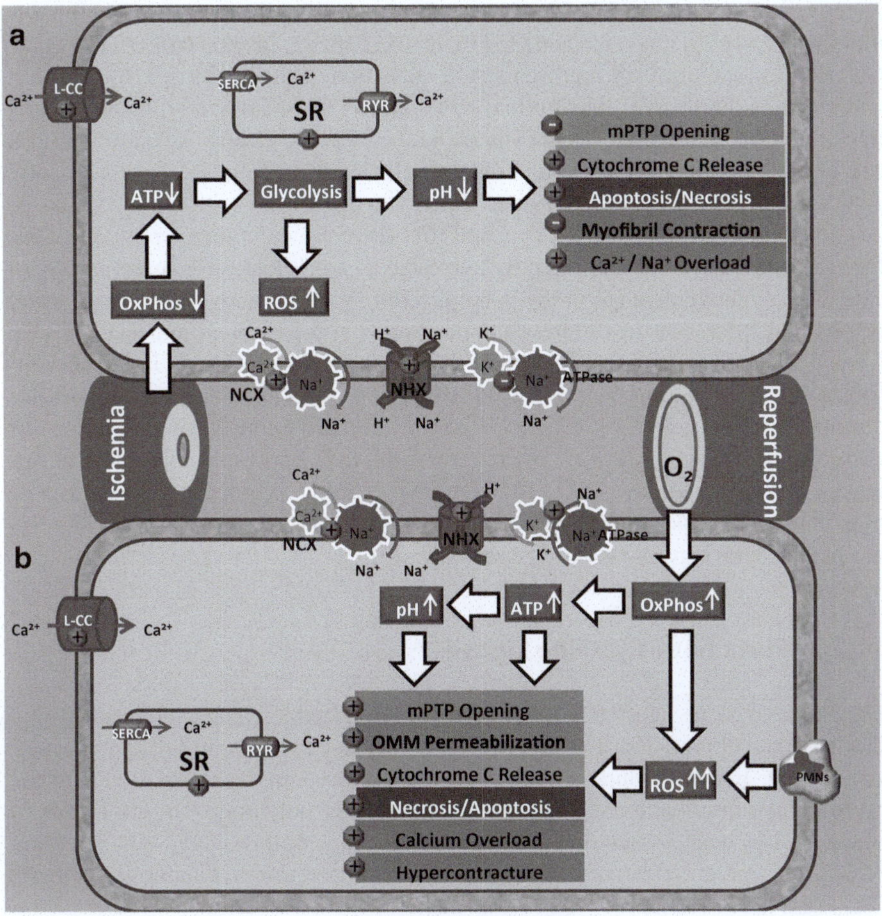

Fig. 11.1 Molecular events during the ischemic-reperfusion injury

enzyme in the cardiovascular system. Multiple studies have shown that the C198T polymorphism of the GPx-1 gene which causes lowered enzyme activity increases the risk of coronary disease and increased thickness of aorta and carotid arteries [11–14]. However, there are contradictory results which refer to the C198T polymorphism stated in the study of Souiden et al. who did not show any association between this polymorphism and coronary heart disease [15]. Glutathione S-transferases (GST) are a family of cytosol and mitochondrial enzymes that catalyze the conjugation of xenobiotics and oxidative compounds with reduced glutathione and therefore prevent cellular injury [16, 17]. The greater susceptibility to oxidative damage is imminent in case of deleterious polymorphisms that lower the expression or activity of GST enzyme family [18, 19]. Many polymorphisms of the GST genes were discovered and evaluated in the context of various heart diseases [20]. The GSTP1 gene located in 11q13.3 has seven exons, and two polymorphisms were well related to the development of coronary artery disease: GSTP1*B polymorphism (rs1695) in exon 5 and GSTP1*C (rs1138272) in exon 6 of GSTP1 gene. GSTP1*B polymorphism represents A-G transition at the position 313 in exon 5 and it causes the Ile>Val substitution, while the GSTP1*C polymorphism is also a transition C-T at position 341 in exon 6 and the effect is Ala>Val substitution [21]. The common feature for these two polymorphisms is that they cause significantly depleted activity of GSTP1 enzyme, which was associated with elevated risk for development of coronary artery disease [16]. There are numerous reports that confirm the relevance of GSTP1*B polymorphism as a risk factor for susceptibility to extended oxidative damage in the pathophysiology of coronary artery disease and myocardial infarction in different ethnic groups [22–25]. Bhat and Gandhi performed a genetic association study and showed that GSTP1*B and GSTP1*C contribute to the five-fold and 5.8 fold elevated risk of coronary artery disease development, respectively [26]. GSTM genes were also evaluated as risk factors for susceptibility to heart diseases. The relevance of GSTM polymorphisms was postulated according to the role of GSTM in detoxification of prooxidants and reduction of oxidative stress detrimental effects. Several studies have associated GSTM polymorphisms with cardiovascular diseases [27, 28].

11.2.2 Glutamate-Cysteine Ligase

Glutamate-cysteine ligase (GCL) is an important enzyme which catalyzes the synthesis of glutathione and it is composed of GCLC catalytic and GCLM modifier subunit [29, 30]. Nakamura et al. showed in their research the association between myocardial infarction and -588T allele of the C-588T polymorphism (rs41303970) of the GCLM gene, which down-regulates the compensatory increased expression of GCLM gene in oxidative stress conditions and decreasing glutathione synthesis [31].

11.2.3 Glutamate Ammonia Ligase

The glutamate ammonia ligase (GLUL) is an important enzyme which catalyzes synthesis of glutamine, important for many metabolic reactions, including glutathione synthesis. The rs10911021 SNP of GLUL gene, which represents a C>T transition down-regulates the gene expression and it was shown to be associated with coronary heart disease among patients with type 2 diabetes mellitus [32]. Shahid et al. reported that this polymorphism is associated not only with an increased risk of coronary events, but with circulatory levels of lipid peroxides, reduced and oxidized glutathione [33].

11.2.4 Superoxide Dismutase

Superoxide dismutase (SOD) is a first line of defense against superoxide anions, and it catalyzes the reaction of superoxide dismutation into peroxide anion. Humans have three SOD isoforms: cytosol Cu^{2+}/Zn^{2+}-SOD1, mitochondrial Mn^{2+}-SOD2 and extracellular Cu^{2+}/Zn^{2+}-SOD3 [34]. Polymorphisms of SOD genes were associated with the initiation and progression of various cardiovascular diseases.

Variations in SOD1 activity are considered to be associated with the extent of cardiac mortality of the general human population. Research of Otaki et al. revealed that rs1041740 polymorphism of SOD1 gene is associated with greater familial risk of cardiac diseases [35]. Souiden et al. also evaluated SOD1 gene and found that another polymorphism Ala16Val of SOD1 gene is connected with increased risk of stenosis in patients with coronary heart disease [15].

The SOD2 is critically important in adequate protection from oxidative stress generated in the respiratory chain functioning [36]. The Ala16Val substitution (rs4880) polymorphism causes decreased transport of SOD2 into the mitochondrial matrix and lessened neutralization of intracellular superoxide anion [37, 38]. Jones et al. demonstrated the connection between rs4880 polymorphism of SOD2 gene, which represents a C>T transition with increased risk of coronary heart disease in diabetic patients [39].

The extracellular SOD3 isoform acts as an extracellular scavenger of superoxide radicals and it is present in plasma, lymph and synovial fluid [40]. It is one of the key enzymes that protect the cardiovascular system from oxidative stress deleterious effects [41]. The rs2284659 polymorphism of SOD3 gene is located in the promotor region and it regulates the degree of gene expression. Mohammedi et al. conducted a study on diabetic patients and concluded that rs2284659 polymorphism greatly influences the plasma redox profile and it lowers protein oxidation products, providing better outcomes of cardiovascular disease [42].

11.2.5 Catalase

Catalase (CAT) is a ubiquitous peroxisomal enzyme that catalyzes breakdown of hydrogen peroxide into water, and it acts in greater amounts of hydrogen peroxide along with the GPx against oxidative injury [43]. The CAT polymorphisms which have lower expression of the CAT gene in common were investigated as risk factors for cardiac disease. The -262C/T polymorphism of the CAT gene promotor causes lower expression of CAT protein and contributes to the development of many pathological conditions, including ischemic heart disease [44]. This polymorphism can also be responsible for in-stent restenosis following invasive coronary revascularization [45]. The -844C/T is another promotor polymorphism of CAT gene, which was evaluated in the context of cardiovascular risk. This polymorphism was proven to be an independent predictor of postoperative occurrence of myocardial infarction among patients who previously underwent heart surgery [46].

11.2.6 NADPH Oxidase

NADPH oxidase represents a multiple subunit enzyme complex which is one of the principal sources of superoxide anion in the vasculature. The critical subunit for oxidase activity is p22phox, and its polymorphism rs4673 was evaluated in cardiovascular diseases [47]. This polymorphism represents C>T transition at position 242 of the CYBA gene that encodes p22phox protein and it causes altered NADPH oxidase activity. Guzik et al. demonstrated that the 242C allele is associated with significantly augmented NADPH activity, which increases prooxidants synthesis and contributed to the elevated risk of coronary atherosclerosis [48].

11.2.7 Myeloperoxidase

Myeloperoxidase (MPO) is a lysosomal enzyme which is abundantly present in neutrophil granulocytes and monocytes and it synthesizes hypochlorite anion which acts as a powerful oxidant compound [49]. The rs2333227 polymorphism represents G>A transition at position −463 of MPO gene and it modulates the binding of SP1 transcription factor. The GG homozygous genotype is associated with upregulated expression of MPO gene and elevated serum activity [50–52]. Farre and Casado correlated the MPO serum activity with the progression of heart failure [53]. Among patients with acute coronary syndromes or ischemic chest pain the increased levels of MPO in circulation were found and they are considered as independent prognostic factors for clinical outcome [54, 55]. Katakami et al. revealed the increased risk of coronary hearts disease in type 2 diabetes patients with MPO rs2333227 polymorphism, along with the polymorphisms of other redox related genes [56].

11.2.8 Nitric Oxide Synthase

Nitric oxide synthase catalyzes the reaction of nitric oxide synthesis from L-arginine, with its endothelial isoform eNOS (NOS3) which reduces circulatory oxidative stress by scavenging superoxide and preventing lipid peroxidation. The significant polymorphism involved in cardiovascular pathology is 894G/T (rs1799983) in exon 7 of NOS3 gene, which boosts intracellular cleavage by proteases and lowers eNOS enzyme activity [57]. This polymorphism is associated with increased cardiopulmonary risk, especially in smokers and patients with renal diseases [58–60].

11.2.9 Angiotensin 1-Receptor

One of the most well characterized polymorphisms of the AT1R gene is A1166C transversion and it causes the enhanced response to the binding of angiotensin II, particularly in homozygous carriers. This polymorphism was previously linked with elevated vasoconstrictor activity which is responsible for the development of arterial hypertension [61]. Beside these recognized effects of A1166C polymorphisms, Cameron et al. showed in their study that A1166CC polymorphism is an independent predictor of increased MPO activity and greater concentration of protein carbonylation products which are formed in oxidative stress conditions [62].

11.3 Genetics of Oxidative Stress in Atrial Fibrillation

The redox metabolism in atrial fibrillation (AF) was in a scope of numerous research studies, including the genetic aspect of the oxidative stress impact. Mihm et al. showed that the activity of the myofibrillar creatine kinase is important for the control of myocyte contractility and that it is significantly sensitive to oxidative stress [63]. If this enzyme is damaged by free radicals and if there is an energy deficit inside myocytes, such conditions lead to the impairment of myocyte's contractile function and contribute to the development of AF.

In the study of Carnes et al. it was revealed that atrial tissue in heart with AF has augmented oxidative stress witnessed by the elevated amount of 3-nitrotyrosine, which is correctable with ascorbate administration [64]. Using the high-throughput cDNA microarray approach in tissue model, Kim et al. demonstrated with high sensitivity and fidelity the important findings, which represent the pattern of altered gene expression in AF, providing the expression hallmark that can be used for future conception of therapeutic strategies [65]. The authors showed the up-regulation of eight pro-oxidative genes which encode following proteins: cytochrome P 450 oxidase, flavin-containing monooxygenase 1, monoamine oxidase B, NADPH oxidase, tyrosine 3-monooxygenase, tyrosinase-related protein 1, ubiquitin-specific protease 8 and xanthine oxidase. Additionally, they revealed the down-regulation of six genes that encode antioxidative defense components: CAT, glutaredoxin, GPx 1, glutathione reductase, heme oxigenase 2 and SOD. Despite the current knowledge, the role of oxidative stress underlying mechanisms in pathogenesis of AF is still ill-defined.

11.4 Genetics of Oxidative Stress in Congestive Heart Failure

There is a substantial interest in the scientific community for the importance of the pathogenic role of oxidative stress in congestive heart failure (CHF). Several researches have brought oxidative stress in relation to the CHF initiation and progression [2, 53, 66]. Membrane associated NADPH oxidases are expressed in the membranes of cardiomyocytes, smooth muscle cells and vascular endothelial cells and they appear to be quantitatively most important source of prooxidants in myocardium of CHF patients [67, 68]. Angiotensin II has the ability to activate NADPH oxidases which then produce the superoxide and peroxide anions by activation of angiotensin type-1 receptor (AT1R) in vascular cells [69, 70]. The mechanism of

NADPH oxidase activation by renin-angiotensin cascade was also demonstrated in studies with experimental animals [71, 72].

Previous researches demonstrate that prooxidant molecules influence the important processes of cardiac remodeling, endothelial homeostasis, NO signaling pathways and cellular proliferation in CHF [2, 53, 66, 73]. Cohn et al. suggested that the process of cardiac remodeling is a determinant of CHF evolution and that it directly influences the survival of such patients [74].

Left ventricular cardiac remodeling is mediated by matrix metalloproteinases (MMPs), which are excessively activated in the presence of increased amounts of prooxidants [75, 76]. Their activity was found to directly correlate with concentrations of 8-isoprostane which is a free-radical peroxide of arachidonic acid, representing a biomarker of oxidative damage [77]. The investigation of Askari et al. revealed that inactivation of plasminogen activator 1 by enhanced oxidative environment leads to the dilatation of the left cardiac ventricle [78].

CHF represents genetically heterogeneous disease, which is still inadequately characterized in terms of inheritance patterns. The study of Katakami et al. demonstrated that multiple oxidative stress-related gene polymorphisms are associated with heart failure among patients with insulin independent diabetes mellitus [56]. These authors elucidated that the accumulation of prooxidant alleles mutually contributes to the cumulative risk for CHD independent of common risk factors, and that it is critical to evaluate collectively all polymorphisms together in order to estimate the individual's risk for disease development.

Kim et al. in their comprehensive study used a candidate SNP analysis for estimation of oxidative stress and injury repair genes on the survival of children with CHF [79]. They showed the association between SNPs of vascular endothelial growth factor A (VEGFA) rs833069 and superoxide dismutase 2 (SOD2) rs2758331 with long-term survival of affected children. The genetic risk is elevated and cumulative when major deleterious alleles are mutually present. On the other hand the protective effect and prolonged survival is observed when minor alleles of VEGFA and SOD2 are present. These findings of Kim et al. are in agreement with previous basic studies that assign VEGFA as a factor of increased cardiovascular permeability, cardiomyocytes damage and decreased survival following acute myocardial infarction [80–82].

The extent of SOD2 expression is critically important for adequate protective effect against free radicals generated in oxidative stress. The rs2758331 polymorphism represents an intronic variant which is responsible for approximately 33% of increased or decreased expression of SOD2 gene, and it directly influences the long-term survival in CHF patients [79, 83]. In their recent study, Kim et al. once more demonstrated that homozygotic children for both VEGFA and SOD2 risk alleles have 16-fold increased death risk following heart surgery, exposing the important effect of genetic variants as modifiers of possible transplantation outcome in CHF patients [84].

11.5 Changes in Regulation of Gene Expression

The cytotoxic effect of oxidative stress is capable to induce cell death by apoptosis or necrosis, and on the other side, its moderate levels influence on signal transduction pathways that control cell growth and posttranslational modifications [85]. In the center of these effect lays an impact on regulation of gene expression through redox-sensitive transcription-regulation network. This modulation is obtained by alterations of transactivation and DNA binding of various transcription factors [86].

The oxidative stress controls many biological processes, through coupling of extracellular and intracellular signalization into alterations of critical gene transcription and posttranslational modifications. These changes come as a result of influence on calcium-dependent signaling as well as on protein kinase and protein phosphatase pathways. The consequences of such processes involve adaptation of cellular function to oxidative environment by mobilization of target genes that enable metabolic changes, repair and survival [87].

In the coronary artery disease the response to oxidative stress leads to the enhanced expression of vascular inflammatory genes repertoire, such as E-selectin, MCP-1 (monocyte chemoattractant protein 1) and VCAM-1 (vascular cell adhesion molecule 1) by means of redox-sensitive signalization and nuclear transcription factors [88]. De Marchi et al. showed that p66Shc protein and protein kinase C (PKC) behave like regulators of intracellular redox homeostasis and extent of oxidative stress [89]. They elucidated the process of redox sensitive signaling cascade dependent upon the adaptor protein p66Shc and PKC which regulate the cytochrome c transport from mitochondria to cytosol, NADPH activity and availability of antioxidative enzymes. This biological mechanism is amplified in obesity by excess glucose levels and it causes the positive feedback loop of oxidative stress progression and damage of cardiovascular tissues [90]. These findings lead to the assumption that the use of antioxidative therapeutics could be used for prevention of oxidative heart damage in obese individuals.

Earlier studies have provided enough evidence for influence of hydrogen-peroxide on different intracellular events. Dhalla et al. demonstrated that hydrogen-peroxide causes the augmented calcium internalization with activation of different chemical reactions during the acute ischemic heart disease [91]. This excess in intracellular calcium concentration following the prooxidant damage of the cell's organelles is another important pathophysiological aspect of oxidative stress in cardiovascular system [92]. Moreover, the hydrogen-peroxide mediated the vascular smooth muscle relaxation via activation of guanylate cyclase and cGMP synthesis [93]. The impact of hydrogen-peroxide on tyrosine phosphorylation mimics the stimulating growth effect of insulin which phosphorylates many proteins in target cells [94]. The MAPKs (mitogen activated protein kinases), phosphorylation of tyrosine and chemotaxis are stimulated by hydrogen-peroxide which acquires these effects by means of PDGF (platelet-derived growth factor) [95]. Torti et al. showed that if cardiomyocytes are exposed to increased hydrogen-peroxide amount, then the expression of alpha-actin, creatine kinase M isoform, myosin light chain 2 and troponin I gene is reduced as evidenced from the depleted levels of corresponding mRNAs [96]. On the other side, a rapid stimulating effect of hydrogen-peroxide on c-fos and c-jun protooncogenes expression was observed in neonatal cardiomyocytes [97]. Temsah et al. demonstrated in their research the inhibitory effect of hydrogen-peroxide on ryanodine receptor on the sarcoplasmic reticulum membrane, calcium-pump ATPase, phospholamban and calsequestrin mRNAs in heart model perfused with xanthine oxidase and hydrogen-peroxide [98].

Beside the fact that oxidative stress influences the rate of transcription of a specific set of genes, the antioxidants can also achieve substantial effects on gene expression. That was proved in a study of Ferran et al. when they showed that an antioxidant pyrrolidinedithiocarbamate can down-regulate the expression of prothrombotic molecules and tumor necrosis factor (TNF) in vascular endothelial cells [99].

The marked progress in the enlightening of the effects of oxidative stress on gene expression in cardiovascular system was made, and the current knowledge serves as a standpoint for future framework in the evaluation of atherogenic effect of certain genes and formulation of novel therapeutic strategies that target redox-sensitive mechanisms.

11.6 Gene Therapy of Cardiac Oxidative Stress

Advances in a novel approach to gene therapy have been involved in an effort to target oxidative stress in myocardial damage, which can occur after ischemic-reperfusion injury, causing the irreversible cardiac tissue damage. The conception of gene therapy that aims to reduce oxidative induced cardiac damage is based on the introduction and expression of genes that would act against the negative consequences of prooxidants [100].

Redox based gene therapy can be performed by two approaches: introduction of prooxidant-depletive genes or genes that modulate the redox-sensitive signaling pathways. Viral vectors are used as a desirable gene delivery mean, because they can transfer target genes into specific cells, including cardiomyocytes. In heart gene therapy the adenovirus is the most suitable viral vector that can transfer DNA sequences into terminally differentiated cells up to the 36 kb in size. One of the most important benefits of adenoviral vectors is the site specific insertion into the host genome [101]. Beside the significant advantages of adenoviral vectors, short-term expression (not more than 1 month), inflammation and cytotoxicity are considered as most important disadvantages [100].

The literature is rich in different studies in which antioxidant genes were introduced in cardiomyocytes in order to reduce the oxidative stress and corresponding heart damage. In the study of Woo et al. authors introduced SOD and CAT by adenoviral vector into heart muscle cells, using the adenoviral vector [102]. They reported highly competent gene transduction by adenovirus and significant protective effects and reduced contractile dysfunction following ischemic-reperfusion injury, owned to the enhanced antioxidative enzyme activity. Yoshida et al. reported that SOD1 nullizygous mice had greater infarct size, attenuated ventricular recovery and higher creatine kinase amount than the wild-type mice [103]. The postischemic myocardial damage was achieved by SOD1 overexpression in transgenic mouse myocardium, leading to the reduction in infarct size and elevated availability of high energy cellular phosphates [104]. SOD2 which is normally present in mitochondria was another subject of antioxidant boosting gene therapy model. The overexpression of SOD2 in the heart by adenoviral vector resulted in the reduction of myocardial infarction size in the left ventricle [105]. These results were additionally confirmed in the study of Chen et al. who found that SOD2 overexpression in mouse myocardium leads to the greater resistance to ischemic-reperfusion injury, based on the lowed lactate dehydrogenase activity and the size of infarct [106]. Li et al. overexpressed SOD3 enzyme in rabbit cardiomyocytes and noticed the cardioprotection against myocardial stunning [107]. In the same study, it was also demonstrated that augmented SOD3 was an effective antioxidant, without a need to simultaneously transduce CAT gene. In their later study, Li et al. demonstrated in the rabbit model that the SOD3 transduction with adenoviral vector reduces myocardial infarction size by 25%by preventing superoxide transfer from one cell to another [108].

The same experimental design with adenoviral gene transduction was used for overexpression of CAT in experimental model of cardiac ischemic injury. Zhu et al. in their research administered the adenoviral vectors carrying CAT into a rabbit's coronary artery, therefore causing the elevated enzyme activity [109]. The increased amount of hydrogen-peroxide generated during the ischemic-reperfusion injury was removed and the adequate myocardial contractility was well-preserved, indicating the usefulness of CAT in the antioxidant therapy by gene transfer in conditions where GPx activity is decreased by ischemic environment.

Human heme oxygenase 1 (hHO-1) was introduced into the myocardium by adenoviral vector in order to test the long term cardioprotection from against ischemic-reperfusion injury [110]. The research showed the reduction of myocardial ischemic injury and the decrease in infarct size, while the inflammation and necrosis were also present in lower quantity, exposing the protective role of hHO-1 and its therapeutic and potentially preventive role of ischemic heart damage.

Weiss et al. created the GPx overexpression model by adenoviral transfer and showed the protection of vascular endothelial cells from homocysteine induced dysfunction [111]. The stabilization of endothelial homeostasis in ischemic cardiac damage was demonstrated in the case of the CAT overexpression in endothelial cells by adenoviral vectors [112, 113].

Beside adenoviral vectors, there were attempts to introduce genes in myocardial cells of interest using the retroviruses for transduction process. In comparison with adenoviral vectors, studies that refer to the use of retroviral vectors are limited due to the fact that retroviruses can induce insertion mutagenesis and that the maximum length of the inserted ssRNA sequence is 9 kb [100]. Retroviral vectors are not suitable for transfection of terminally differentiated cells; however Yang et al. managed to transduce hHO-1 into endothelial cells and to record increased resistance to heme and hydrogen-peroxide cytotoxicity [114]. Sakoda et al. evaluated the transfection with lentivirus and proved that it could be efficient in therapeutic gene introduction into heart muscle cells [115].

Currently there is stagnation in widespread use of cardiac gene therapy that modulates redox metabolism and reduces cardiac oxidative damage, although there are plenty of evidence that beneficial therapeutic and prognostic effects can be achieved. The transfer of antioxidant genes can also be potentially used during the invasive interventions such as balloon angioplasty and stenting, so that endothelial damage could be repaired or prevented. Beforehand the routine administration of gene therapy in cardiac ischemic damage is widely applied in clinical practice it is necessary to refine the vector quality in terms of their tropism, dosing and persistence of transgene effect [116–118].

11.7 Conclusion and Future Perspectives

The oxidative stress genetics has been extensively studied in the past decades, and it revealed the significant weight of heritable factors on the extent of cardiac injury in different pathological states. Using this knowledge, it is possible to lay a solid foundation for future personalized therapy approach based on pharmacogenomics and gene therapy. Moreover, these data are already used as a starting point for analyzing cardiac damage proteomics profile by high-throughput technologies. Beside the treatment strategies, it would be a crucial victory to apply previously stated facts in the preventive measures which represent the goal of the twenty-first century medicine.

References

1. Stocker R, Keaney J (2004) Role of oxidative modifications in atherosclerosis. Physiol Rev 84:1381–1476
2. Giordano F (2005) Oxygen, oxidative stress, hypoxia, and heart failure. J Clin Invest 115:500–508
3. Hare J, Stamler J (2005) NO/redox disequilibrium in the failing heart and cardiovascular system. J Clin Invest 115:509–517
4. Collins FS, Varmus H (2015) A new initiative on precision medicine. N Engl J Med 372:793–795
5. Huggins GS (2015) Genetic susceptibility to oxidative stress and cardiovascular disease. EBioMedicine 2:1864–1865
6. Dominiczak AF (2004) Strategies to reduce oxidative stress in cardiovascular disease. Clin Sci (Lond) 106:219–234
7. Padmanabhan S, Melander O, Hastie C et al (2008) Hypertension and genome-wide association studies: combining high fidelity phenotyping and hypercontrols. J Hypertens 26:1275–1281
8. Park GH, Choe J, Choo HJ et al (2002) Genome-wide expression profiling of 8-chloroadenosine- and 8-chloro-cAMP-treated human neuroblastoma cells using radioactive human cDNA microarray. Exp Mol Med 34:84–193
9. Latronico MV, Catalucci D, Condorelli G (2007) Emerging role of microRNAs in cardiovascular biology. Circ Res 101:1225–1236
10. Beckett GJ, Arthur JR (2005) Selenium and endocrine systems. J Endocrinol 184:455–465
11. Hamanishi T, Furuta H, Kato H et al (2004) Functional variants in the glutathione peroxidase-1 (GPx-1) gene are associated with increased intima-media thickness of carotid arteries and risk of macrovascular diseases in Japanese type 2 diabetic patients. Diabetes 53:2455–2460

12. Tang NP, Wang LS, Yang L et al (2008) Genetic variant in glutathione peroxidase 1 gene is associated with an increased risk of coronary artery disease in a Chinese population. Clin Chim Acta 395:89–93
13. Heslop CL, Tebbutt SJ, Podder M et al (2012) Combined polymorphisms in oxidative stress genes predict coronary artery disease and oxidative stress in coronary angiography patients. Ann Hum Genet 76:435–447
14. Zhang JX, Wang ZM, Zhang JJ et al (2014) Association of glutathione peroxidase-1 (GPx-1) rs1050450 Pro198Leu and Pro197Leu polymorphisms with cardiovascular risk: a meta-analysis of observational studies. J Geriatr Cardiol 11:141–150
15. Souiden Y, Mallouli H, Meskhi S et al (2016) MnSOD and GPx1 polymorphism relationship with coronary heart disease risk and severity. Biol Res 49:22
16. Hayes JD, Flanagan JU, Jowsey IR (2005) Glutathione transferase. Annu Rev Pharmacol Toxicol 45:51–88
17. Li R, Boerwinkle E, Olshan AF et al (2000) Glutathione S-transferase genotype as a susceptibility factor in smoking-related coronary heart disease. Atherosclerosis 149:451–462
18. Turkanoglu A, Demirdogen BC, Demirkaya S et al (2010) Association analysis of GSTT1, GSTM1 genotype polymorphisms and serum total GST activity with ischemic stroke risk. Neurol Sci 31:727–734
19. Doney ASF, Lee S, Leese GP et al (2005) Increased cardiovascular morbidity and mortality in Type 2 diabetes is associated with the Glutathione S-Transferse theta-null genotype: a Go-Darts study. Circulation 111:2927–2934
20. Lo HW, Ali-Osman F (2007) Genetic polymorphism and function of glutathione S-transferase in tumor drug resistance. Curr Opin Pharmacol 7:367–374
21. Ntais C, Polycarpou A, Loannidis JP (2005) Association of GSTM1, GTTT1 and GSTP1 gene polymorphisms with the risk of prostate cancer: a meta-analysis. Cancer Epidemiol Biomark Prev 14:176–181
22. Nomani H, Mozafari H, Ghobadloo SM et al (2011) The association between GSTT1, M1 and P1 polymorphisms with coronary artery disease in Western Iran. Mol Cell Biochem 354:181–187
23. Singh N, Sinha N, Kumar S et al (2011) Glutathione S-transferase gene polymorphism as a susceptibility factor for acute myocardial infarction and smoking in the North Indian population. Cardiology 118:16–21
24. Phulukdaree A, Khan S, Moodley D, Chuturgoon AA (2012) GST polymorphisms and early onset coronary artery disease in young South African Indians. S Afr Med J 102:627–630
25. Yeh HL, Kuo LT, Sung FC et al (2013) GSTM1, GSTT1, GSTP1 and GSTA1 genetic variants are not associated with coronary artery disease in Taiwan. Gene 523:64–69
26. Bhat MA, Gandhi G (2017) Glutathione S-transferase P1 gene polymorphisms and susceptibility to coronary artery disease in a subgroup of north Indian population. J Genet 96:927–932
27. Delles C, Padmanabhan S, Lee WK et al (2008) Glutathione S-transferase variants and hypertension. J Hypertens 26:1343–1352
28. Oniki K, Hori M, Takata K et al (2008) Association between glutathione S-transferase A1, M1 and T1 polymorphisms and hypertension. Pharmacogenet Genomics 18:275–277
29. Meister A, Anderson ME (1983) Glutathione. Annu Rev Biochem 52:711–760
30. Minova HR, Mulcahy RT (1998) An electrophile responsive element (EpRE) regulates gamma-naphthoflavone induction of the human gamma-glutamyl cysteine synthetase regulatory subunit gene. J Biol Chem 273:14683–14689
31. Nakamura S, Kugiyama K, Sugiyama S et al (2002) Polymorphism in the 5′-flanking region of human glutamate-cysteine ligase modifier subunit gene is associated with myocardial infarction. Circulation 105:2968–2973
32. Qi L, Qi Q, Prudente S et al (2013) Association between a genetic variant related to glutamic acid metabolism and coronary heart disease in individuals with type 2 diabetes. JAMA 310:821–828
33. Shahid SU, Shabana HS (2018) The SNP rs10911021 is associated with oxidative stress in coronary heart disease patients from Pakistan. Lipids Health Dis 17:6

34. Fridovich I (1995) Superoxide radical and superoxide dismutases. Annu Rev Biochem 64:97–112
35. Otaki Y, Watanabe T, Nishiyama S et al (2016) The impact of superoxide dismutase-1 genetic variation on cardiovascular and all-cause mortality in a prospective cohort study: The Yamagata (Takahata) Study. PLoS One 11:e0164732
36. Becuwe P, Ennen M, Klotz R et al (2014) Manganese superoxide dismutase in breast cancer: from molecular mechanisms of gene regulation to biological and clinical significance. Free Radic Biol Med 77:139–151
37. Shimoda-Matsubayashi S, Matsumine H, Kobayashi T et al (1996) Structural dimorphism in the mitochondrial targeting sequence in the human manganese superoxide dismutase gene. A predictive evidence for conformational change to influence mitochondrial transport and a study of allelic association in Parkinson's disease. Biochem Biophys Res Commun 226:561–565
38. Sutton A, Khoury H, Prip-Buus C et al (2003) The Ala16Val genetic dimorphism modulates the import of human manganese superoxide dismutase into rat liver mitochondria. Pharmacogenetics 13:145–157
39. Jones DA, Prior SL, Tang TS et al (2010) Association between the rs4880 superoxide dismutase 2 (C>T) gene variant and coronary heart disease in diabetes mellitus. Diabetes Res Clin Pract 90:196–201
40. Stralin P, Karlsson K, Johansson BO, Marklund SL (1995) The interstitium of the human arterial wall contains very large amounts of extracellular superoxide dismutase. Arterioscler Thromb Vasc Biol 15:2032–2036
41. Takatsu H, Tasaki H, Kim HN et al (2001) Overexpression of EC-SOD suppresses endothelial-cell-mediated LDL oxidation. Biochem Biophys Res Commun 285:84–91
42. Mohammedi K, Bellili-Muñoz N, Marklund SL et al (2015) Plasma extracellular superoxide dismutase concentration, allelic variations in the SOD3 gene and risk of myocardial infarction and all-cause mortality in people with type 1 and type 2 diabetes. Cardiovasc Diabetol 14:845
43. Goth L, Rass P, Pay A (2004) Catalase enzyme mutations and their association with diseases. Mol Diagn 8:141–149
44. dos Santos KG, Canani LH, Gross JL et al (2006) The catalase -262C/T promoter polymorphism and diabetic complications in Caucasians with type 2 diabetes. Dis Markers 22:355–359
45. Shuvalova YA, Kaminnyi AI, Meshkov AN et al (2012) Association between polymorphisms of eNOS and GPx-1 genes, activity of free-radical processes and in-stent restenosis. Mol Cell Biochem 370:241–249
46. Podgoreanu MV, White WD, Morris RW et al (2006) Inflammatory gene polymorphism and risk of postoperative myocardial infarction after cardiac surgery. Circulation 114:I275–I281
47. Azumi H, Inoue N, Takeshita S et al (1999) Expression of NADH/NADPH oxidase p22phox in human coronary arteries. Circulation 100:1494–1498
48. Guzik TJ, West NE, Black E et al (2000) Functional effect of the C242T polymorphism in the NAD(P)H oxidase p22phox gene on vascular superoxide production in atherosclerosis. Circulation 102:1744–1747
49. Daugherty A, Dunn JL, Rateri DL, Heinecke JW (1994) Myeloperoxidase, a catalyst for lipoprotein oxidation, is expressed in human atherosclerotic lesions. J Clin Invest 94:437–444
50. Reynolds WF, Chang E, Douer E et al (1997) An allelic association implicates myeloperoxidase in the etiology of acute promyelocytic leukemia. Blood 90:2730–2737
51. Piedrafita FJ, Molander RB, Vansant G et al (1996) An Alu element in the myeloperoxidase promoter contains a composite SP1-thyroid hormone-retinoic acid response element. J Biol Chem 271:14412–14420
52. Katakami N, Kume S, Kaneto H et al (2013) Association of myeloperoxidase G-463A gene polymorphism with diabetic nephropathy in Japanese type 2 diabetic subjects. Endocr J 60:457–471

53. Farre A, Casado S (2001) Heart failure, redox alterations, and endothelial dysfunction. Hypertension 38:1400–1405
54. Baldus S, Heeschen C, Meinertz T et al (2003) Myeloperoxidase serum levels predict risk in patients with acute coronary syndromes. Circulation 108:1440–1445
55. Brennan ML, Penn M, Lente F et al (2003) Prognostic value of myeloperoxidase in patients with chest pain. N Engl J Med 349:1585–1604
56. Katakami N, Kaneto H, Matsuoka TA et al (2014) Accumulation of oxidative stress-related gene polymorphisms and the risk of coronary heart disease events in patients with type 2 diabetes-an 8-year prospective study. Atherosclerosis 235:408–414
57. Wang XL, Sim AS, Wang MX et al (2000) Genotype dependent and cigarette specific effects on endothelial nitric oxide synthase gene expression and enzyme activity. FEBS Lett 471:45–50
58. Tesauro M, Thompson WC, Rogliani P et al (2000) Intracellular processing of endothelial nitric oxide synthase isoforms associated with differences in severity of cardiopulmonary diseases: cleavage of proteins with aspartate vs. glutamate at position 298. Proc Natl Acad Sci U S A 97:2832–2835
59. Veldman BA, Spiering W, Doevendans PA et al (2002) The Glu298Asp polymorphism of the NOS 3 gene as a determinant of the baseline production of nitric oxide. J Hypertens 20:2023–2027
60. Noiri E, Satoh H, Taguchi J et al (2002) Association of eNOS Glu298Asp polymorphism with end-stage renal disease. Hypertension 40:535–540
61. van Geel P, Pinto Y, Voors A et al (2000) Angiotensin II type 1 receptor A116C gene polymorphism is associated with an increased response to angiotensin II in human arteries. Hypertension 35:717–721
62. Cameron VA, Mocatta TJ, Pilbrow AP et al (2006) Angiotensin type-1 receptor A1166C gene polymorphism correlates with oxidative stress levels in human heart failure. Hypertension 47:1155–1161
63. Mihm MJ, Yu F, Carnes CA, Reiser PJ et al (2001) Impaired myofibrillar energetics and oxidative injury during human atrial fibrillation. Circulation 104:174–180
64. Carnes CA, Chung MK, Nakayama T et al (2001) Ascorbate attenuates atrial pacing-induced peroxynitrite formation and electrical remodeling and decreases the incidence of postoperative atrial fibrillation. Circ Res 89:E32–E38
65. Kim YH, Lim DS, Lee JH et al (2003) Gene expression profiling of oxidative stress on atrial fibrillation in humans. Exp Mol Med 35:336–349
66. Mak S, Newton G (2001) The oxidative stress hypothesis of congestive heart failure. Chest 120:2035–2046
67. Griendling K, Minieri C, Ollerenshaw J, Alexander R (1994) Angiotensin II stimulates NADH and NADPH oxidase activity in cultured vascular smooth muscle cells. Circ Res 74:1141–1148
68. Griendling K, Sorescu D, Ushio-Fukai M (2000) NAD(P)H oxidase: role in cardiovascular biology and disease. Circ Res 86:494–501
69. Lassegue U, Sorescu D, Szocs K et al (2001) Novel gp91*phox* homologues in vascular smooth muscle cells: nox1 mediates angiotensin II-sensitive superoxide formation and redox-sensitive signaling pathways. Circ Res 88:888–894
70. Seshiah P, Weber D, Rocic P et al (2002) Angiotensin II stimulation of NAD(P)H oxidase activity: upstream mediators. Circ Res 91:406–413
71. Wang H, Xu S, Johns D et al (2001) Role of NADPH oxidase in the vascular hypertrophic and oxidative stress response to angiotensin II in mice. Circ Res 88:947–953
72. Weber DS, Rocic P, Mellis AM et al (2005) Angiotensin II-induced hypertrophy is potentiated in mice overexpressing p22phox in vascular smooth muscle. Am J Physiol Heart Circ Physiol 288:H37–H42
73. Grieve D, Shah A (2003) Oxidative stress in heart failure: more than just damage. Eur Heart J 24:2161–2163

74. Cohn J, Ferrari R, Sharpe N (2000) Cardiac remodeling-concepts and clinical implications: a consensus paper from an international forum on cardiac remodeling. J Am Coll Cardiol 35:569–582
75. Gopcevic K, Rovcanin B, Kekic D, Radenkovic S (2011) Matrix metalloproteinases and membrane damage markers in sera of patients with acute myocardial infarction. Mol Cell Biochem 350:163–168
76. Gopcevic K, Rovcanin B, Kekic D et al (2017) Gelatinases A and B and antioxidant enzyme activity in the early phase of acute myocardial infarction. Folia Biol (Praha) 63:20–26
77. Kameda K, Matsunaga T, Abe N et al (2003) Correlation of oxidative stress with activity of matrix metalloproteinase in patients with coronary artery disease. Possible role for left ventricular remodelling. Eur Heart J 24:2180–2185
78. Askari A, Brennan ML, Zhou X et al (2003) Myeloperoxidase and plasminogen activator inhibitor 1 play a central role in ventricular remodeling after myocardial infarction. J Exp Med 197:615–624
79. Kim DS, Stanaway IB, Rajagopalan R et al (2012) Results of genome-wide analyses on neurodevelopmental phenotypes at four-year follow-up following cardiac surgery in infancy. PLoS One 7:e45936
80. Weis S, Shintani S, Weber A et al (2004) Src blockade stabilizes a Flk/cadherin complex, reducing edema and tissue injury following myocardial infarction. J Clin Invest 113:885–894
81. Matsudaira K, Maeda K, Okumura N et al (2012) Impact of low levels of vascular endothelial growth factor after myocardial infarction on 6-month clinical outcome. Circ J 76:1509–1516
82. Guo HD, Cui GH, Yang JJ et al (2012) Sustained delivery of VEGF from designer self-assembling peptides improves cardiac function after myocardial infarction. Biochem Biophys Res Commun 424:105–111
83. Bastaki M, Huen K, Manzanillo P et al (2006) Genotype-activity relationship for Mn-superoxide dismutase, glutathione peroxidase 1 and catalase in humans. Pharmacogenet Genomics 16:279–286
84. Kim DS, Kim JH, Burt AA et al (2014) Patient genotypes impact survival after surgery for isolated congenital heart disease. Ann Thorac Surg 98:104–110
85. Schulze-Osthoff K, Bauer MK, Vogt M, Wesselborg S (1997) Oxidative stress and signal transduction. Int J Vitam Nutr Res 67:336–342
86. Allen RG, Tresini M (2000) Oxidative stress and gene regulation. Free Radic Biol Med 28:463–499
87. Suzuki YJ, Forman HJ, Sevanian A (1997) Oxidants as stimulators of signal transduction. Free Radic Biol Med 22:269–285
88. Kunsch C, Medford RM (1999) Oxidative stress as a regulator of gene expression in the vasculature. Circ Res 85:753–766
89. De Marchi E, Baldassari F, Bononi A et al (2013) Oxidative stress in cardiovascular diseases and obesity: role of p66Shc and protein kinase C. Oxidative Med Cell Longev 2013:564961
90. Nishikawa T, Edelstein D, Du XL et al (2000) Normalizing mitochondrial superoxide production blocks three pathways of hyperglycaemic damage. Nature 404:787–790
91. Dhalla NS, Golfman L, Takeda S et al (1999) Evidence for the role of oxidative stress in acute ischemic heart disease: a brief review. Can J Cardiol 15:587–593
92. Dhalla NS, Temsah RM, Netticadan T (2000) Role of oxidative stress in cardiovascular diseases. J Hypertens 18:655–673
93. Moncada S, Higgs A (1993) The L-arginine-nitric oxide pathway. N Engl J Med 329:2002–2012
94. Heffetz D, Bushkin I, Dror R, Zick Y (1990) The insulinomimetic agents H2O2 and vanadate stimulate protein tyrosine phosphorylation in intact cells. J Biol Chem 265:2896–2902
95. Sundaresan M, Yu ZX, Ferrans VJ et al (1995) Requirement for generation of H2O2 for platelet-derived growth factor signal transduction. Science 270:296–299
96. Torti SV, Akimoto H, Lin K et al (1998) Selective inhibition of muscle gene expression by oxidative stress in cardiac cells. J Mol Cell Cardiol 30:1173–1180

97. Peng M, Huang L, Xie ZJ et al (1995) Oxidant-induced activations of nuclear factor-kappa B and activator protein-1 in cardiac myocytes. Cell Mol Biol Res 41:189–197
98. Temsah RM, Netticadan T, Chapman D et al (1999) Alterations in sarcoplasmic reticulum function and gene expression in ischemic-reperfused rat heart. Am J Phys 277:H584–H594
99. Ferran C, Millan MT, Csizmadia V et al (1995) Inhibition of NF-kappa B by pyrrolidine dithiocarbamate blocks endothelial cell activation. Biochem Biophys Res Commun 214:212–223
100. Hingtgen SD, Davisson RL (2001) Gene therapeutic approaches to oxidative stress-induced cardiac disease: principles, progress, and prospects. Antioxid Redox Signal 3:433–449
101. Svensson EC, Marshall DJ, Woodard K et al (1999) Efficient and stable transduction of cardiomyocytes after intramyocardial injection or intracoronary perfusion with recombinant adeno-associated virus vectors. Circulation 99:201–205
102. Woo YJ, Zhang JC, Vijayasarathy C et al (1998) Recombinant adenovirus-mediated cardiac gene transfer of superoxide dismutase and catalase attenuates postischemic contractile dysfunction. Circulation 98:II255–II261
103. Yoshida T, Maulik N, Engelman RM et al (2000) Targeted disruption of the mouse Sod I gene makes the hearts vulnerable to ischemic reperfusion injury. Circ Res 86:264–269
104. Wang P, Chen H, Qin H et al (1998) Overexpression of human copper, zinc-superoxide dismutase (SOD1) prevents postischemic injury. Proc Natl Acad Sci U S A 95:4556–4560
105. Abunasra HJ, Smolenski RT, Morrison K et al (2001) Efficacy of adenoviral gene transfer with manganese superoxide dismutase and endothelial nitric oxide synthase in reducing ischemia and reperfusion injury. Eur J Cardiothorac Surg 20:153–158
106. Chen Z, Siu B, Ho YS et al (1998) Overexpression of Mn-SOD protects against myocardial ischemia/reperfusion injury in transgenic mice. J Mol Cell Cardiol 30:2281–2289
107. Li Q, Bolli R, Qiu Y et al (1998) Gene therapy with extracellular superoxide dismutase attenuates myocardial stunning in conscious rabbits. Circulation 98:1438–1448
108. Li Q, Bolli R, Qiu Y et al (2001) Gene therapy with extracellular superoxide dismutase protects conscious rabbits against myocardial infarction. Circulation 103:1893–1898
109. Zhu HL, Stewart AS, Taylor MD et al (2000) Blocking free radical production via adenoviral gene transfer decreases cardiac ischemia-reperfusion injury. Mol Ther 2:470–475
110. Melo LG, Agrawal R, Zhang L et al (2002) Gene therapy strategy for long-term myocardial protection using adeno-associated virus-mediated delivery of heme oxygenase gene. Circulation 105:602–607
111. Weiss N, Zhang YY, Heydrick S et al (2001) Overexpression of cellular glutathione peroxidase rescues homocyst(e)ine-induced endothelial dysfunction. Proc Natl Acad Sci U S A 98:12503–12508
112. Zanetti M, Katusic ZS, O'Brien T (2002) Adenoviral-mediated overexpression of catalase inhibits endothelial cell proliferation. Am J Physiol Heart Circ Physiol 283:H2620–H2626
113. Erzurum SC, Lemarchand P, Rosenfeld MA et al (1993) Protection of human endothelial cells from oxidant injury by adenovirus mediated transfer of the human catalase cDNA. Nucleic Acids Res 21:1607–1612
114. Yang L, Quan S, Abraham NG (1999) Retrovirus-mediated HO gene transfer into endothelial cells protects against oxidant-induced injury. Am J Phys 277:L127–L133
115. Sakoda T, Kasahara N, Hamamori Y, Kedes L (1999) A high-titer lentiviral production system mediates efficient transduction of differentiated cells including beating cardiac myocytes. J Mol Cell Cardiol 31:2037–2047
116. Hamilton CA, Miller WH, Al-Benna S et al (2005) Corcoran lecture. Cardiovascular genomics and oxidative stress. Hypertension 45:636–642
117. Baker AH (2002) Development and use of gene transfer for treatment of cardiovascular disease. J Card Surg 17:543–548
118. Work LM, Nicklin SA, Baker AH (2003) Targeting gene therapy vectors to the vascular endothelium. Curr Atheroscler Rep 5:163–170

Oxidative Stress in Cardiac Remodeling Post-Ischemia/Reperfusion: Friend or Foe?

12

Emna Abidi, Abdullah Kaplan, George W. Booz, and Fouad A. Zouein

Abbreviations

WHO	World Health Organization
CVD	Cardiovascular disease
Ca^{2+}	Calcium
ROS	Reactive Oxygen Species
I/R	Ischemia Reperfusion
I/RI	Ischemia-Reperfusion Injury
MI	Myocardial Infarction
CHD	Coronary Heart Disease
PCI	Percutaneous Coronary Intervention
CABG	Coronary Artery Bypass Grafting
ATP	Adenosine triphosphate
Na^{+}	Sodium
mPTP	Mitochondrial Permeability Transition Pore
H^{+}	Hydrogen ion
NCX	Na^{+}/Ca^{2+}exchanger
PKC-δ	Protein Kinase C delta
PKC-ε	Protein Kinase C epsilon
PARP	poly (ADP-ribose) polymerase
O_2^{-}	Superoxide anion
XO	Xanthine oxidase
NADPH	Nicotinamide Adenine Dinucleotide Phosphate

E. Abidi · A. Kaplan · F. A. Zouein (✉)
Department of Pharmacology and Toxicology, Faculty of Medicine, American University of Beirut, Beirut, Lebanon

G. W. Booz
Department of Pharmacology and Toxicology, School of Medicine, University of Mississippi Medical Center, Jackson, MS, USA

S. Chakraborti et al. (eds.), *Oxidative Stress in Heart Diseases*,
https://doi.org/10.1007/978-981-13-8273-4_12

NOS	Oxidase Synthase
MPO	Myeloperoxidase
nNOS	neuronal NOS
eNOS	endothelial NOS
iNOS	inducible NOS
NO	Nitric Oxide
$ONOO^-$	peroxynitrite
BH4	tetrahydrobiopterin
H_2O_2	hydrogen peroxide
ETC	Electron Transport Chain
XDH	Xanthine dehydrogenase
IL-1	Interleukine 1
IL-6	Interleukine 6
TNF-α	Tumor Necrosis Factor alpha
PMNs	Polymorphonuclear Lukocytes
HIF-1α	Hypoxia-inducible factor 1-alpha
MIM	Mitochondrial Inner Membrane
MnSOD	Manganese Superoxide Dismutase
H_2O_2	Hydrogen peroxide
MAPKs	Mitogen-activated Protein Kinases
RAF-MEK	Rapidly Accelerated Fibrosarcoma- Mitogen-activated protein kinase kinase pathway
PI3K	PI-3 kinase
HMGB1	High-mobility box 1
TLRs	Toll-Like Receptors
NFκB	Nuclear factor kappa-light-chain-enhancer of activated B cells
PKA	Protein Kinase A
Akt/PKB	Protein kinase B
Bcl-2	B-cell lymphoma 2
MMPs	Matrix metalloproteinases
BK	Big Potassium channels
mitoKATP	Mitochondrial ATP-sensitive K+ channel

12.1 Introduction

Given the heart's high energy demand and function, along with its vital physiological role to the body, a prolonged and non-managed ischemia is detrimental with high risk of morbidity and mortality [1, 2]. Half a century ago, myocardial reperfusion following coronary blood flow obstruction emerged as a promising therapy to rescue the heart from ischemic damage. However, challenging reports emerged since 1970s contradicting the beneficial role of reperfusion on myocardial tissue recovery following ischemia, and highlighting the myocardial ischemia-reperfusion injury (I/RI) concept [3]. Multiple studies, thereafter, exposed the underlying mechanisms behind those findings. Hearse et al [4] were among the first group to report that sudden resumption of metabolic activity to energy-(and oxygen-) starved tissue resulted in a reoxygenation-dependent injury response independent of the hypoxic stress, commonly called "reperfusion injury". I/RI development is multifactorial involving alterations in both mitochondrial and cellular homeostasis, including a shortage in ATP production, alterations in ion gradient homeostasis, excessive inflammation, Ca^{2+} handling dysregulation, and excessive ROS production. In fact, myocardial ROS surge following reperfusion was for long proposed to be the mediator of I/RI [5, 6]. Consistently, a large number of studies have intensively addressed the role of excessive ROS formation during I/R [7]. Of note, ROS is a well-known potent mediator of metabolic disruption, inflammation, necrosis, and cell death in multiple diseases including myocardial injuries [8]. In this chapter we emphasize the importance of ROS-mediated reperfusion injury, and highlight the promising mito-targeted antioxidant therapy. We also examine the paradoxal evidence supporting the beneficial effects of ROS bursts in pre- and postconditioning mechanisms.

12.2 From Permanent Occlusion to Reperfusion: The Bad, the Good, and the Ugly

12.2.1 The Bad: Myocardial Infarction

Coronary blood flow obstruction, commonly termed MI, is characterized by an inadequate blood flow and subsequent nutrient and oxygen deprivation to the affected area. The severity of MI is strongly dependent on the size of the area at risk, the duration of ischemia, and the presence or absence of comorbidities [9]. The onset of MI itself is characterized by multiple life-threatening pathologies, including ventricular fibrillation, atrio-ventricular block [10] and cardiogenic shock [11]. Following hospitalization and stabilization of potentially existing arrhythmias, non-reperfused MI patients undergo adverse remodeling of the myocardium with very poor prognosis and high risk of heart failure development and death. Based on the American Heart Association statistical report, an approximate number of 720,000 Americans are hospitalized either for a first time MI or coronary heart disease (CHD) events with a projection of a median survival of 8.4, 5.6, 7, and 5.5 years for ≥45 year old white males, white females, black males, and black females respectively. Additionally, sudden cardiac death accounts for 13.5% of death certificates with a relatively high lifetime risk for cardiac arrest survivors [12].

12.2.2 The Good: Reperfusion

Given the well-established positive correlation between the duration of ischemia and the extent of myocardial damage, coronary blood flow restoration was an inevitable solution. In the last two decades, researchers have conducted a multitude of studies and reported that the salvage of ischemic cells from inevitable death is only possible by revascularization. Thus, multiple interventions such as percutaneous coronary intervention (PCI), coronary artery bypass grafting (CABG), and pharmacological (thrombolysis) approaches to remove the occluding clot were developed and adopted [13]. Reperfusion has proven to limit the ischemic injury and subsequently the infarct size area. The importance of reperfusion therapy in MI patients was surveyed over the past 20 years and reported a continuous decline of 6-month mortality, along with a further 22% reduction in standardized mortality, from 2010 to 2015 following reperfusion therapy [14] .

12.2.3 The Ugly: Reperfusion Injury

Despite the perpetual improvement of multiple procedures to ensure a rapid, complete, effective, and permanent reopening of the acutely occluded coronary artery, numerous studies revealed that myocardium salvage following blood flow restoration is highly predisposed to another form of injury, known as reperfusion injury [11]. Aside from the reperfusion impact on cardiac remodeling, multiple pathological conditions are known to occur at the onset of blood flow restoration, including arrhythmias, myocardial stunning, and potential microvascular occlusion that could be life-threatening [15].

12.3 Mechanisms of Cellular Cardiac Injury Following I/R

12.3.1 At the Onset of Ischemia

Following coronary artery clotting, cessation of cellular oxygen supply halts mitochondrial membrane polarization, reducing therefore adenosine triphosphate (ATP) formation and increasing mitochondrial ROS production [16]. Subsequently, reduced ATP-dependent Na^+/K^+ pump activity, leads to Na^+ accumulation in the myocyte and lowered mitochondrial resting membrane potential. Na^+ overload within the cell is counter-regulated by the reverse activity of Na^+/Ca^{2+}-exchanger (NCX) that pumps Na^+ out in exchange for Ca^{2+}, resulting eventually in intracellular and intra-mitochondrial Ca^{2+} overload. Concurrently with Ca^{2+} and Na^+ overload, the absence of oxygen supply switches cellular metabolism to anaerobic glycolysis promoting lactate accumulation and cellular acidosis [17]. In summary, ischemia-induced accumulation of intracellular sodium, ROS, and calcium ions, favors, if sustained, the opening of the mitochondrial permeability transition pore (mPTP) [18]. This, together with ATP shortage, determines a loss of contractility, structural disorganization, and apoptotic, necroptotic, and necrotic cell death [19, 20]. However, the acidic conditions during ischemia prevent opening of the mPTP and subsequent cardiomyocyte death (Fig. 12.1a).

12.3.2 At the Onset of Reperfusion

Reperfusion is intended to restore ATP production and reactivate the Na^+/K^+ ATPase to slowly re-establish the sodium gradient, leading to normal cation fluxes and eventually extruding the excess cytosolic and mitochondrial Ca^{2+}. However a massive mitochondrial ROS burst follows reoxygenation during reperfusion, which is further fueled by inflammation, increasing the risk of mPTP opening and cell death [18]. Additionally, persistent high intracellular Ca^{2+} levels observed during the early phase of reperfusion, increase the risk of a damaging myocardial hypercontracture that was otherwise inhibited during acidic ischemia (Fig. 12.1b). Besides, in the setting of ischemic–reperfusion injury, ROS burst is also responsible of the activation of protein kinase C delta (PKC-δ) stimulating its translocation to the mitochondria where it results in cytochrome c release, caspase 3 activation, and a decrease in the activity of pro-survival Akt, as well as poly (ADP-ribose) polymerase (PARP) cleavage in the nucleus. Pharmacological inhibition of PKC-δ is exploited in many therapeutic strategies like preconditioning [21]. In summary, reperfusion-induced cellular damage is largely dependent on ROS burst, Ca^{2+} overload, and mPTP opening [20].

12.4 Myocardial ROS in I/R: Types and Sources

Compelling evidence pointing to the causal interconnection between oxidative stress and I/RI is well established [22]. Oxidative stress is a consequence of the imbalance between ROS production and antioxidant capacity, either because of heightened ROS release and/or an ineffective antioxidant system [23]. Under ischemic conditions, mitochondrial complexes I and III are primarily responsible of the conversion of molecular oxygen to unstable/reactive superoxide (O_2^-) [24]. Cardiomyocytes, containing the highest number of mitochondria, consume a higher level of oxygen than any other cell and subsequently become major ROS producers [25]. As a result, heightened cellular ROS levels ultimately alter cellular homeostasis primarily by damaging proteins, lipids, and nucleic acids [26, 27]. In addition to local ROS production, immune cell infiltration into the myocardium following I/RI contributes substantially to increase ROS levels [28]. Upon reperfusion of the ischemic myocardium, inflammatory reaction is noticeably accelerated. Although inflammation is crucial for myocardial tissue healing, the re-establishment of blood flow to ischemic tissue accelerates and prolongs inflammatory response detrimentally. Among multiple immune cell infiltrations, neutrophils are considered the earliest and the most potent releaser of ROS, followed by macrophages [29]. Interestingly, clinical anti-neutrophil therapies did not succeed in slowing or preventing adverse myocardial remodeling post-MI [30]. These findings imply that local free radical outburst following reperfusion is potentially the main source of ROS-mediated injury during I/R. Xanthine oxidase (OX), nicotinamide adenine dinucleotide phosphate (NADPH) oxidases, mitochondrial electron transport damage and uncoupling, uncoupled nitric oxidase synthase (NOS), and

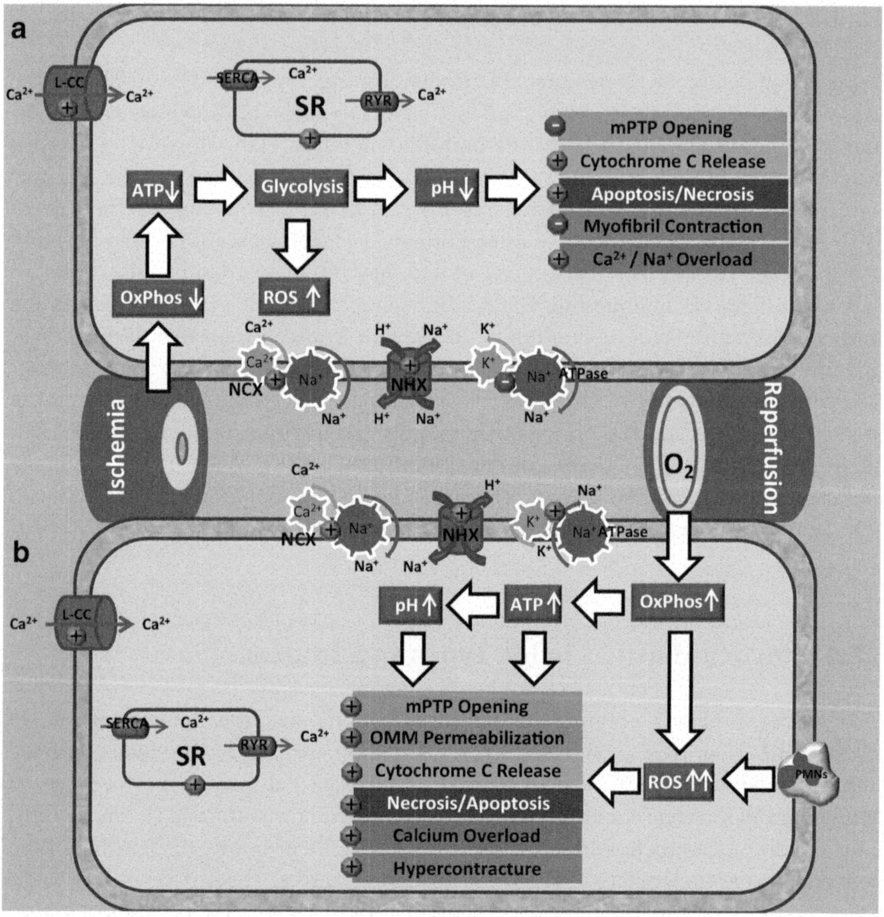

Fig. 12.1 **Schematization of the key components of acute myocardial ischemia reperfusion injury**: (**a**) Loss of oxygen supply in ischemia leads to a loss of ATP production and a switch to anaerobic respiration, resulting in a drop in intracellular pH, accompanied with an increased intracellular and mitochondrial-derived ROS. The ATP consuming Na^+-K^+-pump ceases to function, leading to Na^+ accumulation in the myocyte and the resting membrane potential is lowered. With the development of acidosis, the NHX further increases intracellular Na^+ exacerbating Ca^{2+}overload by forcing the NCX to manage, in a reverse mode the extrusion of Na^+ and the influx of Ca^{2+} into the cell. The sarcolemmal L type voltage-gated Ca^{2+} (L-CC) are activated allowing more Ca^{2+}entry as the resting membrane potential is low. Ca^{2+}pump SERCA2 is now taken up the excess of Ca^{2+}into the SR that releases it subsequently via RYR, leading to contraction and contracture. The acidic conditions during ischemia however prevent the opening of the mPTP and cardiomyocyte hypercontracture. (**b**) During reperfusion ATP production increases leading to the Na^+-K^+-pump reactivation, a slow restoration of both sodium gradient and NCX normal activity extruding the excess of cytosolic Ca^{2+}. An excessive production of ROS accompanies reoxygenation, electron transport chain activation, and immune cells infiltration. ROS burst mediates myocardial reperfusion injury by inducing the opening of the mPTP, causing outer mitochondrial membrane permeabilization, apoptosis, necrosis, acting as a neutrophil and cytokines chemoattractant, mediating dysfunction of the SR and causing myofibril hypercontacture. Restoration of physiological pH

myeloperoxidase (MPO) are the major producers of ROS in reperfused ischemic myocardium [31] and will be discussed in this chapter.

12.4.1 Nitric Oxide Synthases (NOS)

One of the most studied sources of physiological and pathophysiological ROS are the three well-recognized isoforms of NOS enzymes known as neuronal NOS (nNOS), endothelial NOS (eNOS), and inducible NOS (iNOS) that normally produce NO during the oxidation of L-arginine to L-citrulline [32]. While eNOS and nNOS are known to be constitutively expressed in the myocardium, iNOS, although primarily induced in immune cells, is expressed in cardiomyocytes under ischemic conditions [33, 34]. Constitutive myocardial nitric oxide (NO) generation under physiological conditions is essential for physiologic cell signaling [35]. Blood flow and oxygen restoration following reperfusion significantly increase NOS activity and subsequent NO production [36]. Although NO has been reported to be protective against I/R-induced injury in different organs of experimental animals [37] and humans [38], the beneficial effects of NO activity are negated by increased O_2^--mediated peroxynitrite ($ONOO^-$) generation following reperfusion (Table 12.1).

$$NO + O_2^- = ONOO^-$$

Contrarily to NO, $ONOO^-$ is very detrimental to proteins and lipids. $ONOO^-$ can negatively and irreversibly alter the structure and function of NOS by damaging its heme domain and oxidizing the tetrahydrobiopterin (BH4) cofactor [32, 39, 40] ultimately leading to NOS uncoupling an important source of I/R-induced ROS generation [32]. NOS function during I/R depends as well on its structural form; two cellular forms of constitutive NOS exist, the monomer and the homodimer forms. The monomer form is responsible of O_2^- generation in small amounts; the shift towards excessive O_2^- production depends on the homodimer/monomer ratio, intracellular L-arginine supply, and on BH4 oxidation [1].

A decrease in the local BH4/NOS ratio makes the balance, of a stoichiometric relationship between BH4 and eNOS, fall towards increased O_2^- instead of NO [41, 42]. Uncoupled NOS, furthermore, produces more O_2^- that acts as a positive feedback loop leading to further BH4 to BH2 oxidation and the propagation of NOS uncoupling. Besides the described loop, XO [43] and/or NADPH oxidase [44] play an important role in the I/R-induced reduction in BH4 levels by promoting O_2^- generation.Also, an essential factor required for the synthesis of NO by eNOS is

Fig. 12.1 (continued) following reperfusion along with Ca^{2+} overload accentuates mPTP opening leading to an increased infract size, cellular dysfunction, and cell death. *Ca^{2+}* calcium, *Na^+* sodium, *K^+* potassium, *H^+* hydrogen, *O_2* oxygen, *SR* sarcoplasmic reticulum, *SERCA* sarco/endoplasmic reticulum Ca^{2+}-ATPase, *ATP* adenosine triphosphate, *OxPHos* oxidative phosphorylation, *ROS* reactive oxygen species, *mPTP* mitochondrial permeability transition pore, *NCX* $3Na^+/1Ca^{2+}$-exchanger, *NHX* Na^+-H^+-exchanger, *PMNs* polymorphonuclear lukocytes, (+) stimulation, (−) inhibition, ↑ increase, ↓ decrease

Table 12.1 Potential sources of reactive oxygen species in the cardiac tissue exposed to ischemia and reperfusion

Evidence of ROS involvement in I/RI		References
↓Superoxide dismutase activity		[5]
↓Endogenous cellular antioxidant systems		
↓Cellular glutathione-to-glutathione disulfide ratio		
↑ Lipid peroxidation,		
↑O_2^- production at reperfusion		
Oxygen-derived radicals act like mediators of reperfusion injury in isolated heart models in the presence or absence of superoxide		[116–118]
Oxygen-derived free radicals are directly implicated in I/RI		[119–121]
O_2^- was identified as the parent radical that serves as a precursor to the formation of both OH^- and the carbon-centered radical		
Free-radical are highly generated in an intact dog model of I/R		
O_2^- is the parent radical at reperfusion		
Oxygen, nitrogen, and carbon-centered free radicals are generated during I/R in an isolated rabbit and rat heart models		[122–124]
Exogenous administered ROS at the same levels as those observed during reperfusion induced similar calcium overloading, functional depression, and metabolic changes		
Major sources and outcomes of ROS generation in I/RI		
Xanthine oxidase	The time-course of ROS production elicited by I/R in isolated rat hearts is closely correlated with the kinetics of XO substrate accumulation.	[125, 126]
	Increased tissue xanthine and hypoxanthine levels determine the severity of the I/RI	
	Pharmacologic blockade of xanthine oxidase (XO) substrate formation:	
	(−) XO-dependent ROS production	
	(−) Contractile dysfunction that accompanies reperfusion	
	Exogenous administration of hypoxanthine and xanthine :	
	(−) The protective effects of blockade of xanthine oxidase substrate formation	
	Inhibition of XO:	
	↑ Levels of XOR antigen in vascular endothelium of myocardial ischemic tissues	
NADPH oxidase	↑ROS generation	[127–131]
	↑Tissue injury following reperfusion	
	Blunted reperfusion-induced neutrophil accumulation:	
	↓ Tissue injury and/or ROS production	
	The application of a simulated I/R on purified cardiomyocytes in culture:	
	(+) Tissue injury-related responses that are dependent on Nox activity	
	NOX inhibition:	

(continued)

Table 12.1 (continued)

Evidence of ROS involvement in I/RI		References
	↓ ROS production	
	↓ Myocardial infarct size	
	↓Cell death	
	↑Protective effect in isolated buffer (cell free) perfused hearts exposed to I/R	
	Mutant mice deficient in either Nox-1 or Nox-2/ Nox-1, Nox-2 and Nox-1/Nox-2 double knockout mice :	
	↑Protective effect in buffer-perfused Langendorff preparations with I/R hearts models	
	Myocytes release of Nox isoforms:	
	↑ROS generation during I/R	
Mitochondria	***I/RI:***	[125, 132, 133]
	↑Mitochondrial H_2O_2 generation	
	↓Cytochrome c release from the MIM	
	↑Reduction state of cytochrome c	
	Ischemic damage to complex I and III:	
	↑Capacity to generate O_2^- at reperfusion	
	Both associated and/or separated complex I and III isolated from mitochondria obtained from reperfused hearts can generate O_2^-	
Nitric oxide synthase	***I/R:***	[127–131]
	↑Uncoupled NOS	
	↑Myocardial $ONOO^-$ generation by NOS	
	↓Endothelium-dependent vasodilation in porcine coronary arteries	
	In vitro and in vivo models of I/R:	
	BH4 supplementation replenish NOS activity in isolated rat hearts	
	↑↑ uncoupled NOS-derived O_2^- production	
	↑ I/R-induced cardiac inflammation and tissue damage	
	After reperfusion of ischemic heart:	
	↑ Arginase activity	
	↑ O_2^- generation increases	
	↓Arginine levels and NO production decrease	
	I/R+ treatment with a combination of arginine and BH4:	
	↓Infarct size	

I/RI ischemia reperfusion injury, *I/R* ischemia reperfusion, O_2^- superoxide, *ROS* reactive oxygen species, *XO* xantine oxidase, *XOR* xantine oxidase receptor, *Nox* NADPH oxidase, H_2O_2 hydrogen peroxide, *MIM* mitochondrial inner membrane, *NOS* nitric oxide synthase, $ONOO^-$ Peroxynitrite, *BH4* tetrahydrobiopterin, *NO* nitric oxide, (+) stimulation, (−) inhibition, ↑ increase, ↓ decrease arginine, the nitrogen donor, and substrate for arginase I and II [45]. Increased arginase activity leads as well to increased production of O_2^- by NOS, a mechanism called "arginine steal". Finally, it was very early reported that myocardial eNOS actively produces NO during ischemia and reperfusion; however, parallel observations have shown that the enzyme is affected during ischemia. In fact, a prolonged ischemia is accompanied by intracellular acidosis that reversibly or irreversibly inhibits eNOS activity independently of the duration of acidosis [46] (Table 12.1).

12.4.2 Monoamine Oxidase and p66shc

During I/R, mitochondria are responsible of the generation of hydrogen peroxide (H_2O_2) through serotonin oxidization via monoamine oxidase [47]. Serotonin accumulation, as well as increased monoamine oxidase activity, is noted during ischemia and substantially increased following reperfusion [48]. Moreover, mitochondria are also capable of H_2O_2 production using a novel pathway that involves the 66-kDa isoform of the growth factor adaptor protein, p66shc. Ischemic conditions are responsible of translocation of p66shc from the cytosol to the mitochondrial intermembrane space, allowing it to use reducing equivalents from the electron transport chain (ETC) via the oxidation of cytochrome c to make H_2O_2. This reaction acts in a vicious cycle to provide p66shc with increased substrate in the intermembrane space during ischemia [49].

12.4.3 Cellular Xanthine Dehydrogenase vs. Xanthine oxidase

During normoxic physiological conditions, xanthine dehydrogenase (XDH) catalyzes the oxidation of hypoxanthine to xanthine and then to uric acid by coupling the reaction with NAD^+ reduction to yield NADH.

$$\text{hypoxanthine} + \text{NAD}^+ \xrightarrow{\text{XDH}} \text{xanthine} + \text{NADH}$$

$$\text{xanthine} + \text{NAD}^+ \xrightarrow{\text{XDH}} \text{uric acid} + \text{NADH}$$

However, ischemic context enhances the conversion of XDH to XO by the modification of a sulfhydryl group or by proteolytic cleavage. XO is a molybdoflavoenzyme complex that controls, in addition to uric acid production, ROS generation through catalyzing the oxidation of hypoxanthine to xanthine.

$$\text{hypoxanthine} + \text{O}_2 \xrightarrow{\text{XO}} \text{xanthine} + \text{O}_2^-$$

$$\text{xanthine} + \text{O}_2 \xrightarrow{\text{XO}} \text{uric acid} + \text{O}_2^-$$

XO-derived ROS contribute to multiple pathologic conditions including I/RI (Table 12.1). The accumulation of XO following ischemia will increase O_2^- formation. Besides, the oxygen burst at the onset of reperfusion drastically increases O_2^- formation [50] .

12.4.4 Cellular Xanthine Dehydrogenase

Nevertheless, other mechanisms can explain the enhanced superoxide release, independently of XDH to XO conversion. In fact, XDH has an NADH oxidase activity in the presence of acidic conditions (pH 6.5) wherein NADH is oxidized rather than xanthine [33]. XDH is capable of generating superoxide at 4-times the rate of XO. Besides, XDH is the dominant isoform in the early reperfusion period and is most likely a more important source of superoxide than the XO isoform at the onset of the reperfusion.

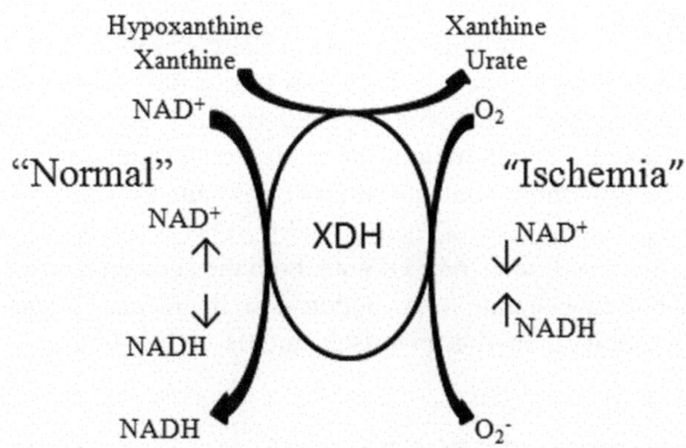

Post-transcriptional regulation of XDH expression is reported during I/R, wherein the hypoxic and inflammatory environments are stimuli associated with increased XDH transcription [51]. On the other hand, XO activity is also regulated at the post-translational level. These modifications have been attributed to O_2 tension that results in phosphorylation of the enzyme by p38 kinase [52]. In addition, along with hypoxic environment, the inflammatory context (mast cell degranulation and macrophage activation), which accompanies I/R, participates via multiple cytokines such as IL-1, IFN-γ, IL-6 and TNF-α to increase XDH/XO mRNA. Another feature of the XO capacity to produce ROS under ischemic conditions is its capacity to act as a nitrate/nitrite reductase (Table 12.1). This enzymatic reaction catalyzes the production of NO by one electron reduction of nitrite, and is optimal under anoxic/hypoxic and acidic conditions [53]. The generated NO, an important substrate for peroxynitrite generation, enhances the oxidative burst in the presence of an ischemia/inflammation loop in the ischemic heart [54] (Table 12.1). Finally, XO participates in leucocyte recruitment upon I/R, followed by neutrophil recruitment and XO-derived ROS secretion [55].

12.4.5 Nicotinamide Adenine Dinucleotide Phosphate (NADPH) Oxidase

The Noxs are a family of 7 isoforms expressed in multiple types of cells, including vascular endothelial cells, smooth muscle cells, fibroblasts, cardiomyocytes, and polymorphonuclear leukocytes (PMNs). Nox isoforms are known as Nox-1 to Nox-5 and dual oxidases (Duox)-1 and -2 [56, 57]. Nox/Duox are considered the major source of ROS in multiple pathological conditions including cardiac remodeling post-MI [58–60] (Table 12.1). The contribution of Nox enzymes to reperfusion injury is documented by multiple studies reporting both an increased expression and/or activity of Nox in ischemic tissue and attenuation of I/R induced injury following Nox inhibition [61, 62]. A large amount of data reported the involvement of multiple factors in the activation of Noxs in I/RI. For example, studies confirmed that hypoxia inhibitory factor-1α (HIF-1α) activation, evoked by the hypoxic state that accompanies ischemia, promote production and activation of Noxs [63]. Among all Nox isoforms, Nox2 is one of the most widely expressed in cardiac cells and, therefore, a prominent ROS producer in myocardial I/R (Table 12.1). The activation of XO and the resulting increase in ROS and intracellular Ca^{2+} levels have been reported to be indispensable for Nox2 activation under ischemic conditions. The stimulation of PKC by XO-generated ROS also contributes to ischemic-evoked Nox2 activation. Furthermore, the inhibition of XO halts ischemic-induced upregulation of HIF-1α proving that Nox2 activation by XO is essential for HIF-1α activation under ischemic conditions [64]. Of note, both the activation of the complement system and increased generation of angiotensin II are also associated with an increase in Nox activity in cardiac post-ischemic tissue [61, 65].

12.4.6 Mitochondrial ETC ROS Production

Mitochondria constitute 33% of the total cardiac myocyte cell volume, highlighting their fundamental role in cardiac function and the high energy demand of the myocardium. The mitochondrial ETC complex is comprised of a series of multi-subunit complexes (complexes I–IV) located in the inner mitochondrial membrane (IMM) and coupled to mobile carriers such as coenzyme Q and cytochrome c. The complexes and cytochrome c contain redox groups (Fe-S clusters and/or heme) that allow for the transfer of electrons along the components of the ETC, generating a proton electrochemical gradient, ultimately promoting ATP production via ATPase [25, 66]. Mitochondria are considered a normal source of ROS that play a crucial role as cell signaling intermediates in order to maintain cellular homeostasis. Under normal physiological conditions, ETC reduces oxygen to water using more than 97% of the entire electron flux through mitochondria. The remaining 2–3% of electrons consistently leak from ETC to form O_2^-. In addition to its important role in signaling, physiological production of O_2^- plays a critical role in multiple crucial cell functions such as metabolism, proliferation, and apoptosis [67].

Following ischemia, the decrease in mitochondrial respiration as well as ATP production, along with complex I/III alterations, increases NADH:NAD^+ ratio and reduces flavin mononucleotide prosthetic group within the NADH dehydrogenase component of complex I. These changes increase the leakage of electrons that form O_2^- via univalent reduction of O_2 and subsequent ROS production beyond physiological levels [24, 25, 68, 69]. Although reduced cytochrome c controls mitochondrial ROS levels by scavenging O_2^-, persistent ischemia increases the oxidized state of cytochrome c contributing further to mitochondrial damage and the accumulation of O_2^- (Table 12.1).

Upon reperfusion, oxygen burst into an already stunned mitochondria drastically increase ROS production to a much higher extent than during ischemia. Additional sites within complex I may contribute to ROS generation. Mitochondrial increase in superoxide production is normally accompanied by an increase in H_2O_2 formation through MnSOD activity within the mitochondrial matrix [70]. Superoxide dismutase enzymes contain either copper, manganese, or nickel metal centers that are reduced or oxidized to convert cellular O_2^- into H_2O_2 (Table 12.1) [71, 72]. H_2O_2 interaction with NO also increases formation of $ONOO^-$. Of note, ROS is able to freely spread within the mitochondrial network mainly through the mPTP and inner membrane ion channels, centralizing therefore cellular damage [25, 73].

12.5 ROS Mediated Adverse Effects in I/R

ROS production during the ischemic, reperfusion, and remodeling phases contribute to cardiac injury post-MI. The extent of injury, however, varies based on the size of the affected myocardium, the magnitude of ROS reactions, and the severity of cardiomyocyte damage. Uncontrolled sustained ROS burst causes modification and denaturation of a multitude of structural and functional molecules leading to irreversible tissue damage. The effect of each ROS, however, depends on its type. OH^- for instance acts instantly right after generation. O_2^- and NO^- radicals on the other hand, are of much lesser reactivity, more specific, and can mediate radical reactions on sites that are distant form their site of production. In the absence of appropriate ROS scavengers, sustained ROS production triggers oxidative vicious cycles that could permanently damage the cells. Of the well-known ROS-mediated cellular damage, lipid peroxidation, protein denaturation, mitochondrial, and DNA damage constitute the basis behind those effects.

12.5.1 DNA Oxidation

OH-mediated hydrogen extraction interferes with cellular DNA, causing purine and/or pyrimidine direct modification and/or fragmentation producing a plethora of DNA lethal lesions [74]. These lesions can induce mutagenesis, crosslinks between DNA strands and proteins, stand breaks, which affect thereafter DNA replication and transcription [75] and ultimately promote a pro-apoptotic and pro-necrotic effect (Table 12.2).

12.5.2 Lipid Peroxidation

Lipid peroxidation is a typical 3 phase oxidative reaction that occurs abundantly during I/R. The alkenes, unsaturated fatty acids and major component of biological membrane's phospholipid bilayers, are very susceptible to hydrogen extraction by ROS. The generated carbon-centered and peroxyl radicals constitute the initial phase of ROS attack followed by an amplification phase also known as the propagation phase [76]. Lipid peroxidation continues with additional similar abstractions until two radical species combine in a termination phase. Reactive aldehydes such as malondialdehyde (MDA), 4-hydroxynonenal, and isoprostane are major end products of this classical oxidation cascade and are known to increase during I/R [77]. Lipid peroxidation byproducts are bioactive and well-involved in the adverse remodeling of I/R. 15-F2t-isoprostane, for example, is reported to induce a dose-dependent vasoconstriction in coronary arteries, promoting therefore cardiac dysfunction following I/R [78] (Table 12.2).

12.5.3 Protein Oxidation

ROS-mediated activation of necrotic and pro-apoptotic proteins determines the severity and the extent of infarct size [79]. For instance, ROS can modify cellular proteins via oxidation and nitration, impairing subsequent myocardial contraction and promoting myocardial stunning following I/R [80]. Similarly to what is observed with lipids, hydrogen extraction by OH^- is a key player in the initiation phase of the oxidative attack on proteins by affecting amino-acid functional groups [81]. Denaturation of proteins by ROS oxidation reactions is due to the cleavage of peptide bonds, functional group cross-linking, and by hydrophobicity alterations of amino acids on protein surfaces [81] (Table 12.2).

Proteins with signaling roles, such as kinases and phosphatases, can also be oxidized by $ONOO^-$, affecting therefore their signaling capacities and impact [82]. By regulating mitogen-activated protein kinases (MAPKs), ROS contribute to cellular responses to mitogens, inflammatory cytokines, and (un)physiological stimuli [83]. Activation of p38 can have either pro- or anti-apoptotic effects, and is exaggerated during I/R. Also, p38 has been reported to play a role in regulating mitochondrial ROS levels and intracellular signaling pathways, as well as controlling mitochondrial events associated with development of I/R-associated damage (Table 12.2). Other signaling pathways that have also been shown to be involved in this regulation include: the RAF-MEK pathway that, in contrast, prevents mitochondrial accumulation of ROS/Ca^{2+} and cell death [83], and the PI-3 kinase (PI3K)/protein kinase C (PKC/AKT) pathway that has a protective role against cellular I/R-induced cell death.

Table 12.2 ROS targets following I/RI

Oxidized target	Effects	References
Lipids	↑Lipid peroxidation.	[77, 134]
	↑Alkene hydrogen abstraction	
	↑Generation of carbon-centered and peroxyl radicals	
	↑Peroxyl and lipid isoforms generation	
	↑Production of MDA ,4-hydroxynonenal, and isoprostane	
	↑Cardiolipin depletion	
Proteins	↑ Physical and chemical modification of myocardial proteins	[79–81, 99, 135–140]
	↑Non-enzymatic modification of cellular proteins	
	↑ Amino acid oxidation and nitration	
	↑ Formation of nitrotyrosine residues on proteins	
	↑ Products of tyrosine oxidation: myocardial 3-nitrotyrosine and dityrosine	
	↑ Particular risk of tyrosine nitration in mitochondrial proteins	
	↑ OH^- hydrogen abstraction in functional groups and backbone α-carbons of all amino acids	
	↑ Cleavage of peptide bonds and cross-linking of functional groups	
	↑Alteration of the hydrophobicity of amino acids on protein surfaces	
	↑Cellular and mitochondrial sulfhydryl groups	
	↑Tyrosine kinases and protein tyrosine phosphatases oxidation	
	↑↑Mitochondrial tyrosine and cysteine residues nitration and oxidation	
	↑↑Reactive aldehydes-mediated electrophilic attack towards nucleophilic amino acids	
	↑4-hydroxynonenal provoked modification and inhibition of the cytochrome oxidase	
	↑ Oxidative activation of necrotic and pro-apoptotic protein	
	↑ MMP-9 cleavage and activation	
DNA	↑ OH^--mediated hydrogen abstraction	[74, 75, 78, 141–143]
	↑ Purine or pyrimidine direct modification and/or fragmentation	
	↑DNA lesions	
	↑DNA bases modification	
	(+) Inter and intra-strand crosslinks	
	↑ DNA–protein crosslinks	
	(+) strand break formation	
	↑Adducts with MDA, and ROS-mediated lipid-peroxidation products	
	↓DNA replication and transcription	

(continued)

Table 12.2 (continued)

Oxidized target	Effects	References
Mitochondria homeostasis	↑Perturbation of the mitochondrial energy production	[93, 96, 99, 144]
	↑Overexuberant liberation of mitochondrial ROS	
	↑Mitochondrial DNA rearrangement and fragmentation	
	↓Mitochondrial enzymes activities	
	↑The susceptibility of mitochondrial DNA to oxidative modification in circulating leukocytes	
	↑Oxidative inactivation of mitochondrial aconitase	
	↑Mitochondrial production of hydroxyl radicals	
	↓Mitochondrial structure and function	
	↑Mitochondrial Ca^{2+} levels	
	↑mPTP opening	
	↑Perforation and lysis	
	↑Mitochondrial depolarization and cell death	

MDA malondialdehyde, *OH⁻* hydroxyl radical, *MMP-9* matrix metalloproteinase 9, *DNA* deoxyribonucleic acid, *ROS* reactive oxygen species, *Ca^{2+}* calcium, *mPTP* mitochondrial permeability transition pore, (+) stimulation, (−) inhibition, ↑ increase, ↓ decrease

ROS are implicated as well in inflammatory signaling, not only by fueling the pro-inflammatory response in a self-perpetuating manner, but also by regulating the process of high-mobility box 1 (HMGB1) protein release that occurs especially in response to cellular damage. HMGB1 is an agonist for Toll-like Receptors (TLRs). Accordingly, TLR4-mediated NFκB activation is recruited for oxidative stress-activated intracellular signaling pathways [84].

Two additional developmental pathways also figure among the most important pathways in this context: the Wnt/s-catenin signaling that is activated by ROS [84] and NOTCH signaling that suppresses ROS production [85]. Nevertheless, this sort of crosslink between intracellular signaling and regulation of mitochondrial ROS production has been demonstrated for p53 [86], protein kinase A (PKA) [87], rapidly accelerated fibrosarcoma (RAF) kinase, protein kinase B (Akt/PKB), and B-cell lymphoma 2 (Bcl-2) [83]. The tyrosine kinase pathway plays a role via p66shc, which acts as a redox enzyme that generates mitochondrial ROS through oxidation of cytochrome c [49]. In addition, oxidative stress leads to alterations in the activation state of different PKCs. This activation provides a protective role in the context of preconditioning by activating the specific PKC-ε isoform [88]. However, activation of PKCδ isoform increases, in a positive-loop manner, ROS generation. Activation of PKCδ by ROS regulates the expression and function of apoptosis-related proteins, and represents a target for caspases leading to cellular death [88].

Activation of enzymes including MMPs and caplains is also pronounced following increased ROS production and pH restoration, and is capable subsequently of degrading crucial functional proteins, such as myosin light chain [89], α-actinin [90], and cardiac troponin [91, 92] (Table 12.2). During ischemia, oxidative stress is also broadly responsible for Na^+/H^+ exchanger (NHX) activation, a mechanism that attempts to restore intracellular pH by increasing cellular Na^+ levels. The Na^+/Ca^{2+} exchanger (NCX) is thus activated, leading to intracellular accumulation of Ca^{2+} and a state of mitochondrial Ca^{2+} overload and depolarization. This phenomenon is exacerbated upon reperfusion with mitochondrial calcium uniporter (CaU) exaggerated opening. ROS-mediated alterations of anion exchanger function leads to pH recovery. Excessive ROS, however, favor opening of the mPTP (Fig. 12.1b) [93], leading to mitochondrial matrix swelling and loss of MOM. This result is fatal due to the pro-apoptotic molecules that are released from the mitochondrial intermembrane space (IMS). Additionally, another type of Ca^{2+} permeable cationic channel is affected by increased ROS production during I/R. The transient receptor potential melastatin 2 (TRPM2) is in fact a ROS sensor [94]. Oxidative stress-mediated activation of TRPM2 results in mitochondrial Na^+ and Ca^{2+} overload, which leads to a disrupted mitochondrial membrane, cytochrome c release, PARP-1 cleavage via induction of caspase-8 activation, and finally apoptotic cell death [95].

It is worthwhile to point out that radical reactions are of a semi-random nature, so they do not necessarily yield irreversible cell damage. For example, the magnitude of the oxidative attack on membranes, proteins, or DNA may not be enough to have an adverse effect on their functions. Besides, if the damaged protein is not of critical functional relevance, normal cell processes, such as phospholipid and protein turnover, can remove the altered biomolecule and the cell will survive. Understanding the contribution of ROS to the development of I/RI may identify additional targets for therapeutic interference. That being said, the understanding of aberrant signaling in this particular pathological condition holds the promise for novel therapeutic approaches that specifically target the regulation of mitochondrial function (Table 12.3).

12.6 General and Mitochondria Targeted Antioxidants Reduce I /R Injury

Mitochondrial Ca^{2+} overload and overexuberant mitochondrial ROS burst, constitute the hallmark of I/R-mediated cardiac cell injury [96, 97]. A massive burst of ROS following reperfusion localizes in mitochondrial regions that progress to swelling and eventually stimulates opening of the mPTP [98, 99]. mPTP opening is

Table 12.3 Summary of selected clinical trials using general ROS scavengers therapeutics for MI

Antioxidant	Effect	Outcome	References
β-carotene	*Harmful*	↑ CAD risk	[145]
Edaravone	*Effective*	↓ Infarct size and reperfusion arrhythmia	[145]
L-carnitine	*Effective*	↓ Level of cardiac MI markers	[146]
Vitamin E	*Effective*	↓ CVD risk	[145]
	Harmful	↑ Increased HF risk	

CAD coronary artery disease, *MI* myocardial infarction, *CVD* cardiovascular disease, *HF* heart failure, ↑ increase, ↓ decrease directly linked to mitochondrial DNA rearrangement and fragmentation, a complete disrupted mitochondrial structure and function (Table 12.2), followed by mitochondrial perforation and lysis [18]. Mitochondria-targeted antioxidant therapy has the ability to salvage I/R-assaulted cardiomyocytes more so than general antioxidants (Table 12.4) at different levels including: (1) preventing excessive detrimental cellular ROS production that is largely and mainly produced by mitochondria with I/R, (2) promoting low and beneficial ROS signaling through protein kinase Cε and its downstream substrates, and (3) preventing harmful ROS signaling through protein kinase Cδ and its downstream effectors. Examples of protective therapies targeting mitochondrial ROS are detailed in Table 12.4.

12.7 The Paradoxal Cardioprotective Effects of ROS

Pre- and postconditioning are manipulations during which short periods or bouts of ischemia are applied by occluding and opening the coronary artery, prior or subsequent to, permanent occlusion [100]. Pharmacological and interventional ischemia pre- and postconditioning has gained immense attention due to its protective effects on cardiac remodeling prior to or following reperfusion [101, 102]. This protection is, however, impeded with application of antioxidants. In fact, unlike excessive and sustained ROS burst that is now proven to be detrimental, low levels of ROS are protective (Table 12.4). A growing body of recent evidence has established that generation of ROS at low levels can serve as a signal mediating physiologic responses. The protective role of preconditioning on the myocardium was first described in 1986 by Murry et al., as a slower ATP depletion rate and smaller infarct size in the heart treated with brief episodes of I/R before prolonged occlusion, followed by reperfusion [101, 102]. Mitochondrial pathways play an important role in promoting the activation of cell survival programs following preconditioning via ROS signaling-dependent mechanisms [103]. A good example of cardioprotective roles of reliable amounts of ROS is the metabolic vasodilator effect of H_2O_2, produced by myocardial mitochondria. H_2O_2 serves as a mediator that couples oxygen consumption to coronary blood flow by acting as an activator of redox- and 4-aminopyridine-sensitive voltage-dependent potassium (Kv) channels in smooth muscle cells [100]. In addition, H_2O_2 that derives from complexes I and III in the endothelial mitochondria's electron transport chain is capable of triggering calcium activated potassium (BKCa) channels in order to enhance acetylcholine- and flow-induced coronary vasodilation [104, 105]. More recently, several methods of preconditioning have been developed including ischemic preconditioning (IPC), exercise preconditioning, and pharmacological preconditioning [106–108]. The opening of mitochondrial ATP-sensitive K^+ (mitoKATP) channel is one of the most important mechanisms activated by preconditioning stimuli (Table 12.4). This activation allows potassium to flow into mitochondria leading to depolarization and matrix alkalization. Subsequently, an increase in ROS production activates downstream survival signaling events through PKC, preventing mPTP opening [106, 109]. Additionally, the generation

Table 12.4 Different mitochondrial ROS targeting strategies

Mitochondrial Target	Antioxidant/ Target name	Description	Effect/outcome	Study model	References
Targeting Defective Mitochondria to Prevent Production of Excessive ROS	**MitoQ**	An ubiquinol with a lipophilic TPP+ modification, which enables mitochondrial delivery of the compound	***Effect:*** Antioxidant activity ↓LDH release ***Outcome:*** (–) Tissue damage (–) Mitochondrial swelling ↓Cytochrome c release (–) Caspase 3 activation ↓Tissue and mitochondrial damage ↓Cardiac disfunction	Langendorff model of rat heart I/R damage	[147]
	SkQ	TPP+ modified ROS scavenger	***Effect:*** (–) Oxidative modification of cytochrom c (–) Peroxidase activity of cytochrom c ***Outcome:*** (–) Cytochrom c/ H_2O_2-induced liposomes permeabilization (–) Apoptosis ↑Protection from oxidation-caused I/RI	Langendorff model of rat heart I/R damage	[148]

(continued)

Table 12.4 (continued)

Mitochondrial Target	Antioxidant/ Target name	Description	Effect/outcome	Study model	References
	SS31 (Bendavia)	A four amino acid synthetic peptide(phenylalanine-darginine-phenylalanine-lysine)	***Effect:*** ↓ ROS levels in the MIM ↓ Mitochondrial ROS production ↓Pathological ROS production in aged mitochondria ***Outcome:*** (−) mPTP opening (−) Hydrophobic cytochrome *c* interactions with cardiolipin in the IMM ↑ Cytochrome *c* to function as an electron carrier ↑Cardioprotection	Multiple animal models of MI and heart failure (mice, rats, guinea pigs, rabbits, and sheep)	[149–151]
	Mitochondrial catalase (mCat)	Mitochondrial enzymatic anti-oxidants	***Effect:*** (−)ROS production ***Outcome:*** ↓Mitochondrial oxidative damage (-)ROS-induced cardiac hypertrophy, fibrosis, and heart failure ↑ Lifespan of mice	Different models of angiotensin II-induced cardiac hypertrophy	[152]

	P110	Slective Drp1 inhibitor	***Effect:*** ↓ROS production ↓ATP production (−) Drp1 enzyme activity (−) Drp1/Fis1 ***Outcome:*** ↓Excessive mitochondrial fission ↓Pathological mitochondrial fragmentation without interfering with normal Drp1 functions ↑ Mitochondrial function ↓Infarct size	Primary cardiomyocytes, *ex vivo*, and an *in vivo* MI models	[153–155]
	iGAPDH	Metabolically inactive or oxidized GAPDH: results from high levels of ROS released by mitochondria	***Effect:*** (−) Nearby GAPDH oxidative post-translational modifications Molecular sensor for detecting and tagging damaged mitochondria ***Outcome:*** ↑ Mitophagy ↑Direct uptake of damaged mitochondria into a lysosomal-like structure	HL1 murine cardiomyocyte cell culture model subjected to simulated I/R	[156]

(continued)

Table 12.4 (continued)

Mitochondrial Target	Antioxidant/ Target name	Description	Effect/outcome	Study model	References
Preventing Harmful ROS Signaling Through Protein Kinase Cδ and its Downstream Effectors	**δV1-1**	Specific PKCδ translocation peptide inhibitor	***Effects:*** (−) PKCδ activation (−) PKCδ translocation to the endoplasmic reticulum ↓ROS generation ***Outcomes:*** ↓Endothelial vascular dysfunction ↑Survival of coronary endothelial cells ↓Cell death	I/R model in cardiac myocytes	[21]
Promote Beneficial ROS Signaling Through Protein Kinase C ε and its Downstream Substrates	**Antimycin A**	Secondary metabolites produced by Streptomyces bacteria	***Effect:*** ↑Physiological ROS production ↑PKCε translocation to the mitochondria ***Outcome:*** ↓Cardiac infarct size	I/R Langendorff model of mice heart	[157]

	ψεRAC	PKCε agonist	***Effect :*** ↑PKCε–HSP90 protein–protein interaction ↑PKCε translocation to the mitochondria ↑Phosphorylation and activity of an intra-mitochondrial PKCε substrate ↑ ALDH2 ↑Physiological ROS generation ***Outcome:*** ↑Cardioprotection following I/RI	I/R model of isolated cardiomyocytes	[158]
Promote Beneficial ROS Signaling through preconditioning mimetic	**Diazoxide**	Pharmacological preconditioning mimetic: mitoKATP channel opener	***Effects:*** ↑ROS generation in a connexin 43-dependent manner ↓Cytosolic LDH release ***Outcome:*** ↑Vasodilation ↑ Hyperglycemic effects	Isolated rat heart model of MI	[159]

(continued)

Table 12.4 (continued)

Mitochondrial Target	Antioxidant/ Target name	Description	Effect/outcome	Study model	References
	Diosgenin and atorvastatin	Pharmacological preconditioning mimetic: mitoKATP channel activators/openers	***Diosgenin effect:*** ↓ROS ↑LDH release ***Diosgenin Outcome:*** Induces a cardioprotective preconditioning effect: ↓Infarct area ***Atorvastatin effect:*** ↓OGD/R-induced ROS levels ↑LDH activity in serum ↓OGD/R-induced mitochondrial Ca^{2+} overload (−) mPTP opening (−) $\Delta\Psi m$ depolarisation (+) mitoKATP channels ***Atorvastatin Outcome:*** ↓ Infarct area (−) NAD^+ release Improves the ultrastructure of myocardium Improves hemodynamic variables	Isolated rat heart model of MI	[160, 161]

	Alda-1	A selective Aldehyde dehydrogenase activator	***Effect :*** ↓H_2O_2 release ↑ALDH2 enzymatic activity ↓Aldehydic load in cardiomyocytes	Oxidative stress-related cardiovascular conditions or models	[162–165]
			Outcomes: (−) down-regulation of cardiac mitochondrial ETC(I and V) (+) Mitochondrial bioenergetic status ↓Ventricular diastolic diameter ↓Posterior wall thickness ↑Cardiac function preservation after I/R (−) MI-induced heart failure ↑↑Mitochondrial function ↑Left ventricular function ↑Viability of iPSC-CMs after ischemia	MI rats models Human ischemic hearts	

TPP+ triphenylphosphonium cation, *I/R* ischemia reperfusion, *LDH* cytosolic enzyme lactate dehydrogenase, *ROS* reactive oxygen species, H_2O_2 hydrogen peroxide, *I/RI* ischemia reperfusion injury, *MIM* mitochondrial inner membrane, *ATP* adenosine triphosphate, *MI* myocardial infarction, *mPTP* mitochondrial permeability transition pore , *Drp1* dynamin related protein 1, *GAPDH* glyceraldehyde-3-phosphate dehydrogenase, *iGAPDH* inactive or oxidized GAPDH, *PKCδ* protein kinase C-delta, *PKCε* protein linase C-epsilon, *HSP90* heat shock protein 90, *OGD/R* oxygen–glucose deprivation/recovery, $\Delta\Psi m$ mitochondrial membrane potential, NAD^+ nicotinamide adenine dinucleotide, *ETC* electron transport chain, *iPSC-CMs* induced human pluripotent stem cell-derived cardiomyocytes, (+) stimulation, (−) inhibition, ↑ increase, ↓ decrease of mild matrix swelling improves ATP synthesis and fatty acid oxidation, conditioning the cells to any potential ischemic injury [109].

Ischemic postconditioning on the other hand was first introduced by Zhao et al. in 2003. This term refers to brief periods of ischemia alternating with brief periods of reflow applied at the onset of reperfusion following sustained ischemia. The timing of post-conditioning interference is crucial given that reperfusion injuries occurs only within several minutes following blood reflow [110]. The basics of preconditioning- and postconditioning-mediated protection are very similar [110, 111]. In fact, similar to preconditioning, the mitoKATP/ROS/PKC axis pathway constitutes the basis of postconditioning protective therapy [110]. However, the degree of protection largely depends on the timing of axes activation following reperfusion [112]. Of note, both pre- and postconditioning share an important effect that underlines their cardioprotective efficacy. In fact, the associated prolongation of cellular acidosis that takes place initially during early reperfusion after ischemia favors inhibition of mPTP opening for a few minutes following reperfusion. Pre- and postconditioning released ROS take advantage of delayed protective pH normalization to induce activation of cell survival programs. Therefore, following pH normalization, an arsenal of downstream effectors that prevent mPTP opening is boosted, to preserve mitochondrial and cellular integrity [113].

Several other signaling pathways are implicated in the infarct-sparing effect of pre- or postconditioning [114]. The Reperfusion Injury Salvage Kinases (RISK) pathway involves the activation of two signaling pathways consisting of pro-survival kinases ERK1/2 and Akt that converge on mitochondria to decrease mPTP opening [115]. The Survivor Activating Factor Enhancement (SAFE) pathway involves the induction of JAK-STAT3 signaling. The relative contribution of the RISK and SAFE pathways to cardiac protection varies with the experimental ischemic protocol, as well as species. Some studies have linked SAFE signaling to the initiation of the RISK pathway, although the mechanism is not defined [115]. Both the RISK and SAFE pathways are activated by ROS.

12.8 Conclusion and Future Direction

Reperfusion of the coronary circulation is necessary to prevent irreversible loss of the myocardium. Yet reperfusion causes further harm to the heart via the generation of ROS, which invariably leads to heart failure and shortened lifespan. These ROS target phospholipids of the cell membrane, various structural, transport, and signaling proteins, and DNA, which may then act synergistically to further ROS generation and damage the heart. MMPs, caspases, and calpains are activated as well, further exacerbating structural damage. Much progress has been made in identifying the sources of ROS, which include NOXs, MAO, uncoupled NOSs, p66shc, xanthine dehydrogenase/oxidase, and mitochondria. Paradoxically, lower levels of ROS may activate a number of signaling mechanisms that tamp down excessive ROS generation by mitochondria as initially revealed in preconditioning experiments. The targets of this manipulation include both direct effects on mitochondria, as well as the upregulation of protective proteins at later time points. For practical reasons, direct preconditioning strategies have little if any translational potential. However, complementary approaches, such as exercise-induced preconditioning and ischemic postconditioning, offer clinical promise. Pharmacological manipulations that specifically target mitochondrial complexes that generate ROS during reperfusion are gaining interest as therapies. Although much progress has been realized in the last decade in understanding the source and implications of ROS as foe in I/R-mediated injury to the heart, the upcoming decade should result in the practical application of therapeutic strategies that are based on the revelation of mechanisms defined by the protective actions of ROS in the myocardium.

References

1. Raedschelders K, Ansley DM, Chen DD (2012) The cellular and molecular origin of reactive oxygen species generation during myocardial ischemia and reperfusion. Pharmacol Ther 133(2):230–255
2. Frank A, Bonney M, Bonney S, Weitzel L, Koeppen M, Eckle T (eds) (2012) Myocardial ischemia reperfusion injury: from basic science to clinical bedside. Seminars in cardiothoracic and vascular anesthesia. SAGE, Los Angeles
3. Reimer KA, Lowe JE, Rasmussen MM, Jennings RB (1977) The wavefront phenomenon of ischemic cell death. 1. Myocardial infarct size vs duration of coronary occlusion in dogs. Circulation 56(5):786–794
4. Hearse D, Humphrey S, Chain E (1973) Abrupt reoxygenation of the anoxic potassium-arrested perfused rat heart: a study of myocardial enzyme release. J Mol Cell Cardiol 5(4):395–407
5. Guarnieri C, Flamigni F, Caldarera C (1980) Role of oxygen in the cellular damage induced by re-oxygenation of hypoxic heart. J Mol Cell Cardiol 12(8):797–808
6. Granger DN, Rutili G, McCord JM (1981) Superoxide radicals in feline intestinal ischemia. Gastroenterology 81(1):22–29
7. Granger DN, Kvietys PR (2015) Reperfusion injury and reactive oxygen species: the evolution of a concept. Redox Biol 6:524–551
8. Davidson SM, Yellon DM, Murphy MP, Duchen MR (2011) Slow calcium waves and redox changes precede mitochondrial permeability transition pore opening in the intact heart during hypoxia and reoxygenation. Cardiovasc Res 93(3):445–453
9. Frangogiannis NG (2015) Pathophysiology of myocardial infarction. Compr Physiol 5(4):1841–1875
10. Spann JF Jr, Moellering RC Jr, Haber E, Wheeler EO (1964) Arrhythmias in acute myocardial infarction: a study utilizing an electrocardiographic monitor for automatic detection and recording of arrhythmias. N Engl J Med 271(9):427–431
11. Bajaj A, Sethi A, Rathor P, Suppogu N, Sethi A (2015) Acute complications of myocardial infarction in the current era: diagnosis and management. J Investig Med 63(7):844–855
12. Benjamin EJ, Virani SS, Callaway CW, Chamberlain AM, Chang AR, Cheng S et al (2018) Heart disease and stroke statistics-2018 Update: a report from the American Heart Association. Circulation 137(12):e67–e492
13. Yellon DM, Hausenloy DJ (2007) Myocardial reperfusion injury. N Engl J Med 357(11):1121–1135
14. Puymirat E, Simon T, Cayla G, Cottin Y, Elbaz M, Coste P et al (2017) Acute myocardial infarction: changes in patient characteristics, management, and 6-month outcomes over a period of 20 years in the FAST-MI program (French Registry of Acute ST-Elevation or Non-ST-Elevation Myocardial Infarction) 1995 to 2015. Circulation 136(20):1908–1919

15. Ferdinandy P, Schulz R, Baxter GF (2007) Interaction of cardiovascular risk factors with myocardial ischemia/reperfusion injury, preconditioning, and postconditioning. Pharmacol Rev 59:418–458
16. Ashraf M, Enthammer M, Haller M, Koziel K, Hermann M, Troppmair J (2012) Intracellular signaling in ischemia/reperfusion injury (IRI): from mechanistic insights to therapeutic options. J Transplant Technol Res 3:002
17. Liu T, O'Rourke B (2013) Regulation of Na^{+}/Ca^{2+} exchanger by pyridine nucleotide redox potential in ventricular myocytes. J Biol Chem M113:496588
18. Seidlmayer LK, Juettner VV, Kettlewell S, Pavlov EV, Blatter LA, Dedkova EN (2015) Distinct mPTP activation mechanisms in ischaemia–reperfusion: contributions of Ca^{2+}, ROS, pH, and inorganic polyphosphate. Cardiovasc Res 106(2):237–248
19. Avkiran M, Marber MS (2002) Na+/H+ exchange inhibitors for cardioprotective therapy: progress, problems and prospects. J Am Coll Cardiol 39(5):747–753
20. Zhou T, Chuang CC, Zuo L (2015) Molecular characterization of reactive oxygen species in myocardial ischemia-reperfusion injury. Biomed Res Int 2015:864946
21. Monti M, Donnini S, Giachetti A, Mochly-Rosen D, Ziche M (2010) δPKC inhibition or εPKC activation repairs endothelial vascular dysfunction by regulating eNOS post-translational modification. J Mol Cell Cardiol 48(4):746–756
22. Vina J, Borras C, Abdelaziz KM, Garcia-Valles R, Gomez-Cabrera MC (2013) The free radical theory of aging revisited: the cell signaling disruption theory of aging. Antioxid Redox Signal 19(8):779–787
23. Ray PD, Huang BW, Tsuji Y (2012) Reactive oxygen species (ROS) homeostasis and redox regulation in cellular signaling. Cell Signal 24(5):981–990
24. Murphy MP (2009) How mitochondria produce reactive oxygen species. Biochem J 417(1):1–13
25. Zorov DB, Juhaszova M, Sollott SJ (2014) Mitochondrial reactive oxygen species (ROS) and ROS-induced ROS release. Physiol Rev 94(3):909–950
26. Chouchani ET, Pell VR, Gaude E, Aksentijević D, Sundier SY, Robb EL et al (2014) Ischaemic accumulation of succinate controls reperfusion injury through mitochondrial ROS. Nature 515(7527):431
27. Chouchani ET, Pell VR, James AM, Work LM, Saeb-Parsy K, Frezza C et al (2016) A unifying mechanism for mitochondrial superoxide production during ischemia-reperfusion injury. Cell Metab 23(2):254–263
28. VanTeeffelen JW (2008) How to prevent leaky vessels during reperfusion? Just keep that glycocalyx sealant in place! Critical care (London, England) 12(4):167
29. Slegtenhorst BR, Dor FJ, Rodriguez H, Voskuil FJ, Tullius SG (2014) Ischemia/reperfusion injury and its consequences on immunity and inflammation. Curr Transplant Rep 1(3):147–154
30. Vinten-Johansen J (2004) Involvement of neutrophils in the pathogenesis of lethal myocardial reperfusion injury. Cardiovasc Res 61(3):481–497
31. Perkins K-AA, Pershad S, Chen Q, McGraw S, Adams JS, Zambrano C et al (2012) The effects of modulating eNOS activity and coupling in ischemia/reperfusion (I/R). Naunyn Schmiedeberg's Arch Pharmacol 385(1):27–38
32. Forstermann U, Munzel T (2006) Endothelial nitric oxide synthase in vascular disease: from marvel to menace. Circulation 113(13):1708–1714
33. Barouch LA, Harrison RW, Skaf MW, Rosas GO, Cappola TP, Kobeissi ZA et al (2002) Nitric oxide regulates the heart by spatial confinement of nitric oxide synthase isoforms. Nature 416(6878):337–339
34. Oess S, Icking A, Fulton D, Govers R, Muller-Esterl W (2006) Subcellular targeting and trafficking of nitric oxide synthases. Biochem J 396(3):401–409
35. Paulus WJ, Bronzwaer JG (2002) Myocardial contractile effects of nitric oxide. Heart Fail Rev 7(4):371–383
36. Fleming I (2010) Molecular mechanisms underlying the activation of eNOS. Pflugers Arch – Eur J Physiol 459(6):793–806

37. Godinez-Rubi M, Rojas-Mayorquin AE, Ortuno-Sahagun D (2013) Nitric oxide donors as neuroprotective agents after an ischemic stroke-related inflammatory reaction. Oxidative Med Cell Longev 2013:297357
38. Roberts BW, Mitchell J, Kilgannon JH, Chansky ME, Trzeciak S (2013) Nitric oxide donor agents for the treatment of ischemia/reperfusion injury in human subjects: a systematic review. Shock (Augusta, Ga) 39(3):229–239
39. Chen W, Druhan LJ, Chen CA, Hemann C, Chen YR, Berka V et al (2010) Peroxynitrite induces destruction of the tetrahydrobiopterin and heme in endothelial nitric oxide synthase: transition from reversible to irreversible enzyme inhibition. Biochemistry 49(14):3129–3137
40. Zhang Y, Tocchetti CG, Krieg T, Moens AL (2012) Oxidative and nitrosative stress in the maintenance of myocardial function. Free Radic Biol Med 53(8):1531–1540
41. Alkaitis MS, Crabtree MJ (2012) Recoupling the cardiac nitric oxide synthases: tetrahydrobiopterin synthesis and recycling. Curr Heart Failure Rep 9(3):200–210
42. Crabtree MJ, Tatham AL, Al-Wakeel Y, Warrick N, Hale AB, Cai S et al (2009) Quantitative regulation of intracellular endothelial nitric-oxide synthase (eNOS) coupling by both tetrahydrobiopterin-eNOS stoichiometry and biopterin redox status: insights from cells with tet-regulated GTP cyclohydrolase I expression. J Biol Chem 284(2):1136–1144
43. De Pascali F, Hemann C, Samons K, Chen C-A, Zweier JL (2014) Hypoxia and reoxygenation induce endothelial nitric oxide synthase uncoupling in endothelial cells through tetrahydrobiopterin depletion and S-glutathionylation. Biochemistry 53(22):3679–3688
44. Siu KL, Lotz C, Ping P, Cai H (2015) Netrin-1 abrogates ischemia/reperfusion-induced cardiac mitochondrial dysfunction via nitric oxide-dependent attenuation of NOX4 activation and recoupling of NOS. J Mol Cell Cardiol 78:174–185
45. Pernow J, Jung C (2013) Arginase as a potential target in the treatment of cardiovascular disease: reversal of arginine steal? Cardiovasc Res 98(3):334–343
46. Giraldez RR, Panda A, Xia Y, Sanders SP, Zweier JL (1997) Decreased nitric-oxide synthase activity causes impaired endothelium-dependent relaxation in the postischemic heart. J Biol Chem 272(34):21420–21426
47. Mialet-Perez J, Bianchi P, Kunduzova O, Parini A (2007) New insights on receptor-dependent and monoamine oxidase-dependent effects of serotonin in the heart. J Neural Transm (Vienna, Austria: 1996) 114(6):823–827
48. Manni ME, Rigacci S, Borchi E, Bargelli V, Miceli C, Giordano C et al (2016) Monoamine oxidase is overactivated in left and right ventricles from ischemic hearts: an intriguing therapeutic target. Oxidative Med Cell Longev 2016:4375418
49. Giorgio M, Migliaccio E, Orsini F, Paolucci D, Moroni M, Contursi C et al (2005) Electron transfer between cytochrome c and p66Shc generates reactive oxygen species that trigger mitochondrial apoptosis. Cell 122(2):221–233
50. George J, Struthers AD (2009) Role of urate, xanthine oxidase and the effects of allopurinol in vascular oxidative stress. Vasc Health Risk Manag 5(1):265–272
51. Wang G, Qian P, Jackson FR, Qian G, Wu G (2008) Sequential activation of JAKs, STATs and xanthine dehydrogenase/oxidase by hypoxia in lung microvascular endothelial cells. Int J Biochem Cell Biol 40(3):461–470
52. Berry CE, Hare JM (2004) Xanthine oxidoreductase and cardiovascular disease: molecular mechanisms and pathophysiological implications. J Physiol 555(Pt 3):589–606
53. Cantu-Medellin N, Kelley EE (2013) Xanthine oxidoreductase-catalyzed reactive species generation: a process in critical need of reevaluation. Redox Biol 1:353–358
54. Maia LB, Pereira V, Mira L, Moura JJ (2015) Nitrite reductase activity of rat and human xanthine oxidase, xanthine dehydrogenase, and aldehyde oxidase: evaluation of their contribution to NO formation in vivo. Biochemistry 54(3):685–710
55. Shafik AN (2013) Febuxostat improves the local and remote organ changes induced by intestinal ischemia/reperfusion in rats. Dig Dis Sci 58(3):650–659
56. Kahles T, Brandes RP (2013) Which NADPH oxidase isoform is relevant for ischemic stroke? The case for nox 2. Antioxid Redox Signal 18(12):1400–1417

57. Brandes RP, Weissmann N, Schroder K (2010) NADPH oxidases in cardiovascular disease. Free Radic Biol Med 49(5):687–706
58. Doerries C, Grote K, Hilfiker-Kleiner D, Luchtefeld M, Schaefer A, Holland SM et al (2007) Critical role of the NAD(P)H oxidase subunit p47phox for left ventricular remodeling/dysfunction and survival after myocardial infarction. Circ Res 100(6):894–903
59. Lambeth JD, Krause KH, Clark RA (2008) NOX enzymes as novel targets for drug development. Semin Immunopathol 30(3):339–363
60. Kleikers PW, Wingler K, Hermans JJ, Diebold I, Altenhofer S, Radermacher KA et al (2012) NADPH oxidases as a source of oxidative stress and molecular target in ischemia/reperfusion injury. J Mol Med (Berlin, Germany) 90(12):1391–1406
61. Simone S, Rascio F, Castellano G, Divella C, Chieti A, Ditonno P et al (2014) Complement-dependent NADPH oxidase enzyme activation in renal ischemia/reperfusion injury. Free Radic Biol Med 74:263–273
62. Gan X, Su G, Zhao W, Huang P, Luo G, Hei Z (2013) The mechanism of sevoflurane preconditioning-induced protections against small intestinal ischemia reperfusion injury is independent of mast cell in rats. Mediat Inflamm 2013:378703
63. Mittal M, Roth M, Konig P, Hofmann S, Dony E, Goyal P et al (2007) Hypoxia-dependent regulation of nonphagocytic NADPH oxidase subunit NOX4 in the pulmonary vasculature. Circ Res 101(3):258–267
64. Nanduri J, Vaddi DR, Khan SA, Wang N, Makarenko V, Semenza GL et al (2015) HIF-1alpha activation by intermittent hypoxia requires NADPH oxidase stimulation by xanthine oxidase. PLoS One 10(3):e0119762
65. Nguyen Dinh Cat A, Montezano AC, Burger D, Touyz RM (2013) Angiotensin II, NADPH oxidase, and redox signaling in the vasculature. Antioxid Redox Signal 19(10):1110–1120
66. Sanderson TH, Reynolds CA, Kumar R, Przyklenk K, Huttemann M (2013) Molecular mechanisms of ischemia-reperfusion injury in brain: pivotal role of the mitochondrial membrane potential in reactive oxygen species generation. Mol Neurobiol 47(1):9–23
67. Cadenas E (2004) Mitochondrial free radical production and cell signaling. Mol Asp Med 25(1-2):17–26
68. Drose S, Brandt U (2012) Molecular mechanisms of superoxide production by the mitochondrial respiratory chain. Adv Exp Med Biol 748:145–169
69. Pasdois P, Parker JE, Griffiths EJ, Halestrap AP (2011) The role of oxidized cytochrome c in regulating mitochondrial reactive oxygen species production and its perturbation in ischaemia. Biochem J 436(2):493–505
70. Chen Q, Lesnefsky EJ (2006) Depletion of cardiolipin and cytochrome c during ischemia increases hydrogen peroxide production from the electron transport chain. Free Radic Biol Med 40(6):976–982
71. Bleier L, Wittig I, Heide H, Steger M, Brandt U, Drose S (2015) Generator-specific targets of mitochondrial reactive oxygen species. Free Radic Biol Med 78:1–10
72. Holzerova E, Prokisch H (2015) Mitochondria: much ado about nothing? How dangerous is reactive oxygen species production? Int J Biochem Cell Biol 63:16–20
73. Aon MA, Cortassa S, Akar FG, Brown DA, Zhou L, O'Rourke B (2009) From mitochondrial dynamics to arrhythmias. Int J Biochem Cell Biol 41(10):1940–1948
74. Jena N, Mishra P (2012) Formation of ring-opened and rearranged products of guanine: mechanisms and biological significance. Free Radic Biol Med 53(1):81–94
75. Niles JC, Wishnok JS, Tannenbaum SR (2006) Peroxynitrite-induced oxidation and nitration products of guanine and 8-oxoguanine: structures and mechanisms of product formation. Nitric Oxide 14(2):109–121
76. Ayala A, Muñoz MF, Argüelles S (2014) Lipid peroxidation: production, metabolism, and signaling mechanisms of malondialdehyde and 4-hydroxy-2-nonenal. Oxidative Med Cell Longev 2014:360438
77. Musiek ES, Yin H, Milne GL, Morrow JD (2005) Recent advances in the biochemistry and clinical relevance of the isoprostane pathway. Lipids 40(10):987–994

78. Xu X, Muller JG, Ye Y, Burrows CJ (2008) DNA– protein cross-links between guanine and lysine depend on the mechanism of oxidation for formation of C5 vs C8 guanosine adducts. J Am Chem Soc 130(2):703–709
79. Krijnen PA, Nijmeijer R, Meijer CJ, Visser CA, Hack CE, Niessen HW (2002) Apoptosis in myocardial ischaemia and infarction. J Clin Pathol 55(11):801–811
80. Kloner RA, Jennings RB (2001) Consequences of brief ischemia: stunning, preconditioning, and their clinical implications: part 2. Circulation 104(25):3158–3167
81. Stadtman E, Levine R (2003) Free radical-mediated oxidation of free amino acids and amino acid residues in proteins. Amino Acids 25(3-4):207–218
82. Peluffo G, Radi R (2007) Biochemistry of protein tyrosine nitration in cardiovascular pathology. Cardiovasc Res 75(2):291–302
83. Kuznetsov AV, Smigelskaite J, Doblander C, Janakiraman M, Hermann M, Wurm M et al (2008) Survival signaling by C-RAF: mitochondrial reactive oxygen species and Ca^{2+} are critical targets. Mol Cell Biol 28(7):2304–2313
84. Lehwald N, Tao GZ, Jang KY, Sorkin M, Knoefel WT, Sylvester KG (2011) Wnt–β-catenin signaling protects against hepatic ischemia and reperfusion injury in mice. Gastroenterology 141(2):707-18. e5
85. Yu HC, Qin HY, He F, Wang L, Fu W, Liu D et al (2011) Canonical notch pathway protects hepatocytes from ischemia/reperfusion injury in mice by repressing reactive oxygen species production through JAK2/STAT3 signaling. Hepatology 54(3):979–988
86. Zhu Y, Prives C (2009) P53 and metabolism: the GAMT connection. Mol Cell 36(3):351–352
87. Acin-Perez R, Salazar E, Kamenetsky M, Buck J, Levin LR, Manfredi G (2009) Cyclic AMP produced inside mitochondria regulates oxidative phosphorylation. Cell Metab 9(3):265–276
88. Brodie C, Blumberg P (2003) Regulation of cell apoptosis by protein kinase c δ. Apoptosis 8(1):19–27
89. Cohen MV, Yang X-M, Downey JM (2007) The pH hypothesis of postconditioning: staccato reperfusion reintroduces oxygen and perpetuates myocardial acidosis. Circulation 115(14):1895–1903
90. Zhang Z, Feng H-Z, Jin J-P (2011) Structure of the NH2-terminal variable region of cardiac troponin T determines its sensitivity to restrictive cleavage in pathophysiological adaptation. Arch Biochem Biophys 515(1-2):37–45
91. Sung MM, Schulz CG, Wang W, Sawicki G, Bautista-López NL, Schulz R (2007) Matrix metalloproteinase-2 degrades the cytoskeletal protein α-actinin in peroxynitrite mediated myocardial injury. J Mol Cell Cardiol 43(4):429–436
92. Hernando V, Inserte J, Sartório CL, Parra VM, Poncelas-Nozal M, Garcia-Dorado D (2010) Calpain translocation and activation as pharmacological targets during myocardial ischemia/reperfusion. J Mol Cell Cardiol 49(2):271–279
93. Perrelli M-G, Pagliaro P, Penna C (2011) Ischemia/reperfusion injury and cardioprotective mechanisms: role of mitochondria and reactive oxygen species. World J Cardiol 3(6):186–200
94. Perraud A-L, Fleig A, Dunn CA, Bagley LA, Launay P, Schmitz C et al (2001) ADP-ribose gating of the calcium-permeable LTRPC2 channel revealed by Nudix motif homology. Nature 411(6837):595
95. Yang K, Chang W, Yang P, Chien C-L, Lai M, Su M et al (2006) Activation of the transient receptor potential M2 channel and poly (ADP-ribose) polymerase is involved in oxidative stress-induced cardiomyocyte death. Cell Death Differ 13(10):1815
96. Camara AK, Bienengraeber M, Stowe DF (2011) Mitochondrial approaches to protect against cardiac ischemia and reperfusion injury. Front Physiol 2:13
97. Ong S-B, Subrayan S, Lim SY, Yellon DM, Davidson SM, Hausenloy DJ (2010) Inhibiting mitochondrial fission protects the heart against ischemia/reperfusion injury. Circulation 121(18):2012–2022
98. Zorov DB, Juhaszova M, Sollott SJ (2006) Mitochondrial ROS-induced ROS release: an update and review. Biochim Biophys Acta (BBA)-Bioenergetics 1757(5–6):509–517
99. Kalogeris T, Bao Y, Korthuis RJ (2014) Mitochondrial reactive oxygen species: a double edged sword in ischemia/reperfusion vs preconditioning. Redox Biol 2:702–714

100. Rogers PA, Chilian WM, Bratz IN, Bryan RM Jr, Dick GM (2007) H_2O_2 activates redox- and 4-aminopyridine-sensitive Kv channels in coronary vascular smooth muscle. Am J Physiol Heart Circ Physiol 292(3):H1404–H1411
101. Ytrehus K, Liu Y, Downey JM (1994) Preconditioning protects ischemic rabbit heart by protein kinase C activation. Am J Phys 266(3 Pt 2):H1145–H1152
102. Murry CE, Jennings RB, Reimer KA (1986) Preconditioning with ischemia: a delay of lethal cell injury in ischemic myocardium. Circulation 74(5):1124–1136
103. Wang Q, Sun AY, Simonyi A, Kalogeris TJ, Miller DK, Sun GY et al (2007) Ethanol preconditioning protects against ischemia/reperfusion-induced brain damage: role of NADPH oxidase-derived ROS. Free Radic Biol Med 43(7):1048–1060
104. Zhang DX, Gutterman DD (2007) Mitochondrial reactive oxygen species-mediated signaling in endothelial cells. Am J Physiol Heart Circ Physiol 292(5):H2023–H2031
105. Xi Q, Cheranov SY, Jaggar JH (2005) Mitochondria-derived reactive oxygen species dilate cerebral arteries by activating Ca^{2+} sparks. Circ Res 97(4):354–362
106. Quindry JC, Hamilton KL (2013) Exercise and cardiac preconditioning against ischemia reperfusion injury. Curr Cardiol Rev 9(3):220–229
107. Ascensao A, Ferreira R, Magalhaes J (2007) Exercise-induced cardioprotection – biochemical, morphological and functional evidence in whole tissue and isolated mitochondria. Int J Cardiol 117(1):16–30
108. Frasier CR, Moore RL, Brown DA (2011) Exercise-induced cardiac preconditioning: how exercise protects your achy-breaky heart. J Appl Physiol (1985) 111(3):905–915
109. Boengler K, Heusch G, Schulz R (2011) Mitochondria in postconditioning. Antioxid Redox Signal 14(5):863–880
110. Ovize M, Baxter GF, Di Lisa F, Ferdinandy P, Garcia-Dorado D, Hausenloy DJ et al (2010) Postconditioning and protection from reperfusion injury: where do we stand? Position paper from the Working Group of Cellular Biology of the Heart of the European Society of Cardiology. Cardiovasc Res 87(3):406–423
111. Skyschally A, van Caster P, Iliodromitis EK, Schulz R, Kremastinos DT, Heusch G (2009) Ischemic postconditioning: experimental models and protocol algorithms. Basic Res Cardiol 104(5):469–483
112. Hausenloy DJ, Wynne AM, Yellon DM (2007) Ischemic preconditioning targets the reperfusion phase. Basic Res Cardiol 102(5):445–452
113. Cohen MV, Yang XM, Downey JM (2008) Acidosis, oxygen, and interference with mitochondrial permeability transition pore formation in the early minutes of reperfusion are critical to postconditioning's success. Basic Res Cardiol 103(5):464–471
114. Rossello X, Yellon DM (2017) The RISK pathway and beyond. Basic Res Cardiol 113(1):2
115. Zouein FA, Altara R, Chen Q, Lesnefsky EJ, Kurdi M, Booz GW (2015) Pivotal importance of STAT3 in protecting the heart from acute and chronic stress: new advancement and unresolved issues. Front Cardiovasc Med 2:36
116. Shlafer M, Kane PF, Kirsh MM (1982) Superoxide dismutase plus catalase enhances the efficacy of hypothermic cardioplegia to protect the globally ischemic, reperfused heart. J Thorac Cardiovasc Surg 83(6):830–839
117. Sies H (1993) Strategies of antioxidant defense. Eur J Biochem 215(2):213–219
118. Ambrosio G, Zweier JL, Jacobus WE, Weisfeldt ML, Flaherty JT (1987) Improvement of postischemic myocardial function and metabolism induced by administration of deferoxamine at the time of reflow: the role of iron in the pathogenesis of reperfusion injury. Circulation 76(4):906–915
119. Arroyo CM, Kramer JH, Dickens BF, Weglicki WB (1987) Identification of free radicals in myocardial ischemia/reperfusion by spin trapping with nitrone DMPO. FEBS Lett 221(1):101–104
120. Arroyo CM, Kramer JH, Leiboff RH, Mergner GW, Dickens BF, Weglicki WB (1987) Spin trapping of oxygen and carbon-centered free radicals in ischemic canine myocardium. Free Radic Biol Med 3(5):313–316

121. Bolli R, Patel BS, Jeroudi MO, Lai EK, McCay PB (1988) Demonstration of free radical generation in "stunned" myocardium of intact dogs with the use of the spin trap alpha-phenyl N-tert-butyl nitrone. J Clin Invest 82(2):476–485
122. Garlick PB, Davies MJ, Hearse DJ, Slater TF (1987) Direct detection of free radicals in the reperfused rat heart using electron spin resonance spectroscopy. Circ Res 61(5):757–760
123. Corretti MC, Koretsune Y, Kusuoka H, Chacko VP, Zweier JL, Marban E (1991) Glycolytic inhibition and calcium overload as consequences of exogenously generated free radicals in rabbit hearts. J Clin Invest 88(3):1014–1025
124. Josephson RA, Silverman HS, Lakatta EG, Stern MD, Zweier JL (1991) Study of the mechanisms of hydrogen peroxide and hydroxyl free radical-induced cellular injury and calcium overload in cardiac myocytes. J Biol Chem 266(4):2354–2361
125. Xia Y, Zweier JL (1995) Substrate control of free radical generation from xanthine oxidase in the postischemic heart. J Biol Chem 270(32):18797–18803
126. Webb A, Bond R, McLean P, Uppal R, Benjamin N, Ahluwalia A (2004) Reduction of nitrite to nitric oxide during ischemia protects against myocardial ischemia-reperfusion damage. Proc Natl Acad Sci U S A 101(37):13683–13688
127. Victorino GP, Ramirez RM, Chong TJ, Curran B, Sadjadi J (2008) Ischemia-reperfusion injury in rats affects hydraulic conductivity in two phases that are temporally and mechanistically separate. Am J Phys Heart Circ Phys 295(5):H2164–H2H71
128. Borchi E, Parri M, Papucci L, Becatti M, Nassi N, Nassi P et al (2009) Role of NADPH oxidase in H9c2 cardiac muscle cells exposed to simulated ischaemia-reperfusion. J Cell Mol Med 13(8b):2724–2735
129. Donoso P, Finkelstein JP, Montecinos L, Said M, Sánchez G, Vittone L et al (2014) Stimulation of NOX2 in isolated hearts reversibly sensitizes RyR2 channels to activation by cytoplasmic calcium. J Mol Cell Cardiol 68:38–46
130. Braunersreuther V, Montecucco F, Ashri M, Pelli G, Galan K, Frias M et al (2013) Role of NADPH oxidase isoforms NOX1, NOX2 and NOX4 in myocardial ischemia/reperfusion injury. J Mol Cell Cardiol 64:99–107
131. Dworakowski R, Anilkumar N, Zhang M, Shah A (2006) Redox signalling involving NADPH oxidase-derived reactive oxygen species. Portland Press Limited, London
132. Chambers DE, Parks DA, Patterson G, Roy R, McCord JM, Yoshida S et al (1985) Xanthine oxidase as a source of free radical damage in myocardial ischemia. J Mol Cell Cardiol 17(2):145–152
133. Webb A, Bond R, McLean P, Uppal R, Benjamin N, Ahluwalia A (2004) Reduction of nitrite to nitric oxide during ischemia protects against myocardial ischemia–reperfusion damage. Proc Natl Acad Sci 101(37):13683–13688
134. Lesnefsky EJ, Chen Q, Moghaddas S, Hassan MO, Tandler B, Hoppel CL (2004) Blockade of electron transport during ischemia protects cardiac mitochondria. J Biol Chem 279(46):47961–47967
135. White MY, Tchen AS, McCarron HC, Hambly BD, Jeremy RW, Cordwell SJ (2006) Proteomics of ischemia and reperfusion injuries in rabbit myocardium with and without intervention by an oxygen-free radical scavenger. Proteomics 6(23):6221–6233
136. Liu B, Tewari AK, Zhang L, Green-Church KB, Zweier JL, Chen YR et al (2009) Proteomic analysis of protein tyrosine nitration after ischemia reperfusion injury: mitochondria as the major target. Biochim Biophys Acta 1794(3):476–485
137. Zhang L, Chen C-L, Kang PT, Garg V, Hu K, Green-Church KB et al (2010) Peroxynitrite-mediated oxidative modifications of complex II: relevance in myocardial infarction. Biochemistry 49(11):2529–2539
138. Viappiani S, Nicolescu AC, Holt A, Sawicki G, Crawford BD, León H et al (2009) Activation and modulation of 72 kDa matrix metalloproteinase-2 by peroxynitrite and glutathione. Biochem Pharmacol 77(5):826–834
139. Fert-Bober J, Leon H, Sawicka J, Basran RS, Devon RM, Schulz R et al (2008) Inhibiting matrix metalloproteinase-2 reduces protein release into coronary effluent from isolated rat hearts during ischemia-reperfusion. Basic Res Cardiol 103(5):431–443

140. Leon H, Baczko I, Sawicki G, Light PE, Schulz R (2008) Inhibition of matrix metalloproteinases prevents peroxynitrite-induced contractile dysfunction in the isolated cardiac myocyte. Br J Pharmacol 153(4):676–683
141. Jena N, Kushwaha P, Mishra P (2008) Reaction of hypochlorous acid with imidazole: formation of 2-chloro-and 2-oxoimidazoles. J Comput Chem 29(1):98–107
142. Shukla P, Jena N, Mishra P (2011) Quantum theoretical study of molecular mechanisms of mutation and cancer-a review. Proc Natl Acad Sci India Sect A-Phys Sci 81:79–98
143. Minko IG, Harbut MB, Kozekov ID, Kozekova A, Jakobs PM, Olson SB et al (2008) Role for DNA polymerase kappa in the processing of N2-N2-guanine interstrand crosslinks. J Biol Chem
144. Loor G, Kondapalli J, Iwase H, Chandel NS, Waypa GB, Guzy RD et al (2011) Mitochondrial oxidant stress triggers cell death in simulated ischemia-reperfusion. Biochim Biophys Acta 1813(7):1382–1394
145. Sugamura K, Keaney JF Jr (2011) Reactive oxygen species in cardiovascular disease. Free Radic Biol Med 51(5):978–992
146. Xue Y-Z, Wang L-X, Liu H-Z, Qi X-W, Wang X-H, Ren H-Z (2007) L-carnitine as an adjunct therapy to percutaneous coronary intervention for non-ST elevation myocardial infarction. Cardiovasc Drugs Ther 21(6):445–448
147. Adlam VJ, Harrison JC, Porteous CM, James AM, Smith RA, Murphy MP et al (2005) Targeting an antioxidant to mitochondria decreases cardiac ischemia-reperfusion injury. FASEB J 19(9):1088–1095
148. Anisimov VN, Egorov MV, Krasilshchikova MS, Lyamzaev KG, Manskikh VN, Moshkin MP et al (2011) Effects of the mitochondria-targeted antioxidant SkQ1 on lifespan of rodents. Aging (Albany NY) 3(11):1110
149. Birk A, Chao W, Bracken C, Warren J, Szeto H (2014) Targeting mitochondrial cardiolipin and the cytochrome c/cardiolipin complex to promote electron transport and optimize mitochondrial ATP synthesis. Br J Pharmacol 171(8):2017–2028
150. Kloner RA, Hale SL, Dai W, Gorman RC, Shuto T, Koomalsingh KJ et al (2012) Reduction of ischemia/reperfusion injury with bendavia, a mitochondria-targeting cytoprotective peptide. J Am Heart Assoc 1(3):e001644
151. Chakrabarti AK, Feeney K, Abueg C, Brown DA, Czyz E, Tendera M et al (2013) Rationale and design of the EMBRACE STEMI study: a phase 2a, randomized, double-blind, placebo-controlled trial to evaluate the safety, tolerability and efficacy of intravenous Bendavia on reperfusion injury in patients treated with standard therapy including primary percutaneous coronary intervention and stenting for ST-segment elevation myocardial infarction. Am Heart J 165(4):509-14. e7
152. Dai D-F, Johnson SC, Villarin JJ, Chin MT, Nieves-Cintrón M, Chen T et al (2011) Mitochondrial oxidative stress mediates angiotensin II–induced cardiac hypertrophy and Gαq overexpression–induced heart failure. Circ Res 108(7):837–846
153. Yeon J-Y, Min S-H, Park H-J, Kim J-W, Lee Y-H, Park S-Y et al (2015) Mdivi-1, mitochondrial fission inhibitor, impairs developmental competence and mitochondrial function of embryos and cells in pigs. J Reprod Dev 61(2):81–89
154. Qi X, Qvit N, Su Y-C, Mochly-Rosen D (2012) Novel Drp1 inhibitor diminishes aberrant mitochondrial fission and neurotoxicity. J Cell Sci 126:789
155. Disatnik MH, Ferreira JC, Campos JC, Gomes KS, Dourado PM, Qi X et al (2013) Acute inhibition of excessive mitochondrial fission after myocardial infarction prevents long-term cardiac dysfunction. J Am Heart Assoc 2(5):e000461
156. Yogalingam G, Hwang S, Ferreira JC, Mochly-Rosen D (2013) Glyceraldehyde-3-phosphate dehydrogenase (GAPDH) phosphorylation by protein kinase C delta (δPKC) inhibits mitochondrial elimination by lysosomal-like structures following ischemia and reoxygenation-induced injury. J Biol Chem M113:466870
157. Kabir AM, Clark JE, Tanno M, Cao X, Hothersall JS, Dashnyam S et al (2006) Cardioprotection initiated by reactive oxygen species is dependent on activation of PKCε. Am J Phys Heart Circ Phys 291(4):H1893–H18H9

158. Budas GR, Churchill EN, Disatnik M-H, Sun L, Mochly-Rosen D (2010) Mitochondrial import of PKCε is mediated by HSP90: a role in cardioprotection from ischaemia and reperfusion injury. Cardiovasc Res 88(1):83–92
159. Heinzel FR, Lisa F (2005) Impairment of diazoxide-induced formation of reactive oxygen species and loss of cardioprotection in connexin 43 deficient mice. Circ Res 97:583–586
160. Badalzadeh R, Yousefi B, Tajaddini A, Ahmadian N (2015) Diosgenin-induced protection against myocardial ischaemia-reperfusion injury is mediated by mitochondrial KATP channels in a rat model. Perfusion 30(7):565–571
161. Zhao Z, Cui W, Zhang H, Gao H, Li X, Wang Y et al (2015) Pre-treatment of a single high-dose of atorvastatin provided cardioprotection in different ischaemia/reperfusion models via activating mitochondrial KATP channel. Eur J Pharmacol 751:89–98
162. Beretta M, Gorren AC, Wenzl MV, Weis R, Russwurm M, Koesling D et al (2009) Characterization of the East Asian variant of aldehyde dehydrogenase-2: bioactivation of nitroglycerin and effects of Alda-1. J Biol Chem M109:014548
163. Gong D, Zhang Y, Zhang H, Gu H, Jiang Q, Hu S (2012) Aldehyde dehydrogenase-2 activation during cardioplegic arrest enhances the cardioprotection against myocardial ischemia–reperfusion injury. Cardiovasc Toxicol 12(4):350–358
164. Gomes KM, Bechara LR, Lima VM, Ribeiro MA, Campos JC, Dourado PM et al (2015) Aldehydic load and aldehyde dehydrogenase 2 profile during the progression of post-myocardial infarction cardiomyopathy: benefits of Alda-1. Int J Cardiol 179:129–138
165. Ebert AD, Kodo K, Liang P, Wu H, Huber BC, Riegler J et al (2014) Characterization of the molecular mechanisms underlying increased ischemic damage in the aldehyde dehydrogenase 2 genetic polymorphism using a human induced pluripotent stem cell model system. Sci Transl Med 6(255):255ra130–255ra130

Redox Aspects of Myocardial Ischemia/ Reperfusion Injury and Cardioprotection

13

Pasquale Pagliaro, Saveria Femminò, and Claudia Penna

13.1 Introduction

It is well known that many cardiovascular diseases (CVDs) are prevalently due to inflammatory processes. Free radical production and inflammatory processes are entangled in a *vicious circle*: the more radicals lead to more inflammation, and inflammation leads to more radicals. These are all well accepted concepts. It is also well accepted that during exercise bursts of free radicals, including reactive oxygen species (ROS), nitric oxide and other reactive nitrogen species (RNS), are produced by the cells of the cardiovascular system. If these concepts are true then a question arises: *why in some cases ROS/RNS are deleterious triggering and exacerbating pathological events, whereas in other circumstances they are beneficial and prevents all CVD?* In this chapter we will see if it is possible to answer to this question. Actually, as we will see, low levels/doses of stressors (*e.g.*, exercise, chemicals/drugs, thermic stimuli, hypoxia, intermittent fasting, toxins and radiation) are known to stimulate a wide range of adaptive responses that may affect the success of subsequent therapeutic interventions for a vast spectrum of disorders. Stressors that trigger adaptive responses also prevent damage in tissues exposed afterward to injurious levels of stressors, including severe psychological stress. Therefore, we can anticipate that ROS/RNS play important role in adaptive mechanisms and they may be both pro- and anti-inflammatory molecules, depending on the context. It is not only a matter of quantity or quality of the reactive species but several factors may influence the outcome, including, for instance, the compartmentalization, the contextual presence of antioxidants and/or activated protective signaling pathways.

Definition of Free Radicals and Reactive Species First of all, let's define biologically relevant free radicals and reactive species. *Free radicals* are atoms or

P. Pagliaro (✉) · S. Femminò · C. Penna
Department of Clinical and Biological Sciences, University of Torino, Torino, Italy
e-mail: pasquale.pagliaro@unito.it

S. Chakraborti et al. (eds.), *Oxidative Stress in Heart Diseases*,
https://doi.org/10.1007/978-981-13-8273-4_13

molecules with an odd (unpaired) number of electrons. They are highly reactive and can be formed when oxygen interacts with certain atoms or group of atoms. *Reactive species* are chemically reactive molecules acting as oxidizing agents and/or they can be easily converted into radicals. In the biological systems, they often contain oxygen or nitrogen molecules. Therefore, reactive species is a collective term that includes not only the oxygen radicals but also some non-radical derivatives (Table 13.1). Thus, all oxygen radicals are reactive species, but not all reactive species are free radicals.

The Role of Reactive Species and Free Radicals in Biological System The role of reactive species and free radicals in several physiological and pathophysiological conditions is becoming clearer. There are no doubts that these species have a prominent role in almost all cardiopathological disorders. Oxygen and nitrogen centered reactive species and radicals, such as superoxide anion, hydrogen peroxide and nitric oxide are known to be important physiological signaling molecules. In particular, these species act on membranes and receptors and can modulate vital intracellular signaling pathways and gene expression. On the other hand, in some

Table 13.1 Some biologically relevant reactive species (formed by and/or acting on living beings)

Reactive Oxygen Species (ROS)	Reactive Nitrogen Species (RNS)
Free radicals	***Free radicals***
Alkoxyl, $RO^{\cdot}$	Nitrogen dioxide, $NO_2^{\cdot}$
Carbon dioxide radical, $CO_2^{\cdot -}$	Nitric oxide, $NO^{\cdot}$
Carbonate, $CO_3^{\cdot -}$	Nitrate radical, $NO_3^{\cdot}$
Hydroperoxyl, $HO_2^{\cdot}$ (protonated superoxide)	
Hydroxyl, $OH^{\cdot}$	
Peroxyl, $RO_2^{\cdot}$	
Singlet $O_2{}^1\Sigma g^+$	
Superoxide, $O_2^{\cdot -}$	
Non-radicals	***Non-radicals***
Hydrogen peroxide, H_2O_2	Alkyl peroxynitrates, RO_2ONO
Hypobromous acid, HOBr	Alkyl peroxynitrites, ROONO
Hypochlorous acid, HOCl	Dinitrogen tetroxide, N_2O_4
Organic peroxides, $ROOH^-$	Dinitrogen trioxide, N_2O_3
Ozone, O_3	Nitronium cation, NO_2^+
Peroxomonocarbonate, $HOOCO_2$	Nitrosyl cation, NO^+
Singlet oxygen ($O_2{}^1\Delta g$)	Nitrous acid, HNO_2
	Nytroxyl, HNO
	Nitroxyl anion, NO^-
	Nitryl chloride, NO_2Cl
	Peroxyacetyl nitrate, $CH_3C(O)OONO_2$
	Peroxynitrate, O_2NOO^-
	Peroxynitrite, $ONOO^-$
	Peroxynitrous acid, ONOOH

circumstances, such as activation of immune cells, ischemia/reperfusion injury (IRI) or drug metabolism, ROS/RNS are associated with cytotoxicity, which in cardiovascular system leads to cardiovascular degenerative disorders. Accordingly, several evidences strength the notion that oxidative signaling plays a pivotal role in cardiovascular physiology. On the other hand, oxidative stress, plays a fundamental role in some cardiovascular pathophysiology.

13.2 Oxidative- or Redox-Signaling Versus Oxidative- or Redox-Stress

The term "redox" refers to oxo-reduction reactions and would be preferable. Nevertheless, from now on we use the terms *"oxidative signaling" or "redox signaling"* with a positive connotation, that is when ROS trigger or mediate cardioprotective actions. Whereas, we use the terms *"oxidative stress" or "redox stress"* to highlight the potential negative effect of ROS. We can use also the terms *nitrosative stress or signaling* when the major players are RNS. Historically, ROS/RNS and free radicals were considered exclusively a cause cellular damage, lacking any physiological function. Indeed, ROS/RNS accumulation and oxidative/nitrosative damage have been linked to multiple pathologies, including neuro- cardio-degenerative diseases, cancer, and diabetes, as well as a cause of acceleration of aging. Thus, ROS/RNS were originally envisioned as a sign of an imperfect system. However, very few, if any, biological systems possess such glaring imperfection. It is likely that the oxidative metabolism is not as ugly as an evil. Nowadays there are no doubt that ROS and RNS are critical for healthy cell function. We should consider that *mitochondria* are the principal source of oxidants in cardiomyocytes and that some cells, especially those of the cardiovascular system, have a high concentration of mitochondria. In particular, cardiomyocytes have the highest mitochondrial density in mammalian cells (35–40% of the volume of cardiomyocytes is occupied by mitochondria). Therefore, the ROS/RNS produced because of the normal function of mitochondria should be considered in their primary role of *signaling molecules*.

In this chapter, we will report the evidence in this field for our strong advances in our knowledge thanks to the work of researchers, physiologists and clinicians. Pointing out that if all free radicals are not all bad, nor antioxidants are all good, we must keep in mind that ROS/RNS:

- are formed as a by-product of the normal metabolism of oxygen and nitrogen;
- are critical for healthy cell function;
- are reactive molecules derived from the most widely diffuse vasodilator, nitric oxide;
- have important roles in physiological homeostasis as major players of cell signaling;
- levels can increase dramatically in some circumstances, such as during exercise;
- may have beneficial effects in various protective and adaptive physiological conditions;

- may have damaging effects in various heart and vascular disorders (coronary heart disease, cardiac arrhythmias, ventricular hypertrophy, and heart failure).

Since ROS/RNS mediated effects are among endogenous mechanisms of either injury or protection, we will mainly analyze the role of endogenous and exogenous antioxidants and their putative role in IRI and in cardioprotection (pre- and postconditioning).

13.2.1 Reactive Oxygen/Nitrogen Species

The chemical reactions between substances are transformations of matter in which the energy stored in the chemical bonds of the starting substances are transferred to the new chemical bonds settled in the final products. Usually, in such transfers *electrons* pass from one energy level to another. In many other reactions the electrons also pass from one atom or molecule to another atom or molecule. The latter are reactions of **ox**idation-**red**uction (or **redox reactions**) (Fig. 13.1).

Oxidation occurs when an electron is *subtracted* from a molecule or an atom.
Reduction occurs when an electron is *accepted* by an atom or molecule.

Indeed, *oxidation and reduction* always take place simultaneously: the electron *subtracted* by a molecule or an atom is *transferred* to another atom or molecule were is *accepted*. Therefore, while substances from which electrons are subtracted are defined ***oxidized,*** substances that accept the electron are said ***reduced***. Actually,

Fig. 13.1 In this reaction the oxidation of an electron donor and the reduction of an acceptor occur

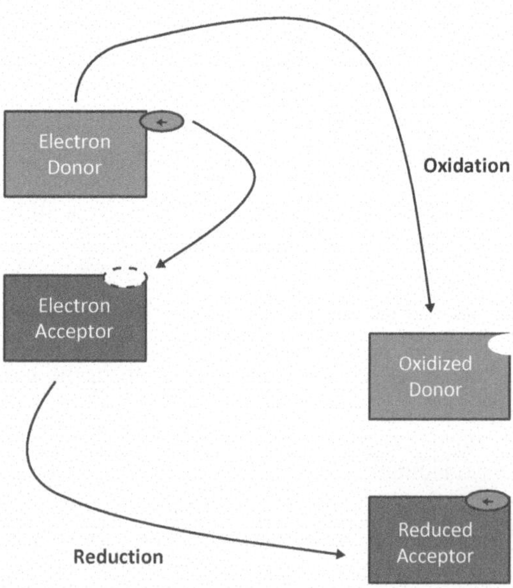

in oxidation-reduction of complex organic molecules there is often a displacement of hydrogen atoms, that is, the electrons traveling together with protons. In nature there are chemical agents called *reactive species* that tend to subtract electrons from another substance. *Therefore,* ***reactive species are subtractors of electrons***. Here we will consider *reactive species containing oxygen and nitrogen*, whose presence in biological materials was discovered about 60 years ago [1].

13.2.2 Reactive Oxygen Species (ROS)

Examples of ROS comprise molecular oxygen, oxygen ions and peroxides. In biological systems ROS are formed as a byproduct of the normal metabolism of oxygen. ROS may be free radicals or non-radicals acting as oxidizing agents. Non-radicals, such as peroxide, for example, may be easily converted to free radicals acquiring unpaired valence electrons and high reactivity. In fact, free radicals are very reactive atoms, molecules, or ions that may easily bonds to other atoms or molecules (covalent bond).

Here we consider the biochemistry of the most represented ROS in biological system. They are the free radicals, superoxide (O_2^-) and hydroxyl ($OH^\cdot$), and the non-radical, hydrogen peroxide (H_2O_2). The other ROS will be briefly described when encountered during the various reactions considered in this Chapter. Since in biological systems, H_2O_2 and $OH^\cdot$ can derive from O_2^- conversion, we analyze first how O_2^- can be formed and then we will see how it can be converted in the other two species, which can be obtained also through other processes.

Superoxide is produced by several processes and reaction in the biological system, including the basic and stimulated production by mitochondria during the normal processes of cell respiration. Although it is biologically quite toxic, it is logical to hypothesize that a substance produced during the *"normal"* life may not be solely deleterious. Indeed, as we will see it may play a pivotal role in cell signaling and survival processes. Actually, also its toxicity turns to be useful because it is exploited by the immune system to kill pathogenic microorganisms and its deficiency may be harmful in this respect.

Superoxide In biological systems O_2^- can be obtained by the reduction of O_2, *via* processes mediated by enzymes such as NAD(P)H oxidases and xanthine oxidase (XO). It can also derive from redox-reactive compounds such as the semi-ubiquinone compound of the electron transport chain. It is the one-electron reduction of molecular oxygen (dioxygen, O_2) that leads to the formation of superoxide anion (O_2^-). Actually, O_2 is considered a diradical containing two unpaired electrons, the addition of a second electron fills one of its two molecular orbitals, leaving a charged ionic species with a single unpaired electron and a negative charge of minus one (−1) (Fig. 13.2). Both O_2 and O_2^- are *paramagnetic* radicals and for this reason they are attracted by magnets. Indeed, a widely used technique for studying paramagnetic species, such as free radicals, is the *electron paramagnetic resonance (EPR) spectroscopy*, which is also called *electron spin resonance spectroscopy* (ESR).

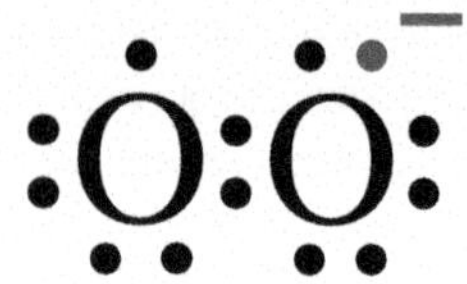

Fig. 13.2 Lewis configuration of superoxide

The magnetic properties of radicals are also considered below.

13.3 Mitochondria Are Important Sources of O_2^- Especially in the Heart

A man of average size consumes about 400,000 ml of O_2 per day at rest. The consumption of oxygen can increase from 5 to 10 times while increasing physical activity. Less than 10% of this O_2 will be used in non-mitochondrial processes, while the rest will be used for mitochondrial respiration. In particular, the heart depends on a continuous oxidative metabolism to meet the continuous demand for ATP and to maintain a redox equilibrium optimal for the contractile function. In fact, the most of cardiac metabolism is aerobic. If we keep in mind that mitochondria make up 40% of the cell mass of cardiomyocytes, we understand how these organelles are central to cardiac cell life, and their energy and redox functions critical not only for the health, but also in diseased and aged heart. Mitochondria can produce O_2^- in the respiratory chain, in particular at complex I and III, as a byproduct of oxidative phosphorylation and by the action of some other enzymes, such as p66shc and MAO; in hypoxic conditions ROS may also derive from complex III [2–4].

13.3.1 ROS Production in Complex I

The flux of O_2^- is in relation with *(a)* the concentration of potential electron donors, *(b)* the concentration of O_2 and *(c)* the velocity of the reaction between the two which follows second-order rate constants. The production of O_2^- by mitochondria *in vivo* and *in vitro* are different. Thereupon, it is not possible to extrapolate O_2^- production by mitochondria *in vivo* from $O_2^{\cdot -}$ generation rates measured in isolated mitochondria. Nevertheless, the production in *vitro* makes easier to understand the modes of operation by mitochondria, which produce O_2^- mainly in complex I in two conditions: (1) *The first mode* is when the mitochondrion has a high proton-motive force (Δp) and a reduced coenzyme Q (CoQ) pool and consequently is not producing ATP; the site of this O_2^- production is uncertain, but may be associated with the CoQ-binding site(s) Therefore, complex I produces large amounts of $O_2^{\cdot -}$ during the so-called reverse electron transport (RET). Succinate-driven RET may lead to $O_2^{\cdot -}$ generation from complex I after an ischemic event, *i.e.* at reperfusion. (2) *The second mode* is when in the mitochondrial matrix there is a high NADH/NAD$^+$ ratio; in this case there is abundant O_2^- generation from the FMN in response to a reduced NADH pool. On the contrary, when mitochondria are actively producing ATP and have a lower Δp and NADH/NAD+ ratio, the O_2^- generation is low and may change in function of ATP production. Therefore, the production of O_2^- within the mitochondrial depends on the $CoQH_2$/CoQ and NADH/NAD+ ratios, the O_2 concentration, and the Δp, which are all extremely variable and not easy to measure *in vivo.*

13.3.2 ROS Production in Complex III

Usually the role of ***Complex III*** is to funnel electrons from the CoQ pool to cytochrome c. It is suggested that at complex III there is an oxygen sensing which is responsible of the paradox of increased ROS during hypoxia. In fact, mitochondrial complex III is considered responsible for cellular oxygen sensing and, therefore, for hypoxia-induced ROS production. The electron donor for superoxide production by complex III of heart mitochondria is considered *ubisemiquinone*. Complex III releases superoxide to both sides of the IMM (inner mitochondrial membrane).

MAO, which is localized to the outer mitochondrial membrane, catalyzes the production of H_2O_2 during catecholamine metabolism. Also the mitochondrial enzymes *aconitase* and *dihydroorotate* can generate superoxide, but their role in ischemia/reperfusion is not clear.

Superoxide is also produced by several enzymes outside the mitochondria, including NAD(P)H oxidases, nitric oxide synthase (NOS) and XO, by the univalent reduction of the so called triplet-state molecular oxygen ($^3O_2$).

13.4 The "NADPH oxidases"

NADPH oxidases is a family of enzymes that is present in several cells and tissues, such as neutrophil, and cells of the cardiovascular system and they have different and specific features in each tissue. For instance, the NADPH oxidases present in vascular cells and neutrophil have a similar structure, but the vascular enzymes produce O_2^- in lesser amounts over longer periods of time [5]. In many cell types these enzymes are the principal producer of ROS. For instance, in the cardiovascular system, Nox1 has been found in vascular smooth muscle, and Nox2 was identified in cardiomyocytes, endothelial cells and fibroblasts. Actually, endothelial cells and cardiomyocytes co-express Nox2 and Nox4. *NADPH oxidases* are each formed of two parts: (1) a *catalytic core*, the cytochrome b558, comprised of gp91phox and p22phox, which is membrane-bound and (2) *cytosolic regulatory subunits* that affect catalytic activity by translocating to and binding the catalytic core. The regulatory subunits comprise the Rac 1/2 GTPase and the proteins p40phox (Ncf4), p41nox (Noxo1), p47phox (Noxo2), p51nox (Noxa1) and p67phox (Noxa2) (Fig. 13.3).

Seven different NOX genes (NOX1 to 5 and DUOX1 and 2) have been described [6]. From Nox1 to Nox4 a p22phox is present, but they have different mechanisms of activation: In the cytosol Nox1 is mainly activated by the subunits Noxa1, Noxo1,

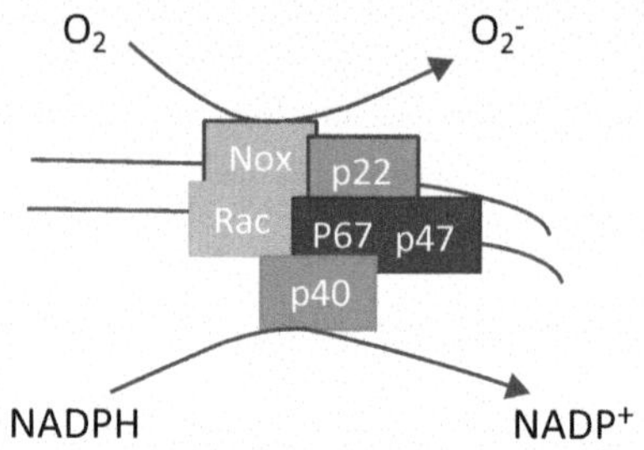

Fig. 13.3 The multisubunit NADPH oxidases are composed of the membrane-bound catalytic subunits NOX (gp91phox in the prototypical phagocyte oxidase) and p22phox, as well as the regulatory subunits p67phox, p47phox, p40phox, and Rac. Enzymatic activity produces superoxide as a byproduct. Several gp91 isoforms have been isolated and characterized as the NOX family of proteins

and Rac1 or 2, which translocate on cell membrane; Nox2 is mainly activated by Noxa2, but may be also activated by Noxa1, which is co-localized with Nox2 in several cell types, including vascular cells. Nox3 is mainly activated by Noxo1 but other cytosolic subunits may be involved. Nox4, Nox5, as well as DUOX1 and 2 activities seems not regulated by cytosolic subunits. Actually, DUOX1 and 2 has an N-terminus containing an extracellular peroxidase-homology domain (PHD) [7] and similarly to Nox5, are mainly regulated by calcium. Importantly, Nox4 and DUOX1 and 2 produce H_2O_2 instead of O_2^- [8]. It has been suggested that the precise subcellular localization of the various Nox proteins may affect ROS produced by the NADPH oxidase. For instance, the NADPH oxidases formed with Nox1 or Nox4 in vascular smooth muscle are located to caveolae and focal adhesions, respectively [9]. In cardiac cells, while Nox2 is located in plasma membrane and its activity is influenced by several factors, including stretch [10, 11]. Nox4 is in intracellular compartments and is constitutively active [12]. The ROS levels produced by Noxs may vary greatly, and may be involved in physiological and pathological processes. In particular, in cardiomyocytes, the steady-state level of ROS produced by Nox2 depends on the amplitude and frequency of cell stretch. Therefore, inotropic changes that depend on the pre-load and heart rate is regulated by a dynamic redox balance that setups cellular Ca^{2+} signaling.

XO produces different amounts of hydrogen peroxide and superoxide during re-oxidation of the enzyme. The role of XO is debated in cardiovascular field. Nevertheless, XO is an enzyme bounding Flavin, which reacts directly with oxygen. Superoxide is not the main product of oxidation by XO but rather the first steps in the over-all reaction result directly in the formation of H_2O_2.

13.5 Superoxide Dismutase as an Example of ROS Scavenger

Intriguingly, aerobic organisms contain not only many enzymes that can produce superoxide, but also various scavengers of ROS. Here we consider only the various isoforms of the enzyme ***superoxide dismutase*** (SOD) that can remove superoxide. Indeed, SOD is a superoxide-scavenging that catalyzes the neutralization of superoxide very rapidly. The enzymes producing superoxide and the various isoforms of SOD are strategically distributed within the cells and in the various organelles. Also other proteins and molecules that can be either oxidized or reduced have SOD-like activity and/or can scavenge O_2^-, as for example hemoglobin, mioglobin, glutathione and nitroxides. This complex system to produce and remove superoxide makes possible the presence of very low steady-state concentration of O_2^- (~10^{-11} M) within the living cells, which can increase or decrease at the occurrence (*i.e.* in physiological conditions) or during pathologic conditions.

The low steady-state concentration is also due to its instability; which is mainly due to its reaction with the cluster [Fe–S] and to its spontaneous SOD-induced $O_2^{\cdot-}$ dismutation to H_2O_2. The instability of O_2^- and its negative charge which obstacle its diffusion through membranes render this oxygen radical a poor signaling factor. Nevertheless, $O_2^{\cdot-}$ can cross the membrane through the anion channels. For this reasons, it has been proposed as probable candidate of cardioprotection triggering in the ischemic conditioning (namely pre- and post-conditioning) conditions (see below and see paragraph on cardioprotection).

The SODs. As above mentioned there are different isoforms of SOD. They are enzymes whose active site requires metals. The different metals used by the enzyme allow do distinguish the major families of SOD, depending on the protein fold and the metal cofactor: the Cu/Zn-SOD (binding both copper and zinc), Fe/Mn-SOD (binding either manganese or iron), and the Ni-SOD, which binds nickel and is contained almost exclusively by prokaryotes. While, the cytosol of the majority of eukaryotic cells contains a Cu/Zn-SOD (SOD 1), mitochondrial matrix contains a Cu/Mn-SOD (SOD 2) isoform. Human mitochondria also contain a Cu/Zn-SOD in the intermembrane space. The importance of $O_2^{\cdot-}$/SOD system in mammalians can be inferred by the observation that mice deprived of mitochondrial SOD (Cu/Mn-SOD) die around 21 days after birth due to cardiomyopathy, neurodegeneration, and lactic acidosis [13]. Yet, mice lacking cytosolic SOD (Cu/Zn-SOD) do not die immediately after birth, but are afflicted by multiple pathologies, such as cataracts, liver cancer, hemolytic anemia, muscle atrophy, thymic involution, and a very rapid drop in female fertility; all together these conditions lead to reduced lifespan [13]. As said, O_2^- can be converted by SODs into $\boldsymbol{H_2O_2}$, which is a non-radical species with many signaling functions. Superoxide can also be converted non-enzymatically into H_2O_2 and singlet oxygen. In the presence of reduced transition metals (*e.g.*, cuprous or ferrous ions), H_2O_2 can be transformed into the highly reactive $^{\cdot}$***OH***. Yet, hydrogen peroxide may be transformed into H_2O by the enzymes glutathione peroxidase and/or catalase. Interestingly by the glutathione peroxidase reaction, GSH is oxidized to GSSG, which can be transformed back to GSH by glutathione reductase in process that transforms NADPH in to $NADP^+$.

13.6 Reactive nitrogen species (RNS)

Reactive nitrogen species (RNS) are reactive molecules derived from nitric oxide (NO). Actually NO is a radical and it should be indicated as ˙NO (we will use both NO or ˙NO (see below). It is usually produced in biological system by the oxidation of one of the terminal guanido-nitrogen atoms of L-arginine [14]. This process is catalyzed by the enzyme nitric oxide synthases (NOSs). Other examples of RNS include peroxynitrite ($ONOO^-$), nitrogen dioxide (NO_2), nitrosonium cation (NO^+), nitroxyl anion (NO^-) and dinitrogen trioxide (N_2O_3) as well as other types of chemically reactive transitional molecules (Table 13.1). Actually, depending on the subcellular and compartmentalized environment, NO can be transformed into various RNS, but some of the biological effects may be due to the transitional formation of S-nitroso-cysteine or S-nitroso-glutathione [15–17]. The systematic name for *"nitric oxide"* is *"nitrogen monoxide"*, in the biological literature, the use of the common name "nitric oxide (NO)" prevails and will be used herein. After the discovery of endogenous nitric oxide generation in mammalian cells in the 1980s, the chemical biology and physiological role of this molecule have been very important research topics for several decades and continues to be so today.

NO Is a Paramagnetic Molecule NO targets radicals and paramagnetic species (e.g. superoxide and dioxygen), transition metals, in metalloenzymes (in particular soluble GC, hemoglobin, and cytochromes), and nuclephiles (e.g. thiols in protein). The reactions can be classified as nitration, nitrosation and nitrosylation. Since nitrosation and nitrosylation of thiols lead to the formation of S-nitrosothiols (RSNOs) and since the exact reaction is unknown this particular reaction is referred to as S-nitros(yl)ation, where the S- suffix is a clear indication of the reaction with a thiol Nitric oxide is very often depicted as ˙NO (with the so-called Lewis dot) showing that it has one unpaired electron (Fig. 13.4).

Therefore, reactive species such as nitric oxide should be represented as ˙*R*, in this case as ˙NO, to indicate the presence of a single unpaired electron. However, for simplicity in this Chapter this convention is not always adopted and we will use "NO" with or without the dot. Indeed, in a biological system the important thing is not that NO may be considered a radical, but that *NO is a paramagnetic molecule*, *i.e.* it is attracted to magnets, whereas most organic molecules repel a magnetic field and thus are called *diamagnetic* (these molecules have all their electrons paired in

Fig. 13.4 The Lewis dot depiction of nitric oxide

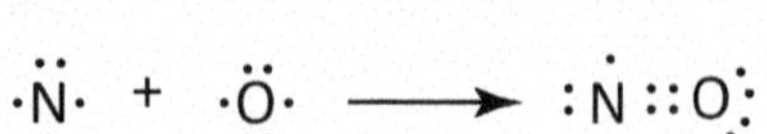

either bonds or in non-bonding orbitals). Paramagnetic and diamagnetic do not react each other or react very slowly in biological conditions. Therefore, NO diffuse easily in the biological medium until running across other paramagnetic species. Thus, for its *paramagnetic properties* NO is an ideal molecule to act as messenger within and between cells. *NO as a paramagnetic molecule is stable.* It has a very low tendency to be reduced or oxidized or even to dimerize in a biological system [18]. *Although NO is a free radical, it is very unreactive.* This characteristic is critical for NO to function as a signaling molecule. Indeed, NO scarcely reacts with diamagnetic species, and most organic molecules are diamagnetic. Therefore, we can say that NO is a paracrine messenger because of the neutral charge of molecules. In biological systems, NO primarily reacts with other paramagnetic molecules as well as with transition metals.

13.6.1 Nitric Oxide Generation

NO is generated by both enzymatic (NOS-dependent and NOS-independent) and non-enzymatic reactions.

13.6.1.1 NOS-Dependent Enzymatic Reactions

Three cytoplasmic nitric oxide synthase (NOS) isoforms have been identified: two constitutive, neuronal (nNOS, NOS I), endothelial (eNOS, NOS III) and one inducible isoform (iNOS, NOS II).

The three NOS isoforms are coded by three distinct genes located at chomosomes 12, 17 and 7. Aminoacid sequence of the three iso-enzymes show less than 59% identification in humans. Nevertheless, aminoacid sequence of each isoform between species has been preserved better (more than 90% for the two cNOSs and more than 80% for iNOS). The principal isoform expressed in the normal vasculature is eNOS. However, the other two isoforms of NOS are also expressed in the majority of the tissues, and expression of all three isoforms is reported to be increased in several chronic diseases. In particular, contrarily to its name, iNOS is expressed *constitutively* in gut and airway epithelium and vascular smooth muscle as well. Moreover, iNOS is transiently expressed in the heart during immune responses in stress conditions and in pathophysiological conditions of the myocardium such as septicemia, heart failure, and during aging. Intriguingly, iNOS in ischemia/reperfusion scenario has been described as protective or deleterious (see Paragraph on cardioprotection). In addition, nNOS, which is mainly expressed in nonadrenergic-noncholinergic nerve endings has been documented to play a role in regulating systemic vascular tone and cardiac contractility as well as basal bronchial epithelium NO production. Thus all three isoforms of NOS could potentially contribute to modulation of vascular tone both in pulmonary and systemic circulation. However, each NOS may have specific spatially and temporarily specific role. As will be seen in the following paragraphs dealing with *NOS isoforms and subcellular NOS compartmentalization*, NO produced by a specific NOS isoform may act as a diffusible messenger within and outside the cells, but generally the main effects are confined in a specific region where NO can target different protein components affecting their function.

Generally, NOS activity is regulated by compartmentalization, availability of substrates and cofactors, endogenous inhibitors, transcriptional, posttranscriptional, and posttranslational modulations.

There is also growing evidence supporting the existence of a *mitochondrial NOS* (mtNOS). It has been suggested that mtNOS is one of the cytoplasmic NOS isoforms, which should be transported to the mitochondria after it has been synthesized in the cytosol. In particular, in the heart nNOS is considered the primary candidate for the cytoplasmic NOS isoform targeted into mitochondria. However, some authors do not support the existence of mtNOS. In brief, it seems that mitochondria do have a NOS, but in a very low amount and the exact nature is not clear [19–21]. Nevertheless, as we will see NO of whatever origin has important effect on mitochondrial function. All isoforms use L-arginine and molecular oxygen as substrates. Moreover, all isoforms need NADPH, tetrahydrobiopterin (BH4), FAD, and FMN as cofactors. The end products are citrulline and NO. It has been proposed a two-step mechanism for the production of NO by NOS. In the first step, L-arginine is converted to NG-hydroxy-L-arginine (L-HOArg) by a reaction requiring one molecule of O_2 and NADPH in the presence of BH4. In the second step, L-HOArg is further oxidized to NO and citrulline.

Under normal conditions, only constitutive NOSs can be detected in heart and coronary vessels using immunohistochemical methods. However vascular smooth muscle, endothelial cells and cardiomyocytes express iNOS upon induction with cytokines and ischemia/reperfusion stimuli in a dose/response fashion [22, 23]. In general, the iNOS enzyme produces much greater amounts of NO (in the micromolar range) than either nNOS or eNOS (in the pico-nanomolar range).

As said NOS activity is determined largely by its site of production. In several cell types eNOS is localized primarily into cell membrane caveolae. In particular, in cardiomyocytes it is localized into caveolae of the sarcolemma and t tubules, where its function is regulated by interaction with scaffolding protein *caveolin-3* and is linked to multiple cell surface receptors, β-adrenergic [24], and bradykinin receptors [25], depressing contractility and downregulating β-adrenergic stimulation. On the other hand, nNOS has been founded in the sarcoplasmic reticulum (SR) where it co-immunoprecipitates with both ryanodine receptors 2 (RyR2) and xanthine oxidoreductase (XOR) nNOS modulates various components of excitation–contraction coupling (ECC), including Ca^{2+} influx through the L-type Ca^{2+} channel (LTCC), Ca^{2+} release from the SR via RyR2, and Ca^{2+} reuptake into the SR via the SR Ca^{2+} ATPase (SERCA2a). Also an inhibitory effect of nNOS on XOR production of O_2^- has been described. Superoxide can then irreversibly increase the open probability of the RyR2 channel and decrease myofilament Ca^{2+} sensitivity. Therefore, the inhibition of XOR may avoid the O_2^- induced reduction in myofilament Ca^{2+} sensitivity, thus contributing to an increase in cell shortening and force of contraction without a necessary change in systolic Ca^{2+}. Therefore, in contrast to eNOS, nNOS has primarily positive inotropic effects in the heart [26].

During pathologic conditions such as ischemic cardiomyopathy or heart failure, the localization of nNOS in subcellular compartments changes. It is translocated from the SR to the sarcolemma and the tight regulation of NOS, which is usually time- and substrate-dependent, is lost. Moreover, gender differences are observed in the role of NOS [27]. The expression of cytosolic iNOS in cardiomyocytes seems to be, in part, responsible for the attenuation of the myocardial inotropic response to β-adrenergic stimulation [28, 29].

13.6.1.2 Uncoupled NOSs

In the absence of the substrate L-arginine and the cofactors 6R)-5,6,7,8-tetrahydro-L biopterin (BH4), NOSs produce superoxide an it is said that it is uncoupled from NO production [30]. The NOS cofactor BH4, is oxidized by ROS produced by NOXs, thus leading to NOS uncoupling [31]. These phenomena, together to the scavenging effects of O_2^-, lead to a vicious cycle which sustain reduced NO levels and impaired endothelium-dependent vasorelaxation. These mechanisms have been also demonstrated *in vivo* in experiments performed on mice overexpressing NOX1. In these mice subjected to Angitensin II induced hypertension, endothelium-dependent relaxation was downregulated as available NO was markedly reduced [32].

13.6.1.3 NOS-Independent Enzymatic and No-Enzymatic NO Production

The production of NO by NOS-independent enzymatic and non-enzymatic reduction of nitrite/nitrate from endogenous and dietary origins is of primary importance during ischemia, a condition characterized by acidotic pH and limited oxygen-dependent NOS activity. Indeed, in a biological system nitrite can be reduced to NO under the acidotic conditions, such as those present in the ischemic tissues. It has been suggested that nitrite, via a mechanism of direct reduction, may represent an alternative source of NO in the ischemic heart, where a limited oxygen-dependent NOS activity has been described [33]. It has also been suggested that nitrite is reduced to NO crossing the capillary territory under normal physiologic conditions [34].

NOS-Independent, Non-enzymatic Production

Nitrite has been proposed as a major physiological source of biologically active NO. Also nitrate, which is in much higher concentrations than nitrite, can be transformed to NO, but it needs a two-step reduction to NO, through nitrite. Therefore, nitrite represent the largest directly-accessible storage pool for NO. AS said low pH and low O_2 tension markedly enhance nitrite reduction, releasing amounts of NO that far exceed those produced by eNOS/nNOS under ischemic conditions; *i.e.* when the activity of these enzymes is reduced. Actually, the reaction is a *nitrite disproportionation*, which increases when H^+ are abundantly available, favoring the release of free NO. Nevertheless, in ischemic cells, this process accounts for only 15–20% *circa* of the total NO produced from nitrite [35]; the remainder is derived from *NOS-independent enzymatic production* [36].

NOS-Independent Enzymatic Production

A number of intracellular proteins display *nitrite reductase activity,* especially in acidotic and ischemic conditions; these include *Cytochrome P450* (a ubiquitous family of enzymes), *Hem-associated globins* (Haemoglobin, myoglobin, neuroglobin and cytoglobin), *Mitochondrial proteins* (components of the oxidative chain: complex III, Cytochrome C and complex IV), *Molybdenum metalloenzymes* (xanthine oxidoreductase, aldehyde oxidase, sulfite oxidase and bacterial nitrate/nitrite reductases), *Carbonic anhydrase* (sixteen isoforms have been described), as well as the *NOS enzymes*. Actually, in anoxia, when the NOSs activity is considered defective, eNOS can use nitrite as a substrate for restoring NO generation [37, 38].

The nitrite reductase activity of the aforementioned mammalian proteins is a well-characterized chemical reaction in many patho-physiological conditions [for reviews see 17, 39–41].

13.6.2 Peroxynitrite ($ONOO^-$)

Among RNS peroxynitrite ($ONOO^-$) anion is the mainly studied oxidant. Therefore, here we will consider principally this RNS. It is a short-lived reactive species that is usually produced by the reaction of $^{\bullet}NO$ and $O_2^{\bullet-}$ radicals at diffusion-controlled rates. The microdomains where peroxynitrite is produced are likely associated with the sources of superoxide (*e.g.* the mitochondrial respiratory chain or the plasma membrane NAD(P)H oxidases). In fact, though $^{\bullet}NO$ is a highly diffusible radical, superoxide as anions hardly diffuse across biomembranes. Therefore, $ONOO^-$ is formed and reacts close to the enzymes producing $O_2^{\bullet-}$ whose activity can be influenced by diffusing $^{\bullet}NO$; thus affecting the final amount of $ONOO^-$ formation. Among peroxynitrite effects are included the activation of matrix metalloproteinases, the DNA oxidative damage and activation of poly(ADP-ribose) polymerase (PARP), as well as the lipid peroxidation, to name only a few [42]. These effects may explain the numerous detrimental effects of peroxynitrite in myocardial ischemia/reperfusion and in other heart diseases. However, it must be noted that $ONOO^-$ does not only trigger direct cytotoxic effects, but it also favors several indirect effects and modulates several protective signaling pathways [43, 44]. Such effects may depend either on a nitrative or an oxidative type of chemistry elicited by peroxynitrite [45]. While the main biological reactions of NO include oxidation leading to formation of nitrite and nitrate, $ONOO^-$ is a reactive oxidant, nitrating and nitrosating agent. Actually, peroxynitrite does not have an impaired electron, thus it is not a radical species. Yet, it may generate nitrate in a complex reaction which include potent oxidants, such as $NO_2/HO\cdot$. In biological systems, reacting with CO_2 peroxynitrite may give rise to other potent oxidants ($NO_2/CO_3^{\cdot-}$) and other intermediates, which may lead to nitrate formation. Moreover, peroxynitrite is not a direct nitrosating agent, but in excess of NO, peroxynitrite and derivative oxidants can be converted to several nitrosating agents, including N_2O_3. The balance between oxidative and nitrosative reactions may explain the different outcome observed with peroxynitrite, which is often harmful, but in some circumstances may be beneficial (see Paragraph on cardioprotection).

13.7 ROS/RNS in Biological System

As we have seen in the previous chapters, ROS and RNS are produced by several cellular redox processes. These reactive species may play a dual role in biological system as they may be either toxic or beneficial compounds. The subtle balance between their two diametrically opposed effects is unquestionably a crucial aspect of life and death. Although ROS/RNS have been involved in numerous pathological conditions, they also play a vital role in several physiological mechanisms and in killing the infecting pathogens [45–47]. However, *inappropriate* generation of ROS/RNS by interacting with biomolecules, including proteins, lipids, enzymes, and DNA, causes damage of cell membranes, misfolding in proteins, oxidative damage of DNA [48, 49]. It is usually said that, at low levels, ROS/RNS may exert beneficial effects but at high concentrations may produce dangerous oxidative stress [50, 51]. Although this may be often true, it is not always true. For this reason, we prefer to use the terms *"appropriate"* and *"inappropriate"* ROS/RNS production for beneficial and detrimental effects, respectively. For example, during exercise an appreciable amount of ROS, such as H_2O_2, O_2^-, NO and ubisemiquinone, are produced, but these are not deleterious and activate signal transduction pathways to induce adaptive homeostasis, including mitochondrial biogenesis [52]. As mentioned and as we will see, also the nature of ROS is not predictive of beneficial or deleterious effects. Therefore, we recommend to use the terms *"appropriate"* and *"inappropriate"* ROS/RNS production.

13.7.1 Deleterious or Beneficial Effects of ROS/RNS

ROS may be protective or may be dangerous. In cardiomyocytes ROS play many physiological influences, including the regulation of mechanical function. Indeed, ROS produced by Noxs and by other sources, including mitochondria, are able to modulate the activity of different protein kinases, which phosphorylate many molecular targets involved in Ca^{2+} signaling and in pro-survival pathways (see Paragraph on cardioprotection). For instance, targets of ROS leading to modulation of intracellular Ca^{2+} concentrations and consequently cardiomyocyte contractility includes membrane channels such as L-TypeCa^{2+} and voltage-dependent Na^+ channels as well as ATPases, such as plasma membrane Ca^{2+} ATPase, and sarco/endoplasmic reticulum Ca^{2+}-ATPase [53, 54]. On the other hand, ROS may contribute to the pathogenesis of several diseases (strong evidence are reported for radiation intoxication and hyperoxic injury). Also to the process of *aging* through oxidative damage determines injury on cells. While the action of O_2^- in determining certain pathological conditions is strong, the role of O_2^- in aging is considered unproven yet. Perhaps, while mice and rats overexpressing CuZnSOD or MnSOD are more resistant to heart attacks and strokes, in other models and organisms (fruit fly Drosophila, yeast and mice), lifespan is shortened and some features of aging cataracts, muscle atrophy, macular degeneration, thymic involution) are accelerated by genetically knocking out CuZnSOD. However, increasing the levels of CuZnSOD, does not seem (except maybe in Drosophila), to steadily increase the lifespan [55]. Indeed, *the most widely accepted opinion is that the oxidative damage (derived among other factors, by* O_2^-*) is only one of several factors that limit the lifespan.*

As said deleterious or beneficial effects of ROS/RNS is not only an issue of *quantity*, but also of *quality* and *compartmentalization*. In fact, also the definition of "*damaging radical*" for some reactive species, such as hydroxyl radical and peroxynitrite is not correct, though rooted in the mind of many people. For instance, if you search on Google "damaging hydroxyl radical" you found about 7.500 results. However, if you search "beneficial hydroxyl radical" you found almost nothing. Yet, it has been clearly demonstrated that *hydroxyl radical formation is a fundamental step of cardioprotection* [56]. Moreover, if you search "damaging peroxynitrite" or "beneficial peroxynitrite" you found about thousands of results for the former and no one for the latter. However, it is well established that *peroxynitrite may play a trigger role in cardioprotection* [57, 58]. On the other hand, nitric oxide, which is very often considered beneficial, and in fact it is very often beneficial, sometime may be detrimental, as for example in septic shock, hemorrhage and negative cardiac contractility [59, 60]. Therefore, the "bad guys" are not always bad and the "good guys" are not always good.

13.7.1.1 Appropriate ROS/RNS Production

The example of "appropriate" ROS/RNS production can be a plethora. For instance, considerable evidence supports the view that redox signaling involving both ROS and RNS is an important contributor to the regulation of coronary physiology and adaptive responses to exercise [53, 61, 62]. For instance, coronary metabolic dilation is mediated by redox-dependent signaling. Indeed, besides the well-known vasodilator effect of NO, another active factor of cardiac metabolic dilation is hydrogen peroxide, which induces dilation by the oxidation of intracellular thiols and is involved in the activation on the p38 MAP kinase [63, 64].

13.7.1.2 Inappropriate ROS/RNS Production

Either high or low levels of ROS/RNS may result in pathological stress to cells and tissues. Usually, small amount of ROS/RNS in the presence of deficient antioxidant defenses damage the cells. Even the exaggerated presence of anti-oxidants may induce a reductive stress. These oxi-reductive stress can have multiple deleterious effects. For instance, considerable evidence supports the view that oxidative damage involving both ROS and RNS is an important contributor to the development of atherosclerosis and/or diabetes [62, 65, 66]. In the context of myocardial ischemia/reperfusion ROS and RNS are emerging as major players in determining either exacerbation of damages or in triggering cardioprotection [49, 67–69]. The following paragraphs will consider these important aspects of ROS/RNS pathophysiology.

13.8 Myocardial Ischemia/Reperfusion Injury and ROS/RNS

Severe hypoxia, acidosis, energy depletion, and ion homeostasis are all typical alterations of ischemic tissue, which lead to cardiac dysfunction and, ultimately, to cell death. Ischemia-reperfusion (I/R) injury occurs when the blood supply to an organ is first reduced and subsequently restored. Of course, the duration of ischemia determines cardiac injury that follows ischemia and reperfusion. The reperfusion may be timing because the sooner it is performed, the greater the amount of saved cardiac tissue. Indeed, a prompt reperfusion has been shown to dramatically reduce the infarct size. However, during the first minutes of reperfusion a large part of the damage takes place: an amplification of ischemic injury or additional damage occur at this stage [70–75]. Intracellular, oxidative/nitrosative stress by ROS/RNS, together to Ca^{2+} overload, inadequate re-synthesis of ATP, and loss of membrane phospholipids have been proposed as contributing to reperfusion injury [76, 77]. Therefore, it is now clear that both ischemia and reperfusion determine the organ damage. As a matter of fact, myocardial reperfusion injury includes arrhythmias, inflammatory responses, microvascular damage, and no reflow phenomenon, as well as transient mechanical dysfunction of the heart or myocardial *stunning* (a transient post-ischemic contractile dysfunction).

It is now clear that, in reperfusion, cell death can occur by apoptosis, autophagy, pyroptosis and necrosis [the reader is redirected to extensive review on this topic [e.g., 72, 78–80]. Nevertheless, in contrast to necrosis, pyroptosis and apoptosis, which are negative phenomena and inevitably lead to cell death, autophagy is not always a negative phenomenon, but under certain conditions, autophagy can be considered a protective mechanism against I/R injury [79, 80]. The vulnerability to I/R injury is likely to be greatly influenced by the autophagic control of protein and organelle quality, such as mitochondria. During reperfusion a burst of ROS/RNS production occurs [72]. Nevertheless, the sources of these ROS/RNS are a matter of controversy. It has been proposed that during ischemia there is a production of ROS/RNS the primes cardiomyocytes for cell death during reperfusion; thus, it is likely that ROS/RNS production during ischemia is a determinant of cardiomyocyte death during the subsequent reperfusion, which is mainly due to reperfusion ROS/RNS burst [72]. The combined effects of ROS/RNS, elevated calcium level and low pH seem mandatory for the opening of the mitochondrial permeability transition pore (mPTP), which plays a critical role in reperfusion damage [72, 81–85]. Post-ischemic reperfusion may result in inappropriate ROS/RNS formation, educed availability of NO, Ca^{2+} overload, and low pH. As said, these modifications together may favor prolonged opening of mPTP, and other processes contributing to cell death, myocardial infarction, stunning, and arrhythmias. It is clear that the prolonged mPTP opening mediates myocardial I/R damage. Indeed, inhibiting mPTP opening at reperfusion protects against cardiac I/R damage. However, short term mPTP opening as proposed as protective mechanism [for Reviews see 67, 86, 87].

Although inappropriate levels of ROS and RNS induce structural modifications of lipids, proteins, and genes that impact on cell function and death, ROS/RNS can activate signaling pathways that contribute to ischemic PreC and PostC. These signaling pathways may protect the heart by inhibiting mPTP opening at reperfusion. The production of ROS during I/R occurs surely in the mitochondrial respiratory chain and by NOX family enzymes and they play a role in I/R injury. Other sources of ROS during I/R include xanthine oxidase (XO) and uncoupled NOS. Although all enzymatic sources are likely to play a certain role in reperfusion injury, priority and emphasis may be given to specific ROS/RNS sources that are predominant in certain tissues. For instance, in the gastrointestinal tract xanthine oxidase may be the predominant source of ROS. However, in the metabolically active organs, such as heart and brain, mitochondria can be the main producers of ROS. Nevertheless, multiple ROS sources contribute for sure to reperfusion injury in most tissues. Evidence exist that ROS produced by one enzymatic source activate and enhance ROS production by a second source. A classic example of this is the so-called ROS-induced ROS release (RIRR) occurring in mitochondria. When mPTP activation occurs by various sources of ROS, intra- and intermitochondrial redox-environment changes leading to RIRR, which is a regenerative cycle of mitochondrial ROS formation and release. This ROS storm contributes to cell damage and death [88]. Besides cell death, serious and often lethal consequence of the acute myocardial infarction (AMI) is the resulting contractile dysfunction, namely *myocardial stunning*. As mentioned in this chapter, ROS/RNS are strong regulators of cardiac function and may have strong deleterious effects on contractility. The imbalanced and high steady state levels of ROS/RNS are involved in the genesis and progression of myocardial stunning [for Reviews see 89–93].

13.9 Cardioprotection and ROS/RNS

As said ROS/RNS balance leads to appropriate or inappropriate redox status in cardioprotection by *preconditioning* and in I/R injury, respectively. This redox balance has been extensively reviewed [17, 49, 94, 95]. Indeed, different cardioprotective strategies, such as pharmacologic and ischemic preconditioning, have been shown to limit mitochondrial dysfunction as demonstrated by a better NADH balance and reduced ROS formation and a reduced mitochondrial Ca^{2+} overload, during ischemia and at reperfusion [96, 97]. Important strategies for reducing the IR injuries are studied, the names of these procedure are Pre- and post-conditioning protocols. These treatments may reduce significantly the contractile dysfunction, reduce infarct size and affect all form of cell death, arrhythmias and endothelial dysfunction. In this scenario of cardioprotection ROS and RNS are two important cardioprotective signaling molecules, which are essential in pre- and post-conditioning processes. The first report on the ischemic preconditioning (PreC) concluded with this phrases "...the multiple anginal episodes that often precede myocardial infarction in man may delay cell death after coronary occlusion, and thereby allow for greater salvage of myocardium through reperfusion therapy" [98]. In this few words are closed the characteristic of PreC.

Nowadays we speak of *"ischemic conditioning"*, which includes several endogenous cardioprotective strategies that share several features. Conditioning protocols can be applied either directly to the heart (ischemic preconditioning or postconditioning) or to another parts of the body, for example a limb, that is remote ischemic preconditioning, applied *before* the ischemic event of the heart, remote ischemic preconditioning, applied *during* the ischemic event or remote ischemic postconditioning, which are brief intermittent ischemia applied to the limb, *after* the ischemic event of the heart. Also pharmacological preconditioning, perconditioning or postconditioning have been described [99]. The Postconditioning (PostC) against AMI was introduced by Vinten johansen group's, and it is characterized by brief cycles of I/R immediately after the global ischemia [100]. In particular, PostC may be performed with one or more brief occlusions of a few seconds (from 5 to 60 s), starting very early in reperfusion, i.e. a few seconds after the end of the infarcting ischemia [100, 101].

In particular, both PreC and PostC attenuate endothelial cell dysfunction by increasing eNOS activity and NO˙ bioavailability in neighboring cells [87, 94, 102–104]. Both pre- or post-conditioning phenomena can be triggered by pharmacological interventions, including exogenous NO-donors, that is pharmacological PreC or pharmacological PostC [105–107]. After that PreC and PostC phenomena have been described, reperfusion injury has been appreciated as a reality. This is important because protection from reperfusion injury is feasible, especially with PostC, which is under the control of the operators. Protective signaling cascades are recruited by both pre- and postconditioning, namely the so-called RISK, SAFE and cGMP/PKG pathways; these may cooperate in inducing protection [for review 104, 108–113]. The cardioprotective signaling pathways are thought to converge on mitochondria (Fig. 13.5).

Fig. 13.5 The main pathways converging on mitochondria for cardioprotection

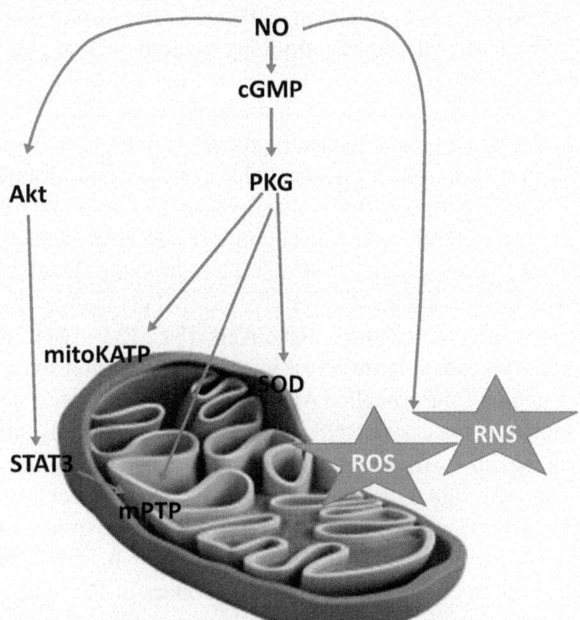

The mitochondria are surely central in cardioprotection. Several mitochondrial proteins have been identified as targets of post-transitional modifications in both pre- and pos-tconditioning. Among these modifications phosphorylation, and nitrosylation are the most studied. In the aforementioned protective pathways, phosphorylative/dephosphorylative and nitrosative processes are largely represented. As mentioned, a pivotal effector of cardioprotection is the inhibition of mPTP opening, which is considered the final target [68, 103, 104] Indeed, as said, the pore is regulated by many factors including ROS/RNS, pH, calcium and potential of mitochondrial membrane, which can be influenced by S-nitrosylation (SNO) of proteins. Besides the decrease of calcium overload by increased reuptake by SNO of SERCA2, also the S-nitrosylation of F1F0-ATPase limits, indirectly, the opening of mPTP, which in turn reduces the breakdown of glycolytic ATP and the fastening of the fall in the potential of mitochondrial membrane. In addition, S-nitrosylation of CyPD (cyclophilline D) [115] and/or of VDAC (voltage dependent anion channel) [68, 69], two putative components of mPTP rich in thiol groups, may occur in cardiopriotection (see below for novel models of mPTP).

13.9.1 Ischemic Preconditioning

Ischemic preconditioning is the protective maneuvers which consists in brief periods (a few minutes) of intermittent ischemia and reperfusion performed before the infarcting ischemia. These maneuvers may trigger two periods of cardioprotection: one that start a few minute after the preconditioning maneuvers and that elapse two-three hours (early preconditioning also known as first window of protection), and a second period of protection, called SWOP (second window of protection) that starts 12–24 h after the preconditioning maneuvers and elapse 48–72 h (the second window of protection is also known as *late preconditioning*) [116–119]. Here we consider the first window of protection, which is obtained immediately before the infarcting ischemia and exerts the most potent cardioprotection against infarct size. Recently it has been suggested that early preconditioning protection is also operative during reperfusion (i.e., in the post-ischemic phase) and limits a large part of the damage not only due to ischemia, but also to reperfusion [78, 120–122] (see also below). Interestingly, PreC can be completely blocked by free radical scavengers, such as mercaptopropionyl glycine (MPG) and/or N-acetyl-cysteine (NAC) given during preconditioning maneuvers [87, 123, 124]. These results confirmed that *redox signaling* is involved in triggering cardioprotection by PreC, that is a clear example of the so-called oxidative paradox (reactive species with a beneficial role). Although, excessive RNS formation during I/R may contribute to reperfusion injury via *nitrative stress* by peroxynitrite, also RNS are important elements in the triggering signal of I-PreC during triggering maneuvers. In fact, it has been reported that I-PreC induces •NO production by activation of differential type of NOS; however controversial papers report that inhibition of NOS abolishes the cardioprotection [125–130]. Even donors of HNO (one electron reduction product of •NO) could induce a preconditioning like effect, which was reversed by NAC [131]. The “classical” protection induced by •NO in I-PreC, is dependent in part by activation of guanylyl cyclase (GC)/cyclic guanosine monophosphate (cGMP)/protein kinase G (PKG), which in turn leads to the opening of the mitoK_{ATP} channel [95]. Sun et al., reported that the I-PreC-induced cardioprotection is due to a *S-nitrosylation signaling* by •NO, rather than to an initial activation of the sGC/cGMP/PKG signaling pathway. In fact, the administration of the sGC inhibitor, ODQ, did not eliminate completely the cardioprotection triggered by I-PreC. Hearts treated with ODQ resulted cardioprotected with a higher S-nitrosylation level of proteins. Therefore, it has been suggested that •NO mediated cardioprotection is regulated mainly by protein S-nitrosylation, at least in some models of cardioprotection [132]. The easy of reversibility and the confirmation of regulated S-nitrosylating and denitrosylating enzymatic and non-enzymatic reactions support the hypothesis that S-nitrosylation regulates the cellular and mitochondrial function through redox-sensible mechanisms [17]. A great part of pre-conditioning protection is due to the limitation of reperfusion injury with a limitation of ROS/RNS stress. This is mainly due to the prevention of mPTP opening in the early phase of reperfusion [112]. In our opinion, the prevention of mPTP opening avoids ROS-induced ROS release and *redox stress*, while *oxidative/nitrosative signaling* may occur and protect the heart via the subsequent intervention of the protective role of the RISK (Reperfusion Injury Salvage Kinase), the SAFE (Survivor Activating Factor Enhancement) and the cGMP/PKG pathways [68, 133], which include phosphorylation of many target proteins. In brief, enzymes that have been shown to take part in these pathways include NOSs, phosphotidylinositol-3-phosphate kinase, PKC, kinase B (PKB or Akt), and extracellular signal-regulated protein kinases. These are a series of kinases that has been termed RISK pathway. SAFE pathway comprise Janus kinase and signal transducer and activator of transcription 3 (JAK-STAT3) and maybe STAT5. Once the heart has been preconditioned, many components of these two pathways are re-activated at reperfusion, leading together to other factors to the prevention of the mPTP formation. Some mitochondria in the cells continue to be functional and do not release pro-apoptotic factors, preventing cell death. However, we do not know how the heart *"remembers"* that it has been preconditioned. Similar mechanisms have been observed with regard to I-PostC, mainly implicating the RISK and SAFE pathways and the prevention of mPTP formation. Pharmacological therapy can thus mimic conditioning by targeting the cells at one of these points at the level of the receptors, the signal transduction pathways, or the mitochondria. Here, we focus on the tens of recent studies reporting *S-nitrosylation* of critical proteins as a pivotal mechanism of cardioprotection by preconditioning [68, 134, 135]. Recently Kohr et al. [136] using two different methods to measure protein oxidation have shown that preconditioning leads to S-nitrosylation of several proteins and that a large majority of these proteins are protected from further oxidation [136]. S-nitrosylation of proteins involved in calcium handling, such as Ca^{2+} channels, phospholamban and SERCA2 have been described [137]. Moreover, multiple S-nitrosylated proteins have been shown by proteomic studies in the presence of PreC [138]. In particular, many of these proteins have been found within mitochondria, including proteins responsible of mitochondrial metabolism (e.g. αKGDH, glycogen phosphorylase, aconitase, glycogen phosphorylase). Other important mitochondrial components that are subjected to S-nitrosylation during the PreC, are the respiratory complexes, including the complex I, which is inhibited when *nitrosylated* [139] or when it is subjected to *nitrosation* by $ONOO^-$ [140, 141]. A mitochondria-selective S-nitrosating agent, MitoSNO, resulted cardioprotective by a mechanism that involves S-nitrosation of complex I and the subsequent slow reactivation of mitochondria at reperfusion, thereby decreasing ROS production [142]. These experiments identify the rapid complex I reactivation as a pivotal pathological aspect of I/R injury and suggest that preventing this reactivation by modification of a cysteine switch is a cardioprotective mechanism and a possible therapeutic strategy. Of note, due to a shielding effect against oxidative/nitrosative stress, if a protein is nitrosylated it is unlike that it can be nitrosated. It may be argued that a rapid reactivation of complex I and the massive ROS production may contribute to the I/R injury, whereas a slow reactivation an a reduced ROS production may be involved in triggering protection. Nevertheless, recent data suggest that ROS derived from mitochondrial p66shc do not contribute to the I/R injury nor they are involved in the cardioprotection byPrecC [143]. Another effect that can be observed in conditioning protection is the inhibition of the so-called F0-F1-ATPase. This can occur by many mechanisms, including the S-nitrosylation, with consequent reduction of ATP consumption of the F0-F1-ATPase working in reserve mode. This typically occurs in I/R of the myocardium [144]. The inhibition of F0-F1-ATPase saves ATP levels and reduces the mitochondrial potential, thereby reducing the driving force for Ca^{2+} uptake into the mitochondria, thus increasing tolerance to I/R challenging [145]. As said above, an important effector of cardioprotection is the inhibition of mPTP opening [68, 103, 114]. In fact, this pore is regulated by ROS, Ca^{2+} and potential of mitochondrial membrane, which are also regulated by S-nitrosylation of critical proteins. Not only the decrease of Ca^{2+} loading by increased reuptake by nitrosylated SERCA2, but also the S-nitrosylation of F0-F1-ATPase reduces indirectly the opening of mPTP, which reduces the breakdown of glycolytic ATP and the acceleration of the fall in the mitochondrial membrane potential. Moreover, CyPD (cyclophilline D) and VDAC (voltage dependent anion channel), which were considered two components of mPTP, are rich in thiol groups, and their S-nitrosylation of may occur in cardioprotection [68, 115]. All together, these data support the view that S-nitrosylation of mitochondrial proteins and proteins involved in calcium handling serves as an important mechanism of preconditioning cardioprotection involving mPTP.

13.9.1.1 The Novel Model of mPTP

Actually, the molecular nature of the mPTP is not clear and continues to be the subject of debate in the specialized/current literature [146–152]. It has been proposed that the pore forms when the dimeric enzyme FoF1-ATP synthase switches from the energy-conserving to the energy-dissipating mode It seems that the pore forms at the interface between the two monomers of the FoF1-ATP synthase [146–149]. It has been also proposed that the subunit *c* of the Fo ATP synthase, which forms the proton-translocating *'c ring'*, represent a critical component of the pore [146]. It seems that the channel is created by the *c ring* itself after that the subunit

F1 has been extruded by a Ca^{2+}-dependent mechanism [147]. Intriguingly, it has been observed that two important component of the peripheral stalk of the ATP synthase, namely the oligomycin sensitivity conferral protein (OSCP) and the *b-subunit*, do not contribute to the mPTP formation. Indeed, cells lacking the membrane domain of *b-subunit* or the entire OSCP display a functional mPTP [151]. Also the *c-subunit* seems not implicated in the pore formation [152]. Therefore, it seems that none of the membrane subunits of the FoF1-ATP synthase that are involved directly in transmembrane proton translocation are involved in forming the mPTP.

13.9.2 Postconditioning

Postconditioning defined as brief (a few seconds) intermittent cycles of ischemia alternating with reperfusion applied immediately after the infarcting ischemic event, has been shown to reduce ischemia/reperfusion damage, in some cases equivalent to that observed with preconditioning. However, cardioprotective modalities of signal transduction also include redox signaling by ROS, S-nitrosylation by NO. and derivative, S-sulfhydration by hydrogen sulfide, and O-linked glycosylation with beta-N-acetylglucosamine [153]. All these modalities can interact and regulate an entire pathway, thus influencing each other. For instance, enzymes can be nitrosylated and/or phosphorylated in specific and different site(s) with consequent increase or decrease of their specific activity. For example, extracellular signal-regulated kinase (ERK) may be S-nitrosylated, thus inhibiting its phosphorylation and activation [154]. Another protein that may undergo NO-mediated S-nitrosylation and phosphorylation is the regulator protein phospholamban, which is involved in the control of cardiac contractility [137, 155]. Both pre and post-conditioning may be triggered by endogenous and exogenous $NO^{\cdot}$ [99]. The relative importance of cGMP/PKG pathway and non-classical processes, such as nitrosylation are under intense investigation. Cardioprotection by ischemic postconditioning (I-PostC) is obtained by short periods of reperfusion intervalled by short periods of ischemia (a few seconds) at be beginning of a reperfusion which follow an infarcting ischemia. Because I-PostC has the advantage that it can be applied after the ischemic insult has occurred, this is therapeutically a more favorable approach than is preconditioning. It requires a complex signaling cascade to be triggered, which includes the opening of mitoK_{ATP} and the activation/inhibition of several enzymes of cardioprotective pathways. With regard to signaling pathways, also for PostC, as for PreC, the greatest attention as focused on the role of the RISK-, the SAFE- and the cGMP/PKG-dependent pathways. Intriguingly, however, I-PostC can be completely blocked by free radical large spectrum scavengers, such as NAC or MPG given during I-PostC maneuvers. However, PostC protection is not abolished if the scavenger is given in reperfusion after the PostC maneuvers have been completed [156, 157]. Even more intricate is the relationship when more selective antioxidant enzymes, such as SOD and catalase, are considered. Indeed, the activity of these enzymes is strongly influenced by pH [49] (see also below) and it is well known that pH changes during ischemia and during reperfusion. Actually, the gradual normalization of intracellular pH in the initial phase of reperfusion plays a role of paramount importance in conditioning strategies. Both in pre- and post-conditioning, acidosis favors redox signaling and the activation of a complex cascade of signal molecules and prevents the opening of mPTP in early post-ischemic phase. A phase in which redox signaling plays a critical role in triggering cardioprotection [68, 87, 158, 159]. In particular, acidosis favors the transient formation of S-nitrosylated protein in postconditioned hearts [159].

13.9.3 Nitric Oxide, Nitration and Nitrosylation May Play a Finely Interconnected Role

It is well known that postconditioning attenuates endothelial cell dysfunction by increasing eNOS activity and •NO bioavailability in neighboring cells [100]. This can be responsible of improved vasodilatation in postconditioned hearts. Moreover, both pre- or post-conditioning protection can be triggered by pharmacological interventions, including the infusion of exogenous •NO-donors, that is pharmacological PreC or pharmacological PostC [95, 133, 160–162]. In an editorial by R. Schulz and P. Ferdinandy [163], which was written as a comment to an interesting article of Sun et al., [132], the authors wonder whether or not "nitric oxide signaling differ in pre- and post-conditioning" and in particular they wonder whether "S-nitrosylation is involved in postconditioning's protection". Above we have seen that S-nitrosylation is involved in ischemic and pharmacological preconditioning. Below, we will see that S-nitrosylation is also involved in ischemic and pharmacological postconditioning. Several studies that used •NO-donors in reperfusion to induce pharmacological PostC revealed an important role of S-nitrosylation of proteins in the mechanisms of protection [139]. We were among the first to show that also I-PostC is mediated by S-nitrosylation of proteins [159]. A finding confirmed by Tong et al., [164], which have shown that several proteins are S-nitrosylated with I-PostC. Due to the abundance of nitrosylated proteins, it is likely that also denitrosylation processes are down-regulated. Indeed, we have shown that PostC, discretely change the activity of antioxidant enzymes in early reperfusion, slightly decreasing SOD and increasing catalase activity [159]. Since SOD may be a de-nitrosylating enzyme [141], these effects may favor S-nitrosylation thus reducing injury due to oxidative-stress. In fact, it has been proposed that the increase in S-nitrosylation could shield critical cysteine residue(s) from further oxidative damage upon reperfusion [49, 164]. Importantly, pro-survival enzyme activation may depend on redox-sensible reactions. For instance, PKC activation can occur *via* S-nitrosylative processes [27] and the activation of PKC plays a central role in sustaining the cardioprotection by postconditioning [156, 157]. The S-nitrosylation of the mitochondrial F0-F1-ATPase described for PreC has also been found in PostC [164]. This is line with interesting findings reported in a recent study, in which Cys294 of the mitochondrial F0-F1-ATPase was found to form a disulfide bond with another cysteine residue in heart failure, whereas the protective cardiac resynchronization therapy led to S-nitrosylation of Cys294 and prevented disulfide formation [165]. It has been find that about 50% of those proteins that were S-nitrosylated by PreC were also S-nitrosylated in PostC [132, 135], suggesting that there might be a common set of proteins targeted by •NO/S-nitrosylation signaling with both PreC and PostC. However, the S-nitrosylation process is not a random reaction, but depends on a number of conditions. In fact, the instantaneous redox state and ultrastructural accessibility of cysteine residue(s) under low-oxygen tension, such as hypoxia, ischemia and postconditioning intermittent ischemia/reperfusion may determine whether a particular thiol in a given protein is subjected to S-nitrosylation [166, 167] During the first minutes of reperfusion usually a typical large burst of ROS occurs in not protected (naïf) hearts. The ROS/RNS burst results in the irreversible oxidation/nitration of a number of important proteins. These proteins are damaged and need to be degraded and re-synthesized to regain normal cell function, otherwise irreversible tissue injury occurs. The shielding effect of S-nitrosylation could be necessary to trigger protection in early reperfusion and to allow sufficient time for the activation of protective signaling. Since S-nitrosylation is a transient readily reversed protein modification, it must be timing. This could be of extreme importance during I-PostC maneuvers. In fact, the ROS/RNS burst is attenuated (not abolished) by I-PostC maneuvers and S-nitrosylation occurring during PostC may shield modified cysteines from more irreversible states of oxidation till the burst of ROS/RNS vanishes. This point of view is in line with the experimental evidence that a delay in performing PostC maneuvers results in a loss of protection [101, 133, 168]. Actually, it has been find that protein nitration may be deleterious in PostC scenario [107, 169]. However, other authors have observed a beneficial effect for this reaction induced by peroxynitrite [170]. We have proposed that tyrosine nitration may be a transient initial effect of I-PostC, which is suddenly followed by the prevalence of protein S-nitrosylation, possibly *via* the so-called *secondary reaction* described above [17]. We have shown in rat hearts that after 7-min of reperfusion I-PostC induces a reduction of the levels of 3-nitrotyrosine formation and a subsequent increase in S-nitrosylation of proteins, which persist for at least the 120 min of reperfusion [17]. In fact, a low level of 3-nitrotyrosine in PostC have often been observed [17, 168, 169], but a prevalent formation of s-nitrosylated proteins have been described [17, 164]. Very recently, we and other authors have shown that protein S-nitrosylation occurs mainly in mitochondria after I-PostC [68, 164]. We have also shown that pharmacological PostC induced by Diazoxide (a drug supposed to promote ROS-signaling through actions on mitoK_{ATP} channels and connexin [103, 171–173] may induce a strong S-nitrosylation of mitochondrial proteins. In another study, the addition of a mitochondria-targeted •NO-donor at the start of reperfusion (*i.e.,* pharmacological PostC) has also been found to be cardioprotective [142]. The •NO-donor used in this study was the so-called MitoSNO, which comprises the •NO-donor SNAP (S-nitroso-N-acetylpenicillamine) conjugated to a triphenylphosphonium (TPP) moiety. The lipophilic TPP allows MitoSNO to pass rapidly through membranes driven by the membrane potential and therefore to accumulate several-hundred-fold within the mitochondria, where it generates •NO and S-nitrosylates thiol proteins [142]. The nitrosylation of proteins by MitoSNO and other donors has been confirmed by other authors both in basal conditions and in the context of postconditioning cardioprotection [49, 142, 173]. It is important to emphasize that phosphorylative pathways may be activate in parallel or in sequence to the nitrosylative processes. For instance, it has been recently reported that the most abundant isoforms of PKG (PKGI) within cardiomyocyte is involved in cardioprotection against I/R injury. However, after cardiomyocyte-specific ablation of the PKGI gene in the mouse heart, it was still possible to protect the hearts with several interventions, including I-PostC or pharmacological PostC with the •NO-donor MitoSNO, *via* S-nitrosylation of mitochondrial proteins [174]. Therefore, the authors concluded that PostC may afford protection either by-passing PKGI or by acting independently or downstream of it. The authors also suggested differences between cGMP/PKGI pathway in myocytes and other cardiac cell types during I-PostC's protection in this *in vivo* study. In fact, they cannot rule out that the exogenous and endogenous •NO may act to protect the heart from I/R injury in a manner that depends on PKG in other cardiac cell types [174]. In fact, PKG pathway has been involved in PostC protection in different models by several authors [158, 159, 169, 174].

13.9.4 Redox Regulation of Transcription in Cardioprotection

The transcription factors hypoxia-inducible factor-1 (HIF-1) and nuclear factor erythroid related factor-2 (Nrf2) are subject to redox regulation and these effects are at the base of adaptive responses triggered by ROS signaling or, alternatively, deleterious effects triggered by redox stress. An appropriate ROS production activate the Nrf2 and HIF-1 pathways which render the organ more resistant to subsequent I/R challenging.

In unstressed cells, Nrf2 is bound to its cytosolic repressor, Kelch-like ECH-associated protein-1 (Keap1), and is largely ubiquitinated and redirect to proteasomal degradation. In the presence of ROS, PKC is activated and can phosphorylate Nrf2 at the Ser40 residue. ROS also oxidize cysteine residues in Keap1, preventing its interaction with Nrf2. Phosphorylate Nrf2 can then easily dissociate from Keap1 and can migrate to the nucleus. In the nucleus, upon binding to antioxidant response elements (AREs) on the genome, Nrf2 transcribes several genes involved in antioxidant defense, mitochondrial biogenesis, and energy regulation. Indeed, several compounds may protect the heart against I/R injury by activating the Nrf2/ARE pathway [175–179]. Moreover, I/R induces de novo Nrf2 protein translation and Nrf2 knockout mice display increased infarct size following I/R and a reduced degree of cardiac protection by I-PreC [180]. It has been also reported that Nrf2 nuclear accumulation occurs together an increased expression of mitochondrial-uncoupling-protein-3 in isolated murine hearts subjected to *ex vivo* I/R challenging [181]. In the presence of ROS signaling also the activation of HIF-1 occurs. Therefore, the inducible subunit of HIF-1, HIF-1α is formed. The latter is regulated by prolyl hydroxylases (PHDs) and may translocate to the nucleus. Under normal oxygen levels, PHDs hydroxylate HIF-1α, allowing von Hippel Lindau to ubiquitinate HIF-1α and to target it for subsequent proteasomal degradation. However, in the presence of ROS, PHDs are inactivated and HIF-1α can translocate to the nucleus. In the nucleus, the constitutively expressed HIF-1β subunit binds to HIF-1α to form the active HIF-1, which binds to hypoxia response elements (HREs) and transcribes several genes for angiogenesis, energy metabolism, and red blood cell production. Whether ROS of mitochondrial origin are necessary for HIF-1 activation and cardioprotection is controversial. However, HIF activation using pharmacological PHD inhibitors such as DMOG results in a level of cardioprotection similar to that obtained with IPC [182]. Likewise, silencing of PHD1 attenuates ex vivo myocardial IR injury [183].

13.10 Conclusions

Oxidative stress is a sign-symbol of the pathophysiology of myocardial I/R injury. In fact, massive ROS generation occurs when oxygen delivery is restored after an ischemic event. While high ROS levels can be deleterious and can kill the cardiac cells, low levels of oxidants can be cardioprotective in therapeutic conditioning approaches. Nevertheless, we have discussed that this is not always the case and it is advisable to talk about *appropriate and inappropriate ROS production.*

Although ROS play an important role in I/R injury, antioxidant therapy fails to prevent I/R related disorders, also because ROS may be necessary for some physiological functions. Thus, the development of the drug that selectively attenuates the pathological oxidative stress without altering the *ROS signaling* in physiological conditions, seems to be an approach still far to be achieved. Perhaps a multiple approach that can limit inflammation on the one hand and production of ROS on the other could be a therapeutic strategy to improve resistance to ischemia and to make the organ more resistant to I/R challenging. Indeed, inflammation and overproduction of ROS are often present in cardiovascular diseases. Moreover, in the presence of inflammation even a modest ROS production can be deleterious. Therefore, compounds that inhibit inflammation on the one hand and activate the protective signaling on the other are promising clinical instruments that may open the window of the hope to defeat I/R injury one day.

Acknowledgments The authors of this chapter are supported for their researches by University of Turin, Italy.

References

1. Commoner B, Townsend J, Pake GE (1954) Free radicals in biological materials. Nature 174:689–691
2. Carpi A, Menabò R, Kaludercic N, Pelicci P, Di Lisa F, Giorgio M (2009) The cardioprotective effects elicited by p66(Shc) ablation demonstrate the crucial role of mitochondrial ROS formation in ischemia/reperfusion injury. Biochim Biophys Acta 1787:774–780
3. Kaludercic N, Mialet-Perez J, Paolocci N, Parini A, Di Lisa F (2014) Monoamine oxidases as sources of oxidants in the heart. J Mol Cell Cardiol 73:34–42

4. Deshwal S, Di Sante M, Di Lisa F, Kaludercic N (2017) Emerging role of monoamine oxidase as a therapeutic target for cardiovascular disease. Curr Opin Pharmacol 33:64–69
5. Lassegue B, Clempus RE (2003) Vascular NAD(P)H oxidases: specific features, expression, and regulation. Am J Physiol Regul Integr Comp Physiol 285:R277–R297
6. Lambeth JD (2004) NOX enzymes and the biology of reactive oxygen. Nat Rev Immunol 4:181–189
7. Donko A, Péterfi Z, Sum A, Leto T, Geiszt M (2005) Dual oxidases. Trans R Soc Lond B Biol Sci 360:2301–2308
8. Martin KD, Frederick LM, vonLoehneysen K, Dinauer MC, Knaus UG (2006) Functional analysis of nox4 reveals unique characteristics compared to other NADPH oxidases. Cell Signal 18:69–82
9. Hilenski LL, Clempus RE, Quinn MT, Lambeth JD, Griendling KK (2004) Distinct subcellular localizations of Nox1 and Nox4 in vascular smooth muscle cells. Arterioscler Thromb Vasc Biol 24:677–683
10. Byrne JA, Grieve DJ, Cave AC, Shah AM (2003) Oxidative stress and heart failure. Arch Mal Coeur Vaiss 96:214–221
11. Byrne JA, Grieve DJ, Bendall JK, Li JM, Gove C, Lambeth JD, Cave AC, Shah AM (2003) Contrasting roles of NADPH oxidase isoforms in pressure-overload versus angiotensin II-induced cardiac hypertrophy. Circ Res 93:802–805
12. Konior A, Schramm A, Czesnikiewicz-Guzik M, Guzik TJ (2014) NADPH oxidases in vascular pathology. Antioxid Redox Signal 20:2794–2814
13. Muller FL, Lustgarten MS, Jang Y, Richardson A, Van Remmen H (2007) Trends in oxidative aging theories. Free Radic Biol Med 43:477–503
14. Palmer RMJ, Rees DD, Ashton DS, Moncada S (1988) L-Arginine is the physiological precursor for the formation of nitric oxide in endothelium dependent relaxation. Biochem Biophys Res Commun 153:1251–1256
15. Stamler JS, Single D, Loscalzo J (1992) Biochemistry of nitric oxide and its redox-activated forms. Science 258:1898–1902
16. Gow AJ, McClelland M, Garner SE, Malcolm S, Ischiropoulos H (1998) The determination of nitrotyrosine residues in proteins. Methods Mol Biol 100:291–299
17. Penna C, Angotti C, Pagliaro P (2014) Protein S-nitrosylation in preconditioning and postconditioning. Exp Biol Med (Maywood) 239:647–662
18. Fukuto JM, Cisneros CJ, Kinkade RL (2013) A comparison of the chemistry associated with the biological signaling and actions of nitroxyl (HNO) and nitric oxide (NO). J Inorg Biochem 118:201–208
19. Lacza Z, Pankotai E, Csordás A, Gero D, Kiss L, Horváth EM, Kollai M, Busija DW, Szabó C (2006) Mitochondrial NO and reactive nitrogen species production: does mtNOS exist? Nitric Oxide 14:162–168
20. Csordás A, Pankotai E, Snipes JA, Cselenyák A, Sárszegi Z, Cziráki A, Gaszner B, Papp L, Benko R, Kiss L, Kovács E, Kollai M, Szabó C, Busija DW, Lacza Z (2007) Human heart mitochondria do not produce physiologically relevant quantities of nitric oxide. Life Sci 80:633–637
21. Zaobornyj T, Ghafourifar P (2012) Strategic localization of heart mitochondrial NOS: a review of the evidence. Am J Physiol Heart Circ Physiol 303:H1283–H1293
22. Bolli R, Dawn B, Tang XL, Qiu Y, Ping P, Xuan YT, Jones WK, Takano H, Guo Y, Zhang J (1998) The nitric oxide hypothesis of late preconditioning. Basic Res Cardiol 93:325–338
23. Guo Y, Sanganalmath SK, Wu W, Zhu X, Huang Y, Tan W, Ildstad ST, Li Q, Bolli R (2012) Identification of inducible nitric oxide synthase in peripheral blood cells as a mediator of myocardial ischemia/reperfusion injury. Basic Res Cardiol 107:253
24. Rastaldo R, Cappello S, Folino A, Berta GN, Sprio AE, Losano G, Samaja M, Pagliaro P (2011) Apelin-13 limits infarct size and improves cardiac postischemic mechanical recovery only if given after ischemia. Am J Physiol Heart Circ Physiol 300:H2308–H2315
25. Zimmet JM, Hare JM (2006) Nitroso-redox interactions in the cardiovascular system. Circulation 114:1531–1544

26. Lim G, Venetucci L, Eisner DA, Casadei B (2008) Does nitric oxide modulate cardiac ryanodine receptor function? Implications for excitation-contraction coupling. Cardiovasc Res 77:256–264
27. Sun J, Picht E, Ginsburg KS, Bers DM, Steenbergen C, Murphy E (2006) Hypercontractile female hearts exhibit increased S-nitrosylation of the L-type Ca^{2+} channel alpha1 subunit and reduced ischemia/reperfusion injury. Circ Res 98:403–411
28. Sarkela TM, Berthiaume J, Elfering S, Gybina AA, Giulivi C (2011) The modulation of oxygen radical production by nitric oxide in mitochondria. J Biol Chem 276:6945–6949
29. Jezek P, Hlavatá L (2005) Mitochondria in homeostasis of reactive oxygen species in cell, tissues, and organism. Int J Biochem Cell Biol 37:2478–2503
30. Xia Y, Tsai AL, Berka V, Zweier JL (1998) Superoxide generation from endothelial nitric-oxide synthase. A Ca^{2+}/calmodulin-dependent and tetrahydrobiopterin regulatory process. J Biol Chem 273:25804–25808
31. Landmesser U, Dikalov S, Price SR, McCann L, Fukai T, Holland SM, Mitch WE, Harrison DG (2003) Oxidation of tetrahydrobiopterin leads to uncoupling of endothelial cell nitric oxide synthase in hypertension. J Clin Invest 111:1201–1209
32. Dikalova AE, Góngora MC, Harrison DG, Lambeth JD, Dikalov S, Griendling KK (2010) Upregulation of Nox1 in vascular smooth muscle leads to impaired endothelium-dependent relaxation via eNOS uncoupling. Am J Physiol Heart Circ Physiol 299:H673–H679
33. Zweier JL, Wang P, Samouilov A, Kuppusamy P (1995) Enzyme-independent formation of nitric oxide in biological tissues. Nat Med 1:804–809
34. Gladwin MT, Shelhamer JH, Schechter AN, Pease-Fye ME, Waclawiw MA, Panza JA, Ognibene FP, Cannon RO (2000) Role of circulating nitrite and S-nitrosohemoglobin in the regulation of regional blood flow in humans. Proc Natl Acad Sci U S A 97:11482–11487
35. Samouilov A, Kuppusamy P, Zweier JL (1998) Evaluation of the magnitude and rate of nitric oxide production from nitrite in biological systems. Arch Biochem Biophys 357:1–7
36. Li H, Cui H, Kundu TK, Alzawahra W, Zweier JL (2008) Nitric oxide production from nitrite occurs primarily in tissues not in the blood: critical role of xanthine oxidase and aldehyde oxidase. J Biol Chem 283:17855–17863
37. Gautier C, van Faassen E, Mikula I, Martasek P, Slama-Schwok A (2006) Endothelial nitric oxide synthase reduces nitrite anions to NO under anoxia. Biochem Biophys Res Commun 341:816–821
38. Vanin AF, Bevers LM, Slama-Schwok A, van Faassen EE (2007) Nitric oxide synthase reduces nitrite to NO under anoxia. Cell Mol Life Sci 64:96–103
39. Tota B, Quintieri AM, Angelone T (2010) The emerging role of nitrite as an endogenous modulator and therapeutic agent of cardiovascular function. Curr Med Chem 17:1915–1925
40. Omar SA, Webb AJ (2014) Nitrite reduction and cardiovascular protection. J Mol Cell Cardiol 73:57–69
41. Maia LB, Moura JJ (2015) Nitrite reduction by molybdoenzymes: a new class of nitric oxide-forming nitrite reductases. J Biol Inorg Chem 20:403–433
42. Pacher P, Liaudet L, Bai P, Mabley JG, Kaminski PM, Virág L, Deb A, Szabó E, Ungvári Z, Wolin MS, Groves JT, Szabó C (2003) Potent metalloporphyrin peroxynitrite decomposition catalyst protects against the development of doxorubicin-induced cardiac dysfunction. Circulation 107:896–904
43. Pacher P, Schulz R, Liaudet L, Szabó C (2005) Nitrosative stress and pharmacological modulation of heart failure. Trends Pharmacol Sci 26:302–310
44. Liaudet L, Vassalli G, Pacher P (2009) Role of peroxynitrite in the redox regulation of cell signal transduction pathways. Front Biosci 14:4809–4814
45. Pacher P, Beckman JS, Liaudet L (2007) Nitric oxide and peroxynitrite in health and disease. Physiol Rev 87:315–424
46. Halliwell B (2007) Biochemistry of oxidative stress. Biochem Soc Trans 35:1147–1150
47. Valko M, Leibfritz D, Moncola J, Cronin MD, Mazur MJ (2007) Free radicals and antioxidants in normal physiological functions and human disease. Int J Biochem Cell Biol 39:44–84

48. Whaley-Connell A, McCullough PA, Sowers JR (2011) The role of oxidative stress in the metabolic syndrome. Rev Cardiovasc Med 12:21–29
49. Tullio F, Angotti C, Perrelli MG, Penna C, Pagliaro P (2013) Redox balance and cardioprotection. Basic Res Cardio 108:392
50. Pham-Huy LA, He H, Pham-Huy C (2008) Free radicals, antioxidants in disease and health. Int J Biomed Sci 4:89–96
51. Sen S, Chakraborty R, Sridhar C, Reddy YSR, De B (2010) Free radicals, antioxidants, diseases and phytomedicines: current status and future prospect. Int J Pharm Sci Rev Res 3:91
52. Davies KJA (2018) Cardiovascular adaptive homeostasis in exercise. Front Physiol 9:369
53. Sag CM, Wagner S, Maier LS (2013) Role of oxidants on calcium and sodium movement in healthy and diseased cardiac myocytes. Free Radic Biol Med 63:338–349
54. Tocchetti CG, Molinaro M, Angelone T, Lionetti V, Madonna R, Mangiacapra F, Moccia F, Penna C, Sartiani L, Quaini F, Pagliaro P (2015) Nitroso-redox balance and modulation of basal myocardial function: an update from the Italian Society of Cardiovascular Research (SIRC). Curr Drug Targets 16:895–903
55. Orr WC, Sohal RS (2003) Does overexpression of Cu, Zn-SOD extend life span in Drosophila melanogaster? Exp Gerontol 38:227–230
56. Garlid AO, Jaburek M, Jacobs JP, Garlid KD (2013) Mitochondrial reactive oxygen species: which ROS signals cardioprotection? Am J Physiol Heart Circ Physiol 305:H960–H968
57. Becker LB (2004) New concepts in reactive oxygen species and cardiovascular reperfusion physiology. Cardiovasc Res 61:461–470
58. Bencsik P, Kupai K, Giricz Z, Görbe A, Pipis J, Murlasits Z, Kocsis GF, Varga-Orvos Z, Puskás LG, Csonka C, Csont T, Ferdinandy P (2010) Role of iNOS and peroxynitrite-matrix metalloproteinase-2 signaling in myocardial late preconditioning in rats. Am J Physiol Heart Circ Physiol 299:H512–H518
59. Vallance P, Moncada S (1993) Role of endogenous nitric oxide in septic shock. New Horiz 1:77–86
60. Rastaldo R, Pagliaro P, Cappello S, Penna C, Mancardi D, Westerhof N, Losano G (2007) Nitric oxide and cardiac function. Life Sci 81:779–793
61. Thannickal VJ, Fanburg BL (2000) Reactive oxygen species in cell signaling. Am J Physiol Lung Cell Mol Physiol 279:L1005–L1028
62. Di Meo S, Reed TT, Venditti P, Victor VM (2016) Role of ROS and RNS sources in physiological and pathological conditions. Oxid Med Cell Longev 2016:1245049
63. Saitoh S, Zhang C, Tune JD, Potter B, Kiyooka T, Rogers PA, Knudson JD, Dick GM, Swafford A, Chilian WM (2006) Hydrogen peroxide: a feed-forward dilator that couples myocardial metabolism to coronary blood flow. Arterioscler Thromb Vasc Biol 26:2614–2621
64. Saitoh S, Kiyooka T, Rocic P, Rogers PA, Zhang C, Swafford A, Dick GM, Viswanathan C, Park Y, Chilian WM (2007) Redox-dependent coronary metabolic dilation. Am J Physiol Heart Circ Physiol 293:H3720–H3725
65. Asmat U, Abad K, Ismail K (2016) Diabetes mellitus and oxidative stress-A concise review. Saudi Pharm J 24:547–553
66. Russo I, Penna C, Musso T, Popara J, Alloatti G, Cavalot F, Pagliaro P (2017) Platelets, diabetes and myocardial ischemia/reperfusion injury. Cardiovasc Diabetol 16:71
67. Perrelli MG, Pagliaro P, Penna C (2011) Ischemia/reperfusion injury and cardioprotective mechanisms: Role of mitochondria and reactive oxygen species. World J Cardiol 3:186–200
68. Penna C, Perrelli MG, Pagliaro P (2013) Mitochondrial pathways, permeability transition pore, and redox signaling in cardioprotection: therapeutic implications. Antioxid Redox Signal 18:556–599
69. Cadenas S (2018) ROS and redox signaling in myocardial ischemia-reperfusion injury and cardioprotection. Free Radic Biol Med 117:76–89
70. Zorov DB, Filburn CR, Klotz LO, Zweier JL, Sollott SJ (2000) Reactive oxygen species (ROS)-induced ROS release: a new phenomenon accompanying induction of the mitochondrial permeability transition in cardiac myocytes. J Exp Med 192:1001–1014

71. Orsini F, Migliaccio E, Moroni M, Contursi C, Raker VA, Piccini D, Martin-Padura I, Pelliccia G, Trinei M, Bono M, Puri C, Tacchetti C, Ferrini M, Mannucci R, Nicoletti I, Lanfrancone L, Giorgio M, Pelicci PG (2004) The life span determinant p66Shc localizes to mitochondria where it associates with mitochondrial heat shock protein 70 and regulates trans-membrane potential. J Biol Chem 279:25689–25695
72. Golstein P, Kroemer G (2007) A multiplicity of cell death pathways. Symposium on apoptotic and non-apoptotic cell death pathways. EMBO Rep 8:829–833
73. Golstein P, Kroemer G (2007) Cell death by necrosis: towards a molecular definition. Trends Biochem Sci 32:7–43
74. Basso E, Petronilli V, Forte MA, Bernardi P (2008) Phosphate is essential for inhibition of the mitochondrial permeability transition pore by cyclosporin A and by cyclophilin D ablation. J Biol Chem 283:26307–26311
75. Piper HM, Abdallah Y, Schäfer C (2004) The first minutes of reperfusion: a window of opportunity for cardioprotection. Cardiovasc Res 61:365–371
76. Bernardi P, Krauskopf A, Basso E, Petronilli V, BlalchyDyson E, Di Lisa F, Forte MA (2006) The mitochondrial permeability transition from in vitro artifact to disease target. FEBS J 273:2077–2099
77. Yellon DM, Hausenloy DJ (2007) Myocardial reperfusion injury. N Engl J Med 357:1121–1135
78. Vinten-Johansen J, Zhao ZQ, Jiang R, Zatta AJ (2005) Myocardial protection in reperfusion with postconditioning. Expert Rev Cardiovasc Ther 3:1035–1045
79. Takagi H, Matsui Y, Sadoshima J (2007) The role of autophagy in mediating cell survival and death during ischemia and reperfusion in the heart. Antioxid Redox Signal 9:1373–1381
80. Przyklenk K, Dong Y, Undyala VV, Whittaker P (2012) Autophagy as a therapeutic target for ischemia-reperfusion injury? Concepts, controversies and challenges. Cardiovasc Res 94:197–205
81. Sheldon JG, Williams SP, Fulton AM, Brindle KM (1996) 31P NMR magnetization transfer study of the control of ATP turnover in Saccharomyces cerevisiae. Proc Natl Acad Sci USA 93:6399–6404
82. Cadenas S, Echtay KS, Harper JA, Jekabsons MB, Buckingham JA, Grau E, Abuin A, Chapman H, Clapham JC, Brand MD (2002) The basal proton conductance of skeletal muscle mitochondria from transgenic mice overexpressing or lacking uncoupling protein-3. J Biol Chem 277:2773–2778
83. Roussel D, Harding M, Runswick MJ, Walker JE, Brand MD (2002) Does any yeast mitochondrial carrier have a native uncoupling protein function? J Bioenerg Biomembr 34:165–176
84. Krauss S, Zhang CY, Scorrano L, Dalgaard LT, St-Pierre J, Grey ST, Lowell BB (2003) Superoxide-mediated activation of uncoupling protein 2 causes pancreatic beta cell dysfunction. J Clin Invest 112:1831–1842
85. Cannon B, Nedergaard J (2004) Brown adipose tissue: function and physiological significance. Physiol Rev 84:277–359
86. Downey JM, Cohen MV (2006) Reducing infarct size in the setting of acute myocardial infarction. Prog Cardiovasc Dis 48:363–371
87. Cohen MV, Downey JM (2011) Ischemic postconditioning: from receptor to end-effector. Antioxid Redox Signal 14:821–831
88. Zorov DB, Juhaszova M, Sollott SJ (2014) Mitochondrial reactive oxygen species (ROS) and ROS-induced ROS release. Physiol Rev 94:909–950
89. Bolli R (1990) Mechanism of myocardial "stunning". Circulation 82:723–738
90. Bolli R (1991) Oxygen-derived free radicals and myocardial reperfusion injury: an overview. Cardiovasc Drugs Ther 5:249–268
91. Bolli R (1992) Myocardial 'stunning' in man. Circulation 86:1671–1691
92. Singal PK, Khaper N, Palace V, Kumar D (1998) The role of oxidative stress in the genesis of heart disease. Cardiovasc Res 40:426–432

93. Stein AB, Tang XL, Guo Y, Xuan YT, Dawn B, Bolli R (2004) Delayed adaptation of the heart to stress: late preconditioning. Stroke 35:2676–2679
94. Schulz R, Kelm M, Heusch G (2004) Nitric oxide in myocardial ischemia/reperfusion injury. Cardiovasc Res 61:402–413
95. Jones SP, Bolli R (2006) The ubiquitous role of nitric oxide in cardioprotection. J Mol Cell Cardiol 40:16–23
96. Riess ML, Camara AK, Novalija E, Chen Q, Rhodes SS, Stowe DF (2002) Anesthetic preconditioning attenuates mitochondrial Ca^{2+} overload during ischemia in Guinea pig intact hearts: reversal by 5-hydroxydecanoic acid. Anesth Analg 95:1540–1546
97. Kevin LG, Camara AK, Riess ML, Novalija E, Stowe DF (2003) Ischemic preconditioning alters real-time measure of O_2 radicals in intact hearts with ischemia and reperfusion. Am J Physiol Heart Circ Physiol 284:H566–H574
98. Murry CE, Jennings R, Reimer KA (1986) Preconditioning with ischemia: a delay of lethal cell injury in ischemic myocardium. Circulation 74:1124–1136
99. Penna C, Granata R, Tocchetti CG, Gallo MP, Alloatti G, Pagliaro P (2015) Endogenous cardioprotective agents: role in pre and postconditioning. Curr Drug Targets 16:843–867
100. Zhao ZQ, Corvera JS, Halkos ME, Kerendi F, Wang NP, Guyton RA, Vinten-Johansen J (2003) Inhibition of myocardial injury by ischemic postconditioning during reperfusion: comparison with ischemic preconditioning. Am J Physiol Heart Circ Physiol 285:H579–H588. Erratum in: Am J Physiol Heart Circ Physiol 286:H477
101. Skyschally A, Van Caster P, Iliodromitis EK, Schulz R, Kremastinos DT, Heusch G (2009) Ischemic postconditioning: experimental models and protocol algorithms. Basic Res Cardiol 104:469–483
102. Yellon DM, Downey JM (2003) Preconditioning the myocardium: from cellular physiology to clinical cardiology. Physiol Rev 83:1113–1151
103. Boengler K, Heusch G, Schulz R (2011) Mitochondria in postconditioning. Antioxid Redox Signal 14:863–880
104. Hausenloy DJ, Lecour S, Yellon DM (2011) Reperfusion injury salvage kinase and survivor activating factor enhancement prosurvival signaling pathways in ischemic postconditioning: two sides of the same coin. Antioxid Redox Signal 14:893–907
105. Siegfried MR, Erhardt J, Rider T, Ma XL, Lefer AM (1992) Cardioprotection and attenuation of endothelial dysfunction by organic nitric oxide donors in myocardial ischemia-reperfusion. J Pharmacol Exp Ther 260:668–675
106. Hill M, Takano H, Tang XL, Kodani E, Shirk G, Bolli R (2001) Nitroglycerin induces late preconditioning against myocardial infarction in conscious rabbits despite development of nitrate tolerance. Circulation 104:694–699
107. Iliodromitis EK, Gaitanaki C, Lazou A, Aggeli IK, Gizas V, Bofilis E, Zoga A, Beis I, Kremastinos DT (2006) Differential activation of mitogen-activated protein kinases in ischemic and nitroglycerin-induced preconditioning. Basic Res Cardiol 101:327–335
108. Garcia-Dorado D, Agulló L, Sartorio CL, Ruiz-Meana M (2009) Myocardial protection against reperfusion injury: the cGMP pathway. Thromb Haemost 101:635–642
109. Hausenloy DJ, Yellon DM (2007) Reperfusion injury salvage kinase signalling: taking a RISK for cardioprotection. Heart Fail Rev 12:217–234
110. Hausenloy DJ, Yellon DM (2008) Preconditioning and postconditioning: new strategies for cardioprotection. Diabetes Obes Metab 10:451–459
111. Hausenloy DJ (2009) Signalling pathways in ischaemic postconditioning. Thromb Haemost 101:626–634
112. Hausenloy DJ, Ong SB, Yellon DM (2009) The mitochondrial permeability transition pore as a target for preconditioning and postconditioning. Basic Res Cardiol 104:189–202
113. Lecour S (2009) Activation of the protective Survivor Activating Factor Enhancement (SAFE) pathway against reperfusion injury: does it go beyond the RISK pathway? J Mol Cell Cardiol 47:32–40
114. Heusch G, Boengler K, Schulz R (2010) Inhibition of mitochondrial permeability transition pore opening: the Holy Grail of cardioprotection. Basic Res Cardiol 105:151–154

115. Nguyen TT, Stevens MV, Kohr M, Steenbergen C, Sack MN, Murphy E (2011) Cysteine 203 of cyclophilin D is critical for cyclophilin D activation of the mitochondrial permeability transition pore. J Biol Chem 286:40184–40192
116. Bolli R (2000) The late phase of preconditioning. Circ Res 87:972–983
117. Pagliaro P, Gattullo D, Rastaldo R, Losano G (2001) Ischemic preconditioning: from the first to the second window of protection. Life Sci 69:1–15
118. Crisafulli A, Melis F, Tocco F, Pittau G, Lorrai L, Gori T, Mancardi D, Concu A, Pagliaro P (2007) Delayed preconditioning-mimetic actions of exercise or nitroglycerin do not affect haemodynamics and exercise performance in trained or sedentary individuals. J Sports Sci 25:1393–1401
119. Hausenloy DJ, Yellon DM (2010) The second window of preconditioning (SWOP) where are we now? Cardiovasc Drugs Ther 24:235–254
120. Zhao ZQ, Vinten-Johansen J (2006) Postconditioning: reduction of reperfusion-induced injury. Cardiovasc Res 70:200–211
121. Ovize M, Baxter GF, Di Lisa F, Ferdinandy P, Garcia-Dorado D, Hausenloy DJ, Heusch G, Vinten-Johansen J, Yellon DM, Schulz R (2010) Working Group of Cellular Biology of Heart of European Society of Cardiology. Postconditioning and protection from reperfusion injury: where do we stand? Position paper from the Working Group of Cellular Biology of the Heart of the European Society of Cardiology. Cardiovasc Res 87:406–423
122. Yang X, Cohen MV, Downey JM (2010) Mechanism of cardioprotection by early ischemic preconditioning. Cardiovasc Drugs Ther 24:225–234
123. Carroll R, Gant VA, Yellon DM (2001) Mitochondrial K(ATP) channel opening protects a human atrial-derived cell line by a mechanism involving free radical generation. Cardiovasc Res 51:691–700
124. Forbes RA, Steenbergen C, Murphy E (2001) Diazoxide-induced cardioprotection requires signaling through a redox-sensitive mechanism. Circ Res 88:802–809
125. Bell RM, Yellon DM (2001) The contribution of endothelial nitric oxide synthase to early ischaemic preconditioning: the lowering of the preconditioning threshold. An investigation in eNOS knockout mice. Cardiovasc Res 52:274–280
126. Bolli R (2001) Cardioprotective function of inducible nitric oxide synthase and role of nitric oxide in myocardial ischemia and preconditioning: an overview of a decade of research. J Mol Cell Cardiol 33:1897–1918
127. Oldenburg O, Qin Q, Krieg T, Yang XM, Philipp S, Critz SD, Cohen MV, Downey JM (2004) Bradykinin induces mitochondrial ROS generation via NO, cGMP, PKG, and mitoKATP channel opening and leads to cardioprotection. Am J Physiol Heart Circ Physiol 286:H468–H476
128. Shiva S, Sack MN, Greer JJ, Duranski M, Ringwood LA, Burwell L, Wang X, MacArthur PH, Shoja A, Raghavachari N, Calvert JW, Brookes PS, Lefer DJ, Gladwin MT (2007) Nitrite augments tolerance to ischemia/reperfusion injury via the modulation of mitochondrial electron transfer. J Exp Med 204:2089–2102
129. Guo Y, Li Q, Wu WJ, Tan W, Zhu X, Mu J, Bolli R (2008) Endothelial nitric oxide synthase is not necessary for the early phase of ischemic preconditioning in the mouse. J Mol Cell Cardiol 44:496–501
130. Simon JN, Duglan D, Casadei B, Carnicer R (2014) Nitric oxide synthase regulation of cardiac excitation-contraction coupling in health and disease. J Mol Cell Cardiol 73:80–91
131. Pagliaro P, Mancardi D, Rastaldo R, Penna C, Gattullo D, Miranda KM, Feelisch M, Wink DA, Kass DA, Paolocci N (2003) Nitroxyl affords thiol-sensitive myocardial protective effects akin to early preconditioning. Free Radic Biol Med 34:33–43
132. Sun J, Aponte AM, Kohr MJ, Tong G, Steenbergen C, Murphy E (2013) Essential role of nitric oxide in acute ischemic preconditioning: S-nitros(yl)ation versus sGC/cGMP/PKG signaling? Free Radic Biol Med 54:105–112
133. Penna C, Mancardi D, Raimondo S, Geuna S, Pagliaro P (2008) The paradigm of postconditioning to protect the heart. J Cell Mol Med 12:435–458

134. Sun J, Morgan M, Shen RF, Steenbergen C, Murphy E (2007) Preconditioning results in S-nitrosylation of proteins involved in regulation of mitochondrial energetic and calcium transport. Circ Res 101:1155–1163
135. Sun J, Kohr MJ, Nguyen T, Aponte AM, Connelly PS, Esfahani SG, Gucek M, Daniels MP, Steenbergen C, Murphy E (2012) Disruption of caveolae blocks ischemic preconditioning-mediated S-nitrosylation of mitochondrial proteins. Antioxid Redox Signal 16:45–56
136. Kohr MJ, Sun J, Aponte A, Wang G, Gucek M, Murphy E, Steenbergen C (2011) Simultaneous measurement of protein oxidation and S-nitrosylation during preconditioning and ischemia/reperfusion injury with resin-assisted capture. Circ Res 108:418–426
137. Angelone T, Quintieri A, Pasqua T, Gentile S, Tota B, Mahata SK, Cerra MC (2012) Phosphodiesterase type-2 and NO-dependent S-nitrosylation mediate the cardioinhibition of the antihypertensive catestatin. Am J Physiol Heart Circ Physiol 302:H431–H442
138. Foster DB, O'Rourke B, Van Eyk JE (2008) What can mitochondrial proteomics tell us about cardioprotection afforded by preconditioning? Expert Rev Proteomics 5:633–636
139. Nadtochiy SM, Burwell LS, Brookes PS (2007) Cardioprotection and mitochondrial S-nitrosation: effects of S-nitroso-2-mercaptopropionyl glycine (SNO-MPG) in cardiac ischemia-reperfusion injury. J Mol Cell Cardiol 42:812–825
140. Galkin A, Moncada S (2007) S-nitrosation of mitochondrial complex I depends on its structural conformation. J Biol Chem 282:37448–37453
141. Sun J, Murphy E (2010) Protein S-nitrosylation and cardioprotection. Circ Res 106(2):285–296
142. Chouchani ET, Methner C, Nadtochiy SM, Logan A, Pell VR, Ding S, James AM, Cochemé HM, Reinhold J, Lilley KS, Partridge L, Fearnley IM, Robinson AJ, Hartley RC, Smith RA, Krieg T, Brookes PS, Murphy MP (2013) Cardioprotection by S-nitrosation of a cysteine switch on mitochondrial complex I. Nat Med 19:753–759
143. Boengler K, Bencsik P, Palóczi J, Kiss K, Pipicz M, Pipis J, Ferdinandy P, Schlüter KD, Schulz R (2017) Lack of contribution of p66shc and its mitochondrial translocation to ischemia-reperfusion injury and cardioprotection by ischemic preconditioning. Front Physiol 8:733
144. Penna C, Pagliaro P, Rastaldo R, Di Pancrazio F, Lippe G, Gattullo D, Mancardi D, Samaja M, Losano G, Mavelli I (2004) F0F1 ATP synthase activity is differently modulated by coronary reactive hyperemia before and after ischemic preconditioning in the goat. Am J Physiol Heart Circ Physiol 287:H2192–H2200
145. Shiva S, Huang Z, Grubina R, Sun J, Ringwood LA, MacArthur PH, Xu X, Murphy E, Darley-Usmar VM, Gladwin MT (2007) Deoxymyoglobin is a nitrite reductase that generates nitric oxide and regulates mitochondrial respiration. Circ Res 100:654–661
146. Bonora M, Bononi A, De Marchi E, Giorgi C, Lebiedzinska M, Marchi S, Patergnani S, Rimessi A, Suski JM, Wojtala A, Wieckowski MR, Kroemer G, Galluzzi L, Pinton P (2013) Role of the c subunit of the FO ATP synthase in mitochondrial permeability transition. Cell Cycle 12:674–683
147. Alavian KN, Beutner G, Lazrove E, Sacchetti S, Park HA, Licznerski P, Li H, Nabili P, Hockensmith K, Graham M, Porter GA Jr, Jonas EA (2014) An uncoupling channel within the c-subunit ring of the F1FO ATP synthase is the mitochondrial permeability transition pore. Proc Natl Acad Sci USA 111:10580–10585
148. Bernardi P, Di Lisa F (2015) The mitochondrial permeability transition pore: molecular nature and role as a target in cardioprotection. J Mol Cell Cardiol 78:100–106
149. Bernardi P, Rasola A, Forte M, Lippe G (2015) The mitochondrial permeability transition pore: channel formation by F-ATP synthase, integration in signal transduction, and role in pathophysiology. Physiol Rev 95:1111–1155
150. Halestrap AP, Richardson AP (2015) The mitochondrial permeability transition: a current perspective on its identity and role in ischaemia/reperfusion injury. J Mol Cell Cardiol 78:129–141
151. He J, Carroll J, Ding S, Fearnley IM, Walker JE (2017) Permeability transition in human mitochondria persists in the absence of peripheral stalk subunits of ATP synthase. Proc Natl Acad Sci USA 114:9086–9091

152. He J, Ford HC, Carroll J, Ding S, Fearnley IM, Walker JE (2017) Persistence of the mitochondrial permeability transition in the absence of subunit c of human ATP synthase. Proc Natl Acad Sci USA 114:3409–3414
153. Pagliaro P, Moro F, Tullio F, Perrelli MG, Penna C (2011) Cardioprotective pathways during reperfusion: focus on redox signaling and other modalities of cell signaling. Antioxid Redox Signal 14:833–850
154. Feng X, Sun T, Bei Y, Ding S, Zheng W, Lu Y, Shen P (2013) S-nitrosylation of ERK inhibits ERK phosphorylation and induces apoptosis. Sci Rep 3:1814
155. Filice E, Angelone T, De Francesco EM, Pellegrino D, Maggiolini M, Cerra MC (2011) Crucial role of phospholamban phosphorylation and S-nitrosylation in the negative lusitropism induced by 17β-estradiol in the male rat heart. Cell Physiol Biochem 28:41–52
156. Penna C, Rastaldo R, Mancardi D, Raimondo S, Cappello S, Gattullo D, Losano G, Pagliaro P (2006) Post-conditioning induced cardioprotection requires signaling through a redox-sensitive mechanism, mitochondrial ATP-sensitive K^+ channel and protein kinase C activation. Basic Res Cardiol 101:180–189
157. Downey JM, Cohen MV (2006) A really radical observation – a comment on Penna et al. in Basic Res Cardiol (2006) 101:180–189. Basic Res Cardiol 101:190–191
158. Inserte J, Barba I, Poncelas-Nozal M, Hernando V, Agulló L, Ruiz-Meana M, Garcia-Dorado D (2011) cGMP/PKG pathway mediates myocardial postconditioning protection in rat hearts by delaying normalization of intracellular acidosis during reperfusion. J Mol Cell Cardiol 50:903–909
159. Penna C Perrelli MG, Tullio F, Moro F, Parisella ML, Merlino A, Pagliaro P (2011) Post-ischemic early acidosis in cardiac postconditioning modifies the activity of antioxidant enzymes, reduces nitration, and favors protein S-nitrosylation. Pflugers Arch 462:219–233. Erratum in: Pflugers Arch. 2014 Dec;466:2339–2341
160. Valen G, Vaage J (2005) Pre- and postconditioning during cardiac surgery. Basic Res Cardiol 100:179–186
161. Gross ER, Gross GJ (2006) Ligand triggers of classical preconditioning and postconditioning. Cardiovasc Res 70:212–221
162. Tissier R, Waintraub X, Couvreur N, Gervais M, Bruneval P, Mandet C, Zini R, Enriquez B, Berdeaux A, Ghaleh B (2007) Pharmacological postconditioning with the phytoestrogen genistein. J Mol Cell Cardiol 42:79–87
163. Schulz R, Ferdinandy P (2013) Does nitric oxide signaling differ in pre- and post-conditioning? Importance of S-nitrosylation vs. protein kinase G activation. Free Radic Biol Med 54:113–115
164. Tong G, Aponte AM, Kohr MJ, Steenbergen C, Murphy E, Sun J (2014) Postconditioning leads to an increase in protein S-nitrosylation. Am J Physiol Heart Circ Physiol 306:H825–H832
165. Wang SB, Foster DB, Rucker J, O'Rourke B, Kass DA, Van Eyk JE (2011) Redox regulation of mitochondrial ATP synthase: implications for cardiac resynchronization therapy. Circ Res 109:750–757
166. Saini HK, Machackova J, Dhalla NS (2004) Role of reactive oxygen species in ischemic preconditioning of subcellular organelles in the heart. Antioxid Redox Signal 6:393–404
167. Foster DB, Ho AS, Rucker J, Garlid AO, Chen L, Sidor A, Garlid KD, O'Rourke B (2012) Mitochondrial ROMK channel is a molecular component of mitoK(ATP). Circ Res 111:446–454
168. Hausenloy DJ, Erik Bøtker H, Condorelli G, Ferdinandy P, Garcia-Dorado D, Heusch G, Lecour S, van Laake LW, Madonna R, Ruiz-Meana M, Schulz R, Sluijter JP, Yellon DM, Ovize M (2013) Translating cardioprotection for patient benefit: position paper from the Working Group of Cellular Biology of the Heart of the European Society of Cardiology. Cardiovasc Res 98:7–27
169. Inserte J, Hernando V, Vilardosa Ú, Abad E, Poncelas-Nozal M, Garcia-Dorado D (2013) Activation of cGMP/protein kinase G pathway in postconditioned myocardium depends on reduced oxidative stress and preserved endothelial nitric oxide synthase coupling. J Am Heart Assoc 2:e005975

170. Kupai K, Csonka C, Fekete V, Odendaal L, van Rooyen J, Marais de W, Csont T, Ferdinandy P (2009) Cholesterol diet-induced hyperlipidemia impairs the cardioprotective effect of postconditioning: role of peroxynitrite. Am J Physiol Heart Circ Physiol 297:H1729–H1735
171. Boengler K, Konietzka I, Buechert A, Heinen Y, Garcia-Dorado D, Heusch G, Schulz R (2007) Loss of ischemic preconditioning's cardioprotection in aged mouse hearts is associated with reduced gap junctional and mitochondrial levels of connexin 43. Am J Physiol Heart Circ Physiol 292:H1764–H1769
172. Sánchez JA, Rodríguez-Sinovas A, Barba I, Miró-Casas E, Fernández-Sanz C, Ruiz-Meana M, Alburquerque-Béjar JJ, García-Dorado D (2013) Activation of RISK and SAFE pathways is not involved in the effects of Cx43 deficiency on tolerance to ischemia-reperfusion injury and preconditioning protection. Basic Res Cardiol 108:351
173. James AM, Sharpley MS, Manas AR, Frerman FE, Hirst J, Smith RA, Murphy MP (2007) Interaction of the mitochondria-targeted antioxidant MitoQ with phospholipid bilayers and ubiquinone oxidoreductases. J Biol Chem 282:14708–14718
174. Methner C, Lukowski R, Grube K, Loga F, Smith RA, Murphy MP, Hofmann F, Krieg T (2013) Protection through postconditioning or a mitochondria-targeted S-nitrosothiol is unaffected by cardiomyocyte-selective ablation of protein kinase G. Basic Res Cardiol 108:337
175. Zhang Y, Sano M, Shinmura K, Tamaki K, Katsumata Y, Matsuhashi T, Morizane S, Ito H, Hishiki T, Endo J, Zhou H, Yuasa S, Kaneda R, Suematsu M, Fukuda K (2010) 4-hydroxy-2-nonenal protects against cardiac ischemiareperfusion injury via the Nrf2-dependent pathway. J Mol Cell Cardiol 49:576–586
176. Ashrafian H, Czibik G, Bellahcene M, Aksentijević D, Smith AC, Mitchell SJ, Dodd MS, Kirwan J, Byrne JJ, Ludwig C, Isackson H, Yavari A, Støttrup NB, Contractor H, Cahill TJ, Sahgal N, Ball DR, Birkler RI, Hargreaves I, Tennant DA, Land J, Lygate CA, Johannsen M, Kharbanda RK, Neubauer S, Redwood C, de Cabo R, Ahmet I, Talan M, Günther UL, Robinson AJ, Viant MR, Pollard PJ, Tyler DJ, Watkins H (2012) Fumarate is cardioprotective via activation of the Nrf2 antioxidant pathway. Cell Metab 15:361–371
177. Katsumata Y, Shinmura K, Sugiura Y, Tohyama S, Matsuhashi T, Ito H, Yan X, Ito K, Yuasa S, Ieda M, Urade Y, Suematsu M, Fukuda K, Sano M (2014) Endogenous prostaglandin D2 and its metabolites protect the heart against ischemia-reperfusion injury by activating Nrf2. Hypertension 63:80–87
178. Yu H, Shi L, Zhao S, Sun Y, Gao Y, Sun Y, Qi G (2016) Triptolide attenuates myocardial ischemia/reperfusion injuries in rats by inducing the activation of Nrf2/HO-1 defense pathway. Cardiovasc Toxicol 16:325–335
179. Farias JG, Carrasco-Pozo C, Carrasco R, Loza R, Sepulveda N, Alvarez P, Quezada M, Quinones J, Molina V, Castillo RL (2017) Polyunsaturated fatty acid induces cardioprotection against ischemia-reperfusion through the inhibition of NFkappaB and induction of Nrf2. Exp Biol Med 242:1104–1114
180. Xu B, Zhang J, Strom J, Lee S, Chen QM (2014) Myocardial ischemic reperfusion induces de novo Nrf2 protein translation. Biochim Biophys Acta 1842:1638–1647
181. Anedda A, Lopez-Bernardo E, Acosta-Iborra B, Saadeh Suleiman M, Landazuri MO, Cadenas S (2013) The transcription factor Nrf2 promotes survival by enhancing the expression of uncoupling protein 3 under conditions of oxidative stress. Free Radic Biol Med 61C:395–407
182. Eckle T, Kohler D, Lehmann R, El Kasmi K, Eltzschig HK (2008) Hypoxia-inducible factor-1 is central to cardioprotection: a new paradigm for ischemic preconditioning. Circulation 118:166–175
183. Adluri RS, Thirunavukkarasu M, Dunna NR, Zhan L, Oriowo B, Takeda K, Sanchez JA, Otani H, Maulik G, Fong GH, Maulik N (2011) Disruption of hypoxiainducible transcription factor-prolyl hydroxylase domain-1 (PHD-1−/−) attenuates ex vivo myocardial ischemia/reperfusion injury through hypoxia-inducible factor- 1alpha transcription factor and its target genes in mice. Antioxid Redox Signal 15:1789–1797

Role of Oxidative Stress in Myocardial Ischemia and Infarction

14

Bodh I. Jugdutt and Bernadine A. Jugdutt

Abbreviations

AMPK	AMP-activated protein kinase
ATP	adenosine triphosphate
BH4	tetrahydrobiopterin
CABG	coronary artery bypass surgery
CaMKII	Ca^{2+}/calmodulin kinase II
cGMP	cyclic guanosine monophosphate
CP	creatine phosphate
DES	drug-eluting stents
DTB	door-to-balloon
ETC	electron transfer chain
FACoA	long-chain fatty-acyl-CoA
FAcarn	long-chain fatty-acyl-carnitine
GIK	glucose-insulin-potassium
GLP-1	anti-diabetic glucagon-like peptide-1
GTP	guanosine triphosphate
H_2O_2	hydrogen peroxide
HOCl	hypochlorous acid
IL	interleukin
I/R	ischemia-reperfusion
IRA	infarct-related artery
LAD	left anterior descending
LC	left circumflex
LV	left ventricular

B. I. Jugdutt (✉) · B. A. Jugdutt
Cardiology Division, Department of Medicine, Faculty of Medicine, Walter MacKenzie Health Sciences Centre, University of Alberta, Edmonton, AB, Canada
e-mail: bjugdutt@ualberta.ca

S. Chakraborti et al. (eds.), *Oxidative Stress in Heart Diseases*,
https://doi.org/10.1007/978-981-13-8273-4_14

MI	myocardial infarction
MPTP	mitochondrial permeability transition pore
MRI	magnetic resonance imaging
MVO2	myocardial oxygen consumption
•NO	nitric oxide
NOO^-	NO-derived peroxynitrite
NOS	nitric oxide synthase
NOX	NADPH oxidase
NSTEMI	non-ST segment elevation MI
O_2	oxygen
OFRs	oxygen free radicals
OXS	oxidative stress
•OH	hydroxyl radical
PCI	percutaneous coronary intervention
PDH	pyruvate dehydrogenase
PKG	phosphokinase G
PPARs	peroxisome proliferator-activated receptors
PPCI	primary PCI
PTP	permeability transition pore
$O_2•^-$	superoxide anion radical
RCT	randomized clinical trial
RIP3	receptor-interacting protein 3
ROS	reactive oxygen species
SERCA	sarcoendoplasmic reticulum (SR) calcium transport ATPase
SOD	superoxide dismutase
SR	sarcoplasmic reticulum
STEMI	ST-segment elevation MI
TAG	triacyl glycerol
TNF	tumor necrosis factor.

14.1 Introduction

Myocardial ischemia, ischemia-reperfusion (I/R) and myocardial infarction (MI) are major causes of morbidity and mortality due to ischemic heart disease in developed and developing countries of the world. Whereas timely restoration of myocardial blood flow saves muscle and has indeed been life-saving, efforts to prevent I/R injury and reperfusion damage have been frustrating for nearly four decades despite recognition that reactive oxygen species (ROS) and the associated oxidative stress (OXS) were the major culprits. A workshop held at the National Institutes of Health, Bethesda, Maryland, U.S.A in the early 1980s underscored the roles of oxygen free radicals (OFRs), also known as ROS, and OXS in the pathophysiology of cardiovascular diseases including myocardial ischemia, I/R and MI. That workshop inspired many investigators, and research over the subsequent 4 decades has generated a wealth of knowledge on the pathobiology, physiology, biochemistry and pharmacology of ROS and OXS that has improved our understanding of their roles in the pathophysiology of those conditions. However, several pharmacological interventions that were developed and tested for preventing and limiting the harmful effects of ROS and OXS during myocardial ischemia, IR and MI in the clinical arena did not prove to be effective for many reasons [1–3]; the search for new approaches that can be translated to the bedside therefore goes on. Ongoing translational research is focused on providing a broader understanding of the biology of OXS, identifying the key players in ROS regulation and dysregulation, and unravelling pathways and targets for intervention. Such data may allow future development of novel pharmacological treatments and strategies for the limitation and prevention of ROS-induced damage during myocardial ischemia, I/R and MI. If these efforts succeed, the clinical and socioeconomic impact will likely be tremendous.

This chapter focuses on some pertinent areas for future drug development targeted at reducing ROS-induced damage during myocardial ischemia, I/R and MI in patients, with the ultimate goals of validation in carefully designed randomized clinical trials (RCTs) and application in the real world.

14.2 Oxygen Supply and Demand in the Pathophysiology of Ischemia and Infarction

14.2.1 A Few Pertinent Pearls from the Basic Sciences

Studies of the biology, physiology and biochemistry of OFRs and OXS have underpinned four points that are highly pertinent in the context of the pathophysiology of myocardial ischemia, I/R and MI. **First**, preservation of the "milieu intérieur" or "homeostasis", proposed by Claude Bernard and Walter Bradford Cannon, respectively, is critical for normal physiological and metabolic processes, and a failure of endogenous homeostatic mechanisms results in dysfunction and disease. **Second**, oxygen (O_2) is vital for aerobic respiration, in which O_2 acts as the final acceptor of electrons in the electron transport chain (ETC) leading to the generation of energy in the form of adenosine triphosphate (ATP); importantly, O_2-driven aerobic respiration produces more ATP per mole of glucose than anaerobic respiration (theoretically 38 versus 2 ATPs; effectively 30–32 versus 2 ATPs). However, O_2 is also highly reactive and, as such, can generate OFRs that can be toxic to tissues and cause damage when the capacity of endogenous protective antioxidants and enzymes is exceeded and homeostasis fails. **Third**, metabolism through the Krebs cycle (also called tricarboxylic acid cycle or citric acid cycle) is vitally important for cells to convert substrates such as carbohydrate (sugars), fat and protein into intermediates of glycolysis or respiration and thereby generate the energy needed to support life. Briefly, at the cellular level, catabolic pathways in the cytosol convert the substrates into metabolites that converge on the Krebs cycle in the mitochondrion through acetyl-CoA (Fig. 14.1). While metabolism of fatty acid yields more ATP per carbon than carbohydrates (total yield 39 versus 30 ATP per 6 carbons from fat and sugar, respectively), it is important to note that these processes are dependent on O_2 supply

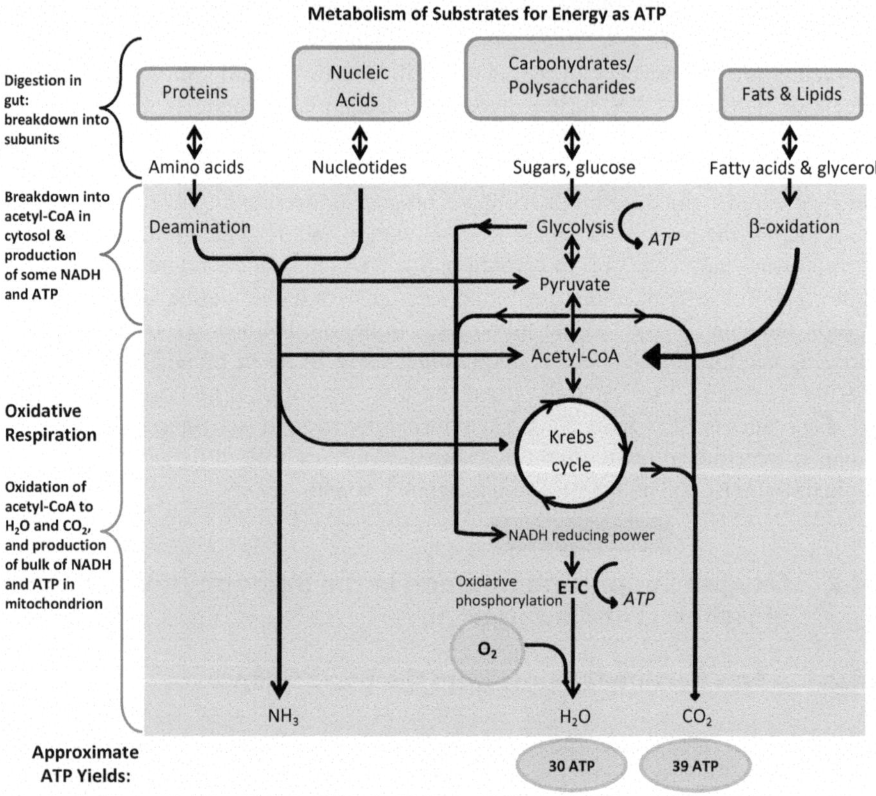

Fig. 14.1 Simplified schematic highlighting the main substrates for oxidative metabolism and energy yield from fat metabolism. Not shown, the net energy field from palminate oxidation is estimated as 110 ATP/mole. *ETC* electron transport chain

for aerobic respiration through the Krebs cycle and ETC in the mitochondrion (Fig. 14.1). **Fourth**, the myocardium depends on high energy supply as ATP in order to maintain normal contractile function and cardiac output to ensure organ perfusion continuously during a lifetime, and the mitochondrion is the "powerhouse" for the continued energy supply.

14.2.2 The Myocardial Oxygen Supply and Demand Equation

The function of the heart is to pump blood containing O_2 and nutrients continuously to all organs throughout life to sustain life. In humans and other mammals, the heart has two pumps (right and left); the right pump directs deoxygenated blood to the lungs for re-oxygenation while the left pump sends the oxygenated blood to all tissues to maintain tissue perfusion, oxygenation and nutrition (Fig. 14.2a). The heart muscle itself needs a continuous supply of blood, nutrients, O_2 and energy in the

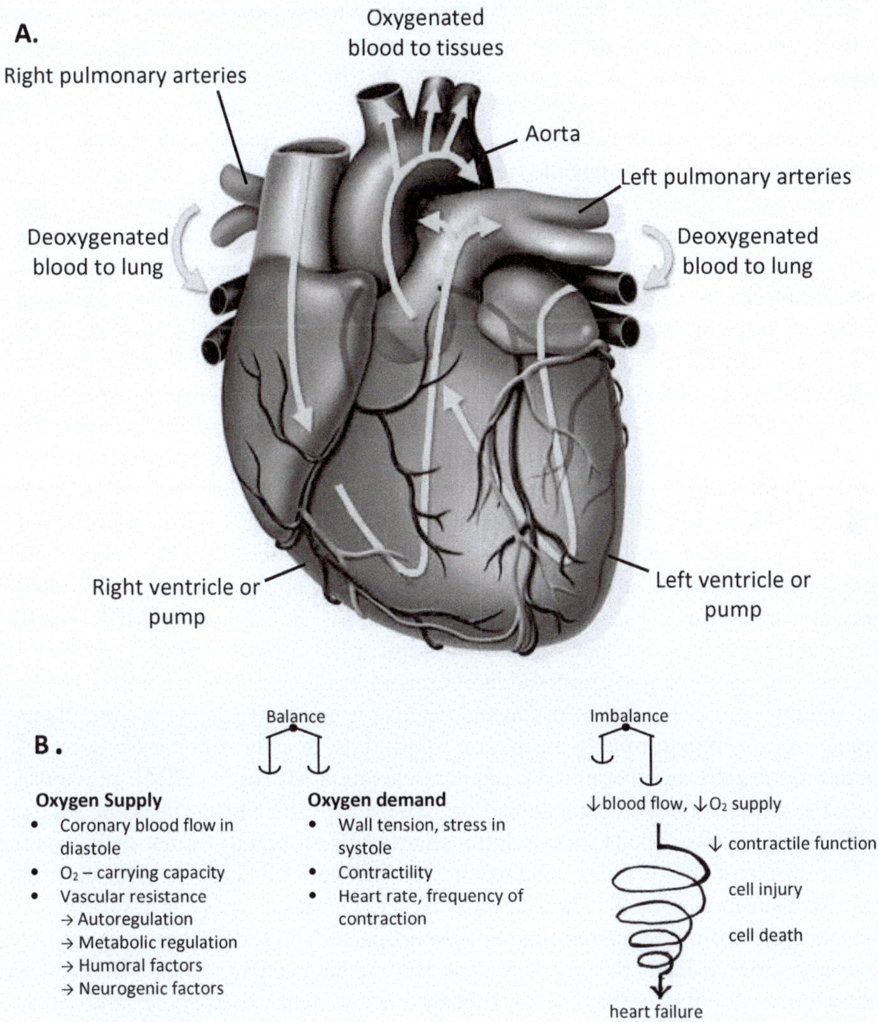

Fig. 14.2 Simplified schematic depicting: (**a**) Oxygenation of blood through the lungs and delivery of oxygenated blood by the heart pumps, respectively. (**b**) Balance of determinants of myocardial oxygen supply and demand and consequences of imbalance. (Adapted from Selwyn [4] and Braunwald [23])

form of ATP (Fig. 14.2a). Just to maintain the cyclic contraction and relaxation during the normal 70–80 beats per minute, the myocardium of the normal resting human heart consumes about 8–15 mL of O_2 per 100 g/min, and that is supplied by a coronary blood flow of about 70–90 mL/min [4]. Several studies have shown that myocardial tension or wall stress, contractility, and frequency of contraction or heart rate account for most of the myocardial oxygen consumption (MVO_2) [4]. When myocardial O_2 demand increases, as during physical activity or sympathetic stimulation, it needs to be balanced by an increase in myocardial blood flow and an increase in oxidative metabolism so as to generate more ATP to balance the O_2 demand and thereby restore homeostasis (Fig. 14.2b). The heart's own O_2 and blood supply comes through its own two large epicardial coronary arteries (right and left) that arborize into intramural arterioles and a collateral microcirculation. Briefly, the right coronary artery supplies the right ventricular myocardium while the main left coronary artery divides further into two large arteries, the left anterior descending (LAD) and left circumflex (LC) branches, to supply the anterior and posterior parts of the left ventricular myocardium (Fig. 14.2a). It should be noted that myocardial blood flow is also under autoregulatory control by multiple factors such as myogenic, neurogenic, and chemical mediators including nitric oxide (NO), serotonin, adenosine diphosphate (ADP), epinephrine and vasopressin; furthermore, patients with coronary atherosclerosis and comorbidities such as hypertension and left ventricular hypertrophy have a blunting of autoregulation of myocardial blood flow [4].

The balance between myocardial O_2 supply and demand in the heart is critical for survival, and failure to maintain this balance, or homeostasis, results in a cascade of myocardial jeopardy, injury, cell death and heart failure (Fig. 14.2b). Clinical studies have established that myocardial ischemia is usually due to restriction of blood flow caused by critical stenoses of epicardial coronary arteries leading to an imbalance between myocardial O2 supply and demand [4]. In contrast, an MI is usually the result of occlusion of one of the major epicardial coronary arteries. At first, it was felt that MI resulted from sudden coronary occlusion by a thrombus [5] but subsequent studies have shown that the occlusion involved atherosclerotic plaque rupture followed by thrombus formation [6–8]; thus, in acute coronary syndromes, including unstable angina, non-ST-segment-elevation MI (or NSTEMI, previously called non-transmural or subendocardial MI) and ST-segment-elevation MI (or STEMI, previously called Q-wave or transmural MI), thrombus formation is a more gradual process [6–8]. Other studies showed that most STEMI patients have an occlusive thrombus [6, 9] whereas most NSTEMI patients have a non-occlusive thrombus [7, 9]; the latter finding suggested that in NSTEMI, whatever coronary flow is preserved leads to only partial imbalance between supply and demand so that necrosis is only confined to the subendocardial layer [9]. Together, these findings revolutionized the approach to therapy and further endorsed the idea of achieving early reperfusion by opening of the infarct-related artery (IRA) with percutaneous coronary intervention (PCI) using a catheter and by applying early anti-platelet, anti-thrombotic and adjunctive therapies to salvage all possible jeopardized ischemic myocardium.

14.2.3 Switch from Aerobic to Anaerobic Metabolism During Myocardial Ischemia

Although anaerobic respiration in the cytosol is less efficient than aerobic respiration in the mitochondrion in terms of the net ATP yield per molecule of glucose, it is still a very effective source of ATP in human skeletal muscles during short bouts of strenuous exercise [10, 11]. As summarized by Hargreaves [11], skeletal muscle uses intra- and extramuscular substrates such as creatine phosphate (CP), glycogen, glucose, lactate and free fatty acids during normal contraction; and during prolonged submaximal exercise, most of the ATP comes from oxidative metabolism of muscle glycogen, blood glucose and free fatty acids. However, during spurts of high-intensity exercise, most of the energy comes from degradation of CP and breakdown of glycogen to lactate, and the rest from oxidative metabolism; lactic acid build-up causes fatigue and an O_2 debt that needs to be repaid during recovery and hyperventilation.

In contrast, the heart is aerobic and cannot work under O_2 debt. While it weighs only about 1 % of the body weight in the adult, it consumes nearly 10 % of the total O_2 consumption. When it is subjected to ischemia and anoxia, it shuts down aerobic metabolism and switches to anaerobic metabolism; however, anaerobic metabolism cannot supply the ATP it needs to maintain viability as the severity of the insult increases. Evidence from preclinical studies suggests that the early contractile dysfunction is mainly due to metabolic dysfunction (Fig. 14.3) rather than changes in cytosolic calcium [12, 13], and the fall in pH plays a critical role [14]. During severe myocardial ischemia, anaerobic metabolism alone cannot meet the ATP demand for continued viability [15, 16]; with severely reduced blood flow and O_2 supply, failure to clear metabolic products leads to a build-up of NADH, lactate, H^+, CO_2, long-chain fatty-acyl-CoA (FACoA) and long-chain fatty-acyl-carnitine (FAcarn) that blunt or block ATP production and contribute to reduced myocardial function [13, 17]. At the same time, coronary venous blood shows end-products of degradation of purine nucleotides such as inosine and hypoxanthine, besides lactate and lysophosphoglycerides. Although disputed, evidence suggests that the acyl esters FACoA and FAcarn act as detergents that cause mitochondrial and plasma membrane dysfunction during prolonged ischemia (Fig. 14.3) [17–20]. Nearly 95% of CoA present in heart cells seems to be localized in the matrix of the mitochondrion [13]. Lysophosphoglyceride formation during ischemia has been implicated in malignant dysrhythmias [21]. During severe ischemia, the release of enzymes such as troponins, creatine kinases and lactic dehydrogenases into venous blood is indicative of and proportional to the degree of myocardial damage and is used in diagnosis [1, 2].

14.2.4 Fatty Acid Metabolism During Myocardial Ischemia

The importance of fatty acid metabolism during myocardial ischemia has been reviewed [22]. Of note, most of those studies used the 'global ischemia' models of the working rat heart and human hearts during open heart surgery with cardioplegia. Normally, the adult heart gets >95% of its ATP from mitochondrial oxidative phosphorylation and the remaining 5% from glycolysis and formation of guanosine triphosphate (GTP) in the Krebs cycle; the myocardial ATP pool is limited (about 5 μmol/g wet wt) and undergoes complete turnover nearly every 10 s [12, 22]. About 50–70% of the acetyl CoA comes from fatty acid β-oxidation and the remaining 30–50% from oxidation of pyruvate derived from glycolysis and oxidation of

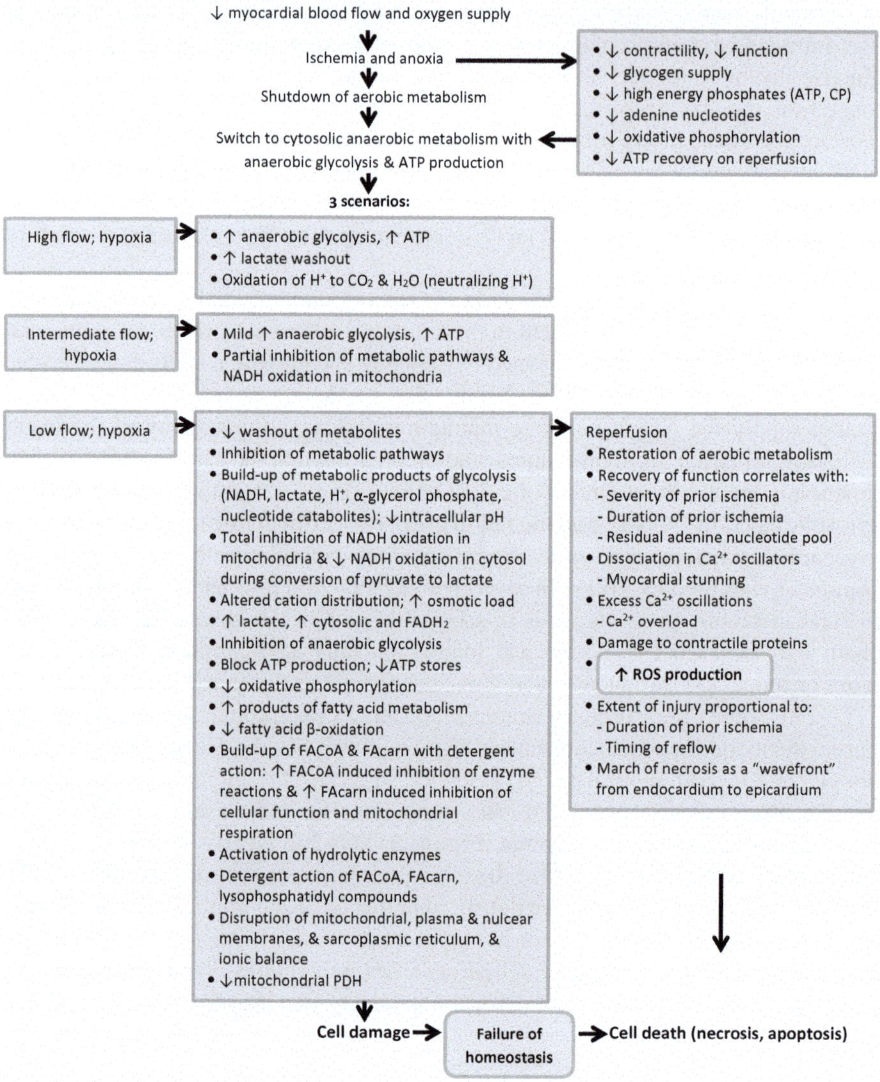

Fig. 14.3 Schematic of metabolic consequences of ischemia due to decreased myocardial blood flow. (Adapted from Neely and Morgan [12], Neely and Feuvray [13], Meissner and Morgan [14], Jennings [18], Reimer and Jennings [58], and Braunwald and Kloner [61])

lactate [12, 22]. As mentioned before, during severe ischemia, the ability to meet the high demand for energy and ATP to maintain normal metabolism, contractility and ionic hemostasis is outstripped; this is despite mitochondrial oxidative phosphorylation, glycolysis and formation of ATP and GTP in the Krebs cycle, and β-oxidation of fatty acids (Fig. 14.3); in contrast, with mild to moderate ischemia, O_2 deprivation leads to reduced ATP output from oxidative phosphorylation and catabolism of fatty acid and pyruvate, unchanged carbohydrate oxidation, swift conversion of pyruvate to lactate and NADH to NAD+, and continued fatty acid β-oxidation [22]. The regulation of fatty acid β-oxidation in the heart is under complex multifactorial control [22, 23]; a source of free fatty acids is from labile stores of triacylglycerol (TAG). Most of the fatty acid for β-oxidation is derived from chylomicron-TAG with a little from very-low-density lipoprotein (VLDL)-TAG, rather than albumin-bound fatty acids [22, 23]. Reduced fatty acid β-oxidation during severe ischemia can contribute to lipotoxicity [13, 22]. Whereas FACoA accumulates mainly in the mitochondrial matrix, FAcarn accumulates in both the mitochondrial matrix and cytosol [13, 22]. Of interest, AMP-activated protein kinase (AMPK) appears to act as a fuel sensor that modulates fatty acid β-oxidation in response to energy demands. In situations where fatty acid β-oxidation increases, this is met by shuttling more of the NADH and FADH2 reducing power to the ETC and ROS formation such as superoxide [22]. A downside to increasing fatty acid usage is a blunting of ATP transfer from the mitochondrial matrix to the cytosol where it is hydrolysed, thereby decreasing contractile power. Various therapies that target fatty acid metabolism have been proposed, including a glucose-insulin-potassium (GIK) cocktail, peroxisome proliferator-activated receptors (PPARs), nicotinic acid, β-adrenergic blockade, sarcolemmal fatty acid uptake inhibitors, mitochondrial fatty acid uptake inhibitors (such as etomoxir, perhexiline and malonyl CoA decarboxylase inhibitors), mitochondrial fatty acid β-oxidation inhibitors (such as trimetazidine and ranolazine), and dichloroacetate (inhibitor of pyruvate dehydrogenase kinase, thereby stimulating mitochondrial pyruvate dehydrogenase). Of these, ranolazine has been proposed as an anti-anginal agent for chronic stable angina.

14.2.5 Salvage of Jeopardized Myocardium and the Open Artery Hypothesis

In order to formulate therapeutic strategies for interrupting the progression of myocardial ischemia, I/R and MI to heart failure in patients, it was essential to understand the determinants of myocardial O_2 consumption (MVO_2) [24]. Seminal clinical and preclinical studies of determinants of MVO_2 in Eugene Braunwald's group during the late 1960s and early 1970s revolutionized contemporary thinking with concepts of salvage of the jeopardized myocardium by pharmacotherapy and early reperfusion [24–28]. In patients with STEMI, they showed that carotid sinus stimulation to increase coronary blood flow reduced ischemic injury [25]. In studies with coronary artery occlusion in the anesthetized dog model, they established that the extent of acute myocardial ischemic injury correlated with subsequent necrosis measured 24 h later [26]. Additionally, they showed that reperfusion of the ischemic zone to increase O_2 supply even as late as 3 h could salvage jeopardized myocardium [27]. Also in the dog model, they showed that by manipulating the determinants of MVO_2, they could increase or decrease the extent of ischemic injury assessed by electrocardiographic ST-segment elevations by various metabolic and pharmacologic interventions [28, 29].

During the same era, cardiovascular surgeons began restoring myocardial blood flow and O_2 supply in the territory of the critically stenotic coronary artery by performing coronary artery bypass grafting (CABG) in patients with stable angina and this procedure was subsequently shown to abolish the ischemic response to atrial pacing stress that increased MVO_2 [30].

In the latter half of the 1970s, intravenous β-adrenergic blockade to improve the balance between O_2 supply and demand in patients with acute STEMI was shown to decrease ischemic injury (assessed by the sum of ST-segment elevations or ΣST by precordial mapping) and myocardial work (assessed by the rate-pressure product, an index of myocardial work and O_2 demand) without signs of heart failure [31]. That study [31] suggested that β-adrenergic blockade might be a promising strategy for limiting infarct size in patients with evolving acute STEMI.

14.2.6 Myocardial Salvage in the Territory Infarct-Related Artery or the "Region at Risk"

In order to test the myocardial salvage concept and assess various potential infarct-limiting pharmacologic agents, more robust studies were needed for the objective assessment of infarct size relative to the "region at risk" in the territory of the infarct-related artery (IRA), rather than just as a percent of the whole left ventricle. This idea was tested in subsequent studies using a conscious canine model; these studies demonstrated that the "region at risk", defined by the post-mortem coronary arteriography technique developed by Fulton [32] and applied by Schaper [33], as well as the collateral blood flow within the "region at risk" after a coronary artery occlusion, assessed by the radio-active microsphere technique, and the subsequent grossly visible infarct size could be accurately mapped [34, 35]. These studies demonstrated a gradient in collateral blood flow across the "region at risk", from the borders to the centre, that correlated with the amount of subsequent myocardial necrosis [34, 35], thereby allowing several interventions to be tested [36–41]. Of note, the OFR scavenger, superoxide dismutase (SOD), was shown to reduce reperfusion injury and infarct size [41]. Subsequently, the degree of mechanical dysfunction as left ventricular dysynergy across the area of necrosis within the region at risk was quantified by two-dimensional echocardiography (2D-Echo) in the conscious dog model [42], thus providing a means of assessing infarct size by non-invasive 2D-Echo imaging at the bedside in subsequent studies of patients with STEMI and preclinical studies in the conscious dog model using various interventions [43–56]. Magnetic resonance imaging (MRI) is now widely used to assess myocardial infarct size and scar size relative to the region at risk in patients with MI in vivo [57].

14.2.7 Reperfusion and Reperfusion Injury

Also during the 1970s, other studies in the canine model demonstrated that coronary occlusion was followed by a "march to necrosis" over time, with transmural extension as a "wavefront" from endocardium to epicardium, that could be aborted by early reperfusion and thereby salvage ischemic myocardium [58], whereas late reperfusion resulted in reperfusion injury associated with further myocardial damage, arrhythmias and intramyocardial hemorrhage [59]. As reviewed by Jennings and Reimer [59], now well-known characteristics of irreversible injury in cardiomyocytes include "severely depressed ATP levels to <10% of control", shutdown of anaerobic glycolysis with increasing levels of metabolites, marked increase in osmolar load, mitochondrial swelling with accumulation of amorphous matrix densities, and sarcolemmal disruption. They again underscored the "wavefront" concept, that in the anesthetized dog model of regional myocardial ischemia, severely ischemic cardiomyocytes were mostly dead by 60 min whereas moderate to mildly ischemic cardiomyocytes in the mid and epicardial myocardium "survived for as long as 6 hours" [59]. They noted that recovery after reperfusion was variable; whereas aerobic metabolism is restored within minutes, adenine nucleotides remain depressed even after 4 days, stunning disappears after 1–2 days, and the preconditioning effect is mostly gone after 2 h [59]. Jennings had previously made the important observations that cell death developed within 20 min of ischemia, and infarcts were still developing between 60 and 120 min after occlusion in the dog model [18]. Using an anesthetized dog model of 40 min of coronary occlusion and 3 days of reperfusion, they found a similar relation between infarct size post-reperfusion and the region at risk defined by in vitro latex injection of the coronary bed [60]. Of note, reperfused infarcts display altered morphology with distinct patchiness [41], suggesting islands of surviving myocardium.

The concept of the stunned myocardium with "prolonged, postischemic ischemic ventricular dysfunction" was reviewed by Braunwald and Kloner [61] before the contribution of OXS and OFRs became apparent. They underscored two important points: first, that brief episodes of non-lethal ischemia produced prolonged myocardial stunning associated with defects in function, metabolism and structure that persisted for days but eventually cleared; second, that repeated episodes of stunning caused chronic postischemic ventricular dysfunction which, if prolonged, could lead to ischemic cardiomyopathy [61].

14.2.8 Clinical Experience with Reperfusion

Taken together, the aforementioned and other early studies laid the foundation for modern reperfusion therapy. From a historical perspective, several concurrent landmark clinical studies between the late 1970s and early 1980s demonstrated that timing was of utmost importance for myocardial salvage which is only optimal after very early reperfusion [1, 2]. Driven by the dream of salvaging as much of the ischemic myocardium in the jeopardized zone as possible, cardiologists marched forward into the era of thrombolytic therapy early after the onset of acute MI to limit infarct size, salvage ventricular function and improve survival [62–66]. No cardiac surgeon would attempt CABG during that time frame. Fortunately in the mid to late 1970s, Andreas Gruentzig pioneered the use of percutaneous transluminal coronary angioplasty (PTCA) or percutaneous coronary intervention (PCI), which was first applied for patients with coronary stenosis as an alternative to CABG [67, 68]. Since unlike CABG, PTCA could be applied to more acute settings, this followed soon after, first for unstable angina [69] and then for acute MI [70–73]. These bold early steps burgeoned into the present generation of invasive and interventional cardiologists whose principal goal is to achieve as prompt and complete recanalization of the IRA as possible [1, 2].

When Marcus de Wood first demonstrated coronary thrombosis in acute MI in the 1980s [6, 7], the finding gave an added boost to the mission of applying very early thrombolytic therapy to open the IRA. Indeed, that led to the salvage of millions of lives worldwide [1, 2]. The subsequent finding that restenosis of arteries opened by PTCA can occur by 6 months later [74] led to the performance of repeat PCIs and use of stents and various generations of drug-eluting stents (DES) and adjunctive therapies to maintain patency [1, 2, 7 5]. Still guided by the idea that "an open artery is by far better than an occluded one" and the well-known dictum "time is muscle," every effort continues to be made to ensure earlier and earlier reperfusion, often by combined timely PCI, coronary artery stenting, thrombolytic therapy, and even thrombus aspiration where applicable, in order to optimize myocardial salvage and limit reperfusion damage in acute MI [1, 2].

Primary PCI (PPCI) is now considered the preferred therapy in STEMI patients for whom lytics are contraindicated and those who are at high risk [76]. Moreover, Simari et al [76] found that, with regard to the ability to salvage jeopardized myocardium, PPCI without prior lytic therapy was equally effective as lytics whereas "immediate adjunctive PTCA" after lytics was associated with increased risk; however, "rescue PTCA" after failed lytics was beneficial whereas "deferred PTCA" several weeks after MI in order to prevent recurrence of ischemia showed neither mortality nor re-infarction benefit [76].

Besides the choice of the ideal PCI approach, the issue of time delays still looms over us and needs to be addressed worldwide but that can be problematic [77–82]. Despite attempts to abbreviate the door-to-balloon (DTB) time by reducing patient transfer time, various other factors and obstacles are involved and have to be addressed (but this is beyond the scope of this chapter). Suffice it to underscore 2 points: first, that De Luca et al found that every 30-min delay in delivery of PPCI increases mortality by 7.5% [83]; and second, that Cannon et al found from an analysis of 4 registers that, in the real world, a DTB time of more than 90 min was associated with the worse prognosis [84].

The advent of the balloon catheter provided Hugenholtz and associates the opportunity to study physiologic responses to episodes of ischemia in patients undergoing PTCA for coronary stenosis. They made 2 pertinent observations; first, they found that 40–60 s of coronary occlusion did not produce persistent dysfunction of ventricular mechanics, myocardial blood flow or lactate metabolism [85]; second, they documented that recovery of diastolic function was delayed after flow and systolic function had recovered [86]. These findings, and others up until two decades ago, led Heusch to comment in a most compelling and provocative article that although "stunning has emerged as an important paradigm of I/R injury" "its clinical importance appears minimal" [87]. However, Heusch also noted that with balloon inflation or occlusion of 4–7 min or more, PTCA does produce stunning that resolves after 24 h [88].

Clearly, even at the most physically accessible centres with readily available PPCI, most patients arrive more than 30 min after the onset of STEMI. The sobering fact remains that, despite impressive improvement in mortality and morbidity after STEMI and acute coronary syndromes over the last four decades as a result of timely reperfusion and adjunctive therapies and adherance to published management guidelines [1, 2, 75], there is still a 7% mortality and 22% morbidity, respectively, at 1 year after prompt reperfusion and PPCI in patients with STEMI [1, 3, 89]. There is therefore a need to target reperfusion injury in patients with STEMI aggressively [90] despite the "trials and tribulations" [3].

14.2.9 Some Known Mechanisms and Potential Targets of Reperfusion Injury

The question that remains is, what else to target?

Extensive research has shed light on several molecular and cellular mechanisms and mediators of I/R injury that could potentially be targeted by interventions to attenuate the injury [91–93] and some of these are depicted in Fig. 14.4. Taken together, the evidence suggests that remodeling of cellular, subcellular, metabolic and molecular processes during ischemia prime the myocardium for further damage during reperfusion. The main changes that set the stage for I/R injury are summarized in Fig. 14.3.

Briefly, the key metabolic, structural and functional changes during ischemia include: (i) aerobic metabolism shutdown with a switch to anaerobic metabolism and glycolysis, and downstream consequences; (ii) depletion of high-energy phosphate stores such as ATP and CP; (iii) H+ overload with reduced intracellular pH, cation redistribution, increased lactate, and osmotic overload; (iv) structural remodeling with cell swelling and disruption of cell, mitochondria, sarcolemma and sarcoplasmic reticulum (SR); (v) decreased mitochondrial pyruvate dehydrogenase (PDH) that persists 30 min post-reperfusion [93–95]; (vi) contractile failure with functional recovery if reperfusion is achieved within 30 min.

Based on the aforementioned reviews [3, 61–64, 91–93], some known key processes that can be targeted during and after reperfusion to reduce or limit I/R damage include (Fig. 14.4): (i) free radical overload with OFRs and oxidants and the resulting OXS; (ii) pro-inflammatory cytokine overload with interleukin (IL)-6 and tumor necrosis factor (TNF)-α and increased inflammation; (iii) metabolic remodeling with H^+ ion, Ca^{2+} ion and osmotic overload; (iv) structural damage to organelles and membranes including the mitochondria and its membranes, the plasmalemma, SR and nuclear membrane; (v) persistent ventricular dysfunction, stunning and no-reflow; and (vi) cell death with necrosis and apoptosis (Fig. 14.5).

Reperfusion after ≥ 15 min of ischemia	Potential Pharmacological Interventions
• Intracellular remodeling	
- oxidative stress (↑ ROS)	- antioxidants (SOD+catalase, glutathione peroxidase); N-acetyl cysteine; mercapto-propionyl-glycine
- ↑ inflammation (↑ ROS)	- inhibitors of inflammation (TGFβ, IL-10)
- Ca^{2+} overload	- Ca^{2+} channel inhibitors, exchangers
- platelet activation	- antiplatelet agents
- leucocyte aggregation	- inhibitors of neutrophil aggregation & activation
- endothelial dysfunction	- vasodilators, adenosine, K-ATP channel opener, NO
- H+ overload	- Na+/H+ exchange or antiporter inhibitor (cariporide)
• Metabolic remodeling	
- depletion of high energy stores	- adenosine, insulin
- ↑ proteolytic activity (calpain)	- protease inhibitors; calpain inhibitors
- ↑ arrhythmias	- anti-arrhythmic agents
• Remodeling of gene expression	- mRNAs; gene therapy (prophylactic); gene transfer (Ec-SOD) - miRNAs (miRNA-126, 133, 144)
• Remodeling of translation mechanisms	- several potential targets
• Remodeling of cardiac function	
- ventricular dysfunction	- post-ischemic conditioning; NO modulation; K-ATP channel opener; NO/cGMP pathway (ANP, GIK); Exenatide; liraglutide; NO; NO-donor & nitrite
- Myofibril desensitization to Ca2+	- inotropes
- stunning	- remote ischemic conditioning; SOD+catalase; ARBs; ACE-Is
- apoptosis & necrosis	- Na+/K+ exchange inhibitor, cariporide - CaMKII inhibition for RIP3-mediated programmed cell necrosis
- no reflow	- ARBs; SOD+catalase; adenosine+lidocaine; endothelin A antagonists; glyceryl trinitrate; verapamil; papaverine, nicorandil; NO donor (nitroprusside); glycoprotein IIb/IIIa receptor antagonist (abciximab); eptifibatide; micro-embolism device
• Remodeling of mitochondrial function	
- PDH inhibition - Mitochondrial depolarization	- PDH stimulation, pyruvate; accelerate recovery of aerobic metabolism (adenosine, insulin) for cardioplegia (insulin induces ↑ PDH, ↓ Lactate, ↑ ATP)
- opening of MPTP	- MTP-131 (Bendavia), Cyclosporin-A, TRO40303, PKCδ - MPTP inhibitors
• Multiple targets	Hypothermia, metopolol, adenosine

Fig. 14.4 Schematic of mechanisms of myocardial ischemia-reperfusion injury and stunning, and potential targets for limitation. (Adapted from: Bolli and Marban [91]; Reffelmann and Kloner [92]; Man, Tymchak and Jugdutt [75]; Ducas, Bartekova and Dhalla [93]; and Hausenloy et al [3]). Abbreviations as defined in the text

14.2.10 Reactive Oxygen Species Overload

The known contributors to ROS or OFR overload during I/R include increased superoxide anion ($O_2\bullet^-$), hydrogen peroxide (H_2O_2), hydroxyl radical (•OH), hypochlorite (HOCl) and NO-derived peroxynitrite (NOO^-) [96–98]. Oxygen molecules are reduced to the ROS superoxide ($O_2\bullet^-$). With post-ischemic reperfusion, ROS

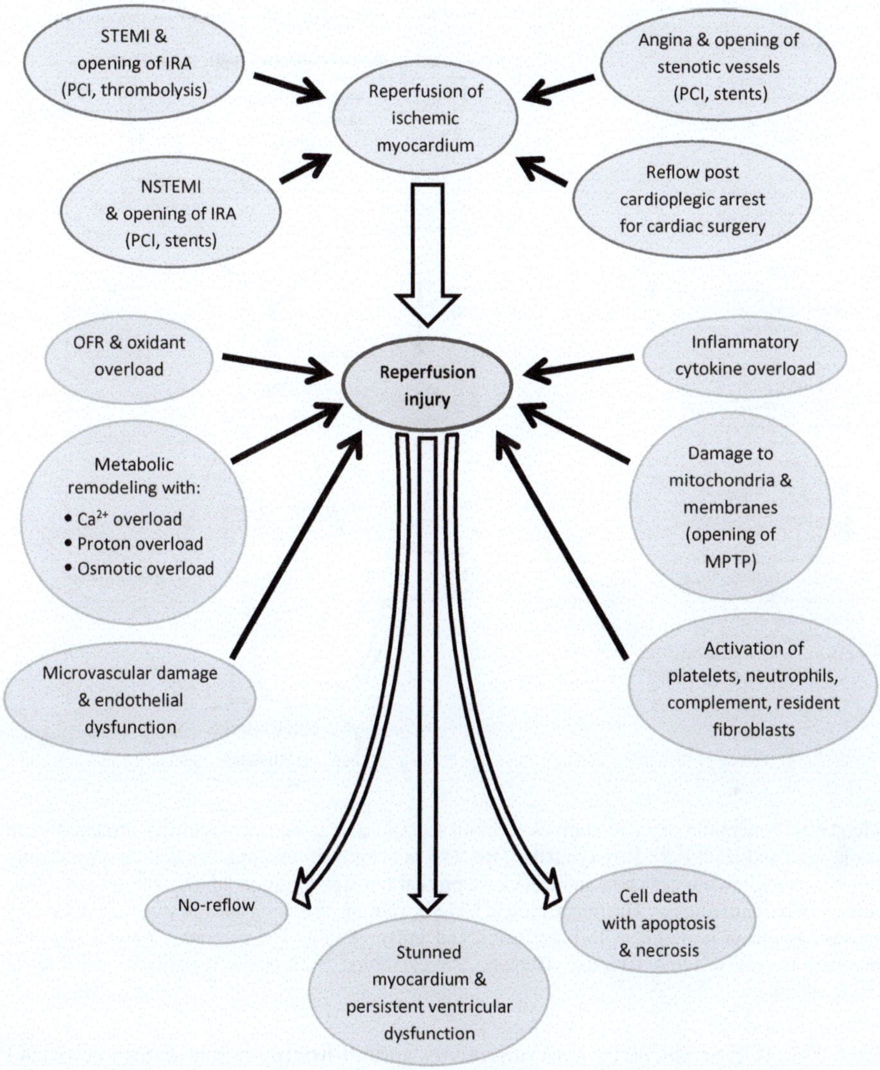

Fig. 14.5 Key processes that can be targeted during ischemia and reperfusion. *Ca^{2+}* calcium ion, *IRA* infarct-related artery, *MPTP* mitochondrial permeability transition pore, *OFR* oxygen free radical, *PCI* percutaneous coronary intervention, *NSTEMI* non-ST segment elevation MI, *STEMI* ST-segment elevation MI

may be produced by both enzymatic and non-enzymatic systems, and in both cardiomyocytes and infiltrating inflammatory cells. The main sources of ROS include enzymes in mitochondrial ETC, plasma membrane, peroxisomes, endoplasmic reticulum and nuclear membrane, and other enzymes such as xanthine oxidase, myeloperoxidase, P450 enzymes, some of which are found in macrophages and neutrophils of the inflammatory reaction, and soluble heme-proteins (Fig. 14.6)

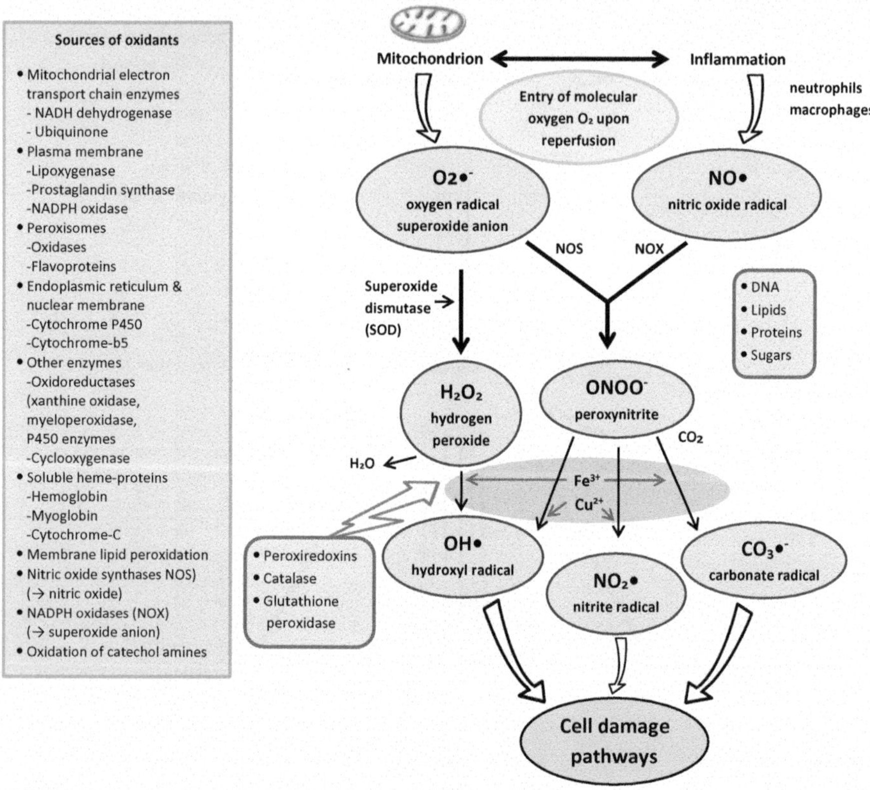

Fig. 14.6 Schematic of main sources of oxidants, superoxide and peroxynitrite. Adapted from Bartesaghi and Radi [98]. Peroxynitrite ($ONOO^-$) is cytotoxic and enhances cell damage during reperfusion following ischemia and can cause protein tyrosine nitration which can serve as a biomarker of oxidative stress. The nitrite radical NO_2• is a strong oxidizing and nitrating agent. Protein tyrosine nitration is pertinent in reperfusion and inflammation. The carbonate radical $CO_3•^-$ is involved mainly in nitro-oxidative damage. Abbreviations: *NOX* NADPH oxidases, *NOS* nitric oxide synthases

[98]. The ROS produced by cardiomyocytes and infiltrating inflammatory cells lead to cellular damage through disruption of membranes and proteins as well as activation of various cell death pathways that trigger apoptosis [99, 100] (Fig. 14.7). In addition, reperfusion itself stimulates neutrophil activation and accumulation of ROS; lipid peroxidation in membranes also contributes ROS. While nitric oxide itself is not reactive, nitric oxide synthases (NOS) generate reactive •NO whereas NADPH oxidases (NOX) generate reactive superoxide which interacts to yield highly reactive peroxynitrite [98]. Peroxynitrite leads to the formation of hydroxyl, nitrite and carbonate radicals which mediate cell damage and tyrosine nitration which serves as a biomarker of OXS [98]. Redox-active transition metals, such as Fe^{3+} and Cu^{2+}, potentiate the oxidant activities and enhance ROS production (Fig. 14.6) [98]. ROS overload, in turn, exerts several harmful effects: (i) it

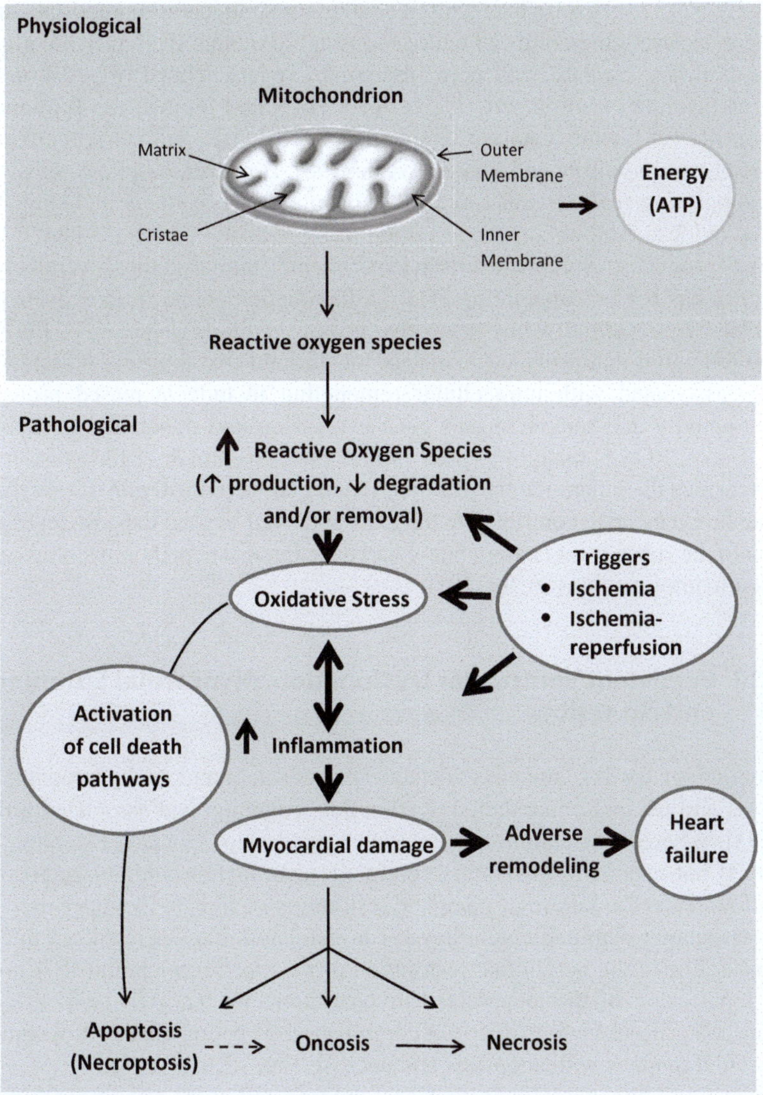

Fig. 14.7 Role of mitochondria in oxidative stress and ischemic myocardial damage. *ATP* adenosine triphosphate

stimulates platelets to release platelet activating factor, thereby attracting more neutrophils that aggravate I/R damage; (ii) it attenuates nitric oxide function and enhances endothelial dysfunction through peroxynitrite formation, thereby aggravating I/R damage; and (iii) it induces blunting of endothelium-dependent vasodilation and enhances endothelin-1-induced vasoconstriction, and thereby contributes to decreased reflow.

Clearly, ROS overload is the result of a failure of endogenous homeostatic mechanisms; the scavenging ability of endogenous anti-oxidants such as SOD, glutathione peroxidases, catalase, and peroxiredoxins is overwhelmed (Fig. 14.6). SOD catalyzes dismutation of superoxide to form H_2O_2 and inhibits the formation of peroxynitrite [98], while catalase, glutathione peroxidases and peroxiredoxins get rid of H_2O_2, peroxynitrite and lipid peroxides [98]. However, despite extensive evidence suggesting that administration of exogenous anti-oxidants including SOD, xanthine oxidase inhibitors (such as allopurinol), N-acetyl cysteine, vitamin E and vitamin C produces favorable results in experimental animal studies, overall clinical experience has been disappointing [1, 2]. Inflammation is known to activate platelets which aggregate into white heads that acquire red tails (made up of fibrin and trapped red cells) that provide the matrix for thrombi that plug the vessels. Other changes associated with intracellular remodeling include increased proteolytic enzyme activity and remodeling of gene expression and translation mechanisms [93]. Of note, ROS is thought to peak in the first 2–10 min after reperfusion [92]; however, since the inflammatory reactions extend far beyond 10 min of reperfusion, they can be expected to contribute to the ROS pool well beyond the post-reperfusion phase into the subsequent healing phase and have important pathophysiological and therapeutic implications [56, 99–101].

14.2.11 Persistent Ventricular Dysfunction, Myocardial Stunning and No-reflow

Remodeling of cardiac function includes the development of hypercontracture, apoptosis and necrosis, endothelial dysfunction, stunning, and associated arrhythmias [93]. Whereas the common belief is that reperfusion can cause stunning with prolonged but eventually reversible contractile dysfunction without necrosis, the "no-reflow" refers to failure of complete restoration of flow in the reperfused zone and is associated with cell death. Several investigators have pointed out the existence of a flaw in the notion that restoration of flow in the epicardial IRA implies 100% reperfusion of the jeopardized myocardium [56, 92, 101–103]. It is now apparent that "no-reflow" is due to a combination of endothelial cell dysfunction and vascular damage with apoptosis and necrosis [56, 92, 100–103].

Kloner and colleagues found that anatomic no-reflow occurs after 90 min of ischemia and is associated with ultrastructural evidence of endothelial damage in the microvasculature of the endocardium [101]. Other pertinent findings included capillary plugging with platelet clumps and fibrin thrombi, and irreversible cardiomyocyte damage with reperfusion after 60 min of ischemia [101]. Becker and colleagues demonstrated, in an anesthetized dog model of coronary occlusion for 90 min followed by reperfusion at repeated intervals between zero and 180 min, that necrosis, determined by electron microscopy, occurs during reperfusion, and the duration of reperfusion was an independent predictor of irreversible injury [103]. In an earlier study, they found that, in the anesthetized dog model of 90 min of coronary occlusion followed by reperfusion for 2 min and 3.5 h, it is myocardium with very low flow during ischemia (measured by the microsphere technique) that shows no-reflow (measured by fluorescent thioflavin-S uptake); they also noted that flow declines progressively in previously well reperfused areas, areas with delayed impairment of flow show "intracapillary erythrocyte stasis" and "intravascular neutrophil accumulation", and areas with early no-reflow show contraction band necrosis [102].

In a dog model of reperfused STEMI, Jugdutt's group [56] showed that reperfusion at 90 min after LAD coronary occlusion was associated with increased pro-inflammatory markers such as inducible-nitric-oxide-synthase (iNOS), cytokines interleukin (IL)-6 and tumor necrosis factor (TNF)-α, and anti-inflammatory markers such as transforming-growth-factor (TGF)-β_1 and IL-10; in addition, they found evidence of increased cardiomyocyte damage (ischemic injury, infarct size, apoptosis, blood flow impairment, and no-reflow), adverse left ventricular (LV) remodeling with LV dilation and dysfunction as well as adverse extracellular matrix (ECM) remodeling with increased expression of matrix metalloproteinase (MMP) -9 and MMP-2, matrix proteases such as secretory-leucocyte-protease-inhibitor (SLPI), secreted-protein-acidic-and-rich-in-cysteine (SPARC), as well as osteopontin (OPN) and ADAM-10 and -17. In the same study, they showed that aging results in age-dependent increases in markers of myocardial and vascular damage, as well as markers of adverse structural and matrix remodeling after STEMI [56]. In addition, they showed that early therapy initiated at the onset of reperfusion with the angiotensin II type 1 receptor blocker (ARB) candesartan attenuated all these changes [56]. While these findings await confirmation in properly designed RCTs, the beneficial effect of an ARB on recovery of ventricular function after IR was also demonstrated in several studies using diverse experimental models [104–111]. The importance of timing of pharmacotherapy to match biology for optimal benefit was emphasized before [112].

14.2.12 Strategies to Quench ROS and Reduce OXS by Exogenous ROS Endogenous Scavengers

The idea that the endogenous SOD pathway is overwhelmed during reperfusion led Becker's group to test therapy with exogenous SOD to boost scavenging ability and quench ROS. While a study with recombinant human SOD given at the time of reperfusion in the dog model of 90 min of LC coronary occlusion showed benefit [41], a subsequent RCT with therapy before PTCA in STEMI patients did not [113]. Kloner and associates also tested SOD combined with catalase, another anti-oxidant in the endogenous enzyme system for ROS homeostasis, and found that treatment given at the time of reperfusion after 120 min of LAD coronary occlusion in the dog model did not reduce infarct size (by triphenyltetrazolium staining) or improve subepicardial contractile function (by sonomicrometry) but did attenuate endocardial microvascular injury and "low reflow" and improve regional blood flow [92, 114]. Interestingly, Bolli et al also showed that SOD plus catalase, given over 15 min before and continued for 30 min after reflow, did block ROS production in the stunned zone and improve functional recovery in the dog model [115]. Additionally, they found that significant amounts of ROS, measured by electron paramagnetic resonance and a spin trap in the venous effluent from the stunned zone (via a catheter positioned in the anterior interventricular vein), persisted for several hours after reflow although the peak was over the first 20 min [115].

The collective findings of benefit with the therapies to quench ROS and alleviate OXS await confirmation in a properly designed RCT. The past studies illustrate that myocardial stunning with no-reflow and necrosis after reperfusion are very real [56, 92, 99–103, 112–115], and not so "rare" as suggested before [116] and are in need of therapy [2, 3, 89, 117]; indeed, the dire consequences may be grave and have a negative impact on mortality [2, 3, 89, 117] as well as morbidity and quality of life in a significant number of patients.

14.2.13 Quenching ROS and Attenuating OXS in Patients with STEMI

Eight pertinent points, in translating experimental data to the bedside, need to be taken into account. First, humans with MI tend to be middle-aged, older adults, or elderly, and aging may blunt the response to therapy as found with an ARB after reperfused STEMI in dogs [56] and with post-ischemic conditioning in mice [118, 119]. Second, age equivalence should be taken into account when assessing therapies in animal models, [112]; for example, a 6-week old mouse or rat would be equivalent to a young human child and not even a young adult [112]. Third, ROS production also occurs during healing phases of acute and chronic inflammation and need to be taken into account [118, 120–123]. Fourth, the rate of reperfusion in animal models is usually more abrupt than in humans undergoing PCI and/or thrombolysis [118]. Fifth, besides SOD and catalase, other endogenous protective mechanisms against ROS exist during I/R such as adenosine [92], opening of ATP-sensitive potassium (K-ATP) channels [116], and release of nitric oxide [124]. Sixth, ROS overload and injury after reperfusion of STEMI in humans may be enhanced by comorbidities such as hyperlipidemia, hypertension and diabetes that aggravate endothelial dysfunction [2, 3, 116]. Seventh, targeting I/R injury and OXS in humans is complicated by the multiple factors, players, mechanisms, mediators, signaling pathways, background drugs, pathologies, approaches, and timings involved [1–3] (Fig. 14.3, 14.4, 14.5, 14.6, and 14.7). Eighth, despite successful PPCI after STEMI and achieving TIMI grade 3 flow in epicardial IRAs, reperfusion at the tissue level is often incomplete in as many as 9–15 % of patients due to a combination of microvascular damage and distal embolization of bits of thrombi and debris from atherosclerotic plaques [125, 126]. Of note, while the latest guidelines do not recommend routine aspiration thrombectomy or devices such as filter wires and umbrella PPCI [2], they recognize their usefulness in some cases so long as steps are taken to prevent systemic emboli [125].

14.2.14 Metabolic Remodeling of Cation Movements

Metabolic remodeling of cation movement and Ca^{2+} overload during I/R deserve further comment. It is known that in normal cardiomyocytes, depolarization is associated with a rapid sequence of activation of sodium channels, increase in intracellular sodium, release of calcium from SR, actin-myosin coupling and binding to troponin, sarcomere shortening, and generation of force and contraction. During subsequent repolarization, sodium/potassium (Na^+/K^+) ATPase and sodium/calcium (Na^+/Ca^{2+}) exchangers restore intracellular homeostasis. Ducas et al summarized the ionic movements during I/R [93]. The switches to anaerobic metabolism during ischemia and back to aerobic metabolism during reperfusion set forth the tandem development of cytosolic H^+ ion or proton overload and acidosis, Ca^{2+} ion overload, and lactate and osmotic overload that trigger a march to cell injury, damage to organelles (mitochondria, SR and sarcolemma), contractile dysfunction and death [93]. The critical role of membrane transporters in the regulation of pH has been reviewed by Karmazyn and associates [127, 128]. H^+ ion overload triggers the sequential activation of the sodium/hydrogen (Na^+/H^+) ion exchanger, exchange of H^+ ions for Na^+ ions with a build-up of intracellular Na^+ ions, activation of the sarcolemmal Na^+/Ca^{2+} ion exchanger, exchange of Na^+ for Ca^{2+} ions, development of intracellular Ca^{2+} overload, stimulation of Ca^{2+} ATPase, depletion of ATP, and adverse structural remodeling of transmembrane pumps and exchangers [93, 127, 128]. The depletion of high energy phosphates during ischemia attenuates the activity of Na^+/K^+ ATPase and its ability to clear excess Na^+ ions during ischemia [93]. Structural remodeling of the transmembrane pumps and exchangers modify the amino-acid and sulfhydryl terminals of proteins, and thereby dampen their ability to shift cations [93]. Damage to the sarcolemmal pumps with reflow facilitates the massive shift of Ca^{2+} from the SR. Evidence suggests that Na^+/K^+ exchange mediates IR injury and its inhibitors are beneficial [127].

14.2.15 Mitochondrial Pyruvate Dehydrogenase and Contractile Dysfunction

Studies to address the persistent contractile failure often seen during reflow after cardioplegic arrest pointed to mitochondrial PDH as a potential therapeutic target. The collective evidence suggests that the activity of mitochondrial PDH, a critical enzyme that converts pyruvate arising from glycolysis (anaerobic metabolism) to form acetyl-CoA for the Krebs cycle through oxidative decarboxylation, is reduced during flow-induced ischemia as well as reperfusion causing contractile dysfunction that can be salvaged by either stimulating PDH or infusing pyruvate. The reason for the high lactate and reduced ATP, frequently associated with low-output failure during reflow after corrective surgery in children with cyanotic heart disease, was traced to impaired aerobic metabolism due to transcriptional downregulation of mitochondrial enzymes such as PDH, as well as cytochrome-c oxidase, succinate cytochrome-c reductase, succinate dehydrogenase, and citrate synthase [95].

Subsequently, the high lactate, associated with contractile dysfunction and low cardiac output during reperfusion after cardioplegic arrest in patients undergoing CABG, was explained by a delayed switch-back from anaerobic to aerobic metabolism and persistent generation of lactate from glycolysis [94].

In the isolated rat heart model, Churchill et al studied the likely mechanism for ischemia-induced PDH inhibition and its partial recovery during reperfusion, and showed that reperfusion-induced translocation of δPKC to the mitochondria leads to phosphorylation and activation of PDH kinase-2 that results in phosphorylation-dependent PDH inhibition [129]. Interestingly, in the transgenic mice model with chronic inhibition of PDH induced by overexpression of PDH kinase-4, an inhibitor of PDH, together with an insulin-resistance profile, Chambers et al found that chronic activation of PDH kinase-4 triggered "transcriptional and post-transcriptional mechanisms that re-program the heart for chronic high rates of fatty acid oxidation" without adverse functional or metabolic sequelae [130]. Taken together, the findings suggest that newer approaches may be found for treating contractile dysfunction during I/R besides pharmacologic stimulation of PDH.

It is interesting to note that the piperazine compound ranolazine, which is approved as an anti-anginal agent, was originally considered for its inhibition of fatty acid oxidation, shift to more energy efficient glucose oxidation with improved ATP/O^2, and attenuation of the build-up of H^+ ions, lactate and fatty acyl intermediates [131]; it was later shown to inhibit β1 and β2 adrenoceptors, inhibit late Na^+ channels, and lower total inward Na^+ flux and the subsequent Ca^{2+} overload; it was also shown to reduce the late inward Ca^{2+} current, inward Na^+/Ca^{2+} exchange current, and the outward repolarizing rectifier K^+ current [132]. These latter effects expanded the application of ranolazine from chronic stable angina to suppression of arrhythmias in patients with ischemic heart disease [132]. The clinical trials with ranolazine [132, 133] illustrate the problems of testing efficacy of new pharmacological agents on top of recommended background therapies (Tables 14.1 and 14.2).

14.2.16 Reperfusion and Mitochondrial Damage

Damage to mitochondria during reperfusion makes them dysfunctional and the main contributors to the damage are protons, Ca^{2+} ions, and osmotic and ROS overload which depend on the severity and duration of prior ischemia. A critical event is the prompt opening of the mitochondrial permeability transition pore (MPTP) of the inner membrane and other membrane channels within minutes of reperfusion; this results in free traffic of proteins across the outer membrane, increased osmotic stress on the outer membrane, eventual membrane rupture and release of mitochondrial proteins and ROS [93]. The damaged mitochondria also leak cytochrome C, proteases and caspases, and trigger cell death pathways. Mitochondrial Ca^{2+} overload leads to uncoupling of oxidative phosphorylation and reduced ATP production (switch from synthesis to hydrolysis), which in turn accentuates altered ion gradients and degradation of cellular enzymes [93, 134, 135]. Opening of the MPTP upon reperfusion has been linked to several events, some of which are connected to

Table 14.1 Pharmacologic agents during acute STEMI and reperfusion

Thrombolytics
Antiplatelet agents
$P2Y_{12}$ inhibitor (prasugrel, ticagrelor), clopidogrel
aspirin
$GPII_b/III_a$ inhibitors
cangrelor
Anticoagulants
unfractionated heparin
bivalirudin
enoxaparin
fondaparinux
Fibrinolytics
fibrin-specific agent (tenecteplase, alteplase, reteplase)
antiplatelet agents (aspirin, clopidogrel, $P2Y_{12}$ inhibitor)
anticoagulants (enoxaparin, unfractionated heparin)

Table 14.2 Maintenance after STEMI [1, 2]

Antithrombotic agents
aspirin
antiplatelet agents
proton pump inhibitors
oral anticoagulants
Beta-blockers
Lipid lowering agents
statins
ezetimide
PCSK9 inhibitors
Nitrates
Calcium antagonists
Angiotensin-converting enzyme (ACE) inhibitors & angiotensin II receptor blockers (ARBs)
Mineralocorticoid/ aldosterone receptor antagonists (MRAs)
eplerenone

altered Ca^{2+} and H^+ homeostasis, including: (i) hypercontracture resulting from high and oscillating Ca^{2+} as ATP supply increases [3, 136]; (ii) dysregulated activation of calpain as the pH is normalized; calpain is a member of the non-lysosomal neutral cysteine protease family that is regulated by the endogenous inhibitor calpastatin [3, 137]; (iii) increased Ca^{2+} overload due to activation of the calcium-dependent enzymes calcineurin and Ca^{2+}/calmodulin kinase II (CaMKII) [3, 137]; calcineurin activation is thought to be due to increased cytosolic Ca^{2+} mediated by dephosphorylation of phospholamban and inhibition of Ca^{2+}-ATPase (SERCA) activity and Ca^{2+} uptake in SR; and CaMKII activation is thought to hasten recovery of pH by phosphorylation of the Na^+/H^+ exchanger, modulation of Ca^{2+} entry into the cell and reactivation of processes suppressed by high H^+ [3, 137]. CaMKII activation and MPTP opening may also be involved in receptor-interacting protein 3 (RIP3)-mediated programmed cell necrosis and apoptosis (necroptosis) after I/R [3, 138]. ROS overload is another important cause of MPTP opening, and quenching of mitochondrial ROS by inhibition of succinate dehydrogenase with malonate has been suggested to prevent MPTP opening and limit infarct size [3, 139]. Since ROS induces endothelial NOS uncoupling at the onset of reperfusion, it has been suggested that the NO-dependent activation of the cGMP/phosphokinase G (PKG) pathway that is involved in postconditioning-induced cardioprotection is due to attenuation of superoxide production at the onset of reperfusion, thereby reducing oxidation of tetrahydrobiopterin (BH4) which is a cofactor in NOS coupling, and limiting NOS uncoupling [140]. Inhibition of mitochondrial ROS-triggered activation of NADPH-oxidase has been shown to attenuate mitochondrial derived superoxide production [141], suggesting that mitochondria might be a focus for anti-oxidant treatment [142].

14.2.17 Central Role of Mitochondria in Oxidative Stress

It is known that the myocardium depends on a continuous supply of high energy ATP in order to maintain normal contractile function and cardiac output to ensure organ perfusion during a lifetime. The ATP is supplied by oxidative phosphorylation of substrates and the ETC in the myocardial mitochondrion. The principal substrate for ATP generation is through mitochondrial β-oxidation of fatty acid, followed by oxidation of glucose, lactate, amino acid and ketones, as well as glycolysis [97, 143]. Besides being the powerhouse for energy production, the mitochondrion also plays a central role in the production of ROS in tissues throughout the body including the myocardium [96, 97, 144] (Fig. 14.7). In addition, the mitochondrion orchestrates the homeostasis of redox and calcium and thereby regulates susceptibility to myocardial injury and programmed cell death or apoptosis. In order to ensure these functions in myocardium, the population of mitochondria in myocardium is abundant, making up about 25% of the volume of cardiomyocytes [20, 145].

14.2.18 The Myocardial Mitochondrion and Homeostasis of Reactive Oxygen Species

Under normal conditions, the production of ROS is tightly regulated in all tissues. In fact, low levels of ROS modulate normal mitochondrial and cellular signaling and function, and maintain redox homeostasis [97]. An imbalance in ROS production and removal via degradation or consumption results in increasing levels of ROS and OXS. As mentioned before, the production of ROS in tissues involves both enzymatic and non-enzymatic systems, and the removal of ROS typically involves endogenous antioxidants, ROS scavengers and other removal systems (Fig. 14.4). However, as mentioned before, ROS removal mechanisms may be overwhelmed during various pathologies, and increasing levels of ROS can be very damaging to all tissues, including the myocardium, especially during injury [97]. During myocardial ischemia, IR, MI and reperfused MI, high ROS levels are very damaging, can exacerbate myocardial injury and contribute to further myocardial damage, dysfunction, adverse cardiac remodeling and heart failure in tandem sequence, setting up a vicious cycle (Fig. 14.7). Unchecked, this cycle leads to increased mortality and morbidity after MI.

Of interest, evidence suggests that mitochondria are programmed for different functions in different cells and tissues [146] so that different phenotypes exist. Which specific substrate is used for myocardial mitochondrial metabolism depends on the specific underlying pathology; here we focus on myocardial ischemia, IR and MI, but other conditions including myocardial hypertrophy, heart failure and diabetic cardiomyopathy differ in substrate use [20, 147, 148].

Mitochondrial oxidative metabolism involves oxidative phosphorylation with oxidation of the substrate, reduction of NAD and FAD, donation of electrons down the ETC and complexes I to IV, utilization of oxygen and generation of ATP [97, 149]. Both genetic and acquired mechanisms such as myocardial ischemia and MI can disrupt mitochondrial oxidative phosphorylation. This disruption of mitochondrial oxidative phosphorylation and ETC not only impairs ATP generation, but is also associated with impairment of intracellular homeostasis of calcium and ROS, which triggers a vicious cycle of decreased contractile function and increased ROS generation, decreased cardiac output and alteration in redox balance, and increased myocardial damage and heart failure [97].

Although the SR is the main organelle that regulates Ca^{2+}, the shift of cytosolic Ca^{2+} into mitochondria normally serves to activate the various enzymes needed for oxidative phosphorylation to generate ATP for contraction; however, during Ca^{2+} overload as with I/R, the mitochondria act as a sink for Ca^{2+} which leads to mitochondrial ROS production and activation of cell death [97]. Evidence exists of cross-talk between mitochondria and the nucleus [149], the SR and other organelles.

Further discussion of other aspects of ROS biology is beyond the scope of this chapter but is reviewed elsewhere. Some of the therapies that were used for I/R in patients [150] are listed in Table 14.3.

Table 14.3 Summary of therapies targeting reperfusion injury in patients with acute myocardial infarction

Ischemic conditioning	
Post conditioning	several positive MI studies
Remote conditioning	several positive MI studies
NO/cGMP pathway	
Atrionatriuretic peptide (ANP)	one positive MI study
Glucose-insulin-potassium (GIK)	inconsistent results
Exenatide	positive MI studies
Nitric oxide/nitrite	inconsistent results
Mitochondria and MPTP	
MTP-131 (Bendavia)	neutral study
Cyclosporine A	inconsistent/neutral study
TRO40303 (mitochondrial targeting drug)	neutral MI study
PKC-δ inhibition	neutral MI study
Multiple targets	
Hypothermia	neutral MI study
Metoprolol	positive MI study
Adenosine	inconsistent MI study

MI myocardial infarction
Adapted from Hausenloy [3]

14.3 2018 Update on Therapies for Quenching ROS and Limiting Reperfusion Damage

The question that remains is, what else and how to target? Hausenloy and colleagues have provided a summary of the therapies that have shown or might show benefit [3]. As previously underscored, the multiplicity of factors, players, mechanisms, mediators, signaling pathways, background drugs and pathologies involved, and the approaches and timings employed [1–3] add to the complexity of the problems of what and when to target. A clear understanding of the underlying mechanisms and their interplay is imperative in designing RCTs. For example, concerning the central issue of cardiomyocyte cell death alone, besides well-known mechanisms of apoptosis and necrosis [99, 100], such as Ca^{2+} overload, osmotic overload, proton overload, metabolic remodeling, OXS, inflammation, and mitochondrial dysfunction (Figs. 14.4 and 14.5), additional mechanisms that contribute to I/R injury and infarct size [3] include: (i) no-reflow due to microvascular injury and dysfunction; (ii) attraction of inflammatory cells into the injured zone due to increased permeability of endothelium; (iii) platelet activation and adhesion to reperfused endothelium due to P-selectin-ligand interactions, GPIIb/IIIa binding to fibrinogen, or fibronectin receptor and platelet-leucocyte conjugate formation [151]; (iv) enhanced inflammation and I/R damage due to inflammasome activation in cardiac fibroblasts [152]; (v) duration of ischemia which modulates mitochondrial permeability transition (calcein) and I/R-induced cardiomyocyte death [153]; and (vi) duration of ischemia and timing of reperfusion, which when made late, after 2 h, does not limit infarct size despite benefits on overall ventricular remodeling [154, 155].

Several interventions were attempted but produced variable results (Table 14.3). First, anti-inflammatory therapy given at the time of reperfusion was neutral [156, 157]. Second, preconditioning was extensively reviewed [90, 158, 159] and studied in patients [3, 150]; ischemic post-conditioning (I-Post) applied after reopening the IRA was shown to reduce infarct size assessed by enzyme release, nuclear imaging and MRI in patients with STEMI [160, 161]. Third, remote ischemic conditioning (RIC) using transient arm or leg ischemia was shown to reduce infarct size in patients with CABG [162], and in STEMI in patients reperfused by PPCI or thrombolysis [163]; long-term outcomes are under study. Fourth, since NO/cyclic guanosine monophosphate (cGMP) signaling is blunted in reperfused MI, therapies targeting NO/cGMP signaling pathway were studied; particulate guanylate cyclase with natriuretic peptide (NP) was shown to reduce infarct size [164] whereas GIK produced mixed results [165]. Fifth, since anti-diabetic glucagon-like peptide-1 (GLP-1) reduces MI size in experimental studies, it was studied in STEMI patients; the GLP-1 analog exenatide reduced I/R injury and MI size including after PPCI [166–168] while the analog liraglutide only improved systolic function in STEMI patients [169]. Sixth, NO and nitrite showed no benefit in STEMI and PPCI patients [170, 171] but reduced MI size in patients with a fully occluded IRA in one study [172]. Seventh, Cyclosporin A (CsA) which inhibits MPTP opening was successful in some animals [173–177] but failed when given before PPCI in STEMI patients [89, 178–180] or during aortic valve surgery [181].

Eighth, the mitochondrial targeting peptide MTP-131 or Bendavia, which reduces ROS production by targeting cardiolipin in the inner mitochondrial membrane, was shown to reduce MI size after I/R in animal models [182, 183] but failed to reduce MI size in STEMI patients when given before PPCI [184, 185]. Ninth, angiotensin II is known to stimulate ROS production and release of NO, activate NADPH oxidase, and generate superoxide and peroxynitrite, in addition to producing several other harmful cardiovascular effects. After I/R, angiotensin II type 1 receptor blockers (ARBs) have been shown to exert several beneficial effects in experimental animal models of I/R [56, 104–111], including attenuation of OXS in heart failure [109], inhibition of apoptosis [56, 110], reduction of infarct size and no-reflow [56], reversal of post-transcriptional modification of δ-subunit of ATP synthase [106], alteration of metabolic, functional and structural proteins [104], improvement of matrix protease balance [105], improvement of pro- and anti-inflammatory cytokine balance [56], and upregulation of beneficial angiotensin II type 2 receptors [107, 108]. Tenth, the β-adrenergic blocker metoprolol given before I/R has been shown to reduce MI size in experimental animals [186], and limit MI size when given in the ambulance before PPCI in patients with STEMI [187, 188]; the results of another study using MRI for MI size are pending [189]. Eleventh, the mitochondrial targeting drug, TRO40303, which targets inhibition of MPTP opening and reduces ROS through binding to the translocator protein TSPO in the outer mitochondrial membrane, was shown to limit MI size in small animal studies [190] but failed to do so in a large animal model [191] and in STEMI patients [192].

Twelfth, protein kinase C-δ inhibition, a known mediator of ischemic preconditioning, was promising in one study of intracoronary delcasertib given before PPCI in STEMI patients [193] but a subsequent study of intravenous delcasertib failed to reduce MI size after STEMI [194]. Thirteenth, adenosine, which is known for its cardioprotective effects, has produced inconsistent results with respect to reduction of MI size during STEMI [3, 195], although a recent metanalysis revealed benefit in heart failure after reperfused STEMI [196]. Fourteenth, with regard to effects of other targets on MI size in reperfused STEMI, hypothermia has produced neutral results [3]. Fifteenth, FXO6, a naturally occurring peptide derived from human fibrin, was found to limit MI size in animal models of I/R injury and was recently shown to reduce the necrotic zone evaluated by MRI when given at the onset of reperfusion in STEMI patients [156]. Sixteenth, another interesting recent target is use of the catalytic antioxidant Mangafodipir as a cardioprotective adjunct during PPCI in STEMI patients; the initial results showed a trend towards benefit [197].

14.4 Future Directions

The collective evidence indicates that targeting OXS during myocardial ischemia, I/R and MI in humans is complicated by the multiple factors, players, mechanisms, mediators, signaling pathways, background drugs, pathologies and co-morbidities, approaches, and timings involved. Over the last four decades, extensive research has generated a wealth of data and identified a host of novel mechanisms, pathways and molecules that can potentially be targeted. The assimilation of the new knowledge is essential for successful translation of data from small to large animals and finally to humans; however, this process has been slow and a common scenario has been frustration in translating the many successes found in animals to the bedside in the real world. Certainly, there have been some successes, but the need for new targets, novel pharmacological agents, improved strategies, and better designed RCTs remains and may require collaboration at a global level in order to minimize the residual mortality and morbidity associated with OXS after reperfused MI.

References

1. O'Gara PT, Kushner FG, Ascheim DD, Casey DE Jr, Chung MK, de Lemos JA et al (2013) 2013 ACCF/AHA guideline for the management of ST-elevation myocardial infarction: executive summary: a report of the American college of cardiology foundation/American Heart Association task force on practice guidelines. J Am Coll Cardiol 61:485–510
2. Ibanez B, James S, Agewall S, Antunes MJ, Bucciarelli-Ducci C, Bueno H et al (2018) 2017 ESC guidelines for the management of acute myocardial infarction in patients presenting with ST-segment elevation: The task force for the management of acute myocardial infarction in patients presenting with ST-segment elevation of the European Society of Cardiology (ESC). Eur Heart J 39:119–177
3. Hausenloy DJ, Botker HE, Engstrom T, Erlinge D, Heusch G, Ibanez B et al (2017) Targeting reperfusion injury in patient with ST-segment elevation myocardial infarction: trials and tribulations. Eur Heart J 38:935–941

4. Selwyn AP (2010) Coronary physiology and pathophysiology. In: Brown DL, Jeremias A (eds) Textbook of cardiac intensive care, 2nd edn. Elsevier, Philadelphia, pp 69–72
5. Herrick JB (1912) Clinical features of sudden obstruction of the coronary arteries. JAMA 9:2015–2022
6. DeWood MA, Spores J, Notske R, Mouser LT, Burroughs R, Golden MS, Lang HT (1980) Prevalence of total coronary occlusion during the early hours of transmural myocardial infarction. N Engl J Med 303:897–902
7. DeWood MA, Stifter WF, Simpson CS, Spores J, Eugster GS, Judge TP, Hinnen ML (1986) Coronary arteriographic findings soon after non-Q-wave myocardial infarction. N Engl J Med 315:417–423
8. Libby P (2001) Current concepts of the pathogenesis of the acute coronary syndromes. Circulation 104:365–372
9. Thygesen K, Alpert JS, White HD, on behalf of the joint ESC/ACCF/AHA/WHF Task force for the redefinition of myocardial infarction (2007) Universal definition of myocardial infarction. Eur Heart J 28:2525–2538
10. Spriet LL (1990) Anaerobic metabolism in human skeletal muscle during short-term, intense activity. Can J Physiol Pharmacol 70:157–165
11. Hargreaves M (2000) Skeletal muscle metabolism during exercise in humans. Clin Exp Pharmacol Physiol 27:225–228
12. Neely JR, Morgan HE (1974) Relationship between carbohydrate metabolism and energy balance of heart muscle. Annu Rev Physiol 36:413–459
13. Neely JR, Feuvray D (1981) Metabolic products and myocardial ischemia. Am J Pathol 102:282–291
14. Meissner A, Morgan JP (1995) Contractile dysfunction and abnormal Ca^{2+} modulation during post-schemic reperfusion in the rat heart. Am J Physiol 268:H100–H111
15. Steenbergen C, Deleeuw G, Rich T, Williamson JR (1977) Effects of acidosis and ischemia on contractility and intracellular pH of rat heart. Circ Res 41:849–858
16. Weissler AM, Kruger FA, Baba N, Scarpelli DG, Leighton RF, Gallimore JK (1968) Role of anaerobic metabolism in the preservation of functional capacity and structure of anoxic myocardium. J Clin Invest 47:403–416
17. Henry PD, Sobel RE, Braunwald E (1974) Protection of hypoxic guinea pig hearts with glucose and insulin. Am J Physiol 226:309–313
18. Jennings RB (1969) Early phase of myocardial ischemic injury and infarction. Am J Cardiol 24:753–765
19. Weglicki WB, Owens K, Urschel CW, Serur JR, Sonnenblick EH (1972) Hydrolysis of myocardial lipids during acidosis and ischemia. In: Dhalla NS (ed) Recent advances in studies on cardiac structure and metabolism, vol 3, pp 781–793
20. Schaper J, Hehrlein F, Schlepper M, Thiedemann K-U (1977) Ultrastructural alterations during ischemia and reperfusion in human hearts during cardiac surgery. J Mol Cell Cardiol 9:175–189
21. Sobel BE, Corr PB, Robison AK, Goldstein RA, Witkowski FX, Klein MS (1978) Accumulation of lysophosphoglycerides with arrhythmogenic properties in ischemic myocardium. J Clin Invest 62:546–553
22. Lopaschuk GD, Ussher JR, Folmes CD, Jaswal JS, Stanley WC (2010) Myocardial fatty acid metabolism in health and disease. Physiol Rev 90:207–258
23. Van der Vusse GJ, van Bilsen M, Glatz JF (2000) Cardiac fatty acid uptake and transport in health and disease. Cardiovasc Res 45:279–293
24. Braunwald E (1969) Thirteenth Bowditch Lecture: the determinants of myocardial oxygen consumption. Physiologist 12:65–93
25. Braunwald E, Epstein SE, Glick G, Wechsler AS, Braunwald NS (1967) Relief of angina pectoris by electrical stimulation of the carotid-sinus nerves. N Engl J Med 277:1278–1283
26. Maroko PR, Kjekshus JK, Sobel BE, Watanabe T, Covell JW, Ross J Jr, Braunwald E (1971) Factors influencing infarct size following experimental coronary artery occlusions. Circulation 43:67–82

27. Maroko PR, Braunwald E (1973) Modification of myocardial infarction size after coronary occlusion. Ann Intern Med 79:720–733
28. Braunwald E (1976) Symposium on protection of the ischemic myocardium. Circulation 53(Suppl 3):I-1–I-2
29. Maroko PR, Braunwald E (1976) Effects of metabolic and pharmacologic interventions on myocardial infarct size following coronary occlusion. Circulation 53(Suppl 3):I-162–I-168
30. Jugdutt BI, Lee SJ, Taylor RF (1976) Abolition of ischemic response to atrial pacing following aortocoronary bypass surgery. Circulation 54:225–229
31. Jugdutt BI, Lee SJ (1978) Intravenous therapy with propranolol in acute myocardial infarction: effects on changes in the S-T segment and hemodynamics. Chest 74:514–521
32. Fulton WFM (1965) The Coronary Arteries. Arteriography, microanatomy, and pathogenesis of obliterative coronary artery disease. Charles C. Thomas, Springfield
33. Schaper W (1971) The collateral circulation of the heart. North-Holland Publishing Co, Amsterdam
34. Jugdutt BI, Hutchins GM, Bulkley BH, Becker LC (1979) Myocardial infarction in the conscious dog: three-dimensional mapping of infarct, collateral flow and region at risk. Circulation 60:1141–1150
35. Jugdutt BI, Becker LC, Hutchins GM (1979) Early changes in collateral blood flow during myocardial infarction in conscious dogs. Am J Physiol 237:H371–H380
36. Jugdutt BI, Hutchins GM, Bulkley BH, Pitt B, Becker LC (1979) Effect of indomethacin on collateral blood flow and infarct size in the conscious dog. Circulation 59:734–743
37. Jugdutt BI, Hutchins GM, Bulkley BH, Becker LC (1980) Salvage of ischemic myocardium by ibuprofen during infarction in the conscious dog. Am J Cardiol 46:74–82
38. Jugdutt BI, Becker LC, Hutchins GM, Bulkley BH, Reid PR, Kallman CH (1981) Effect of intravenous nitroglycerin on collateral blood flow and infarct size in the conscious dog. Circulation 63:17–28
39. Jugdutt BI, Hutchins GM, Bulkley BH, Becker LC (1981) Dissimilar effects of prostacyclin, prostaglandin E1, and prostaglandin E2 on myocardial infarct size after coronary occlusion in conscious dogs. Circ Res 49:685–700
40. Blumenthal DS, Hutchins GM, Jugdutt BI, Becker LC (1981) Salvage of ischemic myocardium by dipyridamole in the conscious dog. Circulation 64:915–923
41. Ambrosio G, Becker LC, Hutchins GM, Weisman HF, Weisfeldt ML (1986) Reduction in experimental infarct size by recombinant human superoxide dismutase: insights into the pathophysiology of reperfusion injury. Circulation 74:1424–1143
42. Lieberman AN, Weiss JL, Jugdutt BI, Becker LC, Bulkley BH, Garrison JG et al (1981) Two-dimensional echocardiography and infarct size: relationship of regional wall motion and thickening to the extent of myocardial infarction in the dog. Circulation 63:739–746
43. Jugdutt BI, Sussex BA, Haraphongse M, Rossall RE (1983) Right ventricular involvement in transmural inferior wall infarction: two-dimensional echocardiographic and clinical correlations. Clin Invest Med 6:261–273
44. Jugdutt BI, Sussex BA, Sivaram CA, Rossall RE (1984) Right ventricular infarction: two-dimensional echocardiographic evaluation. Am Heart J 107:505–518
45. Jugdutt BI, Sussex BA, Warnica JW, Rossall RE (1983) Persistent reduction in left ventricular asynergy in patients with acute myocardial infarction by intravenous infusion of nitroglycerin. Circulation 68:1264–1273
46. Jugdutt BI (1983) Myocardial salvage by intravenous nitroglycerin in conscious dogs: loss of beneficial effect with marked nitroglycerin-induced hypotension. Circulation 68:673–684
47. Jugdutt BI, Warnica JW (1988) Intravenous nitroglycerin therapy to limit myocardial infarct size, expansion, and complications. Effect of timing, dosage, and infarct location. Circulation 78:906–919. Erratum in: Circulation 1989 79:1151
48. Jugdutt BI, Warnica JW (1989) Tolerance with low dose intravenous nitroglycerin therapy in acute myocardial infarction. Am J Cardiol 64:581–587
49. Jugdutt BI, Warnica JW (1987) Improved left ventricular function with metoprolol in acute myocardial infarction. Circulation 76:IV–276. (Abstract)

50. Jugdutt BI, Tymchak WJ, Burton JR (1991) Preservation of left ventricular geometry and function after late reperfusion and intravenous nitroglycerin in acute transmural myocardial infarction. Circulation 84:II–683. (Abstract)
51. Jugdutt BI, Tymchak WJ, Humen DP, Gulamhusein S, Tang SB (1992) Effect of thrombolysis and prolonged captopril and nitroglycerin on infarct size and remodeling in transmural myocardial infarction. J Am Coll Cardiol 19(Suppl A):205A. (Abstract)
52. Jugdutt BI, Menon V, Kumar D, Idikio H (2002) Vascular remodeling during healing after myocardial infarction in the dog model: effects of reperfusion, amlodipine and enalapril. J Am Coll Cardiol 39:1538–1545
53. Jugdutt BI, Idikio H, Uwiera RR (2007) Angiotensin receptor blockade and angiotensin-converting-enzyme inhibition limit adverse remodeling of infarct zone collagens and global diastolic dysfunction during healing after reperfused ST-elevation myocardial infarction. Mol Cell Biochem 303:27–38
54. Palaniyappan A, Uwiera RR, Idikio H, Jugdutt BI (2009) Comparison of vasopeptidase inhibitor omapatrilat and angiotensin receptor blocker candesartan on extracellular matrix, myeloperoxidase, cytokines, and ventricular remodeling during healing after reperfused myocardial infarction. Mol Cell Biochem 321:9–22
55. Jugdutt BI, Palaniyappan A, Uwiera RR, Idikio H (2009) Role of healing-specific-matricellular proteins and matrix metalloproteinases in age-related enhanced early remodeling after reperfused STEMI in dogs. Mol Cell Biochem 322:25–36
56. Jugdutt BI, Jelani A, Palaniyappan A, Idikio H, Uweira RE, Menon V et al (2010) Aging-related early changes in markers of ventricular and matrix remodeling after reperfused ST-segment elevation myocardial infarction in the canine model: effect of early therapy with an angiotensin II type 1 receptor blocker. Circulation 122:341–351
57. Ubachs JFA, Sörensson P, Henrik E, Carlsson M, Jovinge S, Pernow J et al (2012) Myocardium at risk by magnetic resonance imaging: head-to-head comparison of T2-weighted imaging and contrast-enhanced steady-state free precession. Eur Heart J Cardiovasc Imaging 13:1008–1015
58. Reimer KA, Lowe JE, Rasmussen MM, Jennings RB (1977) The wavefront phenomenon of ischemic cell death. 1. Myocardial infarct size vs duration of coronary occlusion in dogs. Circulation 56:786–794
59. Jennings RB, Reimer KA (1991) The cell biology of acute myocardial ischemia. Annu Rev Med 42:225–246
60. Lowe JE, Reimer KA, Jennings RB (1978) Experimental infarct size as a function of the amount of myocardium at risk. Am J Pathol 90:363–379
61. Braunwald E, Kloner RA (1982) The stunned myocardium: prolonged, post ischemic ventricular dysfunction. Circulation 66:1146–1149
62. Rentrop KP, Feit F, Blanke H, Stecy P, Schneider R, Rey M et al (1984) Effects of intracoronary streptokinase and intracoronary nitroglycerin infusion on coronary angiographic patterns and mortality in patients with acute myocardial infarction. N Engl J Med 311:1457–1463
63. Simoons ML, Serruys PW, van den Brand M, Bär F, de Zwaan C, Res J et al (1985) Improved survival after early thrombolysis in acute myocardial infarction. A randomized trial conducted by the Interuniversity Cardiology Institute in the Netherlands. Lancet 2:578–582
64. Kennedy JW, Ritchie JL, Davis KB, Stadius ML, Maynard C, Fritz JK (1985) The Western Washington randomized trial of intracoronary streptokinase trial in acute myocardial infarction. N Engl J Med 312:1073–1078
65. Simoons ML, Serruys PW, van den Brand M, Res J, Verheugt FW, Krauss XH et al (1986) Early thrombolysis in acute myocardial infarction: reduction of infarct size, preservation of left ventricular function and improved survival. J Am Coll Cardiol 7:718–728
66. Serruys PW, Simoons ML, Suryapranata H, Vermeer F, Wijns W, van den Brand M et al (1986) Preservation of global and regional left ventricular function after early thrombolysis in acute myocardial infarction. J Am Coll Cardiol 7:729–742
67. Grüntzig A (1978) Transluminal dilatation of coronary stenosis. Lancet 1:263

68. Grüntzig AR, Senning Å, Siegenthaler WE (1979) Nonoperative dilatation of coronary-artery stenosis: percutaneous transluminal coronary angioplasty. N Engl J Med 301:61–68
69. de Feyter PJ, Suryapranata H, Serruys PW, Beatt K, van Domburg R, van den Brand M et al (1988) Coronary angioplasty for unstable angina: immediate and late results in 200 consecutive patients with identification of risk factors for unfavorable early and late outcome. J Am Coll Cardiol 12:324–333
70. Rentrop P, Blanke H, Karsch KR, Kreutzer H (1979) Initial experience with transluminal recanalization of the recently occluded infarct-related coronary artery in acute myocardial infarction – comparison with conventionally treated patients. Clin Cardiol 2:92–105
71. Meyer J, Merx W, Schweizer P, Erbel R, Kiesslich T, Dörr R et al (1982) Advantages of combining transluminal coronary angioplasty (PTCA) with selective coronary thrombolysis in acute infarction. Circulation 66:905–913
72. Hartzler GO, Rutherford BD, McConahay DR (1984) Percutaneous transluminal coronary angioplasty: application for acute myocardial infarction. Am J Cardiol 53:117C–121C
73. Suryapranata H, Serruys PW, de Feyter PJ, van den Brand M, Beatt K, van Domburg R et al (1988) Coronary angioplasty immediately after thrombolysis in 115 consecutive patients with acute myocardial infarction. Am Heart J 115:519–529
74. Gruentzig AR, King SB III, Schlumff SW (1987) Longterm follow-up after percutaneous transluminal coronary angioplasty.The early Zurich experience. N Engl J Med 316:1127–1132
75. Man J, Tymchak W, Jugdutt BI (2010) Adjunctive pharmacologic treatment for acute myocardial infarction. In: Brown DL, Jeremias A (eds) Textbook of cardiac intensive care, 2nd edn. Elsevier, Philadelphia, pp 1–72
76. Simari RD, Berger PB, Bell MR, Gibbons RJ, Holmes DR Jr (1994) Mayo Clin Proc 69:346–358
77. Welsh RC, Goldstein P, Adgey J, Verheugt F, Bestilny SA, Wallentin L, for the ASSENT 3 plus trial investigators et al (2004) Variations in pre-hospital fibrinolysis process of care: insights from the assessment of the safety and efficacy of a new thrombolytic 3 plus international acute myocardial infarction prehospital care survey. Eur J Emerg Med 11:134–140
78. Welsh RC, Chang W, Goldstein P, Adgey J, Granger CB, Verheugt FW et al (2005) Time to treatment and the impact of a physician on prehospital management of acute ST elevation myocardial infarction: insights from the ASSENT-3 PLUS trial. Heart 91:1400–1406
79. Terkelsen CJ, Lassen JF, Nørgaard BL, Gerdes JC, Poulsen SH, Bendix K et al (2005) Reduction of treatment delay in patients with ST-elevation myocardial infarction: impact of pre-hospital diagnosis and direct referral to primary percutanous coronary intervention. Eur Heart J 26:770–777
80. Crea F, Falk V, Filippatos G, Fox K, Huber K, Kastrati A et al (2008) Management of acute myocardial infarction in patients presenting with persistent ST-segment elevation: the task force on the management of ST-Segment elevation acute myocardial infarction of the European Society of Cardiology. Eur Heart J 29:2909–2945
81. Kushner FG, Hand M, Smith SC, King SB, Anderson JL, Antman EM et al (2009) 2009 focused updates: ACC/AHA guidelines for the management of patients with ST-elevation myocardial infarction (updating the 2004 guideline and 2007 focused update) and ACC/AHA/SCAI guidelines on percutaneous coronary intervention (updating the 2005 guideline and 2007 focused update) a report of the American College of Cardiology Foundation/American Heart Association task force on practice guidelines. J Am Coll Cardiol 54:2205–2241
82. Estevez-Loureiro R, Lopez-Sainz A, de Prado AP, Cuellas C, Santos RC, Alonzo-Orcajo N et al (2014) Timely reperfusion for ST-elevation myocardial infarction: effect of direct transfer to primary angioplasty on time delays and clinical outcomes. World J Cardiol 6:424–433
83. De Luca G, Suryapranata H, Ottervanger JP, Antman EM (2004) Time delay to treatment and mortality in primary angioplasty for acute myocardial infarction: every minute of delay counts. Circulation 109:1223–1225
84. Cannon CP, Gibson CM, Lambrew CT, Shoultz DA, Levy D, French WJ et al (2000) Relationship of symptom-onset-to-balloon time and door-to-balloon time with mortality in patients undergoing angioplasty for acute myocardial infarction. JAMA 283:2941–2947

85. Serruys PW, Wijns W, van den Brand M, Mey S, Slager CJ, Schuurbiers JCH et al (1984) Left ventricular performance, regional blood flow, wall motion and lactate metabolism during transluminal angioplasty. Circulation 70:25–36
86. Wijns WW, Serruys PW, Slager CJ, Grimm J, Krayenbuehl HP, Hugenholtz PG et al (1986) Effect of coronary occlusion during percutaneous transluminal angioplasty in humans on left ventricular chamber stiffness and regional diastolic pressure-radius relations. J Am Coll Cardiol 7:455–463
87. Heusch G (1998) Stunning-great paradigmatic, but little clinical evidence. Basic Res Cardio 93:164–166
88. Sheiban I, Tonni S, Benussi P, Martini A, Trevi GP (1993) Left ventricular dysfunction following transient ischemia induced by transluminal coronary angioplasty. Benficial effects of calcium antagonists against post-ischaemic myocardial stunning. Eur Heart J 14(suppl A):14–21
89. Yingzhong C, Lin C, Chunbin W (2016) Clinical effects of cyclosporine a on reperfusion injury in myocardial infarction: a meta-analysis of randomized controlled trials. SpringerPlus 5:1117–1126
90. Bulluck H, Yellon DM, Hausenloy DJ (2016) Reducing myocardial infarct size: challenges and future opportunities. Heart 102:341–348
91. Bolli R, Marban E (1999) Molecular and cellular mechanisms of myocardial stunning. Physiol Rev 79:609–634
92. Reffelmann T, Kloner RA (2002) The "no-reflow" phenomenon: basic science and clinical correlates. Heart 87:162–168
93. Ducas A, Bartekova M, Dhalla NS (2015) Ischemia-reperfusion injury of the heart: moving forward with our knowledge. J Hear Health 1:1–10. https://doi.org/10.16966/2379-769X.110
94. Rao V, Ivanov J, Weisel RD, Cohen G, Borger MA, Mickle DA (2001) Lactate release during reperfusion predicts low cardiac output syndrome after coronary bypass surgery. Ann Thorac Surg 71:1925–1930
95. Merante F, Mickle DA, Weisel RD, Li RK, Tumiati LC, al RV (1998) Myocardial aerobic metabolism is impaired in a cell culture model of cyanotic heart disease. Am J Physiol 275:H1673–H1681
96. Murphy E, Ardehali H, Balaban RS, DiLisa F, Dorn GW, Kitsis RN et al (2016) American Heart Association council on basic cardiovascular sciences, council on clinical cardiology, and council on functional genomics and translational biology. Mitochondrial function, biology, and role in disease: a scientific statement from the American Heart Association. Circ Res 118:1960–1991
97. Sack MN, Fyhrquist FY, Saijonmaa OJ, Fuster V, Kovacic JC (2017) Basic biology of oxidative stress and the cardiovascular system. Part 1 of 3-Part series. J Am Coll Cardiol 70:196–211
98. Bartesagh S, Radi (2018) Fundamentals on the biochemistry of peroxynitrite and protein tyrosine nitration. Redox Biol 14:618–625
99. Kumar D, Jugdutt BI (2003) Apoptosis and oxidants in the heart. J Lab Clin Med 142:288–297
100. Jugdutt BI, Idikio HA (2005) Apoptosis and oncosis in acute coronary syndromes: assessment and implications. Mol Cell Biochem 270:177–200
101. Kloner RA, Ganote CE, Jennings RB (1974) The "no-reflow" phenomenon after temporary coronary occlusion in dogs. J Clin Invest 54:1496–1508
102. Ambrosio G, Weisman HF, Mannisi JA, Becker LC (1989) Progressive impairment of regional myocardial perfusion after initial restoration of postischemic blood flow. Circulation 80:1846–1861
103. Becker LC, Jeremy RW, Schaper J, Schaper W (1999) Ultrastructural assessment of myocardial necrosis occurring during ischemia and 3-h reperfusion in the dog. Am J Physiol 277:H243–H252
104. Jugdutt BI, Sawicki G (2004) AT1 receptor blockade alters metabolic, functional and structural proteins after reperfused myocardial infarction: detection using proteomics. Mol Cell Biochem 263:179–188. Erratum in: Mol Cell Biochem (2007) 300:269-270

105. Sawicki G, Menon V, Jugdutt BI (2004) Improved balance between TIMP-3 and MMP-9 after regional myocardial ischemia-reperfusion during AT1 receptor blockade. J Card Fail 10:442–449
106. Sawicki G, Jugdutt BI (2007) Valsartan reverses post-translational modifications of the delta-subunit of ATP synthase during in vivo canine reperfused myocardial infarction. Proteomics 7:2100–2110
107. Jugdutt BI, Menon V (2004) AT1 receptor blockade limits myocardial injury and upregulates AT2 receptors during reperfused myocardial infarction. Mol Cell Biochem 260:111–118
108. Jugdutt BI, Menon V (2004) Valsartan-induced cardioprotection involves angiotensin II type 2 receptor upregulation in dog and rat models of in vivo reperfused myocardial infarction. J Card Fail 10:74–82
109. Moe G, Konig A, Liu P, Jugdutt BI (2008) Selective type 1 angiotensin II receptor blockade attenuates oxidative stress and regulates angiotensin II receptors in the canine failing heart. Mol Cell Biochem 17:97–104
110. Moudgil R, Menon V, Xu Y, Musat-Marcu S, Kumar D, Jugdutt BI (2001) Postischemic apoptosis and functional recovery after angiotensin II type 1 receptor blockade in isolated working rat hearts. J Hypertens 19:1121–1129
111. Ford WR, Khan MI, Jugdutt BI (1998) Effect of the novel angiotensin II type 1 receptor antagonist L-158,809 on acute infarct expansion and acute anterior myocardial infarction in the dog. Can J Cardiol 14:73–80
112. Jugdutt BI (2011) Optimizing pharmacotherapy for limiting cardiovascular remodeling a matter of timing therapy to match biology. J Am Coll Cardiol 57:2029–2030
113. Flaherty JT, Pitt B, Gruber JW, Heuser RR, Rothbaum DA, Burwell LR et al (1994) Recombinant human superoxide dismutase (h-SOD) fails to improve recovery of ventricular function in patients undergoing coronary angioplasty for acute myocardial infarction. Circulation 89:1982–1991
114. Przyklenk K, Kloner RA (1989) "Reperfusion injury" by oxygen-derived free radicals? Effect of superoxide dismutase plus catalase, given at the time of reperfusion, on myocardial infarct size, contractile function, coronary microvasculature, and regional myocardial blood flow. Circ Res 674:86–96
115. Bolli R, Jeroudi MO, Patel BS, DuBose CM, Lai EK, Roberts R et al (1989) Direct evidence that oxygen-derived free radicals contribute to postischemic myocardial dysfunction in the intact dog. Proc Natl Acad Sci USA 86:4695–4599
116. Verma S, Fedak PWM, Weisel RD, Butany J, Rao V, Maitland A et al (2002) Fundamentals of reperfusion Injury for the clinical cardiologist. Circulation 105:2332–2336
117. Cung TT, Morel O, Rioufol CG, Garcia-Dorado D, Angoulvant D, Bonnefoy-Cudraz E et al (2015) Cyclosporine before PCI in patients with acute myocardial infarction. N Engl J Med 373:1021–1031
118. Jugdutt BI, Jelani A (2008) Aging and defective healing, adverse remodeling, and blunted post-conditioning in the reperfused wounded heart. J Am Coll Cardiol 51:1399–1403
119. Przyklenk K, Maynard M, Darling CE, Whittaker P (2008) Aging mouse hearts are refractory to infarct size reduction with post-conditioning. J Am Coll Cardiol 51:1393–1398
120. Jugdutt BI, Jelani A (2013) Chapter 27: Aging and markers of adverse remodeling after myocardial infarction. In: Jugdutt BI, Dhalla NS (eds) Cardiac remodeling: molecular mechanisms. Springer, New York, pp 487–512
121. Frangogiannis NG, Smith CW, Entman ML (2002) The inflammatory response in myocardial infarction. Cardiovasc Res 53:31–47
122. Bujak M, Kweon HJ, Chatila K, Li N, Taffet G, Frangogiannis NG (2008) Aging-related defects are associated with adverse cardiac remodeling in a mouse model of reperfused myocardial infarction. J Am Coll Cardiol 51:1384–1392
123. Nahrendorf M, Pittet MJ, Swirski FK (2010) Monocytes: Protagonists of infarct inflammation and repair after myocardial infarction. Circulation 121:2437–2445
124. Jugdutt BI (2002) Nitric oxide and cardioprotection during ischemia-reperfusion. Heart Fail Rev 7:391–405

125. Neumann F-J, Gick M (2018) Direct stenting in ST-segment elevation myocardial infarction: convenient, but not improving outcomes. Eur Heart J 39:2480–2483
126. Mahmood KO, Jolly SS, James S, Dzavik V, Cairns JA, Olivecrona CK et al (2018) Clinical impact of direct stenting and interaction with thrombus aspiration in patients with ST-segment elevation myocardial infarction undergoing percutaneous coronary intervention: thrombectomy trialists collaboration. Eur Heart J 39:2472–2479
127. Karmazyn M, Gan XT, Humphreys RA, Yoshida H, Kusumoto K (1999) The myocardial Na+-H+ exchange structure, regulation, and its role in heart disease. Circ Res 85:777–786
128. Karmazyn M, Moffat MP (1993) Role of Na+/H+ exchange in cardiac physiology and pathophysiology: mediation of myocardial reperfusion injury by the pH paradox. Cardiovasc Res 27:915–924
129. Churchill EN, Murriel CL, Chen CH, Mochly-Rosen D, Szweda LI (2005) Reperfusion-induced translocation of δPKC to cardiac mitochondria prevents pyruvate dehydrogenase reactivation. Circ Res 97:78–85
130. Chambers KT, Leone TC, Sambandam N, Kovacs A, Wagg CS, Lopaschuk GD et al (2011) Chronic inhibition of pyruvate dehydrogenase in heart triggers an adaptive metabolic response. J Biol Chem 286:11155–11162
131. McCormack JC, Barr EL, Wolff AA, Lopaschuk GD (1996) Ranolazine stimulates glucose oxidation in normoxic, ischemic and repefused ischemic hearts. Circulation 93:135–142
132. Vitulano N, Ciaidella P, Gustapane M, Vitulano L, Pedicino A, Pelargonio D (2015) Ranolazine: beyond the treatment of chronic stable angina pectoris. Int J Clin Cardiol 2(4):1–7. ISSN: 2378-2951
133. Codolosa JN, Acharjee S, Figueredo VM (2014) Update on ranolazine in the management of angina. Vasc Health Risk Manag 10:353–362
134. Halestrap AP, Clarke SJ, Javadov SA (2004) Mitochondrial permeability transition pore opening during myocardial reperfusion-a target for cardioprotection. Cardiovasc Res 61:372–385
135. Kornfeld O, Hwang S, Disatnik MH, Chen CH, Qvit N, Mochly-Rosen D (2015) Mitochondrial reactive oxygen species at the heart of the matter: new therapeutic approaches for cardiovascular diseases. Circ Res 116:1783–1799
136. Garcia-Dorado D, Ruiz-Meana M, Inserte J, Rodriguez-Sinovas A, Piper HM (2012) Calcium-mediated cell death during myocardial reperfusion. Cardiovasc Res 94:168–180
137. Inserte J, Hernando V, Garcia-Dorado D (2012) Contribution of calpains to myocardial ischaemia/reperfusion injury. Cardiovasc Res 96:23–31
138. Zhang T, Zhang Y, Cui M, Jin L, Wang Y, Lv F et al (2016) CaMKII is a RIP3 substrate mediating ischemia- and oxidative stress-induced myocardial necroptosis. Nat Med 22:175–182
139. Valls-Lacalle L, Barba I, Miro-Casas E, Alburquerque-Bejar JJ, Ruiz-Meana M, Fuertes-Agudo M et al (2016) Succinate dehydrogenase inhibition with malonate during reperfusion reduces infarct size by preventing mitochondrial permeability transition. Cardiovasc Res 109:374–384
140. Inserte J, Hernando V, Vilardosa U, Abad E, Poncelas-Nozal M, Garcia-Dorado D (2013) Activation of cGMP/protein kinase G pathway in postconditioned myocardium depends on reduced oxidative stress and preserved endothelial nitric oxide synthase coupling. J Am Heart Assoc 2:e005975
141. Dikalova AE, Bikineyeva AT, Budzyn K, Nazarewicz RR, McCann L, Lewis W et al (2010) Therapeutic targeting of mitochondrial superoxide in hypertension. Circ Res 107:106–116
142. O'Connor PM, Gutterman DG (2010) Resurrecting hope for antioxidant treatment of cardiovascular disease. Focus on mitochondria. Circ Res 107:9–11
143. Taegtmeyer H (1994) Energy metabolism of the heart: from basic concepts to clinical applications. Curr Probl Cardiol 19:59–113
144. Chen YR, Zweir JL (2014) Cardiac mitochondria and reactive oxgen species generation. Circ Res 114:424–537
145. Schaper J, Meiser E, Stammler G (1985) Ultrastructural morphometric analysis of myocardium from dogs, rats, hamsters, mice, and from human hearts. Circ Res 56:377–391

146. Johnson DT, Harris RA, French S, Blair PV, You J, Bemis KG et al (2007) Tissue heterogeneity of the mammalian mitochondrial proteome. Am J Physiol Cell Physiol 292:C689–C697
147. Opie LH, Sack MN (2002) Metabolic plasticity and the promotion of cardiac protection in ischemia and ischemic preconditioning. J Mol Cell Cardiol 34:1077–1089
148. Sack MN (2009) Type 2 diabetes, mitochondrial biology and the heart. J Mol Cell Cardiol 46:842–849
149. Marín-García J, Damle S, Jugdutt BI, Moe GW (2012) Nuclear-mitochondrial cross-talk in global myocardial ischemia. A time-course analysis. Mol Cell Biochem 364:225
150. Ibáñez B, Heusch G, Ovize M, Van de Werf F (2015) Evolving therapies for myocardial ischemia/reperfusion injury. J Am Coll Cardiol 65:1454–1471
151. Barrabes JA, Inserte J, Mirabet M, Quiroga A, Hernando V, Figueras J et al (2010) Antagonism of P2Y12 or GPIIb/IIIa receptors reduces plateletmediated myocardial injury after ischaemia and reperfusion in isolated rat hearts. Thromb Haemost 104:128–135
152. Kawaguchi M, Takahashi M, Hata T, Kashima Y, Usui F, Morimoto H et al (2011) Inflammasome activation of cardiac fibroblasts is essential for myocardial ischemia/reperfusion injury. Circulation 123:594–604
153. Ruiz-Meana M, Inserte J, Fernandez-Sanz C, Hernando V, Miro-Casas E, Barba I et al (2011) The role of mitochondrial permeability transition in reperfusion-induced cardiomyocyte death depends on the duration of ischemia. Basic Res Cardiol 106:1259–1268
154. Jugdutt BI (1997) Effect of reperfusion on ventricular mass, topography and function during healing of anterior infarction. Am J Physiol 272:H1205–H1211
155. Jugdutt BI (2012) Ischemia/Infarction. Heart Fail Clin 8:43–51
156. Atar D, Petzelbauer P, Schwitter J, Huber K, Rensing B, Kasprzak JD et al (2009) Effect of intravenous FX06 as an adjunct to primary percutaneous coronary intervention for acute ST-segment elevation myocardial infarction results of the F.I.R.E. Efficacy of FX06 in the prevention of myocardial reperfusion injury trial. J Am Coll Cardiol 53:720–729
157. Armstrong PW, Granger CB, Adams PX, Hamm C, Holmes D Jr, O'Neill WW (2007) Pexelizumab for acute ST-elevation myocardial infarction in patients undergoing primary percutaneous coronary intervention: a randomized controlled trial. JAMA 297:43–51
158. Kloner RA, Jennings RB (2001) Consequences of brief ischemia: stunning, preconditioning, and their clinical implications: part II. Circulation 104:3158–3167
159. Hausenloy DJ, Yellon DM (2013) Myocardial ischemia-reperfusion injury: a neglected therapeutic target. J Clin Invest 123:92–100
160. Staat P, Rioufol G, Piot C, Cottin Y, Cung TT, L'Huillier I et al (2005) Postconditioning the human heart. Circulation 112:2143–2148
161. Thibault H, Piot C, Staat P, Bontemps L, Sportouch C, Rioufol G et al (2008) Long-term benefit of postconditioning. Circulation 117:1037–1044
162. Hausenloy DJ, Candilio L, Evans R, Ariti C, Jenkins DP, Kolvekar S et al (2015) Remote ischemic preconditioning and outcomes of cardiac surgery. N Engl J Med 373:1408–1417
163. Yellon DM, Ackbarkhan AK, Balgobin V, Bulluck H, Deelchand A, Dhuny MR et al (2015) Remote ischemic conditioning reduces myocardial infarct size in STEMI patients treated by thrombolysis. J Am Coll Cardiol 65:2764–2765
164. Kitakaze M, Asakura M, Kim J, Shintani Y, Asanuma H, Hamasaki T et al (2007) Human atrial natriuretic peptide and nicorandil as adjuncts to reperfusion treatment for acute myocardial infarction (J-WIND): two randomised trials. Lancet 370:1483–1493
165. Selker HP, Beshansky JR, Sheehan PR, Massaro JM, Griffith JL, D'Agostino RB et al (2012) Out-of-hospital administration of intravenous glucose-insulin-potassium in patients with suspected acute coronary syndromes: the IMMEDIATE randomized controlled trial. JAMA 307:1925–1933
166. Lonborg J, Vejlstrup N, Kelbaek H, Botker HE, Kim WY, Mathiasen AB et al (2012) Exenatide reduces reperfusion injury in patients with ST-segment elevation myocardial infarction. Eur Heart 33:1491–1499

167. Lonborg J, Kclbaek H, Vejlstrup N, Botker HE, Kim WY, Holmvang L et al (2012) Exenatide reduces final infarct size in patients with ST-segment-elevation myocardial infarction and short-duration of ischemia. Circ Cardiovasc Interv 5:288–295
168. Woo JS, KimW HSJ, Kim JB, Kim SJ, Kim WS et al (2013) Cardioprotective effects of exenatide in patients with ST-segment-elevation myocardial infarction undergoing primary percutaneous coronary intervention: results of exenatide myocardial protection in revascularization study. Arterioscler Thromb Vasc Biol 33:2252–2260
169. Chen WR, Hu SY, Chen YD, Zhang Y, Qian G, Wang J et al (2015) Effects of liraglutide on left ventricular function in patients with ST-segment elevation myocardial infarction undergoing primary percutaneous coronary intervention. Am Heart J 170:845–854
170. Lefer D, Jones S, Steenbergen C, Kukreja R, Guo Y, Tang X-L et al (2014) Sodium nitrite fails to limit myocardial infarct size: results from the CAESAR cardioprotection consortium. FASEB J 28:LB645
171. Siddiqi N, Neil C, Bruce M, MacLennan G, Cotton S, Papadopoulou S et al (2014) Intravenous sodium nitrite in acute ST-elevation myocardial infarction: a randomized controlled trial (NIAMI). Eur Heart J 35:1255–1262
172. Jones DA, Pellaton C, Velmurugan S, Rathod KS, Andiapen M, Antoniou S et al (2015) Randomized phase 2 trial of intracoronary nitrite during acute myocardial infarction. Circ Res 116:437–447
173. Hausenloy DJ, Maddock HL, Baxter GF, Yellon DM (2002) Inhibiting mitochondrial permeability transition pore opening: a new paradigm for myocardial preconditioning? Cardiovasc Res 55:534–543
174. Argaud L, Gateau-Roesch O, Muntean D, Chalabreysse L, Loufouat J, Robert D et al (2005) Specific inhibition of the mitochondrial permeability transition prevents lethal reperfusion injury. J Mol Cell Cardiol 38:367–374
175. Skyschally A, Schulz R, Heusch G (2010) Cyclosporine A at reperfusion reduces infarct size in pigs. Cardiovasc Drugs Ther 24:85–87
176. Karlsson LO, Zhou AX, Larsson E, Astrom-Olsson K, Mansson C, Akyurek LM et al (2010) Cyclosporine does not reduce myocardial infarct size in a porcine ischemia-reperfusion model. J Cardiovasc Pharmacol Ther 15:182–189
177. Lim WY, Messow CM, Berry C (2012) Cyclosporin variably and inconsistently reduces infarct size in experimental models of reperfused myocardial infarction: a systematic review and meta-analysis. Br J Pharmacol 165:2034–2043
178. Piot C, Croisille P, Staat P, Thibault H, Rioufol G, Mewton N et al (2008) Effect of cyclosporine on reperfusion injury in acute myocardial infarction. N Engl J Med 359:473–481
179. Ghaffari S, Kazemi B, Toluey M, Sepehrvand N (2013) The effect of prethrombolytic cyclosporine-A injection on clinical outcome of acute anterior ST-elevation myocardial infarction. Cardiovasc Ther 31:e34–e39
180. Ottani F, Latini R, Staszewsky L, La Vecchia L, Locuratolo N, Sicuro M et al (2016) Cyclosporine A in reperfused myocardial infarction. The multicenter, controlled ,open-label Cycle trial. J Amer Coll Cardiol 67:365–374
181. Chiari P, Angoulvant D, Mewton N, Desebbe O, Obadia JF, Robin J et al (2014) Cyclosporine protects the heart during aortic valve surgery. Anesthesiology 121:232–238
182. Dai W, Shi J, Gupta RC, Sabbah HN, Hale SL, Kloner RA (2014) Bendavia, a mitochondria-targeting peptide, improves postinfarction cardiac function, prevents adverse left ventricular remodeling, and restores mitochondria-related gene expression in rats. J Cardiovasc Pharmacol 64:543–553
183. Shi J, Dai W, Hale SL, Brown DA, Wang M, Han X et al (2015) Bendavia restores mitochondrial energy metabolism gene expression and suppresses cardiac fibrosis in the border zone of the infarcted heart. Life Sci 141:170–178
184. Chakrabarti AK, Feeney K, Abueg C et al (2013) Rationale and design of the EMBRACE STEMI study: a phase 2a, randomized, double-blind, placebo-controlled trial to evaluate the safety, tolerability and efficacy of intravenous Bendavia on reperfusion injury in patients

treated with standard therapy including primary percutaneous coronary intervention and stenting for ST-segment elevation myocardial infarction. Am Heart J 165:509–514

185. Gibson CM, Giugliano RP, Kloner RA, Bode C, Tendera M, Jánosi A et al (2016) EMBRACE STEMI study: a Phase 2a trial to evaluate the safety, tolerability, and efficacy of intravenous MTP-131 on reperfusion injury in patients undergoing primary percutaneous coronary intervention. Eur Heart J 37:1296–1303
186. Ibanez B, Prat-Gonzalez S, Speidl WS, Vilahur G, Pinero A, Cimmino G et al (2007) Early metoprolol administration before coronary reperfusionresults in increased myocardial salvage: analysis of ischemic myocardium at risk using cardiac magnetic resonance. Circulation 115:2909–2916
187. Ibanez B, Macaya C, Sanchez-Brunete V, Pizarro G, Fernandez-Friera L, Mateos A et al (2013) Effect of early metoprolol on infarct size in ST-segment-elevation myocardial infarction patients undergoing primary percutaneous coronary intervention: the effect of metoprolol in cardioprotection during an acute myocardial infarction (METOCARD-CNIC) trial. Circulation 128:1495–1503
188. Pizarro G, Fernandez-Friera L, Fuster V, Fernandez-Jimenez R, Garcia-Ruiz JM, Garcia-Alvarez A et al (2014) Long-term benefit of early pre-reperfusion metoprolol administration in patients with acute myocardial infarction: results from the METOCARD-CNIC trial (Effect of metoprolol in cardioprotection during an acute myocardial infarction). J Am Coll Cardiol 63:2356–2362
189. Roolvink V, Rasoul S, Ottervanger JP, Dambrink JH, Lipsic E, van der Horst IC et al (2014) Rationale and design of a double-blind, multicenter, randomized, placebo-controlled clinical trial of early administration of intravenous beta-blockers in patients with ST-elevation myocardial infarction before primary percutaneous coronary intervention. EARLY beta-blocker administration before primary PCI in patients with ST-elevation myocardial infarction trial. Am Heart J 168:661–666
190. Schaller S, Paradis S, Ngoh GA, Assaly R, Buisson B, Drouot C et al (2010) TRO40303, a new cardioprotective compound, inhibits mitochondrial permeability transition. J Pharmacol Exp Ther 333:696–706
191. Hansson MJ, Llwyd O, Morin D, De PD, Arnoux T, Gouarne C et al (2015) Differences in the profile of protection afforded by TRO40303 and mild hypothermia in models of cardiac ischemia/reperfusion injury. Eur J Pharmacol 760:7–19
192. Atar D, Arheden H, Berdeaux A, Bonnet JL, Carlsson M, Clemmensen P et al (2015) Effect of intravenous TRO40303 as an adjunct to primary percutaneous coronary intervention for acute ST-elevation myocardial infarction: MITOCARE study results. Eur Heart J 36:112–119
193. Bates E, Bode C, Costa M, Gibson CM, Granger C, Green C et al (2008) Intracoronary KAI-9803 as an adjunct to primary percutaneous coronary intervention for acute ST-segment elevation myocardial infarction. Circulation 117:886–896
194. Lincoff AM, Roe M, Aylward P, Galla J, Rynkiewicz A, Guetta V et al (2014) Inhibition of delta-protein kinase C by delcasertib as an adjunct to primary percutaneous coronary intervention for acute anterior ST-segment elevation myocardial infarction: results of the PROTECTION AMI randomized controlled trial. Eur Heart J 35:2516–2523
195. Olafsson B, Forman MB, Puett DW, Pou A, Cates CU, Friesinger GC et al (1987) Reduction of reperfusion injury in the canine preparation by intracoronary adenosine: importance of the endothelium and the no-reflow phenomenon. Circulation 76:1135–1145
196. Bulluck H, Sirker A, Loke YK, Garcia-Dorado D, Hausenloy DJ (2015) Clinical benefit of adenosine as an adjunct to reperfusion in ST-elevation myocardial infarction patients: an updated meta-analysis of randomized controlled trials. Int J Cardiol 202:228–237
197. Karlsson J-E, El-Saadi W, Ali M, Puskar W, Skogvard P, Engvall JE et al (2015) Mangafodipir as a cardioprotective adjunct to reperfusion therapy: a feasibility study in patients with ST-segment elevation myocardial infarction. Eur Heart J 1:39–45

Oxidant Stress in Atherosclerosis: Oxidatively Modified LDL and LOX-1

15

Ajoe John Kattoor and Jawahar L. Mehta

15.1 Introduction

Atherosclerosis affecting medium and large sized arteries is characterized by lipid and inflammatory cell accumulation in the vessel walls [1]. The three major hypotheses that explain the pathogenesis of atherosclerosis are: (a) response to injury, (b) response to retention of lipids, and (c) oxidative modification of lipids [2]. Response to injury considers endothelial dysfunction as the main trigger causing release of cytokines and reactive oxygen species leading to an altered vascular homeostasis. Response to retention assumes that LDL permeates to the sub endothelial space, binds to various proteoglycans, and forms foam cells after being engulfed by macrophages. Oxidative modification hypothesis considers oxidized LDL to be the key molecule that contributes to foam cell formation as macrophages have higher affinity than native LDL. Endothelial injury leads to LDL infiltration and accumulation in the subendothelial space (Fig. 15.1). Pathological states cause native LDL to become oxidized. Oxidized-LDL through its receptors causes an increase in the cell adhesion molecule expression on the endothelial cells. These cell adhesion molecules such as vascular cell adhesion molecule-1 (VCAM-1), P and E-selectins, along with chemoattractant proteins (monocyte chemoattractant proteins-1 (MCP-1), eotaxins and interferon (INF)-γ) promote the recruitment of monocytes and T-lymphoctyes into the subendothelial space. Monocytes differentiation of macrophages occurs in the subendothelial space and expresses various scavenger receptors (SRs) such as LOX-1, SRA and CD36. Macrophages via these SRs engulf modified lipoproteins to form foam cells. These lipid laden macrophages are considered to be the hall mark of early atherosclerotic lesion. Further, the inflammatory cells such as T-lymphocytes and mast cells releases more cytokines, growth factors

A. J. Kattoor · J. L. Mehta (✉)
Division of Cardiology, Central Arkansas Veterans Healthcare System,
University of Arkansas for Medical Sciences, Little Rock, AR, USA
e-mail: MehtaJL@uams.edu

S. Chakraborti et al. (eds.), *Oxidative Stress in Heart Diseases*,
https://doi.org/10.1007/978-981-13-8273-4_15

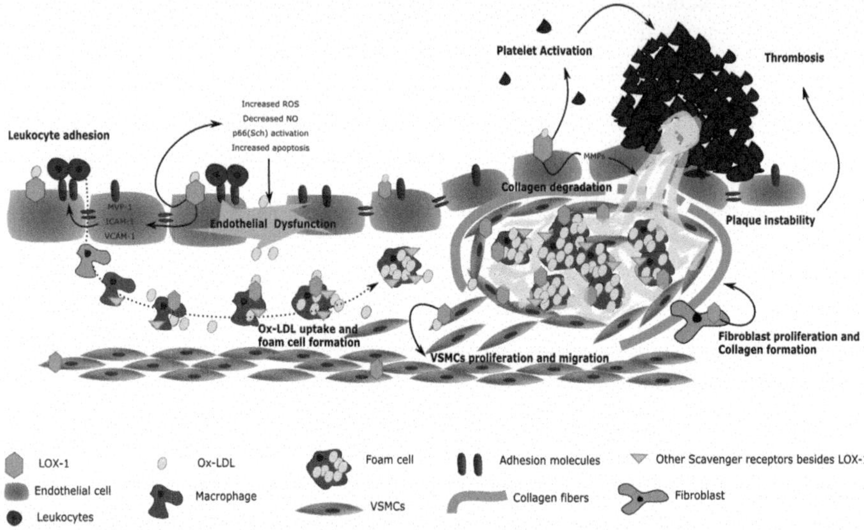

Fig. 15.1 Role of Ox-LDL and LOX-1 in atherogenesis and thrombosis. Reprinted by permission from Rightslink Perimissions Springer Nature Customer Service Center GmbH: Springer Nature, Current Atherosclerosis Reports, Oxidative stress in Atherosclerosis, Kattoor, A.J.; Pothineni, N.V.K.; Palagiri, D.; Mehta, J.L, 2017

and reactive oxygen species (ROS) after migration into the intima. These growth factors and ROS in turn stimulate smooth muscle migration and collagen deposition leading to the formation of an atheromatous plaque. Matrix metalloproteinases (MMPs) which are released in response to oxidative stress degrade the fibrous atheromatous plaque and endothelial cell (ECs) basement membrane thereby causing a physical disruption of the plaque. Physical disruption of plaque may trigger thrombosis of the blood vessel hosting the plaque. In some cases, a healing process takes place leading to further smooth muscle cell and collagen accumulation and converting the fatty atherosclerotic plaque to a fibrous plaque [3].

15.2 Ox-LDL and LOX-1

Oxidized LDL are particles derived from "circulating LDL that have peroxides or their degradation products within it or associated with the particle"[4]. They are generated as a result of oxidative stress. They play an important role in generating chemoattractant proteins and thereby causing leukocyte recruitment into the subendothelial space. Lectin like oxidized LDL receptor-1 (LOX-1) is a 50 kDa transmembrane glycoprotein that was found to be involved in binding, internalization and degradation of ox-LDL [5]. It has a N-terminal cytoplasmic domain, a single transmembrane domain, a neck domain which is extracellular followed by c-type lectin like domain [6]. A variety of cells that play a key role in atherosclerosis such as endothelial cells, macrophages, platelets, fibroblast and smooth muscle cells have been found to harbor LOX-1 [7]. Ox-LDL binds to LOX-1 and causes a rapid internalization complex and activates downstream signaling leading to varied effects in different cell types to promote atherosclerosis (Fig. 15.2). LOX-1 receptors are clustered on the plasma membrane and are located in its lipid rafts. Changes in plasma membrane due to cholesterol depletion can alter the distribution of LOX-1 to a more diffuse one, preventing interaction of the LOX-1 and Ox-LDL in the cell surface [8].

15.3 Ligands and Inducers of LOX-1

Inflammatory cytokines are inducers of LOX-1 expression. TNF-α, IL-1, CRP, INF-γ and lipopolysaccharides and free radicals are able to induce LOX-1 expression in vitro. Hypertension related stimuli such as angiotensin II, endothelin-I, aldosterone and shear stress, hyperglycemic stimuli such as hyperglycemia and advanced glycated end-products, and modified lipoproteins such as ox-LDL, glycoxidized LDL and lysophophatidylcholine have been shown to induce LOX-1 in vitro [9]. In keeping with these concepts developed in vitro, vascular tissues from animals and humans with hypertension, diabetes and hyperlipidemia have high expression of LOX-1. Heart failure, psychological stress, HIV infection and ischemia-reperfusion also can induce LOX-1[10].

15.4 Ox-LDL – LOX-1 Interaction in Atherosclerosis

15.4.1 Endothelial Cells

Ox-LDL enhances expression of E-selectin, P-selectin and vascular cell adhesion molecules (VCAM-1) and intercellular cell adhesion molecules (ICAM-1) through LOX-1 on ECs. Mitogen-activated protein kinase pathway activated by Ox-LDL-LOX-1 interaction generates monocyte chemoattractant protein (MCP-1) expression and increases monocyte adhesion. These molecules leads to leukocyte recruitment and differentiation of monocytes to macrophages in the subendothelial space. Antisense oligodeoxynucleotides to the human LOX-1 gene was shown to suppress ox-LDL mediated upregulation of MCP-1 and monocyte adhesion on human coronary artery ECs [11]. This suggests that ox-LDL acts through LOX-1 on ECs to mediate the expression of chemoattractant proteins and thereby facilitate leukocyte recruitment.

Ox-LDL-LOX-1 interaction leads to activation of both intrinsic and extrinsic apoptotic pathways in ECs. Ox-LDL activates caspase-3 and caspase-9 and inhibits antiapoptotic proteins such as B- cell lymphoma 2 (Bcl-2) and cellular inhibitor of apoptosis protein 1 (c-IAP-1) [12]. On the other hand, Fas ligand mediated apoptosis is enhanced due to the upregulation of Fas on ECs by ox-LDL [13].

Angiotensin converting enzyme (ACE) and endothelin-1 generation on ECs are upregulated by Ox-LDL and they in turn increase LOX-1 expression leading to ox-LDL uptake by ECs [14–16].

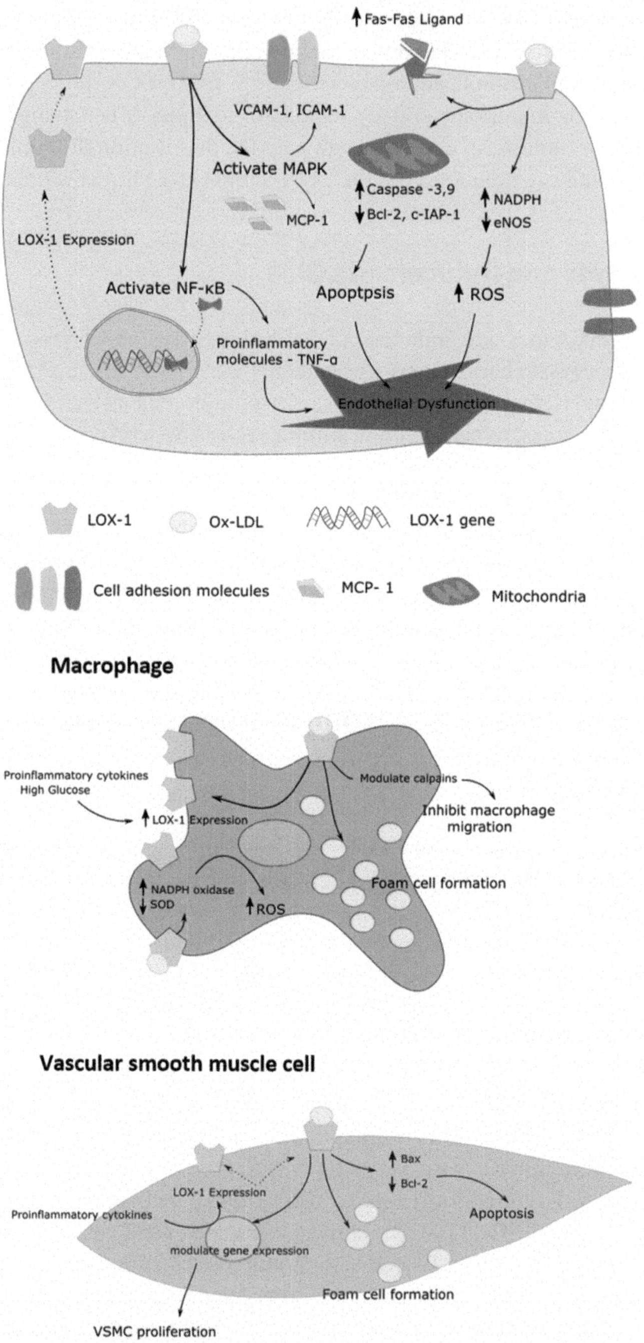

Fig. 15.2 Effects of ox-LDL-LOX-1 interaction in endothelial cells, vascular smooth muscle cells and macrophages. Reprinted by permission from Eureka Science: Bentham Science Publishers Ltd., Current Medicinal Chemistry; Role of Ox-LDL and LOX-1 in Atherogenesis, Kattoor, A.J.; Kanuri, S.H.; Mehta, J.L, 2018

Ox-LDL through LOX-1 contributes to suppression of nitric oxide (NO) generation and increase in ROS generation in ECs. Endothelial nitric oxide synthase (eNOS) is displaced from its caveolar membrane location by ox-LDL and leads to its dysfunction. Moreover, ox-LDL causes arginase II activation which leads to downregulation of eNOS as it competes for their common substrate L-arginine [17]. Increased ROS generation by ox-LDL in ECs causes inactivation of preformed NO. In addition, NADPH oxidase expression is also increased by Ox-LDL contributing to ROS generation [18].

5′ flanking region of the LOX-1 gene has a NF-κB binding site. Ox-LDL though LOX-1 activates NF-κB and increases the expression of LOX-1. This leads to activation of a vicious cycle of increased ox-LDL mediated pro-inflammatory signaling. Activation of NF-κB also increases adhesion molecules and TNF-α expression in ECs [5].

Synthesis of matrix metalloproteinases such as MMP-1, MMP-3 and MMP-9 is increased by ox-LDL through LOX-1. This creates an imbalance between the metalloproteinases and its tissue inhibitor leading to increased degradation of the fibrotic cap of atherosclerotic plaque predisposing to its rupture.

Thus, ox-LDL via LOX-1 induces expression of cell adhesion molecules, contributes to endothelial dysfunction by NO depletion, promotes EC apoptosis, and predisposes to plaque rupture by creating an imbalance in metalloproteinases and collagen synthesis.

It is of note that LOX-1 is expressed on ECs in several-fold larger number than any other SR [19]. Thus LOX-1 is the primary receptor for ox-LDL binding and its uptake and downstream signaling.

15.4.2 Effects on Macrophages

Macrophages are important cells in the atherosclerotic vasculature. Macrophages transform into foam cells. Plaque formation in the early stages of atherosclerosis development represents accumulation of a large number of form cells. Macrophages express a host of SRs, including LOX-1, SR-A and CD36. LOX-1 accounts for about 40% of ox-LDL uptake in macrophages in a proinflammatory environment. In normal circumstances, it accounts for only 5–10% of ox-LDL uptake. Ox-LDL causes a rise in ROS and decreases activity of superoxide dismutase in macrophages. siRNA targeting LOX-1 was shown to impair NADPH oxidase system and MAPK activation in in-vitro studies. High glucose levels, ox-LDL and proinflammatory cytokines also upregulate ox-LDL and leads to lipid accumulation and foam cell formation [20].

In addition, LOX-1 modulates calcium depended proteases – calpains which influence cell migration. Thus ox-LDL through LOX-1 increases macrophage attachment there by contributing to atherosclerosis [21].

15.4.3 Effects on Vascular Smooth Muscle Cells

Vascular smooth muscle cell (VSMC) proliferation and migration are key features of atherosclerosis. SMCs can also transform into foam cells. VSMCs contain a large number of LOX-1 receptors, perhaps LOX-1 density is more on SMCs than on ECs in an inflammatory state. Pro-inflammatory cytokines such as TNF-α, IL-1 and IFN-γ significantly increase the expression of LOX-1 on VSMCs [22]. Recent studies on inhibition of LOX-1 gene by microRNA let-7g suggest that ox-LDL mediated VSMC proliferation and migration were decreased by LOX-1 gene inhibition. Also, LOX-1 suppress miR-141 expression thereby contributing to VSMC proliferation [23]. Ox-LDL also induces growth factors such as insulin-like growth factor (IGF-1), platelet derived growth factor (PDGF) and epidermal growth factor (EGF) [24]. These growth factors also promote VSMC proliferation.

Apoptosis of smooth muscle cells is increased in inflammatory states due to enhanced expression of pro-apoptotic proteins like bcl-2-associated X protein (bax) and suppression of antiapoptotic bcl-2 by ox-LDL. This leads to an instability in the VSMC layer and predisposition of atherosclerotic plaque to rupture [25]. In addition, LOX-1 also causes ox-LDL uptake and lipid accumulation in smooth muscle progenitor cells leading to formation of foam cells [26].

15.4.4 Effects on Platelets

Platelets are important constituents of the clot that leads to acute myocardial ischemia. Platelets also thought to contribute to atherogenesis by release of PDGF and other growth factors and vasoconstrictors [27]. LOX-1 is expressed on the surface of platelets in an activation dependent manner [28]. Anti-LOX-1 antibody has been showed to inhibit ADP-induced platelet aggregation. LOX-1 contributes to thrombus formation by acting on ADP-induced activation of fibrinogen receptors such as alpha(IIb)beta(3) and alpha(2)beta(1) integrins [29]. Endothelin-1 is induced in ECs by LOX-1 and CD40 interaction with activated platelets, and leads to endothelial dysfunction [30]. Recent studies show that generation of CD147 which is an MMP inducer is stimulated by ox-LDL–LOX-1 interaction contributing to plaque instability [31].

15.5 Diagnostic Value of LOX-1

The extracellular domain of LOX-1, thought to be cleaved by ADAM10 metalloproteinase, is called the soluble LOX-1 (sLOX-1). It is being evaluated as a potential biomarker for cardiovascular disease. sLOX-1 levels have been associated with diabetes mellitus type 2, hypertension and smoking [32–34]. Since there are more readily available methods to diagnose these diseases, sLOX-1's utility in this regard is limited. Civelek et al., proposed using sLOX-1 as a predictor for metabolic syndrome; however, a diagnostic cut off is still under research [35].

Hayashida et al, proposed using sLOX-1 as a marker for plaque instability [36]. This was due to earlier peak levels for sLOX-1 compared to troponin T in a cohort of 521 patients with acute coronary syndrome. Coronary sinus blood samples had a higher level of sLOX-1 compared to aortic blood samples in patients with acute coronary syndrome and exertional angina suggesting the origin of sLOX-1 to coronary circulation [37]. Patients with percutaneous coronary intervention related periprocedural myocardial infarction (PCI-RPMI) which is predisposed by plaque vulnerability, was also found to be have a higher sLOX-1 level compared to those with stable angina [38]. Hence, sLOX-1 levels can potentially be used as a predictor of PCI-RPMI.

In a recent cross sectional study, Liu et al explored the feasibility of using LOX-1 as a predictor for in-stent restenosis (ISR) in patients undergoing percutaneous coronary intervention (PCI) [39]. Level of sLOX-1 during early post-PCI period (1–7 days) were compared between patients who developed ISR (41 patients), non-significant lesions (51 patients) and control group (96 patient without coronary artery disease). A significantly higher level of sLOX-1 was present in patients in the ISR group compared to the both non-significant lesion group ($p = 0.005$) and control group ($p < 0.001$) suggesting that sLOX-1 may be used as a predictor for ISR post-PCI.

Further large scale studies are needed before diagnostic and prognostic applications of LOX-1 can be put to clinical practice.

15.6 Therapeutic Implications of LOX-1

Many naturally occurring compounds have been found to modulate LOX-1 and thereby affecting atherogenesis. Gingko biloba extract, curcumin, and bergamot peet have been shown to decrease LOX-1 expression. Resveratrol, berberine, and tanshionone II-A have been shown to influence atherosclerosis by inhibition of ROS generation by ox-LDL [40–43]. Medications such as aspirin, and statins have also been found to reduce ox-LDL mediated expression of LOX-1 and adhesion molecules [44, 45].

Structure based drug design, RNA interference techinques and monoclonal antibodies are currently being studied for its effects in modulating LOX-1. Ox-LDL–LOX-1 interaction was found to be inhibited by the modified phospholipid molecule PLAzPC by binding to ox-LDL binding site of LOX-1 [46]. Through structure based drug design techniques, Thakkar et al were able to screen 5 molecules that could potentially inhibit LOX-1 from a database of 18 million molecular structures. Two of those molecules were found to inhibit the downstream signaling, LOX-1 mRNA expression and ox-LDL update by 80% in ECs [47].

Non-coding RNAs that control gene expression by post transcription effects are known as microRNAs (miRNAs). Let-7g miRNA which binds to the 3′ UTR region of LOX-1 mRNA has been able to block LOX-1 expression and ox-LDL uptake in human aortic SMCs [48]. In a recent study, miR-98 was studied for its effects on LOX-1 expression and foam cell formation in mouse peritoneal macrophages.

Enhancement of miR-98 decreased LOX-1 expression, lipid accumulation and foam cell formation, whereas inhibition of miR-98 had opposite effects. This suggest that miR-98 can be potentially used in modulation of LOX-1 effects in atherosclerosis [49].Use of siRNA has been studied in vitro to regulate LOX-1 expression. Amati et al used antisense *olr-l* (gene encoding LOX-1) to downregulate LOX- mRNA and LOX-1 protein in the aorta of mice using this technique [50].

Multiple trials in mice and cell based models have shown efficacy of LOX-1 antibodies to inhibit ox-LDL-mediated effects. The highly conserved C-type lectin domain of LOX-1 in mammalian species makes it difficult for human use [51]. Chimeric chicken-human antibodies that decrease ox-LDL uptake have been developed after immunizing chicken with recombinant human LOX-1 [52]. Though multiple large molecules have been developed and are currently being studied with regard to inhibition of LOX-1. Further research is needed before these large and small molecules targeting LOX-1 can be used clinically.

15.7 Conclusion

Ox-LDL through its receptor LOX-1 acts on endothelial cells, macrophages, VSMCs and platelets, and influences multiple signaling pathways that contribute to atherosclerosis. Many of the currently used drugs modulate atherosclerosis by their action on LOX-1 receptor. New molecules that modulate LOX-1 are currently under investigation. sLOX-1 is a potential biomarker in acute coronary injury and further studies needs to be done before it can be put to clinical use.

Conflicts of Interest None

Funding/Declarations Self/None

References

1. Hansson GK, Hermansson A (2011) The immune system in atherosclerosis. Nat Immunol 12(3):204–212
2. Yoshimoto R, Fujita Y, Kakino A, Iwamoto S, Takaya T, Sawamura T (2011) The discovery of LOX-1, its ligands and clinical significance. Cardiovasc Drugs Ther 25(5):379–391
3. Libby P (2002) Inflammation in atherosclerosis. Nature 420(6917):868–874
4. Parthasarathy S, Raghavamenon A, Garelnabi MO, Santanam N (2010) Oxidized low-density lipoprotein. Methods Mol Biol (Clifton, N.J.) 610:403
5. Sawamura T, Kume N, Aoyama T, Moriwaki H, Hoshikawa H, Aiba Y et al (1997) An endothelial receptor for oxidized low-density lipoprotein. Nature 386(6620):73
6. Aoyama T, Sawamura T, Furutani Y, Matsuoka R, Yoshida MC, Fujiwara H et al (1999) Structure and chromosomal assignment of the human lectin-like oxidized low-density-lipoprotein receptor-1 (LOX-1) gene. Biochem J 339(Pt 1):177–184
7. Kattoor AJ, Kanuri SH, Mehta JL (2018) Role of Ox-LDL and LOX-1 in atherogenesis. Curr Med Chem

8. Matarazzo S, Quitadamo MC, Mango R, Ciccone S, Novelli G, Biocca S (2012) Cholesterol-lowering drugs inhibit lectin-like oxidized low-density lipoprotein-1 receptor function by membrane raft disruption. Mol Pharmacol 82(2):246–254
9. Pirillo A, Norata GD, Catapano AL (2013) LOX-1, OxLDL, and atherosclerosis. Mediators Inflamm 2013:152786
10. Xu S, Ogura S, Chen J, Little PJ, Moss J, Liu P (2013) LOX-1 in atherosclerosis: biological functions and pharmacological modifiers. Cell Mol Life Sci 70(16):2859–2872
11. Li D, Mehta JL (2000) Antisense to LOX-1 inhibits oxidized LDL-mediated upregulation of monocyte chemoattractant protein-1 and monocyte adhesion to human coronary artery endothelial cells. Circulation 101(25):2889–2895
12. Chen J, Mehta JL, Haider N, Zhang X, Narula J, Li D (2004) Role of caspases in Ox-LDL-induced apoptotic cascade in human coronary artery endothelial cells. Circ Res 94(3):370–376
13. Salvayre R, Auge N, Benoist H, Negre-Salvayre A (2002) Oxidized low-density lipoprotein-induced apoptosis. Biochim Biophys Acta 1585(2-3):213–221
14. Pernow J, Shemyakin A, Böhm F (2012) New perspectives on endothelin-1 in atherosclerosis and diabetes mellitus. Life Sci 91(13–14):507–516
15. Li D, Singh RM, Liu L, Chen H, Singh BM, Kazzaz N et al (2003) Oxidized-LDL through LOX-1 increases the expression of angiotensin converting enzyme in human coronary artery endothelial cells. Cardiovasc Res 57(1):238–243
16. Morawietz H, Rueckschloss U, Niemann B, Duerrschmidt N, Galle J, Hakim K et al (1999) Angiotensin II induces LOX-1, the human endothelial receptor for oxidized low-density lipoprotein. Circulation 100(9):899–902
17. Ryoo S, Bhunia A, Chang F, Shoukas A, Berkowitz DE, Romer LH (2011) OxLDL-dependent activation of arginase II is dependent on the LOX-1 receptor and downstream RhoA signaling. Atherosclerosis 214(2):279–287
18. Rueckschloss U, Galle J, Holtz J, Zerkowski HR, Morawietz H (2001) Induction of NAD(P)H oxidase by oxidized low-density lipoprotein in human endothelial cells: antioxidative potential of hydroxymethylglutaryl coenzyme A reductase inhibitor therapy. Circulation 104(15):1767–1772
19. Khaidakov M, Wang X, Mehta JL (2011) Potential involvement of LOX-1 in functional consequences of endothelial senescence. PLoS One 6(6):e20964
20. Yang H, Bian Y, Zhang H, Gao F, Xiao C, Liang B et al (2015) LOX-1 is implicated in oxidized low-density lipoprotein-induced oxidative stress of macrophages in atherosclerosis. Mol Med Rep 12(4):5335–5341
21. Wang X, Ding Z, Lin J, Guo Z, Mehta JL (2015) LOX-1 in macrophage migration in response to ox-LDL and the involvement of calpains. Biochem Biophys Res Commun 467(1):135–139
22. Eto H, Miyata M, Kume N, Minami M, Itabe H, Orihara K et al (2006) Expression of lectin-like oxidized LDL receptor-1 in smooth muscle cells after vascular injury. Biochem Biophys Res Commun 341(2):591–598
23. Zhang Y, Chen B, Ming L, Qin H, Zheng L, Yue Z et al (2015) MicroRNA-141 inhibits vascular smooth muscle cell proliferation through targeting PAPP-A. Int J Clin Exp Pathol 8(11):14401
24. Goyal T, Mitra S, Khaidakov M, Wang X, Singla S, Ding Z et al (2012) Current concepts of the role of oxidized LDL receptors in atherosclerosis. Curr Atheroscler Rep 14(2):150–159
25. Kataoka H, Kume N, Miyamoto S, Minami M, Morimoto M, Hayashida K et al (2001 Jun) Oxidized LDL modulates Bax/Bcl-2 through the lectinlike Ox-LDL receptor-1 in vascular smooth muscle cells. Arterioscler Thromb Vasc Biol 21(6):955–960
26. Yang H, Mohamed ASS, Zhou S-H (2012) Oxidized low density lipoprotein, stem cells, and atherosclerosis. Lipids Health Dis 11(1):85
27. Yeung J, Li W, Holinstat M (2018) Platelet signaling and disease: targeted therapy for thrombosis and other related diseases. Pharmacol Rev 70(3):526–548
28. Chen M, Kakutani M, Naruko T, Ueda M, Narumiya S, Masaki T et al (2001) Activation-dependent surface expression of LOX-1 in human platelets. Biochem Biophys Res Commun 282(1):153–158

29. Marwali MR, Hu C, Mohandas B, Dandapat A, Deonikar P, Chen J et al (2007) Modulation of ADP-induced platelet activation by aspirin and pravastatin: role of lectin-like oxidized low-density lipoprotein receptor-1, nitric oxide, oxidative stress, and inside-out integrin signaling. J Pharmacol Exp Ther 322(3):1324–1332
30. Sakurai K, Cominacini L, Garbin U, Fratta Pasini A, Sasaki N, Takuwa Y et al (2004) Induction of endothelin-1 production in endothelial cells via co-operative action between CD40 and lectin-like oxidized LDL receptor (LOX-1). J Cardiovasc Pharmacol 44(Suppl 1):173
31. Yang S, Li Y, Du D (2013) Oxidized low-density lipoprotein-induced CD147 expression and its inhibition by high-density lipoprotein on platelets in vitro. Thromb Res 132(6):702–711
32. Yavuzer S, Yavuzer H, Cengiz M, Erman H, Altıparmak MR, Korkmazer B et al (2015) Endothelial damage in white coat hypertension: role of lectin-like oxidized low-density lipoprotein-1. J Hum Hypertens 29(2):92–98
33. Takanabe-Mori R, Ono K, Wada H, Takaya T, Ura S, Yamakage H et al (2013) Lectin-like oxidized low-density lipoprotein receptor-1 plays an important role in vascular inflammation in current smokers. J Atheroscler Thromb 20(6):585–590
34. Tan KCB, Shiu SWM, Wong Y, Leng L, Bucala R (2008) Soluble lectin-like oxidized low density lipoprotein receptor-1 in type 2 diabetes mellitus. J Lipid Res 49(7):1438–1444
35. Civelek S, Kutnu M, Uzun H, Erdenen F, Altunoglu E, Andican G et al (2015) Soluble lectin-like oxidized LDL receptor 1 as a possible mediator of endothelial dysfunction in patients with metabolic syndrome. J Clin Lab Anal 29(3):184–190
36. Hayashida K, Kume N, Murase T, Minami M, Nakagawa D, Inada T et al (2005) Serum soluble lectin-like oxidized low-density lipoprotein receptor-1 levels are elevated in acute coronary syndrome: a novel marker for early diagnosis. Circulation 112(6):812–818
37. Misaka T, Suzuki S, Sakamoto N, Yamaki T, Sugimoto K, Kunii H et al (2014) Significance of soluble lectin-like oxidized LDL receptor-1 levels in systemic and coronary circulation in acute coronary syndrome. Biomed Res Int 2014:649185
38. Balin M, Celik A, Kobat MA, Baydas A (2012) Circulating soluble lectin-like oxidized low-density lipoprotein receptor-1 levels predict percutaneous coronary intervention-related periprocedural myocardial infarction in stable patients undergoing elective native single-vessel PCI. J Thromb Thrombolysis 34(4):483–490
39. Liu J, Liu Y, Jia K, Huo Z, Huo Q, Liu Z et al (2018) Clinical analysis of lectin-like oxidized low-density lipoprotein receptor-1 in patients with in-stent restenosis after percutaneous coronary intervention. Medicine (Baltimore) 97(17):e0366
40. Ou H, Lee W, Lee I-, Chiu T, Tsai K, Lin C, et al. Ginkgo biloba extract attenuates oxLDL-induced oxidative functional damages in endothelial cells. J Appl Physiol 2009;106(5):1674-1685.
41. Kang B, Khan JA, Ryu S, Shekhar R, Seung K, Mehta JL (2010) Curcumin reduces angiotensin II-mediated cardiomyocyte growth via LOX-1 inhibition. J Cardiovasc Pharmacol 55(2):176–183
42. Mollace V, Ragusa S, Sacco I, Muscoli C, Sculco F, Visalli V et al (2008) The protective effect of bergamot oil extract on lecitine-like oxyLDL receptor-1 expression in balloon injury-related neointima formation. J Cardiovasc Pharmacol Ther 13(2):120–129
43. Zeya B, Arjuman A, Chandra NC (2016) Lectin-like oxidized low-density lipoprotein (LDL) receptor (LOX-1): a chameleon receptor for oxidized LDL. Biochemistry 55(32):4437–4444
44. Mehta JL, Chen J, Yu F, Li DY (2004) Aspirin inhibits ox-LDL-mediated LOX-1 expression and metalloproteinase-1 in human coronary endothelial cells. Cardiovasc Res 64(2):243–249
45. Li D, Chen H, Romeo F, Sawamura T, Saldeen T, Mehta JL (2002) Statins modulate oxidized low-density lipoprotein-mediated adhesion molecule expression in human coronary artery endothelial cells: role of LOX-1. J Pharmacol Exp Ther 302(2):601–605
46. Falconi M, Ciccone S, D'Arrigo P, Viani F, Sorge R, Novelli G et al (2013) Design of a novel LOX-1 receptor antagonist mimicking the natural substrate. Biochem Biophys Res Commun 438(2):340–345
47. Thakkar S, Wang X, Khaidakov M, Dai Y, Gokulan K, Mehta JL et al (2015) Structure-based design targeted at LOX-1, a receptor for oxidized low-density lipoprotein. Sci Rep 5:16740

48. Ding Z, Wang X, Khaidakov M, Liu S, Mehta JL (2012) MicroRNA hsa-let-7g targets lectin-like oxidized low-density lipoprotein receptor-1 expression and inhibits apoptosis in human smooth muscle cells. Exp Biol Med (Maywood) 237(9):1093–1100
49. Dai Y, Wu X, Dai D, Li J, Mehta JL (2018) MicroRNA-98 regulates foam cell formation and lipid accumulation through repression of LOX-1. Redox Biol 6(16):255–262
50. Amati F, Diano L, Vecchione L, Norata GD, Koyama Y, Cutuli L et al (2012) LOX-1 inhibition in ApoE KO mice using a schizophyllan-based antisense oligonucleotide therapy. Mol Ther Nucleic Acids 1:e58
51. Sugimoto K, Ishibashi T, Sawamura T, Inoue N, Kamioka M, Uekita H et al (2009) LOX-1-MT1-MMP axis is crucial for RhoA and Rac1 activation induced by oxidized low-density lipoprotein in endothelial cells. Cardiovasc Res 84(1):127–136
52. Pothineni NVK, Karathanasis SK, Ding Z, Arulandu A, Varughese KI, Mehta JL (2017) LOX-1 in atherosclerosis and myocardial ischemia: biology, genetics, and modulation. J Am Coll Cardiol 69(22):2759–2768

Role of Oxidative Stress and Cardiovascular Risk Factors in Ischemic Heart Disease

16

Monika Bartekova, Kristina Ferenczyova, Marek Jelemensky, and Naranjan S Dhalla

16.1 Introduction

Ischemic heart disease (IHD), the most common type of cardiovascular disease (CVD), is a major cause of mortality and disability worldwide. Pathophysiology of IHD is characterized by reduced blood flow in coronary arteries due to narrowing lumen, leading to an insufficient supply of oxygen and substrates to the affected area of the heart. The reduction of the arterial lumen in IHD is mostly caused by the formation of atherosclerotic plaque and development of thrombosis or spasm in the coronary arteries, which results in no-flow ischemia. In clinical situations, the blood supply to the ischemic heart is usually restored by reperfusion upon angioplasty, coronary by-pass surgery or thrombolytic therapy. In addition, the opening of collateral vessels may occur as an endogenous adaptive mechanism in response to ischemia and this may maintain the proper oxygen and nutrition delivery to the ischemic area of the heart. However, if the blood supply to the ischemic heart is not restored by reperfusion or opening of collaterals at the proper time, it may lead to irreversible changes (ischemia-reperfusion, I/R injury) at the subcellular level including malfunction of different intra- as well as extracellular proteins, lipid peroxidation associated with altered membrane permeability, abnormal gene expression, defects in subcellular organelles (such as mitochondria, sarcolemma, sarcoplasmic reticulum or myofibrils), and changes in cardiac performance. Thus,

M. Bartekova (✉) · K. Ferenczyova · M. Jelemensky
Institute for Heart Research, Centre of Experimental Medicine, Slovak Academy of Sciences, Bratislava, Slovak Republic
e-mail: monika.bartekova@savba.sk

N. S. Dhalla
Institute of Cardiovascular Sciences, St. Boniface Hospital Albrechtsen Research Centre, Department of Physiology and Pathophysiology, Max Rady College of Medicine, University of Manitoba, Winnipeg, MB, Canada
e-mail: NSDhalla@sbrc.ca

S. Chakraborti et al. (eds.), *Oxidative Stress in Heart Diseases*,
https://doi.org/10.1007/978-981-13-8273-4_16

reperfusion may even worsen the ischemic injury, accelerate the development of tissue necrosis and induce further deterioration of cardiac function [1, 2].

Oxidative stress is considered to be the main mechanism involved in the development of CVD including I/R injury to the heart [3]. During ischemia there occurs an imbalance between the production of pro-oxidant substances and the capacity of endogenous antioxidants leading to impaired redox homeostasis in cardiac cells [4]. Abundant formation of the reactive oxygen species (ROS), as a major factor associated with oxidative stress, leads to lipid peroxidation, oxidation of thiol groups and modification of phospholipids as well as proteins during ischemia [5]. Consequently, the oxidative stress associated with ischemic insult to the heart leads to altered membrane permeability in cardiac cells and dysfunction of various intra- as well as extracellular proteins.

Cardiovascular risk factors including atherosclerosis, hypertension and thrombosis are known to contribute to the occurrence of ischemic events in the heart, and oxidative stress is believed to play an important role in the genesis of these risk factors. Experimental evidence has indicated a causal role for oxidative stress in the development of hypertension. In fact increased formation of ROS has been shown to promote endothelial dysfunction leading to vascular damage as the major mechanism involved in the pathophysiology of hypertension [6–8]. It has been also documented that increased production of ROS in the vascular wall, primarily by ROS-producing enzymes such as NADPH oxidase, xanthine oxidase or uncoupled endothelial nitric oxide synthase (eNOS), plays a crucial role in the development of atherosclerosis [9–11]. Furthermore, thrombosis as a contributory factor to the genesis of atherosclerosis is also proposed to be associated with oxidative stress and altered redox state in platelets and/or vasculature [10, 12]. Thus, oxidative stress seems to be the major player in the development of cellular damage associated with I/R injury in the heart because it contributes to the genesis of cardiovascular risk factors for the occurrence of IHD as well as participates directly in pathological mechanisms leading to ischemia-induced damage to cardiac cells. The present chapter deals with the current knowledge on the role of oxidative stress and endogenous ROS production in the pathophysiology of cardiac I/R injury. In addition, it is intended to discuss the role of pro-oxidants and the involvement of some of major cardiovascular risk factors for the occurrence and severity of IHD.

16.2 Formation of ROS in the Cell

In general, ROS include oxyradicals such as the superoxide anion (O_2^-) and the hydroxyl radical (OH) as well as oxidants such as hydrogen peroxide (H_2O_2), and hypochlorous acid (HClO). In addition, reactive forms of nitrogen species (RNS), namely nitric oxide (NO^-), peroxynitrite ($ONOO^-$), S-nitrosothiols and dinitrozyl complexes, represent other type of free radicals acting in the body. Under physiological conditions, ROS are produced in the cell at low rate, and play an important role in signaling and defense mechanisms. The life-time of ROS is relatively short, and their rate of production is regulated by multiple antioxidant systems to minimize their damaging effects. ROS are, however, highly reactive and can interact with cell components including lipids, proteins and DNA leading to their oxidation and serious irreversible damages. ROS are produced in the cell by several mechanisms in both physiological and pathological conditions; these include ROS-producing enzymes such as mitochondrial cytochromes, xanthine oxidoreductase and nicotinamide adenine dinucleotide phosphate (NADPH) oxidase [13]. In addition, ROS may also be produced by uncoupled nitric oxide synthase (NOS). It is pointed out that NOS primarily catalyzes the production of nitric oxide (NO); however, when it becomes functionally uncoupled due to reduced levels of co-factor, tetrahydrobiopterin (BH_4), it switches to produce deleterious rather than protective NO, thus acting as a secondary source of ROS in the cell [14, 15]. Finally, ROS are produced in a sequential chain reaction where the generation of one free radical leads to the formation of other radicals [11]. Thus, in pathological conditions, the imbalance between pro-oxidants and anti-oxidant mechanisms results in oxidative stress that could lead the development of various diseases.

NADPH oxidase (NOX) is a membrane-bound complex located primarily in the plasma membrane. This enzyme catalyzes the production of a superoxide radical by transferring electron from NADPH to oxygen and is believed to be a primary source of superoxide anions and H_2O_2 in the vessel wall. In cardiomyocytes, ROS production can be stimulated by norepinephrine via α1-adrenergic receptors while in the vascular smooth muscle cells and endothelial cells, NOX-induced production of ROS can be enhanced by vasoconstrictors such as angiotensin II and endothelin-1 through angiotensin- and endothelin- receptors, respectively [11, 16]. Xanthine oxidoreductase (XOR) is also a major producer of ROS in the cell which appears in two forms, xanthine oxidase (XO) and xanthine dehydrogenase (XDH). Both forms catalyze the oxidation of hypoxanthine to xanthine as well as the oxidation of xanthine to uric acid; however, while XO reduces only oxygen, XDH can reduce either oxygen or NAD^+. XOR generates superoxide via NADH oxidase activity and can also produce NO via nitrate and nitrite reductase activities. It should be noted that XOR has been shown to be involved in the development of CVD including cardiac I/R injury [17]. Other major producers of ROS within the cell are mitochondria where ROS are generated as by-products of the oxidative phosphorylation during ATP production via an electron transport from NADH and $FADH_2$ to oxygen. Such ROS may potentially contribute to the oxidative damage of the cell during I/R, particularly, when the mitochondrial aerobic respiration continues in the absence of oxygen and results in the occurrence of intracellular Ca^{2+}-overload and activation of apoptotic cascades [18, 19].

16.3 Role of Oxidative Stress in Hypertension

Hypertension is a chronic disease, which is characterized by elevation of blood pressure, and is one of the main cardiovascular risk factors which significantly contributes to the development of IHD. The pathogenesis of hypertension is multifactorial in nature as it includes both genetic and non-genetic factors. More than 90% of hypertensive subjects belong to the essential hypertension category, which is known as primary or idiopathic hypertension of unknown origin. Some patients have secondary hypertension, which is most likely a result of the persisting disease with well-known mechanisms, such as narrowing of the arteries, as well as chronic kidney or endocrine disorders. In addition to the environmental factors, other features involved in the genesis of essential hypertension include: increased sympathetic nervous system activity due to psychosocial stress, enhanced production of vasoconstrictors, deficiency of vasodilators such as NO, high sodium intake, inadequate calcium and potassium intake, increased renin secretion leading to increased formation of angiotensin II, increased vessels resistance, metabolic disorders, insulin resistance and obesity, increased activity of vascular growth factors, and altered cellular ion transport [20]. It has been documented that persistent hypertension is a major risk factor for different kinds of CVD including coronary artery as well as peripheral artery disease, stroke, aortic aneurysm or chronic kidney disease [21].

There is increasing evidence that oxidative stress associated with overproduction of ROS is one of the pathological mechanisms for the development of hypertension [6–8, 22]. This disease is tightly connected with altered function of the vasculature, which in turn leads to increased peripheral resistance for determining blood pressure. In vascular smooth muscle and endothelial cells, the primary source of ROS is NADPH oxidase (NOX), which is considered to be involved in the genesis of hypertension. NOX-induced production of ROS in the vascular system may be induced by vasoconstrictors such as angiotensin II (Ang II), endothelin-1 and norepinephrine as well as by aldosterone [11]. It has been reported that the activation of AT1 receptors by Ang II may result in the induction of ROS production via membrane NOX activation in rats [23]. Regarding the role of endothelin-1 in the oxidative stress-mediated hypertension, it was shown that incubation of rat aortic rings with endothelin-1 leads to enhanced superoxide production and vasoconstriction through activation of NOX and uncoupled NOS [24]. Several isoforms of NOX including NOX1, NOX2, as well as NOX4 have been reported to be involved in the ROS production and the development of hypertension. Smooth muscle-specific NOX1 overexpression has been shown to potentiate the Ang II-induced hypertension and vascular smooth muscle hypertrophy in transgenic mice [25]. The role of NOX1 in Ang II-induced hypertension was confirmed in a study in NOX1-deficient mice by reducing the bioavailability of NO [26].

Numerous studies have reported that NOX2 isoform is also involved in Ang II-induced hypertension. In transgenic mice with endothelial-specific overexpression of NOX2, it was demonstrated that Ang II causes a greater increase in ROS production in transgenic compared to wild-type aortic smooth muscle and attenuates the acetylcholine-induced vasorelaxation. Administration of Ang II was discovered to increase blood pressure more in the transgenic compared to the wild-type mice. These results indicated that NOX2 mediated ROS production contributes to Ang II-induced endothelial dysfunction, vascular remodeling and hypertension [27]. It has also been suggested that inhibition of NOX2 abolished increased production of mitochondrial and cytoplasmic superoxide which was stimulated by activation of mitochondrial ATP-sensitive K^+ channels (mitoKATP). In addition, inhibition of reverse electron transport from complex II to complex I of respiratory chain in the mitochondrial matrix which causes increased production of superoxide reduced blood pressure significantly. These results suggest that mitoKATP-mediated mitochondrial superoxide production stimulates cytoplasmic NOX2 and through this mechanism endothelial oxidative stress contributes to the development hypertension [28]. It is pointed out that NOX4 isoform has been also identified to be involved in Ang II-induced ROS formation, most likely via mediating eNOS dysfunction; this results in decreased bioavailability of NO, which in turn contributes to vascular endothelial dysfunction and hypertension [29]. On the other hand, more recently it has been shown that silencing of NOX2, but not NOX1, NOX4, or NOX5, inhibits Ang II-induced superoxide production in both mitochondria and cytoplasm in cultured human aortic endothelial cells. Moreover, depletion of NOX2 in a mouse model inhibited Ang II-induced superoxide production and attenuated Ang II-induced hypertension suggesting NOX2 to be the dominant NOX isoform responsible for ROS production in the development of hypertension [30]. Significant role of the NOX-derived ROS in pathogenesis of hypertension has been discussed in several recent comprehensive reviews [14, 31–33].

Another ROS-producing enzyme suggested to be tightly involved in the development of hypertension is xanthine oxidase (XO). It has been shown that Ang II increases XO protein levels and the XO-dependent superoxide production was prevented by NOX inhibition in cultured endothelial cells. Also, the endothelium-bound XO activity was reduced by losartan and allopurinol in patients with coronary disease. The inhibition of XO with oxypurinol improved endothelium-dependent vasodilation before, but not after losartan or allopurinol therapy in these patients suggesting a contributory role of XO-produced superoxide in endothelial dysfunction [34]. It has been also shown that XO levels are significantly enhanced in the spontaneously hypertensive rats (SHR). In addition, adrenalectomy led to a reduction of XO to normotensive levels and reduced the blood pressure and free radical production in SHR to normotensive levels indicating the involvement of adrenal pathway in this XO-mediated mechanism in hypertension [35]. On the other hand, different studies focused on the role of renal XO activity in hypertension suggested that XO may play a role in end-organ damage in hypertension, but not in the development of hypertension [36, 37].

In addition to NOX and XO, some other ROS producers have been suggested to be involved in the oxidative stress-mediated development of hypertension; these include mitochondrial respiratory chain enzyme complexes. If any damage occurs to the mitochondrial respiratory chain, the mitochondrial respiration become dysfunctional, and the transfer of electrons to O_2 increases the mitochondrial ROS formation subsequently [11, 38]. Reduced capacity of the mitochondrial respiratory chain has been observed in the brain stem of the spontaneously hypertensive rats (SHR) as well as in the Ang II-induced neurogenic hypertension [39]. It has been documented that mitochondrial dysfunction in hypertension results in defective calcium homeostasis and impaired energy production [40, 41]. Hypertension is also associated with structural abnormalities of mitochondria such as decreased mitochondrial mass and density which may result in impaired energy production and accelerated ROS formation due to instability of electron transport chain complexes [41, 42]. Hypertension affects mitochondrial dynamics including decreased mRNA expression of the fusion proteins, mitofusin-1 and -2, as well as increased mitochondrial fragmentation. Stimulation of oxidative stress has been documented in Dahl salt-sensitive rats fed with a high-salt diet, which is considered as a hypertensive model of heart failure [43]. Finally, it has been shown that overexpression of myocardial adenine nucleotide translocase 1 and consequent accelerated mitochondrial ADP/ATP transport attenuates hypertension-induced heart disease, suggesting the improvement in mitochondrial function as a basic principle for new strategies in the treatment of heart disease [44].

ROS producers including those derived from uncoupled NOS have been proposed to be implicated in the development of hypertension. For instance, it has been shown that increased superoxide release by uncoupled eNOS contributes to impaired pulmonary vasodilation in persistent pulmonary hypertension in newborns [45]. In a mouse model of hypertension, it has been documented that hypertensive hearts with diastolic dysfunction, but without systolic dysfunction or cardiac hypertrophy, showed increased oxidized biopterins, NOS-dependent superoxide production, reduced NO formation, and dephosphorylated phospholamban. Moreover, feeding hypertensive mice with tetrahydrobiopterin (BH_4) was observed to improve cardiac BH_4 stores, phosphorylated phospholamban levels, and diastolic dysfunction, suggesting that uncoupled cardiac NOS mediates diastolic dysfunction in hypertension-induced heart failure with preserved ejection fraction [46]. It has been also shown that supplementation with BH_4, a cofactor determining for NOS function, augments endothelium-dependent vasodilation in both normotensive and hypertensive subjects; these findings provide an indirect evidence of uncoupled NOS in the development of hypertension [47]. It appears that there is a crosstalk between the major ROS producing systems such as NADPH oxidase, NOX, XO, mitochondrial respiratory enzymes and eNOS, and that these respond to different ligands including Ang II for ROS overproduction. Thus oxidative stress associated with the development of hypertension is most likely the result of ROS produced by several sources. The main features of the crosstalk of different ROS producers have been reviewed previously [11, 33, 48, 49]. Proposed triggers for the generation of oxidative stress leading to the development of hypertension are shown in Fig. 16.1.

16.4 Role of Oxidative Stress in Atherosclerosis

Atherosclerosis, a chronic inflammatory disease, is characterized by accumulation of lipids and inflammatory cells in the walls of large and medium-sized arteries including coronary arteries. The atherosclerotic plaque is made up of fat, cholesterol, calcium, and other substances found in the blood. The plaque slowly narrows the arterial lumen what in turn limits the flow of oxygen-rich blood to different parts of the body. In fact atherosclerosis is a leading cause of CVD resulting in high rate of mortality in the population [50, 51]. The exact cause for the genesis of atherosclerosis is not known; however, it is proposed that several lifestyle-related risk factors

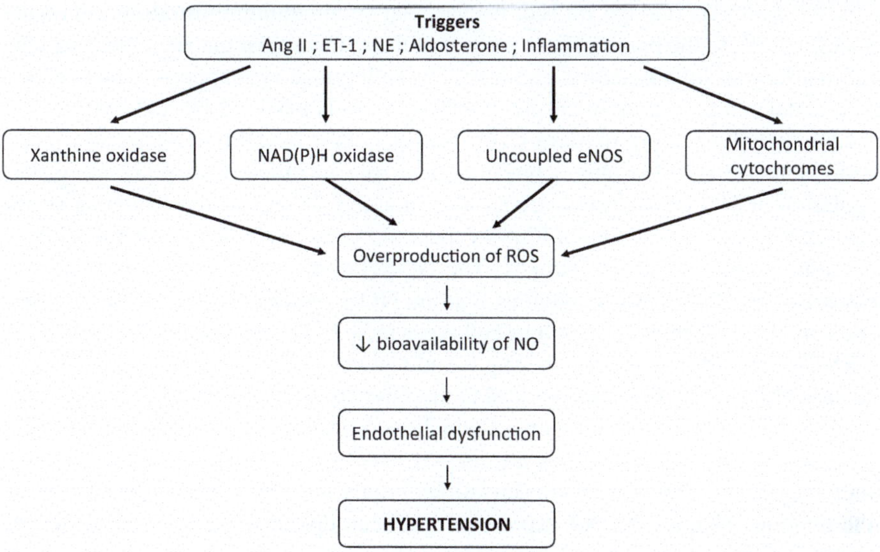

Fig. 16.1 Proposed role of oxidative stress in inducing alterations in vasculature leading to development of hypertension. *Ang II* Angiotensin II, *ET-1* endothelin 1, *NE* norepinephrine, *ROS* reactive oxygen species, *eNOS* endothelial NO synthase, *NAD(P)H oxidase* nicotinamide adenine dinucleotide phosphate oxidase

including metabolic syndrome, diabetes, smoking, obesity and unhealthy diet may contribute to plaque formation. In addition to these factors, oxidative stress induced by generation of excessive amount of ROS has been shown to be a critical and final common mechanism in the development of atherosclerosis. Uncontrolled production of ROS has been implicated in creating vascular injury, which has a devastating effect on vascular homeostasis. Thus, together with high levels of lipoprotein particles (LDL and VLDL), hyperglycemia, insulin resistance and inflammation, oxidative stress is one of the main causes of atherogenesis in the coronary artery system, which leads to ischemic injury to the heart.

For the genesis of atherosclerosis, abundant ROS formation is due to hyperglycemia which is a consequence of obesity or metabolic syndrome, leading to increased oxidative stress in vascular endothelial cells. Thus, high levels of ROS contribute to endothelial dysfunction due to increased intracellular glucose concentration and in fact disturbances resulting from oxidative stress lead to less availability of nitric oxide (NO), DNA damage and oxidation of phospholipids forming membranes in endothelial cells. Another mechanism by which hyperglycemia stimulates oxidative stress is spontaneous oxidation of glucose to produce reactive intermediates (glyoxal and methylglyoxal) while allowing the formation of ROS [52]. Also, the free fatty acids, the concentration of which is elevated in the metabolic syndrome are degraded by β-oxidation to produce acetylcoenzyme A that enters the Krebs cycle. Free radicals can also react with LDL particles to form oxidized lipoprotein particles (oxLDL) that are no longer able to stimulate LDL receptors in liver cells, but penetrate into the walls of the vascular endothelium. Infiltration and retention of oxLDL to the proteoglycan present in the extracellular matrix in the vascular wall triggers an inflammatory response. The inflammatory response is induced by endothelial cell activation releasing of triacyglycerols from oxLDL [53, 54].

Endothelial xanthine oxidase (XO) together with NADPH oxidase and nitric oxide synthase (eNOS) play a physiological role in inflammatory signaling, the regulation of NO production and vascular function. However, the oxidative stress generated by overproduction of ROS by these enzymes may induce endothelial dysfunction, leading to atherosclerosis. The NOX family proteins are unique in the production of ROS and it is obvious that the NOX protein family is important for redox-mediated signaling in various cell types [55]. In vascular smooth muscle cells (VSMCs), growth factors such as Ang II and platelet-derived growth factor (PDGF) increase H_2O_2 via NOX1, thereby lead to their hypertrophy, migration and proliferation [56, 57]. In addition, monocytes which were infiltrated into the vascular wall can produce O_2^- via NOX2. They are also involved in the oxidation of LDL, thereby enhancing further infiltration and activation of macrophages and ROS production in atheromatous plaques. NOX4 expression has been documented to be increased in intimal lesions of coronary arteries for the occurrence of atherosclerosis in humans [58]. Furthermore, a variety of oxidized lipids stimulate NOX4 expression in macrophages [59]. Taken together, vascular NOX proteins are intimately involved in the development and progression of atherosclerosis.

Xanthine oxidase (XO) generates superoxide and hydrogen peroxide by using O_2 as an electron acceptor [60]. The expression and activity of endothelial XO are enhanced by pro-atherosclerotic stimuli such as Ang II [34] and periodically repeated shear stress [61], suggesting a contribution of XO-derived superoxide to atherosclerosis. The activity of both types of endothelial XO [62] and plasma XO [63] is increased in experimental atherosclerosis, as well as in human atherosclerotic plaque [64, 65]. XO inhibitors, such as allopurinol, tungsten [66] and febuxostat [67] have been shown to reduce the development of atherosclerosis in ApoE-KO mice. Normally, mitochondrial oxidative phosphorylation generates physiological levels of superoxide, which is converted to hydrogen peroxide by the manganese-dependent superoxide dismutase (MnSOD/SOD2) and subsequently by glutathione peroxidase 1 (GPX1) to H_2O [68]. Under pathological conditions, mitochondrial oxidative stress can occur because of excessive ROS production or insufficient ROS removal and it has been shown that atherosclerosis in human is associated with mitochondrial oxidative stress [69]. The importance of mitochondrial redox balance is supported by cardiac deletion of mitochondrial SOD2, which causes perinatal lethality in cardiac myopathy and congestive heart failure [70]. It has been also reported that heterozygous SOD2+/− knockout mice on ApoE-KO background show increased ROS levels in the mitochondria and enhanced atherogenesis at arterial branches [71].

Under physiological conditions, eNOS produces NO, which represents a key vasoprotective factor of the endothelium [72, 73]. However, under pathological conditions associated with oxidative stress, dysfunction of eNOS has been shown to reduce the formation of NO. Oxidative stress contributes to endothelial dysfunction

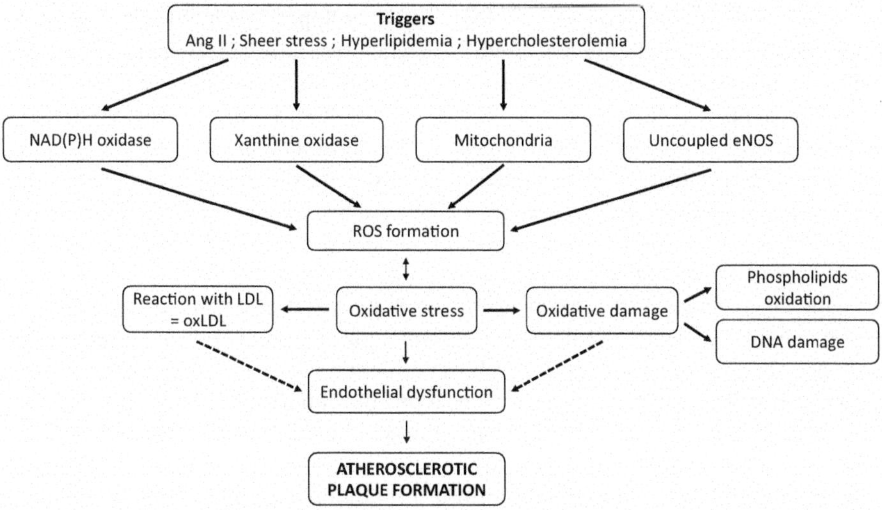

Fig. 16.2 Proposed role of oxidative stress in inducing alterations in vasculature leading to atherosclerotic plaque formation. *Ang II* Angiotensin II, *ROS* reactive oxygen species, *eNOS* endothelial NO synthase, *NAD(P)H oxidase* nicotinamide adenine dinucleotide phosphate oxidase, *oxLDL* oxidized low density lipoproteins

primarily due to the rapid inactivation of NO by excess superoxide production; oxidative stress trigger uncoupling of eNOS such that it produces superoxide at the expense of NO [74]. Mechanistically, deficiency of eNOS cofactor tetrahydrobiopterin (BH_4), eNOS substrate l-arginine, or eNOS S-glutathionylation are likely to be the major causes for eNOS uncoupling [75]. Peroxynitrite and superoxide can oxidize BH_4 leading to BH_4 deficiency and in fact enhanced ROS production from uncoupled eNOS has been shown in mouse [76] and human models of atherosclerosis [77]. Thus, oxidative stress evoked by excessive ROS production by all major ROS-producing enzymes including NOX isoforms, XO, mitochondrial oxidative phosphorylation respiratory chain as well as uncoupled eNOS, widely contributes to the atherosclerotic plaque formation. As in the case of hypertension, targeting oxidative stress with inhibitors of ROS-producing enzymes and antioxidant therapies may help to attenuate the pathological processes associated with atherogenesis, which in turn may prevent the occurrence of different CVD including IHD. Proposed triggers for the generation of oxidative stress leading to the development of atherosclerosis are given in Fig. 16.2.

16.5 Role of Oxidative Stress in Thrombosis

Thrombosis is the formation of a blood clot inside a vessel, leading to reduced blood flow for the circulation. When a blood vessel (either vein or artery) is injured, the body forms a blood clot to prevent blood loss via activation of platelets and fibrin.

However, under certain conditions blood clot may appear even when the vessel is not injured. Thus in pathological conditions when regulatory mechanisms of hemostasis are imbalanced, excessive quantities of thrombin may initiate thrombosis. Moreover, a piece of the clot can break free and begin to circulate as an embolus and lodge somewhere else as an embolism, called thromboembolism. While arterial thrombosis is a critical event in the development of arterial diseases associated with myocardial infarction and stroke, venous thrombosis leads to congestion of the affected part of the body [78]. It is noted that pathogenesis of thrombosis includes some kind of inflammatory process due to trauma, surgery or infection, which cause endothelial damage in the vessel wall. The main mechanism for the initiation of blood coagulation is via the activated tissue factor [78]; however, inflammation as well as other stimuli such as hypercholesterolemia can also lead to altered gene expression in the endothelium leading to a pro-thrombotic state [79]. In this situation, endothelial cells downregulate anti-thrombotic substances such as thrombomodulin, a key modulator of thrombin activity, which in turn may result in sustained activation of thrombin and subsequent pro-thrombotic state [80]. Thus, endothelial injury is almost invariably involved in the thrombus formation in arteries. As high rates of blood flow hinder clot formation, arterial and cardiac clots are rich in platelets, which are required for clot formation in areas under high stress due to blood flow [79].

In addition to inflammation and trauma, oxidative stress has been implicated in the genesis of thrombosis. It should be emphasized that dramatic changes in redox status occur during normal platelet stimulation because platelet aggregation is associated with a burst of oxygen consumption; however, conditions that provoke oxidative stress may also be prothrombotic [12]. ROS derived from both platelets and other vascular sources have been shown to alter platelet responses. Superoxide, which is produced by platelets, is known to augment platelet aggregation responses [12, 81]. Platelet-derived ROS may come from several sources such as NADPH oxidase (NOX), cyclooxygenases, uncoupled eNOS, xanthine oxidase (XO) and mitochondrial respiration, among which NOX is considered to play the crucial role [82–84]. Growing number of studies provide evidence that NOX is implicated in altered platelet activation via superoxide production, and that platelet-associated NOX mediates a thrombogenic phenotype [84, 85]. Platelet NOX is a protein complex consisting of cellular subunits $p47^{phox}$, $p67^{phox}$, and the membrane-bound proteins, $p22^{phox}$ and $gp91^{phox}$, that together with Rac1/2 (a small GTPase) form the active enzyme complex [86]. Recently, it has been documented that expression of $gp91^{phox}$ NOX subunit is increased in pulmonary arteries in a mouse model of pulmonary hypertension, and that NOX-derived superoxide formation in venous thrombosis and endothelial dysfunction play an important role in pulmonary hypertension [87]. It has also been shown that NOX-dependent superoxide release in platelets may be induced by collagen activation and consequently enhance the thrombus formation via increasing the availability of released ADP, which is essential for stable thrombus formation [88]. A key role of NOX-derived ROS in the oxidative stress-mediated platelet activation and thrombosis has been reviewed recently [84].

NOX2 is considered a major isoform of the enzyme contributing to the thrombosis and is expressed in platelets and the megakaryocyte cell-line [84]. NOX2-generated superoxide anion is rapidly converted into the longer lasting H_2O_2, which is the major contributor for the oxidation of lipids and proteins [84]. Interestingly, patients with hereditary deficiency of NOX2 show almost no ROS production by platelets [89]. NOX2 in platelets is activated by several pathways including CD40 ligand signaling, collagen/glycoprotein VI activation or oxidized –LDL/CD36 pathway [84]. Consequently, the downstream signaling pathway of platelet NOX2 activation, in addition to enhanced production of ROS, includes the increase in thromboxane A2 levels, expression of P-selectin and release of intracellular Ca^{2+}. It is pointed out that intracellular Ca^{2+} release modulates granular secretion as an early stage of cell activation, platelet whereas NOX2-derived ROS mediate the oxidation of sulfhydryl groups in glycoprotein Ibα and enhance its ligand-binding function with the von Willebrand factor on endothelial cells to promote platelet aggregation [84]. Finally, platelet NOX2 contributes to the formation of 8-iso-PGF2α, a reliable marker of oxidative stress. Elevated levels of this stable isoprostane have been shown to be associated with platelet activation [90]. In view of the potential involvement of NOX-derived ROS in the thrombogenesis, both NOX inhibitors and superoxide scavengers have been reported to reduce platelet aggregation and thrombus formation [91, 92].

In addition to NOX-derived ROS, other sources such as mitochondrial oxidation may contribute to oxidative stress in platelets. It has been documented that treatment of platelets with thrombin stimulates depolarization of mitochondria and potentiates the generation of H_2O_2; the thrombin-induced apoptosis may also be mediated by the generation of H_2O_2 in platelets [93]. On the other hand, it has been shown that mitochondria-targeted ROS scavenger, Mito-TEMPO, can ameliorate the hyperthermia-impaired platelet aggregation as well as hyperthermia-triggered platelet apoptosis. These data indicate the contributory role of mitochondrial ROS in platelet apoptosis which may be involved in the hyperthermia-induced thrombocytopenia seen during in the combined therapy of various solid tumors. Thus, therapeutic approaches targeting mitochondrial ROS would have potential clinical utility in oxidative damage-mediated platelet-associated disorders [94]. However, the majority of experimental data suggest that NOX seems to be the main ROS producer in the development of thrombosis rather than mitochondrial cytochromes. Finally, it should be mentioned that platelets are equipped with an effective enzymatic antioxidant system consisting of superoxide dismutase (SOD), glutathione peroxidase (GPX), catalase (CAT), glutathione transferase (GST) and glutathione reductase (GSSG-R), which maintain the redox balance in the cell. However, under pathological conditions an imbalance between the formation of ROS and efficiency of endogenous antioxidants may occur, and this may consequently contribute to the pathogenesis of thrombotic disease via enhanced platelet activation mediated by elevated intracellular ROS levels [84]. Proposed triggers for the generation of oxidative stress leading to the development of the thrombosis are shown in Fig. 16.3.

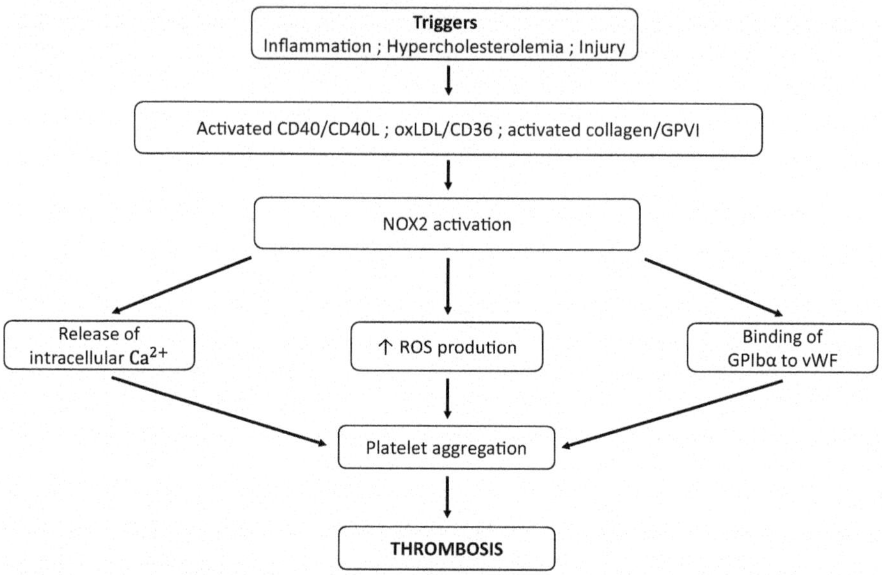

Fig. 16.3 Proposed role of NADPH oxidase 2 in oxidative stress-induced changes in platelets leading to thrombus formation. *ROS* reactive oxygen species, *NOX2* nicotinamide adenine dinucleotide phosphate oxidase 2, *CD40(L)* cluster of differentiation 40(ligand), *CD36* cluster of differentiation 36, *oxLDL* oxidized low density lipoproteins, *GPVI* glycoprotein VI, *GPIba* glycoprotein Ibα, *vWF* von Willebrand factor

16.6 Role of Oxidative Stress in Subcellular Defects in Cardiac I/R Injury

In a healthy heart, the level of constantly generated ROS, mainly in mitochondria, is strictly controlled and maintained low by superoxide dismutase (SOD) and catalase (CAT). During I/R there occurs an imbalance between the production of pro-oxidant substances and the capacity of endogenous antioxidant leading to impaired redox homeostasis in cardiac cells [4]. Maintaining the balance between pro-oxidant and anti-oxidant levels is at risk from the onset of ischemia, but during reperfusion period the central role is played by a mitochondrial disorder [95]. The mitochondrial respiratory chain is the largest source of ROS and RNS, and therefore it is not surprising that during the first few seconds after the onset of reperfusion, a massive production of O_2^-, NO^- and $ONOO^-$ occurs due to the disruption of the mitochondrial membrane potential or the imbalance in the distribution of Ca^{2+} cations [96, 97]. Generation of O_2^- in the mitochondria is primarily due to the electron leakage from complexes I and III of the electron transport chain [98]. This leads to damage in the mitochondrial membrane phospholipid, cardiolipin, and consequent decrease in ATP production [99]. It has been documented that worsened heart function due to I/R is associated with decreased mitochondrial respiration and oxidative phosphorylation, and notably, the observed alterations of cardiac performance and mitochondrial function could be prevented by pretreatment with oxyradical scavenging agents or antioxidants [100]. In addition to mitochondrial damage and lipid peroxidation, oxidative stress during I/R leads to structural and functional alterations of important cellular proteins such as membrane channels proteins, calcium handling proteins or sarcomeric proteins. One of the proposed mechanisms leading to protein alterations and degradation due to oxidative stress in I/R is the activation of endogenous proteinases such as matrix metalloproteinases (MMPs), cathepsins or calpains [101]. Another mechanism of oxidative stress-induced impairment of functionality of different proteins is depressing their phosphorylation, especially in the case of Ca^{2+} handling proteins such as Ca^{2+} pump ATPase (SERCA) [102] and sarcomeric proteins essential for cardiac contraction including troponins T and I, tropomyosin or actin [103]. Finally, oxidative stress during I/R can result in the oxidation or nitrosylation of contractile proteins leading to changes in their structural conformation and/or functionality [103]. Taken together, various ROS-induced post-translational modifications of proteins including oxidation, phosphorylation, and protein cleavage significantly contribute to I/R-induced cardiac dysfunction. The complex view on the role of oxidative stress in the development of I/R injury, could be found in recent reviews where this topic has been discussed comprehensively [2, 41, 97, 104].

In view of the crucial role of oxidative stress in the development of I/R injury to the heart, increasing the antioxidant capacity of the heart as well as administration of exogenous antioxidants have been suggested as therapeutic strategies for preventing the negative consequences of I/R injury. It has been found that overexpression of SOD or CAT protects the heart against I/R injury [105, 106]. Also, increased activity of SOD due to different treatments or preconditioning has been reported to be associated with cardioprotection against I/R [107–112]. In addition, treatment with various exogenous antioxidants such as polyphenols, plant mixtures, antioxidant vitamins or synthetic antioxidants have been observed to evoke cardioprotective effects in I/R injury [107, 113–116]. However, despite numerous experimental studies documenting the positive effects of antioxidant treatments in the prevention of I/R injury in animal models, the effectiveness of antioxidant supplementation in patients with ischemic heart disease is still controversial [117]. Taking together, oxidative stress mediated mitochondrial dysfunction and the following sequence of biochemical events including lipid peroxidation and impaired ATP production, as well as functional alterations of different cellular proteins due to increased ROS production, are the cardinal features of the myocardial I/R injury. Thus, searching for the effective antioxidant therapies for preventing oxidative stress-induced damage due to cardiac I/R injury is still challenging.

16.7 Conclusions

It has become evident that oxidative stress is one of the major players in the genesis of risk factors such as hypertension, atherosclerosis and thrombosis for the occurrence of IHD. Enhanced ROS production by NADPH oxidases, xanthine oxidoreductase or uncoupled NO synthase, as well as ROS produced by mitochondria in the vasculature, mainly in vascular endothelial cells and vascular smooth muscle cells, and in the platelets, has been shown to be associated with these risk factors for IHD. Moreover, oxidative stress plays a significant role in the development of subcellular defects due to I/R injury in the heart. Thus, enhanced production of ROS, leading to an imbalance between their production and degradation by endogenous antioxidant systems, represents a promising target for the prevention of risk factors of IHD as well as the negative consequences of I/R injury in the heart tissue. Hence, interventions leading to decreased activity of ROS producing enzymes or enhanced activity of endogenous antioxidants may lead to decreased occurrence of risk factors of IHD and prevent the heart tissue from I/R injury. Moreover, decreasing oxidative stress may prevent the development of subcellular defects in the myocardium and improve cardiac performance in the ischemia–reperfused hearts. Treatment with exogenous antioxidants might be a promising strategy to reverse the redox imbalance as well as prevent the genesis of IHD and development of I/R injury to the heart.

Acknowledgements The infrastructure support for this project was provided by the St. Boniface Hospital Research Foundation, Winnipeg, Canada.

References

1. Dhalla NS, Saini HK, Tappia PS et al (2007) Potential role and mechanisms of subcellular remodeling in cardiac dysfunction due to ischemic heart disease. J Cardiovasc Med 8(4):238–250
2. Bartekova M, Barancik M, Dhalla NS (2016) Role of oxidative stress in subcellular defects in ischemic heart disease. In: Gelpi RJ, Boveris A, Paderosa JJ (eds) Biochemistry of oxidative stress: physiopathology and clinical aspects, Advances in biochemistry in health and disease, vol 16. Springer, Cham, pp 129–146
3. Dhalla NS, Temsah RM, Netticadan T (2000) Role of oxidative stress in cardiovascular diseases. J Hypertens 18(6):655–673
4. Dhalla NS, Elmoselhi AB, Hata T, Makino N (2000) Status of myocardial antioxidants in ischemia-reperfusion injury. Cardiovasc Res 47(3):446–456
5. Ceconi C, Cargnoni A, Pasini E et al (1991) Evaluation of phospholipid peroxidation as malondialdehyde during myocardial ischemia and reperfusion injury. Am J Phys 260:H1057–H1061
6. Montezano AC, Dulak-Lis M, Tsiropoulou S, Harvey A, Briones AM, Touyz RM (2015) Oxidative stress and human hypertension: vascular mechanisms, biomarkers, and novel therapies. Can J Cardiol 31(5):631–641. https://doi.org/10.1016/j.cjca.2015.02.008
7. Dinh QN, Drummond GR, Sobey CG, Chrissobolis S (2014) Roles of inflammation, oxidative stress, and vascular dysfunction in hypertension. Biomed Res Int 2014:406960. https://doi.org/10.1155/2014/406960
8. Sinha N, Dabla PK (2015) Oxidative stress and antioxidants in hypertension – a current review. Curr Hypertens Rev 11(2):132–142
9. Li H, Horke S, Förstermann U (2014) Vascular oxidative stress, nitric oxide and atherosclerosis. Atherosclerosis 237(1):208–219. https://doi.org/10.1016/j.atherosclerosis.2014.09.001
10. Yang X, Li Y, Li Y et al (2017) Oxidative stress-mediated atherosclerosis: mechanisms and therapies. Front Physiol 8:600. https://doi.org/10.3389/fphys.2017.00600. eCollection 2017

11. Majzunova M, Dovinova I, Barancik M, Chan JY (2013) Redox signaling in pathophysiology of hypertension. Biomed Sci 20:69
12. Freedman JE (2008) Oxidative stress and platelets. Arterioscler Thromb Vasc Biol 28(3):s11–s16. https://doi.org/10.1161/ATVBAHA.107.159178
13. Müller BA, Dhalla NS (2010) Mechanisms of the beneficial actions of ischemic preconditioning on subcellular remodeling in ischemic-reperfused heart. Curr Cardiol Rev 6(4):255–264
14. Montezano AC, Touyz RM (2012) Reactive oxygen species and endothelial function: role of nitric oxide synthase uncoupling and NOX family nicotinamide adenine dinucleotide phosphate oxidases. Basic Clin Pharmacol Toxicol 110(1):87–94
15. Zweier JL, Chen CA, Druhan LJ (2011) S-glutathionylation reshapes our understanding of endothelial nitric oxide synthase uncoupling and nitric oxide/reactive oxygen species-mediated signaling. Antioxid Redox Signal 14(10):1769–1775. https://doi.org/10.1089/ars.2011.3904
16. Balakumar P, Jagadeesh G (2010) Multifarious molecular signaling cascades of cardiac hypertrophy: can the muddy waters be cleared? Pharmacol Res 62(5):365–383
17. Berry CE, Hare JM (2004) Xanthine oxidoreductase and cardiovascular disease: molecular mechanisms and pathophysiological implications. J Physiol 555(Pt 3):589–606
18. Chen Q, Camara AK, Stowe DF et al (2007) Modulation of electron transport protects cardiac mitochondria and decreases myocardial injury during ischemia and reperfusion. Am J Physiol Cell Physiol 292(1):C137–C147
19. McLeod CJ, Aziz A, Hoyt RF Jr et al (2005) Uncoupling proteins 2 and 3 function in concert to augment tolerance to cardiac ischemia. J Biol Chem 280(39):33470–33476
20. Oparil S, Zaman MA, Calhoun DA (2003) Pathogenesis of hypertension. Ann Intern Med 139(9):761–776
21. Pierdomenico SD, Di Nicola M, Esposito AL et al (2009) Prognostic value of different indices of blood pressure variability in hypertensive patients. Am J Hypertens 22(8):842–847. https://doi.org/10.1038/ajh.2009.103
22. Brito R, Castillo G, González J, Valls N, Rodrigo R (2015) Oxidative stress in hypertension: mechanisms and therapeutic opportunities. Exp Clin Endocrinol Diabetes 123(6):325–335. https://doi.org/10.1055/s-0035-1548765
23. Rajagopalan S, Kurz S, Münzel T et al (1996) Angiotensin II-mediated hypertension in the rat increases vascular superoxide production via membrane NADH/NADPH oxidase activation. Contribution to alterations of vasomotor tone. J Clin Invest 97(8):1916–1923
24. Loomis ED, Sullivan JC, Osmond DA, Pollock DM, Pollock JS (2005) Endothelin mediates superoxide production and vasoconstriction through activation of NADPH oxidase and uncoupled nitric-oxide synthase in the rat aorta. J Pharmacol Exp Ther 315(3):1058–1064
25. Dikalova A, Clempus R, Lassègue B et al (2005) NOX1 overexpression potentiates angiotensin II-induced hypertension and vascular smooth muscle hypertrophy in transgenic mice. Circulation 112(17):2668–2676
26. Matsuno K, Yamada H, Iwata K et al (2005) NOX1 is involved in angiotensin II-mediated hypertension: a study in NOX1-deficient mice. Circulation 112(17):2677–2685
27. Murdoch CE, Alom-Ruiz SP, Wang M et al (2011) Role of endothelial NOX2 NADPH oxidase in angiotensin II-induced hypertension and vasomotor dysfunction. Basic Res Cardiol 106(4):527–538. https://doi.org/10.1007/s00395-011-0179-7
28. Nazarewicz RR, Dikalova AE, Bikineyeva A, Dikalov SI (2013) NOX2 as a potential target of mitochondrial superoxide and its role in endothelial oxidative stress. Am J Physiol Heart Circ Physiol 305(8):H1131–H1140. https://doi.org/10.1152/ajpheart.00063
29. Lee DY, Wauquier F, Eid AA et al (2013) NOX4 NADPH oxidase mediates peroxynitrite-dependent uncoupling of endothelial nitric-oxide synthase and fibronectin expression in response to angiotensin II: role of mitochondrial reactive oxygen species. J Biol Chem 288(40):28668–28686. https://doi.org/10.1074/jbc.M113.470971
30. Dikalov SI, Nazarewicz RR, Bikineyeva A et al (2014) Nox2-induced production of mitochondrial superoxide in angiotensin II-mediated endothelial oxidative stress and hypertension. Antioxid Redox Signal 20(2):281–294. https://doi.org/10.1089/ars.2012.4918

31. Montezano AC, Touyz RM (2012) Oxidative stress, Noxs, and hypertension: experimental evidence and clinical controversies. Ann Med 44(Suppl 1):S2–S16. https://doi.org/10.3109/07853890.2011.653393
32. García-Redondo AB, Aguado A, Briones AM, Salaices M (2016) NADPH oxidases and vascular remodeling in cardiovascular diseases. Pharmacol Res 114:110–120. https://doi.org/10.1016/j.phrs.2016.10.015
33. Daiber A, Di Lisa F, Oelze M et al (2017) Crosstalk of mitochondria with NADPH oxidase via reactive oxygen and nitrogen species signalling and its role for vascular function. Br J Pharmacol 174(12):1670–1689. https://doi.org/10.1111/bph.13403
34. Landmesser U, Spiekermann S, Preuss C et al (2007) Angiotensin II induces endothelial xanthine oxidase activation: role for endothelial dysfunction in patients with coronary disease. Arterioscler Thromb Vasc Biol 27(4):943–948
35. DeLano FA, Parks DA, Ruedi JM, Babior BM, Schmid-Schönbein GW (2006) Microvascular display of xanthine oxidase and NADPH oxidase in the spontaneously hypertensive rat. Microcirculation 13(7):551–566
36. Laakso J, Mervaala E, Himberg JJ et al (1998) Increased kidney xanthine oxidoreductase activity in salt-induced experimental hypertension. Hypertension 32(5):902–906
37. Laakso JT, Teräväinen TL, Martelin E, Vaskonen T, Lapatto R (2004) Renal xanthine oxidoreductase activity during development of hypertension in spontaneously hypertensive rats. J Hypertens 22(7):1333–1340
38. Radi R, Cassina A, Hodara R, Quijano C, Castro L (2002) Peroxynitrite reactions and formation in mitochondria. Free Radic Biol Med 33:1451–1464
39. Chan SH, Wu KL, Chang AY, Tai MH, Chan JY (2009) Oxidative impairment of mitochondrial electron transport chain complexes in rostral ventrolateral medulla contributes to neurogenic hypertension. Hypertension 53:217–227
40. Postnov IUV (2001) The role of mitochondrial calcium overload and energy deficiency in pathogenesis of arterial hypertension. Arkh Patol 63:3–10
41. Siasos G, Tsigkou V, Kosmopoulos M et al (2018) Mitochondria and cardiovascular diseases-from pathophysiology to treatment. Ann Transl Med 6(12):256. https://doi.org/10.21037/atm.2018.06.21
42. Cogliati S, Frezza C, Soriano ME et al (2013) Mitochondrial cristae shape determines respiratory chain supercomplexes assembly and respiratory efficiency. Cell 155(1):160–171. https://doi.org/10.1016/j.cell.2013.08.032
43. Rimbaud S, Ruiz M, Piquereau J et al (2011) Resveratrol improves survival, hemodynamics and energetics in a rat model of hypertension leading to heart failure. PLoS One 6(10):e26391. https://doi.org/10.1371/journal.pone.0026391
44. Walther T, Tschöpe C, Sterner-Kock A et al (2007) Accelerated mitochondrial adenosine diphosphate/adenosine triphosphate transport improves hypertension-induced heart disease. Circulation 115(3):333–344
45. Konduri GG, Bakhutashvili I, Eis A, Pritchard K Jr (2007) Oxidant stress from uncoupled nitric oxide synthase impairs vasodilation in fetal lambs with persistent pulmonary hypertension. Am J Physiol Heart Circ Physiol 292(4):H1812–H1820
46. Silberman GA, Fan TH, Liu H et al (2010) Uncoupled cardiac nitric oxide synthase mediates diastolic dysfunction. Circulation 121(4):519–528. https://doi.org/10.1161/CIRCULATIONAHA.109.883777
47. Higashi Y, Sasaki S, Nakagawa K et al (2002) Tetrahydrobiopterin enhances forearm vascular response to acetylcholine in both normotensive and hypertensive individuals. Am J Hypertens 15(4 Pt 1):326–332
48. Dikalov S (2011) Cross talk between mitochondria and NADPH oxidases. Free Radic Biol Med 51:1289–1301
49. Schulz E, Wenzel P, Munzel T, Daiber A (2014) Mitochondrial redox signaling: interaction of mitochondrial reactive oxygen species with other sources of oxidative stress. Antioxid Redox Signal 20:308–324

50. Badimón L, Vilahur G, Padró T (2009) Lipoproteins, platelets and atherothrombosis. Rev Esp Cardiol 62(10):1161–1178
51. Torres N, Guevara-Cruz M, Velázquez-Villegas LA, Tovar AR (2015) Nutrition and atherosclerosis. Arch Med Res 46(5):408–426. https://doi.org/10.1016/j.arcmed.2015.05.010
52. Victor VM, Rocha M, Solá E, Bañuls C, Garcia-Malpartida K, Hernández-Mijares A (2009) Oxidative stress, endothelial dysfunction and atherosclerosis. Curr Pharm Des 15(26):2988–3002
53. Kita T, Kume N, Minami M et al (2001) Role of oxidized LDL in atherosclerosis. Ann N Y Acad Sci 947:199–205
54. Skålén K, Gustafsson M, Rydberg EK et al (2002) Subendothelial retention of atherogenic lipoproteins in early atherosclerosis. Nature 417(6890):750–754
55. Bedard K, Krause KH (2007) The NOX family of ROS-generating NADPH oxidases: physiology and pathophysiology. Physiol Rev 87(1):245–313
56. Lassegue B, Griendling KK (2010) NADPH oxidases: functions and pathologies in the vasculature. Arterioscler Thromb Vasc Biol 30:653–661
57. Griendling KK, Sorescu D, Ushio-Fukai M (2000) NAD(P)H oxidase: role in cardiovascular biology and disease. Circ Res 86:494–501
58. Sorescu D, Weiss D, Lassègue B et al (2002) Superoxide production and expression of nox family proteins in human atherosclerosis. Circulation 105:1429–1435
59. Lee S, Gharavi NM, Honda H et al (2009) A role for NADPH oxidase 4 in the activation of vascular endothelial cells by oxidized phospholipids. Free Radic Biol Med 47:145–151
60. Nishino T, Okamoto K, Eger BT, Pai EF, Nishino T (2008) Mammalian xanthine oxidoreductase – mechanism of transition from xanthine dehydrogenase to xanthine oxidase. FEBS J 275:3278–3289. https://doi.org/10.1111/j.1742-4658.2008.06489.x
61. McNally JS, Davis ME, Giddens DP et al (2003) Role of xanthine oxidoreductase and NAD(P)H oxidase in endothelial superoxide production in response to oscillatory shear stress. Am J Physiol Heart Circ Physiol 285:H2290–H2297. https://doi.org/10.1152/ajpheart.00515.2003
62. Ohara Y, Peterson TE, Harrison DG (1993) Hypercholesterolemia increases endothelial superoxide anion production. J Clin Invest 91:2546–2551. https://doi.org/10.1172/JCI116491
63. White CR, Darley-Usmar V, Berrington WR et al (1996) Circulating plasma xanthine oxidase contributes to vascular dysfunction in hypercholesterolemic rabbits. Proc Natl Acad Sci U S A 93:8745–8749
64. Patetsios P, Song M, Shutze WP et al (2001) Identifiation of uric acid and xanthine oxidase in atherosclerotic plaque. Am J Cardiol 88:188–191
65. Guzik TJ, Sadowski J, Guzik B et al (2006) Coronary artery superoxide production and nox isoform expression in human coronary artery disease. Arterioscler Thromb Vasc Biol 26:333–339. https://doi.org/10.1161/01.ATV.0000196651.64776.51
66. Schröder K, Vecchione C, Jung O et al (2006) Xanthine oxidase inhibitor tungsten prevents the development of atherosclerosis in ApoE knockout mice fed a Western-type diet. Free Radic Biol Med 41:1353–1360. https://doi.org/10.1016/j.freeradbiomed.2006.03.026
67. Nomura J, Busso N, Ives A et al (2014) Xanthine oxidase inhibition by febuxostat attenuates experimental atherosclerosis in mice. Sci Rep 4:4554. https://doi.org/10.1038/srep04554
68. Wang Y, Wang GZ, Rabinovitch PS, Tabas I (2014) Macrophage mitochondrial oxidative stress promotes atherosclerosis and nuclear factor-κB-mediated inflmmation in macrophages. Circ Res 114:421–433. https://doi.org/10.1161/CIRCRESAHA.114.302153
69. Corral-Debrinski M, Shoffner JM, Lott MT, Wallace DC (1992) Association of mitochondrial DNA damage with aging and coronary atherosclerotic heart disease. Mutat Res 275:169–180
70. Nojiri H, Shimizu T, Funakoshi M et al (2006) Oxidative stress causes heart failure with impaired mitochondrial respiration. J Biol Chem 281:33789–33801. https://doi.org/10.1074/jbc.M602118200
71. Ballinger SW, Patterson C, Knight-Lozano CA et al (2002) Mitochondrial integrity and function in atherogenesis. Circulation 106:544–549

72. Li H, Förstermann U (2009) Prevention of atherosclerosis by interference with the vascular nitric oxide system. Curr Pharm Des 15:3133–3145
73. Förstermann U, Sessa WC (2012) Nitric oxide synthases: regulation and function. Eur Heart J 33:829–837. https://doi.org/10.1093/eurheartj/ehr304
74. Li H, Förstermann U (2013) Uncoupling of endothelial NO synthase in atherosclerosis and vascular disease. Curr Opin Pharmacol 13:161–167. https://doi.org/10.1016/j.coph.2013.01.006
75. Förstermann U, Münzel T (2006) Endothelial nitric oxide synthase in vascular disease: from marvel to menace. Circulation 113:1708–1714. https://doi.org/10.1161/CIRCULATIONAHA.105.602532
76. Alp NJ, McAteer MA, Khoo J, Choudhury RP, Channon KM (2004) Increased endothelial tetrahydrobiopterin synthesis by targeted transgenic GTPcyclohydrolase I overexpression reduces endothelial dysfunction and atherosclerosis in ApoE-knockout mice. Arterioscler Thromb Vasc Biol 24:445–450. https://doi.org/10.1161/01.ATV.0000115637.48689.77
77. Antoniades C, Shirodaria C, Crabtree M et al (2007) Altered plasma versus vascular biopterins in human atherosclerosis reveal relationships between endothelial nitric oxide synthase coupling, endothelial function, and inflammation. Circulation 116:2851–2859. https://doi.org/10.1161/CIRCULATIONAHA.107.704155
78. Furie B, Furie BC (2008) Mechanisms of thrombus formation. N Engl J Med 359(9):938–949. https://doi.org/10.1056/NEJMra0801082
79. Kumar V (2015) Robbins and Cotran pathologic basis of disease. Elsevier, Philadelphia, pp 122–130. ISBN 978-1-4557-2613-2614
80. Ito T, Kakihana Y, Maruyama I (2016) Thrombomodulin as an intravascular safeguard against inflammatory and thrombotic diseases. Expert Opin Ther Targets 20(2):151–158. https://doi.org/10.1517/14728222.2016.1086750
81. Handin RI, Karabin R, Boxer GJ (1977) Enhancement of platelet function by superoxide anion. J Clin Invest 59:959–965
82. Wachowicz B, Olas B, Zbikowska HM, Buczynski A (2002) Generation of reactive oxygen species in blood platelets. Platelets 13(3):175–182
83. Stokes KY, Russell JM, Jennings MH, Alexander JS, Granger DN (2007) Platelet-associated NAD(P)H oxidase contributes to the thrombogenic phenotype induced by hypercholesterolemia. Free Radic Biol Med 43:22–30
84. Fuentes E, Gibbins JM, Holbrook LM, Palomo I (2018) NADPH oxidase 2 (NOX2): a key target of oxidative stress-mediated platelet activation and thrombosis. Trends Cardiovasc Med. pii: S1050-1738(18)30044–6. https://doi.org/10.1016/j.tcm.2018.03.001. [Epub ahead of print]
85. Delaney MK, Kim K, Estevez B et al (2016) Differential roles of the NADPH – oxidase 1 and 2 in platelet activation and thrombosis. Arterioscler Thromb Vasc Biol 36(5):846–854
86. Seno T, Inoue N, Gao D et al (2001) Involvement of NADH/NADPH oxidase in human platelet ROS production. Thromb Res 103(5):399–409
87. Brandt M, Giokoglu E, Garlapati V et al (2018) Pulmonary arterial hypertension and endothelial dysfunction is linked to NADPH oxidase-derived superoxide formation in venous thrombosis and pulmonary embolism in mice. Oxidative Med Cell Longev 2018:1860513. https://doi.org/10.1155/2018/1860513. eCollection 2018
88. Krötz F, Sohn HY, Gloe T et al (2002) NAD(P)H oxidase-dependent platelet superoxide anion release increases platelet recruitment. Blood 100(3):917–924
89. Violi F, Pignatelli P (2014) Platelet NOX, a novel target for anti-thrombotic treatment. Thromb Haemost 111(5):817–823. https://doi.org/10.1160/TH13-10-0818
90. Carnevale R, Iuliano L, Nocella C et al (2013) Relationship between platelet and urinary 8-Iso-PGF2alpha levels in subjects with different degrees of NOX2 regulation. J Am Heart Assoc 2(3):e000198
91. Begonja AJ, Gambaryan S, Geiger J et al (2005) Platelet NAD(P)H-oxidase-generated ROS production regulates alphaIIbbeta3-integrin activation independent of the NO/cGMP pathway. Blood 106:2757–2760

92. Begonja AJ, Teichmann L, Geiger J, Gambaryan S, Walter U (2006) Platelet regulation by NO/cGMP signaling and NAD(P)H oxidase-generated ROS. Blood Cells Mol Dis 36(2):166–170
93. Lopez JJ, Salido GM, Go'mez-Arteta E, Rosado JA, Pariente JA (2007) Thrombin induces apoptotic events through the generation of reactive oxygen species in human platelets. J Thromb Haemost 5:1283–1291
94. Wang Z, Cai F, Chen X, Luo M, Hu L, Lu Y (2013) The role of mitochondria-derived reactive oxygen species in hyperthermia-induced platelet apoptosis. PLoS One 8(9):e75044. https://doi.org/10.1371/journal.pone.0075044. eCollection 2013
95. Navarro-Yepes J, Burns M, Anandhan A et al (2014) Oxidative stress, redox signaling, and autophagy: cell death versus survival. Antioxid Redox Signal 21(1):66–85. https://doi.org/10.1089/ars.2014.5837
96. Lucchesi BR (1990) Myocardial ischemia, reperfusion and free radical injury. Am J Cardiol 65(19):14–23. https://doi.org/10.1016/0002-9149(90)90120-P
97. Kurian GA, Rajagopal R, Vedantham S, Rajesh M (2016) The role of oxidative stress in myocardial ischemia and reperfusion injury and remodeling: revisited. Oxidative Med Cell Longev 2016:1656450. https://doi.org/10.1155/2016/1656450
98. Tanaka-Esposito C, Chen Q, Lesnefsky EJ (2012) Blockade of electron transport before ischemia protects mitochondria and decreases myocardial injury during reperfusion in aged rat hearts. Transl Res 160(3):207–216. https://doi.org/10.1016/j.trsl.2012.01.024
99. Paradies G, Petrosillo G, Pistolese M, Ruggiero FM (2002) Reactive oxygen species affect mitochondrial electron transport complex I activity through oxidative cardiolipin damage. Gene 286(1):135–141
100. Makazan Z, Saini HK, Dhalla NS (2007) Role of oxidative stress in alterations of mitochondrial function in ischemic-reperfused hearts. Am J Physiol Heart Circ Physiol 292(4):H1986–H1994
101. Müller AL, Hryshko LV, Dhalla NS (2013) Extracellular and intracellular proteases in cardiac dysfunction due to ischemia-reperfusion injury. Int J Cardiol 164(1):39–47
102. Netticadan T, Temsah R, Osada M, Dhalla NS (1999) Status of Ca^{2+}/calmodulin protein kinase phosphorylation of cardiac SR proteins in ischemia-reperfusion. Am J Phys 277:C384–C391
103. Steinberg SF (2013) Oxidative stress and sarcomeric proteins. Circ Res 112(2):393–405
104. Sinning C, Westermann D, Clemmensen P (2017) Oxidative stress in ischemia and reperfusion: current concepts, novel ideas and future perspectives. Biomark Med 11(11):11031–11040. https://doi.org/10.2217/bmm-2017-0110
105. Chen Z, Siu B, Ho YS et al (1998) Overexpression of MnSOD protects against myocardial ischemia/reperfusion injury in transgenic mice. J Mol Cell Cardiol 30(11):2281–2289
106. Woo YJ, Zhang JC, Vijayasarathy C et al (1998) Recombinant adenovirus-mediated cardiac gene transfer of superoxide dismutase and catalase attenuates postischemic contractile dysfunction. Circulation 98:II 255–II 261
107. Barteková M, Šimončíková P, Fogarassyová M et al (2015) Quercetin improves postischemic recovery of heart function in doxorubicin-treated rats and prevents doxorubicin-induced matrix metalloproteinase-2 activation and apoptosis induction. Int J Mol Sci 16(4):8168–8185
108. Cheng L, Jin Z, Zhao R et al (2015) Resveratrol attenuates inflammation and oxidative stress induced by myocardial ischemia-reperfusion injury: role of Nrf2/ARE pathway. Int J Clin Exp Med 8(7):10420–10428
109. Sun G, Li Y, Ji Z (2015) Atorvastatin attenuates inflammation and oxidative stress induced by ischemia/reperfusion in rat heart via the Nrf2 transcription factor. Int J Clin Exp Med 8(9):14837–14845
110. Qu D, Han J, Ren H et al (2016) Cardioprotective effects of astragalin against myocardial ischemia/reperfusion injury in isolated rat heart. Oxidative Med Cell Longev 2016:8194690
111. Das DK, Engelman RM, Kimura Y (1993) Molecular adaptation of cellular defences following preconditioning of the heart by repeated ischaemia. Cardiovasc Res 27:578–584
112. Hoshida S, Kuzuya T, Fuji H et al (1993) Sublethal ischemia alters myocardial antioxidant activity in canine heart. Am J Phys 264:H33–H39

113. Brosková Z, Drábiková K, Sotníková R et al (2013) Effect of plant polyphenols on ischemia-reperfusion injury of the isolated rat heart and vessels. Phytother Res 27(7):1018–1022
114. Li W, Wu M, Tang L et al (2015) Novel curcumin analogue 14p protects against myocardial ischemia reperfusion injury through Nrf2-activating anti-oxidative activity. Toxicol Appl Pharmacol 282(2):175–183
115. Sethi R, Takeda N, Nagano M, Dhalla NS (2000) Beneficial effects of vitamin E treatment in acute myocardial infarction. J Cardiovasc Pharmacol Ther 5(1):51–58
116. Bartekova M, Barancik M, Ferenczyova K, Dhalla NS (2018) Beneficial effects of N-acetylcysteine and N-mercaptopropionylglycine on ischemia reperfusion injury in the heart. Curr Med Chem 25(3):355–366. https://doi.org/10.2174/0929867324666170608111917
117. Stone PH, Lloyd-Jones DM, Kinlay S et al (2005) Effect of intensive lipid lowering, with or without antioxidant vitamins, compared with moderate lipid lowering on myocardial ischemia in patients with stable coronary artery disease: the vascular basis for the treatment of myocardial ischemia study. Circulation 111(14):1747–1755

Part III

Diabetes Induced Cardiovascular Dysfunction

Oxidative Stress and Labile Zinc in Heart Dysfunction Under Hyperglycemia

17

Belma Turan

17.1 Introduction

The epidemiological and clinical studies, in general aspects, agree that *Diabetes Mellitus* with its complications, including cardiovascular disorders, is one of the major health disorders worldwide [1–4]. Due to statistical document, the prevalence of diabetes worldwide will reach about 8% among populations in 2030 [5]. The incidence of ischemic heart disease is higher with the higher risk of mortality rather than those of nondiabetics [6]. In addition, special cardiomyopathy, named diabetic cardiomyopathy was first defined by Rubler and his co-workers (1972) in diabetic patients, who had heart failure with independent of arterial disfunction [7]. All over the World, diabetes is becoming the most important disorders in humans, including children, mostly due to increases in overweightness and obesity via high carbohydrate and fat diets and lack of daily physical inactivity. Due to the WHO documents (http://www.euro.who.int/), there are about 60 million people with diabetes in the European Countries (about 10.3% of men and 9.6% of women aged 25 years and older). Worldwide documents show that there are considerable numbers of diabetes-related human deaths and these numbers will reach to twofold until 2030 (http://www.eatlas.idf.org/; [8]). In addition, clinical and epidemiological studies emphasize that the type 2 diabetes incidence is increasing in Asian populations, although they have low body mass index with respect to increased body weight [9].

As mentioned in several books, review articles original and review articles, as well as public documents, the cardiac dysfunction can be developed during diabetes and resulted from multiple parameters including, cellular $[Ca^{2+}]_i$ – and $[Zn^{2+}]_i$ -dysregulations, dysregulations in the suborganelle levels (i.e. in Sarco(endoplasmic) reticulum and mitochondrial functions) as well as an uncoupling between them [10] besides glucotoxicity, lipotoxicity and fibrosis in the tissue levels [11–13].

B. Turan (✉)
Department of Biophysics, Faculty of Medicine, Ankara University, Ankara, Turkey
e-mail: belma.turan@medicine.ankara.edu.tr

S. Chakraborti et al. (eds.), *Oxidative Stress in Heart Diseases*,
https://doi.org/10.1007/978-981-13-8273-4_17

Hyperglycemia is recognized as the primary accountable factor in the induction of alterations in vital organs and tissues. The heat is an important hyperglycemia-targeting organ. A nonenzymatic attraction of high glucose to protein, named as glycation, represents one possible mechanism, of which high-level blood glucose in serum/plasma of mammalians can lead to pathophysiological damages [14–18]. However, diabetologists have reported the importance of non-understood factors' contributions to the development of hyperglycemia associated organ damage. Importantly, most of the published data emphasized the important roles and also contributions of a high amount of free radical production associated increased oxidative stress in hyperglycemic cells [19, 20]. The data at cellular levels demonstrate that to hyperglycemia induces important changess in mitochondrial structure and function, such as alterations in xanthine oxidoreductase and nicotinamide adenine dinucleotide phosphate, NAD(P)H oxidase, and induction of uncoupled endothelial nitric oxide synthase (eNOS), which, in turn, stimulate the increases in cellular oxidative stress via increases in both reactive oxygen-nitrogen species (ROS/RNS). All these are underlying the changes in epigenetic mechanisms (such as changes in acetylation in both histone and nonhistone proteins and in DNA methylation level). Furthermore, these alterations, in later periods, contribute to the progression of hyperglycemia associated heart dysfunction (i.e. diabetic cardiomyopathy) (Fig. 17.1).

In this content, experimental studies demonstrated that the Zn^{2+} release during an excitation-contraction cycle could increase in cardiomyocytes under oxidative

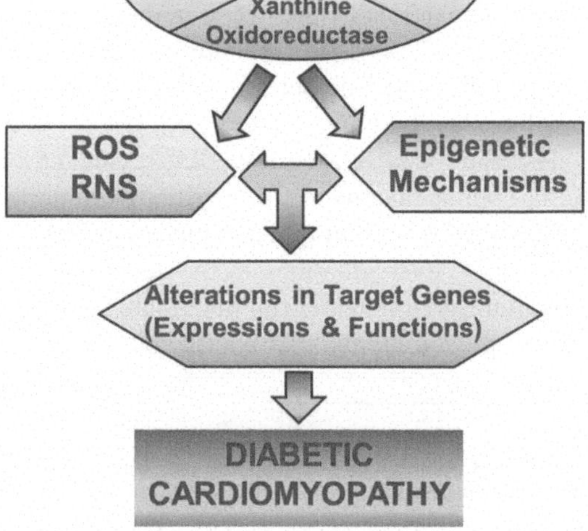

Fig. 17.1 A pathway underlying the hyperglycemia associated subcellular and cellular changes, which are further leading to the induction of special cardiomyopathy in diabetic subjects. Reactive oxygen species (ROS), reactive nitrogen species (RNS), NAD(P)H oxidases; membrane-associated enzymes, catalyze the 1-electron reduction of oxygen using NADH or NADPH, endothelial nitric oxide synthase (eNOS)

stress and hyperglycemia, inducing a marked increase in cytosolic $[Zn^{2+}]_i$ [21], which further could trigger production of pro-oxidants promoting to oxidative damage in cells and tissues [16]. However, early studies demonstrated that acute expose cardiomyocytes to oxidants leads to a significant increase in $[Zn^{2+}]_i$ [22]. Moreover, experimental studies showed the role of both acute and chronic hyperglycemia in the abnormal function of the heart, with parallel induction of oxidative stress and high $[Zn^{2+}]_i$ [18]. All above studies demonstrated that similar to $[Ca^{2+}]_i$, $[Zn^{2+}]_i$ have vital roles in cardiomyocyte function in the mammalian heart [23] and serves up as an important secondary messenger [24]. Therefore, the present review focused on the already knows and present hypothesis on the relation between high $[Zn^{2+}]_i$ and increased oxidative stress in heart dysfunction under acute and chronic hyperglycemia.

17.2 Induction of Oxidative Stress and Redox Dysregulation in Hyperglycemic Heart

The oxidative stress in cells is defined as disturbance in the balance between proper ROS production/use, in the ability of the cell to detoxify oxidative reagents, or and/or impairment in the repair processes [19, 20]. In mammalian cells, ROS are produced physiologically and intracellularly via mitochondria and its production is important for several cellular functions. ROS involves superoxide ($O_2^{\cdot-}$) and hydroxyl radicals ($OH^{\cdot}$), besides oxidizing agents such as hydrogen peroxide (H_2O_2) (see review article [25]).

Under hyperglycemia, there are many different sources to increase oxidative stress, including mitochondrial pathways and arising from the oxidative biochemistry of high glucose [26–33]. The role of mitochondrial pathology is presented in alteration of heart function (cardiomyopathy) in diabetics, however, the experimental and clinical approaches to improve mitochondrial function are not able to solve these complications. Systemic hyperglycemia in humans is one of the major etiological components in the diabetic cardiomyopathy because high cellular glucose can initiate the high amount of ROS and RNS production and/or alter the antioxidant defense mechanisms. However, as mentioned in previous paragraphs, high glucose can directly cause increases in ROS and RNS generation as well as insufficient antioxidant defense in cardiomyocytes. Consequently, resultant imbalance between production and detoxification of oxidants in cardiomyocytes further promote several deleterious effects, including cellular $[Ca^{2+}]_i$-dyshomeostasis [14–17, 34, 35] and dysregulation of redox signalling [11, 12, 28, 36, 37]. Cellular redox dysregulation in the heart tissue underlines the induction of many physiopathological processes in the heart, including atherosclerosis in the vascular system, cardiac hypertrophy and dense fibrosis at both cellular and tissue levels [38]. Briefly, high ROS can induce harmful processes in both cellular and organ functions, at most, though some specific modifications in target molecules/proteins, which in turn undergo redox-sensitive alterations in their function [15, 17, 39, 40].

In the content of ROS, there are both free radicals and non-radical species, which have high chemical reactivity in the cells. In physiological condition, cells have a well-controlled balance system against the highly production of ROS, however, when the existance of unbalanced with different pathological stimuli, cells/tissue/organs get the effect of oxidative stress [31, 41–44]. This imbalance is mostly associated with an imbalance in cellular redox control, which further leads to oxidative stress, [39]. Increases in oxidative stress, via high glucose, is a key factor for the initiation of diabetes associated organ dysfunctions, including cardiac dysfunction. High glucose in cardiac cells underlines several harmful processes at cytosolic, subcellular, and molecular levels, which in turn damage cardiac tissue and depressed cardiac performance [45, 46]. Briefly, high glucose, having either direct, indirect, or both affects the myocardium via affecting both metabolic and nonmetabolic signaling pathways, Ca^{2+}-handling proteins, and several phosphatases in cellular levels as well as serious structural changes in the heart tissue. All these changes contribute to the heart dysfunction in diabetic humans [15, 16, 47, 48].

Since diabetic cardiomyopathy is a complex disorder with the contribution of several internal and external factors, its pathogenesis is not exactly very well understood. As mentioned in the previous paragraph, important changes in the energy metabolism of the heart, characterized with reduced glucose uptake, mitochondrial dysfunctioning, and their consequent events, as directly or indirectly, such as $[Ca^{2+}]_i$-dyshomeostasis in cardiomyocytes, lead to altered excitation-contraction coupling and thereby insufficient contractile activity of the heart [15, 49–51]. Among altered events, a reduction in sarcoplasmic reticulum (SR)-Ca^{2+}-load via abnormal SR-pump activity (SERCA2a) and over photophosphorylation in SR-ryanodine receptors (RyR2) via activations in both protein kinase A (PKA) and protein kinase C (PKC) and calcium-calmodulin kinase II, CaMKII can underline the insufficient heart function in either streptozotocin-diabetic rats or metabolic syndrome rat together with high oxidative stress [15, 52–54]. In those animal models, the contractile dysfunction is also explained with a left ventricular remodeling, although the presence of slowing in action potential duration and reducing in SR Ca^{2+}-release via dysfunctioning of RyR2 [55]. Furthermore, some other studies demonstrated the role of altered redox status in depressed cardiac function in diabetic humans. In those studies, some authors considered that RyR2 has major role in the left ventricular remodeling development in diabetes, while at the cellular level, the slower action potential and reduced SERCA2a expression could only underline the changes in intracellular transient Ca^{2+}-changes (at most kinetics of these changes) in diabetic rat cardiomyocytes [55–57].

17.3 Cross-Talk Between Zinc and Oxidative Stress in Cardiovascular Function

In living organisms, Zn^{2+}, being one of the important transition metal ion, is an important ion for animals and humans. Zn^{2+} is not a redox element and has basically two important roles, such as catalytic activity and structural component [58]. In mammalian cardiomyocytes, Zn^{2+} plays important role for the regulation of protein expression, including inter-molecular interactions, phosphorylation, and oxidation processes [16, 59]. In addition, Zn^{2+} plays important role to stabilize the high level transcription factors, including Zn^{2+}-fingers [60, 61]. In an early study in this field, a very striking example was provided for the correlation between the acute oxidant application and a very rapidly $[Zn^{2+}]_i$ increase in ventricular cardiomyocytes [22], while increased production of oxidants (both ROS and RNS) could trigger rapidly important increases in $[Zn^{2+}]_i$ [18, 21]. In these regards, early studies have reported how the high $[Zn^{2+}]_i$ is associated with mitochondrial dysfunction in neurons, leading to not only reduced ATP production but also induced increase ROS production via activation of peroxisome proliferator-activated receptors, PARP-1 [62, 63].

More particularly, in noncontacting (resting) ventricular cardiomyocytes from STZ-diabetic rats, the high $[Zn^{2+}]_i$ together with high $[Ca^{2+}]_i$ are responsible, in part, responsible from the alterations of the electrical and mechanical function of the cells [16]. However, there are some studies emphasizing the positive role of high $[Zn^{2+}]_i$ in cells. In this regard, Song et al. [64] mentioned that somehow zinc-compounds have an antioxidant action to protect the heart against oxidative stress. Contrary to them, a direct increase in $[Zn^{2+}]_i$ together with high oxidative agents induces important alterations in excitation-contraction coupling activity of cardiomyocytes from left ventricle [21, 22]. In an another study, Lin et al. [65] demonstrated that I/R, and exposed to either ROS, RNS, or both provided important and deleterious increase in the oxygen radicals, resulting important increases in cytosolic $[Zn^{2+}]_i$ via increases in suborganelle Zn^{2+}-release including S(E)R and mitochondria Zn^{2+} release. The studies performed with a Zn^{2+}-chelator (high affinity to bind Zn^{2+}), N,N,N0,N0, tetrakis(2-pyridylmethyl) ethylene diamine pentaethylene (TPEN), demonstrated the deleterious role of high $[Zn^{2+}]_i$, protection myocytes against apoptosis via abolishing the Zn^{2+}-release. In line with these findings, Mato et al. [66] further examined how $[Zn^{2+}]_i$-dysregulation can contribute to oligodendrocyte injury, via AMPA receptors, in parallel to $[Zn^{2+}]_i$ elevation resulting from AMPA receptor activation via promoting Ca^{2+}-dependent cytosolic acidification. A hypothetical pathway is prepared to demonstrate how hyperglycemia associated increased oxidative stress and cellular $[Zn^{2+}]_i$ can correlate to cardiac dysfunction in diabetic mammalians (Fig. 17.2).

Clinical data in this field present some contradictions for the role of high $[Zn^{2+}]_i$ on heart function. As an example, a decrease zinc level in the body determined mostly by serum zinc level and defined as zinc deficiency has been considered as a risk factor for development of high glucose associated cellular/tissue/organ dysfunctions in diabetic humans. The clinical results showed that serum Zn^{2+} level (but not Cu^{2+}) is significantly low in diabetic patients, while supplementations with zinc compounds provided important beneficial effects, including preservation of heart function [67–69]. These benefits with zinc supplementation seem to link its functioning as an antioxidant through participation in activities of superoxide dismutase and thioredoxin as well as its inhibition effect in lipid peroxidation [70]. Furthermore, in cells, intracellular free Zn^{2+} is a potent metallothionein inducer to scavenge the intracellular oxidants [71, 72]. A supporting study was performed by Wang et al.

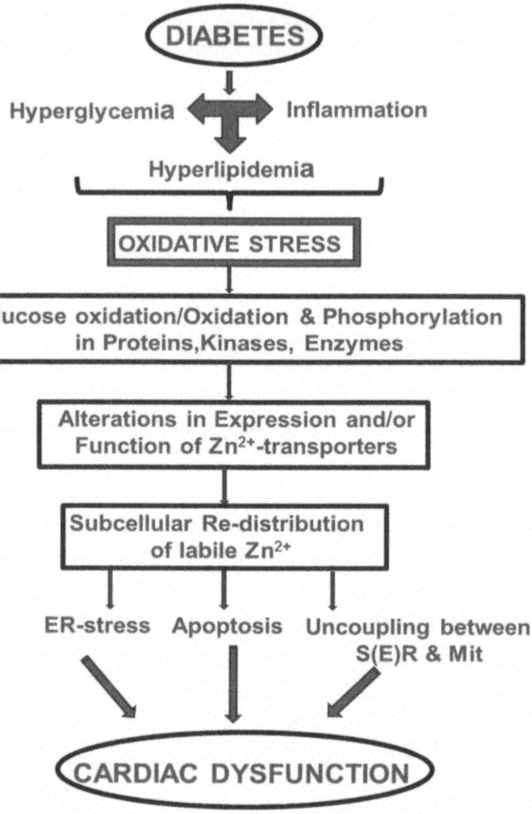

Fig. 17.2 A hypothetical pathway on increases and redistribution of cellular free/labile Zn^{2+} and cardiac dysfunction in diabetic subjects. Oxidative stress in cells from diabetic subject heart increases mainly due to hyperglycemia, hyperlipidemia and inflammation, which leads to oxidation and/or phosphorylation in several proteins, kinases, and enzymes. These changes can further lead to changes in intracellular free/labile Zn^{2+} levels as well as its subcellular distribution via changes in the function of Zn^{2+}–transporter families. Altogether these changes can further induce cardiac dysfunction via induction of apoptosis, ER stress and uncoupling between S(E)R and mitochondria. *S(E)R* Sarco(endo)plasmic reticulum, *Mit* mitochondria [10, 86, 88]

[73]. They demonstrated the important cardioprotective effect of induction of cardiac metallothionein via zinc-supplementation in diabetic cardiomyopathy. Moreover, in a review article by Efeovbokhan et al. [74], they discussed how high $[Zn^{2+}]_i$ could serve as cardio-protectant in congestive heart failure by an antioxidant considering the pharmaco-physiologic potential of $[Zn^{2+}]_i$ at the cellular level.

17.4 Role of Zn^{2+}-Transporters on Cardiomyocyte Function Under High Glucose

Even in early studies with mammalians showed that the cellular transport, localization, and free level of Zn^{2+} are strongly controlled and regulated with severeal ways, including Zn^{2+}-transporters [75–78]. Recently, to examine how the cellular Zn^{2+}-fluxes existing and what types of transporters playing a role for the regulation of cellular Zn^{2+}-distribution, most of the studies are focused on the locations and roles of specific Zn^{2+}-transporters. From these recent efforts, right now, there are the valuable amount of information on the physiological roles of Zn^{2+}-transporters in cellular $[Zn^{2+}]_i$-handling (see review articles [79–82]. Supporting these statements, recent growing evidence documented that any failure to function of $[Zn^{2+}]_i$-homeostasis due to dysfunctioning in Zn^{2+}-transporters results in not only induction but also the progression of many diseases in mammalians [10, 83–90]. However, there are very little documents in literature related with the roles and status of Zn^{2+}-transporters in the pathogenesis of cardiovascular disorders, and therefore, in this section, it will be summarized the already knowns in this area.

The first functional study with Zn^{2+}-transporters was performed in HL-1 embryonic rat atrial cell line by Levy et al. [91]. They demonstrated that ZnT1 interacts with the beta-subunit of voltage-dependent L-type Ca^{2+}-channel via leading to an important decline in expression of its alpha-subunit. However, in early studies, it has been demonstrated that ZnT1 has a role in resistance to Zn^{2+} associated cellular toxicity [92, 93]. Moreover, Mor et al. [94] studied the ZnT1 role in the same cell line in the expression level of T-type Ca^{2+}-channels in the sarcolemma through Ras-ERK signaling. Briefly, those ZnT1 related studies point out its important and broad regulatory roles, partially but directly related to $[Zn^{2+}]_i$-handling [95]. In a later study, the authors mentioned that ZnT1 extrudes Zn^{2+} from mammalian cells via a Zn^{2+}/H^+ exchanger in HEK 293 T cells [96].

The first detailed gene profile analysis to identify the expression levels of Zn^{2+}-transporters in human organs were performed by Yang et al. [97], whereas the mRNA level of *ZnT7* in a number of mouse tissues, including heart tissue, was examined by Kirschke and Huang (2003) [98]. There is no fully demonstration of subcellular localizations of Zn^{2+}-transporters in cardiomyocytes yet although some of them have shown in different organelles in different cell types. For example, the ZIP7 localization was shown into ER in yeast, being responsible for the unfolded-protein-response [99]. Huang et al. (2005) have shown the localization of ZIP7 on Golgi in ovary-cells [100], whereas others demonstrated ZIP7 localization on ER in breast cancer cells [101, 102].

Although Yang et al. (2013) demonstrated the protein expression levels not only in ZIP7 but also in ZnT7 in the human cardiac tissue, the functions and cellular localization of ZIP7 and ZnT7 are not known very well, under both physiological and pathophysiological conditions in the heart [97]. In this regard, Turan's lab demonstrated the ZIP7 and ZnT7 localization on S(E)R and mitochondria in cardiomyocytes from left ventricle of rat heart [10, 86]. The same team also demonstrated the increased ZIP7-phosphorylation in hyperglycemic cardiomyocytes as well as in cardiomyocytes from diabetic rat heart [103]. Following discovery of recombinant FRET-based sensors targeting Zn^{2+}, it has been shown important redistribution of free Zn^{2+} in hyperglycemic cardiomyocytes, which are characterized with high cytosolic $[Zn^{2+}]_i$ and lacking of the proper amount of free Zn^{2+} in the S(E)R. Importantly, less free Zn^{2+} in the S(E)R could underline the ER stress in cardiomyocytes, as well. Furthermore, Turan's lab also demonstrated that the protein expression levels of ZIP7, ZIP14, and ZnT8 were markedly increased with decreased ZIP8 and ZnT7 in hypertrophic rat heart [88]. The relation between the cellular distribution of free Zn^{2+} and ER stress and their coupled-role in cellular dysfunction under pathological conditions have been demonstrated by others, as well [99, 104]. Moreover, with detailed examinations of subcellular localization of ZIP8, ZnT8, and ZIP14 in ventricular H9c2 cell line by using confocal microscopy imaging and calculation of Pearson's coefficients from images demonstrated that ZIP8, ZIP14 and ZnT8 were localized to both sarcolemma and S(E)R [88]. Moreover, in these heart tissue parallel to the different expression levels of these transporters, the increases in the ER stress markers were also confirming the relation between ER stress and redistribution of subcellular free Zn^{2+}. A supporting study on the role of Zn^{2+}-transporters in cardiac function is given by Wen et al. who showed the role of ZIP8 in proper structure and function of left ventricle [89].

Under the light of already documented data, in here, the localization and roles of these Zn^{2+}-transporters in mammalian cardiomyocytes are summarized on a hypothetical map in Fig. 17.3. Overall, already knowns strongly suggest the important role of Zn^{2+}–transporters on the cross-talk between increased $[Zn^{2+}]_i$ and induction of cardiac dysfunction under different pathological conditions, in part, via induction of ER stress. In these regards, the differential distribution in the localizations of Zn^{2+}-transporters as well as their significant role in cross-talk of S(E)R-mitochondria in cardiomyocytes, one can suggest their important contribution to Zn^{2+}-distribution between cellular-compartments of cardiomyocytes, under hyperglycemia, hypertrophy, and/or heart failure. All these can propose a new strategy for prevention/therapy of these pathologies via considering a discovery of new actors for a well-controlled cellular $[Zn^{2+}]$.

17.5 Conclusions

Diabetes, with parallel to obesity and unhealthy feeding and lifestyle in the current populations, is affecting over 200 million humans, and cardiovascular diseases are the major disorders cause the deaths in these populations [2]. Although cardiovascular disorders are associated with several comorbidities, diabetes in humans and experimental animals, as a chronic disease, can induce insufficient ventricular function, being independent of other risk factors including atherosclerosis [16, 17, 105]. The exact mechanisms related with the pathogenesis of hyperglycemia associated events are not clear yet, however, several studies have suggested the important roles of defective subcellular functioning such as S(E)R [15, 21], mitochondria [10, 106] and extracellular matrix [14, 34], at most, due to increased oxidative stress. Recent studies are well established that increases in both ROS and RNS direct and/or

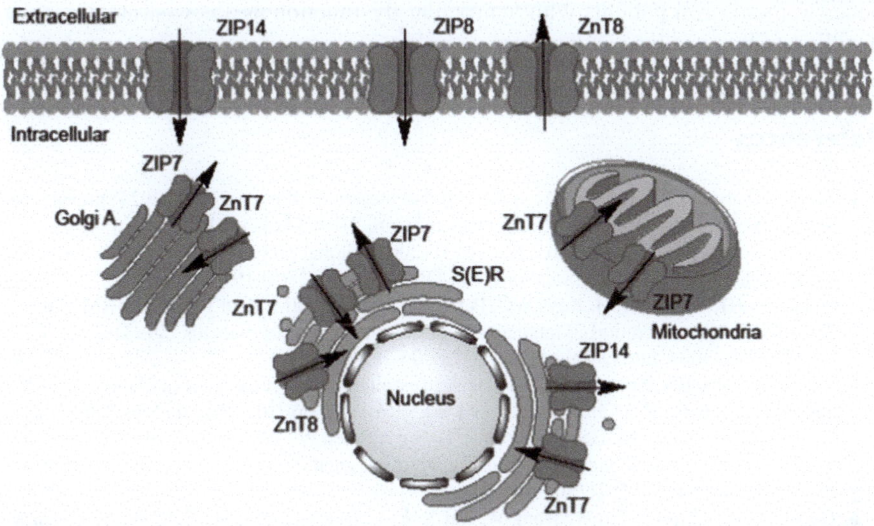

Fig. 17.3 **Cellular and subcellular localization of Zn^{2+}–transporter families, ZIPs and ZnTs and Zn^{2+}–transport directions by these transporters in cardiomyocytes from the mammalian heart**. The primary localization of Zn^{2+}–transporters (blue boxes) and direction of Zn^{2+}-transports (black arrows) is shown on the membranes of cells. Turan's lab examined the localization and protein expression levels in ventricular cells such as ZIP14, ZIP8, ZnT8 along with ZIP7 and ZnT7. The current data obtained in have demonstrated that ZIP8, ZIP14, and ZnT8 are localized to both sarcolemma and S(E)R, while ZIP7 and ZnT7 are localized to S(E)R, Golgi and mitochondria. Their expressions levels can change differentially in the heart under different pathological conditions [10, 86, 88]

indirectly contribute to deleterious changes in cardiomyocytes and cardiac tissues, as well in diabetics [16, 107, 108]. More importantly, it has been shown that, if cardiomyocytes exposed to ROS and/or RNS, the level of $[Zn^{2+}]_i$ in resting cells can increase rapidly through the high amount of Zn^{2+} release from intracellular stores [22, 86]. These events can provide a close relationship between the harmul action of high $[Zn^{2+}]_i$ and increases in oxidative stress in the heart [18, 59, 109]. Zinc, in general aspects, has more than one major biological roles in a mammalian organism, including immune function, oxidative stress, apoptosis, and aging. Therefore, either zinc-deficiency or zinc-excess can be detrimental to heart function in humans. The cellular level of $[Zn^{2+}]_i$ is mainly regulated with several ways, while labile Zn^{2+} has a critical effect in homeostasis of cellular redox state. Considering some aspects of chronic diseases, including diabetes and aging, some similarities to ones in zinc-deficiency status, here are several common clinical outcomes, which affect negatively the immunological and oxidative stress status, and thereby cause the induction of fatal alterations in organ/tissue/cell of mammalians.

Collectively, the summary, given in here, reveal potentially important Zn^{2+}-transporters and their roles on Zn^{2+}-influx into cytosol underlying ventricular dysfunction during myocardial morphogenesis. It can be suggested that these Zn^{2+} -transporters (such as ZIP14, ZIP8, ZnT8, ZIP7, and ZnT7) may be good candidate genes to monitor in patients with ventricular dysfunction as a new strategy for the handling of human chronic diseases.

References

1. King H, Aubert RE, Herman WH (1998) Global burden of diabetes, 1995–2025: prevalence, numerical estimates, and projections. Diabetes Care 21(9):1414–1431
2. Wild S, Roglic G, Green A, Sicree R, King H (2004) Global prevalence of diabetes: estimates for the year 2000 and projections for 2030. Diabetes Care 27(5):1047–1053
3. Halanych JH, Safford MM, Keys WC, Person SD, Shikany JM, Kim YI, Centor RM, Allison JJ (2007) Burden of comorbid medical conditions and quality of diabetes care. Diabetes Care 30(12):2999–3004. https://doi.org/10.2337/dc06-1836
4. Shaw JE, Sicree RA, Zimmet PZ (2010) Global estimates of the prevalence of diabetes for 2010 and 2030. Diabetes Res Clin Pract 87(1):4–14. https://doi.org/10.1016/j.diabres.2009.10.007
5. Stovring H, Andersen M, Beck-Nielsen H, Green A, Vach W (2003) Rising prevalence of diabetes: evidence from a Danish pharmaco-epidemiological database. Lancet 362(9383):537–538. https://doi.org/10.1016/S0140-6736(03)14116-5
6. Huo X, Gao L, Guo L, Xu W, Wang W, Zhi X, Li L, Ren Y, Qi X, Sun Z, Li W, Ji Q, Ran X, Su B, Hao C, Lu J, Guo X, Zhuo H, Zhang D, Pan C, Weng J, Hu D, Yang X, Ji L (2016) Risk of non-fatal cardiovascular diseases in early-onset versus late-onset type 2 diabetes in China: a cross-sectional study. Lancet Diabetes Endocrinol 4(2):115–124. https://doi.org/10.1016/S2213-8587(15)00508-2
7. Rubler S, Dlugash J, Yuceoglu YZ, Kumral T, Branwood AW, Grishman A (1972) New type of cardiomyopathy associated with diabetic glomerulosclerosis. Am J Cardiol 30(6):595–602
8. Monesi L, Baviera M, Marzona I, Avanzini F, Monesi G, Nobili A, Tettamanti M, Cortesi L, Riva E, Fortino I, Bortolotti A, Fontana G, Merlino L, Roncaglioni MC (2012) Prevalence, incidence and mortality of diagnosed diabetes: evidence from an Italian population-based study. Diabet Med 29(3):385–392. https://doi.org/10.1111/j.1464-5491.2011.03446.x
9. Jung HH, Park JI, Jeong JS (2017) Incidence of diabetes and its mortality according to body mass index in South Koreans aged 40–79 years. Clin Epidemiol 9:667–678. https://doi.org/10.2147/CLEP.S146860
10. Tuncay E, Bitirim CV, Olgar Y, Durak A, Rutter GA, Turan B (2018) Zn(2+)-transporters ZIP7 and ZnT7 play important role in progression of cardiac dysfunction via affecting sarco(endo)plasmic reticulum-mitochondria coupling in hyperglycemic cardiomyocytes. Mitochondrion. https://doi.org/10.1016/j.mito.2017.12.011
11. Poornima IG, Parikh P, Shannon RP (2006) Diabetic cardiomyopathy: the search for a unifying hypothesis. Circ Res 98(5):596–605. https://doi.org/10.1161/01.RES.0000207406.94146.c2
12. Fox CS, Coady S, Sorlie PD, D'Agostino RB Sr, Pencina MJ, Vasan RS, Meigs JB, Levy D, Savage PJ (2007) Increasing cardiovascular disease burden due to diabetes mellitus: the Framingham heart study. Circulation 115(12):1544–1550. https://doi.org/10.1161/CIRCULATIONAHA.106.658948
13. Boudina S, Abel ED (2007) Diabetic cardiomyopathy revisited. Circulation 115(25):3213–3223. https://doi.org/10.1161/CIRCULATIONAHA.106.679597
14. Yaras N, Sariahmetoglu M, Bilginoglu A, Aydemir-Koksoy A, Onay-Besikci A, Turan B, Schulz R (2008) Protective action of doxycycline against diabetic cardiomyopathy in rats. Br J Pharmacol 155(8):1174–1184. https://doi.org/10.1038/bjp.2008.373
15. Yaras N, Ugur M, Ozdemir S, Gurdal H, Purali N, Lacampagne A, Vassort G, Turan B (2005) Effects of diabetes on ryanodine receptor Ca release channel (RyR2) and Ca2+ homeostasis in rat heart. Diabetes 54(11):3082–3088

16. Tuncay E, Okatan EN, Vassort G, Turan B (2013) ss-blocker timolol prevents arrhythmogenic Ca(2)(+) release and normalizes Ca(2)(+) and Zn(2)(+) dyshomeostasis in hyperglycemic rat heart. PLoS One 8(7):e71014. https://doi.org/10.1371/journal.pone.0071014
17. Tuncay E, Okatan EN, Toy A, Turan B (2014) Enhancement of cellular antioxidant-defence preserves diastolic dysfunction via regulation of both diastolic Zn2+ and Ca2+ and prevention of RyR2-leak in hyperglycemic cardiomyocytes. Oxidative Med Cell Longev 2014:290381. https://doi.org/10.1155/2014/290381
18. Tuncay E, Turan B (2016) Intracellular Zn(2+) increase in cardiomyocytes induces both electrical and mechanical dysfunction in heart via endogenous generation of reactive nitrogen species. Biol Trace Elem Res 169(2):294–302. https://doi.org/10.1007/s12011-015-0423-3
19. Maritim AC, Sanders RA, Watkins JB 3rd (2003) Diabetes, oxidative stress, and antioxidants: a review. J Biochem Mol Toxicol 17(1):24–38. https://doi.org/10.1002/jbt.10058
20. Baynes JW, Thorpe SR (1999) Role of oxidative stress in diabetic complications: a new perspective on an old paradigm. Diabetes 48(1):1–9
21. Tuncay E, Bilginoglu A, Sozmen NN, Zeydanli EN, Ugur M, Vassort G, Turan B (2011) Intracellular free zinc during cardiac excitation-contraction cycle: calcium and redox dependencies. Cardiovasc Res 89(3):634–642. https://doi.org/10.1093/cvr/cvq352
22. Turan B, Fliss H, Desilets M (1997) Oxidants increase intracellular free Zn2+ concentration in rabbit ventricular myocytes. Am J Phys 272(5 Pt 2):H2095–H2106
23. Atar D, Backx PH, Appel MM, Gao WD, Marban E (1995) Excitation-transcription coupling mediated by zinc influx through voltage-dependent calcium channels. J Biol Chem 270(6):2473–2477
24. Hirano T, Murakami M, Fukada T, Nishida K, Yamasaki S, Suzuki T (2008) Roles of zinc and zinc signaling in immunity: zinc as an intracellular signaling molecule. Adv Immunol 97:149–176. https://doi.org/10.1016/S0065-2776(08)00003-5S0065-2776(08)00003-5. [pii]
25. Vassort G, Turan B (2010) Protective role of antioxidants in diabetes-induced cardiac dysfunction. Cardiovasc Toxicol 10(2):73–86. https://doi.org/10.1007/s12012-010-9064-0
26. Duncan JG, Fong JL, Medeiros DM, Finck BN, Kelly DP (2007) Insulin-resistant heart exhibits a mitochondrial biogenic response driven by the peroxisome proliferator-activated receptor-alpha/PGC-1alpha gene regulatory pathway. Circulation 115(7):909–917. https://doi.org/10.1161/CIRCULATIONAHA.106.662296
27. Shen X, Zheng S, Metreveli NS, Epstein PN (2006) Protection of cardiac mitochondria by overexpression of MnSOD reduces diabetic cardiomyopathy. Diabetes 55(3):798–805
28. Brownlee M (2005) The pathobiology of diabetic complications: a unifying mechanism. Diabetes 54(6):1615–1625
29. Oliveira PJ, Seica R, Coxito PM, Rolo AP, Palmeira CM, Santos MS, Moreno AJ (2003) Enhanced permeability transition explains the reduced calcium uptake in cardiac mitochondria from streptozotocin-induced diabetic rats. FEBS Lett 554(3):511–514
30. Turko IV, Li L, Aulak KS, Stuehr DJ, Chang JY, Murad F (2003) Protein tyrosine nitration in the mitochondria from diabetic mouse heart. Implications to dysfunctional mitochondria in diabetes. J Biol Chem 278(36):33972–33977. https://doi.org/10.1074/jbc.M303734200
31. Marra G, Cotroneo P, Pitocco D, Manto A, Di Leo MA, Ruotolo V, Caputo S, Giardina B, Ghirlanda G, Santini SA (2002) Early increase of oxidative stress and reduced antioxidant defenses in patients with uncomplicated type 1 diabetes: a case for gender difference. Diabetes Care 25(2):370–375
32. Stockklauser-Farber K, Ballhausen T, Laufer A, Rosen P (2000) Influence of diabetes on cardiac nitric oxide synthase expression and activity. Biochim Biophys Acta 1535(1):10–20
33. Pierce GN, Russell JC (1997) Regulation of intracellular Ca2+ in the heart during diabetes. Cardiovasc Res 34(1):41–47
34. Aydemir-Koksoy A, Bilginoglu A, Sariahmetoglu M, Schulz R, Turan B (2010) Antioxidant treatment protects diabetic rats from cardiac dysfunction by preserving contractile protein targets of oxidative stress. J Nutr Biochem 21(9):827–833. https://doi.org/10.1016/j.jnutbio.2009.06.006

35. Ayaz M, Ozdemir S, Ugur M, Vassort G, Turan B (2004) Effects of selenium on altered mechanical and electrical cardiac activities of diabetic rat. Arch Biochem Biophys 426(1):83–90. https://doi.org/10.1016/j.abb.2004.03.030
36. Shah AM, Channon KM (2004) Free radicals and redox signalling in cardiovascular disease. Heart 90(5):486–487
37. Klatt P, Lamas S (2000) Regulation of protein function by S-glutathiolation in response to oxidative and nitrosative stress. Eur J Biochem 267(16):4928–4944
38. Finkel T (1999) Signal transduction by reactive oxygen species in non-phagocytic cells. J Leukoc Biol 65(3):337–340
39. Gao WD, Liu Y, Marban E (1996) Selective effects of oxygen free radicals on excitation-contraction coupling in ventricular muscle. Implications for the mechanism of stunned myocardium. Circulation 94(10):2597–2604
40. Schilling JD (2015) The mitochondria in diabetic heart failure: from pathogenesis to therapeutic promise. Antioxid Redox Signal 22(17):1515–1526. https://doi.org/10.1089/ars.2015.6294
41. Kannel WB, Hjortland M, Castelli WP (1974) Role of diabetes in congestive heart failure: the Framingham study. Am J Cardiol 34(1):29–34
42. Vanden Hoek TL, Becker LB, Shao Z, Li C, Schumacker PT (1998) Reactive oxygen species released from mitochondria during brief hypoxia induce preconditioning in cardiomyocytes. J Biol Chem 273(29):18092–18098
43. Hassouna A, Loubani M, Matata BM, Fowler A, Standen NB, Galinanes M (2006) Mitochondrial dysfunction as the cause of the failure to precondition the diabetic human myocardium. Cardiovasc Res 69(2):450–458. https://doi.org/10.1016/j.cardiores.2005.11.004
44. Forstermann U (2008) Oxidative stress in vascular disease: causes, defense mechanisms and potential therapies. Nat Clin Pract Cardiovasc Med 5(6):338–349. https://doi.org/10.1038/ncpcardio1211
45. Ceriello A (2008) Possible role of oxidative stress in the pathogenesis of hypertension. Diabetes Care 31(Suppl 2):S181–S184. https://doi.org/10.2337/dc08-s245
46. Johansen JS, Harris AK, Rychly DJ, Ergul A (2005) Oxidative stress and the use of antioxidants in diabetes: linking basic science to clinical practice. Cardiovasc Diabetol 4:5. https://doi.org/10.1186/1475-2840-4-5
47. Ayaz M, Can B, Ozdemir S, Turan B (2002) Protective effect of selenium treatment on diabetes-induced myocardial structural alterations. Biol Trace Elem Res 89(3):215–226
48. Ozdemir S, Ugur M, Gurdal H, Turan B (2005) Treatment with AT(1) receptor blocker restores diabetes-induced alterations in intracellular Ca(2+) transients and contractile function of rat myocardium. Arch Biochem Biophys 435(1):166–174. https://doi.org/10.1016/j.abb.2004.11.027
49. Choi KM, Zhong Y, Hoit BD, Grupp IL, Hahn H, Dilly KW, Guatimosim S, Lederer WJ, Matlib MA (2002) Defective intracellular Ca(2+) signaling contributes to cardiomyopathy in Type 1 diabetic rats. Am J Phys Heart Circ Phys 283(4):H1398–H1408. https://doi.org/10.1152/ajpheart.00313.2002
50. Ishikawa T, Kajiwara H, Kurihara S (1999) Alterations in contractile properties and Ca2+ handling in streptozotocin-induced diabetic rat myocardium. Am J Phys 277(6 Pt 2):H2185–H2194
51. Ren J, Davidoff AJ (1997) Diabetes rapidly induces contractile dysfunctions in isolated ventricular myocytes. Am J Phys 272(1 Pt 2):H148–H158. https://doi.org/10.1152/ajpheart.1997.272.1.H148
52. Bidasee KR, Nallani K, Besch HR Jr, Dincer UD (2003) Streptozotocin-induced diabetes increases disulfide bond formation on cardiac ryanodine receptor (RyR2). J Pharmacol Exp Ther 305(3):989–998. https://doi.org/10.1124/jpet.102.046201
53. Pereira L, Matthes J, Schuster I, Valdivia HH, Herzig S, Richard S, Gomez AM (2006) Mechanisms of [Ca2+]i transient decrease in cardiomyopathy of db/db type 2 diabetic mice. Diabetes 55(3):608–615

54. Yaras N, Tuncay E, Purali N, Sahinoglu B, Vassort G, Turan B (2007) Sex-related effects on diabetes-induced alterations in calcium release in the rat heart. Am J Phys Heart Circ Phys 293(6):H3584–H3592. https://doi.org/10.1152/ajpheart.00619.2007
55. Zhang L, Cannell MB, Phillips AR, Cooper GJ, Ward ML (2008) Altered calcium homeostasis does not explain the contractile deficit of diabetic cardiomyopathy. Diabetes 57(8):2158–2166. https://doi.org/10.2337/db08-0140
56. Hertelendi Z, Toth A, Borbely A, Galajda Z, van der Velden J, Stienen GJ, Edes I, Papp Z (2008) Oxidation of myofilament protein sulfhydryl groups reduces the contractile force and its Ca2+ sensitivity in human cardiomyocytes. Antioxid Redox Signal 10(7):1175–1184. https://doi.org/10.1089/ars.2007.2014
57. Turan B, Desilets M, Acan LN, Hotomaroglu O, Vannier C, Vassort G (1996) Oxidative effects of selenite on rat ventricular contractility and Ca movements. Cardiovasc Res 32(2):351–361
58. Vallee BL, Auld DS (1992) Functional zinc-binding motifs in enzymes and DNA-binding proteins. Faraday Discuss 93:47–65
59. Degirmenci S, Olgar Y, Durak A, Tuncay E, Turan B (2018) Cytosolic increased labile Zn(2+) contributes to arrhythmogenic action potentials in left ventricular cardiomyocytes through protein thiol oxidation and cellular ATP depletion. J Trace Elem Med Biol 48:202–212. https://doi.org/10.1016/j.jtemb.2018.04.014
60. Tupler R, Perini G, Green MR (2001) Expressing the human genome. Nature 409(6822):832–833. https://doi.org/10.1038/35057011
61. Andreini C, Banci L, Bertini I, Rosato A (2006) Counting the zinc-proteins encoded in the human genome. J Proteome Res 5(1):196–201. https://doi.org/10.1021/pr050361j
62. Dineley KE, Richards LL, Votyakova TV, Reynolds IJ (2005) Zinc causes loss of membrane potential and elevates reactive oxygen species in rat brain mitochondria. Mitochondrion 5(1):55–65
63. Gazaryan IG, Krasinskaya IP, Kristal BS, Brown AM (2007) Zinc irreversibly damages major enzymes of energy production and antioxidant defense prior to mitochondrial permeability transition. J Biol Chem 282(33):24373–24380. https://doi.org/10.1074/jbc.M611376200
64. Song Y, Wang J, Li XK, Cai L (2005) Zinc and the diabetic heart. Biometals : an international journal on the role of metal ions in biology. Biochem Med 18(4):325–332. https://doi.org/10.1007/s10534-005-3689-7
65. Lin CL, Tseng HC, Chen RF, Chen WP, Su MJ, Fang KM, Wu ML (2011) Intracellular zinc release-activated ERK-dependent GSK-3beta-p53 and Noxa-Mcl-1 signaling are both involved in cardiac ischemic-reperfusion injury. Cell Death Differ 18(10):1651–1663. https://doi.org/10.1038/cdd.2011.80
66. Mato S, Sanchez-Gomez MV, Bernal-Chico A, Matute C (2013) Cytosolic zinc accumulation contributes to excitotoxic oligodendroglial death. Glia 61(5):750–764. https://doi.org/10.1002/glia.22470
67. Maret W, Sandstead HH (2006) Zinc requirements and the risks and benefits of zinc supplementation. J Trace Elem Med Biol 20(1):3–18. https://doi.org/10.1016/j.jtemb.2006.01.006
68. Roussel AM, Kerkeni A, Zouari N, Mahjoub S, Matheau JM, Anderson RA (2003) Antioxidant effects of zinc supplementation in Tunisians with type 2 diabetes mellitus. J Am Coll Nutr 22(4):316–321
69. Kumar SD, Vijaya M, Samy RP, Dheen ST, Ren M, Watt F, Kang YJ, Bay BH, Tay SS (2012) Zinc supplementation prevents cardiomyocyte apoptosis and congenital heart defects in embryos of diabetic mice. Free Radic Biol Med 53(8):1595–1606. https://doi.org/10.1016/j.freeradbiomed.2012.07.008
70. Prasad AS, Bao B, Beck FW, Kucuk O, Sarkar FH (2004) Antioxidant effect of zinc in humans. Free Radic Biol Med 37(8):1182–1190. https://doi.org/10.1016/j.freeradbiomed.2004.07.007
71. Song Y, Wang J, Li Y, Du Y, Arteel GE, Saari JT, Kang YJ, Cai L (2005) Cardiac metallothionein synthesis in streptozotocin-induced diabetic mice, and its protection against diabetes-induced cardiac injury. Am J Pathol 167(1):17–26. https://doi.org/10.1016/S0002-9440(10)62949-5

72. Ali MM, Frei E, Straub J, Breuer A, Wiessler M (2002) Induction of metallothionein by zinc protects from daunorubicin toxicity in rats. Toxicology 179(1–2):85–93
73. Wang J, Song Y, Elsherif L, Song Z, Zhou G, Prabhu SD, Saari JT, Cai L (2006) Cardiac metallothionein induction plays the major role in the prevention of diabetic cardiomyopathy by zinc supplementation. Circulation 113(4):544–554. https://doi.org/10.1161/CIRCULATIONAHA.105.537894
74. Efeovbokhan N, Bhattacharya SK, Ahokas RA, Sun Y, Guntaka RV, Gerling IC, Weber KT (2014) Zinc and the prooxidant heart failure phenotype. J Cardiovasc Pharmacol 64(4):393–400. https://doi.org/10.1097/FJC.0000000000000125
75. Eide DJ (2000) Metal ion transport in eukaryotic microorganisms: insights from Saccharomyces cerevisiae. Adv Microb Physiol 43:1–38
76. Liuzzi JP, Cousins RJ (2004) Mammalian zinc transporters. Annu Rev Nutr 24:151–172. https://doi.org/10.1146/annurev.nutr.24.012003.132402
77. Shiroi A, Yoshikawa M, Yokota H, Fukui H, Ishizaka S, Tatsumi K, Takahashi Y (2002) Identification of insulin-producing cells derived from embryonic stem cells by zinc-chelating dithizone. Stem Cells 20(4):284–292. https://doi.org/10.1634/stemcells.20-4-284
78. Ye B, Maret W, Vallee BL (2001) Zinc metallothionein imported into liver mitochondria modulates respiration. Proc Natl Acad Sci U S A 98(5):2317–2322. https://doi.org/10.1073/pnas.041619198
79. Kambe T, Tsuji T, Hashimoto A, Itsumura N (2015) The physiological, biochemical, and molecular roles of zinc transporters in zinc homeostasis and metabolism. Physiol Rev 95(3):749–784. https://doi.org/10.1152/physrev.00035.2014
80. Kambe T, Matsunaga M, Takeda TA (2017) Understanding the contribution of zinc transporters in the function of the early secretory pathway. Int J Mol Sci 18(10):pii: E2179. https://doi.org/10.3390/ijms18102179
81. Turan B, Tuncay E (2017) Impact of labile zinc on heart function: from physiology to pathophysiology. Int J Mol Sci 18(11):pii: E2395. https://doi.org/10.3390/ijms18112395
82. Fukada T, Kambe T (2018) Welcome to the world of zinc signaling. Int J Mol Sci 19(3). https://doi.org/10.3390/ijms19030785
83. Beharier O, Etzion Y, Levi S, Mor M, Mor M, Dror S, Kahn J, Katz A, Moran A (2010) The involvement of ZnT-1, a new modulator of cardiac L-type calcium channels, in [corrected] atrial tachycardia remodeling. [corrected]. Ann N Y Acad Sci 1188:87–95. https://doi.org/10.1111/j.1749-6632.2009.05087.x
84. Zhao L, Oliver E, Maratou K, Atanur SS, Dubois OD, Cotroneo E, Chen CN, Wang L, Arce C, Chabosseau PL, Ponsa-Cobas J, Frid MG, Moyon B, Webster Z, Aldashev A, Ferrer J, Rutter GA, Stenmark KR, Aitman TJ, Wilkins MR (2015) The zinc transporter ZIP12 regulates the pulmonary vascular response to chronic hypoxia. Nature 524(7565):356–360. https://doi.org/10.1038/nature14620
85. Tepaamorndech S, Kirschke CP, Pedersen TL, Keyes WR, Newman JW, Huang L (2016) Zinc transporter 7 deficiency affects lipid synthesis in adipocytes by inhibiting insulin-dependent Akt activation and glucose uptake. FEBS J 283(2):378–394. https://doi.org/10.1111/febs.13582
86. Tuncay E, Bitirim VC, Durak A, Carrat GRJ, Taylor KM, Rutter GA, Turan B (2017) Hyperglycemia-induced changes in ZIP7 and ZnT7 expression cause Zn(2+) release from the sarco(endo)plasmic reticulum and mediate ER stress in the heart. Diabetes 66(5):1346–1358. https://doi.org/10.2337/db16-1099
87. Lin W, Vann DR, Doulias PT, Wang T, Landesberg G, Li X, Ricciotti E, Scalia R, He M, Hand NJ, Rader DJ (2017) Hepatic metal ion transporter ZIP8 regulates manganese homeostasis and manganese-dependent enzyme activity. J Clin Invest 127(6):2407–2417. https://doi.org/10.1172/JCI90896
88. Olgar Y, Durak A, Tuncay E, Bitirim CV, Ozcinar E, Inan MB, Tokcaer-Keskin Z, Akcali KC, Akar AR, Turan B (2018) Increased free Zn(2+) correlates induction of sarco(endo)plasmic reticulum stress via altered expression levels of Zn(2+) -transporters in heart failure. J Cell Mol Med 22(3):1944–1956. https://doi.org/10.1111/jcmm.13480

89. Lin W, Li D, Cheng L, Li L, Liu F, Hand NJ, Epstein JA, Rader DJ (2018) Zinc transporter Slc39a8 is essential for cardiac ventricular compaction. J Clin Invest 128(2):826–833. https://doi.org/10.1172/JCI96993
90. Huang L, Tepaamorndech S, Kirschke CP, Newman JW, Keyes WR, Pedersen TL, Dumnil J (2018) Aberrant fatty acid metabolism in skeletal muscle contributes to insulin resistance in zinc transporter 7 (znt7)-knockout mice. J Biol Chem 293(20):7549–7563. https://doi.org/10.1074/jbc.M117.817692
91. Levy S, Beharier O, Etzion Y, Mor M, Buzaglo L, Shaltiel L, Gheber LA, Kahn J, Muslin AJ, Katz A, Gitler D, Moran A (2009) Molecular basis for zinc transporter 1 action as an endogenous inhibitor of L-type calcium channels. J Biol Chem 284(47):32434–32443. https://doi.org/10.1074/jbc.M109.058842
92. Palmiter RD, Huang L (2004) Efflux and compartmentalization of zinc by members of the SLC30 family of solute carriers. Pflugers Arch 447(5):744–751. https://doi.org/10.1007/s00424-003-1070-7
93. Tsuda M, Imaizumi K, Katayama T, Kitagawa K, Wanaka A, Tohyama M, Takagi T (1997) Expression of zinc transporter gene, ZnT-1, is induced after transient forebrain ischemia in the gerbil. J Neurosci 17(17):6678–6684
94. Mor M, Beharier O, Levy S, Kahn J, Dror S, Blumenthal D, Gheber LA, Peretz A, Katz A, Moran A, Etzion Y (2012) ZnT-1 enhances the activity and surface expression of T-type calcium channels through activation of Ras-ERK signaling. Am J Phys Cell Physiol 303(2):C192–C203. https://doi.org/10.1152/ajpcell.00427.2011
95. Beharier O, Dror S, Levy S, Kahn J, Mor M, Etzion S, Gitler D, Katz A, Muslin AJ, Moran A, Etzion Y (2012) ZnT-1 protects HL-1 cells from simulated ischemia-reperfusion through activation of Ras-ERK signaling. J Mol Med 90(2):127–138. https://doi.org/10.1007/s00109-011-0845-0
96. Shusterman E, Beharier O, Shiri L, Zarivach R, Etzion Y, Campbell CR, Lee IH, Okabayashi K, Dinudom A, Cook DI, Katz A, Moran A (2014) ZnT-1 extrudes zinc from mammalian cells functioning as a Zn(2+)/H(+) exchanger. Metallomics Integr Biometal Sci 6(9):1656–1663. https://doi.org/10.1039/c4mt00108g
97. Yang J, Zhang Y, Cui X, Yao W, Yu X, Cen P, Hodges SE, Fisher WE, Brunicardi FC, Chen C, Yao Q, Li M (2013) Gene profile identifies zinc transporters differentially expressed in normal human organs and human pancreatic cancer. Curr Mol Med 13(3):401–409
98. Kirschke CP, Huang L (2003) ZnT7, a novel mammalian zinc transporter, accumulates zinc in the Golgi apparatus. J Biol Chem 278(6):4096–4102. https://doi.org/10.1074/jbc.M207644200
99. Ellis CD, Wang F, MacDiarmid CW, Clark S, Lyons T, Eide DJ (2004) Zinc and the Msc2 zinc transporter protein are required for endoplasmic reticulum function. J Cell Biol 166(3):325–335. https://doi.org/10.1083/jcb.200401157
100. Huang L, Kirschke CP, Zhang Y, Yu YY (2005) The ZIP7 gene (Slc39a7) encodes a zinc transporter involved in zinc homeostasis of the Golgi apparatus. J Biol Chem 280(15):15456–15463. https://doi.org/10.1074/jbc.M412188200
101. Taylor KM, Vichova P, Jordan N, Hiscox S, Hendley R, Nicholson RI (2008) ZIP7-mediated intracellular zinc transport contributes to aberrant growth factor signaling in antihormone-resistant breast cancer cells. Endocrinology 149(10):4912–4920. https://doi.org/10.1210/en.2008-0351
102. Hogstrand C, Kille P, Nicholson RI, Taylor KM (2009) Zinc transporters and cancer: a potential role for ZIP7 as a hub for tyrosine kinase activation. Trends Mol Med 15(3):101–111. https://doi.org/10.1016/j.molmed.2009.01.004
103. Tuncay E, Bitirim VC, Durak A, Carrat GR, Taylor K, Rutter GA, Turan B (2017) Hyperglycemia-induced changes in ZIP7 and ZnT7 expression cause Zn_2+ release from the Sarco(endo)plasmic reticulum and mediate ER-stress in the heart. Diabetes 66(5):1346–1358. https://doi.org/10.2337/db16-1099
104. Harding HP, Ron D (2002) Endoplasmic reticulum stress and the development of diabetes: a review. Diabetes 51(Suppl 3):S455–S461

105. Go AS, Mozaffarian D, Roger VL, Benjamin EJ, Berry JD, Borden WB, Bravata DM, Dai S, Ford ES, Fox CS, Franco S, Fullerton HJ, Gillespie C, Hailpern SM, Heit JA, Howard VJ, Huffman MD, Kissela BM, Kittner SJ, Lackland DT, Lichtman JH, Lisabeth LD, Magid D, Marcus GM, Marelli A, Matchar DB, McGuire DK, Mohler ER, Moy CS, Mussolino ME, Nichol G, Paynter NP, Schreiner PJ, Sorlie PD, Stein J, Turan TN, Virani SS, Wong ND, Woo D, Turner MB, American Heart Association Statistics C, Stroke Statistics S (2013) Executive summary: heart disease and stroke statistics--2013 update: a report from the American Heart Association. Circulation 127(1):143–152. https://doi.org/10.1161/CIR.0b013e318282ab8f
106. Dhalla NS, Takeda S, Elimban V (2013) Mechanisms of the beneficial effects of vitamin B6 and pyridoxal 5-phosphate on cardiac performance in ischemic heart disease. Clin Chem Lab Med 51(3):535–543. https://doi.org/10.1515/cclm-2012-0553
107. Malaiyandi LM, Dineley KE, Reynolds IJ (2004) Divergent consequences arise from metallothionein overexpression in astrocytes: zinc buffering and oxidant-induced zinc release. Glia 45(4):346–353. https://doi.org/10.1002/glia.10332
108. Ayaz M, Turan B (2006) Selenium prevents diabetes-induced alterations in [Zn_2+]i and metallothionein level of rat heart via restoration of cell redox cycle. Am J Phys Heart Circ Phys 290(3):H1071–H1080. https://doi.org/10.1152/ajpheart.00754.2005
109. Billur D, Tuncay E, Okatan EN, Olgar Y, Durak AT, Degirmenci S, Can B, Turan B (2016) Interplay between cytosolic free Zn_2+ and mitochondrion morphological changes in rat ventricular cardiomyocytes. Biol Trace Elem Res 174(1):177–188. https://doi.org/10.1007/s12011-016-0704-5

Histone Deacetylases and Oxidative Stress: Role in Diabetic Cardiomyopathy

18

Bhoomika M. Patel

18.1 Diabetic Cardiomyopathy

Healthcare research has come a long way from initial discovery of diabetes in 1550 BC with several breakthroughs including the isolation of insulin by Best and Banting in 1922. Although discovery of insulin was a major leap forward in history of diabetes, diabetic complications were not known for several years. Five decades after insulin discovery i.e. in 1972, first evidence of diabetic cardiomyopathy was reported by Rubler et al. [1]. They examined post-mortem findings of 27 patients with proven case of diabetic glomerulosclerosis and they documented that out of 27 patients, 04 patients had cardiomegaly, left ventricular hypertrophy and congestive cardiac failure with no known cause and they hypothesized the cardiac disease is secondary to diabetic microangiopathy [1]. Since this first discovery, it has been more than 45 years now, and research in diabetic cardiomyopathy has advanced to a large extent. Several new targets and molecular mechanisms have been identified not only for cardiovascular diseases [2, 3] but also for diabetic cardiomyopathy [4, 5]. The major factors contributing to diabetic cardiomyopathy are hyperglycemia and insulin resistance, oxidative stress, abnormal free fatty acid metabolism, advanced glycation end products, renin-angiotensin system etc. [6]. A brief summary of mechanisms of diabetic cardiomyopathy has been provided in Fig. 18.1 and the molecular signalling pathways affecting diabetic cardiomyopathy are depicted in Fig. 18.2.

Currently, there is no specific treatment for diabetic cardiomyopathy; however it involves use of anti-diabetic agents, anti-hypertensives, lipid lowering drugs, and management of heart failure depending upon the stage and extent of cardiomyopathy. Metformin [7, 8], glucagon like peptide-1 (GLP-1) agonist [9, 10], Dipeptidyl peptidase 4 (DPP-4) inhibitors [11, 12] and sodium-glucose cotransporter 2 (SGLT-2) inhibitors [13, 14] have been reported to exhibit beneficial role in

B. M. Patel (✉)
Department of Pharmacology, Institute of Pharmacy, Nirma University, Ahmedabad, India

S. Chakraborti et al. (eds.), *Oxidative Stress in Heart Diseases*,
https://doi.org/10.1007/978-981-13-8273-4_18

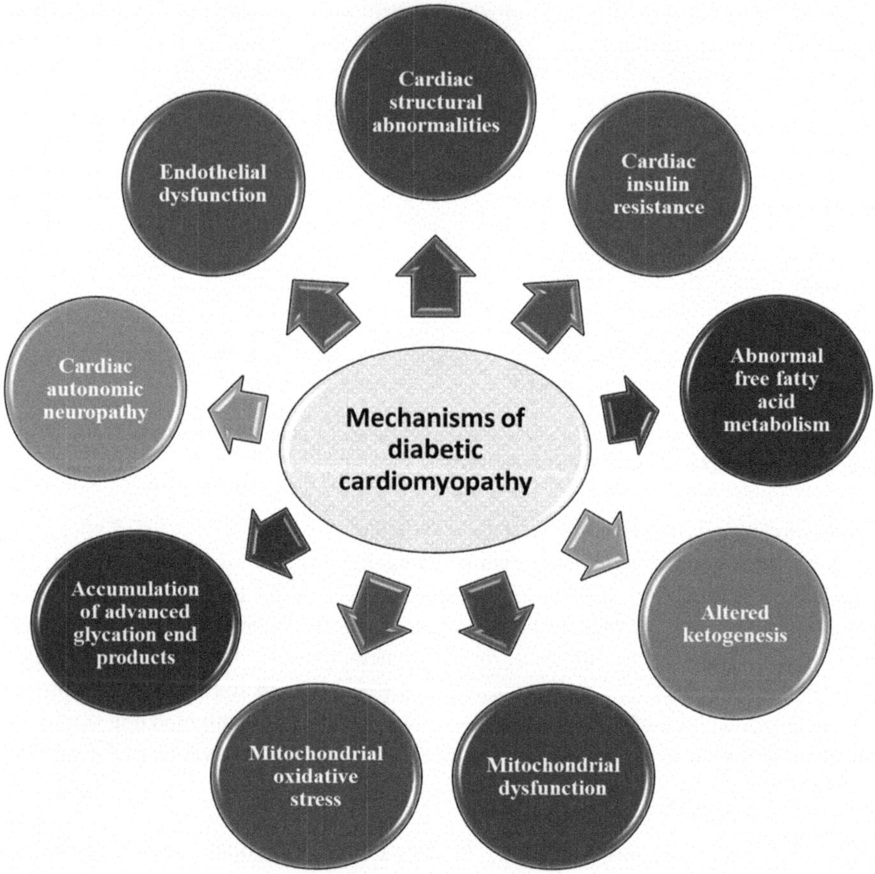

Fig. 18.1 Mechanisms leading to diabetic cardiomyopathy

cardiomyopathy. Thiazolidinediones are reported to worsen the cardiovascular complications of diabetes [15]. Amongst various anti-hypertensive agents, spironolactone [16, 17], β-blockers [18], renin inhibitors [19], angiotensin II receptor blockers [20–22], calcium channel blockers [23] and angiotensin converting enzyme (ACE) inhibitors [24–26] have been documented to exhibit cardioprotective actions in diabetes. Some agents like buspirone is also reported to be beneficial owing to its effect on serotonin [27]; statins are reported to reduce cardiovascular mortality in diabetics [28, 29].

18.2 Oxidative Stress and Diabetic Cardiomyopathy

Oxidative stress plays a major role in diabetic cardiomyopathy. During diabetes, hyperglycemia and insulin resistance produces increase in nicotinamide adenine dinucleotide and flavin adenine dinucleotide flux to the mitochondrial respiratory

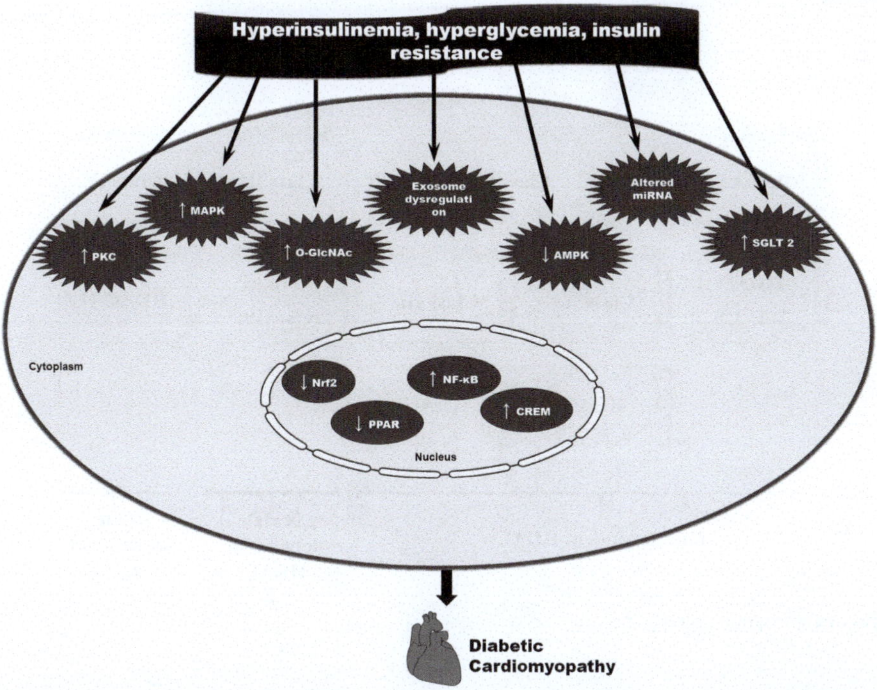

Fig. 18.2 Molecular mechanisms and signaling pathways which are altered during diabetes and lead to cardiomyopathy. *PKC* Protein kinase C, *MAPK* Mitogen activated protein kinase, *O-GlcNAc* O-linked N-acetylglucosamine, *AMPK* AMP-activated protein kinase, *SGLT-2* sodium–glucose cotransporter-2, *Nrf2* nuclear factor erythroid 2–related factor 2, *PPAR* peroxisome proliferator-activated receptor, *NF-κB* nuclear factor kappa B, *CREM* cyclic adenosine 5′-monophosphatere-sponsive element modulator

chain [30]. This leads to causes hyperpolarization of the inner mitochondrial membrane alongwith inhibition of electron transport in complex II and produces excessive production of reactive oxygen species (ROS). In addition to this, uncoupling of nitric oxide synthase along with excessive xanthine oxidase, and microsomal P-450 enzyme activity are additional sources of ROS. ROS increases formation of advanced glycation end products, enhances polyol pathway and inhibits e-nitric oxide synthase (e-NOS) activity leading to cardiomyopathy [31]. Increased nicotinamide adenine dinucleotide phosphate oxidase, which is mediated through renin angiotensin aldosterone system (RAAS), also increases formation of ROS and it is established that RAAS is implicated in diabetic cardiomyopathy [32, 33]. Targeting mitochondrial oxidative stress have provided some evidences pertaining to prevention of diabetic cardiomyopathy. A free radical scavenger d-Arg-2′, 6′-dimethyltyrosine-Lys-Phe-NH2 (SS31) is stated to prevent cardiac hypertrophy, diastolic dysfunction, cardiac hypertrophy and promotes oxidative phosphorylation [34, 35]. Similar effects are exerted by several other anti-oxidants viz. coenzyme Q10 [36], edavarone [37], vitamin E [38], resveratrol [39, 40], taxifolin [41], procyanidin B2

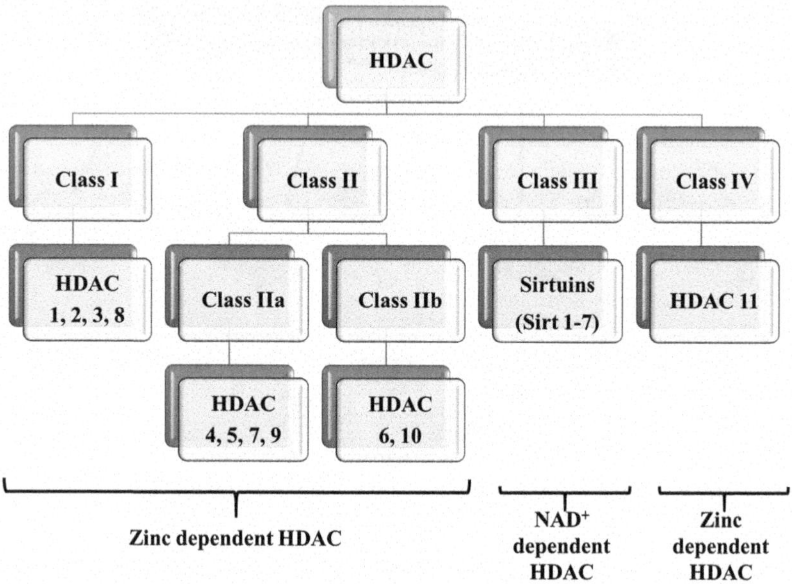

Fig. 18.3 HDAC and their classes

[42], luteolin [43], tempol [44], L-glutamine [45], wogonin [46], curcumin [47], β-casomorphin-7 [48], rutin [49, 50], α-tocopherol [51], α-lipoic acid [52], phlorizin [53], N-acetyl cysteine [54], apocynin [55] and genistein [56].

18.3 Histone Deacetylases (HDACs)

Histone deacetylases (HDACs) are group of key enzymes involved in epigenetic regulation of genes by removing acetyl group from lysine residue of histone proteins and conversely histone acetyl transferases (HAT) are involved in acetylation of histone proteins. 18 types of HDACs have been identified till date which are divided into four classes as depicted in Fig. 18.3.

HDACs have been implicated in several disorders including cancer [57, 58], inflammatory bowel diseases [59], psychiatric disorders [60], Alzheimer's disease [61], AIDS [62], ischemic stroke [63], kidney diseases [64] and cardiovascular diseases [65] including cardiac hypertrophy [66]. Furthermore, HDACs have also been known to regulate several important genes which are related to diabetes [67, 68].

18.4 Histone Modifications in Diabetic Cardiomyopathy

Epigenetic modifications involving alterations in histone acetylation and deacetylation are described to be involved in diabetes and its complications [69, 70]. Such epigenetic modifications play a key role in diabetic cardiomyopathy by regulating the pathophysiological and molecular signaling pathways. Evidences support the role of HDAC in pathophysiology of diabetic cardiomyopathy and several studies involving use of HDAC inhibitors have shown to exhibit protective effect which is discussed in the upcoming sections.

18.4.1 Histone Acetylation in Diabetic Cardiomyopathy

Documented evidences suggest that there is upregulation of epigenetic mechanisms in diabetic heart as depicted from increase in H3K9 and H3K23 acetylation alongwith H3K4 dimethylation, and phosphorylation at serine 10 in the hearts of diabetic rats with renal failure [71]. In spontaneously diabetic Goto-Kakizaki rat with myocardial infarction, it was found that phosphorylation and acetylation of histone proteins regulate Akt – FOXO3a, Sirt1 – p53 and the MAPK p38 pathways and play a role in cardiac remodeling [72]. Certain *in vitro* evidences also suggest role of altered histone acetylation in diabetic cardiac complications. In one study, endothelial cells when subjected to high glucose concentration produced a significant increase in p300 levels resulting in increased histone acetylation in the promoter regions of genes pertaining to extracellular matrix. Additionally, when p300 was inhibited, there was prevention of cardiomyocyte hypertrophy, thus, suggesting a role of histone acetylation in regulation of gene expression [73, 74]. On similar lines, H9C2 rat cardiomyocyte cells when incubated with high glucose, depicted increase in association of p53 with HDAC1 along with decrease in association of acetylated histone-4 with the insulin like growth factor (IGF)-1R promoter [75].

18.4.2 HDAC in Diabetic Cardiomyopathy

O-linked β-N-acetylglucosamine (O-GlcNAc) is one of the cytosolic molecule which causes acylation of serine and threonine residues of proteins and regulates various cell functions including cardiac actions [76]. It has been reported that in diabetic hearts, there is an increase in O-GlcNAc levels which is also co-related to diabetic cardiomyopathy [77]. Decreased levels of HDAC1, HDAC2 along with increase in HDAC3 mRNA expression has been documented in the heart of diabetic animals [78]. Moreover, it was suggested that the diabetic heart functions were compromised due to decrease in physical association of O-GlcNAc with mSin3A/HDAC1/2 and this was prevented by physical exercise [78]. On similar lines, a significant increase in cardiac HDAC3 activity has been reported in OVE26 diabetic animals in which HDAC3 down-regulated expression of one MAP kinase phosphatase, DUSP5 by deacetylating histone H3 at the primer region of DUSP5 gene. This led to pro-hypertrophic effect in the diabetic animals [79]. In another report, specific cardiac Hdac3a knock-out mice depicted cardiac hypertrophy along with altered glucose metabolism due to downregulation of GLUT4 and upregulation of pyruvate dehydrogenase kinase 4 (PDK4) again confirming the role of HDAC3 in cardiomyopathy and diabetic states [80].

Poly adenosine diphosphate ribose polymerase 1 (PARP1) is reported to induce oxidative stress and play a key role in diabetic cardiomyopathy [81]. Documented evidences have suggested that wild rats fed with high galactose exhibited increased p300 transcript levels along with increased levels of deacetylated lysine while $PARP^{-/-}$ mice fed a galactose enriched diet illustrated similar p300 levels as compared to wild control mice [82]. Recently, Wu et al. [83] reported that ischemic reperfusion injury produced a significant increase in HDAC activity with attenuated ratio of Ac-H3/H3 and Ac-H4/H4 in diabetic rats induced by single dose of streptozotocin which was attributed to decrease in phosphorylated levels of Akt.

18.4.3 Sirtuins in Diabetic Cardiomyopathy

Silent Information Regulator 2 (SIR2) proteins i.e. Sirtuins belong to class III HDACs which are NAD+ dependent enzymes [84]. Sirtuins are reported to play important role in diabetic cardiomyopathy through modulation of several pathways. SIRT1 is evidenced to upregulate and phosphorylate ERK1/2 at Thr202/tyr204 residue which causes expression of scaffold protein Homer1α and this Homer1α acts as Ca^{+2} dependent endogenous scavenger of ROS [85]. Similarly, SIRT1 overexpression produce decrease in superoxide generation and increases activity of superoxide dismutase enzyme in ischemic reperfusion injury in diabetic animals [86]. Moreover, SIRT1 is reported to down regulate pro-apoptotic substances during ischemic reperfusion injury and thereby inhibits apoptosis and reduce oxidative stress induced cardiomyopathy [87]. In diabetes, increase in glucose level increase P300 which activates transforming growth factor-β (TGF-β) and SIRT1 inhibits P300 and thereby causes reduction in the activity of TGF-β and prevents heart failure in diabetes [88]. Autophagy, which is an intracellular process of protein digestion is documented to be suppressed in diabetes, specifically cardiac autophagy [89]. Upregulation of fibroblast growth factor 21 (FGF – 21) is stated to increase autophagy mediated by SIRT1, thereby prevent diabetes induced cardiac fibrosis [90].

18.4.4 Therapeutic Opportunity by Modulation of HDACs

Various evidences have suggested the role of HAT and HDACs in regulation of imperative genes which are involved in the pathophysiology diabetic cardiomyopathy. Hence, it is worthwhile to explore targeting of these established cascades for as a therapeutic approach. In an animal model of type 1 diabetes induced by streptozotocin in ICR mice, HDAC inhibitor, sodium butyrate when given in drinking water, produced inhibition of interstitial fibrosis and myocyte hypertrophy in the heart of diabetic mice and preserved ventricular function and prevented cardiac remodeling [91]. Sodium butyrate produced an increase in myocardial superoxide dismutase, glucose transporter-1 (GLUT-1) and GLUT-4 protein levels and attenuated myocardial apoptosis [91]. Similarly, magnesium valproate in type 1 diabetes in rats induced by streptozotocin prevented diabetes induced dyslipidemia, cardiac hypertrophy and fibrosis [92]. Although the authors attributed this effect to estrogen modulatory effects of magnesium valproate, magnesium valproate is previously reported to exhibit HDAC inhibitory actions [93]. Furthermore, magnesium valproate was documented to control diabetic dyslipidemia, left ventricular hypertrophy, left ventricular dysfunction and cardiac oxidative stress in an animal model of type 2 diabetes mellitus induced cardiovascular complications by administering streptozotocin in 2 day old pups of wistar rats [94].

Peroxisome proliferator-activated receptors (PPARs) are described to regulate cardiac glucose and lipid homeostasis which in turn are thought to be regulated by HDACs owing to their anti-inflammatory effects. MPT0E014, a pan HDAC inhibitor is stated to reduce dyslipidemia and myocardial inflammation in nicotinamide and streptozotocin induced diabetic animals [95]. MPT0E014 increased the protein expression of PPARα and PPARδ while it reduced the expression of PPARγ suggesting the link between PPAR and HDAC [95]. RGFP966, selective HDAC3 inhibitor and valproic acid, pan HDAC inhibitor are documented to prevent cardiac hypertrophy, cardiac fibrosis, cardiac collagen accumulation and preserved cardiac function in OVE 26 type 1 diabetic mice [79]. DUSP5 is a nuclear ERK1/2-specific phosphatase which regulates cardiac hypertetoph and RGFP966 suppressed diabetes induced ERK1/2 activation in nucleus and increase DUSP5 expression [79]. In another study involving type 2 diabetes, MPT0E014 is shown to exhibit cardioprotective effects via increased protein expression of phosphorylated 5′ adenosine monophosphate-activated protein kinase (AMPK) α2, GLUT-4, and insulin receptor substrate-1 (IRS), alongwith decrease in expression of p-mTOR-S2448 and poly adenosine diphosphate ribose polymerase 1 (PARP1), TNF-α and IL-6 [96].

In a recent study, trichostatin A is reported to prevent myocardial injury in type 1 diabetic rats subjected to ischemic reperfusion injury [83]. Trichostatin A reduced cardiac biomarkers, infarct size, improved hemodynamics, diminished apoptosis and activated Akt/FoXo3 pathway. The authors also studied effect of trichostatin A on H9C2 cardiomyocytes subjected to high glucose concentrations and found that reduced overindulgence of mitochondrial membrane potential, protected the integrity of mitochondrial permeability transition pore (mPTP), and decreased cell apoptosis and these effects were revered when Akt inhibitor was given confirming the role of Akt/FoXo3 pathway [83]. In another *in vitro* study using HUVEC cell line, trichostatin A is reported to prevent glucose mediated upregulation of PARP, fibronectin and p300 mRNA levels [82].

Nuclear factor-like 2 (Nrf2) transcription factor mediates the induction of antioxidant and cytoprotective genes and thereby controls oxidative stress and is reported to play a role in several diseases including diabetes [97]. Furthermore, histone acetyltransferase P300/CREB-binding protein (CBP) mediated acetylation and HDAC mediated deacetylation of lysine residures regulate the Nrf2 activity and its nucleo-cytoplasmic localization [98, 99]. Sulforaphane is an inhibitor of Keap1-mediated degradation of Nrf2 and also HDAC inhibitor and in diabetic mice it is reported to preserve cardiac hemodynamics, control cardiac hypertrophy, reduce atrial natriuretic peptide, inflammation, fibrosis and oxidative stress [100].

Certain drugs have been reported to exhibit cardioprotective actions via modulation of SIRT expression. Garlic produces improvement in SIRT-3 and superoxide dismutase activity in STZ induced diabetic animals and thereby prevented cardiac oxidative stress [101]. Similarly, resveratrol in animals models of diabetes prevents diabetes induced changes in SIRT1 expression in type 1 diabetic heart and SIRT1, 2,5 and SIRT-5 expression in type 2 diabetic heart [102]. Further, L-arginine exhibits its cardioprotective effects in diabetic hearts owing to its SIRT1 modulatory action [103].

18.5 Conclusions

Research in diabetes has come a long way after the discovery of insulin. Several pathways responsible for diabetic complications are known and molecular mechanisms are being explored. Given the fact that diabetes is associated with environmental factors modulate genetic factors, epigenetic mechanisms in diabetic cardiomyopathy is one of the key regulating mechanisms. Histone modification is one such epigenetic mechanism. However, limited research is being carried out in this direction to understand the role of HAT and HDAC in diabetic cardiomyopathy. Moreover, there is a requirement of research to determine the specific isoforms of HDAC which are dysregulated during diabetic cardiomypathy. Some studies have been carried out with respect to HDAC inhibitors in diabetic cardiomyopathy but again are limited. Since sirtuins are one class of HDAC which behave differently than other HDACs, future studies should be targeted in developing novel scaffolds which serve as modulators of different class of HDACs. This will help in future for developing effective therapies for hyperglycemia induced cardiovascular complication and improve patients' quality of life.

References

1. Rubler S, Dlugash J, Yuceoglu YZ, Kumral T, Branwood AW, Grishman A (1972) New type of cardiomyopathy associated with diabetic glomerulosclerosis. Am J Cardiol 30:595–602
2. Raghunathan S, Patel BM (2013) Therapeutic implications of small interfering RNA in cardiovascular diseases. Fundam Clin Pharmacol 27(1):1–20
3. Rawal H, Patel BM (2018) Opioids in cardiovascular disease: therapeutic options. J Cardiovasc Pharmacol Ther 23(4):279–291
4. Goyal BR, Mehta AA (2013) Diabetic cardiomyopathy: pathophysiological mechanisms and cardiac dysfunction. Hum Exp Toxicol 32(6):571–590
5. Patel BM, Mehta AA (2013) The choice of anti-hypertensive agents in diabetic subjects. Diab Vasc Dis Res 10(50):385–396
6. Jia G, Hill MA, Sowers JR (2018) Diabetic cardiomyopathy: an update of mechanisms contributing to this clinical entity. Circ Res 122(4):624–638
7. Wong AK, Symon R, AlZadjali MA, Ang DS, Ogston S, Choy A et al (2012) The effect of metformin on insulin resistance and exercise parameters in patients with heart failure. Eur J Heart Fail 14:1303–1310
8. Xie Z, Lau K, Eby B, Lozano P, He C, Pennington B et al (2011) Improvement of cardiac functions by chronic metformin treatment is associated with enhanced cardiac autophagy in diabetic OVE26 mice. Diabetes 60:1770–1778

9. Younce CW, Burmeister MA, Ayala JE (2013) Exendin-4 attenuates high glucose-induced cardiomyocyte apoptosis via inhibition of endoplasmic reticulum stress and activation of SERCA2a. Am J Phys Cell Physiol 304:C508–C518
10. Doehner W, Frenneaux M, Anker SD (2014) Metabolic impairment in heart failure: the myocardial and systemic perspective. J Am Coll Cardiol 64:1388–1400
11. Witteles RM, Keu KV, Quon A, Tavana H, Fowler MB (2012) Dipeptidyl peptidase 4 inhibition increases myocardial glucose uptake in nonischemic cardiomyopathy. J Card Fail 18:804–809
12. Bostick B, Habibi J, Ma L, Aroor A, Rehmer N, Hayden MR et al (2014) Dipeptidyl peptidase inhibition prevents diastolic dysfunction and reduces myocardial fibrosis in a mouse model of Western diet induced obesity. Metabolism 63:1000–1011
13. Inzucchi SE, Zinman B, Wanner C, Ferrari R, Fitchett D, Hantel S et al (2015) SGLT-2 inhibitors and cardiovascular risk: proposed pathways and review of ongoing outcome trials. Diab Vasc Dis Res 12:90–100
14. Zinman B, Wanner C, Lachin JM, Fitchett D, Bluhmki E, Hantel S et al (2015) Empagliflozin, cardiovascular outcomes, and mortality in type 2 diabetes. N Engl J Med 373:2117–2128
15. Caglayan E, Stauber B, Collins AR, Lyon CJ, Yin F, Liu J et al (2008) Differential roles of cardiomyocyte and macrophage peroxisome proliferator-activated receptor gamma in cardiac fibrosis. Diabetes 57:2470–2479
16. Goyal BR, Solanki N, Goyal RK, Mehta AA (2009) Investigation into the cardiac effects of spironolactone in the experimental model of type 1 diabetes. J Cardiovasc Pharmacol 54(6):502–509
17. Patel BM, Kakadiya J, Goyal RK, Mehta AA (2013) Effect of spironolactone on cardiovascular complications associated with type-2 diabetes in rats. Exp Clin Endocrinol Diabetes 121(08):441–447
18. Sharma V, McNeill JH (2011) Parallel effects of β-adrenoceptor blockade on cardiac function and fatty acid oxidation in the diabetic heart: confronting the maze. World J Cardiol 3:281–302
19. Thomas CM, Yong QC, Seqqat R, Chandel N, Feldman DL, Baker KM et al (2013) Direct renin inhibition prevents cardiac dysfunction in a diabetic mouse model: comparison with an angiotensin receptor antagonist and angiotensin-converting enzyme inhibitor. Clin Sci (Lond) 124:529–541
20. Machackova J, Liu X, Lukas A, Dhalla NS (2004) Renin-angiotensin blockade attenuates cardiac myofibrillar remodelling in chronic diabetes. Mol Cell Biochem 261:271–278
21. Goyal BR, Parmar K, Goyal RK, Mehta AA (2011) Beneficial role of telmisartan on cardiovascular complications associated with STZ-induced type-2 diabetic rats. Pharmacol Rep 63(4):956–966
22. Goyal BR, Mesariya P, Goyal RK, Mehta AA (2008) Effect of telmisartan on cardiovascular complications associated with STZ-diabetic rats. Mol Cell Biochem 314(1–2):123–131
23. Mohamad HE, Askar ME, Hafez MM (2011) Management of cardiac fibrosis in diabetic rats: the role of peroxisome proliferator activated receptor gamma (PPAR-gamma) and calcium channel blockers (CCBs). Diabetol Metab Syndr 3:4
24. Symeonides P, Koulouris S, Vratsista E, Triantafyllou K, Ioannidis G, Thalassinos N et al (2007) Both ramipril and telmisartan reverse indices of early diabetic cardiomyopathy: a comparative study. Eur J Echocardiogr 8:480–486
25. Patel BM, Bhadada SV (2014) Type 2 diabetes induced cardiovascular complications: comparative evaluation of spironolactone, atenolol, metoprolol, ramipril and perindopril. Clin Exp Hypertens 36(5):340–347
26. Goyal BR, Patel MM, Bhadada SV (2011) Comparative evaluation of spironolactone, atenolol, metoprolol, ramipril and perindopril on diabetes induced cardiovascular complications in type 1 diabetes in rats. Int J Diabetes Metab 19(1):11–18
27. Raghunathan S, Tank P, Bhadada SV, Patel BM (2014). Evaluation of Buspirone on Streptozotocin induced Type 1 diabetes and its associated complications. Biomed Res Int 2014:9. https://doi.org/10.1155/2014/948427. Article ID 948427

28. Chen YH, Feng B, Chen ZW (2012) Statins for primary prevention of cardiovascular and cerebrovascular events in diabetic patients without established cardiovascular diseases: a meta-analysis. Exp Clin Endocrinol Diabetes 120:116–120
29. Dai QM, Lu J, Liu NF (2011) Fluvastatin attenuates myocardial interstitial fibrosis and cardiac dysfunction in diabetic rats by inhibiting over-expression of connective tissue growth factor. Chin Med J 124:89–94
30. Teshima Y, Takahashi N, Nishio S, Saito S, Kondo H, Fukui A et al (2014) Production of reactive oxygen species in the diabetic heart. Roles of mitochondria and NADPH oxidase. Circ J 78:300–306
31. Giacco F, Brownlee M (2010) Oxidative stress and diabetic complications. Circ Res 107:1058–1070
32. Thomas CM, Yong QC, Rosa RM, Seqqat R, Gopal S, Casarini DE et al (2014) Cardiac-specific suppression of NF- κB signaling prevents diabetic cardiomyopathy via inhibition of the renin-angiotensin system. Am J Physiol Heart Circ Physiol 307:H1036–H1045
33. Patel BM, Mehta AA (2012) Aldosterone and angiotensin: role in diabetes and cardiovascular diseases. Eur J Pharmacol 697(1–3):1–12
34. Szeto HH (2014) First-in-class cardiolipin-protective compound as a therapeutic agent to restore mitochondrial bioenergetics. Br J Pharmacol 171:2029–2050
35. Xu YJ, Tappia PS, Neki NS, Dhalla NS (2014) Prevention of diabetes-induced cardiovascular complications upon treatment with antioxidants. Heart Fail Rev 19:113–121
36. Huynh K, Kiriazis H, Du XJ, Love JE, Jandeleit-Dahm KA, Forbes JM et al (2012) Coenzyme Q10 attenuates diastolic dysfunction, cardiomyocyte hypertrophy and cardiac fibrosis in the db/db mouse model of type 2 diabetes. Diabetologia 55:1544–1553
37. Ji L, Liu Y, Zhang Y, Chang W, Gong J, Wei S et al (2016) The antioxidant edaravone prevents cardiac dysfunction by suppressing oxidative stress in type 1 diabetic rats and in high-glucose-induced injured H9c2 cardiomyoblasts. Can J Physiol Pharmacol 94(9):996–1006
38. Abdel-Raheem MH, Salim SU, Mosad E, Al-Rifaay A, Salama HS, Hasan-Ali H (2015) Antiapoptotic and antioxidant effects of carvedilol and vitamin E protect against diabetic nephropathy and cardiomyopathy in diabetic Wistar albino rats. Horm Metab Res 47(2):97–106
39. Mohammadshahi M, Haidari F, Soufi FG (2014) Chronic resveratrol administration improves diabetic cardiomyopathy in part by reducing oxidative stress. Cardiol J 21(1):39–46
40. Wu H, Li GN, Xie J, Li R, Chen QH, Chen JZ et al (2016) Resveratrol ameliorates myocardial fibrosis by inhibiting ROS/ERK/TGF-β/periostin pathway in STZ-induced diabetic mice. BMC Cardiovasc Disord 16:5
41. Sun X, Chen RC, Yang ZH, Sun GB, Wang M, Ma XJ et al (2014) Taxifolin prevents diabetic cardiomyopathy in vivo and in vitro by inhibition of oxidative stress and cell apoptosis. Food Chem Toxicol 63:221–232
42. Luan SS, Yu F, Li BY, Qin RJ, Li XL, Cai Q et al (2014) Quantitative proteomics study of protective effects of grape seed procyanidin B2 on diabetic cardiomyopathy in db/db mice. Biosci Biotechnol Biochem 78(9):1577–1583
43. Wang G, Li W, Lu X, Bao P, Zhao X (2012) Luteolin ameliorates cardiac failure in type I diabetic cardiomyopathy. J Diabetes Complicat 26(4):259–265
44. Taye A, Abouzied MM, Mohafez OM (2013) Tempol ameliorates cardiac fibrosis in streptozotocin-induced diabetic rats: role of oxidative stress in diabetic cardiomyopathy. Naunyn Schmiedeberg's Arch Pharmacol 386(12):1071–1080
45. Badole SL, Jangam GB, Chaudhari SM, Ghule AE, Zanwar AA (2014) L-glutamine supplementation prevents the development of experimental diabetic cardiomyopathy in streptozotocin-nicotinamide induced diabetic rats. PLoS One 9(3):e92697
46. Khan S, Zhang D, Zhang Y, Li M, Wang C (2016) Wogonin attenuates diabetic cardiomyopathy through its anti-inflammatory and anti-oxidative properties. Mol Cell Endocrinol 428:101–108

47. Soetikno V, Sari FR, Sukumaran V, Lakshmanan AP, Mito S, Harima M et al (2012) Curcumin prevents diabetic cardiomyopathy in streptozotocin-induced diabetic rats: possible involvement of PKC-MAPK signaling pathway. Eur J Pharm Sci 47(3):604–614
48. Han DN, Zhang DH, Wang LP, Zhang YS (2013) Protective effect of β-casomorphin-7 on cardiomyopathy of streptozotocin-induced diabetic rats via inhibition of hyperglycemia and oxidative stress. Peptides 44:120–126
49. Guimaraes JF, Muzio BP, Rosa CM, Nascimento AF, Sugizaki MM, Fernandes AA et al (2015) Rutin administration attenuates myocardial dysfunction in diabetic rats. Cardiovasc Diabetol 14:90
50. Saklani R, Gupta SK, Mohanty IR, Kumar B, Srivastava S, Mathur R (2016) Cardioprotective effects of rutin via alteration in TNF-α, CRP, and BNP levels coupled with antioxidant effect in STZ-induced diabetic rats. Mol Cell Biochem 420(1–2):65–72
51. Celik T, Yuksel C, Iyisoy A (2010) Alpha tocopherol use in the management of diabetic cardiomyopathy: lessons learned from randomized clinical trials. J Diabetes Complicat 24(4):286–288
52. Hegazy SK, Tolba OA, Mostafa TM, Eid MA, El-Afify DR (2013) Alpha-lipoic acid improves subclinical left ventricular dysfunction in asymptomatic patients with type 1 diabetes. Rev Diabet Stud 10(1):58–67
53. Cai Q, Li B, Yu F, Lu W, Zhang Z, Yin M et al (2013) Investigation of the protective effects of phlorizin on diabetic cardiomyopathy in db/db mice by quantitative proteomics. J Diabetes Res 2013:263845
54. Su W, Zhang Y, Zhang Q, Xu J, Zhan L, Zhu Q et al (2016) N-acetylcysteine attenuates myocardial dysfunction and postischemic injury by restoring caveolin-3/eNOS signaling in diabetic rats. Cardiovasc Diabetol 15(1):146
55. Rosa CM, Gimenes R, Campos DH, Guirado GN, Gimenes C, Fernandes AA et al (2016) Apocynin influence on oxidative stress and cardiac remodeling of spontaneously hypertensive rats with diabetes mellitus. Cardiovasc Diabetol 15(1):126
56. Gupta SK, Dongare S, Mathur R, Mohanty IR, Srivastava S, Mathur S et al (2015) Genistein ameliorates cardiac inflammation and oxidative stress in streptozotocin-induced diabetic cardiomyopathy in rats. Mol Cell Biochem 408(1–2):63–72
57. McClure JJ, Li X, Chou CJ (2018) Advances and challenges of HDAC inhibitors in cancer therapeutics. Adv Cancer Res 138:183–211
58. Patel M, Patel BM (2018) Repurposing of sodium valproate in colon cancer associated with diabetes mellitus: role of HDAC inhibition. Eur J Pharm Sci 121:188–199
59. Felice C, Lewis A, Armuzzi A, Lindsay JO, Silver A (2015) Review article: selective histone deacetylase isoforms as potential therapeutic targets in inflammatory bowel diseases. Aliment Pharmacol Ther 41(1):26–38
60. Ganguly S, Seth S (2018) A translational perspective on histone acetylation modulators in psychiatric disorders. Psychopharmacology (Berl). https://doi.org/10.1007/s00213-018-4947-z. [Epub ahead of print]
61. Dubey H, Gulati K, Ray A (2018) Recent studies on cellular and molecular mechanisms in Alzheimer's disease: focus on epigenetic factors and histone deacetylase. Rev Neurosci 29(3):241–260
62. Sachweh MC, Drummond CJ, Higgins M, Campbell J, Laín S (2013) Incompatible effects of p53 and HDAC inhibition on p21 expression and cell cycle progression. Cell Death Dis 4:e533
63. Kim HJ, Rowe M, Ren M, Hong JS, Chen PS, Chuang DM (2007) Histone deacetylase inhibitors exhibit anti-inflammatory and neuroprotective effects in a rat permanent ischemic model of stroke: multiple mechanisms of action. J Pharmacol Exp Ther 321:892–901
64. Fontecha-Barriuso M, Martin-Sanchez D, Ruiz-Andres O, Poveda J, Sanchez-Niño MD, Valiño-Rivas L et al (2018) Targeting epigenetic DNA and histone modifications to treat kidney disease. Nephrol Dial Transplant. https://doi.org/10.1093/ndt/gfy009. [Epub ahead of print]

65. Yoon S, Eom GH (2016) HDAC and HDAC inhibitor: from cancer to cardiovascular diseases. Chonnam Med J 52(1):1–11
66. Patel BM (2018) Sodium butyrate controls cardiac hypertrophy in experimental models of rats. Cardiovasc Toxicol 18(1):1–8
67. Gray S, De Meyts P (2005) Role of histone and transcription factor acetylation in diabetes pathogenesis. Diabetes Metab Res Rev 21(5):416–433
68. Miao F, Gonzalo I, Lanting L, Natarajan R (2004) In vivo chromatin remodeling events leading to inflammatory gene transcription under diabetic conditions. J Biol Chem 279:18091–18097
69. Khullar M, Cheema BS, Raut SK (2017) Emerging evidence of epigenetic modifications in vascular complication of diabetes. Front Endocrinol (Lausanne) 8:237
70. Al-Haddad R, Karnib N, Assaad RA, Bilen Y, Emmanuel N, Ghanem A et al (2016) Epigenetic changes in diabetes. Neurosci Lett 625:64–69
71. Gaikwad A, Sayyed S, Lichtnekert J, Tikoo K, Anders H (2010) Renal failure increases cardiac histone h3 acetylation, dimethylation, and phosphorylation and the induction of cardiomyopathy-related genes in type 2 diabetes. Am J Pathol 176(3):1079–1083
72. Vahtola E, Louhelainen M, Forste'n H, Merasto S, Raivio J, Kaheinen P et al (2010) Sirtuin1-p53, forkhead box O3a, p38 and post-infarct cardiac remodeling in the spontaneously diabetic Goto-Kakizaki rat. Cardiovasc Diabetol 27(9):5
73. Chen S, Feng B, George B, Chakrabarti R, Chen M, Chakrabarti S (2010) Transcriptional coactivator p300 regulates glucose induced gene expression in endothelial cells. Am J Physiol Endocrinol Metab 298:E127–E137
74. Feng B, Chen S, Chiu J, George B, Chakrabarti S (2008) Regulation of cardiomyocyte hypertrophy in diabetes at the transcriptional level. Am J Physiol Endocrinol Metab 294:E1119–E1126
75. Yu XY, Geng YJ, Liang JL, Lin QX, Lin SG, Zhang S et al (2010) High levels of glucose induce apoptosis in cardiomyocyte via epigenetic regulation of the insulin-like growth factor receptor. Exp Cell Res 316(17):2903–2909
76. Wang Z, Gucek M, Hart GW (2008) Cross-talk between GlcNAcylation and phosphorylation: site-specific phosphorylation dynamics in response to globally elevated O-GlcNAc. Proc Natl Acad Sci U S A 105:13793–13798
77. Marsh SA, Collins HE, Chatham JC (2014) Protein O-GlcNAcylation and cardiovascular (patho) physiology. J Biol Chem 289:34449–34456
78. Cox EJ, Marsh SA (2013) Exercise and diabetes have opposite effects on the assembly and O-GlcNAc modification of the mSin3A/HDAC1/2 complex in the heart. Cardiovasc Diabetol 12:101
79. Xu Z, Tong Q, Zhang Z, Wang S, Zheng Y, Liu Q et al (2017) Inhibition of HDAC3 prevents diabetic cardiomyopathy in OVE26 mice via epigenetic regulation of DUSP5-ERK1/2 pathway. Clin Sci 131(15):1841–1857
80. Montgomery RL, Potthoff MJ, Haberland M, Qi X, Matsuzaki S, Humphries KM et al (2008) Maintenance of cardiac energy metabolism by histone deacetylase 3 in mice. J Clin Invest 118(11):3588–3597
81. Pacher P, Liaudet L, Soriano FG, Mabley JG, Szabo E, Szabo C (2002) The role of poly(ADPribose) polymerase activation in the development of myocardial and endothelial dysfunction in diabetes. Diabetes 51(2):514–521
82. Chiu J, Farhangkhoee H, Xu BY, Chen S, George B, Chakrabarti S (2008) PARP mediates structural alterations in diabetic cardiomyopathy. J Mol Cell Cardiol 45(3):385–393
83. Wu Y, Leng Y, Meng Q, Xue R, Zhao B, Zhan L et al (2017) Suppression of excessive histone deacetylases activity in diabetic hearts attenuates myocardial ischemia/reperfusion injury via mitochondria apoptosis pathway. J Diabetes Res 2017:8208065
84. Winnik S, Auwerx J, Sinclair DA, Matter CM (2015) Protective effects of sirtuins in cardiovascular diseases: from bench to bedside. Eur Heart J 36(48):3404–3412
85. Duan J, Yin Y, Wei G, Cui J, Zhang E, Guan Y et al (2015) Chikusetsu saponin IVa confers cardioprotection via SIRT1/ ERK1/2 and Homer1a pathway. Sci Rep 5:18123

86. Koka S, Aluri HS, Xi L, Lesnefsky EJ, Kukreja RC (2014) Chronic inhibition of phosphodiesterase 5 with tadalafil attenuates mitochondrial dysfunction in type 2 diabetic hearts: role of NO/SIRT1/PGC-1a signaling. Am J Physiol Heart Circ Physiol 306(11):H1558–H1568
87. Becatti M, Taddei N, Cecchi C, Nassi N, Nassi PA, Fiorillo C (2012) SIRT1 modulates MAPK pathways in ischemic–reperfused cardiomyocytes. Cell Mol Life Sci 69(13):2245–2260
88. Bugyei-Twum A, Advani A, Advani SL, Zhang Y, Thai K, Kelly DJ et al (2014) High glucose induces Smad activation via the transcriptional coregulator p300 and contributes to cardiac fibrosis and hypertrophy. Cardiovasc Diabetol 13(1):1
89. He C, Zhu H, Li H, Zou MH, Xie Z (2013) Dissociation of Bcl-2–Beclin1 complex by activated AMPK enhances cardiac autophagy and protects against cardiomyocyte apoptosis in diabetes. Diabetes 62(4):1270–1281
90. Zhang J, Cheng Y, Gu J, Wang S, Zhou S, Wang Y et al (2016) Fenofibrate increases cardiac autophagy via FGF21/SIRT1 and prevents fibrosis and inflammation in the hearts of Type 1 diabetic mice. Clin Sci 130(8):625–641
91. Chen Y, Du J, Zhao YT, Zhang L, Lv G, Zhuang S et al (2015) Histone deacetylase (HDAC) inhibition improves myocardial function and prevents cardiac remodeling in diabetic mice. Cardiovasc Diabetol 14:99
92. Rabadiya S, Bhadada S, Dudhrejiya A, Vaishnav D, Patel BM (2018) Magnesium valproate ameliorates type 1 diabetes and cardiomyopathy in diabetic rats through estrogen receptors. Biomed Pharmacother 97:919–927
93. Raghunathan S, Goyal RK, Patel BM (2017) Selective inhibition of HDAC2 by magnesium valproate attenuates cardiac hypertrophy. Can J Physiol Pharmacol 95(3):260–267
94. Patel BM, Raghunathan S, Porwal U (2014) Cardioprotective effects of magnesium valproate in type 2 diabetes mellitus. Eur J Pharmacol 728:128–134
95. Lee T-I, Kao Y-H, Tsai W-C, Chung C-C, Chen Y-C, Chen Y-J (2016) HDAC inhibition modulates cardiac PPARs and fatty acid metabolism in diabetic cardiomyopathy. PPAR Res 2016:5938740
96. Lee TI, Bai KJ, Chen YC, Lee TW, Chung CC, Tsai WC et al (2018) Histone deacetylase inhibition of cardiac autophagy in rats on a high fat diet with low dose streptozotocin-induced type 2 diabetes mellitus. Mol Med Rep 17(1):594–601
97. Uruno A, Yagishita Y, Yamamoto M (2015) The Keap1-Nrf2 system and diabetes mellitus. Arch Biochem Biophys 566:76–84
98. Kawai Y, Garduno L, Theodore M, Yang J, Arinze IJ (2011) Acetylation deacetylation of the transcription factor Nrf2 (nuclear factor erythroid 2-related factor 2) regulates its transcriptional activity and nucleocytoplasmic localization. J Biol Chem 286:7629–7640
99. Lee DY, Lee CI, Lin TE, Lim SH, Zhou J, Tseng YC et al (2012) Role of histone deacetylases in transcription factor regulation and cell cycle modulation in endothelial cells in response to disturbed flow. Proc Natl Acad Sci U S A 109:1967–1972
100. Ichikawa T, Li J, Meyer CJ, Janicki JS, Hannink M, Cui T (2009) Dihydro-CDDO trifluoroethyl amide (dh404), a novel Nrf2 activator, suppresses oxidative stress in cardiomyocytes. PLoS One 4:e8391
101. Sultana MR, Bagul PK, Katare PB, Anwar Mohammed S, Padiya R, Banerjee SK (2016) Garlic activates SIRT-3 to prevent cardiac oxidative stress and mitochondrial dysfunction in diabetes. Life Sci 164:42–51
102. Bagul PK, Dinda AK, Banerjee SK (2015) Effect of resveratrol on sirtuins expression and cardiac complications in diabetes. Biochem Biophys Res Commun 468(1–2):221–227
103. Rizk SM, El-Maraghy SA, Nassar NN (2014) A novel role for SIRT-1 in L-arginine protection against STZ induced myocardial fibrosis in rats. PLoS One 9(12):e114560

Regulating Inflammatory Cytokines in the Diabetic Heart

19

Santosh K. Yadav, Tyler N. Kambis, and Paras K. Mishra

19.1 Introduction

Diabetes mellitus (DM) is a disease where patients have increased levels of blood glucose (fasting ≥126 mg/dL, non-fasting ≥200 mg/dL) and elevated levels (≥6.5%) of glycosylated hemoglobin (HbA1C) [1]. Prevalence of DM is very high (around 425 million people) and it is rapidly increasing (estimated to be 628.6 million people in 2045) [2] in the world. DM increases the risk of heart failure [3, 4] and mortality [5] compared to non-diabetics. Although men are more prone to DM than women are, the risk of DM-induced stroke and heart failure is almost similar in men and women [6–8]. There are two major types of DM: type-1 DM (T1DM) and type-2 DM (T2DM). More people (95%) have T2DM than T1DM (5%) [9]. T1DM has very high blood glucose levels (>500 mg/DL) due to lack of insulin production by pancreatic beta cells. Insulin treatment to T1DM patients that lowers the blood glucose levels causes high fluctuations in the glucose levels. Therefore, T1DM patients are challenged to adapt to this high fluctuations in the blood glucose levels. In the heart, cardiomyocytes require constant supply of energy in the form of ATP to maintain the contraction-relaxation cycle of the heart. In the T1DM heart, cardiomyocytes are always in stress condition to adapt to the high glucose fluctuations, especially after insulin injection. Thus, T1DM is more detrimental to the heart than T2DM, where the blood glucose levels are moderately high (around 250–350 mg/dL). In T2DM, pancreatic beta cells produce insulin but insulin signaling is

S. K. Yadav · T. N. Kambis
Department of Cellular and Integrative Physiology, University of Nebraska Medical Center, Omaha, NE, USA

P. K. Mishra (✉)
Department of Cellular and Integrative Physiology, University of Nebraska Medical Center, Omaha, NE, USA

Department of Anesthesiology, University of Nebraska Medical Center, Omaha, NE, USA
e-mail: paraskumar.mishra@unmc.edu

S. Chakraborti et al. (eds.), *Oxidative Stress in Heart Diseases*,
https://doi.org/10.1007/978-981-13-8273-4_19

defective that impairs cellular glucose uptake resulting in hyperglycemia. Because insulin is not completely used in glucose uptake, both insulin and glucose levels are high in T2DM. However, due to presence of insulin in T2DM, blood glucose levels do not increase to a very high level as ocuurs in T1DM. Therefore, blood-glucose reducing drugs such as metformin do not cause glucose fluctuations to a very high degree resulting in less adaptive stress to the cardiomyocytes in the T2DM heart. Thus, T1DM is more severe than T2DM for developing diabetic cardiomyopathy (DMCM). If T2DM remains untreated, it may lead to T1DM because constant hyperglycemia triggers pancreatic beta cells to work incessantly without rest leading to their death and subsequent devlopment of T1DM phenotype. Turning of T2DM into T1DM phenotype is a severe condition, where less insulin is produced and the available insulin is unable to transport glucose efficiently into the cell. Thus, these patients need treatment with both insulin injection (to maintain insulin level) and metformin (to improve glucose transport to the cell). Cardiomyocytes of these double DM (T1DM and T2DM) patients are deprived of glucose (due to less insulin production) and in adaptive stress (due to high glucose fluctuation when treated with insulin), which make them more prone to DMCM. Tight glycemic control in the DM patients were unable to decrease the risk of heart failure in clinical trials [10]. Therefore, novel therapeutic approaches are warranted to mitigate DM-induced heart failure. Thus, it is important to investigate the causes of cardiac remodeling and heart failure in the DM heart. One of the hallmarks of the DM heart is increased inflammation [11]. Thus, investigating the causes of increased inflammation and its signaling that mediates myocardial cell death, cardiac remodeling and heart failure is important for developing a novel therapeutic strategy for DMCM and DM-induced heart failure (Fig. 19.1).

19.2 Diabetic Cardiomyopathy

Diabetic cardiomyopathy (DMCM) is a DM-induced cardiac muscle disorder. It was discovered in 1972, when the postmortem of patients with proved record of diabetic glomerulosclerosis were examined for the evidence of primary myocardial disease. Four DM patients who were not suffering from hypertension, significant obstruction of coronary arteries or valvular disease demonstrated cardiomegaly and congestive heart failure [12]. It was concluded that the main cause of the heart failure in these patients was DM. Thus, DM is an independent cause of cardiomyopathy leading to heart failure. DM increases the risk of heart failure when compared to the age and gender-matched non-diabetic patients [4]. In the age group below 54 years, it is found that the risk of heart failure for DM patients is ~nine fold higher compared to the non-DM patients [3]. DMCM is a complex disease with unknown etiology and it may be caused by several factors, including inflammation [13].

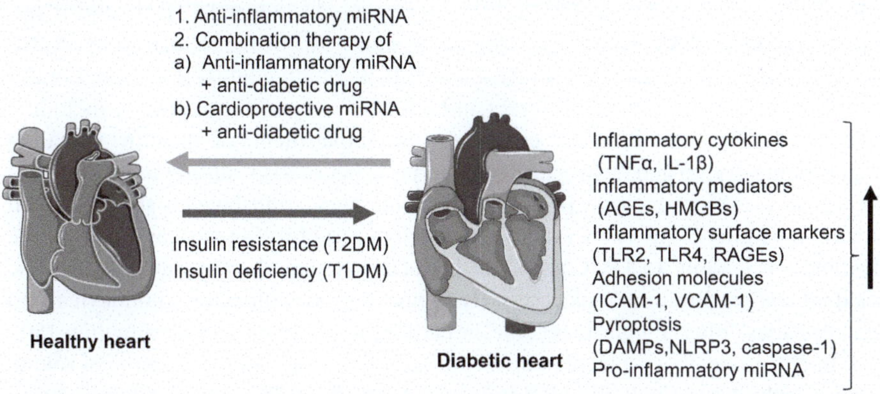

Fig. 19.1 Schematic showing the inflammatory changes in the diabetic heart and the potential therapeutic strategies to ameliorate diabetic heart failure. Due to insulin deficiency (T1DM) or resistance (T2DM), the heart undergoes pathological remodeling leading to diastolic dysfunction. The heart geometry changes to round shape and it increases its size (hypertrophy). The diabetic heart has increased inflammation that promotes inflammatory cell death pathways, pyroptosis. Since glucose-lowering drugs were unable to mitigate diabetic cardiomyopathy, an alternative and a novel therapeutic approach would be treatment with either an anti-inflammatory miRNA, a combination treatment of anti-inflammatory miRNA and anti-diabetic drug, or a combination of cardioprotective miRNA and anti-diabetic drug. Abbreviations: *TNFα* Tumor necrosis factor-α, *IL-1β* Intgerleukin-1β, *AGE*s Advanced glycation end products, *HMGB* members of high mobility group protein, *TLR* Toll-like receptor, *RAGE*s Receptor of AGEs, *ICAM* Intracellular adhesion molecule-1, *VCAM-1* Vascular adhesion molecule-1, *DAMP*s Death associated molecular patterns, *NLRP3* Nod-like receptor family pyrin domain containing-3

19.3 Role of Inflammation and Their Regulators in Diabetic Cardiomyopathy

Inflammation plays a crucial role in cardiovascular disease including ischemic heart disease [14] and hypertension-induced heart failure [15]. It promotes adverse cardiac remodeling [16, 17]. Increased inflammation destroys pancreatic β-cells leading to T1DM phenotype [18–21]. In the DM hearts, pro-inflammatory cytokines and chemokines including TNFα, IL-6, IL-1β, IL-18 MCP-1, adiponectin, and inflammatory mediators such as HMGBs (members of high mobility group proteins), AGEs (advanced glycation end products), and lipoprotein- associated phospholipase A2 are upregulated [22–30]. Activation of various inflammatory cytokines upregulates their cell surface receptors such as TLR2 (toll like receptor- 2), TLR4 and RAGEs (receptor for advanced glycation end products), and promotes NFkB signaling [31–33]. DM also induces different types of cell adhesion molecules such as vascular cell adhesion molecule-1 (VCAM-1) and intracellular adhesion molecule-1 (ICAM-1) and upregulates C-reactive protein, which further promotes inflammation [34–37]. Excessive inflammation in the DM heart could be due to increased damage-associated molecular pattern (DAMPs) released after myocardial cell death [38, 39]. DAMPs and activation of NFkB signaling promote pathological remodeling in the DM heart [40, 41]. In streptozotocin-induced T1DM, NFkB signaling induces RAS (renin-angiotensin system) to instigate cardiomyopathy. Notably, cardiac-specific overexpression of IkB-α, which inhibits RAS, ameliorates DMCM [42]. Activation of different inflammatory signaling pathways including NFkB, MAPK and PI3/Akt contribute to impairment of insulin signaling (T2DM phenotype) and destruction of pancreatic beta cells resulting in insulin deficiency, a T1DM phenotype [43–46]. These changes contribute to structural and functional remodeling such as myocardial hypertrophy, fibrosis, and contractile dysfunction leading to DMCM [47–49]. Recent studies demonstrate that cardiac resident and infiltrating leukocytes (lymphocytes, granulocytes, monocytes and macrophages) play a significant role in cardiac remodeling and dysfunction [50–52]. In DM mice, macrophages-dependent secretion of IL-1β plays a crucial role in cardiac arrhythmia. Notably, inhibition of NLRP3 (nod-like receptor family pyrin domain containing-3)-dependent inflammasome formation and IL-1β receptor blockade alleviated DM-induced cardiac arrhythmias [53]. NLRP3-induced inflammasome formation promotes DMCM [11, 51]. Recently, we have reviewed the role of immune-metabolism in the diabetic heart and elaborated the role of non-coding RNAs in the regulation of inflammation and DMCM [54]. MicroRNAs (miRNAs) are a class of non-coding regulatory RNAs that has could be a promising therapeutic target for cardiovascular disease [55], including DMCM [56]. They also play important roles in the regulation of inflammation and cardiac dysfunction in the diabetic heart [57–62]. Thus, inflammation has an important role in DM-induced cardiac remodeling and dysfunction (Table 19.1).

19.4 Potential Therapeutic Targets/Candidates for Inflammation-Induced Diabetic Cardiomyopathy

A number of empirical evidences support that inflammatory cytokines and related signaling pathways are upregulated in the DM heart [24, 49] [53, 63]. Therefore, inhibition of inflammation could be a promising approach to mitigate DMCM. Blockers of NFkB and IL-1β have been used for the treatment of T2DM and details are elaborated in a review article by Donath and Shoelson [21]. Insulin has an independent anti-inflammatory effect in newly diagnosed T2DM [64]. Drugs that improve insulin sensitization in T2DM or insulin level in T1DM reduces inflammation and they have cardioprotective effects [64, 65]. Metformin, a commonly used drug for T2DM, activates AMPK and protects the heart against pathological cardiac remodeling [66, 67]. Metformin may also exert anti-inflammatory effects by inhibiting NFkB signaling [68]. However, the use of metformin is restricted for the diabetic patients with cardiac dysfunction because of its lactic acidosis property [69]. Other insulin-sensitizing agents such as thiazolidinedione (TZD), PPAR agonist, pioglitazone and rosiglitazone have anti- inflammatory effects and they have been used to prevent cardiac dysfunction in T2DM animals as well as human patients [65, 70]. We know that activation of RAAS (renin-angiotensin-aldosterone systems) that releases angiotensin-II (Ang-II) contributes to DMCM and are

Table 19.1 Different regulators of inflammation and their potential targets

Regulators	Targets	References
Drug		
Fenofibrate	IL-1β and IL-6	[90]
Simvastatin	IL-6 and TNFα	[91, 92]
Candesartan	TNFα	[93]
Enalapril and losartan	IL-1 and IL-6	[94, 95]
Rosiglitazone and pioglitazone	Indirectly IL-6	[96]
Statin	IL-1	[97]
Pravastatin	IL-8	[98]
microRNA (miRNA)		
miR-124	Indirectly targets IL-6 and TNF-alpha	[99]
miR-203	Targets TNF-alpha	[100]
miR-155	Targets TNF-alpha	[101]
miR-223	Indirectly targets NF-kB	[102]
miR-146	Targets IL-6, IL-8, and TNF-alpha	[103]
Long-noncoding RNA (lnc-RNA)		
MALAT1	Targets NF-kB	[104]
EPS	Indirectly targets NF-kB	[105]
Lethe	Targets NF-kB	[106]
NKILA	Indirectly targets NF-kB	[107]
lincRNA-p21	Indirectly targets NF-kB	[108]
MIRT2	Indirectly targets NF-kB	[109]
Lnc-IL7R	Represses IL-6 and IL-8	[110]

associated with increased inflammation. Thus, targeting RAAS and Ang-II by antagonist could be a cardioprotective approach for DMCM [71, 72]. In DM, the levels of free fatty acids are increased that derange the cardio-metabolic function leading to DMCM [54]. Statin that reduces circulating free fatty acids [73] could be a cardioprotective drug for DM patients. Although anti-inflammatory and insulin sensitizing drugs have cardioprotective effects, they often have side effects on vital organs including the heart [74, 75]. Thus, regulating the levels of inflammation by endogenous regulators such as non-coding RNAs could be an alternative approach to mitigate DMCM and DM-induced heart failure [58, 76–83].

19.5 Future Perspectives

DMCM is a major emerging heath concern in the world. With increasing numbers of diabetic individuals and silent progression of DM, the risk of DMCM is very high. Although different anti-diabetic drugs including insulin, metformin, glucagon-like peptide (GLP-1) antagonist, RAAS inhibitor, or a combination of these drugs have been used to reduce the levels of blood glucose in the diabetic patients, however, none of them could cure DM and some of them have detrimental effects on the heart [84, 85]. Several meta-analysis of randomized control trials using anti-diabetic drugs revealed no effects or harmful effects on the heart except SGLT-2 (sodium-glucose cotransporter-2) – inhibitors, which has cardioprotective effects [86–89]. Future studies focusing on regulation of inflammation by non-coding RNAs and a combination therapy using different glucose-lowering drug (s) with non-coding RNA would be an important strategy to ameliorate DMCM and consequent heart failure.

Acknowledgements This work is supported, in parts, by the National Institutes of Health grants HL-113281 and HL-116205 to Paras K. Mishra.

References

1. Muntner P et al (2018) A comparison of the 2017 American College of Cardiology/American Heart Association blood pressure guideline and the 2017 American Diabetes Association Diabetes and Hypertension Position Statement for U.S. Adults with diabetes. Diabetes Care 41:2322–2329
2. Cho NH et al (2018) IDF diabetes atlas: global estimates of diabetes prevalence for 2017 and projections for 2045. Diabetes Res Clin Pract 138:271–281
3. Nichols GA et al (2004) The incidence of congestive heart failure in type 2 diabetes: an update. Diabetes Care 27(8):1879–1884
4. Fox CS (2010) Cardiovascular disease risk factors, type 2 diabetes mellitus, and the Framingham Heart Study. Trends Cardiovasc Med 20(3):90–95
5. Haffner SM et al (1998) Mortality from coronary heart disease in subjects with type 2 diabetes and in nondiabetic subjects with and without prior myocardial infarction. N Engl J Med 339(4):229–234
6. Nordstrom A et al (2016) Higher prevalence of type 2 diabetes in men than in women is associated with differences in visceral fat mass. J Clin Endocrinol Metab 101(10):3740–3746
7. Kautzky-Willer A, Harreiter J, Pacini G (2016) Sex and gender differences in risk, pathophysiology and complications of type 2 diabetes mellitus. Endocr Rev 37(3):278–316
8. Ballotari P et al (2017) Sex differences in the effect of type 2 diabetes on major cardiovascular diseases: results from a population-based study in Italy. Int J Endocrinol 2017:6039356
9. Tripathi BK, Srivastava AK (2006) Diabetes mellitus: complications and therapeutics. Med Sci Monit 12(7):RA130–RA147
10. Castagno D et al (2011) Intensive glycemic control has no impact on the risk of heart failure in type 2 diabetic patients: evidence from a 37,229 patient meta-analysis. Am Heart J 162(5):938–948 e2
11. Sharma A et al (2018) Oxidative stress and NLRP3-inflammasome activity as significant drivers of diabetic cardiovascular complications: therapeutic implications. Front Physiol 9:114
12. Rubler S et al (1972) New type of cardiomyopathy associated with diabetic glomerulosclerosis. Am J Cardiol 30(6):595–602
13. Chavali V, Tyagi SC, Mishra PK (2013) Predictors and prevention of diabetic cardiomyopathy. Diabetes Metab Syndr Obes 6:151–160
14. Halade GV, Kain V, Serhan CN (2018) Immune responsive resolvin D1 programs myocardial infarction-induced cardiorenal syndrome in heart failure. FASEB J 32(7):3717–3729
15. Pan JA et al (2018) Extracellular volume by cardiac magnetic resonance is associated with biomarkers of inflammation in hypertensive heart disease. J Hypertens 37:65–72
16. Lee SJ et al (2018) Angiopoietin-2 exacerbates cardiac hypoxia and inflammation after myocardial infarction. J Clin Invest 128:5018–5033
17. Kimura T et al (2018) Tenascin-C accelerates adverse ventricular remodeling after myocardial infarction by modulating macrophage polarization. Cardiovasc Res 115:614–624

18. Cnop M et al (2005) Mechanisms of pancreatic beta-cell death in type 1 and type 2 diabetes: many differences, few similarities. Diabetes 54(Suppl 2):S97–S107
19. Wilcox NS et al (2016) Life and death of beta cells in type 1 diabetes: a comprehensive review. J Autoimmun 71:51–58
20. Pugliese A (2016) Insulitis in the pathogenesis of type 1 diabetes. Pediatr Diabetes 17(Suppl 22):31–36
21. Donath MY, Shoelson SE (2011) Type 2 diabetes as an inflammatory disease. Nat Rev Immunol 11(2):98–107
22. Yu XY et al (2011) Hyperglycemic myocardial damage is mediated by proinflammatory cytokine: macrophage migration inhibitory factor. PLoS One 6(1):e16239
23. Younce CW, Wang K, Kolattukudy PE (2010) Hyperglycaemia-induced cardiomyocyte death is mediated via MCP-1 production and induction of a novel zinc-finger protein MCPIP. Cardiovasc Res 87(4):665–674
24. Grundy SM (2016) Metabolic syndrome update. Trends Cardiovasc Med 26(4):364–373
25. Devaraj S, Jialal I (2000) Low-density lipoprotein postsecretory modification, monocyte function, and circulating adhesion molecules in type 2 diabetic patients with and without macrovascular complications: the effect of alpha-tocopherol supplementation. Circulation 102(2):191–196
26. Frieler RA, Mortensen RM (2015) Immune cell and other noncardiomyocyte regulation of cardiac hypertrophy and remodeling. Circulation 131(11):1019–1030
27. Mann DL (2015) Innate immunity and the failing heart: the cytokine hypothesis revisited. Circ Res 116(7):1254–1268
28. Devaraj S et al (2006) Increased monocytic activity and biomarkers of inflammation in patients with type 1 diabetes. Diabetes 55(3):774–779
29. Volz HC et al (2010) HMGB1: the missing link between diabetes mellitus and heart failure. Basic Res Cardiol 105(6):805–820
30. Yan SF et al (2003) Glycation, inflammation, and RAGE: a scaffold for the macrovascular complications of diabetes and beyond. Circ Res 93(12):1159–1169
31. Yao D, Brownlee M (2010) Hyperglycemia-induced reactive oxygen species increase expression of the receptor for advanced glycation end products (RAGE) and RAGE ligands. Diabetes 59(1):249–255
32. Rojas A et al (2013) The receptor for advanced glycation end-products: a complex signaling scenario for a promiscuous receptor. Cell Signal 25(3):609–614
33. Devaraj S et al (2008) Increased toll-like receptor (TLR) 2 and TLR4 expression in monocytes from patients with type 1 diabetes: further evidence of a proinflammatory state. J Clin Endocrinol Metab 93(2):578–583
34. Mano Y et al (2011) Overexpression of human C-reactive protein exacerbates left ventricular remodeling in diabetic cardiomyopathy. Circ J 75(7):1717–1727
35. Pradhan AD et al (2001) C-reactive protein, interleukin 6, and risk of developing type 2 diabetes mellitus. JAMA 286(3):327–334
36. Matsumoto K et al (2001) Serum concentrations of soluble vascular cell adhesion molecule-1 and E-selectin are elevated in insulin-resistant patients with type 2 diabetes. Diabetes Care 24(9):1697–1698
37. Fasching P, Waldhausl W, Wagner OF (1996) Elevated circulating adhesion molecules in NIDDM--potential mediators in diabetic macroangiopathy. Diabetologia 39(10):1242–1244
38. Zhang W et al (2015) Necrotic myocardial cells release damage-associated molecular patterns that provoke fibroblast activation in vitro and trigger myocardial inflammation and fibrosis in vivo. J Am Heart Assoc 4(6):e001993
39. Hofmann MA et al (1999) Peripheral blood mononuclear cells isolated from patients with diabetic nephropathy show increased activation of the oxidative-stress sensitive transcription factor NF-kappaB. Diabetologia 42(2):222–232
40. Baker RG, Hayden MS, Ghosh S (2011) NF-kappaB, inflammation, and metabolic disease. Cell Metab 13(1):11–22

41. Shah MS, Brownlee M (2016) Molecular and cellular mechanisms of cardiovascular disorders in diabetes. Circ Res 118(11):1808–1829
42. Thomas CM et al (2014) Cardiac-specific suppression of NF-kappaB signaling prevents diabetic cardiomyopathy via inhibition of the renin-angiotensin system. Am J Physiol Heart Circ Physiol 307(7):H1036–H1045
43. Pan Y et al (2014) Inhibition of JNK phosphorylation by a novel curcumin analog prevents high glucose-induced inflammation and apoptosis in cardiomyocytes and the development of diabetic cardiomyopathy. Diabetes 63(10):3497–3511
44. Kaneto H et al (2004) Involvement of oxidative stress and the JNK pathway in glucose toxicity. Rev Diabet Stud 1(4):165–174
45. Li G et al (2007) Tumor necrosis factor-alpha induces insulin resistance in endothelial cells via a p38 mitogen-activated protein kinase-dependent pathway. Endocrinology 148(7):3356–3363
46. Donath MY et al (2003) Inflammatory mediators and islet beta-cell failure: a link between type 1 and type 2 diabetes. J Mol Med (Berl) 81(8):455–470
47. Frati G et al (2017) An overview of the inflammatory signalling mechanisms in the myocardium underlying the development of diabetic cardiomyopathy. Cardiovasc Res 113(4):378–388
48. Poornima IG, Parikh P, Shannon RP (2006) Diabetic cardiomyopathy: the search for a unifying hypothesis. Circ Res 98(5):596–605
49. Nishida K, Otsu K (2017) Inflammation and metabolic cardiomyopathy. Cardiovasc Res 113(4):389–398
50. Westermann D et al (2007) Contributions of inflammation and cardiac matrix metalloproteinase activity to cardiac failure in diabetic cardiomyopathy: the role of angiotensin type 1 receptor antagonism. Diabetes 56(3):641–646
51. Luo B et al (2017) NLRP3 inflammasome as a molecular marker in diabetic cardiomyopathy. Front Physiol 8:519
52. Dick SA, Epelman S (2016) Chronic heart failure and inflammation: what do we really know? Circ Res 119(1):159–176
53. Monnerat G et al (2016) Macrophage-dependent IL-1beta production induces cardiac arrhythmias in diabetic mice. Nat Commun 7:13344
54. Mishra PK et al (2017) Diabetic cardiomyopathy: an immunometabolic perspective. Front Endocrinol (Lausanne) 8:72
55. Mishra PK et al (2009) MicroRNAs as a therapeutic target for cardiovascular diseases. J Cell Mol Med 13(4):778–789
56. Tyagi AC, Sen U, Mishra PK (2011) Synergy of microRNA and stem cell: a novel therapeutic approach for diabetes mellitus and cardiovascular diseases. Curr Diabetes Rev 7(6):367–376
57. Zhang Y et al (2017) Emerging roles for microRNAs in diabetic microvascular disease: novel targets for therapy. Endocr Rev 38(2):145–168
58. Shantikumar S, Caporali A, Emanueli C (2012) Role of microRNAs in diabetes and its cardiovascular complications. Cardiovasc Res 93(4):583–593
59. Liang YZ et al (2018) Identification of stress-related microRNA biomarkers in type 2 diabetes mellitus: a systematic review and meta-analysis. J Diabetes
60. Zhang Y et al (2016) Deletion of interleukin-6 alleviated interstitial fibrosis in streptozotocin-induced diabetic cardiomyopathy of mice through affecting TGFbeta1 and miR-29 pathways. Sci Rep 6:23010
61. Brennan E et al (2017) Protective effect of let-7 miRNA family in regulating inflammation in diabetes-associated atherosclerosis. Diabetes 66(8):2266–2277
62. Wellen KE, Hotamisligil GS (2005) Inflammation, stress, and diabetes. J Clin Invest 115(5):1111–1119
63. Williams LJ, Nye BG, Wende AR (2017) Diabetes-related cardiac dysfunction. Endocrinol Metab (Seoul) 32(2):171–179
64. Mao XM et al (2009) Independent anti-inflammatory effect of insulin in newly diagnosed type 2 diabetes. Diabetes Metab Res Rev 25(5):435–441

65. Paneni F, Luscher TF (2017) Cardiovascular protection in the treatment of type 2 diabetes: a review of clinical trial results across drug classes. Am J Med 130(6S):S18–S29
66. Gundewar S et al (2009) Activation of AMP-activated protein kinase by metformin improves left ventricular function and survival in heart failure. Circ Res 104(3):403–411
67. Xie Z et al (2011) Improvement of cardiac functions by chronic metformin treatment is associated with enhanced cardiac autophagy in diabetic OVE26 mice. Diabetes 60(6):1770–1778
68. Caballero AE et al (2004) The differential effects of metformin on markers of endothelial activation and inflammation in subjects with impaired glucose tolerance: a placebo-controlled, randomized clinical trial. J Clin Endocrinol Metab 89(8):3943–3948
69. Nesti L, Natali A (2017) Metformin effects on the heart and the cardiovascular system: a review of experimental and clinical data. Nutr Metab Cardiovasc Dis 27(8):657–669
70. Pollack RM et al (2016) Anti-inflammatory agents in the treatment of diabetes and its vascular complications. Diabetes Care 39(Suppl 2):S244–S252
71. Shekelle PG et al (2003) Efficacy of angiotensin-converting enzyme inhibitors and beta-blockers in the management of left ventricular systolic dysfunction according to race, gender, and diabetic status: a meta-analysis of major clinical trials. J Am Coll Cardiol 41(9):1529–1538
72. Zaman AK et al (2004) Salutary effects of attenuation of angiotensin II on coronary perivascular fibrosis associated with insulin resistance and obesity. J Mol Cell Cardiol 37(2):525–535
73. Sahebkar A et al (2016) Statin therapy and plasma free fatty acids: a systematic review and meta-analysis of controlled clinical trials. Br J Clin Pharmacol 81(5):807–818
74. Palazzuoli A et al (2018) Clinical impact of oral antidiabetic medications in heart failure patients. Heart Fail Rev 23(3):325–335
75. Kumar R, Kerins DM, Walther T (2016) Cardiovascular safety of anti-diabetic drugs. Eur Heart J Cardiovasc Pharmacother 2(1):32–43
76. Reddy MA et al (2014) Regulation of inflammatory phenotype in macrophages by a diabetes-induced long noncoding RNA. Diabetes 63(12):4249–4261
77. Assmann TS et al (2017) MicroRNA expression profiles and type 1 diabetes mellitus: systematic review and bioinformatic analysis. Endocr Connect 6(8):773–790
78. Ventriglia G et al (2015) MicroRNAs: novel players in the dialogue between pancreatic islets and immune system in autoimmune diabetes. Biomed Res Int 2015:749734
79. Sebastiani G et al (2015) MicroRNA-124a is hyperexpressed in type 2 diabetic human pancreatic islets and negatively regulates insulin secretion. Acta Diabetol 52(3):523–530
80. Lew JK et al (2017) Exercise mediated protection of diabetic heart through modulation of microRNA mediated molecular pathways. Cardiovasc Diabetol 16(1):10
81. Nandi SS et al (2016) Lack of miR-133a decreases contractility of diabetic hearts: a role for novel cross talk between tyrosine aminotransferase and tyrosine hydroxylase. Diabetes 65(10):3075–3090
82. Guo R, Nair S (2017) Role of microRNA in diabetic cardiomyopathy: from mechanism to intervention. Biochim Biophys Acta Mol basis Dis 1863(8):2070–2077
83. Moura J, Borsheim E, Carvalho E (2014) The role of microRNAs in diabetic complications-special emphasis on wound healing. Genes (Basel) 5(4):926–956
84. Younk LM, Lamos EM, Davis SN (2016) Cardiovascular effects of anti-diabetes drugs. Expert Opin Drug Saf 15(9):1239–1257
85. Brown NJ (2012) Cardiovascular effects of antidiabetic agents: focus on blood pressure effects of incretin-based therapies. J Am Soc Hypertens 6(3):163–168
86. Rosano GM, Vitale C, Seferovic P (2017) Heart failure in patients with diabetes mellitus. Card Fail Rev 3(1):52–55
87. Standl E, Schnell O, McGuire DK (2016) Heart failure considerations of Antihyperglycemic medications for type 2 diabetes. Circ Res 118(11):1830–1843
88. Zheng SL et al (2018) Association between use of sodium-glucose cotransporter 2 inhibitors, glucagon-like peptide 1 agonists, and dipeptidyl peptidase 4 inhibitors with all-cause mortality in patients with type 2 diabetes: a systematic review and meta-analysis. JAMA 319(15):1580–1591

89. El Masri D, Ghosh S, Jaber LA (2018) Safety and efficacy of sodium-glucose cotransporter 2 (SGLT2) inhibitors in type 1 diabetes: a systematic review and meta-analysis. Diabetes Res Clin Pract 137:83–92
90. Okopień B et al (2005) Monocyte suppressing action of fenofibrate. Pharmacol Rep 57(3):367–372
91. Sakoda K et al (2006) Simvastatin decreases IL-6 and IL-8 production in epithelial cells. J Dent Res 85(6):520–523
92. Henrich D et al (2007) High dosage of simvastatin reduces TNF-α-induced apoptosis of endothelial progenitor cells but fails to prevent apoptosis induced by IL-1β in vitro. J Surg Res 142(1):13–19
93. Tsutamoto T et al (2000) Angiotensin II type 1 receptor antagonist decreases plasma levels of tumor necrosis factor alpha, interleukin-6 and soluble adhesion molecules in patients with chronic heart failure. J Am Coll Cardiol 35(3):714–721
94. Trevelyan J et al (2004) Effect of enalapril and losartan on cytokines in patients with stable angina pectoris awaiting coronary artery bypass grafting and their interaction with polymorphisms in the interleukin-6 gene. Am J Cardiol 94(5):564–569
95. Andrzejczak D, Górska D, Czarnecka EJPrP (2007) Influence of enalapril, quinapril and losartan on lipopolysaccharide (LPS)-induced serum concentrations of TNF-alpha, IL-1 beta, IL-6 in spontaneously hypertensive rats (SHR). Pharmacol Rep 59(4):437–446
96. Polyzos SA, Kountouras J, Zavos C (2010) Adiponectin as a potential therapeutic agent for nonalcoholic steatohepatitis. Hepatol Res 40(4):446–447
97. Luan Z et al (2003) Statins inhibit secretion of metalloproteinases-1,-2,-3, and-9 from vascular smooth muscle cells and macrophages. Arterioscler Thromb Vasc Biol 23(5):769–775
98. Rosenson RS, Tangney CC, Casey LC (1999) Inhibition of proinflammatory cytokine production by pravastatin. Lancet 353(9157):983–984
99. Sun Y et al (2013) MicroRNA-124 mediates the cholinergic anti-inflammatory action through inhibiting the production of pro-inflammatory cytokines. Cell Res 23(11):1270
100. Sonkoly E, Ståhle M, Pivarcsi A (2008) MicroRNAs and immunity: novel players in the regulation of normal immune function and inflammation. In: Seminars in cancer biology. Elsevier, London
101. Boesch-Saadatmandi C et al (2011) Effect of quercetin and its metabolites isorhamnetin and quercetin-3-glucuronide on inflammatory gene expression: role of miR-155. J Nutr Biochem 22(3):293–299
102. Huang L et al (2016) MicroRNA-223 promotes tumor progression in lung cancer A549 cells via activation of the NF-κB signaling pathway. Oncol Res 24(6):405–413
103. Curtale G et al (2013) Negative regulation of toll-like receptor 4 signaling by IL-10–dependent microRNA-146b. Proc Natl Acad Sci U S A 110(28):11499–11504
104. Puthanveetil P et al (2015) Long non-coding RNA MALAT 1 regulates hyperglycaemia induced inflammatory process in the endothelial cells. J Cell Mol Med 19(6):1418–1425
105. Atianand MK et al (2016) A long noncoding RNA lincRNA-EPS acts as a transcriptional brake to restrain inflammation. Cell 165(7):1672–1685
106. Rapicavoli NA et al (2013) A mammalian pseudogene lncRNA at the interface of inflammation and anti-inflammatory therapeutics. elife 2:e00762
107. Zhao G et al (2016) The long noncoding RNA MALAT1 regulates the lipopolysaccharide-induced inflammatory response through its interaction with NF-κB. FEBS Lett 590(17):2884–2895
108. Fitzgerald KA, Caffrey DR (2014) Long noncoding RNAs in innate and adaptive immunity. Curr Opin Immunol 26:140–146
109. Du M et al (2017) The LPS-inducible lncRNA Mirt2 is a negative regulator of inflammation. Nat Commun 8(1):2049
110. Cui H et al (2014) The human long noncoding RNA lnc-IL 7 R regulates the inflammatory response. Eur J Immunol 44(7):2085–2095

Advanced Glycation End Products: A Potential Contributor of Oxidative Stress for Cardio-Vascular Problems in Diabetes

20

Savita Bansal, Pawan Kumar Kare, Ashok Kumar Tripathi, and Sri Venkata Madhu

Abbreviations

8-OHdG	8-hydroxy-2-deoxy Guanosine
AGEs	Advanced glycation end products
AOPP	Advanced oxidation protein products
CVD	Cardiovascular diseases
eNOS	Endothelial nitric oxide synthase
GO	Glyoxal
H_2O_2	Hydrogen peroxide
HbA_{1C}	Glycated haemoglobin
HDL	High density lipoproteins
HOCl	Hypochlorous acid
iNOS	Inducible nitric oxide synthase
LDL	Low density lipoprotein
MAPK	Mitogen-activated protein kinase
MCP 1	Monocyte-chemotactic protein-1
MDA	Malondialdehyde
MGO	Methylglyoxal
MGO	Methylglyoxal

S. Bansal (✉)
Department of Biochemistry, Institute of Home Economics, University of Delhi, Delhi, India

P. K. Kare
Department of Biochemistry, All India Institute of Medical Sciences (AIIMS), Bhopal, India

A. K. Tripathi
Department of Biochemistry, University College of Medical Sciences, University of Delhi, Delhi, India

S. V. Madhu
Department of Medicine, Centre for Diabetes, Endocrinology & Metabolism, University College of Medical Sciences & Guru Teg Bahadur Hospital, Delhi, India

S. Chakraborti et al. (eds.), *Oxidative Stress in Heart Diseases*,
https://doi.org/10.1007/978-981-13-8273-4_20

MOLD	Methyl glyoxal lysine dimer
NADPH	Nicotinamide adenine dinucleotide phosphate oxidase
NF-κB	nuclear factor-kappa B
NO	Nitric oxide
NOS	Nitric oxide synthase
$ONOO^-$	Peroxynitrite
OS	Oxidative stress
PARP	Poly ADP ribose polymerase
PCO	Protein carbonyls
PKC	Protein kinase C
PON1	Paraoxonase
RAGE	Receptor for advanced glycation end products
ROS	Reactive oxygen species
RS	Reactive species
T2DM	Type 2 diabetes mellitus
VCAM 1	Vascular cell adhesion molecules

20.1 Introduction

Increasing incidence of cardio-vascular complications are the major health concern leading to morbidity and mortality in type 2 diabetes mellitus (T2DM). Cardio-vascular complications are more severe and rapidly progressive in diabetic patients compare to non-diabetic [1]. Hyperglycemia is the primary casual factor leading to pathophysiological alteration in T2DM. Numerous hyperglycemic related mechanisms such as aldose reductase-mediated polyol pathway, hexosamine pathway, protein kinase C (PKC) activation, poly ADP ribose polymerase (PARP) activation and formation of advanced glycation end products (AGEs) are hypothesized to account vascular complications in T2DM [2–5]. Among them, enhanced formation and accumulation of AGEs is one of the important pathway associated with hyperglycemia-mediated detrimental effects in T2DM [6, 7]. AGEs are resulting from non-enzymatic reaction between reducing sugars and free-amino group containing molecules such as proteins, lipids and nucleic acid [8]. AGEs link the hyperglycemia with development of cardio-vascular complications in diabetes as serum level of AGEs in T2DM patients with cardio-vascular diseases (CVD) are higher compared to those without CVD [9–11]. Presence of AGEs has also been noticed in coronary arteries of diabetic patients suggesting the involvement of AGEs in development of accelerated arterial diseases [12]. A follow-up study carried out for 18 years demonstrated that increased levels of AGEs are associated with cardio-vascular mortality in Finnish T2DM women [13].

Of the many factors by which AGEs may induce cardio-vascular complications; oxidative stress (OS) and endothelial dysfunction are considered important. The existence of increased OS is based on decreased antioxidant capacity and increased reactive oxygen species (ROS) generation. High level of ROS leads to number of oxidation and peroxidation reactions at biomolecules (carbohydrates, lipids, proteins and DNA) affecting their functions and properties. Involvement of OS in development and progression of diabetes as well as in atherosclerosis has been noticed in number of studies [14–16].

AGEs mediate their effects at cellular as well as at extracellular level through different pathways including modification of cellular and extracellular matrix proteins, disrupting matrix-matrix and matrix-cell interactions contributing to their pro-fibrotic action [17, 18]. At cellular level they mainly show their effect by interacting with their cell surface receptor namely receptor of advanced glycation end products (RAGE) [19]. Ligation of AGEs with its receptor elicits several intracellular signaling cascades leading to increased cytosolic and mitochondrial reactive oxygen species generation, oxidative stress development, activation of nuclear factor-kappa B (NF-κB), pro-inflammatory and pro-coagulant pathways, smooth and fibroblast proliferation, which are the key factors linking the AGE-RAGE system with diabetic associated cardio-vascular complications [20–23].

This chapter has focused on the role of AGEs, emerges as a crucial mediator of enhanced ROS and oxidative stress-mediated detrimental effects, which can link the hyperglycemia with onset of cardio-vascular problems in T2DM. Also, mentioning the anti-AGEs strategies that may be considered as an ideal candidate for future interventions in amelioration of diabetic associated vascular complications.

20.2 Hyperglycemia, Diabetes and Cardiovascular Diseases

It is estimated that cardiovascular diseases (CVD) are the main cause of morbidity and represent 31% of all global deaths [1]. There are number of risk factors associated with CVD and among them diabetes is one of the important factor [24]. Prevalence of T2DM is increasing at an alarming rate that further enhancing a frequency of CVD. Cardiovascular problems basically seen in diabetes include myocardial infarction, angina, peripheral artery diseases (PAD), stroke, congestive heart failure etc.

Diabetes mellitus (DM) is a chronic condition characterised by non-production of insulin or when body cannot utilize insulin effectively; resulting in a build-up of glucose inside the blood. Exposure of high blood glucose level for a longer time may affect the blood vessels, heart, eyes, kidneys, and nerves. Accumulating data from experimental, pathological, epidemiological and clinical studies have shown that persistent hyperglycemia in diabetes mellitus is the key factor affecting the diabetic vasculature resulting in micro- and macro-vascular complication [25, 26]. Numerous hyperglycemic related mechanisms are hypothesized to account vascular complications in T2DM, which include polyol pathway, protein kinase C (PKC) activation, hexosamine pathway, poly ADP ribose polymerase (PARP) activation and formation of advanced glycation end products (AGEs) [2–5]. Among them, increased formation and accumulation of AGEs is one of the important pathway associated with hyperglycemia-mediated detrimental effects in T2DM [6].

20.3 Advanced Glycation End Products

20.3.1 Formation of Advanced Glycation End Products

Advanced glycation end products (AGEs) are complex group of compounds that are formed by different pathways namely Maillard reaction, polyol pathway, oxidation and peroxidation reaction of biomolecules as shown in Fig. 20.1. Given these differing pathways, diversified AGEs are formed in term of their structure and properties. Maillard reaction describing AGEs formation under hyperglycemic conditions is one of the important pathway and it was first described in 1912 as "browning reaction" due to the associated yellow-brown color change in AGEs [8]. It's mainly resulting from non-enzymatic reaction of reducing sugar with free amino groups present at proteins, lipids and DNA. Products of Maillard reaction has gained attention in recent years due the association of AGEs with certain pathological conditions including diabetes mellitus, cardio-vascular diseases, retinopathy, neuropathy, liver cirrhosis, Alzheimer's diseases as well as during aging process also [7, 27–29].

Maillard reaction is different from enzymatically occur N-/O-linked glycosylation in a cell during the co- and post-translation modifications of proteins required for their normal functioning. Whereas, Maillard reaction is non-enzymatic in nature and occurs spontaneously, which disturb the structure, conformation and functions of affected molecules and alter their cellular properties [30, 31]. Under physiological conditions glycation occur at a very low rate and their products are normally

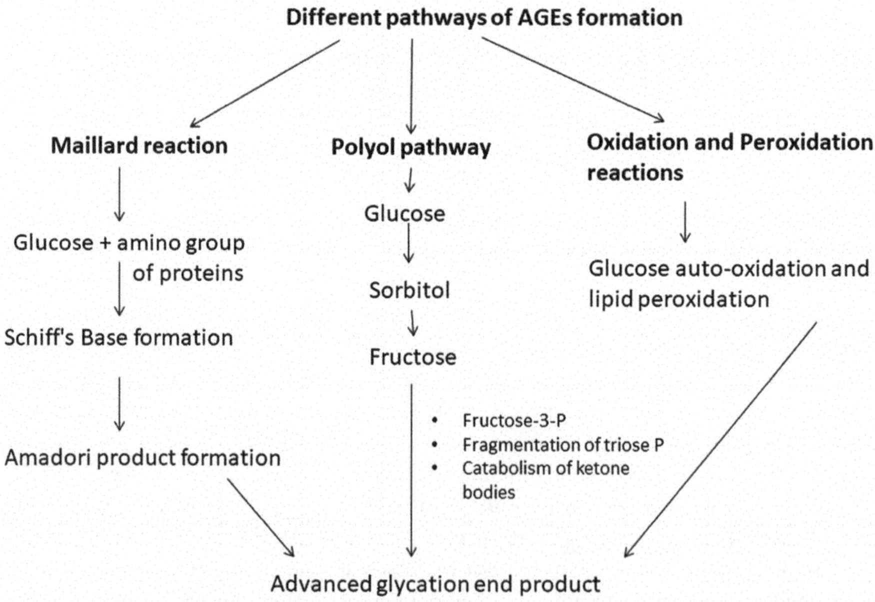

Fig. 20.1 Different pathway of AGEs formation

excreted out from body by different mechanism but this glycation reaction become more rapid and intense under hyperglycemic conditions. Also, during the ageing process their formation is increases, mentioning that glycation has both physiological and pathophysiological significance [32–34]. Formation of AGEs mainly occurs at three stages during Maillard reaction:

1. During the first phase there is a covalent binding of aldehyde or ketone groups of reducing sugars to free amino groups of proteins, lipids and nucleic acids, in a non-enzymatic way forming a reversible intermediates known as Schiff base. The initiation of this step depends on glucose concentration and takes place within hours.
2. In second stage of reaction, Schiff base undergoes rearrangement over a period of time to form ketoamine glycation product, called Amadori's product. These products are more stable compounds, but the reaction is still reversible. The best Amadori's product that was described and identified is HbA1c (glycated haemoglobin) and it is a useful marker of glycemic control but they do not consider under the category of AGEs. Amadori's product also undergo the process of degradation over a period of time to form other highly reactive dicarbonyl intermediates like 3-deoxy-glucosone, glyoxal and methyl-glyoxal that can react again with free amino groups present at proteins or other molecules to form glycation products.
3. During the third phase of reaction, intermediate glycation products of second phase undergo complex series of rearrangement to form irreversible AGEs, characterised by associated yellow-brown color. These chemical rearrangements include dehydration, oxidation and fragmentation reactions. Process of rearrangement is normally slow and spontaneous, often takes month to years for their formation. It is also important to note that this process is enhanced under certain conditions such as oxidative stress, presence of metal ions and catalyst, which increases the post-Amadori formation of AGEs. Once they formed, results in a highly stable structure that is cross-linked and accumulate inside and outside the cell that affect the cell functioning and properties [6]. AGEs are diversified in their structure and some of them also have fluorescent properties.

Apart from Maillard reaction, other important pathways of AGEs formation are auto-oxidation of reducing sugar and peroxidation reaction of lipid molecules that may leads to formation of dicarbonyl derivatives [35]. Also, during glycolysis numbers of reactive intermediates are formed that have a potential to interact with proteins and leading to fast intracellular Maillard reaction resulting in AGE formation [36, 37]. Important reactive intermediates of glycolysis, which are involved in AGE formation include dicarbonyl compounds, glyoxal (GO), methylglyoxal (MGO), 3-deoxyglucosone (3-DG) etc. Also, some of these intermediates GO and MGO can be formed by auto-oxidation of glucose and peroxidation of glycolipids that react with arginine or lysine residues of proteins leading to AGEs formation [38, 39]. Degradation of nucleic acid also represents a source for AGEs formation by releasing a free ribose and results in pentosidine production [40]. Pentosidine can also be formed from glycoxidation of Amadori products or oxidation of arabinose [41]. Pentosidine is one of the important AGEs having a fluorescent property and measured inside the body using different methods. Polyol pathway is also an important mechanism of AGEs formation *in vivo* conditions. During this pathway glucose is converted to sorbitol with the help of enzyme aldose reductase. Next sorbitol leads to fructose formation in presence of sorbitol dehydrogenase. Then fructose metabolites are converted to α-oxoaldehydes that interact with monoacids and results in AGEs formation [42].

Beside the endogenously formed AGEs by different mechanism, they can also originate from exogenous sources such as tobacco, smoke, and diet [43]. Exogenous nature of AGEs are confirmed by various clinical studies showing high level of AGEs in smokers and in individuals on high AGE diets and having a high inflammation and OS in them [44, 45]. Also, evidences from animal studies have shown that on exposure to exogenous AGEs, it leads to renal and vascular complications in them [46, 47].

20.3.2 Type of Advanced Glycation End Products

Because of different pathways of AGEs formation, they comprise a large number of chemical structure and properties like N-carboxyl-methyl-lysine (CML), pyrraline, pentosidine, cross-linked AGEs include GOLD (glyoxal-derived lysine dimer imidazolium salt), MOLD (methyl glyoxal derived lysine dimer-imidazolium salt), DOLD (3-deoxyglucosone-derived lysine dimer-imidazolium salt), etc. [48, 49]. Some AGEs also have fluorescence properties that can be used as a surrogate marker for their identification.

20.4 Cardio-Vascular Disease and Advanced Glycation End Products

Hyperglycemia-mediated formation of AGEs is increasingly evidenced for progression of cardio-vascular complications in T2DM by clinical studies showing high levels of serum AGEs in T2DM patients with CVD compared to those without CVD. High level of AGEs remain significant in T2DM patients with complications compared to those without complication even after adjustment of confounding factors such as duration of diabetes, sex, age by Bonferroni adjustment [9, 50, 51]. Clinical studies have shown that AGEs level increased from 20 to 30% in T2DM without complications and 40–100% with complications having a CVD or microvascular complications [52–55]. Statistical analysis like logistic regression model have shown that higher serum AGEs level will increases the chances of development of cardiovascular complications in T2DM. Accumulation of AGEs has also been identified in atherosclerotic plaque and within myocardium fibers in various studies. Also, gradual deposits and increase in serum level of AGEs have been reported with the severity of atherosclerosis in diabetic patients [56, 57]. Nin et al.

(2011) have shown that level of serum AGEs can act as important marker or predictor of CVD mortality [58]. Koyama et al. (2007) have noticed the relation of pentosidine with severity of heart failure in 141 patients and mentioned that pentosidine as independent risk factor to predict adverse clinical outcome [59].

Important physiological changes that are observed in T2DM patients with cardiovascular complications and high level of AGEs include arterial stiffening, myocardial relaxation abnormalities, atherosclerotic plaque formation and its destabilization, endothelial dysfunction, neointimal proliferation, increased level of oxidised LDL etc. [60–62]. Significant progress has been made in revealing the mechanism leading to AGEs-mediated cardio-vascular changes in T2DM. The multiple mechanisms in relation to AGEs for the progression of CVD in T2DM seem to share a common pathway of prolonged exposure to increased ROS which promote oxidative stress development, vasoconstriction, inflammation, and prothrombotic gene expression in diabetic cardiovascular cells. Also, AGEs modification result in cross-linking of proteins leading to vascular and myocardial stiffness that disturb the structural integrity and biological functions of affected proteins and these changes are found to be associated with isolated systolic hypertension and diastolic heart failure [63].

AGEs induce their detrimental effects at cellular and extracellular level by different mechanism including:

1. Receptor-mediated mechanisms where AGE-RAGE interaction at cell-surface activates the cascade of cell-signalling that results in moderation of gene expression which affect the cell properties and its functioning.
2. Formation and build-up of AGEs in extracellular matrix which leads to cross-linking and decrease the elasticity of vessels.
3. Glycation of proteins, lipids, and nucleic acid leading to deterioration of their structure and functions.

20.5 AGEs-RAGE Interaction

A number of AGEs-binding proteins or receptors have been identified that can be present as soluble form or at the cell surface. Important AGEs binding proteins are: AGE-R1 (oligosaccharyltransferase-48), AGE-R2 (80KD- phosphoprotein), AGE-R3 (galactin-3) and receptor of advanced glycation end products (RAGE) [19, 64–66]. Interaction of AGEs with these proteins or receptors results in both positive and negative impact. During their positive impact, binding of some receptors may leads to clearance of AGEs from the body and hence, lessen the detrimental effects of AGEs. However, interaction of some receptors with AGEs leads to activation of cell-signaling that results in moderation of gene expression, inflammation and affect the cellular properties. The pathological effects of AGEs are largely mediated by cell surface receptors; RAGE (receptor for AGEs) and it is the best characterised receptor for AGEs. RAGE is an approximately 45-kDa protein and member of the immunoglobulin superfamily present on cell-surface of endothelium, vascular smooth muscle cells and invading mononuclear phagocytes [62]. Presences of

RAGE on multiple tissues suggest its potential relevance in activation of pathway that affects micro- and macro-vascular system in diabetes and associated complications. RAGE is a multiligand receptor and AGEs is one of the important ligand.

Enhanced expression of RAGE is found to be associated with certain pathological conditions including diabetes and its associated complications [19]. In diabetes, AGE-RAGE interaction results in activation of diverse downstream pathways including generation of reactive oxygen species (ROS), cytokine production, adhesion molecule expression, endothelial-1, plasmin activator inhibitor 1, production of growth factors such as TNF-α, chemoattraction of inflammatory cells, smooth and fibroblast proliferation, that link its association with diabetic complications [20, 67, 68]. During the activation of above mentioned downstream pathways, AGE-RAGE interaction leads to stimulation of signaling molecules including ERK1/2, p21RAS, MAP kinases, NF-κB, cdc42/rac, and JAK/Stat which modulate the cell properties that possibly promotes a pro-inflammatory and pro-coagulant gene pathways, contributing to cardiovascular complication in diabetes [69, 70].

20.6 Effect of Non-enzymatic Glycation of Biomolecules

Non-enzymatic glycation of free amino groups present at proteins, lipids, and nucleic acid with reducing sugars under hyperglycemic conditions affect the normal functioning of molecules by deterioration of their structure, properties and enzymatic-activity. Proteins of cell and extracellular matrix (ECM) undergo the process of glycation under hyperglycemic conditions. Due to the slow turnover rate and longer half-life of ECM proteins they are more prone to glycation and cross-linking. Modification of extracellular matrix proteins interferes with their normal functioning affecting the cell-matrix and matrix-matrix interactions, which disturb the signalling between matrix and cells contributing to profibrotic action and vascular complications. Formation and accumulation of glycated molecules in ECM also result in cross-links formation, which leads to decreased elasticity, increased stiffness and narrowing of vessels associated with atherosclerosis [61, 71]. During intracellular glycation of proteins not only the glucose but its glycolytic intermediates present inside the cell such as glucose-6-phidphate, glyceraldehyde-3-phosphate, dihydroxyacetone-phosphate, dicarbonyl compounds mainly glyoxal (GO), methyglyoxal (MGO) play an important role in non-enzymatic glycation reaction. Intracellular AGEs are also implicated in activating intracellular signalling pathways that can contribute to diabetic vascular complications [72]. Under hyperglycemic conditions endothelial cell proteins such as fibroblast growth factor undergoes non-enzymatic glycation reaction that affect the vascular properties of cell by affecting mitogenic and eNOS activity [73]. Under hyperglycemic conditions mitochondrial proteins are also undergo the process of glycation which are associated with increased in superoxide formation by mitochondria [74]. In addition, process of glycation also affects the activity of certain anti-oxidant enzymes namely glutathione reductase and glutathione peroxidase, which further enhance the oxidative stress in diabetes [75]. Therefore, glycation of proteins may have deleterious effects by different mechanism under hyperglycemic conditions in diabetes.

Cardiovascular complications are also associated with increased incidence of low-density lipoprotein (LDL) oxidation. It is believed that oxidation of LDL in arterial wall is one of the important casual factor leads to initiation and progression atherosclerosis by foam cell formation. Paraoxonase (PON1) is an enzyme associated with high density lipoproteins (HDL) play an important role in preventing the LDL and cell membranes from their oxidation [76]. Paraoxonase also decreases the oxidised LDL- mediated stimulation of monocyte-chemotactic protein-1 (MCP-1) from endothelial cells and prevent monocyte-endothelial cell interaction during foam cell formation [77]. Such protection provided by PON1 is mainly related to its hydrolysing activity on activated phospholipids and lipid peroxide products. As, PON1 have a peroxidase-like and homocysteine-thiolactonase activity that may be responsible for its anti-atherogenic properties and its protective effect against lipoprotein oxidation [78]. PON1 activity is found to be decreased in diabetes mellitus as well as in cardiovascular patients [77, 79]. It has been postulated that decreased serum PON1 activity associated with diabetes may playing a role in development of premature atherosclerosis and thereby CVD. Several factors may responsible for affecting the PON1 activity in diabetes. *In vitro* studies have shown that glycation of HDL and PON1 is one of the factor that can inhibit the PON1 activity under hyperglycemic conditions [80, 81]. AGEs are also found to be negatively associated with PON1 activity or anti-oxidative capacity of HDL in clinical studies [82].

Also, AGEs-mediated atherosclerotic mechanism involves the impairing of LDL removal by trapping them in sub-endothelium as well as by decreasing the recognition of AGEs-modified LDL through their receptor [83]. Glycation of LDL particle in the phospholipid component and apolipoprotein B is responsible for impairment of hepatic receptor-mediated uptake and its removal. It has been found that glycated LDL are more susceptible to cross-linking with collagen on arterial wall in comparison to non-glycated-LDL. Glycated-LDL is also not able to enter the cell and accumulates outside, which leads to their increased retention in aortic wall and recognition by macrophages. As a result of this, increased localization of AGE-LDL in vessels occurs and results in foam cell formation through macrophage recognition and ingestion [84, 85]. Therefore, glycated-LDL has impact towards cardiovascular complications by atheroma formation under hyperglycemic conditions compare to unmodified LDL.

20.7 AGEs and Reactive Species Production

Reactive oxygen species (ROS) are chemically unstable and extremely reactive free radicals with unpaired electrons such as superoxide anion (O_2^-), hydroxyl radical (OH^-) and lipid radicals. Although some ROS such as hydrogen peroxide (H_2O_2), peroxynitrite ($ONOO^-$) and hypochlorous acid (HOCl) aren't free radicals but due to their oxidizing effect they also lead to oxidative stress. These reactive species have important biological functions at physiological level and required for cellular activities, strengthens synaptic plasticity, maintain the chemical balance and hormone level etc. but excess generation of them may lead to impaired homeostasis and associated physiology. Also, ROS helps to clear the invading pathogens as well as generate the immune response against them [86]. In spite of having their role in different pathways, overproduction of reactive species leads to cell injury and many pathological conditions. Antioxidant defence system is present inside the body including certain enzymes such as superoxide dismutase (SOD), glutathione peroxidase (GPX) and catalase to neutralize the excess ROS [87]. Oxidative stress generation is mainly result from disturbance of equilibrium between production of free radicals and anti-oxidant defence system which leads to tissue injuries. ROS are normally produced by xanthine oxidase (XOS), nicotinamide adenine dinucleotide phosphate oxidase (NADPH), lipoxygenase cytochrome p450, cyclooxygenase (COX), mitochondrial respiration and due to uncoupling of nitric oxide synthase in vascular cells [88].

Among them NADPH oxidase is one of the important enzyme for ROS production within the vascular compartment under certain pathological conditions and infection. In vascular compartment neutrophils are the main producer of ROS by activating NADPH oxidase. Neutrophils (PMN) are the predominant leukocytes and their functions are altered in diabetes with respect to excess superoxide production [89–91]. Therefore, hyperlglycemic conditions leads to activation of neutrophils which leads to exaggerated inflammation and tissue damage. Possible pathway of neutrophils activation and their altered function in diabetes is the presence of AGE receptor (RAGE) on their surface [92]. AGE-RAGE interaction on neutrophil cells under hyperglycemic conditions may leads to activation of certain signalling pathways those results in excess ROS generation. In *vitro* studies have shown that on incubating PMN, monocytes, HUVEC (human unbilical vein endothelial cell), THP1 (human monocyte leukemia cell line) and cardiac myocyte with different concentration of AGEs exhibited significant increase in intracellular and extracellular O_2^- production [93–95]. Therefore, AGE-RAGE interaction trigger the intracellular ROS generation through NADPH oxidase activation that may have significant impact on cellular properties.

Also, under hyperglycemic conditions increased production of ROS occur through mitochondrial electron transport chain which results in activation of protein kinase c (PKC) which further increases the O_2^- generation. Protein kinase C activation leads to up-regulation of NADPH oxidase, thromboxane production and impaired NO release. Increased ROS production by mitochondria triggers the inflammatory cascades responsible for pathogenesis of cardiovascular complications in diabetes. Also, ROS-associated pathways affect the coronary circulation leading to myocardial hypertrophy and fibrosis. Therefore, hyperglycemia-induced ROS generation is associated with the development and progression of vascular dysfunction and unveils the pathophysiology of CVD in diabetes [74, 96].

Apart from ROS, overproduction of reactive nitrogen intermediates (RNI) such as peroxynitrite is also involved in generation of oxidative stress [97]. Nitric oxide is a labile free radical that can react with O_2 leads to formation of potent oxidant peroxynitrite ($ONOO^-$), which on decay leads to formation of another strong oxidant hydroxyl radical (OH^-) [98]. Increased nitrite production is mainly found to be associated with enhanced expression of iNOS (inducible NOS) isoform of NOS enzymes. Research data have shown that on incubation of different kind of cells such as neutrophils, glowmerular mesangial cells from SV40 transgenic mice, murine macrophages and C6 glioma cells with different concentration of AGEs leads increased production of nitrite through iNOS by AGE-RAGE interaction [95, 99–101]. Hence, AGE-RAGE-mediated increased formation of ROS and RNI by activation of NADPH oxidase and iNOS can lead to oxidative damage in diabetic patients by different mechanism, and this may be one of the important pathway associated with AGE-mediated pathology. Possible signal transduction mechanism involved in increased production of reaction species generation through AGE-RAGE interaction may involve induction of certain signaling molecules such as extracellular-signal-regulated-kinase (ERK), Phospholipase A2, phophoinositide3-kinase activation, and p38 mitogen-activated protein kinase (MAPK) [102, 103].

20.8 AGEs, ROS, Oxidative Stress and CVD

Primary causal factors for cardiovascular complications in diabetes are oxidative stress, inflammatory disease, and endothelial dysfunction as evidenced by *in vitro* and epidemiological studies [11, 104, 105]. Diabetic patients are exposed to high level of oxidative stress, which is based on increased reactive species (RS) generation and decreased anti-oxidant defence mechanism of body. Increased RS generation cause a change in activity of signal transduction pathway and concentration of certain transcription factors such as TNFα and NF-κB, which on turn promote the ROS production, inflammatory adhesion molecules and cytokines [26, 106].

Enhanced generation of reactive species by AGEs can also modify the cellular components namely proteins, lipids and nucleic acid and generate structural changes in them, which may affect the cellular properties and its functions. Chronic exposure of biomolecule to high level of ROS may lead to oxidation, peroxidation and glyoxidation reactions that result in formation of protein carbonyl (PCO), oxidation of thiol group, lipid peroxidation (MDA), generation of advanced oxidation protein products (AOPP), and 8-OHdG. Increased level of these oxidative markers are observed in diabetic patients [10, 107, 108]. Oxidative injury to biomolecules has also been reported in presence of high AGEs concentration in tissue and body fluid of diabetic patients [11, 14]. Positive correlation between the AGEs and various oxidative stress markers revealed the contributory effect of enhanced level of AGEs towards increased oxidative burden with diabetic pathology.

Lipid peroxidation of cell components plays an important role in pathogenesis of cardiovascular complications in diabetes [109]. Among the lipid peroxidation products, serum MDA (melondialdehyde) level is a sensitive marker used in monitoring of oxidative stress status. MDA is a decay product of the peroxidation of polyunsaturated fatty acid such as arachidonic, eicosapentaenoic and docosahexaevoic acid and its concentration is high in diabetic individuals compared to healthy ones [110, 111]. Melondialdehyde play an important role in modification of LDL leading to oxidised-LDL (ox-LDL) formation [112]. Experimental models and various clinical studies reported that ox-LDL play a significant role in pathogenesis of atherosclerosis [113]. Mechanism behind the association of ox-LDL with CVD involves the uptake of ox-LDL by macrophage through scavenger receptor pathway, which results in cholesterol ester-rich foam cell formation and endothelial dysfunction [114, 115]. Also, the presence of ox-LDL is also noticed in atherosclerotic plaques by using certain immune-histochemical staining [116].

Oxidation of proteins by RS also represents an ideal candidate in the onset of vascular-complication in diabetes. Some of the important biomarkers of protein oxidation are PCO and AOPP. Advanced oxidation protein products are formed in presence of chlorinated oxidants such as hypochlorous acid and chloramines. AOPP are the dityrosine-containing cross-linked protein-structure and represent an important marker to measure the degree of protein oxidation [107, 117]. Elevated biomarkers of protein oxidation are also found to be associated with diabetes mellitus during correlation analysis of clinical studies and are present in high level in diabetic individuals compared to healthy individuals and more so in diabetic-vascular complications [118, 122]. Sarkar et al. (2010) have also reported that content of protein carbonyl have association with insulin resistance in diabetes [108]. Oxidative modification of proteins due to addition of dityrosine and carbonyl molecules may leads to loss of structural, functional and catalytic properties of affected proteins. Such type of oxidative modification in proteins may have impact on development and progression of diabetic-vascular complications.

Enhanced production of ROS also affects the functional properties of DNA which results in oxidative injury to DNA including modification of DNA bases in the form of single- and double- DNA strand breaks, 8-hydroxydeoxyguanosine (8-OHdG) formation, generation of apurinic sites, damage to deoxyribose sugar, DNA-protein cross-linkage, and damage to DNA repair system [119, 120]. The most studied and frequently detected oxidised marker of DNA lesion is 8-OHdG, an oxidised nucleoside of DNA. This modified nucleoside form of DNA is considered as a novel biomarker of DNA modification under oxidative stress in vivo conditions. Increased levels of 8-OHdG have been detected in tissue and body fluids including mononuclear cells, pancreatic islet, mitochondrial DNA and urine of diabetic and CVD patients [121–124]. Research evidence suggests that 8-OHdG lesions of DNA may leads to somatic mutations those results in smooth muscle proliferation and associated with pathogenesis of atherosclerotic plaque formation [125, 126]. Also, it is found that 8-OHdG is strongly mutagenic which may leads to increased frequency of spontaneous transverion mutation G.C $\rightarrow$ A.T in repair deficient cells [127].

20.9 AGEs and Endothelial Cells Dysfunction

Impairment of endothelial functions is one of the most important factors for development and progression of diabetic-vascular complications. Endothelial dysfunction is mainly associated with reduced anti-coagulant properties, less nitric oxide (NO) production, increased ROS generation, enhanced expression of adhesion molecule, chemokine and cytokine release from endothelium [128]. These conditions lead to inflammation, vasoconstriction, oxidative stress, myofibroblast migration, and proliferation inside the endothelial vessel, all of which play an important role in development and progression of vascular complications and results in atherosclerosis [104]. AGEs may be one of the important factor affecting the endothelial function in diabetes and thereby leading to CVD. Receptors of AGEs are also present on the endothelial cell surface and their presence on these cells suggests a potential relevance of AGE-RAGE interaction with endothelial dysfunction associated with pathogenesis of diabetic-vascular complications.

Role of AGEs in endothelial dysfunction has been observed in human diabetes, as serum AGEs level in T2DM patients is negatively associated with extent of endothelium-dependent vasodilation [129]. Nitric oxide (NO) is the most important endogenous vasodilator with anti-inflammatory, anti-proliferative effects [106, 130]. Because of these properties NO is recognised as an endogenous anti-atherogenic factor. Also, NO plays an important role by acting as a mediator or regulator of various biological processes associated with nervous, immune, and cardio-vascular systems [131]. Decreased NO bioavailability through decreased NO production or increased NO inactivation affect the process of vasodilation and induces endothelial dysfunction. In cardiovascular system, NO is involved in vascular smooth muscle cell relaxation leading to arterial vasodilation and increases the blood flow. Increased ROS production by AGEs is one of the reason for inactivation of NO as well their conversion to peroxynitrite form, that act as free radical and responsible for affecting the integrity of endothelial cells.

Nitric oxide production inside the endothelial is mainly carried out by endothelial NO synthase (eNOS). Formation of AGEs inside the endothelial cells has been found to be associated with reduced expression of eNOS and increased eNOS-mRNA degradation [130]. Also, peroxynitrite form of NO is responsible for inhibition of eNOS activity by depletion of tetrahydrobiopterin (BH4) that act as cofactor for eNOS [132]. Intracellular formation of AGEs in endothelial cells by non-enzymatic glycation of certain proteins including basic fibroblast growth factor under hyperglycemic conditions also leads to altered vascular functions such as reduced mitogenic activity of endothelial cells [73]. Apart from above mentioned effects of AGEs on NO properties, AGEs are also responsible for enhance production of asymmetric dimethylarginine (ADMA) that is an endogenous inhibitor of eNOS by AGE-RAGE interaction on endothelial cells [133]. ADMA is now considered as one of the strongest marker of cardiovascular diseases progression and revealed the importance of AGE–RAGE axis in decreased production or impaired bioavailability of NO in cardiovascular problems in diabetes. Therefore, reduced eNOS activity and inactivation of NO by AGEs may have an important impact in the pathogenesis of diabetes associated atherosclerosis by affecting vasodilating properties of endothelial cells.

Cardiovascular complications are further characterised by adhesion of monocyte to endothelial cells and transmigration into the sub-endothelial space. AGE-RAGE interaction under hyperglycemic conditions is found to enhance this process through activation of pro-inflammatory molecules such as NF- κB. Activation of NF-κB results in over expression of inflammatory genes and adhesion molecules such as VCAM-1 that facilitates the monocyte adhesion to endothelial cells [134, 135]. At adhesion site monocyte and oxidised lipid leads to foam cells formation which results in fatty streak formation in vessel wall. Subsequently, these fatty streaks are converted into advanced lesions that become unstable and rupture over a period of time which, trigger the thrombo-embolic events and result in the clinical manifestations of cardiovascular disease [136].

Increased levels of AGEs are also associated with platelet activation, aggregation and stimulate pro-coagulant activity by increasing the expression of tissue factor which is responsible for thrombus formation and main initiator of coagulation cascade [137]. Available research data has also shown the involvement of AGEs in endothelial cell damage through up-regulation of protease-activated receptor-1and −2 by potentiates thrombin or factor-Xa [13, 138]. In addition of it, AGE-RAGE interactions also inhibit the prostacyclin production and induce the generation of plasminogen activator inhibitor-1 in endothelial cells [139]. These data suggest that, AGEs may have potential to cause platelet aggregation and fibrin stabilization, thereby contributing to the development and progression of vascular injury in diabetes.

In diabetes, there is also decreased in endothelial progenitor cell (EPC) function and mobilization, which could contribute to high risk for cardiovascular complications [140]. AGEs enhance apoptosis and suppress the migration and tube formation of late EPC by down-regulation of Akt and cyclooxygenase-2 through interaction with RAGE [141]. Moreover, AGEs modification of fibronectin by glycation of Arg-Gly-Asp motif leads to impairment of vascular repair by inhibiting EPC adhesion, migration and spreading [141].

Ligation of AGE with its receptor elicits several intracellular signaling cascades leading to cytosolic and mitochondrial ROS generation and enhanced production of ROS can affects the functional properties of biomolecules by binding with them. DNA is one of major target of endogenous oxidants that leads to oxidative injury of DNA that may result in several type of damages such as modification of DNA bases in the form of 8-OHdG, single- and double-DNA strand breaks, generation of apurinic sites by loss of purines, damage to deoxyribose sugar, DNA-protein cross-linkage, and damage to DNA repair system [119, 120]. Presence of oxidative damage to DNA has been reported in presence of high AGEs concentration in tissue and body fluid of diabetic patients [124, 142]. *In vitro* and *in vivo* reports have shown that DNA damage also associated with pathogenesis of atherosclerosis [122, 123, 126]. Further, DNA damage along with activation of various apoptotic trigger, mediated through AGE-RAGE interaction may also induces endothelial cell apoptosis; that is a prominent feature of atherosclcrosis, and has been implicated in the pathophysiology of vascular complications in diabetes [143].

20.10 Anti-AGEs Therapy

Since AGEs play a significant role in development and progression of diabetic associated vascular complications, they present a promising target for therapeutic interventions. Various pathways or agents associated with inhibition of AGEs formation or attenuate their pathological effects are considered as an ideal candidate for pharmaceutical intervention in the amelioration of AGE-mediated diabetic vascular complications. Therapies that can work against AGE-associated problems include the diverse pathways such as inhibition of Amadori products formation (e.g aminoguanidine, pyridoxamine, LR-90), cross link breaker (e.g alagebrium), decreasing AGE-RAGE interaction (e.g anti-RAGE antibody, SiRNA against RAGE), detoxifying the dicarbonyls intermediates and interrupting signalling pathways that are associated with AGE-mediated effects [144–149]. However, Food and Drug Administration doesn't approve any agents that have significant role in AGE-modification and its associated effects to date, although AGE-associated medications are in clinical and preclinical testing.

Also, oxidative stress play a significant role in development of vascular complications, therefore anti-oxidants are thought to have beneficial effect on AGE-mediated MVC. Haidara et al., concluded that administration of antioxidants might have a cardio-protective effect in experimental setting against endothelial dysfunction and provide an ideal candidate of reducing cardiovascular complication in diabetes [150]. Many anti-oxidants such as N-acetylcysteine, and alpha–lipoic acid have found to inhibit the enhanced VCAM-1 expression by blocking the induction of specific DNA-binding activity for NF-κB at VACM-1 promoter of AGE-treated endothelial cells [151, 152]. Naturally occurring polyphenols like resveratrol (phytoalexin present in red wine and grape juice) which is known for its beneficial effect on cardiovascular disease have shown to restore the endothelial functions in T2DM thereby, suggesting the potential for new treatment lines to promote vascular health under hyperglycemic conditions [153].

20.11 Conclusion

Enhanced formation of AGEs under hyperglycemic conditions is an important biochemical abnormality that accompanies the development of cardiovascular complications in diabetes through increased ROS generation, inflammation, glycation of biomolecules, enhanced oxidative stress development, cellular proliferation, and others that may possibly exacerbate damaging effects on cardiac function. Also, AGEs affect the endothelial function through alteration of its vasodilating and adhesive properties by activating the RAGE-mediated intracellular signaling pathway. Therefore, detection of AGEs levels may be a useful marker in monitoring and tailoring the treatment of diabetes and its associated complications. Agents that can decrease the AGEs formation or attenuate their detrimental effects may be considered as ideal candidates for pharmaceutical intervention in the amelioration of diabetes and its associated vascular-complications. Henceforth, anti-AGEs strategies acting synergistically with conventional approaches may play an important role in improvement and upgrading of currently available therapeutic options for vascular complications in diabetes.

Acknowledgments This work was supported by the Indian Council of Medical Research, Government of India, New Delhi and the Council of Scientific and Industrial Research, New Delhi, India.

References

1. Beckman JA, Creager MA, Libby P (2002) Diabetes and atherosclerosis: epidemiology, pathophysiology and management. JAMA 287:2570–2581
2. Bonnefont-Rousselot D (2002) Glucose and reactive oxygen species. Curr Opin Clin Nutr Metab Care 5:561–568
3. Spitaler MM, Graier WF (2002) Vascular targets of redox signaling in diabetes mellitus. Diabetologia 45:476–494
4. Kaneko M, Bucciarelli L, Hwang YC et al (2005) Aldose reductase and AGE-RAGE pathways: key players in myocardial ischemic injury. Ann N Y Acad Sci 1043:702–709
5. Huebschmann AG, Regensteiner JG, Vlassara H et al (2006) Diabetes and advanced glycoxidation end products. Diabetes Care 29:1420–1432
6. Goh SY, Cooper ME (2008) The role of advanced glycation end products in progression and complications of diabetes. J Clin Endocrinol Metab 93:1143–1152
7. Basta G, Schmidt AM, De Caterina R (2004) Advanced glycation end products and vascular inflammation: implications for accelerated atherosclerosis in diabetes. Cardiovasc Res 63:582–592
8. John WG, Lamb EJ (1993) The Maillard or browning reaction in diabetes. Eye 7:230–237
9. Kiuchi K, Nejima J, Takano T et al (2001) Increased serum concentrations of advanced glycation end products: a marker of coronary artery disease activity in type 2 diabetic patients. Heart 85:87–91
10. Bansal S, Chawla D, Siddarth M et al (2013) A study on serum advanced glycation end products and its association with oxidative stress and paraoxonase activity in type 2 diabetic patients with vascular complications. Clin Biochem 46:109–114
11. Kaneda H, Taguchi J, Ogasawara K (2002) Increased level of advanced oxidation protein products in patients with coronary artery disease. Atherosclerosis 162:221–225
12. Schleicher ED, Wagner E, Nerlich AG (1999) Increased accumulation of the glycoxidation product N-(carboxymethyl) lysine in human tissues in diabetes and ageing. J Clin Invest 99:457–468
13. Kilhovd BK, Juutilainen A, Lehto S et al (2007) Increased serum levels of advanced glycation end products predict total, cardiovascular and coronary mortality in women with type 2 diabetes: a population-based 18 year follow-up study. Diabetologia 50:1409–1417
14. Pan HZ, Zhang H, Chang D et al (2008) The change of oxidative stress products in diabetes mellitus and diabetic retinopathy. Br J Ophthalmol 92:548–551
15. Hayashi T, Mori T, Yamashita C et al (2008) Regulation of oxidative stress and cardioprotection in diabetes mellitus. Curr Cardiol Rev 4(4):251–258

16. Moris D, Spartalis M, Spartalis E et al (2017) The role of reactive oxygen species in the pathophysiology of cardiovascular diseases and the clinical significance of myocardial redox. Ann Transl Med 5(16):326
17. Haitoglou CS, Tsilibary EC, Brownlee M et al (1992) Altered cellular interactions between endothelial cells and nonenzymatically glucosylated laminin/type IV collagen. J Biol Chem 267:12404–12407
18. Throckmorton DC, Brogden AP, Min B et al (1995) PDGF and TGF-beta mediate collagen production by mesangial cells exposed to advanced glycosylation end products. Kidney Int 48:111–117
19. Yan SF, Ramasamy R, Schmidt AM (2010) The RAGE axis: a fundamental mechanism signaling danger to the vulnerable vasculature. Circ Res 106:842–853
20. Goldin A, Beckman JA, Schmidt AM et al (2006) Advanced glycation end products: sparking the development of diabetic vascular injury. Circulation 114:597–605
21. Coughlan MT, Thorburn DR, Penfold SA et al (2009) RAGE-induced cytosolic ROS promote mitochondrial superoxide generation in diabetes. J Am Soc Nephrol 20:742–752
22. Fukami K, Yamagishi S, Okuda S (2014) Role of AGEs-RAGE system in cardiovascular disease. Curr Pharm Des 20(14):2395–2402
23. Schmidt AM, Stern D (2000) Atherosclerosis and diabetes: the RAGE connection. Curr Atheroscler Rep 2:430–436
24. Stirban AO, Tschoepe D (2008) Cardiovascular complications in diabetes. Diabetes Care 31:s215–ss21
25. Hanssen KF (1997) Blood glucose control and microvascular and macrovascular complications. Diabetes 46:S101–S1S3
26. Brownlee M (2001) Biochemistry and molecular cell biology of diabetic complications. Nature 414:813–820
27. Sebekova K, Kupcova V, Schinzel R et al (2002) Markedly elevated levels of plasma advanced glycation end products in patients with liver cirrhosis-amelioration by liver transplantation. J Hepatol 36:66–71
28. Boehm BO, Schilling S, Rosinger S et al (2004) Elevated serum levels of N (epsilon)-carboxymethyl- lysine, an advanced glycation end product, are associated with proliferative diabetic retinopathy and macular oedema. Diabetologia 47:1376–1379
29. Wada R, Yagihashi S (2005) Role of advanced glycation end products and their receptors in development of diabetic neuropathy. Ann N Y Acad Sci 1043:598–604
30. Khalifah RG, Baynes JW, Hudson BG (1999) Amadorins: novel post-Amadori inhibitors of advanced glycation reactions. Biochem Biophys Res Commun 257:251–258
31. Jakus V, Rietbrock N (2004) Advanced glycation end-products and the progress of diabetic vascular complications. Physiol Res 53:131–142
32. Monnier VM et al (1990) Nonenzymatic glycosylation, the Maillard reaction and the aging process. J Gerontol 45:B105–BB11
33. Singh R, Barden A, Mori T et al (2001) Advanced glycation end-products: a review. Diabetologia 44:129–146
34. Aronson D, Rayfield EJ (2002) How hyperglycemia promotes atherosclerosis: molecular mechanisms. Cardiovasc Diabetol 1:1
35. Lyons T, Jenkins AJ (1997) Glycation, oxidation and lipoxidation in the development of the complications of diabetes mellitus: a carbonyl stress hypothesis. Diabetes Rev 5:365–391
36. Miyata T, van Ypersele DS, Kurokawa K et al (1999) Alterations in nonenzymatic biochemistry in uremia: origin and significance of "carbonyl stress" in long term uremic complications. Kidney Int 55:389–399
37. Thornalley PJ (2005) Dicarbonyl intermediates in the maillard reaction. Ann N Y Acad Sci 1043:111–117
38. Kilhovd BK, Giardino I, Torjesen PA et al (2003) Increased serum levels of the specific AGE-compound methylglyoxal-derived hydroimidazolone in patients with type 2 diabetes. Metabolism 52:163–167

39. Turk Z (2010) Glycotoxines, carbonyl stress and relevance to diabetes and its complications. Physiol Res 59:147–156
40. Booth AA, Khalifah RG, Todd P et al (1997) In vitro kinetic studies of formation of antigenic advanced glycation end products (AGEs) novel inhibition of post-Amadori glycation pathways. J Biol Chem 272:5430–5437
41. Miyata T, Ueda Y, Yamada Y et al (1998) Accumulation of carbonyls accelerates the formation of pentosidine, an advanced glycation end product: carbonyl stress in uremia. J Am Soc Nephrol 9:2349–2356
42. Lorenzi M (2007) The polyol pathway as a mechanism for diabetic retinopathy: attractive, elusive and resilient. Exp Diabetes Res 2007:61038
43. Koschinsky T, He CJ, Mitsuhashi T et al (1997) Orally absorbed reactive glycation products (glycotoxins): an environmental risk factor in diabetic nephropathy. Proc Natl Acad Sci 94:6474–6479
44. Nicholl ID, Stitt AW, Moore JE et al (1998) Increased levels of advanced glycation endproducts in the lenses and blood vessels of cigarette smokers. Mol Med 4:594–601
45. Vlassara H, Cai W, Crandall J et al (2002) Inflammatory mediators are induced by dietary glycotoxins, a major risk factor for diabetic angiopathy. Proc Natl Acad Sci 99:15596–15601
46. Zheng F, He C, Cai W et al (2002) Prevention of diabetic nephropathy in mice by a diet low in glycoxidation products. Diabetes Metab Res Rev 18:224–237
47. Uribarri J, Tuttle KR (2006) Advanced glycation end products and nephrotoxicity of high-protein diets. Clin J Am Soc Nephrol 1:1293–1299
48. Ulrich P, Cerami A (2001) Protein glycation, diabetes, and aging. Recent Prog Horm Res 56:1–21
49. Ahmed N, Thornalley PJ (2003) Quantitative screening of protein biomarkers of early glycation, advanced glycation, oxidation and nitrosation in cellular and extracellular proteins by tandem mass spectrometry multiple reaction monitoring. Biochem Soc Trans 31:1417–1422
50. Lapolla A, Piarulli F, Sartore G et al (2007) Advanced glycation end products and antioxidant status in type 2 diabetic patients with and without peripheral artery disease. Diabetes Care 30:670–676
51. Anitha B, Sampathkumar R, Balasubramanyam M et al (2008) Advanced glycation index and its association with severity of diabetic retinopathy in type 2 diabetic subjects. J Diabet Complicat 22:261–266
52. Sharp PS, Rainbow S, Mukherjee S (2003) Serum levels of low molecular weight advanced glycation end products in diabetic subjects. Diabet Med 20:575–579
53. Tan KC, Chow WS, Tam S et al (2004) Association between acute phase reactants and advanced glycation end products in type 2 diabetes. Diabetes Care 27:223–228
54. Kilhovd BK, Berg TJ, Birkeland KI et al (1999) Serum levels of advanced glycation end products are increased in patients with type 2 diabetes and coronary heart disease. Diabetes Care 22:1543–1548
55. Thomas MC, Tsalamandris C, MacIsaac R et al (2004) Low molecular-weight AGEs are associated with GFR and anemia in patients with type 2 diabetes. Kidney Int 66:1167–1172
56. Yoshida N, Okumura K, Aso Y (2005) High serum pentosidine concentrations are associated with increased arterial stiffness and thickness in patients with type 2 diabetes. Metabolism 54:345–350
57. Willemsen S, Hartog JW, Hummel YM et al (2011) Tissue advanced glycation end products are associated with diastolic function and aerobic exercise capacity in diabetic heart failure patients. Eur J Heart Fail 13:76–82
58. Nin JW, Jorsal A, Ferreira I et al (2011) Higher plasma levels of advanced glycation end products are associated with incident cardiovascular disease and all-cause mortality in type 1 diabetes. Diabetes Care 34:442–447
59. Koyama Y, Takeishi Y, Arimoto T et al (2007) High serum level of pentosidine, an advanced glycation end product (AGE), is a risk factor of patients with heart failure. J Card Fail 13:199–206

60. Zhou H, Tan K, Shiu S et al (2007) Increased serum advanced glycation end products are associated with impairment in HDL antioxidative capacity in diabetic nephropathy. Nephrol Dial Transplant 23:927–933
61. Kass DA, Shapiro EP, Kawaguchi M et al (2001) Improved arterial compliance by a novel advanced glycation end-product crosslink breaker. Circulation 104:1464–1470
62. Schmidt AM, Yan SD, Yan SF et al (2001) The multiligand receptor RAGE as a progression factor amplifying immune and inflammatory responses. J Clin Invest 108:949–955
63. Reddy GK (2004) AGE-related cross-linking of collagen is associated with aortic wall matrix stiffness in the pathogenesis of drug-induced diabetes in rats. Microvasc Res 68:132–142
64. Cai W, Torreggiani M, Zhu L et al (2010) AGER1 regulates endothelial cell NADPH oxidase dependent oxidant stress via PKC-delta: implications for vascular disease. Am J Physiol Cell Physiol 298:C624–C634
65. Li YM, Mitsuhashi T, Wojciechowicz D et al (1996) Molecular identity and cellular distribution of advanced glycation end product receptors: relationship of p60 to OST-48 and p90 to 80K-H membrane proteins. Proc Natl Acad Sci U S A 93:11047–11052
66. Iacobini C, Amadio L, Oddi G et al (2003) Role of galectin-3 in diabetic nephropathy. J Am Soc Nephrol 14:S264–SS70
67. Bierhaus A, Humpert PM, Morcos M et al (2005) Understanding RAGE, the receptor for advanced glycation end products. J Mol Med 83:876–886
68. Bierhaus A, Schiekofer S, Schwaninger M et al (2001) Diabetes-associated sustained activation of the transcription factor nuclear factor-kappa B. Diabetes 50:2792–2808
69. Basta G (2008) Receptor for advanced glycation end products and atherosclerosis: from basic mechanisms to clinical implications. Atherosclerosis 196:9–21
70. Huttunen HL, Fages C, Rauvala H (1999) Receptor for advanced glycation endproducts (RAGE)-mediated neurite outgrowth and activation of NF-κB require the cytoplasmic domain of the receptor of but different downstream signaling pathways. J Biol Chem 274:19919–19924
71. McNulty M, Mahmud A, Feely J (2007) Advanced glycation endproducts and arterial stiffness in hypertension. Am J Hypertens 20:242–247
72. Brownlee M (2005) The pathobiology of diabetic complications: a unifying mechanism. Diabetes 54:1615–1625
73. Giardino I, Edelstein D, Brownlee M (1994) Nonenzymatic glycosylation in vitro and in bovine endothelial cells alters basic fibroblast growth factor activity. A model for intracellular glycosylation in diabetes. J Clin Invest 94:110–117
74. Rosca MG, Mustata TG, Kinter MT et al (2005) Glycation of mitochondrial proteins from diabetic rat kidney is associated with excess superoxide formation. Am J Physiol Renal Physiol 289:F420–FF43
75. Wu L, Juurlink BH (2002) Increased methylglyoxal and oxidative stress in hypertensive rat vascular smooth muscle cells. Hypertension 39:809–814
76. Watson AD, Berliner JA, Hama SY et al (1995) Protective effect of high-density lipoprotein associated paraoxonase-inhibition of the biological activity of minimally oxidised low-density lipoprotein. J Clin Invest 96:2882–2891
77. Mackness MI, Durrington PN, Mackness B (2004) The role of paraoxonase 1 activity in cardiovascular disease: potential for therapeutic intervention. Am J Cardiovasc Drugs 4:211–217
78. Kinumi T, Ogawa Y, Kimata J et al (2005) Proteomic characterization of oxidative dysfunction in human umbilical vein endothelial cells (HUVEC) induced by exposure to oxidized LDL. Free Radic Res 39:1335–1344
79. Karabina SA, Lehner AN, Frank E et al (2005) Oxidative inactivation of paraoxonase – implications in diabetes mellitus and atherosclerosis. Biochim Biophys Acta 1725:213–221
80. Ferretti G, Bacchetti T, Marchionni C et al (2001) Effect of glycation of high density lipoproteins on their physiochemical properties and on paraoxonase activity. Acta Diabetol 38:163–169

81. Zhou H, Tan K, Shiu S, Wong Y (2007) Increased serum advanced glycation end products are associated with impairment in HDL antioxidative capacity in diabetic nephropathy. Nephrol Dial Transplant 23:927–933
82. Brownlee M, Vlassara H, Cerami A (1985) Nonenzymatic glycosylation products on collagen covalently trap low-density lipoprotein. Diabetes 34:938–941
83. Bucala R, Makita Z, Vega G (1994) Modification of low density lipoprotein by advanced glycation end products contributes to the dyslipidemia of diabetes and renal insufficiency. Proc Natl Acad Sci 91:9441–9445
84. Hodgkinson CP, Laxton RC, Patel K et al (2008) Advanced glycation end-product of low density lipoprotein activates the toll-like 4 receptor pathway implications for diabetic atherosclerosis. Arterioscler Thromb Vasc Biol 28:2275–2281
85. Rabbani N, Chittari MV, Bodmer CW et al (2010) Apolipoprotein B100 of LDL cholesterol in patients with type 2 diabetes and effect of metformin. Diabetes 59:1038–1045
86. Betteridge DJ (2000) What is oxidative stress? Metabolism 49:3–8
87. D'Autréaux B, Toledano MB (2007) ROS as signaling molecules: mechanisms that generate specificity in ROS homeostasis. Nat Rev Mol Cell Biol 8(10):813–824
88. Tibaut M, Petrovič D (2016) Oxidative stress genes, antioxidants and coronary artery disease in type 2 diabetes mellitus. Cardiovasc Hematol Agents Med Chem 14:23–38
89. Collison KS, Parhar RS, Saleh SS et al (2002) RAGE-mediated neutrophil dysfunction is evoked by advanced glycation end products (AGEs). J Leukoc Biol 171:433–444
90. Wong RKM, Pettit AI, Quinn PA et al (2003) Advanced glycation end products stimulate an enhanced neutrophil respiratory burst mediated through the activation of cytosolic phospholipase A2 and generation of arachidonic acid. Circulation 108:1858–1864
91. Ding Y, Kantarci A, Hasturk H et al (2007) Activation of RAGE induces elevated O_2 generation by mononuclear phagocytes in diabetes. J Leukoc Biol 81(2):520–527
92. Zhang FL, Gao HQ, Shen L (2007) Inhibitory effect of GSPE on RAGE expression induced by advanced glycation end products in endothelial cells. J Cardiovasc Pharmacol 50:434–440
93. Ayilavarapu S, Kantarci A, Fredman G et al (2010) Diabetes induced oxidative stress is mediated by Ca^{2+}- independent phospholipase A_2 in neutrophils. J Immunol 184:1507–1515
94. Zhang M, Ay LK, Anilkumar N et al (2006) Glycated proteins stimulate reactive oxygen species production in cardiac myocytes: involvement of Nox2 (gp91phox)-containing NADPH oxidase. Circulation 113:1235–1243
95. Bansal S, Siddarth M, Chawla D et al (2012) Advanced glycation end products enhance reactive oxygen and nitrogen species generation in neutrophils in vitro. Mol Cell Biochem 361:289–296
96. Hulsmans M, Van Dooren E, Holvoet P (2012) Mitochondrial reactive oxygen species and risk of atherosclerosis. Curr Atheroscler Rep 14(3):264–276
97. Pitocco D, Zaccardi F, Stasio ED et al (2010) Oxidative stress, nitric oxide, and diabetes. Rev Diabet Stud 7:15–25
98. Bechman JS, Beckman TV, Chen J et al (1990) Apparent hydroxyl radical production by peroxynitrite: implication for endothelial injury from nitric oxide and superoxide. Proc Natl Acad Sci U S A 87:1620–1624
99. Chang PC, Chen TH, Chang CJ et al (2004) Advanced glycosylation end products induce inducible nitric oxide synthase (iNOS) expression via a p38 MAPK dependent pathway. Kidney Int 65:1664–1675
100. Rojas A, Caveda L, Romay C et al (1996) Effects of advanced glycosylation end products on the induction of nitric oxide synthase in murine macrophages. Biochem Biophys Res Commun 225:358–362
101. Lin CH, Lin YF, Chang MC et al (2001) Advanced glycosylation end products induce nitric oxide synthase expression in C6 glioma cells: involvement of a p38 MAP kinase-dependent mechanism. Life Sci 69:2503–2515. 34
102. Toure F, Zahm JM, Garnotel R (2008) Receptor for advanced glycation end-products (RAGE) modulates neutrophil adhesion and migration on glycoxidated extracellular matrix. Biochem J 416:255–261

103. Rizvi AA (2009) Cytokine biomarkers, endothelial inflammation, and atherosclerosis in the metabolic syndrome: emerging concepts. Am J Med Sci 55:659–665
104. Higashi Y, Noma K, Yoshizumi M et al (2009) Endothelial function and oxidative stress in cardiovascular diseases. Circ J 73:411–418
105. Hadi HAR, Al Suwaidi J (2007) Endothelial dysfunction in diabetes mellitus. Vasc Health Risk Manag 3:853–876
106. Pitocco D, Tesauro M, Alessandro R et al (2013) Oxidative stress in diabetes: implications for vascular and other complications. Int J Mol Sci 14(11):21525–21550
107. Piwowar A, Knapik-Krdecka M, Szczecinska J et al (2008) Plasma glycooxidation protein products in type 2 diabetic patients with nephropathy. Diabetes Metab Res Rev 24:549–553
108. Sarkar P, Kaushik K, Mondal MC et al (2010) Elevated level of carbonyl compounds correlates with insulin resistance in type 2 diabetes. Ann Acad Med Singap 39:909–912
109. Soliman GZ (2008) Blood lipid peroxidation (superoxide dismutase, malondialdehyde, glutathione) levels in Egyptian type 2 diabetic patients. Singap Med J 49:129–136
110. Pasaoglu H, Sancak B, Burkan N (2004) Lipid peroxidation and resistance to oxidation in patients with type 2 diabetes mellitus. Tohuku J Exp Med 203:211–218
111. Palinski W, Ord VA, Plump AS et al (1994) ApoE-deficient mice are a model of lipoprotein oxidation in atherogenesis. Demonstration of oxidation-specific epitopes in lesions and high titers of autoantibodies to malondialdehyde-lysine in serum. Arterioscler Thromb 14:605–616
112. Holvoet P, Vanhaecke J, Janssens S et al (1998) Oxidized LDL and malondialdehyde-modified LDL in patients with acute coronary syndromes and stable coronary artery disease. Circulation 98:1487–1494
113. Paudel KR, Lee UW, Kim DW (2016) Chungtaeje on, a Korean fermented tea, prevents the risk of atherosclerosis in rats fed a high-fat atherogenic diet. J Integr Med 14(2):134–142
114. Goyal MT, Mehta JL (2011) Oxidized LDL, LOX-1 and atherosclerosis. Cardiovasc Drugs Ther 25(5):419–429
115. Paudel KR, Karki R, Kim DW (2016) Cepharanthine inhibits in vitro VSMC proliferation and migration and vascular inflammatory responses mediated by RAW264.7. Toxicol in Vitro 34:16–25
116. Sano FT, Umeda T, Hashimoto H (1998) Oxidative stress measurement by in vivo electron spin resonance spectroscopy in rats with streptozotocin-induced diabetes. Diabetologia 41(11):1355–1360
117. Witko-Sarsat V, Friedlander M, Nguyen-Khoa T et al (1998) Advanced oxidation protein products as novel mediator of inflammation and monocyte activation in chronic renal failure. J Immunol 161(2524):32
118. Cakatay U (2005) Protien oxidation parameters in type 2 diabetic patients with good and poor glycemic control. Diabetes Metab 31:551–557
119. Beckman KB, Ames BN (1997) Oxidative decay of DNA. J Biol Chem 272:19633–19636
120. Suantawee T, Cheng H, Adisakwattana S (2016) Protective effect of cyanidin against glucose- and methylglyoxal-induced protein glycation and oxidative DNA damage. Int J Biol Macromol 93:814–821
121. Leinonen J, Lehtimaki T, Toyokuni S et al (1997) New biomarker evidence of oxidative DNA damage in patients with non-insulin-dependent diabetes mellitus. FEBS Lett 417:150–152
122. Nishikawa T, Sasahara T, Kiritoshi S et al (2003) Evaluation of urinary 8-hydroxydeoxyguanosine as a novel biomarker of macrovascular complications in type 2 diabetes. Diabetes Care 26:1507–1512
123. Guerra SD, Lupi R, Marselli L et al (2005) Functional and molecular defects of pancreatic islets in human type 2 diabetes. Diabetes 54:727–735
124. Al-Aubaidy HA, Jelinek HF (2011) Oxidative DNA damage and obesity in type 2 diabetes mellitus. Eur J Endocrinol 164:899–904
125. Ballinger SW, Patterson C, Yan CN (2000) Hydrogen peroxide and peroxynitrite-induced mitochondrial DNA damage and dysfunction in vascular endothelial and smooth muscle cells. Circ Res 86:960–966

126. Andreassi M, Botto N (2003) Coronary atherosclerosis and somatic mutations: and overview of the contributive factors for oxidative DNA damage. Mutat Res 543:67–86
127. Wang D, Kreutzer DA, Essigmann JM (1998) Mutagenicity and repair of oxidative DNA damage: insights from studies using defined lesions. Mutat Res 400:99–115
128. Avogaro A, Albiero M, Menegazzo L et al (2011) Endothelial dysfunction in diabetes. Diabetes Care 34:S285–S290
129. Tan KC, Chow WS, Ai VHG et al (2002) Advanced glycation end products and endothelial dysfunction in type 2 diabetes. Diabetes Care 25:1055–1059
130. Soro-Paavonen A, Zhang WZ, Venardos K et al (2010) Advanced glycation end-products induce vascular dysfunction via resistance to nitric oxide and suppression of endothelial nitric oxide synthase. J Hypertens 28:780–788
131. Sears CE, Bryant SM, Ashley EA et al (2003) Cardiac neuronal nitric oxide synthase isoform regulates myocardial contraction and calcium handling. Circ Res 92:e52–ee9
132. Chen W, Druhan LJ, Chen CA et al (2010) Peroxynitrite induces destruction of the tetrahydrobiopterin and heme in endothelial nitric oxide synthase: transition from reversible to irreversible enzyme inhibition. Biochemistry (Mosc) 49(14):3129–3137
133. Ishibashi Y, Matsui T, Ueda S et al (2014) Irbesartan inhibits advanced glycation end product-induced increase in asymmetric dimethylarginine level in mesangial cells through its anti-oxidative properties. Int J Cardiol 176:1120–1122
134. Price DT, Loscalzo J (1999) Cellular adhesion molecules and atherogenesis. Am J Med 107:85–97
135. Shih PT, Brennan ML, Vora DK et al (1999) Blocking very late antigen-4 integrin decreases leukocyte entry and fatty streak formation in mice fed an atherogenic diet. Circ Res 84:345–351
136. Eriksson JG, Forsen T, Tuomilehto J et al (2001) Early growth and coronary heart disease in later life: longitudinal study. Br Med J 322:949–953
137. Takenaka K, Yamagishi S, Matsui T et al (2006) Role of advanced glycation end products (AGEs) in thrombogenic abnormalities in diabetes. Curr Neurovasc Res 3:73–77
138. Ishibashi Y, Matsui T, Fukami K et al (2015) Rivaroxaban inhibits oxidative and inflammatory reactions in advanced glycation end product-exposed tubular cells by blocking thrombin/protease-activated receptor-2 system. Thromb Res 135:770–773
139. Yamagishi S, Fujimori H, Yonekura H et al (1998) Advanced glycation end products inhibit prostacyclin production and induce plasminogen activator inhibitor-1 in human microvascular endothelial cells. Diabetologia 41:1435–1441
140. Chen Q, Dong L, Wang L (2009) Advanced glycation end products impair function of late endothelial progenitor cells through effects on protein kinase Akt and cyclooxygenase-2. Biochem Biophys Res Commun 381:192–197
141. Bhatwadekar AD, Glenn JV, Li G et al (2008) Advanced glycation of fibronectin impairs vascular repair by endothelial progenitor cells: implications for vasodegeneration in diabetic retinopathy. Invest Ophthalmol Vis Sci 49:1232–1241
142. Botto N, Rizza A, Colombo MG (2001) Evidence for DNA damage in patients with coronary artery disease. Mutat Res 493:23–30
143. Oever IAMVD, Raterman HG, Nurmohamed MT et al (2010) Endothelial dysfunction, inflammation, and apoptosis in diabetes mellitus. Mediat Inflamm 2010:792393
144. Figarola JL, Scott S, Loera S et al (2003) LR-90 a new advanced glycation endproduct inhibitor prevents progression of diabetic nephropathy in streptozotocin-diabetic rats. Diabetologia 46:1140–1152
145. Forbes JM, Yee LT, Thallas V et al (2004) Advanced glycation end product interventions reduce diabetes-accelerated atherosclerosis. Diabetes 53:1813–1823
146. Zieman S, Kass D (2004) Advanced glycation end product cross-linking: pathophysiologic role and therapeutic target in cardiovascular disease. Congest Heart Fail 10:144–149
147. Thornalley PJ (2003) Use of aminoguanidine (Pimagedine) to prevent the formation of advanced glycation endproducts. Arch Biochem Biophys 419:31–40

148. Kaida Y, Fukami K, Matsui T et al (2013) DNA aptamer raised against AGEs blocks the progression of experimental diabetic nephropathy. Diabetes 62:3241–3250
149. Liu H, Zheng F, Zhu L et al (2005) The immune defense protein lysozyme ameliorates acute vascular injury and atherosclerosis in hyperlipidemic mice. Am J Pathol 169:303–313
150. Haidara MA, Yassin HZ, Rateb M et al (2006) Role of oxidative stress in development of cardiovascular complications in diabetes mellitus. Curr Vasc Pharmacol 4:215–227
151. Kunt T, Forst T, Wilhelm A et al (1999) Alpha-lipoic acid reduces expression of vascular cell adhesion molecule-1 and endothelial adhesion of human monocytes after stimulation with advanced glycation end products. Clin Sci (Lond) 96:75–82
152. Zhou ZH, Jiang JL, Peng J et al (2002) Reversal of tolerance to nitroglycerin with N acetylcysteine or captopril: a role of calcitonin gene related peptide. Eur J Pharmacol 439:129–134
153. Zunino SJ, Storms DH (2009) Resveratrol alters proliferative responses and apoptosis in human activated B lymphocytes in vitro. J Nutr 139:1603–1608

Role of iNOS in Insulin Resistance and Endothelial Dysfunction

21

Hobby Aggarwal, Babu Nageswararao Kanuri, and Madhu Dikshit

21.1 Introduction

21.1.1 Insulin Resistance

Insulin resistance (IR) or prediabetic state involves constellation of systemic and tissue specific changes such as altered glucose metabolism, dyslipidemia, sub-acute inflammation, denovo hepatic lipogenesis, gluconeogenesis, tissue hypoxia etc. [1] It occurs due to the inability of insulin to promote glucose disposal to metabolic and vascular tissues as well as the inhibition of its suppressive effects on hepatic glucose production [2]. Being a progressive multisystem disorder, IR linked with diabetes or obesity can lead to myocardial infarction (MI), peripheral vascular disease (Atherosclerosis) and cerebral stroke distressing millions of people all over the globe [3]. According to Centers for Disease Control and Prevention (CDC) estimates, obesity prevalence has increased from 13% in 1962 to 36% in 2010 [4]. As per recent 2017 statistics, 424.9 million people are affected by Type-2-diabetes worldwide and if these trends continue, 628.6 million people will have diabetes by 2035 [5].

H. Aggarwal
Pharmacology Division, CSIR-Central Drug Research Institute, Lucknow, India

B. N. Kanuri
Pharmacology Division, CSIR-Central Drug Research Institute, Lucknow, India

Division of Endocrinology, Diabetes and Metabolism, University of Cincinnati, Cincinnati, OH, USA

M. Dikshit (✉)
Translational Health Science and Technology (THSTI), Department of Biotechnology, Faridabad, India

S. Chakraborti et al. (eds.), *Oxidative Stress in Heart Diseases*,
https://doi.org/10.1007/978-981-13-8273-4_21

21.1.2 IR and Oxidative/Nitosative Stress

Oxidative stress is a resultant of imbalance between the production of pro and antioxidant defenses leading to potential tissue damage in various disease pathologies. Reactive oxygen and nitrogen species (RONS) released from phagocytes and other cells in the gastrointestinal and respiratory tracts at physiologically optimal concentrations act as primary defense mechanism against microorganisms, while higher levels can lead to bio-molecular damage upon the constituents of living organisms [6–8]. Insulin resistance linked with obesity, Type-2-diabetes and cardiovascular disorders is generally associated with chronic systemic oxidative stress along with insulin signals disruption and adipo-cytokines dysregulation [9]. The cause-effect relationship between hyperglycemia, hyperlipidemia and hyperinsulinemia associated with IR and oxidative stress is interlinked as increased RONS has been shown to induce IR in isolated cellular and tissue systems and insulin resistant tissues have shown more oxidative damage [9, 10].

21.1.3 NO/NOS Isoforms/iNOS

Nitric oxide (NO), a highly diffusible gaseous molecule, is an important reactive nitrogen species (RNS) with a short half-life in blood [11]. Being a potent cell-signaling and vasodilator molecule with divergent biological effects, its production is tightly regulated under both physiological and pathophysiological conditions in almost every cell type. The discovery that endothelial derived relaxing factor is NO and it acts as a potential signaling molecule in various tissues led to its recognition as molecule of the year in 1992 and Nobel Prize in Physiology to Robert Furchgott, Louis Ignarro and Ferid Murad in 1998. Various findings in different pathological conditions reveal its pleotropic effects with deeper impact on metabolism [12].

Biological synthesis of NO is tightly regulated from transcriptional to posttranslational levels in multiple tiers through endogenous oxidoreductases called nitric oxide synthases (NOS). Three distinct NOS isoforms identified in mammals are: neuronal (nNOS or NOS1), inducible (iNOS or NOS2) and endothelial (eNOS or NOS3) [13]. Generation of NO from NOS enzymes majorly involves the oxidation of substrate L-arginine to L-citrulline in the presence of NADPH and O_2. Availability of cofactors like FAD, FMN and tetrahydrobiopterin (BH_4) is crucial for the aforementioned NO synthesis. The active NOS enzyme is a homodimer and homodimerization depends on the availability of BH_4 [14]. L-Arginine can also be metabolized by arginase enzyme to urea and L-ornithine, later being the synthesis precursor of polyamines. Intracellular levels of L-arginine and reduced biopterin thus could be a limiting factor for NO biosynthesis [15]. Table 21.1 describes about the cellular distribution of different NOS isoforms.

iNOS is a high output Ca^{2+} independent NOS induced by bacterial endotoxins (LPS), inflammatory cytokines, oxidative stress and nutrient overload [18, 19]. NO released through iNOS stimulation is many folds higher than that produced via activation of constitutive NOS (eNOS and nNOS) isoforms [20, 21]. It is widely known

Table 21.1 Cellular distribution of NOS/iNOS

Isoform	Distribution
NOS1	Neurons (mainly), liver, heart, skeletal muscle, kidney, gastrointestinal tract, vascular smooth muscle cells (VSMCs) and neutrophils
NOS2	Macrophages, mast cells, neutrophils, respiratory tract, vasculature, liver, skeletal muscle, VSMCs, kidney, neurons, colon
NOS3	Vasculature (mainly), respiratory tract, gastrointestinal tract, liver, skeletal muscle, heart, adipose tissue, eosinophils, lymphocytes [16, 17]

to exert biphasic responses through its variable cellular concentration gradients. At physiological concentrations, NO acts as anti-inflammatory, antithrombotic, anti-oxidant etc., by reacting with hydroxyl radical, superoxide anion and hydrogen peroxide all of which are toxic to the cells by themselves and neutralizing them [22]. Whereas, higher amounts can lead to irreversible tissue destruction via peroxynitrite formation, lipid peroxidation, oxidative modification of critical cysteine residues of proteins, nitration of protein tyrosine residues and DNA damage [19].

Studies on both animals and humans have shown that eNOS activity and NO bioavailability is decreased in metabolic disorders such as diabetes, obesity and metabolic syndrome [23, 24]. This can be due to reduced expression/activity of constitutive NOS (eNOS & nNOS) and/or enhanced iNOS expression leading to impaired eNOS activity as well as interaction of available NO with reactive species [25]. Recently, iNOS is also known to be expressed in metabolic tissues involved in the pathogenesis of IR, diabetes and obesity [26]. iNOS-derived NO affects insulin signaling pathway, peroxisome proliferator–activated receptorγ (PPARγ) activity, circulating adiponectin levels etc., thereby modulating genes involved in the glucose and lipids metabolism [27]. iNOS expression, on the other hand, is regulated by Protein kinase B (Akt), mitogen-activated protein kinase (MAPK) p42/44 and extracellular signal-regulated kinase (ERK1/2) in cardiovascular tissues [28]. Literature suggests that NO bioavailability in the vasculature is significantly decreased during insulin resistant states. Decreased NO levels in the endothelium due to generation of high levels of toxic free radicals such as superoxide (O_2^-) and peroxynitrite ($ONOO^-$) supports the hypothesis that IR in vascular endothelium contributes to the accelerated progression of cardiovascular complications such as atherosclerosis and MI [29]. Deficiency in the NO bioavailability, increased level of pro-thrombotic and pro-inflammatory markers and reactive oxygen species (ROS) are thus critical factors for endothelial dysfunction mediated by MAPK/ERK pathway. We and others have hypothesized that reactive oxygen nitrogen species (RONS) play a central role in IR and vascular dysfunction by providing a potential unifying mechanism in the progression of IR, obesity and associated cardiovascular complications at both physiological and pathophysiological levels. Glucotoxicity and lipotoxicity generate inflammatory milieu contributing to vascular damage, thereby linking IR with endothelial dysfunction through different mechanisms [30, 31]. Further sections of this chapter discuss in detail the pathological importance of iNOS in IR and endothelial dysfunction with an overview depicted in Fig. 21.1.

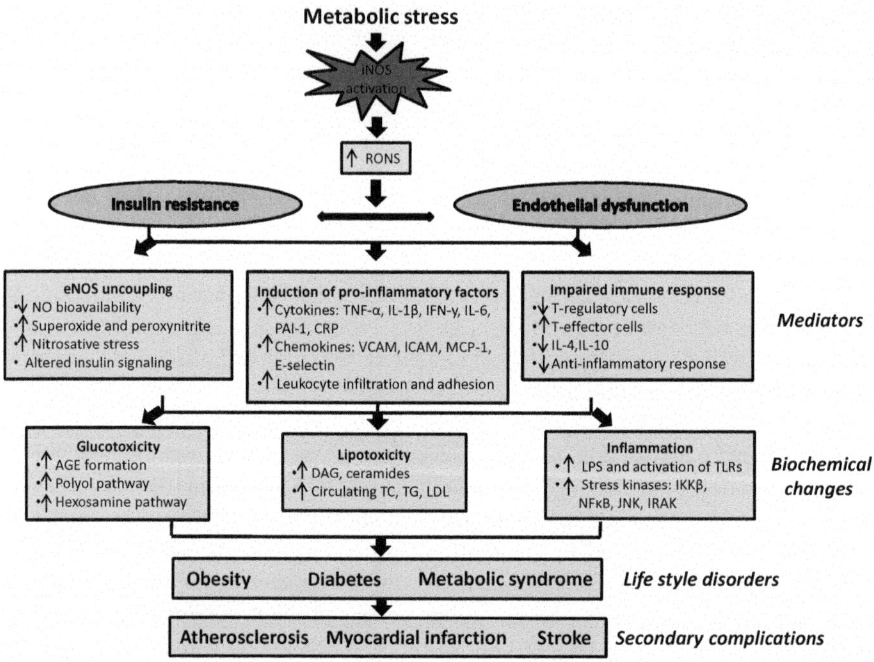

Fig. 21.1 Role of iNOS in insulin resistance and endothelial dysfunction
Metabolic stress causes induction of iNOS with increased production of reactive oxygen and nitrogen species which leads to insulin resistance and endothelial dysfunction due to eNOS uncoupling, induction of pro-inflammatory chemokines and cytokines and impaired immune response. This results in increased glucose, lipids and inflammation leading to cardiovascular complications like obesity, diabetes and metabolic syndrome including secondary complications like atherosclerosis, myocardial infarction and stroke

21.2 Role of iNOS in Insulin Resistance

In general, the levels of circulatory blood glucose are determined by an intricate balance between the rate of intestinal glucose absorption, hepatic glucose production and metabolism by major peripheral tissues such as adipose tissue and skeletal muscle. Upon insulin stimulation, glucose uptake is significantly increased in adipose tissue and skeletal muscle, and glucose oxidation is increased in the heart, while its production in the liver is inhibited. Insulin-induced cell growth and differentiation promotes substrate storage in fat, liver and muscle, which also serves as a regulator in maintaining blood glucose levels [32]. Obesity and Type-2-diabetes are associated with nutrient excess conditions exhibited by increased circulating glucose and lipids [33]. Systemic IR is characterized by increased accumulation of lipid metabolites (FA-CoA, DAG and ceramides) due to the imbalance between the synthesis of fatty acids and their oxidation [34].

As described earlier, iNOS demonstrates its biphasic effects by protecting us from infections on one hand and causing nitrosative stress on other end. In IR involving inflammation and metabolic nutrient stress, iNOS plays an important role in the inactivation of insulin receptor β/IRS-1 and Akt [19, 35] and its expression is regulated by IKKβ-NFκB axis [36]. Different experiments involving rodents have demonstrated augmented iNOS expression in key peripheral metabolic tissues like liver, skeletal muscle, adipose tissue, pancreatic β cells and heart, along with non-metabolic tissues such as aorta, spleen and kidneys during IR, Obesity and Diabetes [26, 37–41]. Adipocytes and subcutaneous adipose tissue of obese humans have more iNOS gene expression when compared with non-obese humans [42]. Elevated serum NOx levels (an indicator of systemic NO synthesis) [43] in obese adolescents strongly correlated with increased body adiposity [44]. Nitrosative stress due to iNOS induction leads to post translational protein modifications including nitrosylation, glutathionylation and nitration. Elevated RONS levels due to nitrosative stress was observed in plasma [45], skeletal muscle [46] and vasculature [47] in animals and patients with diabetes and obesity. LPS induced metabolic endotoxemia condition is associated with iNOS induction. Increased $ONOO^-$ formation causing tyrosine nitration of IRS-1 instead of insulin-dependent tyrosine phosphorylation is found to impair its function. LPS also impairs glucose uptake by the skeletal muscle by causing direct inflammatory insult to the myocytes as well as iNOS dependent cardiovascular dysfunction. Disruption of iNOS protected the mice from LPS-induced defects in cardiac output [48]. iNOS induced IRS-1 degradation was also observed in C2C12 myotubes treated with NO donor (GSNO) or by ectopic iNOS expression [49]. iNOS induction can also result in impairment of c-Jun N-terminal kinase/Stress-activated protein kinases (JNK/SAPK) axis stimulated ER function, thereby causing prolonged ER stress which finally, leads to insulin resistance and obesity [50]. Excessive fat also causes iNOS-induced mitochondrial dysfunction leading to ER stress and decreased adiponectin synthesis in cultured adipocytes [51]. Absence of iNOS completely restored the HFD induced altered insulin signaling in skeletal muscle. But in liver and adipose tissue, IR and ER stress can be caused by both iNOS-dependent and independent mechanisms. Disruption of iNOS was still associated with ER stress and altered insulin signaling in liver and adipose tissue. Insulin signaling was improved when the ER stress was blocked pharmacologically; demonstrating tissue-specific regulation of insulin signaling by iNOS in diabetes and obesity [52]. Thus, iNOS mediated nitrosative modification of key insulin signaling proteins in metabolic tissues is an important mechanism for impaired tissue insulin signaling [53]. Figure 21.2 depicts the downstream insulin signaling affected due to iNOS induction during metabolic stress leading to IR and endothelial dysfunction.

Pathological induction of iNOS during insulin resistant states is also associated with enhanced levels of inflammatory mediators such as TNF-α, IL-6, IL-1β and IFN-γ, which are key participants in chronic low grade inflammation during obesity-linked diabetes and cardiovascular diseases [54]. iNOS, through nitrosative stress or via inflammatory milieu linked to its activation in insulin targeted metabolic tissues; impair energy metabolism thereby participating in iNOS mediated tissue and systemic IR [53]. Pro-inflammatory cytokines like TNF-α can induce serine but not tyrosine mediated phosphorylation of IRS-1 which affects the translocation of

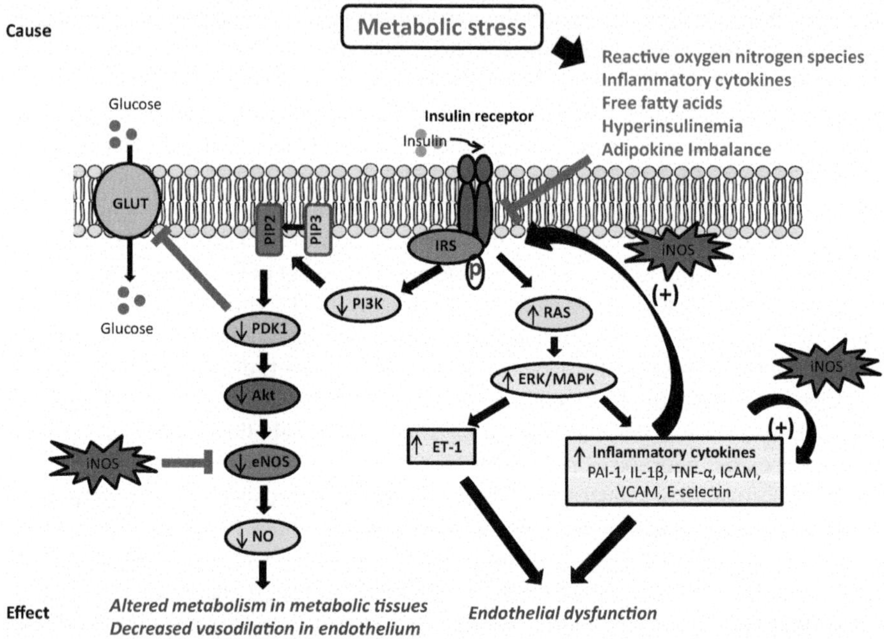

Fig. 21.2 Altered insulin signaling by iNOS leading to insulin resistance and endothelial dysfunction
iNOS induction by metabolic stress leads to increased generation of oxidative and nitrosative stress and pro-inflammatory cytokines with increased lipids, hyperinsulinemia and adipo-cytokines imbalance. This causes decrease in tyrosine phosphorylation of IRS-1/2 with increased serine phosphorylation leading to decreased PI3K and Akt which ultimately decreases the eNOS phosphorylation and activation and production of NO by eNOS leading to altered insulin signaling and metabolism in major metabolic organs with decreased vasodilation in the endothelium. iNOS can also cause direct uncoupling of eNOS by increased production of iNOS-mediated NO with decreased substrate availability leading to IR and endothelial dysfunction. iNOS induction can also leads to activation of RAS/MAPK pathway with increased production of Endothelin-1, pro-inflammatory cytokines and adhesion molecules leading to increased vasoconstriction of endothelium and endothelial dysfunction. Red arrows refers to inhibition

Glut-4 to cell membrane [54, 55], leading to impaired glucose homeostasis. Hepatic IR and mild hyperglycemia in liver specific iNOS transgenic mice indicates a positive correlation between iNOS induced tissue IR and obesity [56]. Studies involving ob/ob mice treated with selective iNOS inhibitor demonstrated reversal in fasting hyperglycemia [39]. iNOS knockout (KO) mice on 30% fructose in drinking water for 8 weeks showed attenuated hepatic steatosis and tissue inflammation, signifying the importance of iNOS in tissue and systemic IR [57]. On the other hand, increased circulating lipids due to systemic IR also induced viscous inflammatory milieu through iNOS activation [19]. Decreased fatty acid transport for the oxidation in mitochondria favors the activation of serine/threonine kinases (eg c-JNK, PKCθ) impairing the insulin signaling pathways including IR, IRS-1/-2 and Akt [58]. In addition, sphingolipids and ceramides produced from free fatty acids inhibit the phosphorylation of Akt, leading to reduced glucose uptake [59]. Literature also suggests that increased serum free fatty acids can attenuate the total expression of GLUT-4 leading to decreased glucose uptake, thereby impairing the glucose homeostasis [60]. Alterations in activities of insulin signaling pathway proteins due to the above mentioned protein modifications lead to dysregulated metabolic homeostasis resulting in metabolic complications such as hyperinsulinemia, hyperglycemia, hyperlipidemia etc. Absence of iNOS reverted lipid induced IR complications and improved the hepatic glucose metabolism [19]. In contrast, dietary administration of iNOS specific inhibitor, L-N^{ω} Nitroarginine to rats suggested an increased fat deposition as well as augmented circulatory and hepatic lipids [61]. Futhermore, significant reduction in skeletal muscle glucose uptake was observed in aged but not in young adult iNOS KO mice during HE clamp [62]. Moreover, iNOS KO mice on 45% HFD for 18 weeks demonstrated marked protection against diet induced systemic IR though they demonstrated increased body weight, elevated fat depots and fasting hyperglycemia [26]. Gut microbiome is a microbial organ which helps in maintaining the energy homeostasis and metabolism, along with insulin sensitivity [63]. Obesity increases the circulating endotoxins leading to inflammation and altered intestinal immunity along with changes in gut microbiome and gut barrier functions. Lipopolysaccharide (LPS) present in gram negative bacterial cell wall promotes inflammation and activates pathogen-associated molecular pattern (PAMP) responses via TLR4 [64] and induces iNOS expression and nitrosative modifications of insulin signaling proteins and IR [65] that is reverted by the disruption of iNOS [66]. NO derived from iNOS induction is also responsible for the impaired insulin release and β-cells destruction in Type-1-diabetes by causing hypoxic injury, apoptosis or necrosis leading to β-cells death [67] and also decreased insulin release in later stages of Type-2-diabetes through decreased expression of key insulin signaling proteins in the pancreas leading to β-cell failure [68].

Nutrient excess and hypoxic conditions can also lead to iNOS induction and excessive NO production by the resident macrophages in adipose tissue in response to inflammatory cytokines by inhibiting differentiation of preadipocytes into adipocytes, promoting tissue fibrosis due to mitochondrial dysfunction and by inducing hypoxia-inducible factor-1α [69]. There is also an increase in gene expression of TNF-α and iNOS in adipose tissue macrophages with decreased expression of *arginase-1* and *IL-10* causing polarization of M2 macrophages to M1 phenotype leading to increased inflammation in obesity and diabetes [70]. However, disruption of iNOS in myeloid cells of mice was not able to protect from HFD-induced IR and obesity suggesting the more important role of iNOS inhibition in tissues other than myeloid cells in IR [71].

As described above, previous reports on obesity/IR and metabolic syndrome (MS) using diet induced obesity (DIO) models have suggested the crucial role of iNOS during systemic IR in dyslipidemic mice models. To understand further the importance of iNOS in IR and obesity, studies were also performed on iNOS KO mice using different diets [low fat diet (LFD) to high fat diet (HFD)], assay protocols, etc. The results suggested the biphasic role of iNOS in IR and insulin signaling as revealed from both protective and detrimental effects, which could be due to the variations in the different protocols used for assessments along with differences in dietary fat content [26, 27, 71]. Further, afore mentioned studies on KO mice have not assessed different parameters for systemic and tissue IR in major metabolic tissues after feeding different diets with diverse regimens. Moreover, parameters of energy homeostasis as well as expression of genes responsible for glucose and lipid homeostasis were also not assessed [26, 71]. As the previous studies performed do not provide complete understanding of role of iNOS in IR, different experiments were conducted using iNOS KO mice in our lab [30]. The results concluded that though iNOS participates in IR and obesity (unpublished) when fed with high fat diets for longer time points (45% HFD for 20 weeks), it is protective against IR at basal conditions (chow and 10% LFD). After 5 weeks of 10% LFD feeding, iNOS KO mice demonstrated altered body glucose and lipid homeostasis as absence of iNOS caused systemic IR (Fig. 21.3). This resulted from tissue specific IR observed in liver and adipose tissue sparing the skeletal muscle. iNOS KO demonstrated hyperinsulinemia, altered glucose tolerance, increased body fat, hepatic gluconeogenesis, hepatic lipids, changes in expression of glucose

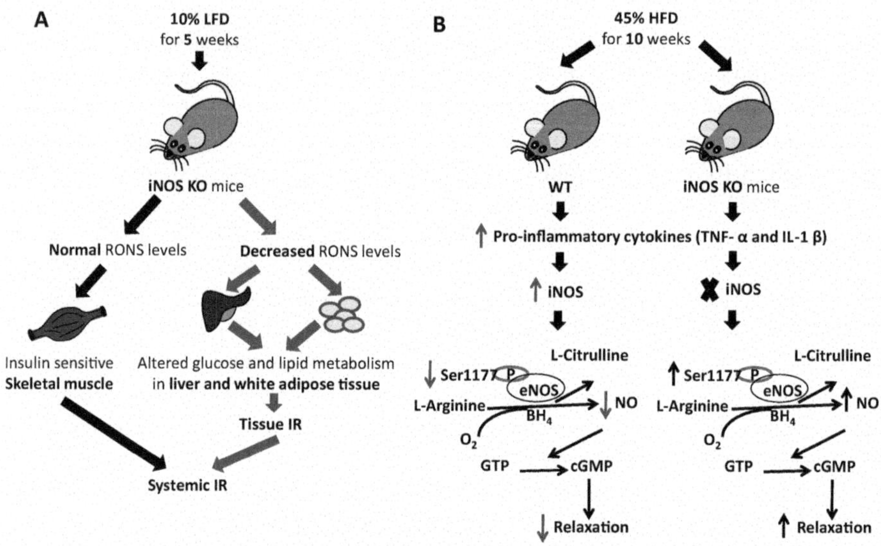

Fig. 21.3 Tissue specific role of iNOS in insulin resistance and endothelial dysfunction (**a**) iNOS KO mice fed with 10% LFD for 5 weeks leads to altered glucose and lipid homeostasis in liver and adipose tissue with decreased RONS levels on one hand; whereas skeletal muscle remained insulin sensitive with normal RONS levels on the other hand. iNOS KO mice displayed systemic insulin resistance and portrays the important role of iNOS in maintaining the normal energy homeostasis of the body with tissue specific response. (**b**) Mice fed with 45% HFD for 10 weeks (both WT and iNOS KO) displayed increase in pro-inflammatory cytokines like TNF-α and IL-1β. The metabolic insult causes increase in iNOS expression in WT mice which leads to decrease in phosphorylation of eNOS along with decreased NO production and vasorelaxation. In absence of iNOS, phosphorylation of eNOS is increased leading to increased NO production and relaxation in the vasculature, thereby protecting the iNOS KO mice from endothelial dysfunction

transporters and its functioning, and altered insulin signaling in liver and adipose tissue but not in muscle (Fig. 21.3). Tissue specific changes in iNOS KO is a consequence of decreased NO bioavailability from attenuated expression of eNOS and nNOS in absence of iNOS [30].

21.3 Endothelium Functionality and iNOS

The wall of a blood vessel is composed of 3 layers namely (i) internal layer of endothelial cells (ii) medial layer of vascular smooth muscle cells (VSMCs), and (iii) outer layer called tunica externa/adventia. The blood in the vessel is separated from the surrounding tissues by a monolayer of endothelial cells covering the internal surface of all the vessels in the body, including conduit and resistance vessels, pre-capillary arterioles and capillaries [72]. Larger conduit vessels such as aorta and carotid/coronary/brachial/femoral arteries, requires a healthy endothelium with smooth, quiescent surface to prevent the activation of clotting factors and pro-inflammatory cytokines and chemokines, accumulation of lipid particles and to inhibit the adhesion of immune cells like platelets and monocytes. Endothelial cells regulate different processes such as maintenance of systemic blood flow and blood pressure in the resistance vessels, transportation and distribution of nutrients like glucose and fats and hormones like insulin, and removal of waste products of metabolism in the pre-capillary arterioles. Vascular smooth muscle forms the middle layer and is responsible for the aortic constriction and relaxation [73]. Outer layer *i.e.* tunica externa/adventia consists of connective elements (fibroblasts, collagen fibers), perivascular adipose tissue, and nerve endings, all of which are responsible for the immune cell functions and inflammatory cell trafficking between the blood vessels and tissues [74]. In addition to the classical insulin dependent metabolic tissues such as liver, skeletal muscle and adipose tissue, it has recently emerged that IR can also manifest in endothelium, where insulin stimulates the release of NO [75] resulting in vascular smooth muscle relaxation [76]. Endothelium was initially considered to be an inert lining but has now received a lot of attention and is recognized as an organ regulating multitude of processes critical to vascular function [77]. NO activates the cGMP pathway in the VSMCs mediating the vasodilation [78].

Endothelial dysfunction is manifested as a systemic pathological event characterized by an imbalance between the vasodilating and the vasoconstricting factors, pro-inflammatory and anti-inflammatory mediators, growth promotion and growth inhibition, and coagulation and fibrinolysis, otherwise, whose equilibrium is tightly regulated in healthy endothelium [79]. NO inhibits adherence and aggregation of platelets and leucocytes to the endothelium surface, suppresses vasoconstriction and inhibits the proliferation of VSMCs. Endothelial dysfunction is basically an outcome of reduced NO bioavailability and is an important indicator for vascular health [29].

Endothelial dysfunction plays a major role in the pathogenesis of IR and its initiation is an important predictor of IR linked diseases such as obesity, diabetes and cardiovascular complications (Fig. 21.1). Loss of insulin signaling in the endothelium causes vascular IR, which results in reduced vasodilation, leading to altered macro and microcirculation of multiple vascular beds projecting diabetes as a vascular disease also along with metabolic disease with decreasing microvascular perfusion and nutritive capillary recruitment (Fig. 21.2) [80]. Reduction in capillary network and microcirculatory blood flow to metabolically active organs lead to impaired insulin-stimulated glucose and lipid homeostasis with decreased surface area for nutrient exchange. Along with disturbed vascular tone, IR/diabetes has been characterized by increased oxidative stress, lipid deposition, inflammation and platelet hyperactivity [81]. In HFD model of obesity, mice fed with 45% HFD for 10 weeks demonstrated a significant decrease in acetylcholine (ACh) induced vasodilation causing endothelial dysfunction (Fig. 21.3) along with elevated body weight and fat, augmented circulatory and hepatic lipids, and hyperinsulinemia. Gene expression analysis revealed that HFD fed aorta with impaired endothelial function also showed augmented expression of iNOS, $p47^{phox}$ and NOX4 (unpublished results from our lab) [82]. In subsequent experiments, mice lacking global iNOS (iNOS KO); demonstrated dyslipidemia induced IR though there was significant improvement in HFD induced endothelial dysfunction (unpublished results from our lab) (Fig. 21.3). Vascular damage due to oxidative stress triggers an inflammatory reaction, which stimulates the release of chemoattractants and cytokines that further aggravates IR and endothelial dysfunction [83]. In fact, IR and endothelial dysfunction are characterized by increase in markers of inflammation. Another study demonstrated that HFD fed iNOS KO mice were protected against the development of insulin resistance and endothelial dysfunction but remained hypertensive with increased vascular ROS [38]. Repeated positive feedback loops aggravating IR increase the susceptibility of patients with metabolic disorders to cardiovascular complications such as atherosclerosis, myocardial infarction (MI) and stroke. Under pathogenic conditions like inflammation and excess caloric intake with physical inactivity, NO bioavailability is decreased with impaired endothelial functionality which leads to induction of IR and dyslipidemia. The enhancement of vascular damage by augmented lipid flux through activation of metabolite sensitive pathways in metabolic organs demystified the correlation between IR and macrovascular compilcations [84].

Initially, iNOS induction compensates for the loss of functional eNOS by producing NO [85, 86] but excess NO and ROS leads to increased tissue dysfunction leading to plaque formation during atherosclerosis [87–91]. *FoxO* transcription factors (encoded by *FoxO1, FoxO3a and FoxO4)*, the downstream messengers of Akt, impair insulin signaling by acting as negative regulators of Akt mediated NO release via inhibition of eNOS expression. Disruption of all three isoforms of *FoxO* in the endothelium increased the NO bioavailability with favorable effects and protection against development of vascular dysfunction and atherosclerosis [92]. Vasoprotective PI3K-Akt pathway activates eNOS under physiological conditions [93], but during insulin resistant states, the balance is skewed towards MAPK/ERK pathway which mediates inflammation, vasoconstriction and VSMCs proliferation [94]. Activation of MAPK/ERK pathway also produces endothelin-1 (ET-1), which is a potent vasoconstrictor contributing to hypertension and endothelial dysfunction [95] with enhanced release of inflammatory markers (e.g., PAI-1, ICAM-1, VCAM-1, and E-selectin) promoting atherosclerosis [96].

Hyperglycemia leads to the conversion of glucose into sorbitol which increases the accumulation of ROS [97], leading to endothelial dysfunction due to impaired vasorelaxation and altered insulin sensitivity [98]. It also leads to O-Glc-N-acylation of IRS-1, impairing the activation of PI3-K and reduces glucose uptake, as well as O-Glc-N-acylation of eNOS at the Akt phosphorylation residues, leading to its inactivation [99]. Increased oxidative stress activates IKKβ-NF-κB axis with iNOS induction and increased pro-inflammatory markers, IL-1β, PAI-1, TGFβ; and TNF-ũFC; [100]. Peroxynitrite formation due to increased ROS enhances endothelial dysfunction by directly uncoupling and inactivating the eNOS. Uridine diphosphate N-acetyl glucosamine in overloaded glycolytic pathway caused by hyperglycemia also reduces eNOS phoshorylation at Ser1177, impairing vascular function [101]. The overall effects of these mechanisms are increased oxidative stress, apoptosis and vascular permeability [102]. FFA stimulates DAG-PKC and adversely affects Akt function due to IRS 1/2 inactivation [103] and enhanced NADPH oxidase induced ROS production [104] promoting release of pro-inflammatory cytokines such as PAI-1, IL-6, CCL-2, VEGF, TGF-1β, MCP-1, IL-1β and CRP, inhibiting NO production by decreasing eNOS expression. ROS also activates NF-κB, which increases ET-1 expression and adhesion molecules ICAM-1 and VCAM-1, thus enhancing the cardiovascular risk even in healthy subjects [105]. Expression of adhesion molecules on endothelial cells promote their contact with monocytes, and differentiate them into macrophages which after loading with lipoproteins turn into foam cells, mobilizing immune cells to build atherosclerotic plaque with impaired insulin signaling [106].

Augmented iNOS expression in cardiovascular tissues during diabetes, impart endothelium damage in the large blood vessels leading to stroke, myocardial ischemia (MI), heart failure and dilated ischemic cardiomyopathy [107–110]. iNOS uncoupling leads to the production of ROS which can induce myofibrillar oxidation contributing to contractile dysfunction [109]. Similarly, iNOS induction and uncoupling is linked to the development of left ventricular dilation, hypertrophy and congestive heart failure (CHF). Acute MI following plaque rupture in a coronary vessel is due to the formation of a thrombus and acute reduction in blood supply. On the other hand, vascular injury is exacerbated when blood supply is re-established. Increased TNF-α expression in myocardial ischemia-reperfusion (I/R) causes induction of superoxide producing systems; xanthine oxidase (XO) and NADPH Oxidase (NOX) with enhanced production of O_2^- which leads to coronary endothelial dysfunction [111]. Absence of NO causes endothelial dysfunction leading to decreased tissue perfusion, myocardial ischemia and vascular remodeling. Whereas, higher levels of NO as observed in failing myocardium, leads to loss of myocytes and decreased contractility suggesting the complex role of NO in CHF [112].

Disruption of iNOS, delays the contractile dysfunction with improved β-adrenergic responses associated with hypertrophy, decreased myocardial apoptosis and improved survival [113]. On the contrary, some studies have demonstrated that iNOS disruption had no effect on severe CHF after MI and I/R injury [114], as well as elevated BP [38]. Mice lacking iNOS showed no signs of improvement in early left ventricular remodeling after myocardial infarction (MI), whereas, significant improvement was observed during late remodeling of myocardium [115]. Dilated cardiomyopathy with increased iNOS derived NO exerts negative inotropic effect on the myocardium with altered systemic hemodynamics and endothelium dependent-impaired coronary vessel relaxation leading to heart failure [116, 117]. Cardiomyocyte-specific overexpression of iNOS increased the occurrence of malignant arrhythmia associated with atrioventricular block, ventricular tachycardia and sudden cardiac death [118]. Interestingly, iNOS offers cardioprotection during late phase of ischemic preconditioning (PC) [119, 120]. It also offers protection from occurrence of abdominal aortic aneurysms in females [121]. iNOS along with eNOS and Akt protects the myocardium from high fructose induced IR [120]. These diverse effects of iNOS-derived NO suggest the Ying-Yang effect of NO/iNOS in cardiovascular diseases.

In diabetes, vascular endothelium of brain becomes thicker with increased permeability and loss of vascular tone [122] impairing the cerebral circulation, thus, increasing the risk of ischemic stroke and cognitive function disruption [123, 124]. In stroke, disrupted cerebral blood flow causes vascular thrombosis and oxygen deficiency leading to ischemia as well as behavioral and functional defects which were reverted by intravenous nitrite infusion [125, 126]. Excessive NO released due to iNOS activation during infections increases the blood brain barrier (BBB) permeability resulting in cytotoxicity in brain [127, 128]. Resident macrophages from the ischemic endothelium and brain parenchyma cause tissue injury and cerebral ischemia due to reduced blood flow, release of inflammatory mediators and activation of leukocytes [129]. Reperfusion of the occluded vessel generates ROS either by reperfusion with oxygenated blood or production within brain and immune cells which further stimulates ischemic cells to secrete inflammatory cytokines, chemokines and adhesion molecules. Various cytotoxic agents like matrix metalloproteinases (MMPs), NO and ROS are activated further causing more cellular and extracellular matrix damage with the disruption of BBB [130]. This further potentiate brain tissue injury and blood along with its components enter the brain [131] leading to decreased tissue perfusion and post-ischemic inflammation [132]. iNOS expression is increased in the brain in the astrocytes, microglia and leukocytes in transient global ischemia [133] along with increased iNOS mRNA, iNOS activity and NO production [134]. NADPH oxidase, cyclooxygenase (COX), xanthine oxidase and xanthine dehydrogenase are the major enzymes responsible for superoxide generation whereas monoamine oxidase (MAO) and myeloperoxidase (MPO) are responsible for the generation of hypochlorous acid and H_2O_2. Superoxide generation can causes direct injury or reacts with NO to generate peroxynitrite in the ischemic brain [135]. Disruption of iNOS genetically or pharmacologically leads to reduced infarct size [136, 137].

Decreased T-cell responsiveness and prolonged peripheral lymphopenia has been observed in stroke patients [138]. Mice with ischemic brain injury due to middle cerebral artery occlusion demonstrated decreased blood supply and changes in splenic B, T and natural killer (NK) cells. Also, there was increased apoptosis of splenocytes, atrophy of spleen and expansion of regulatory T cell (Treg) correlating with decreased lymphocyte counts [139]. The adaptive immune response to the ischemic brain injury depends on the extent of injury and can be skewed towards damaging [T helper cell (Th)1/Th17] or protective (Th2) phenotypes [140]. Also, IL-17-secreting γδT cells contribute to ischemic injury and they do not undergo classical antigen-dependent T-cell activation [141]. Tregs can have a protective effect in ischemic injury to the brain by secreting IL-10 and downregulating post-ischemic inflammation and uniquely expressing nuclear receptor PPAR-γ [142]; whereas, effector T lymphocytes can contribute to focal ischemic injury [143]. However, during reperfusion phase of ischemic brain injury, the presence of intravascular Treg can cause vascular endothelial dysfunction and thrombosis and might contribute to ischemic brain injury [144].

There is an important two way relationship between the insulin resistance and endothelial dysfunction due to diverse physiological, cellular and molecular mechanisms interacting with each other in metabolic and vascular tissues. Parallel physiological and pathophysiological insulin signaling mechanisms in the vascular and metabolic tissues and cross-talk among them, different pathways leading to IR, inflammatory and insulin signaling crosstalk, link between blood flow and glucose metabolism can contribute to both insulin resistance and endothelial dysfunction simultaneously. Thus, instead of pharmacotherapies with single and specific agents, a combination of therapeutic strategies targeting multiple mechanisms is more likely to serve beneficial effects in metabolic and cardiovascular disorders [145].

21.4 Use of L-Arginine, Tetrahydrobiopterin and NO Donors in IR

Reduced NO availability is one of the key factors in the pathogenesis of IR and in associated pathologies such as obesity, diabetes and cardiovascular risk [23, 24]. Search for pharmacotherapies to enhance NO bioavailability is thus an important area of research. L-Arginine, is the substrate for NOS (nNOS, iNOS, eNOS) and NOS catalysis generates equimolar amount of NO and L-citrulline. NO maintains endothelium functionality and also cardiovascular health [146]. Administration of L-arginine has been shown to improve vascular dilation and preserve endothelial functionality [147], enhanced peripheral and liver insulin sensitivity, adipokines release, insulin secretion and tissue oxygenation, as well as alleviated oxidative stress in diabetes [148–150]. On the other hand, iNOS inhibition disrupts hepatic oxygenation and microcirculation, promotes clot formation and ROS generation [151, 152]. However, chronic iNOS inhibition reduces NO levels and improved liver function by increasing eNOS expression and by restoring the disrupted hemodynamics [153]. The natural sources of L-arginine which are recommended to be included in diet are present in both animal and plant derived foods. Animal sources comprise of dairy products, pork, beef, poultry, gelatin and seafood like fish etc. and the plant sources include grains, wheat germ and flour, oatmeal, peanuts, nuts, seeds, chickpeas and soybeans [154].

Tetrahydrobiopterin (BH_4), a cofactor to support NO synthesis [155]; gets oxidised to BH_2 under oxidative stress during obesity and diabetes, leading to a decrease in its intracellular bioavailability and eNOS uncoupling. This leads to superoxide generation which forms $ONOO^-$, adversely affecting endothelial function by oxidation of membrane lipids [156, 157]. Administration of BH_4 improves insulin sensitivity in diabetic animals [158]. Sepiapterin administration increases BH_4, total nitrite, angiogenesis and inhibits nitrosative stress in the infarcted heart of WT mice but not in iNOS deficient mice. This could be used to activate the salvage pathway for BH_4 synthesis and increase NO availability [159]. Intestinal microflora is the source of BH_4 suggesting that BH_4 producing bacteria might be exploited to alleviate cardiovascular diseases [160].

S-Nitrosothiols (RSNOs), the short lived (half-life ~1 h) unstable compounds, act as NO donors by releasing NO upon decomposition [161]. S-nitrosohemoglobin and S-nitrosoglutathione are the few forms of S-nitrosothiols [162]. Rapid inactivation of NO due to scavenging reactions involving hemoglobin, myoglobin and others contribute to its short half-life, making the products of NO metabolism *i.e.* nitrate and nitrite as its reservoirs that can be used for vasorelaxation [163, 164]. Nitroglycerine and sodium nitroprusside are the classical NO donors which are being used for clinical management of cardiovascular disorders such as coronary artery disease and congestive heart failure [165]. Enhanced oxidative and nitrosative stress due to iNOS derived NO production degrading the key insulin signaling proteins, when treated with NO donor indicated the importance of narrow therapeutic window while selecting the doses of different NO donors for pharmacological manipulation of cardiovascular complications [49, 56, 166, 167].

21.5 Conclusions & Future Perspectives

Biphasic effects of NO depend on the concentration gradient in circulation and different organs and tissues. It is widely accepted that constitutive NOS (eNOS and nNOS) maintains homeostasis by regulating diverse metabolic processes, cellular proliferation, apoptosis etc. While the inducible NOS i.e. iNOS, by generating higher amont of NO (~1000 folds more than normal) cause oxidative and nitrosative stress, inflammation and altered endothelium function; leading to diabetes, obesity, atherosclerosis, myocardial infarction and stroke. Surprisingly, recent studies have demonstrated that absence of iNOS also leads to IR and obesity under normal fed conditions, suggesting an optimal level of NO is required to prevent insulin resistance and endothelial dysfunction. It is therefore important to understand the intracellular/extracellular requirements of NO levels generated by three NOS isoforms for the regulation of diverse metabolic processes in different tissues/cell compartments. As NO availability is low in diabetes and obesity, use of L-arginine, tetrahydrobiopterin and NO donors have been claimed to counterbalance metabolic derailment by enhancing the NO bioavailability. Intriguingly, use of iNOS inhibitors have also been suggested for the treatment of IR, as iNOS induction is often found pathogenic. All the results published till date, signify importance of NO/NOS, however establishing an appropriate and optimal therapeutic window, is a challenge. Future studies should therefore be aimed to identify optimal requirements of NO/NOS in various tissues for the management of metabolic disorders and preserving endothelial function to alleviate cardiovascular risk.

References

1. Kahn BB, Flier JS (2000) Obesity and insulin resistance. J Clin Invest 106:473–481
2. Ginsberg HN (2000) Insulin resistance and cardiovascular disease. J Clin Invest 106:453–458
3. Bloomgarden ZT (2007) Insulin resistance, dyslipidemia, and cardiovascular disease. Diabetes Care 30:2164–2170
4. Bhupathiraju SN, Hu FB (2016) Epidemiology of obesity and diabetes and their cardiovascular complications. Circ Res 118:1723–1735
5. International Diabetes Federation (2017) IDF DIABETES ATLAS Eighth edition. At <file:///C:/Users/HOBBY/Downloads/IDF_DA_8e-EN-final.pdf>
6. Fang FC (2004) Antimicrobial reactive oxygen and nitrogen species: concepts and controversies. Nat Rev Microbiol 2:820–832
7. Bashan N, Kovsan J, Kachko I, Ovadia H, Rudich A (2009) Positive and negative regulation of insulin signaling by reactive oxygen and nitrogen species. Physiol Rev 89:27–71
8. Kanuri BN et al (2018) Glucose and lipid metabolism alterations in liver and adipose tissue pre-dispose p47^{phox} knockout mice to systemic insulin resistance. Free Radic Res 52:568–582
9. Houstis N, Rosen ED, Lander ES (2006) Reactive oxygen species have a causal role in multiple forms of insulin resistance. Nature 440:944–948
10. Berdichevsky A, Guarente L, Bose A (2010) Acute oxidative stress can reverse insulin resistance by inactivation of cytoplasmic JNK. J Biol Chem 285:21581–21589
11. Kelm M (1999) Nitric oxide metabolism and breakdown. Biochim Biophys Acta Bioenerg 1411:273–289
12. Levine AB, Punihaole D, Levine TB (2012) Characterization of the role of nitric oxide and its clinical applications. Cardiology 122:55–68
13. Pautz A et al (2010) Regulation of the expression of inducible nitric oxide synthase. Nitric Oxide 23:75–93
14. Tzeng E, Billiar TR, Robbins PD, Loftus M, Stuehr DJ (1995) Expression of human inducible nitric oxide synthase in a tetrahydrobiopterin (H4B)-deficient cell line: H4B promotes assembly of enzyme subunits into an active dimer. Proc Natl Acad Sci U S A 92:11771–11775
15. Mori M, Gotoh T (2000) Regulation of nitric oxide production by arginine metabolic enzymes. Biochem Biophys Res Commun 275:715–719
16. Saluja R et al (2011) Molecular and biochemical characterization of nitric oxide synthase isoforms and their intracellular distribution in human peripheral blood mononuclear cells. Biochim Biophys Acta, Mol Cell Res 1813:1700–1707
17. Villanueva C, Giulivi C (2007) Subcellular and cellular locations of nitric-oxide synthase isoforms as determinants of health and disease. Free Radic Biol Med. https://doi.org/10.1016/j.freeradbiomed.2010.04.004
18. Rana M et al (2015) Turmerone enriched standardized Curcuma longa extract alleviates LPS induced inflammation and cytokine production by regulating TLR4–IRAK1–ROS–MAPK–NFκB axis. J Funct Foods 16:152–163
19. Charbonneau A, Marette A (2010) Inducible nitric oxide synthase induction underlies lipid-induced hepatic insulin resistance in mice potential role of tyrosine nitration of insulin signaling proteins. Diabetes 59:861–871
20. Nathan C, Xie QW (1994) Nitric oxide synthases: roles, tolls, and controls. Cell 78:915–918

21. Michel T, Feron O (1997) Nitric oxide synthases: which, where, how, and why? J Clin Invest 100:2146–2152
22. Kim Y-M, Bergonia H, Lancaster JR (1995) Nitrogen oxide-induced autoprotection in isolated rat hepatocytes. FEBS Lett 374:228–232
23. Kim F et al (2008) Vascular inflammation, insulin resistance, and reduced nitric oxide production precede the onset of peripheral insulin resistance. Arterioscler Thromb Vasc Biol 28:1982–1988
24. Gruber H-J et al (2008) Obesity reduces the bioavailability of nitric oxide in juveniles. Int J Obes 32:826–831
25. Honing MLH, Morrison PJ, Banga JD, Stroes ESG, Rabelink TJ (1998) Nitric oxide availability in diabetes mellitus. Diabetes/Metab Rev 14:241–249
26. Perreault M, Marette A (2001) Targeted disruption of inducible nitric oxide synthase protects against obesity-linked insulin resistance in muscle. Nat Med 7:1138–1143
27. Dallaire P et al (2008) Obese mice lacking inducible nitric oxide synthase are sensitized to the metabolic actions of peroxisome proliferator-activated receptor-gamma agonism. Diabetes 57:1999–2011
28. Sudar E et al (2011) Regulation of inducible nitric oxide synthase activity/expression in rat hearts from ghrelin-treated rats. J Physiol Biochem 67:195–204
29. Duncan ER et al (2008) Effect of endothelium-specific insulin resistance on endothelial function in vivo. Diabetes 57:3307–3314
30. Kanuri BN et al (2017) Altered glucose and lipid homeostasis in liver and adipose tissue predispose inducible NOS knockout mice to insulin resistance. Sci Rep 7:41009
31. Eriksson JW (2007) Metabolic stress in insulin's target cells leads to ROS accumulation – a hypothetical common pathway causing insulin resistance. FEBS Lett 581:3734–3742
32. Saltiel AR, Kahn CR (2001) Insulin signalling and the regulation of glucose and lipid metabolism. Nature 414:799–806
33. Bray GA, Popkin BM (1998) Dietary fat intake does affect obesity! Am J Clin Nutr 68:1157–1173
34. Liu L et al (2007) Upregulation of myocellular DGAT1 augments triglyceride synthesis in skeletal muscle and protects against fat-induced insulin resistance. J Clin Invest 117:1679–1689
35. Liu Z et al (2015) High-fat diet induces hepatic insulin resistance and impairment of synaptic plasticity. PLoS One 10:e0128274
36. Xie QW, Kashiwabara Y, Nathan C (1994) Role of transcription factor NF-kappa B/Rel in induction of nitric oxide synthase. J Biol Chem 269:4705–4708
37. Shimabukuro M, Ohneda M, Lee Y, Unger RH (1997) Role of nitric oxide in obesity-induced beta cell disease. J Clin Invest 100:290–295
38. Noronha BT, Li J-M, Wheatcroft SB, Shah AM, Kearney MT (2005) Inducible nitric oxide synthase has divergent effects on vascular and metabolic function in obesity. Diabetes 54:1082–1089
39. Fujimoto M et al (2005) A role for iNOS in fasting hyperglycemia and impaired insulin signaling in the liver of obese diabetic mice. Diabetes 54:1340–1348
40. Lumeng CN, DeYoung SM, Bodzin JL, Saltiel AR (2006) Increased inflammatory properties of adipose tissue macrophages recruited during diet-induced obesity. Diabetes 56:16–23
41. Tannous M et al (1999) Evidence for iNOS-dependent peroxynitrite production in diabetic platelets. Diabetologia 42:539–544
42. Engeli S et al (2004) Regulation of the nitric oxide system in human adipose tissue. J Lipid Res 45:1640–1648
43. Meininger CJ et al (2004) GTP cyclohydrolase I gene transfer reverses tetrahydrobiopterin deficiency and increases nitric oxide synthesis in endothelial cells and isolated vessels from diabetic rats. FASEB J 18:1900–1902
44. Choi JW et al (2001) Increases in nitric oxide concentrations correlate strongly with body fat in obese humans. Clin Chem 47:1106–1109

45. Mercuri F et al (2001) Detection of nitrotyrosine in the diabetic plasma: evidence of oxidative stress. Diabetologia 44:834–838
46. Torres SH, De Sanctis JB, de L Briceño M, Hernández N, Finol HJ (2004) Inflammation and nitric oxide production in skeletal muscle of type 2 diabetic patients. J Endocrinol 181:419–427
47. Frisbee JC, Maier KG, Stepp DW (2002) Oxidant stress-induced increase in myogenic activation of skeletal muscle resistance arteries in obese Zucker rats. Am J Physiol Circ Physiol 283:H2160–H2168
48. House LM et al (2015) Tissue inflammation and nitric oxide-mediated alterations in cardiovascular function are major determinants of endotoxin-induced insulin resistance. Cardiovasc Diabetol 14:56
49. Sugita H et al (2005) Inducible nitric-oxide synthase and NO donor induce insulin receptor substrate-1 degradation in skeletal muscle cells. J Biol Chem 280:14203–14211
50. Uehara T et al (2006) S-Nitrosylated protein-disulphide isomerase links protein misfolding to neurodegeneration. Nature 441:513–517
51. Jeon MJ et al (2012) Mitochondrial dysfunction and activation of iNOS are responsible for the palmitate-induced decrease in adiponectin synthesis in 3T3L1 adipocytes. Exp Mol Med 44:562
52. Zanotto TM et al (2017) Blocking iNOS and endoplasmic reticulum stress synergistically improves insulin resistance in mice. Mol Metab 6:206–218
53. Pilon G et al (2010) Endotoxin mediated-iNOS induction causes insulin resistance via ONOO− induced tyrosine nitration of IRS-1 in skeletal muscle. PLoS One 5:e15912
54. Hotamisligil GS, Shargill NS, Spiegelman BM (1993) Adipose expression of tumor necrosis factor-α: direct role in obesity-linked insulin resistance. Science 87:91. At <https://www.scopus.com/record/display.uri?eid=2-s2.0-0027459878&origin=inward&txGid=85277588BE5D03F4BA6E74E1DC1E1918.wsnAw8kcdt7IPYLO0V48gA%253a2>
55. Hotamisligil GS, Murray DL, Choy LN, Spiegelman BM (1994) Tumor necrosis factor alpha inhibits signaling from the insulin receptor. Proc Natl Acad Sci U S A 91:4854–4858
56. Shinozaki S et al (2011) Liver-specific inducible nitric-oxide synthase expression is sufficient to cause hepatic insulin resistance and mild hyperglycemia in mice. J Biol Chem 286:34959–34975
57. Spruss A, Kanuri G, Uebel K, Bischoff SC, Bergheim I (2011) Role of the inducible nitric oxide synthase in the onset of fructose-induced steatosis in mice. Antioxid Redox Signal 14:2121–2135
58. Morino K, Petersen KF, Shulman GI (2006) Molecular mechanisms of insulin resistance in humans and their potential links with mitochondrial dysfunction. Diabetes 55:S9–S15
59. Turinsky J, O'Sullivan DM, Bayly BP (1990) 1,2-Diacylglycerol and ceramide levels in insulin-resistant tissues of the rat in vivo. J Biol Chem 265:16880–16885
60. Armoni M, Harel C, Bar-Yoseph F, Milo S, Karnieli E (2005) Free fatty acids repress the GLUT4 gene expression in cardiac muscle via novel response elements. J Biol Chem 280:34786–34795
61. Khedara A, Goto T, Morishima M, Kayashita J, Kato N (1999) Elevated body fat in rats by the dietary nitric oxide synthase inhibitor, L -N $^{\omega}$ nitroarginine. Biosci Biotechnol Biochem 63:698–702
62. Cha H-N et al (2010) Lack of inducible nitric oxide synthase does not prevent aging-associated insulin resistance. Exp Gerontol 45:711–718
63. Cani PD, Delzenne NM (2009) The role of the gut microbiota in energy metabolism and metabolic disease. Curr Pharm Des 15:1546–1558
64. Singh A et al (2016) The IRAK-ERK-p67phox-Nox-2 axis mediates TLR4, 2-induced ROS production for IL-1β transcription and processing in monocytes. Cell Mol Immunol 13:745–763
65. Sugita H et al (2002) Inducible nitric oxide synthase plays a role in LPS-induced hyperglycemia and insulin resistance. Am J Physiol Metab 282:E386–E394

66. Ropelle ER et al (2013) Targeted disruption of inducible nitric oxide synthase protects against aging, S-nitrosation, and insulin resistance in muscle of male mice. Diabetes 62:466–470
67. Ko S-H et al (2008) Inducible nitric oxide synthase-nitric oxide plays an important role in acute and severe hypoxic injury to pancreatic beta cells. Transplantation 85:323–330
68. Tanioka T et al (2011) Inducible nitric-oxide synthase and nitric oxide donor decrease insulin receptor substrate-2 protein expression by promoting proteasome-dependent degradation in pancreatic β-cells. J Biol Chem 286:29388–29396
69. Jang JE et al (2016) Nitric oxide produced by macrophages inhibits adipocyte differentiation and promotes profibrogenic responses in preadipocytes to induce adipose tissue fibrosis. Diabetes 65:2516–2528
70. Lumeng CN, Bodzin JL, Saltiel AR (2007) Obesity induces a phenotypic switch in adipose tissue macrophage polarization. J Clin Invest 117:175–184
71. Lu M et al (2010) Inducible nitric oxide synthase deficiency in myeloid cells does not prevent diet-induced insulin resistance. Mol Endocrinol 24:1413–1422
72. Epstein FH, Vane JR, Änggård EE, Botting RM (1990) Regulatory functions of the vascular endothelium. N Engl J Med 323:27–36
73. Furchgott RF (1999) Endothelium-derived relaxing factor: discovery, early studies, and identification as nitric oxide. Biosci Rep 19:235–251
74. Miao C-Y, Li Z-Y (2012) The role of perivascular adipose tissue in vascular smooth muscle cell growth. Br J Pharmacol 165:643–658
75. Muniyappa R, Montagnani M, Koh KK, Quon MJ (2007) Cardiovascular actions of insulin. Endocr Rev 28:463–491
76. Raghavan SA, Dikshit M (2004) Vascular regulation by the l-arginine metabolites, nitric oxide and agmatine. Pharmacol Res 49:397–414
77. Kearney MT, Duncan ER, Kahn M, Wheatcroft SB (2008) Insulin resistance and endothelial cell dysfunction: studies in mammalian models. Exp Physiol 93:158–163
78. Soskić SS et al (2011) Regulation of inducible nitric oxide synthase (iNOS) and its potential role in insulin resistance, diabetes and heart failure. Open Cardiovasc Med J 5:153–163
79. Deanfield J, Donald A, Ferri C, Giannattasio C, Halcox J, Halligan S, Lerman A, Mancia G, Oliver JJ, Pessina AC, Rizzoni D, Rossi GP, Salvetti A, Schiffrin EL, Taddei S, Webb DJ (2005) Endothelial function and dysfunction. Part I: methodological issues for assessment in the different vascular beds: a statement by the Working Group… – PubMed – NCBI. J Hypertens 23:7–17
80. Cai H, Harrison DG (2011) Endothelial dysfunction in cardiovascular diseases: the role of oxidant stress. At <http://circres.ahajournals.org/content/87/10/. http://www.lww.com/reprints. http://circres.ahajournals.org/Downloadedfrom>
81. Beckman JA, Creager MA, Libby P (2002) Diabetes and atherosclerosis. JAMA 287:2570
82. Priya P, Jitendra K, Sanjay R, Kanuri BN, Sachin K, Kumaravelu J, Dikshit (2017) Inducible nitric oxide synthase deficiency preserves vascular function despite systemic insulin resistance in diet-induced mouse model of obesity. In Molecular medicines for lifestyle diseases: emerging targets and approaches, vol 82
83. Cersosimo E, DeFronzo RA (2006) Insulin resistance and endothelial dysfunction: the road map to cardiovascular diseases. Diabetes Metab Res Rev 22:423–436
84. Brownlee M (2005) The pathobiology of diabetic complications: a unifying mechanism. Diabetes 54:1615–1625
85. Yogo K et al (2000) Different vasculoprotective roles of NO synthase isoforms in vascular lesion formation in mice. Arterioscler Thromb Vasc Biol 20:E96–E100
86. Koglin J, Glysing-Jensen T, Mudgett JS, Russell ME (1998) Exacerbated transplant arteriosclerosis in inducible nitric oxide-deficient mice. Circulation 97:2059–2065
87. Hansson GK et al (1994) Arterial smooth muscle cells express nitric oxide synthase in response to endothelial injury. J Exp Med 180:733–738
88. Kibbe M, Billiar T, Tzeng E (1999) Inducible nitric oxide synthase and vascular injury. Cardiovasc Res 43:650–657

89. Miyoshi T et al (2006) Deficiency of inducible NO synthase reduces advanced but not early atherosclerosis in apolipoprotein E-deficient mice. Life Sci 79:525–531
90. Chyu KY et al (1999) Decreased neointimal thickening after arterial wall injury in inducible nitric oxide synthase knockout mice. Circ Res 85:1192–1198
91. Kuhlencordt PJ, Chen J, Han F, Astern J, Huang PL (2001) Genetic deficiency of inducible nitric oxide synthase reduces atherosclerosis and lowers plasma lipid peroxides in apolipoprotein E-knockout mice. Circulation 103:3099–3104
92. Tsuchiya K et al (2012) FoxOs integrate pleiotropic actions of insulin in vascular endothelium to protect mice from atherosclerosis. Cell Metab 15:372–381
93. Dimmeler S et al (1999) Activation of nitric oxide synthase in endothelial cells by Akt-dependent phosphorylation. Nature 399:601–605
94. Hsueh WA, Law RE (1999) Insulin signaling in the arterial wall. Am J Cardiol 84:21–24
95. Marasciulo FL, Montagnani M, Potenza MA (2006) Endothelin-1: the yin and yang on vascular function. Curr Med Chem 13:1655–1665
96. Montagnani M et al (2002) Inhibition of phosphatidylinositol 3-kinase enhances mitogenic actions of insulin in endothelial cells. J Biol Chem 277:1794–1799
97. Goldin A, Beckman JA, Schmidt AM, Creager MA (2006) Advanced glycation end products: sparking the development of diabetic vascular injury. Circulation 114:597–605
98. Paneni F, Beckman JA, Creager MA, Cosentino F (2013) Diabetes and vascular disease: pathophysiology, clinical consequences, and medical therapy: part I. Eur Heart J 34:2436–2443
99. Janus A, Szahidewicz-Krupska E, Mazur G, Doroszko A (2016) Insulin resistance and endothelial dysfunction constitute a common therapeutic target in cardiometabolic disorders. Mediat Inflamm 2016:3634948
100. Kaiser N et al (1993) Differential regulation of glucose transport and transporters by glucose in vascular endothelial and smooth muscle cells. Diabetes 42:80–89
101. Pieper GM, Riaz-ul-Haq (1997) Activation of nuclear factor-kappaB in cultured endothelial cells by increased glucose concentration: prevention by calphostin C. J Cardiovasc Pharmacol 30:528–532
102. Williams B, Gallacher B, Patel H, Orme C (1997) Glucose-induced protein kinase C activation regulates vascular permeability factor mRNA expression and peptide production by human vascular smooth muscle cells in vitro. Diabetes 46:1497–1503
103. Li H, Li H, Bao Y, Zhang X, Yu Y (2011) Free fatty acids induce endothelial dysfunction and activate protein kinase C and nuclear factor-κB pathway in rat aorta. Int J Cardiol 152:218–224
104. Inoguchi T et al (2000) High glucose level and free fatty acid stimulate reactive oxygen species production through protein kinase C--dependent activation of NAD(P)H oxidase in cultured vascular cells. Diabetes 49:1939–1945
105. Mathew M, Tay E, Cusi K (2010) Elevated plasma free fatty acids increase cardiovascular risk by inducing plasma biomarkers of endothelial activation, myeloperoxidase and PAI-1 in healthy subjects. Cardiovasc Diabetol 9:9
106. Baker RG, Hayden MS, Ghosh S (2011) NF-κB, inflammation, and metabolic disease. Cell Metab 13:11–22
107. Paulus WJ, Frantz S, Kelly RA (2001) Nitric oxide and cardiac contractility in human heart failure: time for reappraisal. Circulation 104:2260–2262
108. Haywood GA et al (1996) Expression of inducible nitric oxide synthase in human heart failure. Circulation 93:1087–1094
109. Heusch P et al (2010) Increased inducible nitric oxide synthase and arginase II expression in heart failure: no net nitrite/nitrate production and protein S -nitrosylation. Am J Physiol Circ Physiol 299:H446–H453
110. Vejlstrup NG et al (1998) Inducible nitric oxide synthase (iNOS) in the human heart: expression and localization in congestive heart failure. J Mol Cell Cardiol 30:1215–1223
111. Zhang C et al (2005) TNF- contributes to endothelial dysfunction in ischemia/reperfusion injury. Arterioscler Thromb Vasc Biol 26:475–480

112. Drexler H, Hornig B (1999) Endothelial dysfunction in human disease. J Mol Cell Cardiol 31:51–60
113. Liu Y-H et al (2005) Role of inducible nitric oxide synthase in cardiac function and remodeling in mice with heart failure due to myocardial infarction. Am J Physiol Heart Circ Physiol 289:H2616–H2623
114. Xi L, Jarrett NC, Hess ML, Kukreja RC (1999) Myocardial ischemia/reperfusion injury in the inducible nitric oxide synthase knockout mice. Life Sci 65:935–945
115. Sam F et al (2001) Mice lacking inducible nitric oxide synthase have improved left ventricular contractile function and reduced apoptotic cell death late after myocardial infarction. Circ Res 89:351–356
116. Rösen P, Ballhausen T, Stockklauser K (1996) Impairment of endothelium dependent relaxation in the diabetic rat heart: mechanisms and implications. Diabetes Res Clin Pract 31:S143–S155
117. Nitenberg A et al (1993) Impairment of coronary vascular reserve and ACh-induced coronary vasodilation in diabetic patients with angiographically normal coronary arteries and normal left ventricular systolic function. Diabetes 42:1017–1025
118. Mungrue IN et al (2002) Cardiomyocyte overexpression of iNOS in mice results in peroxynitrite generation, heart block, and sudden death. J Clin Invest 109:735–743
119. Guo Y et al (1999) The late phase of ischemic preconditioning is abrogated by targeted disruption of the inducible NO synthase gene. Proc Natl Acad Sci U S A 96:11507–11512
120. Prakash P et al (2011) Atorvastatin protects against ischemia-reperfusion injury in fructose-induced insulin resistant rats. Cardiovasc Drugs Ther 25:285–297
121. Lee JK, Borhani M, Ennis TL, Upchurch GR, Thompson RW (2001) Experimental abdominal aortic aneurysms in mice lacking expression of inducible nitric oxide synthase. Arterioscler Thromb Vasc Biol 21:1393–1401
122. Prakash R et al (2012) Enhanced cerebral but not peripheral angiogenesis in the Goto-Kakizaki model of type 2 diabetes involves VEGF and peroxynitrite signaling. Diabetes 61:1533–1542
123. Ergul A, Kelly-Cobbs A, Abdalla M, Fagan SC (2012) Cerebrovascular complications of diabetes: focus on stroke. Endocr Metab Immune Disord Drug Targets 12:148–158
124. Last D et al (2007) Global and regional effects of type 2 diabetes on brain tissue volumes and cerebral vasoreactivity. Diabetes Care 30:1193–1199
125. Moskowitz MA, Lo EH, Iadecola C (2010) The science of stroke: mechanisms in search of treatments. Neuron 67:181–198
126. Calvert JW, Lefer DJ (2010) Clinical translation of nitrite therapy for cardiovascular diseases. Nitric Oxide 22:91–97
127. Shukla A, Dikshit M, Srimal RC (1995) Nitric oxide modulates blood-brain barrier permeability during infections with an inactivated bacterium. Neuroreport 6:1629–1632
128. Faraci FM, Brian JE (1994) Nitric oxide and the cerebral circulation. Stroke 25:692–703
129. Zheng Z, Yenari MA (2004) Post-ischemic inflammation: molecular mechanisms and therapeutic implications. Neurol Res 26:884–892
130. Danton GH, Dietrich WD (2003) Inflammatory mechanisms after ischemia and stroke. J Neuropathol Exp Neurol 62:127–136
131. Rosenberg GA (1999) Ischemic brain edema. Prog Cardiovasc Dis 42:209–216
132. Dirnagl U, Iadecola C, Moskowitz MA (1999) Pathobiology of ischaemic stroke: an integrated view. Trends Neurosci 22:391–397
133. Endoh M, Maiese K, Wagner J (1994) Expression of the inducible form of nitric oxide synthase by reactive astrocytes after transient global ischemia. Brain Res 651:92–100
134. Iadecola C, Zhang F, Xu S, Casey R, Ross ME (1995) Inducible nitric oxide synthase gene expression in brain following cerebral ischemia. J Cereb Blood Flow Metab 15:378–384
135. Chan PH (2001) Reactive oxygen radicals in signaling and damage in the ischemic brain. J Cereb Blood Flow Metab 21:2–14
136. Iadecola C, Zhang F, Xu X (1995) Inhibition of inducible nitric oxide synthase ameliorates cerebral ischemic damage. Am J Physiol Integr Comp Physiol 268:R286–R292

137. Zhao X, Haensel C, Araki E, Ross ME, Iadecola C (2000) Gene-dosing effect and persistence of reduction in ischemic brain injury in mice lacking inducible nitric oxide synthase. Brain Res 872:215–218
138. Członkowska A, Cyrta B, Korlak J (1979) Immunological observations on patients with acute cerebral vascular disease. J Neurol Sci 43:455–464
139. Prass K et al (2003) Stroke-induced immunodeficiency promotes spontaneous bacterial infections and is mediated by sympathetic activation reversal by poststroke T helper cell type 1–like immunostimulation. J Exp Med 198:725–736
140. Korhonen P et al (2015) Immunomodulation by interleukin-33 is protective in stroke through modulation of inflammation. Brain Behav Immun 49:322–336
141. Benakis C et al (2016) Commensal microbiota affects ischemic stroke outcome by regulating intestinal $\gamma\delta$ T cells. Nat Med 22:516–523
142. Hum PD et al (2007) T- and B-cell-deficient mice with experimental stroke have reduced lesion size and inflammation. J Cereb Blood Flow Metab 27:1798–1805
143. Gan Y et al (2014) Ischemic neurons recruit natural killer cells that accelerate brain infarction. Proc Natl Acad Sci 111:2704–2709
144. Liesz A et al (2009) Regulatory T cells are key cerebroprotective immunomodulators in acute experimental stroke. Nat Med 15:192–199
145. Kim J, Montagnani M, Koh KK, Quon MJ (2006) Reciprocal relationships between insulin resistance and endothelial dysfunction: molecular and pathophysiological mechanisms. Circulation 113:1888–1904
146. Jain M, Barthwal MK, Haq W, Katti SB, Dikshit M (2012) Synthesis and pharmacological evaluation of novel arginine analogs as potential inhibitors of acetylcholine-induced relaxation in rat thoracic aortic rings. Chem Biol Drug Des 79:459–469
147. Jun T, Wennmalm A (1994) NO-dependent and -independent elevation of plasma levels of insulin and glucose in rats by L-arginine. Br J Pharmacol 113:345–348
148. Piatti PM et al (2001) Long-term oral L-arginine administration improves peripheral and hepatic insulin sensitivity in type 2 diabetic patients. Diabetes Care 24:875–880
149. Bogdanski P et al (2012) Effect of 3-month L-arginine supplementation on insulin resistance and tumor necrosis factor activity in patients with visceral obesity. At <https://www.europeanreview.org/wp/wp-content/uploads/1081.pdf>
150. Lucotti P et al (2006) Beneficial effects of a long-term oral l -arginine treatment added to a hypocaloric diet and exercise training program in obese, insulin-resistant type 2 diabetic patients. Am J Physiol Metab 291:E906–E912
151. La Mura V et al (2014) Liver sinusoidal endothelial dysfunction after LPS administration: a role for inducible-nitric oxide synthase. J Hepatol 61:1321–1327
152. Darley-Usmar V, Wiseman H, Halliwell B (1995) Nitric oxide and oxygen radicals: a question of balance. FEBS Lett 369:131–135
153. Wei C-L, Hon W-M, Lee K-H, Khoo H-E (2005) Chronic administration of aminoguanidine reduces vascular nitric oxide production and attenuates liver damage in bile duct-ligated rats. Liver Int 25:647–656
154. King DE, Mainous AG III, Geesey ME (2008) Variation in L-arginine intake according to demographic characteristics and cardiovascular risk. Nutr Res 28:21
155. Tsutsui M et al (2015) Significance of nitric oxide synthases: lessons from triple nitric oxide synthases null mice. J Pharmacol Sci 127:42–52
156. Sansbury BE, Hill BG (2014) Regulation of obesity and insulin resistance by nitric oxide. Free Radic Biol Med 73C:383–399
157. Rafikov R et al (2011) eNOS activation and NO function: structural motifs responsible for the posttranslational control of endothelial nitric oxide synthase activity. J Endocrinol 210:271–284
158. Abudukadier A et al (2013) Tetrahydrobiopterin has a glucose-lowering effect by suppressing hepatic gluconeogenesis in an endothelial nitric oxide synthase–dependent manner in diabetic mice. Diabetes 62:3033–3043

159. Shimazu T et al (2011) Sepiapterin enhances angiogenesis and functional recovery in mice after myocardial infarction. Am J Physiol Circ Physiol 301:H2061–H2072
160. Varadharaj S et al (2017) Role of dietary antioxidants in the preservation of vascular function and the modulation of health and disease. Front Cardiovasc Med 4:64
161. Yang Y, Loscalzo J (2005) S-nitrosoprotein formation and localization in endothelial cells. Proc Natl Acad Sci 102:117–122
162. Daaka Y (2012) S-nitrosylation-regulated GPCR signaling. Biochim Biophys Acta 1820:743–751
163. Al-Ani B et al (2006) The release of nitric oxide from S-nitrosothiols promotes angiogenesis. PLoS One 1:e25
164. Kleinbongard P et al (2003) Plasma nitrite reflects constitutive nitric oxide synthase activity in mammals. Free Radic Biol Med 35:790–796
165. Moncada S, Rees DD, Schulz R, Palmer RM (1991) Development and mechanism of a specific supersensitivity to nitrovasodilators after inhibition of vascular nitric oxide synthesis in vivo. Proc Natl Acad Sci U S A 88:2166–2170
166. Park J-H et al (2002) Nitric oxide (NO) pretreatment increases cytokine-induced NO production in cultured rat hepatocytes by suppressing GTP cyclohydrolase I feedback inhibitory protein level and promoting inducible NO synthase dimerization. J Biol Chem 277:47073–47079
167. Carvalho-Filho MA et al (2005) S-nitrosation of the insulin receptor, insulin receptor substrate 1, and protein kinase B/Akt. Diabetes 54:959–967

Oxidative Stress Mechanisms in Type 2 Diabetes Induced Coronary Heart Disease

22

Keerthi Kupsal and Surekha Rani Hanumanth

22.1 Introduction

Type 2 Diabetes (T2D) is a majorly prevalent disorder with approximately 415 million people affected worldwide and this number is expected to rise to 642 million by the year 2040 [1, 2]. T2D represents a group of metabolic disorders characterized by chronic hyperglycemic condition resulting from insulin resistance or abnormal insulin secretion and/or insulin sensitivity.

Type 2 Diabetes in long term is a root cause for several macro and micro vascular complications such as cardiovascular diseases (CVDs), nephropathy, neuropathy & retinopathy etc. [3]. Epidemiological studies on diabetes mellitus have shown that confounding risk factors like age, gender, duration of diabetes and ethnic background are important factors for progressing to diabetic complications [4]. It is estimated that 10%–20% of diabetics progress to macro or microvascular complications and nearly 65% of the diabetes associated deaths are caused by heart diseases and stroke [5]. Despite the morbidity and mortality associated with diabetes related comorbidities, CVDs remain the leading cause of death in T2D [6, 7].

Cardiovascular diseases refer to the group of disorders affecting heart or blood vessels including coronary heart disease (CHD), peripheral arterial disease, stroke, heart failure, cardiomyopathy etc. In 2017, CVDs contributed to about 45% of all the deaths among non-communicable diseases. T2D accounts for two to four fold higher risk of developing CHD, stroke and two to eight fold risk of heart failure in adult diabetic patients [8].

Coronary Heart Disease is one of the major cardiovascular complications of chronic diabetes and is a leading cause of early deaths in diabetic patients. T2D is an established risk factor for CHD and has a higher cardiovascular morbidity and mortality compared to non-diabetic subjects [9]. Adverse effects of hyperglycemia,

K. Kupsal · S. R. Hanumanth (✉)
Department of Genetics, Osmania University, Hyderabad, Telangana, India

S. Chakraborti et al. (eds.), *Oxidative Stress in Heart Diseases*,
https://doi.org/10.1007/978-981-13-8273-4_22

insulin resistance, abnormal carbohydrate metabolism, impaired fasting glucose, impaired glucose tolerance, endothelial dysfunction and oxidative stress, acting on a substrate of genetic susceptibility enhances the risk for development of cardiovascular complications in type 2 diabetic individuals [10, 11].

Several reports have been published on decreased antioxidant capacity in plasma of diabetic patients [12–14]. Hence, depletion of cellular oxidant defence system and increased levels of reactive oxygen species leads to oxidative stress in T2D patients [15]. Diabetes induced oxidative stress has adverse effects on cardiovascular function including reduced nitric oxide bioavailability, increased inflammation, and modification of lipoproteins mediated by several genetic factors. Hence, oxidative stress is the unifying pathophysiological mechanism underlying the development of coronary atherosclerosis in T2D.

In this chapter, we highlight the oxidative stress pathways mediated by insulin and free fatty acids and genetics of oxidative stress related genes in onset and development of CHD in type 2 diabetic individuals (Fig. 22.1).

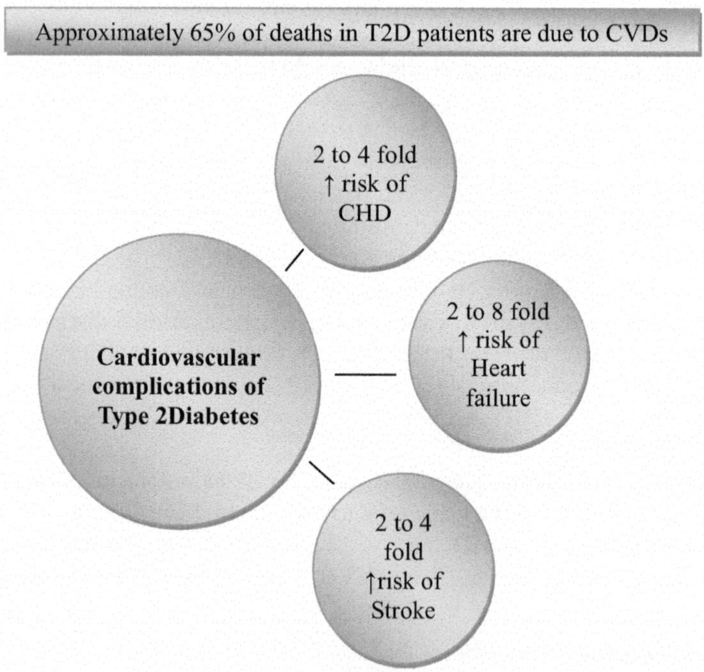

Fig. 22.1 Risk of CVDs in Type 2 Diabetic patients

22.2 Pathophysiology of Coronary Heart Disease in Type 2 Diabetes

Several metabolic and molecular abnormalities in type 2 diabetes including insulin resistance, prolonged hyperglycemia, a combination of dyslipidemia, arterial hypertension and genetic susceptibility activates glucose auto-oxidation, glucosamine pathways, formation of AGEs, and oxidative phosphorylation pathways predisposing T2D individuals to pro-atherogenic effects through diverse mechanisms such as oxidative stress, inflammation etc.

CHD is characterized by endothelial and vascular dysfunction leading to accelerated atherosclerosis, which is a key event in developing CHD. ROS are involved in the progression of endothelial cell dysfunction, accompanied by inactivation of endothelial nitric oxide synthase (eNOS) and decreased NO levels. NO also plays a key role in oxidative stress mainly through the production of reactive oxygen species (ROS). Nitric oxide is a critical regulator of vascular tone in endothelial cells and adequate levels of NO have to be produced in order to maintain the normal vascular physiology. Endothelial dysfunction in diabetes mellitus induces alteration in the signalling pathways that lead to reduced eNOS activation in the endothelium.

NO produced in endothelium modulates vascular dilator tone, maintains vascular homeostasis and vascular integrity by inhibiting the platelet aggregation, leukocyte endothelium adhesion and vascular smooth muscle cell proliferation and migration. NO produced in cardiac smooth muscle regulates cardiac contractility. Conversely, diminished NO availability promotes vascular inflammation and induces the expression of proinflammatory transcription factor nuclear factor kappa B (NF-κB), and induce endothelial dysfunction and increases the entry of modified circulating lipoproteins into the vessel wall. Endothelial dysfunction also results in upregulation of leucocyte adhesion molecules such as intercellular adhesion molecule-1 (ICAM-1) and vascular adhesion molecule-1 (VCAM-1), facilitating transendothelial migration of leukocytes into the tunica intima mediated by a chemoattractant gradient.

In the intima, inflammatory mediators, like M-CSF aid in the detection and internalization of modified lipoprotein particles by macrophages and monocytes ultimately forming foam cells and initiates atherosclerotic lesion formation. During this process, lymphocytes and resident cells of vascular wall secrete various inflammatory cytokines and growth factors that promote proliferation & migration of VSMCs into the intima and penetration through the elastic lamina and collagenous matrix evolving into atheromatous plaque.

Moreover, T2D enhances protein-kinase-C (PKC) activity, NF-κB production and free radical synthesis in VSMCs in the presence of high glucose concentrations [16, 17]. Hence, migration of VSMCs into nascent atherosclerotic lesions leads to the replication and production of extra cellular matrix, which are the major steps in mature lesion formation [18]. VSMCs undergo accelerated apoptosis in the atherosclerotic lesions and diabetic patients tend to have less number of VSMCs in the lesion [19]. Additionally, there is an enhanced production of cytokines in T2D which diminishes vascular smooth muscle synthesis of collagen and elaborates the production of matrix metalloproteinases (MMPs). Enhanced levels of MMPs, increased synthesis and reduced break down by tissue inhibitors of MMPs (TIMPs) results in extracellular matrix degradation [20].

Apoptosis of VSMCs and increased production of MMPs are key determinants of plaque stability. They increase the tendency for plaque destabilization, distinctive of extensive inflammatory infiltrate, reduced VSMCs & collagen, increased neovascularisation, a substantial lipid core, and a thin fibrous cap. This narrows the vessel lumen by more than 50% ultimately leading to atherosclerotic plaque rupture due to metalloproteases, cathepsins, and collagenases secreted from activated macrophages in the plaque [21, 22]. Plaque rupture activates clotting cascade and triggers thrombus formation leading to CHD.

22.3 Concept of Oxidative Stress

Reactive Oxygen Species acts as signal transduction molecules involving in cell protection while excess production of ROS generates oxidative stress. Oxidative stress is the state of imbalance between the production of ROS and buffering capacity of antioxidants. Anti-oxidants (AOX) counteracts the damaging effects caused by oxidants/free radicals and neutralise their effects by providing electrons to the free radicals.

Reactive oxygen species (ROS) is a term which encompasses all highly reactive, oxygen-containing molecules, including free radicals. A free radical is an unstable and highly reactive chemical species/atom/molecule possessing one or more unpaired electrons with a negative charge and to have a balanced charge it captures an electron from a balanced atom, making another free radical, thus initiating a chain reaction.

The cascade of free radical chain reactions within the body results in damage to membrane lipids, nucleic acids, proteins, enzymes, and other small molecules which are very lethal and finally culminates to cell death. All these consequences lead to a wide range of diseases in humans most notably T2D, cardiovascular diseases etc.

Multiple sources of ROS inducing oxidative stress in type 2 diabetics include several enzymatic and non-enzymatic pathways. Oxidative stress acts as a mediator of insulin resistance and progression to glucose intolerance in type 2 diabetes and subsequently contributes to micro and macrovascular complications including coronary atherosclerotic disease (Fig. 22.2).

22.3.1 Generation of Reactive Oxygen Species

Reactive Oxygen Species are a number of reactive molecules, free radicals and ions generated from molecular oxygen (O_2) or formed as by-products generated from endogenous and exogenous sources. Potential endogenous sources include lysosomes, mitochondria, phagocytes, endoplasmic reticulum, peroxisomes, and

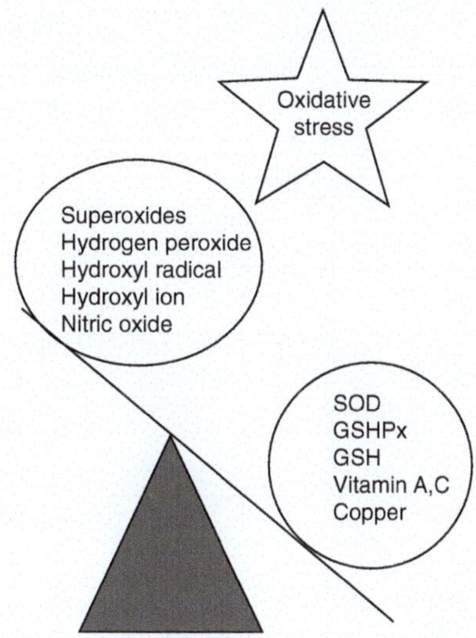

Fig. 22.2 Imbalance between ROS and anti-oxidants

inflammatory cell activation [23]. Exogenous sources include cigarette smoke, alcohol, pollutants and environmental agents such as carcinogens, various xenobiotics, ultrasound, microwave radiation etc. [24, 25].

The reduction of oxygen through the addition of electrons leads to the formation of number of ROS including superoxide (O_2^-), hydrogen peroxide (H_2O_2), hydroxyl radical ($^{\bullet}OH$), hydroxyl ion (OH^-), and nitric oxide (NO).

A primary ROS, superoxide (O_2^-), is formed by one-electron reduction of molecular oxygen while hydrogen peroxide (H_2O_2) is produced by direct two electron reduction of molecular oxygen by family of superoxide dismutase (SOD) enzymes or by reduction of superoxide through dismutation. Hydroxyl radical ($^{\bullet}OH$) arises from electron exchange between superoxide and hydrogen peroxide via the Harber–Weiss reaction or it is also generated by the reduction of hydrogen peroxide by the Fenton reaction. When $O_2\cdot$ and NO are synthesized within a few cell diameters, they will combine to form peroxynitrite ($^{\bullet}ONOO$) by a diffusion-limited reaction.

Under regulated conditions, O_2^- and H_2O_2 mediate redox signalling pathways responsible for physiological processes including cell growth, cell differentiation, metabolism etc. However, excess production of ROS generates oxidative stress, damages DNA, protein, and lipids leading to various pathophysiological conditions [26] (Fig. 22.3).

22.3.2 Role of Anti-Oxidants

Anti-oxidants are beneficial compounds capable of stabilizing, or deactivating, free radicals before they attack cells. ROS production in vivo is regulated by anti-oxidant defense mechanisms which include both enzymatic and non-enzymatic strategies to counteract the effects of ROS.

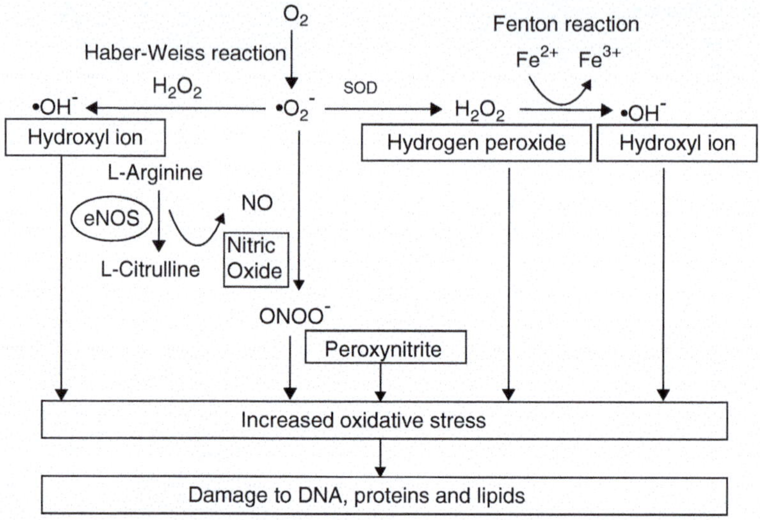

Fig. 22.3 Reactions underlying derivation of ROS

Endogenously synthesized anti-oxidants are one of the products of body's metabolism which may be either enzymatic or non-enzymatic. Enzymatic antioxidants include superoxide dismutase (SOD), glutathione peroxidase (GSHPx), catalase (CAT), glutathione reductase (GR), Thioredoxin (Trx) and Glucose 6 Phosphate Dehydrogenase (G6PDH). SOD plays an important role in the first line of defence against ROS. SOD enzymes are metalloproteins that convert the superoxide to hydrogen peroxide, which is then transformed into water by catalase in lysosomes or by GSHPx in the mitochondria. GSHPx eliminates H_2O_2 by using GR for another substrate, and generates water. CAT enzymes are localised to peroxisome and catalyze the conversion of H_2O_2 to O_2 and H_2O. Gpx enzymes are present in cytosol, mitochondria and plasma membrane and catalyze the conversion of H_2O_2 and lipid peroxides to water and lipid alcohols [27]. Trx induces the formation of a disulfide bond by reducing the oxidized cysteine residues on proteins which is then further reduced by thioredoxin reductase and NADPH [28, 29].

Studies have shown that SOD, catalase and Gpx activities were reduced in diabetic patients due to excessive glycation [30]. It has also been found that cardiac-specific expressions of GPx levels were reduced in diabetic patients [31]. Trx activity was directly suppressed to high glucose exposure invitro and leads to an excessive injury response in an ischemia-reperfusion injury model which suggests the cardioprotective role of Trx in T2D [32].

Non-enzymatic anti-oxidants play a second line of defence against ROS. These include α-lipoic acid, coenzyme Q10 (CoQ10), uric acid, albumin, taurine, tetrahydrobiopterin (BH4), and N-acetylcysteine (NAC) etc.

Table 22.1 Endogenous and non-endogenous anti-oxidants

Endogenously synthesized		Non-endogenous	
Non-enzymatic anti-oxidants	Enzymatic anti-oxidants	Dietary Substances	Trace elements & minerals
Uric acid	Superoxide dismutase (SOD)	Vitamin A	Copper
Bilirubin	Glutathione peroxidise (GSHPx)	Vitamin C	Zinc
Glutathione (GSH)	Catalase (CAT)	Vitamin E	Selenium
Coenzyme Q10 (CoQ10)	Glutathione reductase (GR)	Carotenoids	Manganese
N-Acetylcysteine (NAC)	Thioredoxin (Trx)	Bioflavonoids	Ferrous
Melatonin	Glucose 6 phosphate dehydrogenase (G6PDH)	Polyphenols	
α-Lipoic acid		Folic acid	
Tetrahydrobiopterin (BH4)			
Taurine			
Albumin			

Non-endogenously synthesized anti-oxidants include dietary supplements such as vitamins A, C & E, carotenoids, several bioflavonoids, polyphenols, cofactors like folic acid, anti-oxidant trace elements & minerals (copper, zinc, manganese, selenium and ferrous) etc.

Vitamin C, an essential nutrient regulates intracellular glutathione recovery [33] and is necessary as a co-antioxidant for vitamin E function [34]. Vit C also down-regulates NADPH oxidase, suppresses NF-κβ activation and prevents oxidation of tetrahydrobiopterin – a cofactor of NO synthase [34]. It also increases monocyte adhesion to endothelium. Vitamin E, inhibits lipid peroxidation, downregulates NADPH oxidase [34], reverses endothelial dysfunction [35], decreases monocyte-endothelial cell adhesion [36] and inhibits smooth muscle proliferation etc. Few studies have reported decreased plasma concentrations of vitamin A [37, 38], carotenoids [37], vitamin C [39] & vitamin E [38, 39] in type 2 diabetic subjects compared to controls (Table 22.1).

22.4 Factors Generating ROS in T2D and Enhancing the Risk of CHD

Risk factors for CHD in type 2 diabetes include hyperglycemia and non-glycemic factors like hyperinsulinemia, diabetic dyslipidemia, free fatty acids, endothelial dysfunction etc.

22.4.1 Hyperglycemia

In Type 2 Diabetes, under hyperglycemic conditions, glucose in plasma forms advanced glycation end products (AGEs) by undergoing non-enzymatic reaction. AGEs bind to the receptors for advanced glycation end products (RAGE) on the surface of endothelial cell lining blood vessels and lead to the intracellular generation of ROS, namely superoxides by NADPH oxidase. As a consequence, superoxides activate NFκB, which results in the transcriptional activation of inflammatory genes and development of coronary atherosclerosis.

Hyperglycemia also increases the production of diacylglycerol, a lipid second messenger which causes membrane alteration and activation of Protein kinase-C (PKC) which thereby inhibits the activity of PI3/AKT, subsequent phosphorylation of NOS which results in decreased NO production [40]. Activation of genes relevant to inflammation and decreased NO production leads to pro-atherogenesis in T2D individuals.

Uncontrolled hyperglycemia as evidenced by HbA1c levels has an enhanced propensity to systemic atherosclerosis and severity of CAD [41]. Epidemiology of Diabetes Interventions and Complications [EDIC] and Multiple Risk Factor Intervention Trial (MR-FIT), suggested that cardiovascular complications of T2D are majorly due to high levels of plasma glucose [42, 43].

22.4.2 Hyperinsulinemia

T2D often combined with basal hyperinsulinemia leads to functional changes in blood vessels by impaired NO production by receptor mediated resistance resulting in decreased vasodilation. Structural changes due to hyperinsulinemia occur by pro-atherogenic response mediated by mitogen activated protein kinase (MAPK) pathways which lead to imbalance in homeostatic regulation of vascular function, enhanced oxidative stress, inflammation and subsequent pre-atherosclerotic events in T2D individuals [44].

22.4.3 Free Fatty Acids

In insulin resistance state, decreased insulin function and lack of insulin inhibits lipolysis which leads to increased FFAs generation and lowers lipoprotein lipase activity [45]. Levels of FFAs are also increased due to excessive liberation from adipocytes and diminished uptake by skeletal muscle in T2D patients [46]. FFA which enter adipocytes are rapidly converted into fatty acyl-CoA and stored as triglycerides. Hence, increased mitochondrial oxidation could lead to enhanced mitochondrial superoxide generation in type 2 diabetes subjects.

Certain saturated fatty acids such as palmitate showed the generation of ROS and activates NFκB, expresses monocyte chemotactic factor, serum amyloid A3 (SAA3) and monocyte chemotactic protein-1 (MCP-1) which play an important role in the formation of atherosclerotic plaque [47].

22.4.4 Endothelial Dysfunction

Endothelium is a passive lining for all blood vessels and acts as an interface between circulating blood and VSMCs and plays a key role in the regulation of blood flow, arterial tone, maintenance of vessel wall permeability and vascular homeostasis etc. Endothelial dysfunction suggests the impairment of the capability of endothelium in properly maintaining the vascular homeostasis. Altered signalling pathways in endothelium leads to improper production of NO and its reduced bioavailability to maintain vascular homeostasis, increased synthesis of vasoconstrictor prostanoids and endothelin [48, 49]. Hence, in T2D patients, endothelial dysfunction precedes the development of coronary atherosclerosis and is an indicating factor for progression to CAD.

22.5 Oxidative Stress Mediated Pathways in Type 2 Diabetes

Under diabetic conditions several sources of ROS include glucose and insulin which mediates their effects through altered coagulation, insufficient vasodilation, elevated free fatty acids, insulin resistance etc. by activating numerous pathways that link T2D with CHD are discussed as follows:

22.5.1 eNOS Activation and Vasodilation by PI3K/Akt Pathway

Insulin is one of the important stimuli for eNOS activation in type 2 diabetes which binds to its receptor on endothelial cells leading to phosphorylation of insulin receptor substrate-1 [IRS-1] and subsequent phosphorylation and activation of eNOS and production of NO via PI3 kinase/Akt [50]. In type 2 diabetic patients, hyperglycemia and increase in free fatty acids inhibit PI3K/Akt pathway. Inhibition of PI3 kinase/Akt may reduce the expression of atherothrombic factors in addition decreases the production of protective molecules, including NO. In T2D, activation of PKCβ leads to the activation of NFκB, blocks insulin signalling and reduces the synthesis of NO resulting in reduced vasodilation, abnormal supply of oxygen to cells, increased oxidative stress and atherogenic plaque formation finally leading to coronary heart disease.

Few studies support that mutation in IRS-1, decreases insulin-stimulated eNOS phosphorylation and eNOS gene expression in cultured endothelial cells. Further, specific knockout studies in mice with endothelial specific insulin receptor showed decreased eNOS expression and impaired vasodilator function [51]. Studies on animal models of insulin resistance displayed defects in the PI3 kinase/Akt system and impaired NO bioavailability [52] (Fig. 22.4).

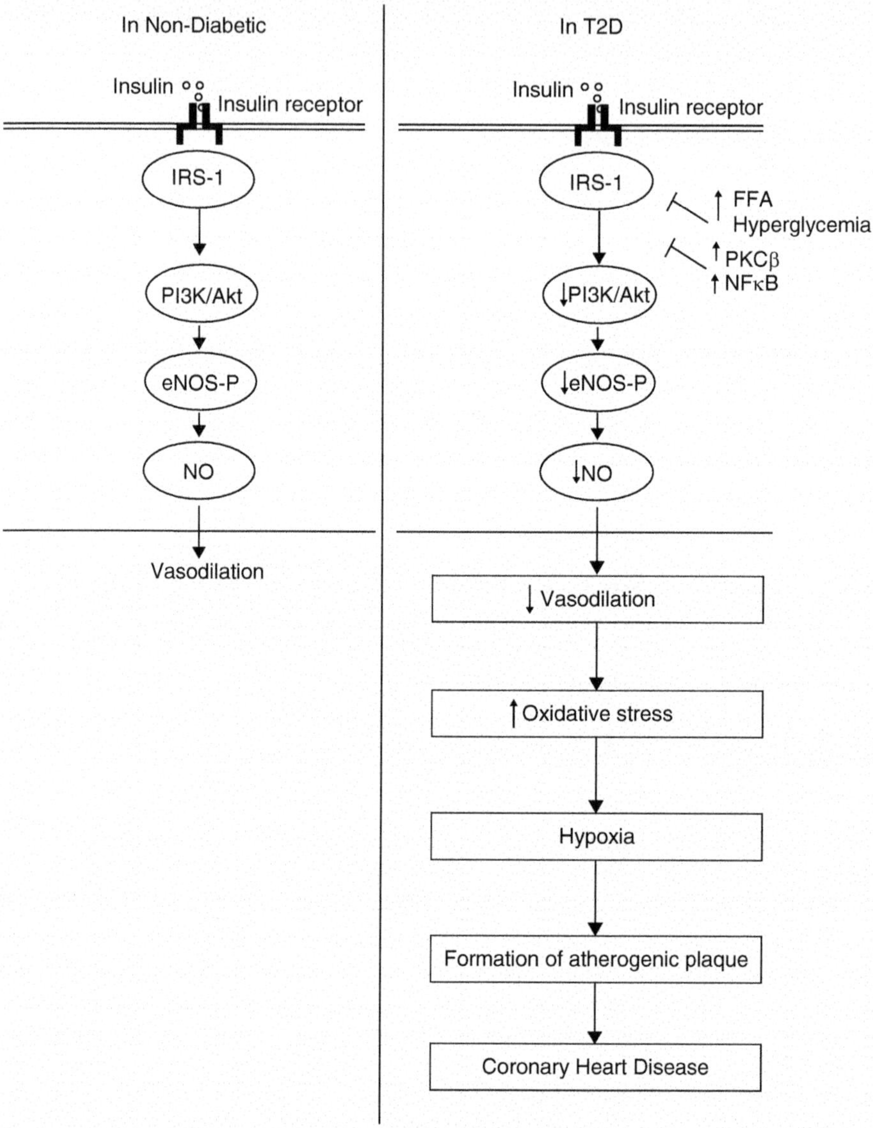

Fig. 22.4 Insulin mediated activation of eNOS by PI3K/Akt pathway

22.5.2 Vasoconstriction and Expression of Pro-Atherogenic Molecules by MAPK-Pathway

Under normal conditions, activation of insulin receptor results in the balance of both processes vasodilation and vasoconstriction via PIK3K/Akt and MAPK pathways to regulate the intermediate metabolic requirements. In type 2 diabetes, in addition to vasodilation, insulin also promotes vasoconstriction and expresses pro-atherogenic molecules in the endothelium through the production of endothelin-1 (ET-1) and PAI-1, ICAM-1, VCAM-1 and E-selectin molecules through MAPK-dependent pathways [53].

Binding of insulin to its receptor initiates a cascade of autophosphorylation events, activates IRS-1 and IRS-2 docking proteins and concurrently activates Src homology containing (Shc) protein. Binding of SH2 domain of Grb2 to phosphorylated tyrosine residues of IRS-1 or Shc activates the preassociated guanosine triphosphate (GTP) exchange factor Sos. Sos inturn activates Ras (rat sarcoma), a small GTP binding protein which subsequently binds and activates the serine-threonine protein kinase Raf (rapidly growing fibrosarcoma). Raf activates MAPK/extra-cellular signal-regulated kinase (MEK) which then activates ERK1/2, also known as p44/42 mitogen-activated protein kinase (MAPK). Activation of MAPK leads to insulin-stimulated production of ET-1, decreased vasodilation, enhanced vasoconstriction and a series of events are followed that lead to coronary heart disease (Fig. 22.5).

22.5.3 NADPH Oxidase Mediated Superoxide Generation in Skeletal Muscle

Under diabetic conditions various factors like AGEs, high glucose, insulin, and angiotensin II activates NADPH oxidase. Membrane bound NADPH oxidases are the principal sources of ROS in diabetes and play a major role in the development of atheroclerosis. NADPH oxidase consists of membrane bound flavocytochrome b558 forming subunits such as gp91 phox (Nox2)/Nox1/Nox4 and p22 phox and the catalytic site of the oxidase and cytosolic components p47 phox and p67 phox. Membrane-bound and cytosolic subunits are called catalytic and regulatory subunits, respectively. In endothelial and smooth muscle cells, Nox 1 and Nox 4 are abundantly expressed. P22phox expression of NADPH oxidases was significantly increased in rat and human diabetic arteries [54, 55].

In skeletal muscles of diabetic patients, NADPH oxidases specifically decrease serine phosphorylation of Akt and GLUT4 translocation impairing the insulin signalling and ameliorate insulin resistance which leads to the production of superoxides [56]. Therefore it is suggested that increased expression of p22 phox may contribute to the development of oxidative stress in diabetic individuals (Fig. 22.6).

22.5.4 Pathway Mediated by Free Fatty Acids

Increased FFAs bind to toll-like receptors (TLRs) and degrades inhibitory complex IkBa by IKKb-kinase and thereby activates NF-kB. The activated NF-kB triggers inflammation due to upregulation of IL-6 and TNF-α. FFA binding to toll like receptors also lead to phosphorylation of insulin receptor substrate-1 (IRS-1) by c-Jun

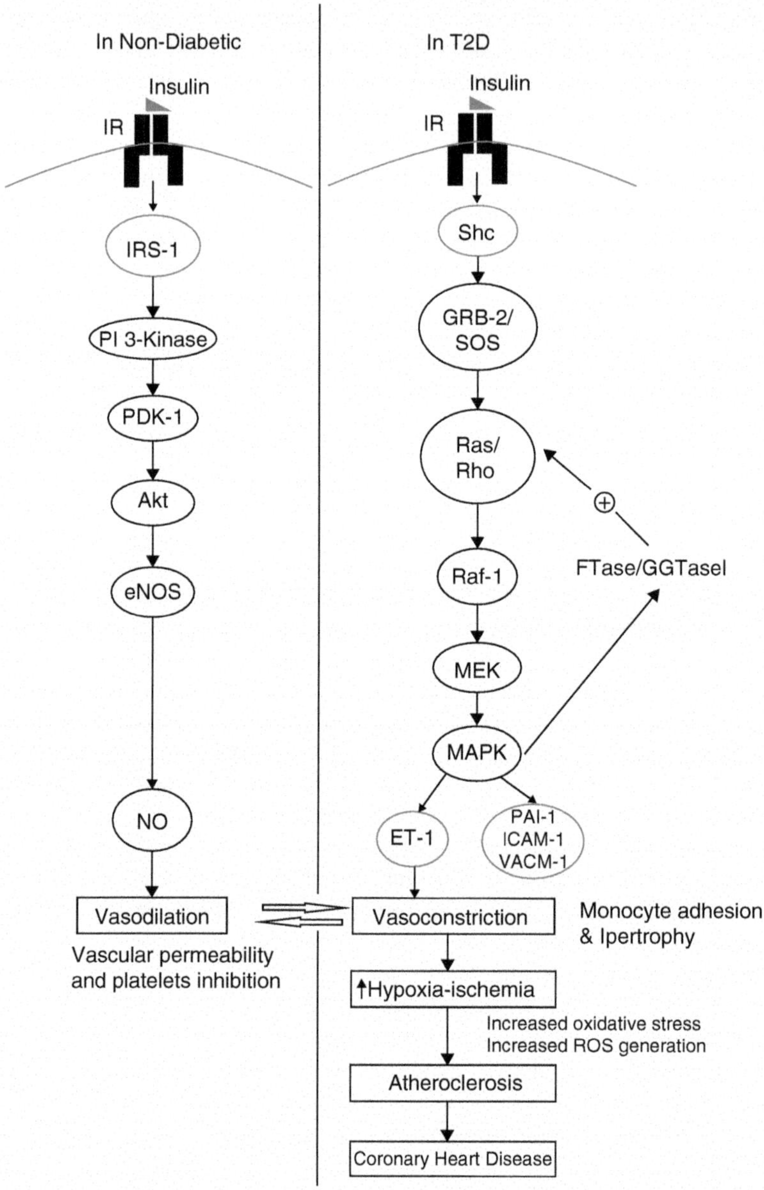

Fig. 22.5 Insulin mediated vasoconstriction and generation of pro-atherogenic molecules

amino-terminal kinase (JNK) and PKC and activates downstream targets PI3-kinase and Akt. All these molecular events results in the down-regulation of the glucose transporter GLUT-4 and occurrence of insulin resistance, a major characteristic feature in type 2 diabetes and markedly increases the risk for CVDs. Thus, downregulation of PI3K/Akt pathway leads to eNOS inhibition and decreased NO production

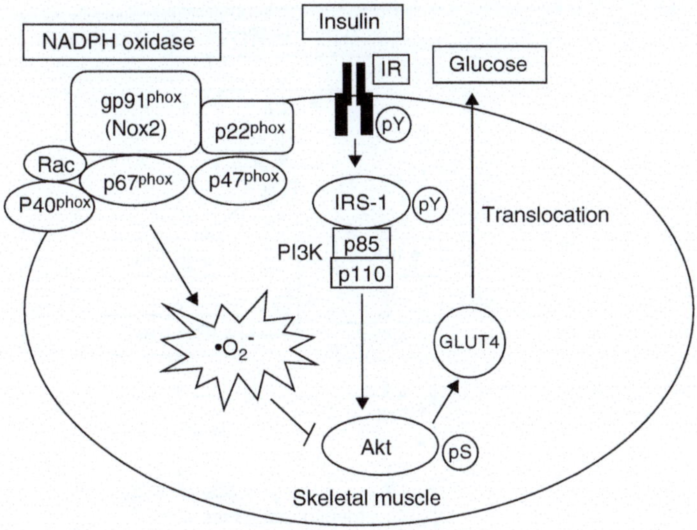

Fig. 22.6 NADPH oxidase mediated superoxide generation

which initates atherosclerosis. Reduced bioavailability of NO triggers inflammatory pathways and increases cytokine production. TNF- α and IL-1 increase NF-kB activity and expression of adhesion molecules and also stimulates the expression of CRP which down-regulates eNOS and increases the production of adhesion molecules and endothelin-1.

In adipocytes, insulin resistance increases the release of free fatty acids from stored triglycerides. Due to lack of insulin stimulation of malonyl CoA production and increased oxidation of free fatty acids in the aortic endothelial cells of type 2 diabetic patients, generates increased superoxide production by the mitochondrial electron transport chain. This activates enhanced production of advanced glycation end products, increased PKC activation, and enhanced N-acetyl glucosamine ($GLcNA_C$) activity.

Free fatty acid induced superoxide production in turn activates pro-inflammatory signals and also inactivates two important anti-atherogenic enzymes, prostacyclin synthase (PGI_2) and eNOS. Studies from non-diabetic rodent models of insulin resistance suggest that inactivation of prostacyclin synthase and eNOS enzymes prevented by inhibition of free fatty acid release from adipose tissue by the rate-limiting enzyme, carnitinepalmitoyltransferase I, for fatty acid oxidation in mitochondria and by reduction of superoxide levels [57]. Excessive superoxide production contributes to mitochondrial damage and precedes the development of atherosclerosis in type 2 diabetic subjects (Fig. 22.7).

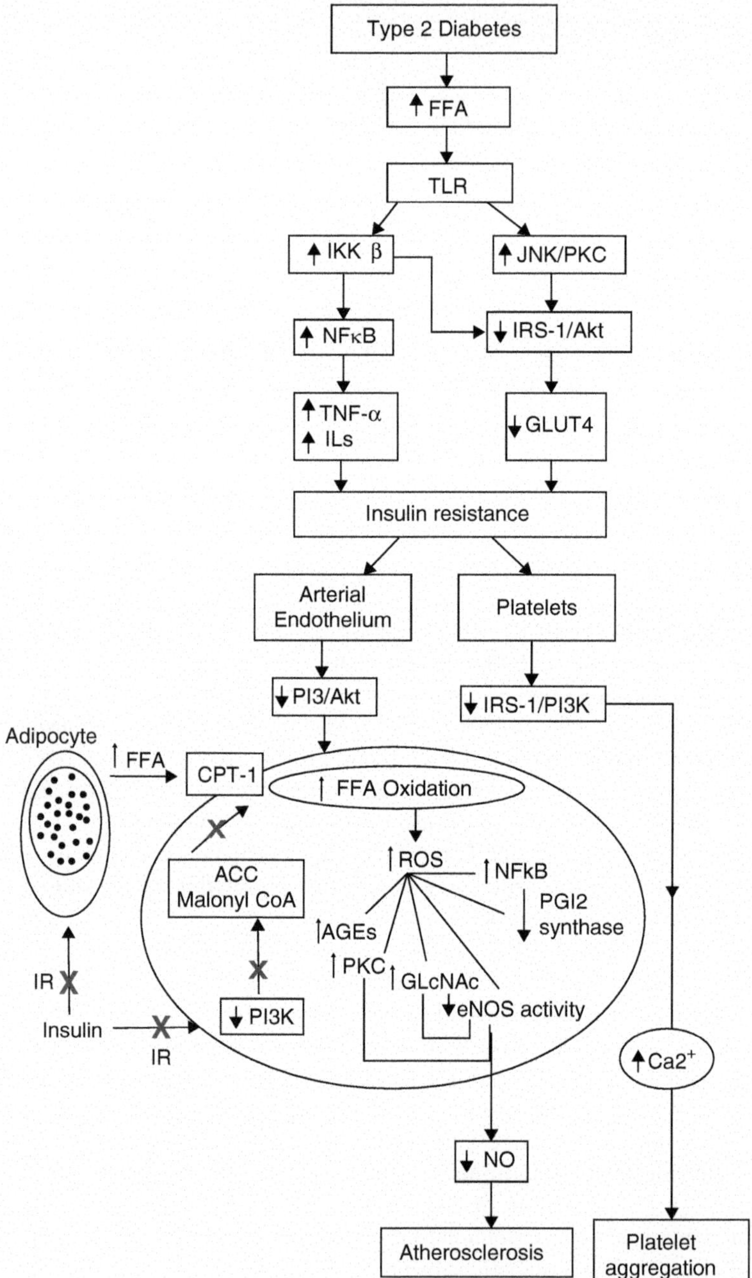

Fig. 22.7 Atherosclerosis progression by free fatty acids and insulin resistance

22.6 Genetics of Oxidative Stress Related Genes and Susceptibility to CHD in T2D

Both type 2 diabetes and cardiovascular diseases constitute the paradigm of common complex genetic traits in their pathophysiology. Whether the genetic risk factors predispose type 2 diabetic individuals to cardiovascular complications remain uncertain. Few studies suggest that risk of CHD is influenced by genetic factors in diabetics compared to non-diabetics. Indeed, vulnerability to oxidative stress in type 2 diabetic patients include functional polymorphic variants in several oxidative stress related genes and anti-oxidant defence systems [58]. Few studies also suggest that the architecture of genetic susceptibility to CHD in oxidative stress related genes might be different in diabetic patients than that in general population [59]. Several studies have reported the association of single nucleotide polymorphisms in oxidative stress related genes and anti-oxidant defense systems in T2D patients with cardiovascular complications [60–62]. Hence, the SNPs of some of the oxidant and antioxidant genes is summarized for identification of genetic markers to improve the risk prediction of CHD in diabetic individuals.

22.6.1 P22phox

One of the major sources of ROS is NADPH oxidase which is a membrane-associated complex. P22phox is one of the essential subunits required for its oxidase activity. It is one of the predominant sources of superoxide production in vascular cells. Hyperglycemia exaggerates superoxide (O^{2-}) production from mitochondrial respiration and leads to the activation of NADPH oxidase and redox sensitive signaling pathways which accelerates the process of atherogenesis in T2D patients.

Human P22phox, encoded by CYBA gene is located on chromosome 16q24. It is composed of six exons spanning 8.5 kb. The three polymorphisms namely C242T (rs4673), A640G (rs1049255) and −930 A/G (rs9632581) in CYBA gene are of considerable interest in research as they have important functional effects.

CYBA C242T polymorphism is located in exon 4 and results in a switch in amino acid [His/Tyr] at residue 72 located in the putative heme-binding sites [63]. Because the histidine residue is considered to be a candidate for the ligand of the heme prosthetic group of cytochrome-b, it has been suggested that this polymorphism is directly associated with the loss of oxidative function and a decreased production of ROS and oxidative stress in the vasculature [64]. Another very common functional polymorphism −930 A/G is located at the position −930 from the ATC codon. The functional effects of this polymorphism resulted in higher transcriptional activity of CYBA gene due to potential binding site for C/EBP transcription factor.

Third polymorphism in CYBA gene, A640G is located in the 3′ untranslated region [65]. It has been assumed that A640G modifies the stability of mRNA and translational activity of CYBA through the interaction with other regions of mRNA. This polymorphism has been found to have effect on ROS generation [66, 67]. All the three polymorphisms of CYBA have been widely investigated for their association with T2D and CHD and found that these may be an indicative for the risk of oxidative stress and subclinical coronary atherosclerosis in type 2 diabetes [68].

22.6.2 Thioredoxin Interacting Protein

Oxidative stress in T2D is linked to enhanced production of superoxides in mitochondria through PKC-dependent activation of membranous NADPH oxidase. The two major intracellular thiol reducing mechanisms are the interacting glutathione and thioredoxin (TRX) systems. TXNIP, an up-regulated gene of oxidative stress, is a binding protein of TRX, a redox protein that neutralises ROS by reducing thiols especially insulin disulfides and limits the damage from oxidative stress.

TXNIP acts as an oxidative stress modulator by inhibiting antioxidant activity of TRX and interacting with antioxidant transcription factors such as Nrf2 transcription factor 8 and activates NLPR3 inflammasome 9. TXNIP also sensitizes cardiomyocytes to ROS-induced apoptosis. Experimental studies suggested that hyperglycemia could induce overexpression of TXNIP and decreased TRX activity which can stimulate ROS production and might trigger hyperglycemia mediated oxidative stress pathways and pre-atherosclerosis in T2D patients [69].

TXNIP gene is located on chromosome 1q21.1. It is a 46-kDa ubiquitously expressed protein which contains 391 amino acid residues. Genetic variations in TXNIP might predispose individuals at inherited risk for developing T2D and CHD. The two SNPs rs7211 and rs7212 are widely studied in T2D and CHD. Both SNPs are located in the 3′ region and alters the gene expression regulating mechanisms like modulating RNA stability and microRNA binding etc. Several studies have shown that increased TXNIP expression was significantly associated with these polymorphisms in T2D and CHD patients [70, 71].

22.6.3 Thioredoxin Reductase 2

Thioredoxin (TRX) system is one of the major thiol-dependent antioxidant systems through its disulfide reductase activity regulating protein dithiol/disulfide balance. The mitochondrial TRX system is composed of NADPH, TRX reductase (TRXR), and TRX. TRXR catalyzes disulfide reduction in TRX with NADPH as a cofactor.

TRXRD2 is a key enzyme in antioxidant defence system predominantly localized to mitochondria and controls the cellular redox environment tby binding to TXNIP and inhibits the reducing activity of TRX leading to mitochondrial dysfunction [72]. TRXRD2 plays an essential role in maintaining vascular endothelial cell function, increases NO bioavailability, and thereby reduces oxidative stress and prevents atherosclerosis development [73, 74]. In T2D, there is an enhanced mitochondrial production of cellular ROS under redox conditions which is central to the pathogenesis and progression of CHD [75].

Thioredoxin reductase 2 (TXNRD2) gene is localized on chromosome 22q11.21 with 18 exons. There are few studies reported on polymorphisms of TXNRD2 in diabetes induced CHD. Kariz et al. has reported that T2D individuals with CC + TT genotypes of T50964C [rs1548357] polymorphism conferred protective role for MI in T2D subjects of Slovenian origin [76].

22.6.4 Myeloperoxidase

Myeloperoxidase (MPO) is a member of the heme peroxidase superfamily, produced primarily by neutrophils and monocytes. Growing evidences suggest that leukocytes are the main source of reactive oxygen and halogens that play a major role in the development of vascular damage and oxidative stress [77, 78]. MPO is a highly oxidant enzyme, catalyzes the production of hypochlorous acid, tyrosyl radical, and reactive nitrogen free radicals and is a key biomarker of vascular inflammation. These highly reactive compounds are involved in oxidative modification of proteins, lipids and DNA.

Hyperglycemia increases MPO activity and stimulates the production of hydrogen peroxide. MPO modifies LDL into atherogenic form, generates dysfunctional HDL, promotes formation of foam cell, rupture of plaque, and accelerates progression of atherosclerosis. MPO has been proposed to be a novel risk indicator for future coronary events in healthy people and a prognostic marker for CAD in Diabetes mellitus patients. Purushothaman et al. has demonstrated that there was an approximately three-fold increased MPO protein expression and was strongly associated with high-risk plaque features in diabetic patients compared to non-diabetic patients [79].

MPO gene is located on chromosome 17q23-q24, and its expression is regulated by nutrilites [80]. Several single nucleotide polymorphisms have been identified in the promoter and coding regions of MPO gene but -463G/A (rs2333227) and -129G/A polymorphisms are widely studied. The MPO -463G/A polymorphism was associated with T2D and an increased risk of CAD in Asians [81].

22.6.5 Poly (ADP-Ribose) Polymerase-1

Poly(ADP-ribose) polymerase-1 (PARP) is the nuclear DNA repair enzyme with multiple regulatory functions. Increased oxidative stress activates PARP enzyme which depletes its substrate, NAD+, slowing the rate of glycolysis, electron transport, ATP formation and also inhibits GAPDH by poly (ADP-ribosy)lation. Hyperactivation of PARP represents tissue damage in various pathological conditions associated with oxidative stress. These processes result in acute endothelial dysfunction in diabetic blood vessels, which importantly contributes to the development of coronary heart disease.

PARP-1 belongs to PARP enzyme family and is a 116-kDa protein that consists of three main domains: the N-terminal DNA-binding domain containing two zinc fingers, the automodification domain, and the C-terminal catalytic domain. PARP-1 through its second zinc finger domain, binds to both single and double stranded DNA breaks, forms homodimers and catalyzes the cleavage of nicotinamide adenine dinucleotide (NAD+) into nicotinamide and ADP-ribose and uses the latter to synthesize branched nucleic acid-like polymers of poly(ADP-ribose) covalently attached to nuclear acceptor proteins. Due to its high negative charge, covalently attached ADP-ribose polymer affects the function of target proteins. Studies have shown that high glucose induced oxidative stress leads to single stranded breaks and PARP activation in murine and human endothelial cells [82].

PARP-1 gene is located on chromosome 1q41–42. A single nucleotide polymorphism Val762Ala (rs1136410), is a non-synonymous A-to-G transition at codon 762 results in conversion of valine to alanine in the catalytic domain of PARP-1 [83]. In a study by Xue et al., PARP-1 rs1136410 SNP has conferred protection against CAD through modulation of PARP activity in Chinese Han population [84]. Few studies conducted on rat and mouse models have proved that PARP activation leads to ET-1 upregulation, downstream effects in diabetes [85] whereas PARP inhibition improved endothelial dysfunction [86] and was protective against progression of cardiovascular complications in diabetes [87].

22.7 Conclusion

Prevention and management of diabetic cardiovascular complications is critical due to its worldwide morbidity and mortality. Understanding the etiological link between T2D & CHD by identification of oxidative stress related molecular and genetic mechanisms may unveil novel strategies and develop new therapeutics to reduce cardiovascular morbidity & mortality in T2D patients. These findings may also provide the basis in developing more sensitive, clinical and genetic markers of oxidative stress for prognosis and risk prediction of CHD in type 2 diabetic patients and also suggest the importance of antioxidant therapy to delay the onset and progression of diabetes induced CHD.

References

1. Beckman JA, Paneni F, Cosentino F, Creager MA (2013) Clinical update. Diabetes and vascular disease: pathophysiology, clinical consequences, and medical therapy: part II. Eur Heart J 34(31):2444–2456
2. Ogurtsova K, Fernandes JDR, Huang Y, Linnenkamp U, Guariguata L, Cho NH et al (2017) IDF diabetes atlas: global estimates for the prevalence of diabetes for 2015 and 2040. Diabetes Res Clin Pract 128:40–50
3. Uppu RM, Parinandi NL (2011) Insulin sensitization and resistance interrelationship revisited with a quantitative molecular model approach. J Diabetes Metab Disord 2(6):106e
4. Li YW, Aronow WS (2011) Diabetes mellitus and cardio vascular disease. J Clin Exp Cardiol 2:114

5. Nathan DM, Cleary PA, Backlund JY et al (2005) Intensive diabetes treatment and cardiovascular disease in patients with type 1diabetes. N Engl J Med 353(25):2643–2653
6. Haffner SM, Lehto S, Rönnemaa T, Pyörälä K, Laakso M (1998) Mortality from coronary heart disease in subjects with type 2 diabetes and in nondiabetic subjects with and without prior myocardial infarction. N Engl J Med 339:229–234
7. Sarwar N, Gao P, Seshasai SR, Gobin R, Kaptoge S et al (2010) Diabetes mellitus, fasting blood glucose concentration, and risk of vascular disease: a collaborative meta analysis of 102 prospective studies. Lancet 375(9733):2215–2222
8. Kannel WB, Feinleib M, McNamara PM, Garrison RJ, Castelli WP (1979) An investigation of coronary heart disease in families. The Framingham offspring study. Am J Epidemiol 110(3):281–290
9. Gu K, Cowie CC, Harris MI (1999) Diabetes and decline in heart disease mortality in US adults. JAMA 281(14):1291–1297
10. Kirpichnikov D, Sowers JR (2001) Diabetes mellitus and diabetes-associated vascular disease. Trends Endocrinol Metab 12(5):225–230
11. Avogaro A, Fadini GP, Gallo A, Pagnin E, de Kreutzenberg S (2006) Endothelial dysfunction in type 2 diabetes mellitus. Nutr Metab Cardiovasc Dis 16(Suppl. 1):S39–S45
12. Sundaram RK, Bhaskar A, Vijayalingam S, Viswanatthan M, Mohan R, Shanmugasundaram KR (1996) Antioxidant status and lipid peroxidation in type II diabetes with and without complications. Clin Sci 90:255–260
13. Nourooz-Zadeh J, Rahimi A, Tajaddini-Sarmadi J et al (1997) Relationships between plasma measures of oxidative stress and metabolic control in NIDDM. Diabetologia 40:647–653
14. Maxwell SRJ, Thomason H, Sandler D et al (1997) Antioxidant status in patients with uncomplicated insulin-dependent and non insulin-dependent diabetes mellitus. Eur J Clin Investig 27:484–490
15. Wali U, Jogana MU, Zarummai AL, Saidu Y (2011) Antioxidant vitamins and trace elements status of diabetics in Sokoto, Nigeria. Niger J Basic Appl Sci 19(1):130–134
16. Inoguchi T, Li P, Umeda F, Yu HY, Kakimoto M, Imamura M, Nawata H (2000) High glucose level and free fatty acid stimulate reactive oxygen species production through protein kinase C--dependent activation of NAD(P)H oxidase in cultured vascular cells. Diabetes 49(11):1939–1945
17. Griffin ME, Marcucci MJ, Cline GW, Bell K, Barucci N, Lee D, Shulman GI (1999) Free fatty acid-induced insulin resistance is associated with activation of protein kinase C theta and alterations in the insulin signaling cascade. Diabetes 48(6):1270–1274
18. Suzuki LA, Poot M, Gerrity RG, Bornfeldt KE (2001) Diabetes accelerates smooth muscle accumulation in lesions of atherosclerosis: lack of direct growth-promoting effects of high glucose levels. Diabetes 50(4):851–860
19. Fukumoto H, Naito Z, Asano G, Aramaki T (1998) Immunohistochemical and morphometric evaluations of coronary atherosclerotic plaques associated with myocardial infarction and diabetes mellitus. J Atheroscler Thromb 5(1):29–35
20. Mehta JL, Rasouli N, Sinha AK, Molavi B (2006) Oxidative stress in diabetes: a mechanistic overview of its effects on atherogenesis and myocardial dysfunction. Int J Biochem Cell Biol 38:794–803
21. Hussain MJ, Peakman M, Gallati H et al (1996) Elevated serum levels of macrophage-derived cytokines precede and accompany the onset of IDDM. Diabetologia 39:60–69
22. Uemura S, Matsushita H, Li W, Glassford AJ, Asagami T, Lee KH, Tsao PS (2001) Diabetes mellitus enhances vascular matrix metalloproteinase activity: role of oxidative stress. Circ Res 88(12):1291–1298
23. Inoue M, Sato EF, Nishikawa M, Park AM, Kira Y, Imada I et al (2003) Mitochondrial generation of reactive oxygen species and its role in aerobic life. Curr Med Chem 10:2495–2505
24. Valko M, Rhodes CJ, Moncol J, Izakovik M, Mazure M (2006) Free radicals, metals and antioxidants in oxidative stress-induced cancer. Chem Biol Interact 160:1–40
25. Sies H (1997) Oxidative stress: oxidants and antioxidants. Exp Physiol 82:291–295

26. Higashi Y, Noma K, Yoshizumi M, Kihara Y (2009) Endothelial function and oxidative stress in cardiovascular diseases. Circ J 73:411–418
27. Arthur JR (2000) The glutathione peroxidases. Cell Mol Life Sci 57:1825–1835
28. Zschauer TC, Matsushima S, Altschmied J, Shao D, Sadoshima J, Haendeler J (2013) Interacting with thioredoxin-1--disease or no disease? Antioxid Redox Signal 18:1053–1062
29. Yamawaki H, Berk BC (2005) Thioredoxin: a multifunctional antioxidant enzyme in kidney, heart and vessels. Curr Opin Nephrol Hypertens 14:149–153
30. Ghattas MH, Abo-Elmatty DM (2012) Association of polymorphic markers of the catalase and superoxide dismutase genes with type 2 diabetes mellitus. DNA Cell Biol 31:1598–1603
31. Sindhu RK, Koo JR, Roberts CK, Vaziri ND (2004) Dysregulation of hepatic superoxide dismutase, catalase and glutathione peroxidase in diabetes: response to insulin and antioxidant therapies. Clin Exp Hypertens 26:43–53
32. Luan R, Liu S, Yin T, Lau WB, Wang Q, Guo WY, Wang HC, Tao L (2009) High glucose sensitizes adult cardiomyocytes to ischaemia/reperfusion injury through nitrative thioredoxin inactivation. Cardiovasc Res 83:294–302
33. Meister A (1994) Glutathione-ascorbic acid antioxidant system in animals. J Biol Chem 269(13):9397–9400
34. Rodrigo R, Libuy M, Feliú F, Hasson D (2013) Molecular basis of cardioprotective effect of antioxidant vitamins in myocardial infarction. Biomed Res Int 2013:1–15
35. Motoyama T, Kawano H, Kugiyama K, Hirashima O, Ohgushi M, Tsunoda R, Moriyama Y, Miyao Y, Yoshimura M, Ogawa H, Yasue H (1998) Vitamin E administration improves impairment of endothelium-dependent vasodilation in patients with coronary spastic angina. J Am Coll Cardiol 32(6):1672–1679
36. Islam KN, Devaraj S, Jialal I (1998) Tocopherol enrichment of monocytes decreases agonist-induced adhesion to human endothelial cells. Circulation 98(21):2255–2261
37. Polidori MC, Mecocci P, Stahl W, Parente B, Cecchetti R, Cherubini A, Cao P, Sies H, Senin U (2000) Plasma levels of lipophilic antioxidants in very old patients with type 2 diabetes. Diabetes Metab Res Rev 16(1):15–19
38. Merzouk S, Hichami A, Madani S, Merzouk H, Berrouiguet AY, Prost J, Moutairou K, Chabane-Sari N, Khan NA (2003) Antioxidant status and levels of different vitamins determined by high performance liquid chromatography in diabetic subjects with multiple complications. Gen Physiol Biophys 22:15–27
39. Odum EP, Ejilemele AA, Wakwe VC (2012) Antioxidant status of type 2 diabetic patients in Port Harcourt, Nigeria. Niger J Clin Pract 15(1):55–58
40. Newton AC (2009) Lipid activation of protein kinases. J Lipid Res 50(Suppl):S266–S271
41. Cameron J, Cruikshank JK (2007) Glucose, insulin, diabetes and mechanisms of arterial dysfunction. Clin Exp Pharmacol Physiol 34:677–682
42. Vinik A, Flemmer M (2002) Diabetes and macrovascular disease. J Diabetes Complicat 16:235–245
43. Epidemiology of Diabetes Interventions and Complications (EDIC) Research Group (1999) Effect of intensive diabetes treatment on carotid artery wall thickness in the epidemiology of diabetes interventions and complications. Diabetes 48(2):383–390
44. Muniyappa R, Iantorno M, Quon MJ (2008) An integrated view of insulin resistance and endothelial dysfunction. Endocrinol Metab Clin N Am 37(3):685–711
45. Surapon T (2015) Oxidative stress, insulin resistance, dyslipidemia and type 2 diabetes mellitus. World J Diabetes 6(3):456–480
46. Paniagua JA (2016) Nutrition, insulin resistance and dysfunctional adipose tissue determine the different components of metabolic syndrome. World J Diabetes 7(19):483–514
47. Han CY, Umemoto T, Omer M, Den Hartigh LJ, Chiba T, LeBoeuf R, Buller CL, Sweet IR, Pennathur S, Abel ED, Chait A (2012) NADPH oxidase-derived reactive oxygen species increases expression of monocyte chemotactic factor genes in cultured adipocytes. J Biol Chem 287(13):10379–10393

48. Luft FC (2002) Proinflammatory effects of angiotensin II and endothelin: targets for progression of cardiovascular and renal diseases. Curr Opin Nephrol Hypertens 11(1):59–66
49. O'Driscoll G, Green D, Rankin J et al (1997) Improvement in endothelial function by angiotensin converting enzyme inhibition in insulin dependent diabetes mellitus. J Clin Invest 100:678–684
50. Boucher J, Kleinridders A, Ronald Kahn C (2014) Insulin receptor signaling in normal and insulin-resistant states. Cold Spring Harb Perspect Biol 6(1):a009191
51. Sengupta A, Patel PA, Yuldasheva NY, Mughal RS, Galloway S, Viswambharan H, Walker AMN, Aziz A, Smith J, Ali N, Mercer BN, Imrie H, Sukumar P, Wheatcroft SB, Kearney MT, Cubbon RM (2018) Endothelial insulin receptor restoration rescues vascular function in male insulin receptor Haploin sufficient mice. Endocrinology 159(8):2917–2925
52. Jiang ZY, Lin Y, Clemont A et al (1999) Characterization of selective resistance to insulin signaling in the vasculature of obese Zucker (fa/fa) rats. J Clin Invest 104(4):447–457
53. Montagnani M, Golovchenko I, Kim I, Koh GY, Goalstone ML, Mundhekar AN, Johansen M, Kucik DF, Quon MJ, Draznin B (2002) Inhibition of phosphatidylinositol 3-kinase enhances mitogenic actions of insulin in endothelial cells. J Biol Chem 277(3):1794–1799
54. Guzik TJ, Mussa S, Gastaldi D et al (2002) Mechanisms of increased vascular superoxide production in human diabetesmellitus: role of NAD(P)H oxidase and endothelial nitric oxide synthase. Circulation 105(14):1656–1662
55. Kim YK, Lee M-S, Son SM et al (2002) Vascular NADH oxidase is involved in impaired endothelium-dependent vasodilation in OLETF rats, a model of type 2 diabetes. Diabetes 51(2):522–527
56. Sowers JR (2004) Insulin resistance and hypertension. Am J Physiol Heart Circ Physiol 286:H1597–H1602
57. Du X, Edelstein D, Obici S, Higham N, Zou MH, Brownlee M (2006) Insulin resistance reduces arterial prostacyclin synthase and eNOS activities by increasing endothelial fatty acid oxidation. J Clin Invest 116:1071–1080
58. Banerjee M, Vats P (2014) Reactive metabolites and antioxidant gene polymorphisms in type 2 diabetes mellitus. Indian J Hum Genet 20(1):10–19
59. Qi L, Parast L, Cai T, Powers C, Gervin EV, Hauser TH, Hu FB, Doria A (2011) Genetic susceptibility to coronary heart disease in type 2 diabetes. J Am Coll Cardiol 58(25):2675–2682
60. Katakami N, Kaneto H, Matsuoka TA, Takahara M, Imamura K, Ishibashi F et al (2010) Accumulation of gene polymorphisms related to oxidative stress is associated with myocardial infarction in Japanese type 2 diabetic patients. Atherosclerosis 212:534–538
61. Dos Santos KG, Canani LH, Gross JL, Tschiedel B, Souto KE, Roisenberg I (2006) The catalase—262C/T promoter polymorphism and diabetic complications in Caucasians with type 2 diabetes. Dis Markers 22(5–6):355–359
62. Jones DA, Prior SL, Tang TS, Bain SC, Hurel SJ, Humphries SE et al (2010) Association between the rs4880 superoxide dismutase 2 (C>T) gene variant and coronary heart disease in diabetes mellitus. Diabetes Res Clin Pract 90(2):196–201
63. Dinauer MC, Pierce EA, Bruns GA, Curnutte JT, Orkin SH (1990) Human neutrophil cytochrome b light chain (p22-phox).Gene structure, chromosomal location, and mutations in cytochrome-negative autosomal recessive chronic granulomatous disease. J Clin Investig 86:1729–1737
64. Whitehead AS, FitzGerald GA (2001) Twenty-first century phox: not yet ready for widespread screening. Circulation 103:7–9
65. De Boer M, De Klein A, Hossle JP, Seger R, Corbeel L et al (1992) Cytochrome b558-negative, autosomal recessive chronic granulomatous disease: two new mutations in the cytochrome b558 light chain of the NADPH oxidase (p22-phox). Am J Hum Genet 51:1127–1135
66. Bedard K, Attar H, Bonnefont J, Jaquet V, Borel C et al (2009) Three common polymorphisms in the CYBA gene form a haplotype associated with decreased ROS generation. Hum Mutat 30:1123–1133

67. Schirmer M, Hoffmann M, Kaya E, Tzvetkov M, Brockmöller J (2007) Genetic polymorphisms of NAD(P)H oxidase: variation in subunit expression and enzyme activity. Pharmacogenomics 8:297–304
68. Moreno MU, San Jose G, Fortuno A, Miguel-Carrasco JL, Beloqui O, Diez J, Zalba G (2011) The A640G CYBA polymorphism associates with subclinical atherosclerosis in. Diabetes 1(3):1467–1474
69. Schulze PC, Yoshioka J, Takahashi T et al (2004) Hyperglycemia promotes oxidative stress through inhibition of thioredoxin function by thioredoxin-interacting protein. J Biol Chem 279:30369–30374
70. Ferreira NE, Omae S, Pereira A, Rodrigues MV, Miyakawa AA, Campos LC, Santos PC, Dallan LA, Martinez TL, Santos RD, Mill JG, Krieger JE, Pereira AC (2012) Thioredoxin interacting protein genetic variation is associated with diabetes and hypertension in the Brazilian general population. Atherosclerosis 221(1):131–136
71. Wang X-b, Han Y-d, Zhang S, Cui N-h, Liu Z-j, Huang Z-l, Li C, Zheng F (2016) Associations of polymorphisms in *TXNIP* and gene–environment interactions with the risk of coronary artery disease in a Chinese Han population. J Cell Mol Med 20(12):2362–2373
72. Saxena G, Chen J, Shalev A (2010) Intracellular shuttling and mitochondrial function of thioredoxin-interacting protein. J Biol Chem 285(6):3997–4005
73. Zhang H, Luo Y, Zhang W, He Y, Dai S, Zhang R et al (2007) Endothelial-specific expression of mitochondrial thioredoxin improves endothelial cell function and reduces atherosclerotic lesions. Am J Pathol 170(3):1108–1120
74. Dai S, He Y, Zhang H, Yu L, Wan T, Xu Z et al (2009) Endothelial specific expression of mitochondrial thioredoxin promotes ischemia-mediated arteriogenesis and angiogenesis. Arterioscler Thromb Vasc Biol 29(4):495–502
75. Aon MA, Stanley BA, Sivakumaran V, Kembro JM, O'Rourke B, Paolocci N et al (2012) Glutathione/thioredoxin systems modulate mitochondrial H2O2 emission: an experimental computational study. J Gen Physiol 139(6):479–491
76. Kariž S, Mankoč S, Petrovič D (2015) Association of thioredoxin reductase 2 (TXNRD2) gene polymorphisms with myocardial infarction in Slovene patients with type 2 diabetes mellitus. Diabetes Res Clin Pract 108(2):323–328
77. Jacobi J, Sela S, Cohen HI, Chezar J, Kristal B (2006) Priming of polymorphonuclear leukocytes: a culprit in the initiation of endothelial cell injury. Am J Physiol Heart Circ Physiol 290(5):H2051–H2058
78. Shurtz-Swirski R, Sela S, Herskovits AT, Shasha SM, Shapiro G, Nasser L et al (2001) Involvement of peripheral polymorphonuclear leukocytes in oxidative stress and inflammation in type 2 diabetic patients. Diabetes Care 24(1):104–110
79. Purushothaman R, Purushothaman M, Alviar CL, Tarricone A, Vasquez M et al (2013) Increased myeloperoxidase expression is associated with increase in intra-plaque hemorrhage, iron content, inflammation and neovascularization in diabetic atherosclerosis: implications for plaque progression. J Cardiovasc Dis Diagn 1:118
80. Denzler KL, Levin WJ, Lee JJ, Lee NA (1997) The murine eosinophil peroxidase maps to chromosome 11. Mamm Genome 8:381–382
81. Ergen H, Karagedik ZEK, Isbir T (2014) An association between MPO-463G/a polymorphism and type 2 diabetes. Folia Biol (Praha) 60(3):108–112
82. Garcia Soriano F, Virág L, Jagtap P, Szabó É, Mabley JG, Liaudet L, Marton A, Hoyt DG, Murthy KG, Salzman AL, Southan GJ, Szabo C (2001) Diabetic endothelial dysfunction: the role of poly (ADPribose) polymerase activation. Nat Med 7:108–113
83. Yu H, Ma H, Yin M, Wei Q (2012) Association between PARP-1 V762A polymorphism and cancer susceptibility: a meta-analysis. Genet Epidemiol 36:56–65
84. Wang X-b, Cui N-h, Zhang S, Shu-renGuo, Liu Z-j, Ming L (2017) PARP-1 Variant Rs1136410 confers protection against coronary artery disease in a Chinese Hanpopulation: a two-stage case-control study involving 5643 subjects. Front Physiol 8:916

85. Chiu J, Xu BY, Chen S, Feng B, Chakrabarti S (2008) Oxidative stress-induced, poly(ADP-ribose) polymerase-dependent upregulation of ET-1 expression in chronic diabetic complications. Can J Physiol Pharmacol 86(6):365–372
86. Soriano FG, Pacher P, Mabley J, Liaudet L, Szabo C (2001) Rapid reversal of the diabetic endothelial dysfunction by pharmacological inhibition of poly(ADP-ribose) polymerase. Circ Res 89:684–691
87. Jagtap P, Szabo C (2005) Poly(ADP-ribose) polymerase and the therapeutic effects of its inhibitors. Nat Rev Drug Discov 4:421–440

Part IV

Hypertension and Heart Failure

Role of Oxidative Stress in the Pathophysiology of Arterial Hypertension and Heart Failure

23

Teresa Sousa, Marta Reina-Couto, and Pedro Gomes

23.1 Introduction

23.1.1 Reactive Oxygen Species (ROS) and Oxidative Stress: General Considerations

Oxygen is essential for cellular respiration and energy production by aerobic organisms. However, the partial reduction of oxygen by several metabolic pathways will inevitably result in the production of reactive oxygen species (ROS). In turn, these short-lived and highly reactive oxygen metabolites have the ability to attack cellular macromolecules, including lipids, proteins and nucleic acids, causing oxidative damage. Under normal conditions, intracellular ROS concentrations are maintained within a balanced, steady-state range, by integrated enzymatic and nonenzymatic antioxidant systems, which not only protect cells from the detrimental effects of

T. Sousa
Department of Biomedicine – Unit of Pharmacology and Therapeutics, Faculty of Medicine, University of Porto, Porto, Portugal

Center for Drug Discovery and Innovative Medicines (MedInUP), Porto, Portugal
e-mail: tsousa@med.up.pt

M. Reina-Couto
Department of Biomedicine – Unit of Pharmacology and Therapeutics, Faculty of Medicine, University of Porto, Porto, Portugal

Center for Drug Discovery and Innovative Medicines (MedInUP), Porto, Portugal

Department of Intensive Medicine, Centro Hospitalar São João, Porto, Portugal

P. Gomes (✉)
Department of Biomedicine – Unit of Pharmacology and Therapeutics, Faculty of Medicine, University of Porto, Porto, Portugal

CNC – Center for Neuroscience and Cell Biology, University of Coimbra, Coimbra, Portugal
e-mail: pgomes@med.up.pt

S. Chakraborti et al. (eds.), *Oxidative Stress in Heart Diseases*,
https://doi.org/10.1007/978-981-13-8273-4_23

ROS but also allow the activation of redox signalling pathways that regulate physiological functions. Disruption of redox equilibrium by persistently elevated ROS concentrations leads to oxidative stress and subsequent dysfunctional signalling and macromolecular damage [1–3]. Oxidative stress is widely recognized as an important contributor to ageing, which is characterized by a progressive decline in biological functions and in the organism's ability to adapt to metabolic stress over time, being also aetiologically involved in the pathogenesis of a wide variety of disease processes, namely arterial hypertension, atherosclerosis, heart failure (HF), diabetic neuropathy, renal diseases, neurological diseases, as well as cancer [4, 5]. In contrast to the well documented role of oxidative stress in cardiac diseases, less is known regarding the role of reductive stress in these processes. Reductive stress is the counterpart of oxidative stress and is defined as an aberrant increase in reducing equivalents, leading to decreased ROS levels. Nevertheless, it is becoming increasingly clear that the biological extremes of the redox spectrum play critical roles in disease pathogenesis [6]. In addition to ROS, there is another class of chemically reactive molecules collectively designated as reactive nitrogen species (RNS), which include various nitric oxide (·NO)-derived compounds. RNS have been recognized as playing important functions in diverse physiological and pathological redox signalling processes [7, 8]. Similarly to ROS, excessive amounts of RNS have been implicated in cell injury and death by inducing nitrosative stress.

23.1.2 Main Types of ROS and RNS

ROS can be divided in two main groups: i) free radicals [e.g. superoxide anion ($O_2^{\cdot-}$), hydroxyl radical (HO·), peroxyl radical (ROO•)], which are unstable and highly reactive species due to the presence of one or more unpaired electrons; and ii) non-radical oxidants [e.g. singlet oxygen ($^1O_2$), hydrogen peroxide (H_2O_2), hypochlorous acid (HOCl)], that have generally more specific reactivity and higher stability. RNS include ·NO and nitrogen dioxide radicals (·NO_2) and also non-radicals such as peroxynitrite ($ONOO^-$), nitrous acid (HNO_2), peroxynitrous acid (ONOOH) and alkyl peroxynitrites (ROONO) (Table 23.1). Among biological ROS and RNS, $O_2^{\cdot-}$, H_2O_2, ·NO and $ONOO^-$ appear to play a prominent role in vascular, cardiac, renal and neuronal regulation and dysregulation, thus representing important targets for strategies aiming to reduce redox dysfunction in cardiovascular diseases [9, 10].

23.1.3 Mechanisms of ROS Generation in Cardiovascular Diseases

A key consideration for ROS/RNS chemistry and biology is the subcellular compartment where a particular species is generated, as discrete microenvironments can determine which targets will be preferentially attacked. ROS are derived from both endogenous and exogenous sources (Fig. 23.1). Intracellular compartments capable of ROS generation include mitochondria, the endoplasmic reticulum, peroxisomes, nuclei, the cytosol, and plasma membrane enzymatic systems. ROS can also be

Table 23.1 Major types of ROS and RNS in living systems and corresponding properties

	Symbol	Half-life (s)	Properties
ROS – *free radicals*			
Superoxide anion	$O_2^{\cdot -}$	10^{-6}	Low reactivity in aqueous solution; signalling function
Hydroxyl	$HO^{\cdot}$	10^{-9}	Most reactive oxygen radical; reacts almost immediately with every molecule in living cells; diffuses a short distance
Peroxyl	$ROO^{\cdot}$	10^{-2}	Weak oxidant; high diffusibility
ROS – *non-radical oxidants*			
Singlet oxygen	$^1O_2$	10^{-6}	Powerful oxidizing agent
Hydrogen peroxide	H_2O_2	10^{-5}	Weak oxidizing and reducing agent; diffuses across membranes; signalling role
Hypochlorous acid	HOCl	n/a	Strong reactive species; released by neutrophils
RNS – *free radicals*			
Nitric oxide	$^{\cdot}NO$	10^{-3}	Can yield potent oxidants during pathological states; endogenous signalling molecule
RNS – *non-radical oxidants*			
Peroxynitrite	$ONOO^-$	10^{-2}	Highly reactive intermediate of $O_2^{\cdot -}$ and $^{\cdot}NO$; permeates cell membranes

n/a no data available

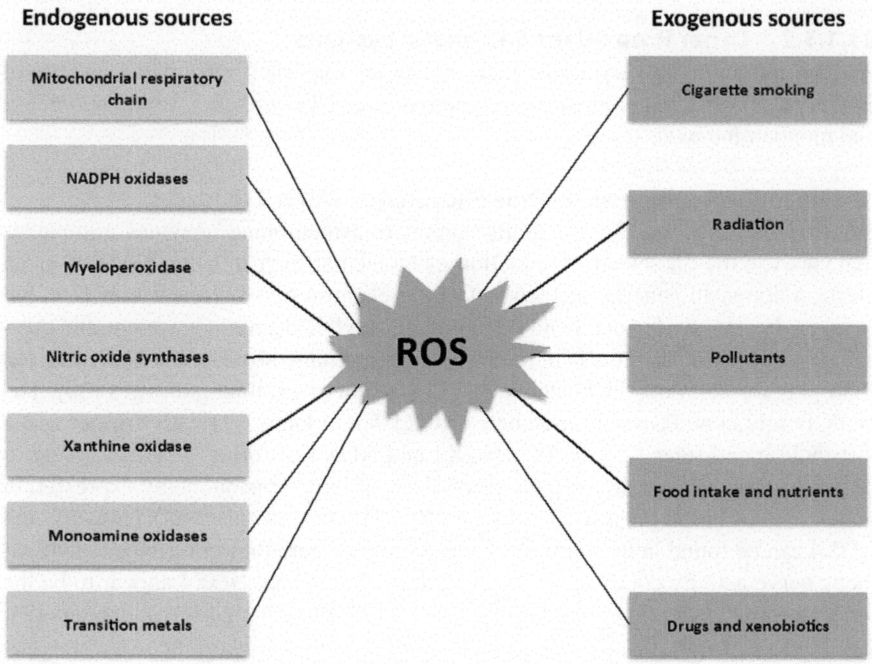

Fig. 23.1 Major endogenous and exogenous sources leading to ROS production

produced in response to external sources, including pollution, alcohol, tobacco smoke, heavy metals, UV radiation.

23.1.3.1 Mitochondria

Mitochondria play a key role in energy metabolism in many tissues. More than 90% of the oxygen consumed by aerobic organisms is utilized by the mitochondrial electron transport chain (ETC), which generates ATP in a process coupled to the reduction of cellular oxygen to water. The mitochondrial respiratory chain complexes are also an important source of ROS within most mammalian cells [11–13]. In fact, about 1–4% of the oxygen used in these reactions is converted to $O_2^{\cdot-}$ and H_2O_2, which may have deleterious consequences to mitochondria if not adequately detoxified [14]. ROS formation in the mitochondria is regulated by the respiratory rate and by the antioxidant enzyme manganese superoxide dismutase (MnSOD) [12]. The mitochondrial respiratory chain appears to be a major source of oxidative stress in some experimental forms of arterial hypertension (e.g. mineralocorticoid hypertension, angiotensin II-induced hypertension) and the inhibition of mitochondrial ROS production has a significant blood pressure-lowering effect in these models [15, 16]. In HF there is also evidence of abnormal ROS production from mitochondrial respiratory chain. Furthermore, the scavenging of mitochondrial ROS has been shown to prevent or reverse HF and to eliminate sudden cardiac death in an animal model of non-ischemic HF that displays important features of human HF (e.g. prolonged QT interval, high incidence of spontaneous sudden cardiac death due to ventricular tachycardia/fibrillation) [17–19].

23.1.3.2 Other Prooxidant Enzymatic Systems

Besides mitochondrial oxidases, there are other important enzymatic sources of ROS, such as NADPH oxidases, myeloperoxidase, ·NO synthases, xanthine oxidase and monoamine oxidases (Fig. 23.1).

Nicotinamide Adenine Nucleotide Phosphate (NADPH) Oxidases

NADPH oxidases (NOX) are multi-subunit transmembrane enzymes complexes that catalyze the one-electron reduction of molecular oxygen using NADPH as an electron donor. In general, the product of the electron transfer reaction is $O_2^{\cdot-}$, but H_2O_2 is also rapidly formed from dismutation of NOX-derived $O_2^{\cdot-}$ due to the presence of superoxide dismutase in the cells or by spontaneous reaction. NOX-derived ROS play a role in host defence and also in various signalling pathways [20]. The NOX family contains seven members (NOX1-5 and Duox1-2) with distinct tissue distribution and roles [20]. NOX1, NOX2 and NOX4 isoforms enzymes appear to be particularly relevant in the pathophysiology of hypertension, being expressed in major sites of blood pressure regulation [20, 21]. For example, NOX1, NOX2 and NOX4 can be found in the central nervous system, where they contribute to sympathetic nerve activity control [21]. In the kidney, NOX2 and NOX4 appear to be the main isoforms regulating renal function and contributing to end-organ damage [21, 22]. These isoforms are also important determinants of vascular tone in several vascular beds, including the renal afferent arteriole, which is critical for the regulation of renal haemodynamics [23–25]. Endothelial function can be regulated by NOX2, which contributes to impaired vasodilation, or by NOX4, which improves endothelial-dependent vasodilation. NOX1 and NOX4 are also involved in vascular smooth muscle cell growth and migration [20, 23, 24]. Of note, recent studies suggest that NOX5, an isoform that is found in humans but absent in rodents, is also implicated in the pathogenesis of cardiovascular diseases, such as hypertension and atherosclerosis [26]. For example, renal proximal tubular cells from human hypertensive subjects appear to express NOX5 in a greater extent than the other isoforms [27]. Furthermore, in mice expressing human NOX5 in podocytes, the renal function becomes impaired and blood pressure increases [26]. NOX5 expression was also shown in human carotid artery atherosclerotic plaques and to be induced in macrophages exposed to a proinflammatory and prooxidant environment [28].

NOX2 and NOX4, the two isoforms expressed in the heart, appear to be especially relevant in HF [29, 30]. NOX2 contributes to angiotensin II-induced cardiac hypertrophy, atrial fibrillation, myocyte death under stress conditions and post-myocardial infarction remodelling. The inactivation of NOX2 was shown to attenuate ventricular dilatation and contractile dysfunction in experimental models of myocardial infarction. NOX2 deletion also abolished angiotensin II-induced cardiac hypertrophy but was not able to prevent the development of HF caused by severe pressure overload [29, 30]. The role of NOX4 in the heart is more controversial, with both protective and detrimental effects reported. For example, mice lacking cardiac NOX4 display either reduced or aggravated maladaptive remodelling in different models of pressure-overload-induced HF [29, 30]. In what concerns to ischemia-reperfusion injury, it appears that both NOX2 and NOX4 contribute to increased ROS production and damage, as evidenced by the reduced myocardial infarct size/area at risk and lower $O_2^{\cdot-}$ production in NOX2 knockout or NOX4 knockout mice subjected to ischemia-reperfusion injury. However, double knockout of NOX2 and NOX4 exacerbates ischemia-reperfusion injury, probably because low levels of ROS generated by these enzymes are necessary to activate adaptive mechanisms that protect the heart against ischemia-reperfusion injury [31].

Myeloperoxidase (MPO)

MPO, a haem-containing enzyme secreted by activated neutrophils and monocytes under inflammatory conditions, produces several oxidizing molecules that can affect lipids and proteins [32]. MPO uses H_2O_2 to produce other ROS/RNS, such as HOCl, chloramines, tyrosyl radicals and nitrogen dioxides [32]. Although MPO-derived ROS have a major role as bactericidal agents, they can also cause tissue damage in the heart, vessels, kidney and brain. Vascular tone and endothelial bioavailability of ·NO appear to be significantly affected by MPO. Interestingly, the MPO G463A polymorphism was associated with an increased risk of hypertension [33]. MPO contributes to vascular and myocardial dysfunction, being significantly increased in acute coronary syndromes and HF [34–36]. Higher MPO values were reported to be associated with increasing likelihood of more advanced HF in chronic systolic HF patients and to predict future adverse clinical events [37].

NO Synthases

The ·NO synthases (NOS) are a family composed of three enzyme isoforms (neuronal NOS, nNOS; inducible NOS, iNOS; endothelial NOS, eNOS) [38]. NOS are the endogenous sources of ·NO in mammalian cells, in a reaction that converts L-arginine to L-citrulline [38]. ·NO exerts a wide array of regulatory functions on the cardiovascular system, including regulation of vascular tone, blood pressure, cardiomyocyte contractility, sympathetic outflow, smooth muscle cell proliferation, renal renin release and natriuresis [39–41]. However, under conditions of limited bioavailability of the cofactor tetrahydrobiopterin (BH_4) or the substrate L-arginine, NOS become unstable and reduces molecular oxygen to $O_2^{·-}$ instead of producing ·NO. This NOS uncoupling is more often described for eNOS and is triggered by oxidative/nitrosative stress [42]. There is evidence that eNOS dysregulation and consequent endothelial dysfunction occur both in hypertension and HF [43, 44]. Treatment with BH_4, which contributes to eNOS recoupling, prevented or attenuated hypertension in spontaneously hypertensive rats [37]. It was also shown to reverse cardiac hypertrophy and fibrosis and to improve chamber and myocyte function in mice with heart disease induced by pressure overload [10, 45].

Xanthine Oxidase

The enzyme xanthine oxidoreductase displays two interchangeable forms, xanthine dehydrogenase (XDH) and xanthine oxidase (XO), that participate in the metabolism of purines by catalyzing the conversion of hypoxanthine to xanthine and xanthine to uric acid [38, 46]. XDH uses NAD^+ as the preferential electron acceptor while XO reduces molecular oxygen in a reaction that generates $O_2^{·-}$ and H_2O_2 [38, 46]. The XO form predominates in oxidative stress conditions and may contribute to endothelial dysfunction due to its localization in the luminal surface of vascular endothelium [38, 46]. Although XO is capable of generating ROS, both XDH and XO generate uric acid which has antioxidant properties, such as the ability to scavenge $ONOO^-$ and $HO^•$, to prevent oxidative inactivation of endothelium enzymes and to stabilize vitamin C [47, 48]. In contrast, uric acid may also exhibit prooxidant and proinflammatory effects. Indeed, increased uric acid levels have been associated with cardiovascular disease [49, 50]. However, it is still unclear whether these effects reflect direct deleterious actions of uric acid or, alternatively, oxidative damage caused by XO-derived ROS.

XO appears to contribute to the pathophysiology of arterial hypertension in SHR, as evidenced by the significant reduction of blood pressure induced by the treatment with XO inhibitors [10, 51]. In humans, some studies have shown a blood pressure-lowering effect of XO inhibition in adolescents with newly diagnosed essential hypertension and an improvement of cardiovascular outcomes in adults with hypertension [52, 53].

In what concerns to heart diseases, XO inhibition was reported to improve left ventricle contractility and myocardial efficiency in an animal model of HF and to attenuate adverse left ventricular remodelling in experimental myocardial infarction

[19]. XO expression and activity was also shown to be increased in coronary arteries from patients with coronary artery disease, contributing to the augmented production of $O_2^{\cdot-}$ [54]. The inhibition of XO with oxypurinol also improved myocardial contractility in patients with ischemic cardiomyopathy [55]. However, other studies failed to demonstrate clinical benefits of oxypurinol treatment in unselected patients with moderate-to-severe HF or in high-risk HF patients with reduced left ventricular ejection fraction and hyperuricemia [56, 57].

Monoamine Oxidases (MAO)

MAO-A and MAO-B are flavoenzymes predominantly located at the outer membrane of mitochondria, being responsible for the oxidative degradation of neurotransmitters (catecholamines, serotonin) and biogenic amines in a process that generates H_2O_2, ammonia and an aldehyde intermediate. All of these products are potentially deleterious, especially for mitochondria. Pathological stimuli such as neurohormonal and/or chronic hemodynamical stress, inflammation and ischemia-reperfusion can increase the availability of MAO substrates, thus augmenting H_2O_2-induced mitochondrial dysfunction in cardiovascular tissues/organs and leading to endothelial dysfunction and HF [9, 58, 59]. In experimental models of hypertension (induced by angiotensin II) and inflammation (induced by lipopolysaccharide), the expression of both MAO isoforms increased in endothelial cells and MAO inhibition attenuated ROS production and restored endothelial-dependent vasodilation [59]. MAO are also important sources of ROS in the heart. There are several important cardiac targets for MAO-derived ROS, besides mitochondria. These include sphingosine kinase-1, an enzyme involved in cell survival, whose inhibition may contribute to cardiomyocyte apoptosis, as well as the contractile proteins, actin and tropomyosin, whose oxidation correlates with ventricular dysfunction, and matrix metalloproteinases, whose activation induces extracellular matrix remodelling. The signalling pathways activated by MAO-derived H_2O_2 depend on the availability of MAO substrates and H_2O_2 concentration in tissues. Lower amounts of H_2O_2 trigger hypertrophy, cell proliferation and matrix remodelling, while higher concentrations lead to mitochondrial dysfunction, apoptosis or necrosis. MAO inhibition appears to be protective in ischemia-reperfusion injury and pressure overload-induced HF [59, 60].

23.1.4 Major Endogenous Antioxidant Systems

All living organisms have adapted and developed an endogenous antioxidant defence system, composed of enzymatic and nonenzymatic antioxidants, that is usually effective in neutralizing deleterious effects of ROS (Fig. 23.2). However, when the antioxidant systems are overwhelmed, as observed in most pathological conditions, oxidative stress ensues. Below we provide an overview of the major antioxidant systems with relevance to cardiovascular diseases.

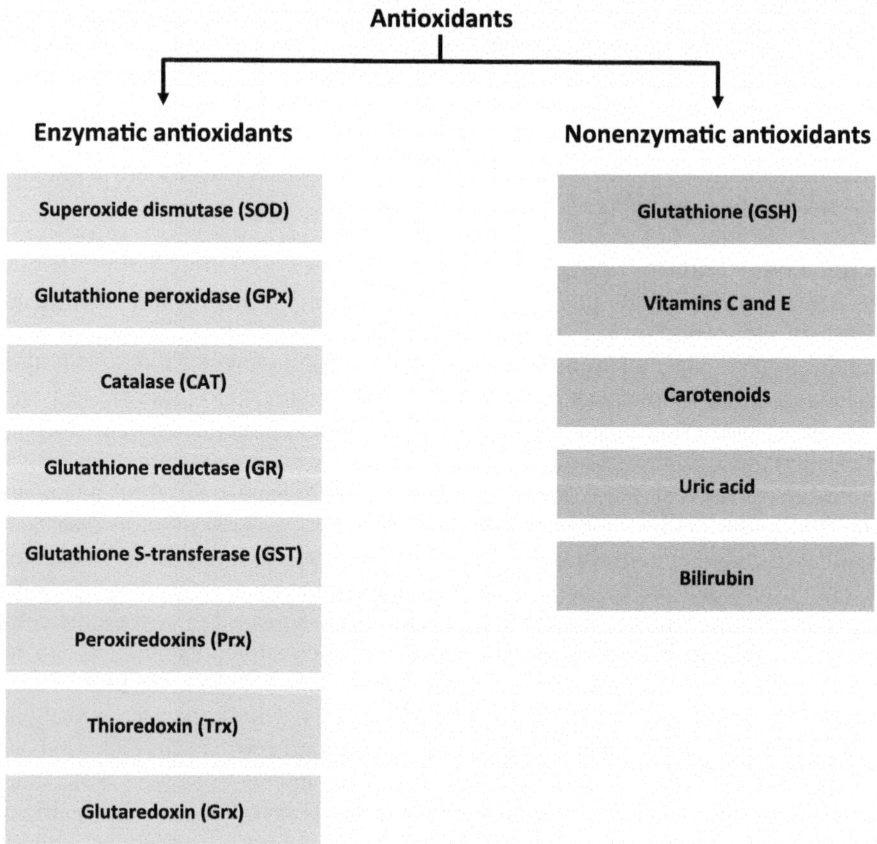

Fig. 23.2 Major enzymatic and nonenzymatic antioxidants

23.1.4.1 Enzymatic Antioxidants

Superoxide Dismutases

Superoxide dismutase (SOD) enzymes consist of three isoforms in mammals: the cytoplasmic Cu/ZnSOD (SOD1), the mitochondrial MnSOD (SOD2), and the extracellular Cu/ZnSOD (SOD3), all of which require catalytic metals (Cu or Mn) for their activity [61]. They are considered the major antioxidant defences against $O_2^{\cdot -}$, being responsible for its dismutation to H_2O_2 and molecular oxygen, which limits the potentially harmful effects of this radical species [61].

Catalase and Glutathione Peroxidase

H_2O_2 produced by the action of SODs or oxidases, such as XO, can be further decomposed to water and oxygen. This is achieved primarily by catalase in the peroxisomes and by glutathione peroxidase (GPx) enzymes in the cytosol and mitochondria. Catalase exists as a tetramer composed of 4 identical monomers, each of which contains a haem group at the active site. Degradation of H_2O_2 is accomplished via the conversion between 2 conformations of catalase-ferricatalase and compound I. GPx are selenium-containing enzymes whose activity is dependent on the amount of reduced glutathione (GSH) available [62]. Besides neutralizing H_2O_2, GPx also degrades lipid hydroperoxides to lipid alcohols. These reactions lead to the oxidation of GSH to oxidized glutathione (GSSG). Catalase and GPx are differentially required for the clearance of high-levels or low-levels of H_2O_2, respectively [63].

Other Enzymatic Defences

In addition to the antioxidant enzymatic systems mentioned above, cells also express other specialized enzymes with direct and/or indirect antioxidant functions. Glutathione reductase (GR) regenerates GSH from GSSG in the presence of NADPH. Glutathione-S-transferase (GST) catalyzes the conjugation of GSH with reactive electrophiles and detoxifies some carbonyl-, peroxide- and epoxide-containing metabolites produced within the cell in oxidative stress conditions. Peroxiredoxins (Prx) are selenium-independent enzymes that decompose H_2O_2, organic hydroperoxides and peroxynitrite, and thioredoxin (Trx) and glutaredoxin (Grx) systems include various enzymes that regulate the thiol-disulphide state of proteins and modulate their structure and activity [10].

23.1.4.2 Nonenzymatic Antioxidants

Nonenzymatic antioxidants, such as GSH, ascorbic acid (vitamin C) and α-tocopherol (vitamin E) play a key role in protecting the cells from oxidative damage and are considered as the second line of defence against active radicals. GSH is termed the master antioxidant given its electron-donating capacity that renders GSH a potent antioxidant *per se*, besides acting as an important cofactor for GPx and other enzymes. Vitamins E and C are among the major dietary antioxidants. Vitamin E, found in lipoproteins, cell membranes and extracellular fluids, terminates lipid peroxidation processes and converts $O_2^{\bullet-}$ and $HO^{\bullet}$ to less reactive forms. Vitamin C is a water-soluble antioxidant that can directly scavenge ROS and lipid hydroperoxides. Carotenoids, such as β-carotene, are lipid soluble antioxidants that function as efficient quenchers of $^1O_2$ but may also scavenge $ROO^{\bullet}$ radicals. Uric acid is a highly abundant aqueous antioxidant, considered to be the main contributor for the antioxidant capacity in the plasma. It has the ability to quench $HO^{\bullet}$ and $ONOO^-$ and may prevent lipid peroxidation, but may also exert prooxidant effects once inside the cells. Bilirubin, the end-product of haem catabolism, has chainbreaking antioxidant properties. Plasma albumin, the predominant plasma protein, is also an antioxidant and can scavenge MPO-derived chlorinated reactive species and $ROO^{\bullet}$ radicals [10].

23.1.5 The Dual "Faces" of ROS

It has long been accepted that elevated ROS levels can cause damage to macromolecules and have been implicated in a vast array of pathologies. More recently, it has become apparent that ROS also serve as signalling molecules to regulate biological and physiological processes and that dysregulated ROS signalling may contribute to a host of human diseases [3]. Downstream of ROS production, several signalling pathways are activated, including protein kinases [mitogen activated protein kinases (MAPKs), protein tyrosine kinases (PTKs), protein kinases B and C] and transcription factors (NF-κB, Nrf2) [64]. Nevertheless, our understanding of the signalling "face" of ROS is still in its infancy, as ROS can often act upstream and/or downstream within a given pathway and sometimes in opposing ways (i.e. inhibitory or stimulatory).

23.2 Evidence for Redox Changes in Experimental and Human Hypertension

23.2.1 Links Between Oxidative Stress and Hypertension

Arterial hypertension, currently defined as systolic blood pressure values ≥140 mmHg and/or diastolic blood pressure ≥90 mm Hg, is a multifactorial, complex disorder, involving many organ systems and constitutes a major risk factor for cardiovascular disease and premature mortality throughout the world [65]. Major pathophysiological mechanisms implicated in the development of hypertension include central nervous system dysregulation and increased activity of sympathetic nervous system, altered renal function with increased renal sodium and water retention and increased peripheral vascular resistance (Fig. 23.3) [51, 66]. The renin-angiotensin-aldosterone system (RAAS) also plays a central role in the regulation of arterial pressure by renal and extrarenal mechanisms (e.g. regulation of sodium homeostasis, autopotentiation of vasoconstrictor responses, vascular hypertrophy, regulation of sympathetic output, facilitation of sympathetic neurotransmitter release, promotion of oxidative stress and inflammation), being intimately involved in hypertension pathophysiology [67–71].

Oxidative stress has emerged as a unifying hypothesis for explaining these diverse mechanisms. Evidence gathered over the last two decades in both experimental models and humans suggests that hypertension arises from increased production of ROS and/or reduced antioxidant capacity in the cardiovascular, renal and central nervous systems [21, 42, 51].

By using animal models of genetic and drug-induced hypertension, we and others have demonstrated increased ROS levels and prooxidant activity, altered antioxidant defences and increased ROS-mediated damage, both at peripheral and central sites of cardiovascular regulation [72–78]. These studies have also underlined the

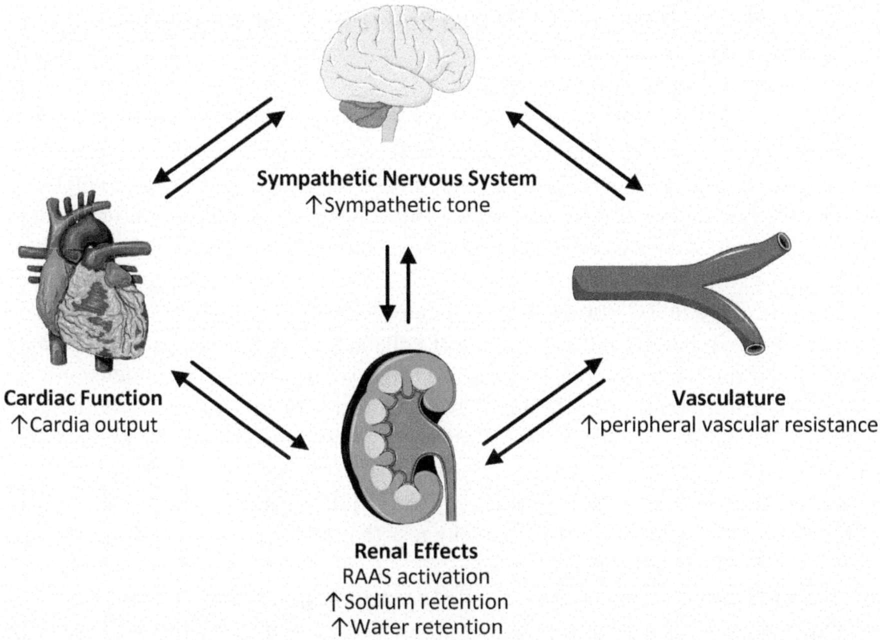

Fig. 23.3 Organs and mechanisms involved in the development and maintenance of arterial hypertension

importance of the kidney in the pathogenesis of hypertension and identified the renal medulla as a major target for angiotensin II-induced redox dysfunction in hypertension [72, 73]. Similarly to what happens in animals, there is also evidence of redox dysfunction in human hypertensive patients, although the association is less consistent and results vary depending on the biological marker of oxidative stress being investigated. The release of $O_2^{\cdot-}$ from peripheral polymorphonuclear leucocytes is increased in hypertensive patients in comparison with normotensive subjects [79]. Plasma H_2O_2 production is augmented in hypertensive patients and, among normotensive subjects, those with a family history of hypertension also exhibit a higher H_2O_2 production [80]. Increased levels of byproducts of protein, lipid and DNA oxidative damage, such as malondialdehyde, 8-isoprostanes, 8-oxo-2′-deoxyguanosine, oxidized low density lipoproteins, carbonyl groups and nitrotyrosine, have also been found in biofluids (i.e., plasma, serum and urine) and blood cells of hypertensive patients [81–83]. Furthermore, both enzymatic and nonenzymatic antioxidant defences appear to be reduced in human hypertension [81, 82, 84, 85]. Despite the vast number of studies reporting a close association between oxidative stress and hypertension, there is still an ongoing debate whether oxidative stress is a cause or a consequence of the disorder [86–88].

23.2.2 Oxidative Stress as Either a Cause or a Consequence of Hypertension

A large body of literature supports the hypothesis that oxidative stress is a major driver of arterial hypertension. In rats, the induction of oxidative stress through the administration of a common environmental heavy metal pollutant (lead), a glutathione synthesis inhibitor (buthionine sulfoximine-BSO) or a SOD inhibitor (sodium diethyldithiocarbamate-DETC), as well as the intrarenal or intrathecal infusion of H_2O_2, lead to increases in blood pressure [72, 89, 90]. The genetic manipulation of enzymes involved in ROS production or metabolism also modifies blood pressure in mice [91–93]. In addition, the exposure of cells and tissues to exogenous oxidants recapitulates molecular events implicated in the pathogenesis of hypertension [72, 94]. Also of importance are the facts that experimental hypertension can be prevented or attenuated by the administration of some antioxidants or inhibitors of ROS production [95–98] and that redox dysregulation, both at systemic and tissue level, precedes the rise in blood pressure [99, 100]. Collectively, these observations in preclinical models of hypertension suggest that oxidative stress plays a causal role in the development of hypertension.

Nevertheless, other authors have failed to demonstrate a direct involvement of oxidative stress in the pathogenesis of hypertension since the administration of antioxidants or inhibitors of ROS generation did not prevent or attenuate experimental hypertension [10]. Indeed, if oxidative stress is causally related to human hypertension, then antioxidants should be able to reduce blood pressure and oxidative damage. However, the majority of clinical trials did not find any blood pressure-lowering effects of antioxidants. One of the largest studies observed no improvement in blood pressure after a 5-year treatment with a combination of vitamin C, vitamin E, and β-carotene versus placebo in subjects thought to be at high risk of cardiovascular disease [101]. Likewise, a recent study found no beneficial effects against major cardiovascular events, including hypertension, after more than a decade of treatment with a multivitamin supplement versus placebo in a population of US male physicians [102]. Furthermore, a meta-analysis failed to reveal a clear benefit after antioxidant supplementation in cardiovascular mortality [103].

There is also evidence that lowering blood pressure *per se* leads to a reduction in oxidative stress and improvement in vascular function [10, 88]. Several antihypertensive drugs with different mechanisms of action, such as angiotensin-converting enzyme inhibitors, angiotensin II receptor antagonists, beta-blockers and calcium channel blockers, have been shown to attenuate oxidative stress markers in experimental and human hypertension [104, 105]. In light of these observations, some authors suggest that oxidative stress may be rather a consequence than a cause of hypertension. However, some of these antihypertensive agents have direct antioxidant properties and others block the RAAS, whose downstream effects are known to be mediated by ROS [10].

23.2.3 Pharmacological Interventions Aimed to Reduce Blood Pressure with Antioxidant Therapies

The rationale for reducing oxidative stress as a therapeutic strategy against hypertension stems from population-based observational studies showing an inverse correlation between plasma antioxidant concentrations, obtained by dietary intake, with blood pressure and cardiovascular risk factors [106]. However, in contrast with preclinical data, no significant improvement in blood pressure has been observed in the vast majority of studies after treatment with single or combination antioxidant therapy in subjects thought to be at high risk of cardiovascular disease (as discussed above in Sect. 23.2.2). A number of potential explanations for the failure of antioxidant supplementation in the chronic suppression of cardiovascular disease in humans have been put forward, including errors in trial design, choice of antioxidants, patient cohorts included in trials, the pathophysiological complexity of ROS/RNS signalling in humans with comorbidities, among others [107]. In what concerns the antioxidants, it is possible that the dose administered and duration of clinical trials were insufficient or agents examined were ineffective and nonspecific. Most antioxidant therapies that have been tested were not chosen because they were proved to be the best antioxidants, but rather because of their easy availability. It is also conceivable that the antioxidants administered failed to target the source of free radicals, particularly if ROS are generated in intracellular organelles and compartments, due to relatively poor uptake of antioxidants by target organs or the interference with other substances that, in some cases, reduce the antihypertensive effects. It is critical to remember that the lack of benefits seen in clinical trials to date does not rule out the essential role of oxidative stress in hypertension and other cardiovascular disorders. Rather, these results highlight the importance of evaluating optimal antioxidant therapies, the ideal cohort of patients to study, and the appropriate trial duration for the future improvement of antioxidant therapy.

23.3 Oxidative Stress in Heart Failure

23.3.1 The Heart, Metabolic Demand and ROS Production

The mammalian heart is the organ with the highest metabolic demand, consuming a large amount of cellular ATP to maintain the contraction-relaxation cycle. Under physiological conditions, this tremendous energy requirement is fulfilled by the high mitochondrial content of cardiomyocytes [19, 108–110]. Mitochondria ensure the production of more ATP through oxidative phosphorylation, whereby the mitochondrial ETC generates a proton gradient that drives ATP synthesis by ATP synthase. Since this process is sustained by O_2, which functions as the final electron acceptor in the ETC, it is not surprising that the heart needs a continuous, as well as adjustable, high supply of O_2 to maintain its function and viability [19, 108, 109]. Normally, most of the O_2 consumed in oxidative phosphorylation is reduced to water. However, electron leakage from the ETC also occurs, thus resulting in the formation of a small amount of ROS, namely $O_2^{\cdot-}$ and H_2O_2, which can be detoxified by endogenous antioxidant enzymes [19, 108–110]. There are several other ROS-producing enzymes in the heart, namely NOXs, XO, uncoupled NOS, MAOs and MPO, that are present in several cell types such as cardiomyocytes, endothelial cells, vascular smooth muscle cells, fibroblasts, neutrophils, monocytes and macrophages [9, 19, 109–111].

Although large amounts of ROS are markedly detrimental, there is evidence that low-to-moderate ROS concentrations in the heart are involved in physiological processes and beneficial adaptive signalling in response to acute changes in workload or brief ischemic episodes [108, 110, 112]. For example, ROS contribute to cardiomyogenesis of embryonic stem cells and proliferation of neonatal cardiac cells [113, 114]. It has also been reported that H_2O_2 derived from dismutation of $O_2^{\cdot-}$ generated by myocardial ETC is involved in coronary dilation, thus linking myocardial oxygen consumption to coronary blood flow [115, 116]. In addition, an increase in mitochondrial-derived ROS appears to mediate the acute inotropic response of cardiomyocytes to β-adrenergic receptor stimulation, being part of the homeostatic physiological signalling in the heart [117]. Importantly, mitochondrial and NOX-derived ROS seem to participate in the protective adaptive responses to moderate hypoxia, through the redox regulation of cardiomyocyte hypoxia-inducible factor activation, and in myocardial ischemic preconditioning, a protective phenomenon triggered by transient ischemic episodes and responsible for enhanced heart resistance to prolonged ischemia-reperfusion scenarios [108, 112, 118, 119].

23.3.2 Role of Oxidative Stress in the Pathophysiology of Heart Failure

HF is a complex clinical syndrome derived from structural and/or functional abnormalities in the heart, leading to impaired ventricular filling or ejection [9, 120]. Cardiac dysfunction triggers compensatory haemodynamic and neurohormonal responses attempting to maintain proper tissue perfusion, but these ultimately become maladaptive and deleterious [121]. Typical symptoms of this syndrome include shortness of breath, ankle swelling, fatigue, tiredness and reduced tolerance to exercise [120]. HF is usually a chronic, progressive and terminal illness, associated to poor quality of life for the patient due to the increase in symptoms frequency, severity and distress along disease course. Its prevalence in developed countries ranges from 1–2% in adults but can increase to values equal or higher than 10% in people with more than 70 years old, posing an enormous economic burden on healthcare systems. Of note, HF is the most frequent diagnosis responsible for hospitalization among patients aged 65 years or older [122, 123]. HF aetiologies include those related with diseased myocardium (e.g. ischemic heart disease; toxic damage due to alcoholism, drugs of abuse, medications such as cytostatics, heavy metals or radiation; immune-mediated and inflammatory damage caused by infections or auto-immune conditions; metabolic derangements such as thyroid diseases and pheocromocytoma; infiltration related with malignancy or other diseases such as amyloidosis; genetic disorders), with abnormal loading conditions (e.g. arterial hypertension; valve and myocardial structural defects; pericardial and endomyocardial pathologies; high output states such as severe anaemia; volume overload caused by renal failure) or with arrhythmias [120].

Despite therapeutic advances, chronic HF often decompensates, leading to the rapid aggravation of symptoms and/or signs of HF and thus requiring hospitalization [120, 124]. The term acute HF frequently refers to this state of acute decompensation of chronic HF but may also represent new-onset HF ("*de novo*" HF) resulting, for example, from acute myocardial dysfunction due to ischemic, inflammatory or toxic insults, acute valve insufficiency or cardiac tamponade (a condition characterized by heart compression and dysfunction as a consequence of pericardial accumulation of fluid, pus, blood, clots or gas due to blunt or penetrating trauma, accidental cardiac perforation following catheterization, infection, cancer and aortic aneurysm rupture) [120, 125].

As mentioned previously, low-to-moderate amounts of ROS contribute to physiological and beneficial adaptive responses in the heart. However, when prooxidant and antioxidant systems are imbalanced, leading to a prevailing prooxidant status, macromolecular damage and harmful signalling may occur and contribute to the genesis and progression of HF [9, 19, 112]. In the heart, there are many processes or targets that can be adversely affected by ROS (Fig. 23.4), namely cardiac contractility, myocardial remodelling, cardiomyocyte apoptosis, mitochondria and

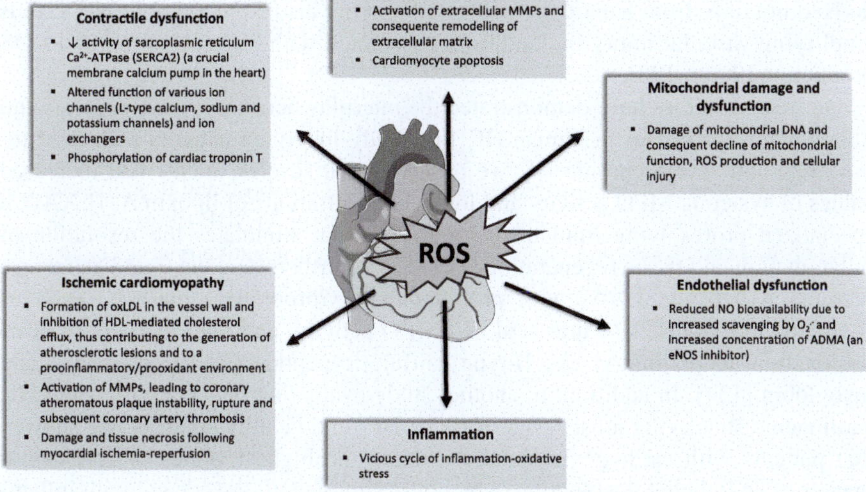

Fig. 23.4 Adverse effects of ROS in the heart. *ADMA* asymmetric dimethylarginine, *eNOS* endothelial nitric oxide synthase, *DNA* deoxyribonucleic acid, *HDL* high-density lipoprotein, *MMPs* matrix metalloproteinases, *oxLDL* oxidized low-density lipoprotein, *ROS* reactive oxygen species

endothelium [9, 19, 109, 112, 126]. ROS also contribute to ischemic cardiomyopathy by promoting the formation of oxidized low-density lipoprotein (oxLDL), which plays a central role in the pathogenesis of atherosclerosis [9, 109]. Furthermore, the redox sensitive alteration of apolipoprotein A-I, the major protein constituent in high-density lipoproteins (HDL), inhibits the efflux of cholesterol, contributing to atherosclerotic lesions formation and to a prooxidant and proinflammatory environment [127]. The activation of matrix metalloproteinases by ROS is also involved in coronary atheromatous plaque instability, rupture and subsequent coronary artery thrombosis [9, 109]. Of note, after a significant myocardial ischemic insult, the restoration of oxygen supply during the reperfusion phase is responsible for the generation of high amounts of ROS, which contribute to extensive damage and tissue necrosis in the heart [9, 109, 127].

Inflammation plays a central role in the development and progression of chronic HF, regardless of aetiologies [128, 129]. It is also considered an important precipitator and prognostic factor in acute HF [130]. Oxidative stress and inflammation are closely interconnected, contributing to the pathophysiology of HF [9, 109, 131]. Several transcription factors that regulate the expression of proinflammatory cytokines are activated under oxidative stress conditions [9, 109, 131]. In turn, proinflammatory cytokines induce the generation of ROS, thus creating a potential vicious cycle of oxidation and inflammation [9, 131, 132]. Moreover, the production of large amounts of ROS is a feature of activated inflammatory cells, and MPO, a major effector enzyme of neutrophils that is released into the extracellular space during leukocyte activation, also functions as a link between oxidative stress and inflammation [9, 34, 36]. This enzyme uses H_2O_2 as a substrate to produce HOCl, which is a potent prooxidant and proinflammatory molecule. Importantly, MPO has the ability to bind and infiltrate in the vascular wall and to utilize H_2O_2 derived not only from leukocyte oxidative burst but also from vascular NOX, thus amplifying vascular injury in conditions associated with higher than normal ROS production [9, 36, 133].

Our recent studies have demonstrated the interplay between oxidative stress and inflammatory processes in human HF. In a study involving patients with mild-to-moderate and severe chronic HF, we observed that severe patients had increased values of systemic MPO activity and lower concentrations of lipoxin A_4 (LXA_4), a specialized proresolving lipid mediator (SPM) that stimulates the resolution of inflammation and tissue regeneration [121, 134]. Furthermore, we found an inverse correlation between LXA_4 with proinflammatory/prooxidant markers, such as C-reactive protein (CRP), uric acid and MPO activity, and with markers of heart dysfunction and/or injury, like B-type natriuretic peptide (BNP), troponin I and myoglobin [134]. In addition, in another study evaluating patients with acute HF, cardiogenic shock (the most severe form of HF) and healthy controls, we showed that patients with cardiogenic shock exhibited the highest values of endocan, a marker of endothelial dysfunction, which was significantly associated with inflammatory status [135, 136]. Among the controls and patients evaluated, serum nitrotyrosine, a marker of oxidative/nitrosative stress, was significantly and positively correlated with CRP and high-sensitivity-troponin I, which are markers of inflammation and myocardial damage, respectively [135]. We also observed that resolvin E1 (RvE1), another mediator of inflammation resolution, increased in line with acute HF severity and was significantly associated with inflammatory/oxidant status and endothelial dysfunction [136].

Noteworthy, LXs and Rvs, besides possessing proresolving and anti-inflammatory properties, have also been shown to exert several protective effects on redox status that may be particularly relevant in the context of HF. These include the blockade of NOX enzymes in endothelial cells and macrophages, inhibition of ROS generation by leukocytes and vascular smooth muscle cells, blockade of angiotensin II-, thrombin- or tumor necrosis factor-α (TNF-α)-induced ROS production in endothelial cells, increased SOD activity and reduced malondialdehyde (MDA) content in the heart, induction of haem oxygenase-1 in endothelial cells and cardiomyocytes and upregulation of nuclear factor erythroid-2 related factor 2 (Nrf2) in cardiomyocytes [121]. Thus, strategies targeting inflammation or promoting its resolution will likely attenuate oxidative stress, and vice-versa, in patients with HF.

23.3.3 Biomarkers of Oxidative Stress in Human Heart Failure

Human HF was recently divided into 3 categories according to left ventricular ejection fraction (LVEF): reduced (HFrEF), preserved (HFpEF) or mid-range (HFmrEF) [120]. This definition only comprises the clinical manifestation of an underlying structural and/or functional cardiac abnormality resulting from a myriad of insults, of which the ischemic is the most prevalent in HFrEF. Thus, in clinical trials it is difficult to understand oxidative stress as a cause or consequence of the disease because HF prevails in older ages and it remains underdiagnosed and untreated [137, 138]. It is well established that ageing is associated with increased ROS accumulation, lipid peroxidation and mitophagy as well as atherosclerosis, diabetes and obesity, major risk factors for ischemic heart disease and HF. Despite the association between oxidative stress with clinical outcome in patients with coronary artery disease, no redox biomarker is currently in routine clinical use, in part because they are not specific for individual disease processes [139, 140]. The question remains whether plasma oxidation products reflect systemic or vascular redox state or other biological processes, as well as what is their value for independent stratification and therapeutic management, critical issues to consider them as "biomarkers". Nevertheless, and surmounting the difficulties associated with the short half-life, limited diffusion and requirement of invasive biopsies to quantify ROS in human tissues, indirect indexes of oxidative stress are gaining increasing acceptance among established biomarkers of HF [141]. These biomarkers can be grouped into three main categories, namely prooxidant enzymes, products of oxidized macromolecules and antioxidant defences.

23.3.3.1 Prooxidant Enzymes

Myeloperoxidase contributes to endothelial dysfunction and mediates dysregulation of vascular tone [10]. The plasma concentration of MPO is elevated in HF patients compared to controls and its systemic activity is also increased in severe chronic HF compared to mild to moderate HF [134, 142]. Treatment with the inodilator levosimendan seems to reduce the concentration of plasma MPO by decreasing its release from neutrophils in patients with acute decompensation of chronic HF [143]. MPO was selected among others as an incremental prognostic biomarker in a multimarker risk strategy of stratification for cardiovascular death or HF in patients with acute myocardial infarction [144] and it also seems to differentiate forms of acute HF with cardiorenal syndrome [145]. These results, along with its predictive value for cardiovascular morbidity and mortality observed in other relatively large prospective studies, its therapeutic implications and the feasibility of its commercial assays, make MPO one of the most promising redox biomarkers for clinical application [146, 147].

Although NOXs are important cardiovascular ROS sources, available data about the involvement of NOXs in human HF is scarce. One study has described increased NOX2 expression and raised NOX activity in myocardial tissue from human failing hearts compared with non-failing controls, but information is lacking regarding NOX2 association with prognosis or treatment, with the exception of a study reporting a downregulatory effect of mediterranean diet on soluble NOX2-derived peptide values in patients with atrial fibrillation [148]. Nevertheless, compelling evidence suggests that redox protective effects of RAAS inhibitors, which are part of HF pharmacological treatment, are due to the prevention of vascular and phagocytic NOX activation [24, 120, 137].

The ·NO-generating enzyme eNOS has some limitations as a redox biomarker in humans, not only for its localization (in the vessel wall and cardiomyocytes) but also because the complex regulation of the biosynthesis of its cofactor, BH_4, makes hard to estimate the ratio of reduced to oxidized forms (BH_4/BH_2) and consequently to calculate eNOS uncoupling, which is responsible for generating $O_2^{\cdot-}$ instead of NO. The administration of BH_4 does not improve vascular oxidative stress in patients with coronary disease [149] but indirect strategies like folates [150], statins [9] or polyphenols [151] could do so, thus reinforcing the interest of this pathway for future research in HF. Of note, in high-risk diabetic patients, the cardioprotection and reduction of risk of re-infarction and all-cause mortality afforded by metformin, which is no longer contra-indicated in HF, seems to be related, at least in part, with increased ·NO bioavailability [152]. Also, the superiority of ticagrelor vs. clopidogrel in reducing cardiovascular events can be explained by the higher ·NO concentrations triggered by ticagrelor, compared to clopidogrel, through an adenosine-mediated pathway that activates eNOS [153, 154].

23.3.3.2 Products of Oxidized Macromolecules

Lipid peroxidation results from ROS attack to polyunsaturated fatty acids (PUFA) in cell membranes. End-products of lipid peroxidation, including isoprostanes and MDA, affect membrane fluidity, inactivate receptors and enzymes attached to it, and even threaten cell viability. This lipid susceptibility to ROS has attracted considerable attention to the evaluation of lipid peroxides as biomarkers of oxidative stress.

Isoprostanes are produced by ROS-induced peroxidation of arachidonic acid and then released by phospholipases [155]. The most stable and thus most commonly quantified are F_2-isoprostanes, which can be assessed in tissues and biological fluids. In HF, isoprostane levels in plasma, urine and pericardial fluid correlate with disease severity and ventricular dilatation [156, 157]. Recent works are hypothesizing that they could be used in a precocious strategy to identify populations with sub-clinical increased cardiovascular risk. In addition, they could also be used to monitor the protective effect of diets (e.g. low-sodium diet), as well as dietary adequacy, in patients with HF [158, 159].

MDA, another product of lipid peroxidation, is routinely evaluated by the thiobarbituric acid-reactive substances (TBARS) assay. There is evidence of increased systemic and intraplatelet production of TBARS in patients with acute or chronic HF [160]. Furthermore, a reduction in TBARS levels was observed in HF patients after treatment with a beta-blocker, short-term inotropic support and vitamin C, but not with the addition of an angiotensin II receptor antagonist to angiotensin converting enzyme inhibitor therapy [161, 162]. MDA appears to contribute to the formation of OxLDL [163] which have been proposed to be an useful predictor of mortality in patients with CHF [164]. MDA or MDA-modified LDL are being evaluated in device studies in advanced HF to monitor oxidative stress in patients under device therapy (implantable cardioverter defibrillator, continuous-flow left ventricular assist device) [165, 166].

Oxidative posttranslational modifications of cellular proteins by means of tyrosine nitration, protein carbonylation, and S-glutathionylation, can accurately reflect oxidative stress in HF patients. One of the most emblematic examples of protein oxidation in HF is myocardial sarco/endoplasmic reticulum Ca^{2+}-ATPase (SERCA) inactivation by nitration, which may contribute to reduced contractility and progression of HF [167]. Furthermore, ceruloplasmin tyrosine nitration with consequent antioxidant reduced activity is associated with reduced survival in patients with HF [168]. Protein nitration by peroxynitrite and haem peroxidase can result in gain of function or inactivation of different proteins in plasma, vessel wall and myocardium that link nitrosative stress to cardiovascular disease and, for that reason, nitrotyrosine is also emerging as a good candidate for a marker of cardiovascular risk [169].

Protein carbonyls can result from oxidation of amino acid side chains, reaction with lipid peroxidation products and glycation/glycoxidation of Lys amino groups. They are very stable and represent a good mirror of protein oxidation Increased carbonyls were found in diaphragm biopsies from patients with end-stage HF, probably resulting from increased Nox2-derived ROS and imbalanced antioxidant enzymes [170]. The inodilator levosimendan prevented the increase in MDA, protein carbonyls and nitrotyrosine in hospitalized patients with worsening HF. These results point to a cardioprotective effect of this drug and thus its wider use in advanced CHF patients has been hypothesized [171]. Additionally, there is evidence that a polymorphism in angiotensin II type 1 receptor can predict the formation of carbonyls in HF patients, suggesting that angiotensin signalling contributes to oxidative stress in HF [172].

8-hydroxy-2′-deoxyguanosine (8-OHdG) results from oxidative DNA damage and its levels can be quantified in urine. In fact, 8-OHdG was demonstrated to be higher with increasing HF severity and correlated with left ventricular ejection fraction in patients with chronic HF [173]. It also seems to be a tool to evaluate beta-blocker responsiveness in chronic HF patients or even to diagnose subclinical left ventricular diastolic dysfunction in hypertensive patients [174, 175], but more data is needed and/or combination with other biomarkers in a multipanel strategy.

23.3.3.3 Antioxidant Defences

Antioxidant enzymes (e.g. catalase, GPx, SOD) can be measured in blood samples but their values are hard to interpret and these studies have low reproducibility or therapeutic/prognostic implications [176–178] On the other hand, there has been an enthusiastic exploration of non-enzymatic antioxidants, such as biopyrrins (oxidative metabolites of bilirubin) and albumin since urinary levels of biopyrrins have been shown to be associated with HF severity [179] and oxidative stress has been proposed as a cause for the development of hypoalbuminemia in ischemic HF [180]. Nevertheless, the disappointing results of studies evaluating the effects of antioxidant administration in HF patients, particularly the failure of vitamins C and E to improve prognosis and the deleterious effects observed in HOPE and HOPE TOO trials, restrained the enthusiasm in this area [181–183].

23.3.3.4 Other Oxidative Stress Markers

Uric acid is an end-product of purine metabolism in humans derived from XO that catalyses its conversion from hypoxanthine. Although it is one of the most abundant aqueous antioxidants in plasma, it can also exert prooxidant effects [10]. Uric acid is frequently accepted as a biomarker of HF [141] but remains a controversial issue because causality in its relationship with cardiovascular disease remains uncertain. Although affected by renal function and diuretic use, there is enough evidence demonstrating that it can work as an independent and simple, albeit nonspecific, predictor of excessive oxidative stress and of adverse prognosis in HF [184].

23.4 Concluding Remarks

A vast body of literature accumulated over the past decades has firmly implicated oxidative stress in the pathogenesis and progression of cardiovascular diseases, including hypertension and HF, as well as associated risk factors and comorbidities. Key molecular events in hypertension and HF, such as oxidative modification of lipids and proteins, endothelial cell activation and inflammation, are facilitated by oxidative stress. More recently, the role of redox signalling and specific molecular targets have also been appreciated. Despite the significant progress in understanding the pathophysiology of these conditions and the promising results in pre-clinical animal models, clinical trials of antioxidant approaches to prevent cardiovascular mortality and morbidity have been, so far, disappointing. Several hypotheses have been put forward, including the failure to appreciate the complexity of the effects of

ROS or inappropriate antioxidant selection or dosage, which warrants future research on new compounds with improved properties. Finally, more human data is required to provide clinical relevance and determine the potential for clinical translation. Nevertheless, several studies indicate that oxidative stress biomarkers may be useful for risk stratification and to monitor the protective effects of pharmacological treatment, diets or devices in human HF.

Acknowledgements We apologize to all colleagues whose work has not been discussed or cited owing to space limitations. TS is currently supported by FEDER funds via COMPETE (Portugal 2020) and by national funds through the Portuguese Foundation for Science and Technology (FCT) (project grant PTDC/MEC-CAR/32188/2017; SFRH/BPD/112005/2015). PG is funded by FEDER, Centro2020 Regional Operational Programme: CENTRO-01-0145-FEDER-000012-HealthyAging2020, COMPETE 2020-Operational Programme for Competitiveness & Internationalization; and FCT (strategic project UID/NEU/04539/2013; SFRH/BPD/111815/2015).

References

1. Dickinson BC, Chang CJ (2011) Chemistry and biology of reactive oxygen species in signaling or stress responses. Nat Chem Biol 7(8):504–511
2. Droge W (2002) Free radicals in the physiological control of cell function. Physiol Rev 82(1):47–95
3. Schieber M, Chandel NS (2014) ROS function in redox signaling and oxidative stress. Curr Biol 24(10):R453–R462
4. Alfadda AA, Sallam RM (2012) Reactive oxygen species in health and disease. J Biomed Biotechnol 2012:936486
5. Valko M et al (2007) Free radicals and antioxidants in normal physiological functions and human disease. Int J Biochem Cell Biol 39(1):44–84
6. Brewer AC et al (2013) Reductive stress linked to small HSPs, G6PD, and Nrf2 pathways in heart disease. Antioxid Redox Signal 18(9):1114–1127
7. Martinez MC, Andriantsitohaina R (2009) Reactive nitrogen species: molecular mechanisms and potential significance in health and disease. Antioxid Redox Signal 11(3):669–702
8. Turko IV, Murad F (2002) Protein nitration in cardiovascular diseases. Pharmacol Rev 54(4):619–634
9. Costa S et al (2016) Statins and oxidative stress in chronic heart failure. Rev Port Cardiol 35(1):41–57
10. Sousa T et al (2012) Lipid peroxidation and antioxidants in arterial hypertension. In: Catala A (ed) Lipid peroxidation. IntechOpen, Rijeka, pp 345–392
11. Chen YR, Zweier JL (2014) Cardiac mitochondria and reactive oxygen species generation. Circ Res 114(3):524–537
12. Murphy MP (2009) How mitochondria produce reactive oxygen species. Biochem J 417(1):1–13
13. Addabbo F, Montagnani M, Goligorsky MS (2009) Mitochondria and reactive oxygen species. Hypertension 53(6):885–892
14. Chance B, Sies H, Boveris A (1979) Hydroperoxide metabolism in mammalian organs. Physiol Rev 59(3):527–605
15. Nazarewicz RR et al (2013) Nox2 as a potential target of mitochondrial superoxide and its role in endothelial oxidative stress. Am J Physiol Heart Circ Physiol 305(8):H1131–H1140
16. Zhang A et al (2011) Relative contributions of mitochondria and NADPH oxidase to deoxycorticosterone acetate-salt hypertension in mice. Kidney Int 80(1):51–60

17. Dey S et al (2018) Mitochondrial ROS drive sudden cardiac death and chronic proteome remodeling in heart failure. Circ Res 123(3):356–371
18. Moris D et al (2017) The role of reactive oxygen species in myocardial redox signaling and regulation. Ann Transl Med 5(16):324
19. Tsutsui H, Kinugawa S, Matsushima S (2011) Oxidative stress and heart failure. Am J Physiol Heart Circ Physiol 301(6):H2181–H2190
20. Bedard K, Krause KH (2007) The NOX family of ROS-generating NADPH oxidases: physiology and pathophysiology. Physiol Rev 87(1):245–313
21. Datla SR, Griendling KK (2010) Reactive oxygen species, NADPH oxidases, and hypertension. Hypertension 56(3):325–330
22. Nistala R, Whaley-Connell A, Sowers JR (2008) Redox control of renal function and hypertension. Antioxid Redox Signal 10(12):2047–2089
23. Brandes RP, Takac I, Schroder K (2011) No superoxide--no stress?: Nox4, the good NADPH oxidase! Arterioscler Thromb Vasc Biol 31(6):1255–1257
24. Brown DI, Griendling KK (2009) Nox proteins in signal transduction. Free Radic Biol Med 47(9):1239–1253
25. Sedeek M et al (2013) NADPH oxidases, reactive oxygen species, and the kidney: friend and foe. J Am Soc Nephrol 24(10):1512–1518
26. Montezano AC et al (2015) Redox signaling, Nox5 and vascular remodeling in hypertension. Curr Opin Nephrol Hypertens 24(5):425–433
27. Yu P et al (2014) Unique role of NADPH oxidase 5 in oxidative stress in human renal proximal tubule cells. Redox Biol 2:570–579
28. Manea A et al (2015) Human monocytes and macrophages express NADPH oxidase 5; a potential source of reactive oxygen species in atherosclerosis. Biochem Biophys Res Commun 461(1):172–179
29. Munzel T et al (2017) Impact of oxidative stress on the heart and vasculature: part 2 of a 3-part series. J Am Coll Cardiol 70(2):212–229
30. Zhang M et al (2013) NADPH oxidases in heart failure: poachers or gamekeepers? Antioxid Redox Signal 18(9):1024–1041
31. Matsushima S et al (2013) Broad suppression of NADPH oxidase activity exacerbates ischemia/reperfusion injury through inadvertent downregulation of hypoxia-inducible factor-1alpha and upregulation of peroxisome proliferator-activated receptor-alpha. Circ Res 112(8):1135–1149
32. Davies MJ (2011) Myeloperoxidase-derived oxidation: mechanisms of biological damage and its prevention. J Clin Biochem Nutr 48(1):8–19
33. Liu YC et al (2013) Genetic polymorphisms of myeloperoxidase and their effect on hypertension. Blood Press 22(5):282–289
34. Anatoliotakis N et al (2013) Myeloperoxidase: expressing inflammation and oxidative stress in cardiovascular disease. Curr Top Med Chem 13(2):115–138
35. Baldus S et al (2006) Heparins increase endothelial nitric oxide bioavailability by liberating vessel-immobilized myeloperoxidase. Circulation 113(15):1871–1878
36. Nussbaum C et al (2013) Myeloperoxidase: a leukocyte-derived protagonist of inflammation and cardiovascular disease. Antioxid Redox Signal 18(6):692–713
37. Tang WH et al (2007) Prognostic value and echocardiographic determinants of plasma myeloperoxidase levels in chronic heart failure. J Am Coll Cardiol 49(24):2364–2370
38. Forstermann U, Sessa WC (2012) Nitric oxide synthases: regulation and function. Eur Heart J 33(7):829–837
39. Gewaltig MT, Kojda G (2002) Vasoprotection by nitric oxide: mechanisms and therapeutic potential. Cardiovasc Res 55(2):250–260
40. Horita S et al (2014) Regulatory roles of nitric oxide and angiotensin II on renal tubular transport. World J Nephrol 3(4):295–301
41. Massion PB et al (2003) Nitric oxide and cardiac function: ten years after, and continuing. Circ Res 93(5):388–398

42. Briones AM, Touyz RM (2010) Oxidative stress and hypertension: current concepts. Curr Hypertens Rep 12(2):135–142
43. Li H et al (2006) Reversal of endothelial nitric oxide synthase uncoupling and up-regulation of endothelial nitric oxide synthase expression lowers blood pressure in hypertensive rats. J Am Coll Cardiol 47(12):2536–2544
44. Yamamoto E et al (2015) The pivotal role of eNOS uncoupling in vascular endothelial dysfunction in patients with heart failure with preserved ejection fraction. Int J Cardiol 190:335–337
45. Moens AL et al (2008) Reversal of cardiac hypertrophy and fibrosis from pressure overload by tetrahydrobiopterin: efficacy of recoupling nitric oxide synthase as a therapeutic strategy. Circulation 117(20):2626–2636
46. Berry CE, Hare JM (2004) Xanthine oxidoreductase and cardiovascular disease: molecular mechanisms and pathophysiological implications. J Physiol 555(Pt 3):589–606
47. Kuzkaya N et al (2005) Interactions of peroxynitrite with uric acid in the presence of ascorbate and thiols: implications for uncoupling endothelial nitric oxide synthase. Biochem Pharmacol 70(3):343–354
48. Hooper DC et al (1998) Uric acid, a natural scavenger of peroxynitrite, in experimental allergic encephalomyelitis and multiple sclerosis. Proc Natl Acad Sci U S A 95(2):675–680
49. Johnson RJ et al (2003) Is there a pathogenetic role for uric acid in hypertension and cardiovascular and renal disease? Hypertension 41(6):1183–1190
50. Niskanen LK et al (2004) Uric acid level as a risk factor for cardiovascular and all-cause mortality in middle-aged men: a prospective cohort study. Arch Intern Med 164(14):1546–1551
51. Loperena R, Harrison DG (2017) Oxidative stress and hypertensive diseases. Med Clin North Am 101(1):169–193
52. Feig DI, Soletsky B, Johnson RJ (2008) Effect of allopurinol on blood pressure of adolescents with newly diagnosed essential hypertension: a randomized trial. JAMA 300(8):924–932
53. MacIsaac RL et al (2016) Allopurinol and cardiovascular outcomes in adults with hypertension. Hypertension 67(3):535–540
54. Guzik TJ et al (2006) Coronary artery superoxide production and NOX isoform expression in human coronary artery disease. Arterioscler Thromb Vasc Biol 26(2):333–339
55. Baldus S et al (2006) Inhibition of xanthine oxidase improves myocardial contractility in patients with ischemic cardiomyopathy. Free Radic Biol Med 41(8):1282–1288
56. Givertz MM et al (2015) Effects of xanthine oxidase inhibition in hyperuricemic heart failure patients: the xanthine oxidase inhibition for hyperuricemic heart failure patients (EXACT-HF) study. Circulation 131(20):1763–1771
57. Hare JM et al (2008) Impact of oxypurinol in patients with symptomatic heart failure. Results of the OPT-CHF study. J Am Coll Cardiol 51(24):2301–2309
58. Casas AI et al (2015) Reactive oxygen-related diseases: therapeutic targets and emerging clinical indications. Antioxid Redox Signal 23(14):1171–1185
59. Deshwal S et al (2017) Emerging role of monoamine oxidase as a therapeutic target for cardiovascular disease. Curr Opin Pharmacol 33:64–69
60. Kaludercic N et al (2011) Monoamine oxidases (MAO) in the pathogenesis of heart failure and ischemia/reperfusion injury. Biochim Biophys Acta 1813(7):1323–1332
61. Fukai T, Ushio-Fukai M (2011) Superoxide dismutases: role in redox signaling, vascular function, and diseases. Antioxid Redox Signal 15(6):1583–1606
62. Lubos E, Loscalzo J, Handy DE (2011) Glutathione peroxidase-1 in health and disease: from molecular mechanisms to therapeutic opportunities. Antioxid Redox Signal 15(7):1957–1997
63. Wassmann S, Wassmann K, Nickenig G (2004) Modulation of oxidant and antioxidant enzyme expression and function in vascular cells. Hypertension 44(4):381–386
64. Morgan MJ, Liu ZG (2011) Crosstalk of reactive oxygen species and NF-kappaB signaling. Cell Res 21(1):103–115
65. Williams B et al (2018) ESC/ESH guidelines for the management of arterial hypertension. Eur Heart J 39(33):3021–3104

66. Coffman TM (2011) Under pressure: the search for the essential mechanisms of hypertension. Nat Med 17(11):1402–1409
67. Davisson RL, Zimmerman MC (2010) Angiotensin II, oxidant signaling, and hypertension: down to a T? Hypertension 55(2):228–230
68. Kobori H et al (2007) The intrarenal renin-angiotensin system: from physiology to the pathobiology of hypertension and kidney disease. Pharmacol Rev 59(3):251–287
69. Reckelhoff JF, Romero JC (2003) Role of oxidative stress in angiotensin-induced hypertension. Am J Phys Regul Integr Comp Phys 284(4):R893–R912
70. Weir MR, Dzau VJ (1999) The renin-angiotensin-aldosterone system: a specific target for hypertension management. Am J Hypertens 12(12 Pt 3):205S–213S
71. Zimmerman MC et al (2004) Hypertension caused by angiotensin II infusion involves increased superoxide production in the central nervous system. Circ Res 95(2):210–216
72. Makino A et al (2003) Increased renal medullary H2O2 leads to hypertension. Hypertension 42(1):25–30
73. Sousa T et al (2012) Role of H(2)O(2) in hypertension, renin-angiotensin system activation and renal medullary disfunction caused by angiotensin II. Br J Pharmacol 166(8):2386–2401
74. Gomes P et al (2009) Aging increases oxidative stress and renal expression of oxidant and antioxidant enzymes that are associated with an increased trend in systolic blood pressure. Oxidative Med Cell Longev 2(3):138–145
75. Simao S et al (2011) Age-related changes in renal expression of oxidant and antioxidant enzymes and oxidative stress markers in male SHR and WKY rats. Exp Gerontol 46(6):468–474
76. Gomes P et al (2013) Loss of oxidative stress tolerance in hypertension is linked to reduced catalase activity and increased c-Jun NH2-terminal kinase activation. Free Radic Biol Med 56:112–122
77. Ulker S et al (2003) Impaired activities of antioxidant enzymes elicit endothelial dysfunction in spontaneous hypertensive rats despite enhanced vascular nitric oxide generation. Cardiovasc Res 59(2):488–500
78. Chan SH et al (2009) Oxidative impairment of mitochondrial electron transport chain complexes in rostral ventrolateral medulla contributes to neurogenic hypertension. Hypertension 53(2):217–227
79. Kristal B et al (1998) Participation of peripheral polymorphonuclear leukocytes in the oxidative stress and inflammation in patients with essential hypertension. Am J Hypertens 11(8 Pt 1):921–928
80. Lacy F et al (2000) Plasma hydrogen peroxide production in human essential hypertension: role of heredity, gender, and ethnicity. Hypertension 36(5):878–884
81. Zhou L et al (2006) Reduction in extracellular superoxide dismutase activity in African-American patients with hypertension. Free Radic Biol Med 41(9):1384–1391
82. Redon J et al (2003) Antioxidant activities and oxidative stress byproducts in human hypertension. Hypertension 41(5):1096–1101
83. Kedziora-Kornatowska K et al (2004) The markers of oxidative stress and activity of the antioxidant system in the blood of elderly patients with essential arterial hypertension. Cell Mol Biol Lett 9(4A):635–641
84. Wen Y et al (1996) Lipid peroxidation and antioxidant vitamins C and E in hypertensive patients. Ir J Med Sci 165(3):210–212
85. Pedro-Botet J et al (2000) Decreased endogenous antioxidant enzymatic status in essential hypertension. J Hum Hypertens 14(6):343–345
86. Ward NC, Croft KD (2006) Hypertension and oxidative stress. Clin Exp Pharmacol Physiol 33(9):872–876
87. Ceriello A (2008) Possible role of oxidative stress in the pathogenesis of hypertension. Diabetes Care 31(Suppl 2):S181–S184
88. Grossman E (2008) Does increased oxidative stress cause hypertension? Diabetes Care 31(Suppl 2):S185–S189

89. Lin HH et al (2003) Hydrogen peroxide increases the activity of rat sympathetic preganglionic neurons in vivo and in vitro. Neuroscience 121(3):641–647
90. Vaziri ND et al (2000) Induction of oxidative stress by glutathione depletion causes severe hypertension in normal rats. Hypertension 36(1):142–146
91. Dikalova A et al (2005) Nox1 overexpression potentiates angiotensin II-induced hypertension and vascular smooth muscle hypertrophy in transgenic mice. Circulation 112(17):2668–2676
92. Godin N et al (2010) Catalase overexpression prevents hypertension and tubular apoptosis in angiotensinogen transgenic mice. Kidney Int 77(12):1086–1097
93. Gavazzi G et al (2006) Decreased blood pressure in NOX1-deficient mice. FEBS Lett 580(2):497–504
94. Simao S et al (2011) H2 O2 stimulates Cl- /HCO 3- exchanger activity through oxidation of thiol groups in immortalized SHR renal proximal tubular epithelial cells. J Cell Biochem 112(12):3660–3665
95. Baumer AT et al (2007) The NAD(P)H oxidase inhibitor apocynin improves endothelial NO/superoxide balance and lowers effectively blood pressure in spontaneously hypertensive rats: comparison to calcium channel blockade. Clin Exp Hypertens 29(5):287–299
96. Beswick RA et al (2001) NADH/NADPH oxidase and enhanced superoxide production in the mineralocorticoid hypertensive rat. Hypertension 38(5):1107–1111
97. Sousa T et al (2008) Role of superoxide and hydrogen peroxide in hypertension induced by an antagonist of adenosine receptors. Eur J Pharmacol 588(2–3):267–276
98. Zhang Y et al (2004) The antioxidant tempol prevents and partially reverses dexamethasone-induced hypertension in the rat. Am J Hypertens 17(3):260–265
99. Nabha L et al (2005) Vascular oxidative stress precedes high blood pressure in spontaneously hypertensive rats. Clin Exp Hypertens 27(1):71–82
100. Wilcox CS (2005) Oxidative stress and nitric oxide deficiency in the kidney: a critical link to hypertension? Am J Phys Regul Integr Comp Phys 289(4):R913–R935
101. Heart Protection Study Collaborative, G (2002) MRC/BHF heart protection study of antioxidant vitamin supplementation in 20,536 high-risk individuals: a randomised placebo-controlled trial. Lancet 360(9326):23–33
102. Sesso HD et al (2012) Multivitamins in the prevention of cardiovascular disease in men: the physicians' health study II randomized controlled trial. JAMA 308(17):1751–1760
103. Vivekananthan DP et al (2003) Use of antioxidant vitamins for the prevention of cardiovascular disease: meta-analysis of randomised trials. Lancet 361(9374):2017–2023
104. Baykal Y et al (2003) Effects of antihypertensive agents, alpha receptor blockers, beta blockers, angiotensin-converting enzyme inhibitors, angiotensin receptor blockers and calcium channel blockers, on oxidative stress. J Hypertens 21(6):1207–1211
105. de Cavanagh EM et al (2010) Vascular structure and oxidative stress in salt-loaded spontaneously hypertensive rats: effects of losartan and atenolol. Am J Hypertens 23(12):1318–1325
106. Dauchet L et al (2006) Fruit and vegetable consumption and risk of coronary heart disease: a meta-analysis of cohort studies. J Nutr 136(10):2588–2593
107. Steinhubl SR (2008) Why have antioxidants failed in clinical trials? Am J Cardiol 101(10A):14D–19D
108. Brown DA et al (2017) Expert consensus document: mitochondrial function as a therapeutic target in heart failure. Nat Rev Cardiol 14(4):238–250
109. Giordano FJ (2005) Oxygen, oxidative stress, hypoxia, and heart failure. J Clin Invest 115(3):500–508
110. Kalogeris T, Bao Y, Korthuis RJ (2014) Mitochondrial reactive oxygen species: a double edged sword in ischemia/reperfusion vs preconditioning. Redox Biol 2:702–714
111. Haworth RA, Potter KT, Russell DC (2010) Role of arachidonic acid, lipoxygenase, and mitochondrial depolarization in reperfusion arrhythmias. Am J Physiol Heart Circ Physiol 299(1):H165–H174
112. Santos CX et al (2011) Redox signaling in cardiac myocytes. Free Radic Biol Med 50(7):777–793

113. Buggisch M et al (2007) Stimulation of ES-cell-derived cardiomyogenesis and neonatal cardiac cell proliferation by reactive oxygen species and NADPH oxidase. J Cell Sci 120(Pt 5):885–894
114. Sauer H et al (2000) Role of reactive oxygen species and phosphatidylinositol 3-kinase in cardiomyocyte differentiation of embryonic stem cells. FEBS Lett 476(3):218–223
115. Saitoh S et al (2007) Redox-dependent coronary metabolic dilation. Am J Physiol Heart Circ Physiol 293(6):H3720–H3725
116. Saitoh S et al (2006) Hydrogen peroxide: a feed-forward dilator that couples myocardial metabolism to coronary blood flow. Arterioscler Thromb Vasc Biol 26(12):2614–2621
117. Andersson DC et al (2011) Mitochondrial production of reactive oxygen species contributes to the beta-adrenergic stimulation of mouse cardiomycytes. J Physiol 589(Pt 7):1791–1801
118. Loor G, Schumacker PT (2008) Role of hypoxia-inducible factor in cell survival during myocardial ischemia-reperfusion. Cell Death Differ 15(4):686–690
119. Zhang M et al (2010) NADPH oxidase-4 mediates protection against chronic load-induced stress in mouse hearts by enhancing angiogenesis. Proc Natl Acad Sci U S A 107(42):18121–18126
120. Ponikowski P et al (2016) ESC guidelines for the diagnosis and treatment of acute and chronic heart failure: the task force for the diagnosis and treatment of acute and chronic heart failure of the European Society of Cardiology (ESC). Developed with the special contribution of the Heart Failure Association (HFA) of the ESC. Eur J Heart Fail 18(8):891–975
121. Reina-Couto M et al (2016) Resolving inflammation in heart failure: novel protective lipid mediators. Curr Drug Targets 17(10):1206–1223
122. Braunwald E (2015) The war against heart failure: the lancet lecture. Lancet 385(9970):812–824
123. Guha K, McDonagh T (2013) Heart failure epidemiology: European perspective. Curr Cardiol Rev 9(2):123–127
124. Ramani GV, Uber PA, Mehra MR (2010) Chronic heart failure: contemporary diagnosis and management. Mayo Clin Proc 85(2):180–195
125. Spodick DH (2003) Acute cardiac tamponade. N Engl J Med 349(7):684–690
126. von Haehling S et al (2010) Elevated levels of asymmetric dimethylarginine in chronic heart failure: a pathophysiologic link between oxygen radical load and impaired vasodilator capacity and the therapeutic effect of allopurinol. Clin Pharmacol Ther 88(4):506–512
127. Kanaan GN, Harper ME (2017) Cellular redox dysfunction in the development of cardiovascular diseases. Biochim Biophys Acta Gen Subj 1861(11 Pt A):2822–2829
128. Heymans S et al (2009) Inflammation as a therapeutic target in heart failure? A scientific statement from the translational research Committee of the Heart Failure Association of the European Society of Cardiology. Eur J Heart Fail 11(2):119–129
129. Hofmann U, Frantz S (2013) How can we cure a heart "in flame"? A translational view on inflammation in heart failure. Basic Res Cardiol 108(4):356
130. Mueller C et al (2006) Inflammation and long-term mortality in acute congestive heart failure. Am Heart J 151(4):845–850
131. Khaper N et al (2010) Targeting the vicious inflammation-oxidative stress cycle for the management of heart failure. Antioxid Redox Signal 13(7):1033–1049
132. Chen X et al (2008) Role of reactive oxygen species in tumor necrosis factor-alpha induced endothelial dysfunction. Curr Hypertens Rev 4(4):245–255
133. Zhang C et al (2003) Interaction of myeloperoxidase with vascular NAD(P)H oxidase-derived reactive oxygen species in vasculature: implications for vascular diseases. Am J Physiol Heart Circ Physiol 285(6):H2563–H2572
134. Reina-Couto M et al (2014) Impaired resolution of inflammation in human chronic heart failure. Eur J Clin Investig 44(6):527–538
135. Reina-Couto M et al (2018) Endocan as a new biomarker of severity in acute heart failure. Eur J Heart Fail 20:P459
136. Reina-Couto M et al (2018) Inflammation resolution mediators in acute heart failure. J Hypertens 36(e-Supplement 1):e211

137. Munzel T et al (2015) Pathophysiological role of oxidative stress in systolic and diastolic heart failure and its therapeutic implications. Eur Heart J 36(38):2555–2564
138. van Riet EE et al (2014) Prevalence of unrecognized heart failure in older persons with shortness of breath on exertion. Eur J Heart Fail 16(7):772–777
139. Karimi Galougahi K et al (2015) Redox biomarkers in cardiovascular medicine. Eur Heart J 36(25):1576–1582. 1582a-b
140. Patel RS et al (2016) Novel biomarker of oxidative stress is associated with risk of death in patients with coronary artery disease. Circulation 133(4):361–369
141. Braunwald E (2008) Biomarkers in heart failure. N Engl J Med 358(20):2148–2159
142. Tang WH et al (2006) Plasma myeloperoxidase levels in patients with chronic heart failure. Am J Cardiol 98(6):796–799
143. Adam M et al (2015) Levosimendan displays anti-inflammatory effects and decreases MPO bioavailability in patients with severe heart failure. Sci Rep 5:9704
144. O'Donoghue ML et al (2016) Multimarker risk stratification in patients with acute myocardial infarction. J Am Heart Assoc 5(5):e002586
145. Virzi GM et al (2018) Levels of proinflammatory cytokines, oxidative stress, and tissue damage markers in patients with acute heart failure with and without cardiorenal syndrome type 1. Cardiorenal Med 8(4):321–331
146. Kataoka Y et al (2014) Myeloperoxidase levels predict accelerated progression of coronary atherosclerosis in diabetic patients: insights from intravascular ultrasound. Atherosclerosis 232(2):377–383
147. Meuwese MC et al (2007) Serum myeloperoxidase levels are associated with the future risk of coronary artery disease in apparently healthy individuals: the EPIC-norfolk prospective population study. J Am Coll Cardiol 50(2):159–165
148. Pastori D et al (2015) Does mediterranean diet reduce cardiovascular events and oxidative stress in atrial fibrillation? Antioxid Redox Signal 23(8):682–687
149. Cunnington C et al (2012) Systemic and vascular oxidation limits the efficacy of oral tetrahydrobiopterin treatment in patients with coronary artery disease. Circulation 125(11):1356–1366
150. Shirodaria C et al (2007) Global improvement of vascular function and redox state with low-dose folic acid: implications for folate therapy in patients with coronary artery disease. Circulation 115(17):2262–2270
151. Santos CN et al (2018) Pure polyphenols applications for cardiac health and disease. Curr Pharm Des 24(19):2137–2156
152. Driver C et al (2018) Cardioprotective effects of metformin. J Cardiovasc Pharmacol 72(2):121–127
153. Alemayehu M et al (2017) Effect of ticagrelor versus clopidogrel on vascular reactivity. J Am Coll Cardiol 69(17):2246–2248
154. Nanhwan MK et al (2014) Chronic treatment with ticagrelor limits myocardial infarct size: an adenosine and cyclooxygenase-2-dependent effect. Arterioscler Thromb Vasc Biol 34(9):2078–2085
155. Montuschi P, Barnes PJ, Roberts LJ 2nd (2004) Isoprostanes: markers and mediators of oxidative stress. FASEB J 18(15):1791–1800
156. Mallat Z et al (1998) Elevated levels of 8-iso-prostaglandin F2alpha in pericardial fluid of patients with heart failure: a potential role for in vivo oxidant stress in ventricular dilatation and progression to heart failure. Circulation 97(16):1536–1539
157. Polidori MC et al (2004) Increased F2 isoprostane plasma levels in patients with congestive heart failure are correlated with antioxidant status and disease severity. J Card Fail 10(4):334–338
158. Davies SS, Roberts LJ 2nd (2011) F2-isoprostanes as an indicator and risk factor for coronary heart disease. Free Radic Biol Med 50(5):559–566
159. Hummel SL et al (2012) Low-sodium dietary approaches to stop hypertension diet reduces blood pressure, arterial stiffness, and oxidative stress in hypertensive heart failure with preserved ejection fraction. Hypertension 60(5):1200–1206

160. de Meirelles LR et al (2011) Platelet activation, oxidative stress and overexpression of inducible nitric oxide synthase in moderate heart failure. Clin Exp Pharmacol Physiol 38(10):705–710
161. Ellis GR et al (2002) Addition of candesartan to angiotensin converting enzyme inhibitor therapy in patients with chronic heart failure does not reduce levels of oxidative stress. Eur J Heart Fail 4(2):193–199
162. White M et al (2006) Increased systemic inflammation and oxidative stress in patients with worsening congestive heart failure: improvement after short-term inotropic support. Clin Sci (Lond) 110(4):483–489
163. Amaki T et al (2004) Circulating malondialdehyde modified LDL is a biochemical risk marker for coronary artery disease. Heart 90(10):1211–1213
164. Charach G et al (2015) Usefulness of antibodies to oxidized low-density lipoproteins as predictors of morbidity and prognosis in heart failure patients aged >/=65 years. Am J Cardiol 116(9):1379–1384
165. Kato M et al (2017) Stretching exercises improve vascular endothelial dysfunction through attenuation of oxidative stress in chronic heart failure patients with an implantable cardioverter defibrillator. J Cardiopulm Rehabil Prev 37(2):130–138
166. Mondal NK et al (2016) Systemic inflammatory response syndrome in end-stage heart failure patients following continuous-flow left ventricular assist device implantation: differences in plasma redox status and leukocyte activation. Artif Organs 40(5):434–443
167. Lokuta AJ et al (2005) Increased nitration of sarcoplasmic reticulum Ca2+-ATPase in human heart failure. Circulation 111(8):988–995
168. Cabassi A et al (2014) Low serum ferroxidase I activity is associated with mortality in heart failure and related to both peroxynitrite-induced cysteine oxidation and tyrosine nitration of ceruloplasmin. Circ Res 114(11):1723–1732
169. Peluffo G, Radi R (2007) Biochemistry of protein tyrosine nitration in cardiovascular pathology. Cardiovasc Res 75(2):291–302
170. Ahn B et al (2016) Diaphragm abnormalities in patients with end-stage heart failure: NADPH oxidase upregulation and protein oxidation. Front Physiol 7:686
171. Parissis JT et al (2007) Effects of Levosimendan on circulating markers of oxidative and nitrosative stress in patients with advanced heart failure. Atherosclerosis 195(2):e210–e215
172. Cameron VA et al (2006) Angiotensin type-1 receptor A1166C gene polymorphism correlates with oxidative stress levels in human heart failure. Hypertension 47(6):1155–1161
173. Kobayashi S et al (2011) Urinary 8-hydroxy-2′-deoxyguanosine reflects symptomatic status and severity of systolic dysfunction in patients with chronic heart failure. Eur J Heart Fail 13(1):29–36
174. Masugata H et al (2013) Association between oxidative stress assessed by urinary 8-hydroxydeoxyguanosine and the cardiac function in hypertensive patients without overt heart disease. Clin Exp Hypertens 35(5):308–312
175. Susa T et al (2012) Urinary 8-hydroxy-2′-deoxyguanosine as a novel biomarker for predicting cardiac events and evaluating the effectiveness of carvedilol treatment in patients with chronic systolic heart failure. Circ J 76(1):117–126
176. Blankenberg S et al (2003) Glutathione peroxidase 1 activity and cardiovascular events in patients with coronary artery disease. N Engl J Med 349(17):1605–1613
177. Caruso R et al (2007) Pre-operative redox state affects 1-month survival in patients with advanced heart failure undergoing left ventricular assist device implantation. J Heart Lung Transplant 26(11):1177–1181
178. Sam F et al (2005) Increased reactive oxygen species production and functional alterations in antioxidant enzymes in human failing myocardium. J Card Fail 11(6):473–480
179. Hokamaki J et al (2004) Urinary biopyrrins levels are elevated in relation to severity of heart failure. J Am Coll Cardiol 43(10):1880–1885
180. Ellidag HY et al (2014) Oxidative stress and ischemia-modified albumin in chronic ischemic heart failure. Redox Rep 19(3):118–123

181. Lonn E et al (2005) Effects of long-term vitamin E supplementation on cardiovascular events and cancer: a randomized controlled trial. JAMA 293(11):1338–1347
182. Thomson MJ, Frenneaux MP, Kaski JC (2009) Antioxidant treatment for heart failure: friend or foe? QJM 102(5):305–310
183. Yamauchi Y et al (2017) Is serum uric acid independently associated with left ventricular mass index, ejection fraction, and B-type natriuretic peptide among female and male cardiac patients? Int Heart J 58(4):562–569
184. Wannamethee SG et al (2018) Serum uric acid as a potential marker for heart failure risk in men on antihypertensive treatment: the British regional heart study. Int J Cardiol 252:187–192

Pulmonary Hypertension and the Right Ventricle: The Roles of Mitochondrial Reactive Oxygen Species in Causing Further Right Ventricular Mitochondrial Changes

24

Gerald J. Maarman

24.1 Pulmonary Hypertension

Pulmonary hypertension (PH) is defined as mean pulmonary arterial pressure $\geq$ 25 mm Hg [1], and associates with diseases such as congenital heart disease and chronic obstructive pulmonary disease [2]. The global prevalence of PH is largely unknown, mostly due to the broad PH classification and the lack of national registries in countries such as those on the African continent [3, 4]. Irrespective, a recent report [3] suggests the prevalence of PH as approximately 1% of the global population and up to 10% in older patients. The most commonly reported causes of PH includes left -sided heart and lung diseases, and 80% of these patients are resident in developing countries [3]. In these countries, PH also associates with HIV, rheumatic heart disease, schistosomiasis and congenital heart disease in younger patients [3]. Considering the high burden of HIV and rheumatic heart disease in countries like South Africa [5], it is highly likely that PH is not as rare as once believed [6, 7]. In addition, the lack of PH registries [8] suggests that the prevalence of PH may currently be underestimated. It is nonetheless the consensus that PH is a global health concern that requires considerable research and intervention [3].

The underlying mechanisms of PH are not well understood, but three major molecular pathways have been shown to contribute to the pathogenesis of PH [9]. These pathways include endothelin mediated proliferation, and prostacyclin and nitric oxide pathways of vascular modulation [9]. In brief, these pathways cause vasoconstriction of the pulmonary arterioles and excessive proliferation of

G. J. Maarman (✉)
Hatter Institute for Cardiovascular Research in Africa (HICRA) and MRC Inter-University, Cape Heart Group, Department of Medicine, Faculty of Health Sciences, University of Cape Town, Cape Town, South Africa

Cardiovascular Research Group, Division of Medical Physiology, Department of Biomedical Sciences, Stellenbosch University, Tygerberg, South Africa
e-mail: gmaarman@sun.ac.za

S. Chakraborti et al. (eds.), *Oxidative Stress in Heart Diseases*,
https://doi.org/10.1007/978-981-13-8273-4_24

pulmonary arterial smooth muscle cells [9]. The pulmonary vasculature undergoes intimal thickening and concentric hypertrophy, and perivascular fibrosis of distal pulmonary arterioles that leads to the generation of plexogenic lesions in severe PH [9]. All these pulmonary vascular changes cause obliteration of the pulmonary arteries with subsequent elevation of mean pulmonary artery pressure [9].

24.2 The Right Ventricle in PH

Due to the anatomical relationship between the pulmonary circulatory system and the heart, elevated pressures caused by PH directly affects the right ventricle (RV) [10]. Normally, the RV has an anatomical design that is reflective of the relatively low pressure of the pulmonary circulation [11, 12], as opposed to the high pressure of systemic circulation. Thus, as opposed to the ellipsoidal *shape of the left ventricle* [13], *the RV* is smaller in diameter, has a thin wall and crescent shape [2, 14]. RV output is significantly lower than *left ventricular* output and the timing and pattern of ejection differ significantly from that of the left ventricle [12]. Pressure in the pulmonary artery is low during diastole and the hemodynamic effects of alterations in RV compliance are different for the RV compared to the left ventricle. Subsequently, the hydraulic impedance is lower and compliance higher [12]. Similarly to the LV [13], the RV is sensitive to fluctuations in ventricular wall stress secondary to increased pressure or volume overload [10]. As previously described [2], PH elevates pulmonary vascular resistance and pulmonary vascular compliance. Consequently, the RV afterload becomes elevated, RV dilation occurs [2], and the RV transitions from adaptive (hypertrophy) to maladaptive (dilation) ventricular remodelling [13].

24.3 RV Remodelling Underlined by Mitochondrial ROS and Oxidative Stress

The aforementioned RV changes associate with a number of molecular processes including altered bioenergetics, neuro-hormonal changes, alterations in ion channels and contractile proteins and increased production of reactive oxygen species (ROS) [2]. Biological sources of ROS during PH includes the nicotinamide adenine dinucleotide phosphate *oxidase (NOX), myeloperoxidase, xanthine oxidase, lipoxygenase, cyclooxygenase, uncoupled endothelial nitric oxide, cytochrome-P450 and cardiac* mitochondria [15, 16]. Considering the abundance of mitochondria in the human body [17], they are considered major biological sources of ROS, especially during PH and RV remodelling [18, 19] (Fig. 24.1).

Mitochondria are known as the powerhouses of the heart and they employ metabolic pathways to generate energy for the maintenance of cardiac processes [20]. Energy production occurs via the electron transfer system that comprises a series of complexes (one to four) [21], of which complex-1 and 3 are considered mitochondrial sites for ROS production [22]. Electron transfer from complex-1 to complex-3 results in the production of the highly reactive superoxide anion on the matrix side of the mitochondrial membrane [23] (Fig. 24.2).

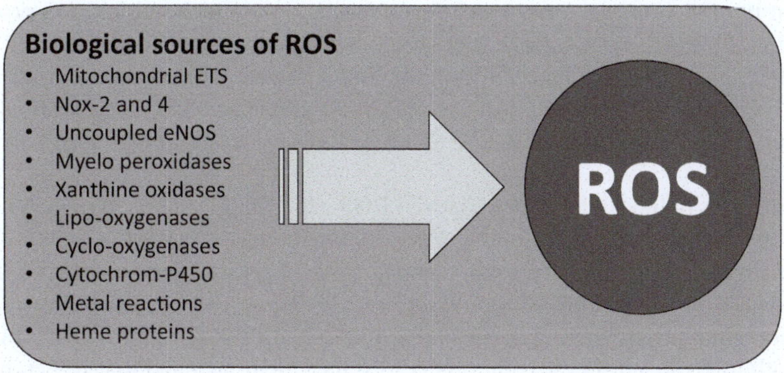

Fig. 24.1 A representation of biological sources of ROS during PH, including mitochondrial electron transfer system (ETS), nicotinamide adenine dinucleotide phosphate oxidase (NOX), uncoupled endothelial nitric oxide (eNOS), myeloperoxidase, xanthine oxidase, lipoxygenase, cyclooxygenase, cytochrome-P450, metal reactions and heme-proteins

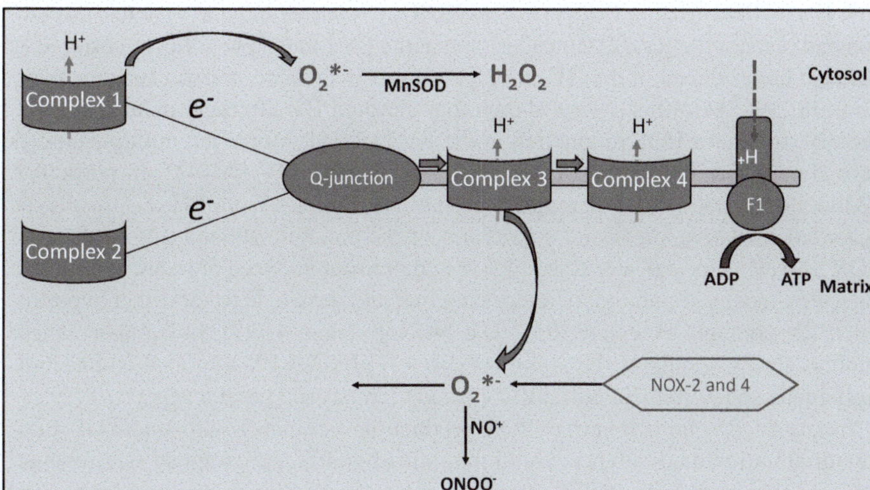

Fig. 24.2 This is a depiction of how the mitochondria uses multiple mechanisms to generate ROS

Superoxide has three fates. First, it can cross the mitochondrial membrane into the cytosol where it reacts with nitrogen to form toxic peroxynitrite [23]. Second, superoxide can undergo conversion to hydrogen peroxide, a reaction which is catalysed by manganese-dependent superoxide dismutase [23]. Third, zinc/copper-dependent superoxide dismutase catalyses the conversion of superoxide to hydrogen peroxide or hydroxyl radical [23]. The hydroxyl radical is highly active and estimated to last milliseconds in the cellular space, whereas hydrogen peroxide is further converted to water as a by-product of reactions catalysed by catalase [23] or glutathione peroxidase [24]. It should also be noted that xanthine oxidase could be present in mitochondrial spaces and thus contribute to mitochondrial ROS production [25, 26]. Furthermore, there is crosstalk between the mitochondria and NOX-1 to 5, and upon translocation of NOX to the mitochondrial matrix, it generates ROS [27]. Therefore, mitochondria can use multiple mechanisms [28] in order to generate ROS [29, 30] within the RV under pathologic conditions caused by PH (Fig. 24.2).

In PH, it is believed that mitochondrial ROS can also be generated in the pulmonary vasculature by these aforementioned mitochondrial processes [31, 32]. In hypoxia-induced PH, lung expression and activity of NOX-2 and 4 is increased and exacerbates mitochondrial ROS production [33]. These ROS can be released into systemic circulation where it causes systemic oxidative stress in blood plasma. In line with this, our group [34] and others [35] have demonstrated that in PH, blood plasma display elevated oxidative stress [34]. These findings are in support of Iridova et al. (2002) who demonstrated that PH associates with elevated plasma malonic dialdehyde, a proxy for oxidative stress [36]. Subsequent reports confirmed the presence of excessive oxidative stress in the plasma or systemic circulation of patients with PH [37].

It is also likely that in PH, (1) excess ROS are released into the systemic circulation and reaches the RV. (2) Considering that the RV has its own mitochondria, ROS can also be produced in the RV, in response to PH-induced pressure/volume overload [38, 39]. Akin this, it was shown that elevated RV afterload induced by PH, directly stimulates ROS production in the RV [15, 16]. Moreover, multiple studies have shown that PH increases oxidative stress in the RV [40–42]. In particular, Redout et al. (2007) have demonstrated that RV samples from PH rats, displayed increased oxidative stress and expression of the catalytic subunit gp91 (phox) of NOX as well as its activator Rac1 [42]. Another group induced pressure overload in the RV by means of pulmonary artery banding, and were able to show that hypertrophied RV contains excessive ROS, and NOX-1, 2 and 4 [43]. In summary, these findings strongly support the argument that PH elevates RV afterload and accordingly increase RV mitochondrial ROS production beyond basal levels.

Increased RV mitochondrial ROS production associates with impaired mitochondrial antioxidant defence. Naturally, mitochondria are equipped with copper/zinc and manganese dependent superoxide dismutase, glutathione peroxidase and catalase [44]. Accumulating evidence suggests that PH correlates with reduced or impaired activities of these mitochondrial antioxidant enzymes [32, 36, 40]. A previous study conducted in our group also demonstrated that apart from increased oxidative stress, PH is associated with impaired activities of these mitochondrial antioxidant enzymes [34]. Therefore, an important point to raise is that PH excessively increases RV mitochondrial ROS production that impairs RV mitochondrial antioxidant enzyme activity. This imbalance between RV mitochondrial ROS production and adequate antioxidant defence contributes to excessive RV oxidative stress in PH.

RV mitochondrial ROS have also been implicated as inducers of ventricular fibrosis [45] that contribute to RV remodelling [46] in PH [47–50]. First, mitochondrial ROS production can stimulate resident RV fibroblasts and upregulate transforming growth factor beta-1 mediated pro-fibrotic responses [51]. This stimulates the synthesis of matrix proteins, extracellular matrix and collagen deposition that eventually results in RV fibrosis [51]. Second, RV-mitochondrial ROS stimulates RV fibrosis through the ROS-sensitive transcription factors nuclear factor kappa-B and/or activator protein-1 [51]. These can increase the expression of collagen, fibronectin and osteopontin, enhance metallopeptidase inhibitor-1 and 2 activity and decrease matrix metallopeptidase-1 activity [51]. This results in less collagen degradation that favours collagen synthesis culminating in RV fibrosis. Therefore, RV mitochondrial ROS production not only augments RV fibrosis but also simulates remodelling of the extracellular matrix. These RV mitochondrial ROS therefore, contributes to pathologic RV remodelling in PH.

24.3.1 Effects of Mitochondrial ROS on RV Mitochondrial Gene Expression & Dynamics

The aforementioned changes that RV mitochondrial ROS induce, are further underpinned by changes in mitochondrial gene expression [52, 53]. In fact, in various models of PH, RV mitochondrial ROS production correlates with downregulation of crucial genes encoding proteins, which are important for proper RV function [54]. Enache et al. (2013) induced PH with monocrotaline and studied RV mitochondrial changes 4 weeks after the monocrotaline injection [55]. At week two, the RV did not display any changes in the expression of mitochondrial genes sirtuin-1, peroxisome proliferator-activated receptor gamma coactivator-*1* alpha or citrate synthase [55]. However, the expression of all these genes was reduced after 4 weeks of PH. Sirtuin-1 is known to deacetylate peroxisome proliferator-activated receptor gamma coactivator-*1* alpha and regulate its activity [56]. These data [55] suggest that prolonged PH reduces sirtuin-1 that in turn reduces peroxisome proliferator-activated receptor gamma coactivator-*1* alpha activity. Peroxisome proliferator-activated receptor gamma coactivator-*1* alpha enables mitochondrial biogenesis [57]. Therefore, there is a link between excessive RV mitochondrial ROS production in PH and downregulated RV mitochondrial gene expression [58].

Another important component to the mitochondrial changes observed in RV, is mitochondrial fission and fusion, processes responsible for mitochondrial dynamics [59, 60]. This is relevant as increased RV mitochondrial ROS production is linked with changes in mitochondrial dynamics [59, 61]. Alterations in mitochondrial shape can influence RV function, and mitochondrial fission and fusion embody these shape changes (described as mitochondrial dynamics) [59, 60]. Mitochondrial dynamics are generally controlled by mitochondrial fusion and fission proteins (protein optic atrophy-1, mitofusin-1 and 2), and essential for RV mitochondria function [59, 60]. To date, no studies have investigated the role of mitochondrial fission and fusion the RV in PH. However, in left ventricular hypertrophy and failure, mitochondrial protein optic atrophy-1 is decreased and mitochondria fragmented [62]. In other models, partial deficiency in protein optic atrophy-1 reduced mitochondrial number and function [63], and increased susceptibility to left ventricular hypertrophy and cardiac dysfunction [64]. On the other hand, ablation of cardiac mitofusin-1 and mitofusin-2 fragments ventricular mitochondria and impairs mitochondrial respiration [65]. Mitofusin-2 is also downregulated in left ventricular hypertrophy and its overexpression reverses left ventricular hypertrophy [66, 67]. With the absence of such data for RV remodelling, future studies could investigate whether there is a link between RV mitochondrial ROS production and RV mitochondrial fission and fusion in a model of PH.

24.3.2 Effects of ROS on RV Mitochondrial Metabolism

We've now established that RV mitochondrial ROS impairs RV mitochondrial gene expression. However, these RV mitochondrial ROS also impairs RV mitochondrial energy substrate utilization by causing a metabolic shift [68, 69]. In the healthy heart, mitochondrial energy substrate utilization mostly comprises fatty acid oxidation, and to a lesser extent glycolysis and glucose oxidation [68, 69]. However, with ventricular hypertrophy and eventually ventricular failure, there is a metabolic shift from fatty acid oxidation toward higher glycolysis and glucose oxidation rates [68, 69].

As previously reviewed [2], in PH the RV displays a metabolic shift as the cardiomyocytes have increased glycolysis. In the monocrotaline and pulmonary artery banding animal models [70] of RV remodelling, RV compensated/adaptive hypertrophy associates with enhanced glucose oxidation, increased expression of glucose transporter-1 and pyruvate dehydrogenase [71]. These findings are consistent with the existing knowledge that ventricular remodelling correlates with a metabolic shift from fatty acid oxidation to glycolysis [68, 69]. These changes also coincide with the upregulation of glucose uptake in RV hypertrophy [72–74] as a means to increase glycolysis. This is required in order to maintain energy homeostasis and has been adequately demonstrated in PH [72, 74]. In prolonged PH, reduced RV mitochondrial complex-1 linked oxidative phosphorylation was observed when glucose-oxidation substrates were used to stimulate electron transfer through complex-1 and 2 [55, 75]. In particular, the reduced oxidative phosphorylation after the glutamate titration, could suggest that PH causes changes to the enzyme, glutamate dehydrogenase [29]. Future studies could investigate the effects of PH on glutamate dehydrogenase activity in the RV. Succinate dehydrogenase (complex-2) is the only mitochondrial complex that forms part of the tri-carboxylic acid cycle [29]. Therefore, the unaffected succinate dehydrogenase suggests that PH does not affect RV complex-2 activity. These findings support the notion that in maladaptive RV remodelling as in PH, RV mitochondrial metabolism shifts from fatty acid oxidation to glucose oxidation. However, the latter is also reduced in the RV during far progressed PH. Unfortunately, these glucose mediated metabolic pathways do not produce sufficient energy to maintain ventricular function in PH, and thus contributes to impaired RV function in PH [14, 76]. This aforementioned evidence supports the concept that RV mitochondrial ROS produced during PH can alter RV mitochondrial energy production pathways.

24.3.3 Effects of Mitochondrial ROS on RV Mitochondrial Respiration

Another aspect of the adverse effects RV mitochondrial ROS have is impaired RV mitochondrial respiration. In recent years, few studies have identified a key role for mitochondria in the pathogenesis of PH [77–80]. These studies demonstrated that PH is associated with altered RV mitochondrial respiration [77–80]. During PH-induced compensated RV hypertrophy, the RV displays increased RV mitochondrial ROS production and elevated complex-1 and 2 linked respiration [42]. Whereas, during the decompensated stage, the RV displays reduced mitochondrial complex activity [55]. This is an important observation, as increased ROS have been shown to cause structural damage to mitochondrial complexes [81] that impairs mitochondrial respiration. Accordingly, in prolonged PH, the RV has been shown to display reduced complex-1 linked respiration [55]. This while complex-2 linked respiration remained unchanged, showing only a trend towards a decrease [55]. These findings were corroborated by another group who later demonstrated that in prolonged PH, the RV displayed reduced complex-1, and unchanged complex-2 linked respiration [75, 82]. These findings suggest that in PH, RV-mitochondrial complex-1 linked respiration [75] and most likely due to the impact of excessive RV-mitochondrial ROS.

24.4 Conclusions

This chapter discussed the roles and impact of RV mitochondrial ROS on RV mitochondrial gene expression and dynamics, RV mitochondrial respiration and RV structure. The potential role of mitochondria-targeted antioxidants in protecting the RV in PH was briefly discussed. The chapter highlighted that RV mitochondrial ROS cause changes in RV mitochondrial gene/protein expression, dynamics, metabolism and respiration. RV mitochondrial ROS also activates RV fibrosis and extracellular matrix changes and the evidence suggests that altogether, these processes contribute to RV remodelling observed in PH.

Acknowledgements I would like to thank the National Research Foundation of South Africa and the Hatter Institute for Cardiovascular Research in Africa (HICRA), for their continued support. Special thanks for Prof. Sajal Chakraborti for the invitation to contribute a chapter to this highly important book series.

Conflict of Interest The author has no conflict of interest to declare.

References

1. Galie N et al (2016) 2015 ESC/ERS guidelines for the diagnosis and treatment of pulmonary hypertension the joint task force for the diagnosis and treatment of pulmonary hypertension of the European Society of Cardiology (ESC) and the European Respiratory Society (ERS) endorsed by: Association for European Paediatric and Congenital Cardiology (AEPC), International Society for Heart and Lung Transplantation (ISHLT). Eur Heart J 37(1):67–119
2. Maarman GJ et al (2017) Novel putative pharmacological therapies to protect the right ventricle in pulmonary hypertension: a review of current literature. Br J Pharmacol 174(7):497–511
3. Hoeper MM et al (2016) A global view of pulmonary hypertension. Lancet Respir Med 4(4):306–322
4. Mocumbi AO, Thienemann F, Sliwa K (2015) A global perspective on the epidemiology of pulmonary hypertension. Can J Cardiol 31(4):375–381
5. Kharsany AB, Karim QA (2016) HIV infection and aids in Sub-Saharan Africa: current status, challenges and opportunities. Open AIDS J 10:34–48
6. Sliwa K et al (2008) Spectrum of heart disease and risk factors in a black urban population in South Africa (the heart of Sowe to study): a cohort study. Lancet 371(9616):915–922
7. Thienemann F et al (2016) The causes, treatment, and outcome of pulmonary hypertension in Africa: insights from the pan African pulmonary hypertension cohort (PAPUCO) registry. Int J Cardiol 221:205–211
8. Gomberg-Maitland M, Michelakis ED (2010) A global pulmonary arterial hypertension registry: is it needed? Is it feasible? Pulmonary vascular disease: the global perspective. Chest 137(6):95s–101s
9. Maron BA, Loscalzo J (2013) Pulmonary hypertension: pathophysiology and signaling pathways. Handb Exp Pharmacol 218:31–58
10. Noordegraaf AV, Westerhof BE, Westerhof N (2017) Y the relationship between the right ventricle and its load in pulmonary hypertension. J Am Coll Cardiol 69(2):236–243
11. Naeije R, Brimioulle S, Dewachter L (2014) Biomechanics of the right ventricle in health and disease (2013 grover conference series). Pulm Circ 4(3):395–406
12. Penny DJ, Redington AN (2016) Function of the left and right ventricles and the interactions between them. Pediatr Crit Care Med 17(8):S112–S118
13. Opie LH et al (2006) Controversies in ventricular remodelling. Lancet 367(9507):356–367
14. Vonk-Noordegraaf A et al (2013) Right heart adaptation to pulmonary arterial hypertension physiology and pathobiology. J Am Coll Cardiol 62(25):D22–D33
15. Fulton DJR et al (2017) Reactive oxygen and nitrogen species in the development of pulmonary hypertension. Antioxidants 6(3):E54
16. Jernigan NL et al (2017) Contribution of reactive oxygen species to the pathogenesis of pulmonary arterial hypertension. PLoS One 12(6):e0180455
17. Robin ED, Wong R (1988) Mitochondrial-DNA molecules and virtual number of mitochondria per cell in mammalian-cells. J Cell Physiol 136(3):507–513
18. Zorov DB, Juhaszova M, Sollott SJ (2014) Mitochondrial reactive oxygen species (ROS) and ROS-induced ros release. Physiol Rev 94(3):909–950
19. Zorov DB et al (2000) Reactive oxygen species (ROS)-induced ROS release: a new phenomenon accompanying induction of the mitochondrial permeability transition in cardiac myocytes. J Exp Med 192(7):1001–1014
20. Murphy E et al (2016) Mitochondrial function, biology, and role in disease. A scientific statement from the American Heart Association. Circ Res 118(12):1960–1991
21. Rosca MG, Hoppel CL (2013) Mitochondrial dysfunction in heart failure. Heart Fail Rev 18(5):607–622. https://doi.org/10.1007/s10741-012-9340-0
22. Murphy MP (2009) How mitochondria produce reactive oxygen species. Biochem J 417:1–13
23. Camello-Almaraz C et al (2006) Mitochondrial reactive oxygen species and Ca^{2+} signaling. Am J Phys Cell Phys 291(5):C1082–C1088

24. Birben E et al (2012) Oxidative stress and antioxidant defense. World Allergy Organ J 5(1):9–19
25. Kelley EE et al (2010) Hydrogen peroxide is the major oxidant product of xanthine oxidase. Free Radic Biol Med 48(4):493–498
26. Vergeade A et al (2012) Xanthine oxidase contributes to mitochondrial ROS generation in an experimental model of cocaine-induced diastolic dysfunction. J Cardiovasc Pharmacol 60(6):538–543
27. Dikalov S et al (2011) Crosstalk between mitochondrial ROS and NADPH oxidases in cardiovascular and degenerative diseases: application of mitochondria-targeted antioxidants. Free Radic Biol Med 51:S85–S86
28. Zorov DB et al (2001) Mechanisms of "reactive oxygen species (ROS)-induced ROS release" during induction of the mitochondrial permeability transition (MPT). Biophys J 80(1):500a–500a
29. Gnaiger E (2014) Mitochondrial pathways and respiratory control, an introduction to oxphos analysis, 4th edn. OROBOROS MiPNet Publications, Innsbruck
30. Grivennikova VG, Vinogradov AD (2013) Mitochondrial production of reactive oxygen species. Biochemistry (Mosc) 78(13):1490–1511
31. Maarman GJ (2017) Natural antioxidants as potential therapy, and a promising role for melatonin against pulmonary hypertension. Adv Exp Med Biol 967:161–178
32. Wong CM et al (2013) Reactive oxygen species and antioxidants in pulmonary hypertension. Antioxid Redox Signal 18(14):1789–1796
33. Adesina SE et al (2015) Targeting mitochondrial reactive oxygen species to modulate hypoxia-induced pulmonary hypertension. Free Radic Biol Med 87:36–47
34. Maarman G et al (2015) Melatonin as a preventive and curative therapy against pulmonary hypertension. J Pineal Res 59(3):343–353
35. Zuo L et al (2014) Molecular characterization of reactive oxygen species in systemic and pulmonary hypertension. Am J Hypertens 27(5):643–650
36. Irodova NL et al (2002) Oxidative stress in patients with primary pulmonary hypertension. Bull Exp Biol Med 133(6):580–582
37. Spiekermann S, Schenk K, Hoeper MM (2009) Increased xanthine oxidase activity in idiopathic pulmonary arterial hypertension. Eur Respir J 34(1):276
38. Zhang S et al (2015) Oxidative stress and nitric oxide signaling related biomarkers in patients with pulmonary hypertension: a case control study. BMC Pulm Med 15:50
39. Reis GS et al (2013) Oxidative-stress biomarkers in patients with pulmonary hypertension. Pulm Circ 3(4):856–861
40. Leichsenring-Silva F et al (2011) Association of the time course of pulmonary arterial hypertension with changes in oxidative stress in the left ventricle. Clin Exp Pharmacol Physiol 38(12):804–810
41. Redout EM et al (2010) Antioxidant treatment attenuates pulmonary arterial hypertension-induced heart failure. Am J Phys Heart Circ Phys 298(3):H1038–H1047
42. Redout EM et al (2007) Right-ventricular failure is associated with increased mitochondrial complex ii activity and production of reactive oxygen species. Cardiovasc Res 75(4):770–781
43. Lu X et al (2011) Elevated oxidative stress and endothelial dysfunction in right coronary artery of right ventricular hypertrophy. J Appl Physiol 110(6):1674–1681
44. Rabilloud T et al (2001) The mitochondrial antioxidant defence system and its response to oxidative stress. Proteomics 1(9):1105–1110
45. Lijnen PJ, van Pelt JF, Fagard RH (2012) Stimulation of reactive oxygen species and collagen synthesis by angiotensin II in cardiac fibroblasts. Cardiovasc Ther 30(1):e1–e8
46. Moris D et al (2017) The role of reactive oxygen species in the pathophysiology of cardiovascular diseases and the clinical significance of myocardial redox. Ann Transl Med 5(16):326
47. Aguero J et al (2014) Characterization of right ventricular remodeling and failure in a chronic pulmonary hypertension model. Am J Phys Heart Circ Phys 307(8):H1204–H1215
48. Naeije R, Manes A (2014) The right ventricle in pulmonary arterial hypertension. Eur Respir Rev 23(134):476–487

49. Rain S et al (2016) Right ventricular myocardial stiffness in experimental pulmonary arterial hypertension relative contribution of fibrosis and myofibril stiffness. Circ Heart Fail 9(7):e002636
50. Ryan JJ, Archer SL (2014) The right ventricle in pulmonary arterial hypertension disorders of metabolism, angiogenesis and adrenergic signaling in right ventricular failure. Circ Res 115(1):176–188
51. Richter K, Kietzmann T (2016) Reactive oxygen species and fibrosis: further evidence of a significant liaison. Cell Tissue Res 365(3):591–605
52. Buermans HPJ et al (2005) Microarray analysis reveals pivotal divergent MRNA expression profiles early in the development of either compensated ventricular hypertrophy or heart failure. Physiol Genomics 21(3):314–323
53. Pearce SF et al (2017) Regulation of mammalian mitochondrial gene expression: recent advances. Trends Biochem Sci 42(8):625–639
54. Gomez-Arroyo J et al (2013) Metabolic gene remodeling and mitochondrial dysfunction in failing right ventricular hypertrophy secondary to pulmonary arterial hypertension. Circ Heart Fail 6(1):136–144
55. Enache I et al (2013) Skeletal muscle mitochondrial dysfunction precedes right ventricular impairment in experimental pulmonary hypertension. Mol Cell Biochem 373(1–2):161–170
56. Cantó C, Auwerx J (2009) Pgc-1alpha, sirt1 and AMPK, an energy sensing network that controls energy expenditure. Curr Opin Lipidol 20(2):98–105
57. Rowe GC, Jiang A, Arany Z (2010) Pgc-1 coactivators in cardiac development and disease. Circ Res 107(7):825–838
58. Larsen S et al (2012) Biomarkers of mitochondrial content in skeletal muscle of healthy young human subjects. J Physiol 590(14):3349–3360
59. Ong SB et al (2015) Mitochondrial fusion and fission proteins as novel therapeutic targets for treating cardiovascular disease. Eur J Pharmacol 763:104–114
60. Sharp WW, Archer SL (2015) Mitochondrial dynamics in cardiovascular disease: fission and fusion foretell form and function. J Mol Med 93(3):225–228
61. Song MS et al (2014) Super-suppression of mitochondrial reactive oxygen species signaling impairs compensatory autophagy in primary mitophagic cardiomyopathy. Circ Res 115(3):348–353
62. Chen L et al (2009) Mitochondrial opa1, apoptosis, and heart failure. Cardiovasc Res 84(1):91–99
63. Chen L et al (2012) Opa1 mutation and late-onset cardiomyopathy: mitochondrial dysfunction and MTDNA instability. J Am Heart Assoc 1(5):e003012
64. Piquereau J et al (2012) Down-regulation of opa1 alters mouse mitochondrial morphology, PTP function, and cardiac adaptation to pressure overload. Cardiovasc Res 94(3):408–417
65. Chen Y, Liu YQ, Dorn GW (2011) Mitochondrial fusion is essential for organelle function and cardiac homeostasis. Circ Res 109(12):1327–1U36
66. Fang L et al (2007) Down-regulation of mitofusin-2 expression in cardiac hypertrophy in vitro and in vivo. Life Sci 80(23):2154–2160
67. Yu HY et al (2011) Mitofusin 2 inhibits angiotensin II-induced myocardial hypertrophy. J Cardiovasc Pharmacol Ther 16(2):205–211
68. Lopaschuk GD et al (2010) Myocardial fatty acid metabolism in health and disease. Physiol Rev 90(1):207–258
69. Stanley WC, Recchia FA, Lopaschuk GD (2005) Myocardial substrate metabolism in the normal and failing heart. Physiol Rev 85(3):1093–1129
70. Maarman G et al (2013) A comprehensive review: the evolution of animal models in pulmonary hypertension research; are we there yet? Pulm Circ 3(4):739–756
71. Piao L et al (2010) The inhibition of pyruvate dehydrogenase kinase improves impaired cardiac function and electrical remodeling in two models of right ventricular hypertrophy: resuscitating the hibernating right ventricle. J Mol Med 88(1):47–60

72. Lundgrin EL et al (2013) Fasting 2-deoxy-2-[18f]fluoro-d-glucose positron emission tomography to detect metabolic changes in pulmonary arterial hypertension hearts over 1 year. Ann Am Thorac Soc 10(1):1–9
73. Marsboom G et al (2012) Lung (18)f-fluorodeoxyglucose positron emission tomography for diagnosis and monitoring of pulmonary arterial hypertension. Am J Respir Crit Care Med 185(6):670–679
74. Oikawa M et al (2005) Increased [18f]fluorodeoxyglucose accumulation in right ventricular free wall in patients with pulmonary hypertension and the effect of epoprostenol. J Am Coll Cardiol 45(11):1849–1855
75. Daicho T et al (2009) Possible involvement of mitochondrial energy-producing ability in the development of right ventricular failure in monocrotaline-induced pulmonary hypertensive rats. J Pharmacol Sci 111(1):33–43
76. Vonk Noordegraaf A, Galie N (2011) The role of the right ventricle in pulmonary arterial hypertension. Eur Respir Rev 20(122):243–253
77. Marsboom G et al (2012) Dynamin-related protein 1-mediated mitochondrial mitotic fission permits hyperproliferation of vascular smooth muscle cells and offers a novel therapeutic target in pulmonary hypertension. Circ Res 110(11):1484–1497
78. Ryan J et al (2015) Mitochondrial dynamics in pulmonary arterial hypertension. J Mol Med 93(3):229–242
79. Ryan JJ, Archer SL (2015) Emerging concepts in the molecular basis of pulmonary arterial hypertension part I: metabolic plasticity and mitochondrial dynamics in the pulmonary circulation and right ventricle in pulmonary arterial hypertension. Circulation 131(19):1691–1702
80. Sharma H, Ladner AO, Dhillon NK (2017) Mitochondrial dynamics and HIV/opioids mediated pulmonary endothelial dysfunction: implications in HIV-related pulmonary arterial hypertension. Am J Respir Crit Care Med 195
81. Schwall CT, Greenwood VL, Alder NN (2012) The stability and activity of respiratory complex II is cardiolipin-dependent. Biochim Biophys Acta, Bioenerg 1817(9):1588–1596
82. Tang Z et al (2002) Heart and breast muscle mitochondrial dysfunction in pulmonary hypertension syndrome in broilers (Gallus domesticus). Comp Biochem Physiol A Mol Integr Physiol 132(3):527–540

Mitochondrial Dysfunction and Oxidative Stress: Focusing on Cardiac Hypertrophy and Heart Failure

25

Parmeshwar B. Katare, Hina L. Nizami, and Sanjay K. Banerjee

Abbreviations

ATP	Adenosine triphosphate
DAMPs	Damage associated molecular patterns
NAD	Nicotinamide adenine dinucleotide
NOX	NADPH oxidases
OXPHOS	Oxidative phosphorylation system
PRRs	Pathogen recognition receptors
RNS	Reactive nitrogen species
ROS	Reactive oxygen species
TCA	Tricarboxylic acid cycle
TFAM	Mitochondrial transcription factor A
TLR	Toll like receptor
XDH	Xanthine dehydrogenase
XOD	Xanthine oxidase

25.1 Introduction

Mitochondria plays a pivotal role in the cell by producing energy required for cellular functions. Mitochondria are crucial for generating adenosine triphosphate (ATP) via oxidative phosphorylation system (OXPHOS) [1]. In addition to that, mitochondria is also important for several biological functions such as production of heat, calcium homeostasis, stress signalling and defence responses [2]. Heart

P. B. Katare · H. L. Nizami · S. K. Banerjee (✉)
Drug Discovery Research Centre (DDRC), Translational Health Science and Technology Institute (THSTI), Faridabad, Haryana, India
e-mail: skbanerjee@thsti.res.in

S. Chakraborti et al. (eds.), *Oxidative Stress in Heart Diseases*,
https://doi.org/10.1007/978-981-13-8273-4_25

being the most energy consuming organ in the body, primarily depends on mitochondria for steady supply of ATP.

Heart failure is one of the leading causes of death in the industrialized countries [3]. A complex clinical syndrome, essentially marked by compromised cardiac function and chronic heart failure is often preceded by cardiac hypertrophy. Cardiac hypertrophy is the enlargement of the cardiomyocytes, which is initially an adaptive response to increased pressure or volume load in heart [4]. Prolonged haemodynamic load, however, results in pathological cardiac hypertrophy involving reactivation of fetal gene program, myocardial fibrosis and increased heart weight indices, eventually leading to heart failure [5]. Though the exact reasons for this transition from cardiac hypertrophy to failure remain elusive, changes in signalling modules such as cardiac contractile machinery, calcium ion homeostasis, mitochondrial dysfunction and oxidative stress have been speculated to play critical roles in the development and progression of heart failure [6].

Mitochondria, occupying about 30% mass of a cardiomyocyte [7], produce >95% ATP required by the heart [8]. The energy requirements of the heart are met primarily by fatty acid oxidation that takes place in the mitochondrial matrix, followed by electron transport in the inner mitochondrial membrane that leads to generation of ATP coupled to reduction of oxygen to water. Apart from being the single major source of energy required for continuous contraction and relaxation, mitochondria are also involved in regulation of calcium handling, redox equilibrium and cell death [9]. About 2% of the oxygen consumed by the mitochondria gets converted to reactive oxygen species (ROS) as by-products of the respiratory chain [10]. Dysfunctional mitochondria may not only be deficient in oxidative phosphorylation-induced energy production, but also generate increased levels of ROS.

ROS are highly reactive chemical species formed from molecular oxygen, which include superoxide, hydroxyl free radicals and non-radical hydrogen peroxide [11]. Under physiological conditions, redox equilibrium exists in the cell due to detoxification of ROS by the cellular antioxidant response. This antioxidant system is composed of enzymes such as superoxide dismutase (SOD), glutathione peroxidase (GSHPx) and catalase, and other non-enzymatic components. Cellular stress that triggers increased ROS generation exceeding the antioxidant defence results in a state called oxidative stress [12]. Oxidative stress observed in the cardiomyocytes of failing hearts may arise from several sources, including the mitochondria, NADPH oxidase (NOX), xanthine oxidase, and uncoupled nitric oxide synthases (NOS) [13, 14].

Mitochondrial dysfunction and oxidative stress share a bidirectional association, and have been implicated in cardiac hypertrophy and failure [15, 16]. Mitochondria, apart from being a producer of ROS, can themselves be attacked by these reactive intermediates of oxygen [15]. While oxidative stress induces cellular damage and death directly by attacking biomolecules such as membrane phospholipids and nucleic acids, and indirectly by activating maladaptive signalling cascades, mitochondrial dysfunction is implicated in the transition from compensatory hypertrophy to heart failure [16, 17]. Mitochondrial dysfunction, oxidative stress and their association, thus, assume importance in the biology, diagnostic approaches and therapy of cardiac hypertrophy and heart failure. Compared to other organs, heart is very sensitive to mitochondrial defects. Thus, mitochondrial diseases preferentially affect the heart. Similarly, impaired cardiac conditions due to defects in either OXPHOS or ETC are associated with mitochondrial dysfunction [18]. Any redox imbalance in cells leads to ROS/RNS-dependent modifications of key proteins which are involved in electron coupling and mitochondrial oxidative phosphorylation. This affects the function of proteins which contribute to cellular injury or death [18, 19]. A malicious cycle of these complex interactions could thus lead to decompensation of the failing heart.

Current treatments used in cardiac hypertrophy and failure are inadequate. However, new evidence has suggested changes in mitochondria plays a crucial role in progression and severity of the disease. Here in the present book chapter, we are going discuss the central role of mitochondria in developing cardiac hypertrophy and failure and future strategy to develop novel therapy.

25.2 Cardiomyocytes, Heart and Cardiovascular System

Cardiovascular system, as the name suggests, is composed of heart and blood vessels that form a body-spanning network for perfusion of tissues with blood. The cardiovascular system developed close to 600 million years ago, and evolved in complexity from a single chamber in primitive organisms such as Drosophila to a compartmentalized system in higher animals [20]. Despite the progressing structural complexity of the mammalian cardiovascular system, its primary function remains the same and its importance is underscored by the impact of the growing burden of cardiovascular diseases on not only morbidity and mortality but general quality of life.

Blood transports nutrients, gases, hormones and metabolites to the target tissues in the body through a network of arteries, veins and capillaries. Heart functions as the incessant pump that pushes blood through this network, by continuous contraction and relaxation. Located between the lungs slightly left to the sternum, and covered by a double-layered membrane called pericardium, a normal adult heart weighs 200–425 grams and beats about 100,000 times daily, pumping about 7571 litres of blood [21]. The heart is composed of two atria and two ventricles; the atrium and ventricle on the right side pump deoxygenated blood to the pulmonary circulation, whereas the chambers on the left side pump oxygenated blood into the arteries supplying the body. The left ventricle is the largest and strongest chamber of the heart that – with walls that are only about half an inch in thickness – pumps blood through the aortic valve into the body. Contractile activity of the heart, composed of systolic and diastolic phases, is regulated intrinsically at the level of electrical conduction through AV node, Purkinje fibres and SA node, and extrinsically by the autonomic nervous system [21].

Heart, the first organ formed in developing foetus, is built up of cardiomyocytes supported by a matrix of fibroblasts and nourished by vascular network of coronaries [22]. Cardiomyocytes are terminally differentiated contractile cells that constitute 80% of the volume of the myocardium despite being only 20% of the total number of cells. Capillaries in the myocardium are finely aligned with the cardiomyocytes in a one-to-one ratio [18]. Adult cardiomyocytes are binucleate cells containing abundant myofibrils arranged in sarcomeric contractile units and abundant mitochondria for energy generation. The mitochondria of cardiomyocytes have distinct structural features such as large numbers of tightly packed cristae formed from invaginations of the inner mitochondrial membrane to fulfil their high energy output [23]. Adjacent cardiomyocytes are connected end-to-end by the intercalated discs, and side-to-side by desmosomes. Contraction and relaxation of cardiomyocytes during each cardiac cycle is under fine regulation of cyclic increases and decreases in intracellular Ca2+ initiated by membrane depolarization and sustained by Ca2+ release and re-uptake by the sarcoplasmic reticulum. Keeping in view their low proliferative capacity, adult cardiomyocytes undergo hypertrophic growth to compensate for the increased workload in conditions of stress [24].

25.3 Mitochondrial Dysfunction and Oxidative Stress: Emphasis on the Heart

25.3.1 Mitochondria: Essential Organelle of Cardiomyocytes

Mitochondria are double membrane-bound, rod-shaped and semi-autonomous organelles. Thought to have originated from an ancient symbiotic event wherein a nucleated cell engulfed an aerobic prokaryote, mitochondria still retain some characteristics of modern prokaryotes such as their own circular genome, a membrane laced with electron transport proteins, and the ability to divide by simple fission independent of the cell [25, 26]. While the outer mitochondrial membrane is sufficiently porous to allow passage of ions and small proteins, the inner mitochondrial membrane exhibits restricted permeability similar to that of cell membrane. The inner membrane surrounds the mitochondrial matrix that is the site for citric acid cycle, the electrons produced from which travel through the electron transport chain in this membrane. The protein complexes of the ETC push protons into the intermembrane space to create a gradient that utilised to the synthesis of ATP, coupled with the reduction of oxygen to water at the end of ETC, thus the name oxidative phosphorylation for the process. Majority of the mitochondrial proteins, such as enzymes required for the citric acid cycle, the proteins involved in DNA replication and transcription, and ribosomal proteins, are synthesised from nuclear genes. The 16.5 kb mitochondrial genome encodes for 13 OXPHOS subunits, 22 tRNAs and two rRNA subunits, transcribed from the light-strand promoter and the heavy strand promoter, under the control of the mitochondrial transcription factor A (TFAM) [26].

Mitochondria in the cardiomyocytes show distinct subcellular distribution in the sarcolemmal, perinuclear and intrafibrillar regions, and are relatively firmly fixed [7]. Mitochondrial fission, fusion, and autophagy machinery in adult cardiomyocytes regulates the energy generation and structural integrity of the organelle, and is under strict physiological control. Altered expression of proteins that regulate mitochondrial dynamics, such as biogenesis, fragmentation and hyperplasia, has been reported in humans as well as animal models of heart failure [7, 27]. Since fatty acid oxidation that takes place in the mitochondrial matrix is the primary source of energy for an adult heart, mitochondria are the central organelles that regulate energy metabolism in the heart and correlate with its function and oxygen consumption [28]. Events involved in cardiac contraction and relaxation, such as release of actin from myosin and sequestration of calcium during diastole, utilize about 90% of the ATP produced in the heart [29]. Abnormalities such as impaired ETC complex activity, depleted mtDNA, and disrupted import of proteins into the mitochondria directly affect cardiac oxidative phosphorylation. Mutations in the mtDNA are associated with many inherited familial cardiomyopathies, as expression of mutant proteins in the mitochondria disturbs the energy homeostasis [30].

25.3.2 Mitochondrial Dysfunction and Oxidative Stress: Implication in Cardiac Disease Pathology

Mitochondrial dysfunction, characterized by a loss of efficiency in the electron transport chain and reduced synthesis of high-energy molecules such as ATP, is a characteristic of aging, and essentially, of all chronic diseases [31]. Mitochondrial abnormalities in cardiac hypertrophy and failure, apart from reduced capacity to generate ATP, are directly linked to cardiomyocyte damage and, therefore, link to disease progression. Dysfunctional mitochondria are a major source of reactive oxygen species (ROS) production, which can induce cellular damage. Abnormal mitochondria and ROS can trigger programmed cell death through the release of cytochrome c into the cytosolic compartment and activation of caspases [32]. Mitochondrial dysfunction and oxidative stress are both linked to aberrant cellular calcium homeostasis, vascular smooth muscle pathology, myofibrillar disruption, and altered cell differentiation, all implicated in the etiology of cardiac hypertrophy and failure [33].

ROS is a by-product of cellular metabolism of oxygen and performs an important role in cellular signalling and homeostasis [12]. In normal physiology, ROS exist in a delicate balance with the cellular antioxidants, such as MnSOD and catalase, to maintain redox equilibrium in the cell. Any alteration in this equilibrium that exhausts the antioxidant's capacity to counter balance ROS may lead to oxidative stress in the cell. Oxidative stress causes cell injury and death through nucleic acid damage and oxidation of several biomolecules including amino acids, fatty acids and enzymes [11]. Mitochondria, being the major contributor of ROS, play an important role in this equilibrium. In cardiomyocytes, approximately 90% of the antioxidant activity is performed by MnSOD, located in the mitochondria [34]. Oxidative damage to mitochondria, compromises energy production, and progression of mitochondrial dysfunction contributes to further cardiac structural and functional defects, thus driving a vicious cycle of cardiac damage and failure [35].

25.3.3 Oxidative Stress-Induced Mitochondrial Dysfunction

During the normal functioning of cell, 2% of oxygen is converted to superoxide due to incomplete conversion to water in electron transport chain [36]. This superoxide is detoxified by endogenous antioxidants in the cell such as MnSOD, catalase, GSH etc. However, during the time of perturbed redox balance, ROS production in cells may be increased, which cause damage to cell as well as mitochondria. Increased superoxide may react with membrane phospholipids and produce highly reactive malonaldehyde (MDA) or 4-hydroxy-2-nonenal (HNE), which may further damage the cellular lipids, proteins and nucleic acids [37].

Cellular metabolic performance directly correlates with the cellular antioxidant response and NADPH level [38]. Increased ROS and lipid peroxidation level has shown to be increased in tissues and pericardial fluid of heart failure patients, which directly correlates with the contractile dysfunction [36]. Apart from this, protein carbonyls, an oxidation products of important amino acids such as cysteine and lysine, have been shown to be increased in myofilament oxidation and systolic failure in end stage human heart failure [18]. In the damaged mitochondria, the blocked electron transport chain leads to further accumulation of excess free radicals. Mitochondrial DNA is located very proximity to this source of ROS and may get damaged, introducing mutations in the mtDNA. Increased accumulation of mutated mtDNA in the cell, may further lead to chronic innate inflammatory response in cells [39].

25.3.4 Mitochondrial Dysfunction-Induced Oxidative Stress

Mitochondria are the adapted organelles in the cell. It has been believed that, mitochondria are a prokaryotic cell which was incorporated in mammalian cell long back and got symbiotically associated in the cell permanently. Mitochondria are the energy power house of the cell. It continuously burns energy sources to produce energy rich molecules ATP. Therefore, dysfunction of this important organelle in heart directly enhances ROS generation, which could be very critical to cause cardiac hypertrophy and failure. There are many causes of mitochondrial dysfunction, which have been discussed below.

25.4 Sources of ROS in Cells

All sources of ROS in cell as depicted in Fig. 25.1 have been reported to be involved in cardiac hypertrophy and failure [40, 41]. Different sources of ROS activated by different kind of pathological conditions play important role in the generation of unfavourable condition and maladaptive changes during stress condition.

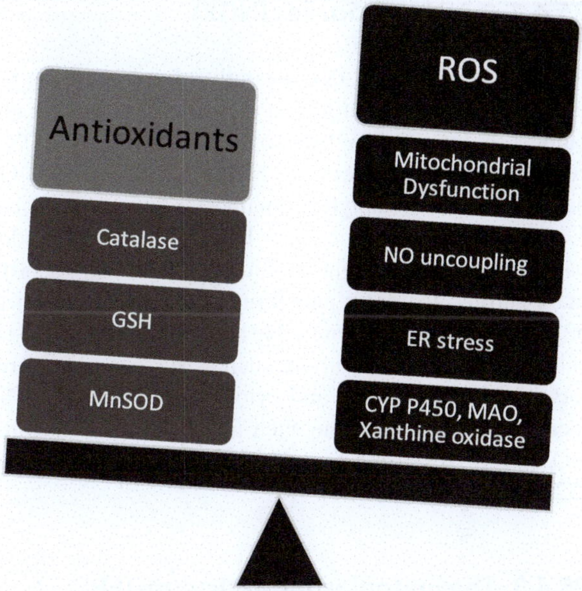

Fig. 25.1 Oxidative stress is defined as an imbalance between endogenous antioxidants and reactive oxygen species (ROS)

25.4.1 Mitochondria

Maintaining a proper balance between oxidants and antioxidants in the mitochondria is very crucial for healthy redox balance in cell [33]. In the mitochondria, ROS is generated due to incomplete reduction of O2 to water, leading to generation of $O2^-$ and peroxide molecules. During mitochondrial respiration, O2 may escape through electron transport chain to generate highly reactive oxygen species [42]. Mitochondrial enzyme such as, mono-amino oxidase (MAO) can also contribute to the generation of ROS. Elevated ROS generation in cell have been reported in many cardiac diseases. Recently we reported an increased ROS generation and lipid peroxidation in rat model of cardiac hypertrophy and diabetic cardiomyopathy [43, 44]. This elevation in mitochondrial ROS generation can be very detrimental in damaged cells and play a vital role in disease progression from cardiac hypertrophy to failure [45]. However, mitochondria possess their own defence system to protect from over production of ROS. Mitochondrial antioxidants are crucial for survival, as genetic deletion of mitochondrial antioxidant gene TRX reductase 2 is embryonically lethal due to cardiac dysfunction [46]. In vitro, inhibition of TRX reductase 2 in isolated mitochondria, results in increased H2O2 generation [47]. Similarly, genetic deletion of mitochondrial Mn-SOD leads to fatal dilated cardiomyopathy. On the other hand, mice overexpressing mitochondrial-targeted catalase found to be protected from cardiac disease as well as showed a prolonged life span [48].

25.4.2 NADPH Oxidases (NOX)

NADPH oxidases (NOX), a family of enzymes are a major source of ROS, and implicated in many cardiovascular diseases. Among all sources of ROS, NOX are unique, as the primary function of NOX is only ROS production. NOX family is composed of seven members NOX1-NOX5, and DUOX1 and DUOX2. Each one has different core catalytic domain. Cardiomyocytes mostly express NOX1, 2, 4 and 5, and generate low amount of ROS to regulate of redox balance, redox signalling pathway, cell differentiation and proliferation [49, 50]. NOX2 and NOX4 are the main cardiac isoforms and form a heterodimer with a 22 kD subunit termed as p22phox. NOX2 is an inducible sarcolemmal enzyme that can be induced by many cellular hypertrophic stimuli, such as sympathetic β1 receptor agonist, growth factors, mechanical forces and cytokines. On the other hand, NOX4 is intracellular membrane bound enzyme such as ER and is constitutively expressed [51]. Both, NOX2 and NOX4 are found to be upregulated in cardiac hypertrophy and failure [52, 53].

25.4.3 The Endoplasmic Reticulum (ER)

Structurally endoplasmic reticulum (ER) is composed of lumen, which is responsible for most of the biochemical protein modifications. The ER lumen has an extreme oxidising environment which is required for oxidation and folding of proteins. Two enzymes, lysyl oxidase (LOX) and prolyl oxidase (PHD), use oxygen (O2) in the process of oxidation of Lysine and proline, respectively. ER also plays a vital role in introducing disulfide bonds into nascent proteins. In the first step, enzyme ER oxidase 1 (Ero1) is oxidized by molecular O2 to generate H2O2. In second step, protein disulfide isomerase (PDI) transfers the disulfides from Ero1 to nascent proteins, which is necessary for normal protein folding. Thus, ER stress can activate oxidases and generate H2O2. It is reported that, prolonged ER stress is associated with cardio myocyte apoptosis and heart failure [54].

25.4.4 Monoamine Oxidases (MAO)

Monoamine oxidase (MAO) performs the oxidative deamination of catecholamines. It is principally located in the mitochondria and makes it a very important source of ROS during pathological conditions. The reaction of oxidative deamination results in the generation of H2O2, which is an important source of ROS in the mitochondria. It has been reported that, MAO is an important source of oxidative stress in mouse model of pressure overload and contributes significantly towards cardiac dysfunction [55, 56]. The mitochondrial location of MAO is also crucial in its mechanism of oxidative stress in heart.

25.4.5 Uncoupled NO Synthases (NOS)

Cardiomyocytes can express both endothelial and neuronal nitric oxide synthase (NOS). Under certain conditions, such as inflammation and cellular stress condition, inducible nitric oxide synthase (iNOS) is over-expressed. During many pathological conditions, such as hypertension, and cardiac hypertrophy, NO may get uncoupled and results in increased generation of ROS and oxidative stress in the cardiomyocytes [57].

25.4.6 Cytochrome P450 Oxidase

Cytochrome P450 (CYP P450) family enzymes are involved in the metabolism of drugs in the body. CYP P450 oxidase enzymes oxidise its substrate and could be an important source of oxidative stress under certain circumstances. CYP P450 oxidase is known to be upregulated in heart diseases such as cardiac hypertrophy. CYP P450 knockout mice attenuated cardiac dysfunction when they were crossed with a dilated cardiomyopathy mouse model [58].

25.4.7 Xanthine/Xanthine Oxidase

Xanthine dehydrogenase (XDH) and xanthine oxidase (XOD) are a single gene protein that may exist in two different but interconvertible forms [59]. XDH acts on hypoxanthine or xanthine in presence of NAD, a cofactor and produce NADH, whereas, XOD acts on same substrate but utilizes O2 as a cofactor to produce superoxide anion ($O2^-$) and uric acid. During the pathological condition of ischemia and hypoxia, xanthine dehydrogenase (XDH) is converted into xanthine oxidase (XOD) which then utilizes O2 to produce superoxide and ROS [60]. Although, XOD appears to be an important source of ROS in cardiac pathological conditions like ischemia, its role in the development of cardiac hypertrophy remains to be unleashed [61]. It should be taken into account that, XOD is expressed in very low amount in human hearts. Therefore, extent of XOD mediated ROS contribution to human cardiac pathologies remains elusive [62].

25.5 Causes of Mitochondrial Dysfunction

There are different causes of mitochondrial dysfunction as mentioned below (Fig. 25.2).

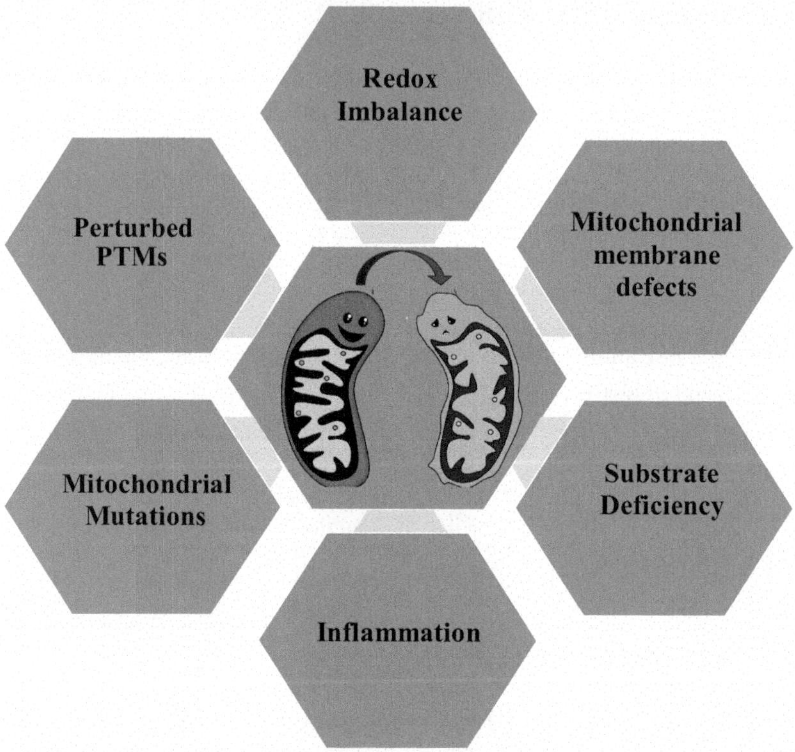

Fig. 25.2 Causes of mitochondrial dysfunction

25.5.1 Nutritional and Substrate Deficiency

Nutritional or substrate deficiency in the mitochondria may arise due to defect in the mitochondria or in the cell itself. This will directly affect energy production and cellular homeostasis. Apart from lack of substrate, sometimes metabolic enzyme deficiency in the ETC cycle may also contribute towards mitochondrial dysfunction.

It has been reported that, deficiency of coenzyme Q or L carnitine may lead to mitochondrial dysfunction. Data has shown that, supplementation of deficient nutrient can restore the mitochondrial function to normal [31].

25.5.2 Redox Imbalance Due to Raised Oxidative Stress

Reactive oxygen species play an important role in protection against infectious agents. ROS is present in the cell always in equilibrium with antioxidants. Disturbed redox balance due to over production of ROS may lead to toxic effects at its site of origin i.e. mitochondria and other organelles in cell.

It has been shown that, increased ROS is associated with mitochondrial dysfunction in cardiac hypertrophy and failure [63, 64]. Therefore, attenuating the increased cellular ROS using antioxidants becomes an attractive target for the treatment of mitochondrial dysfunction.

25.5.3 Post-Translational Modification of Mitochondrial Proteins

In our laboratory, recently Pankaj et al. reported the association between acetylation of proteins and mitochondrial dysfunction in diabetic heart [65]. Sirtuins are the group of enzymes, which regulates the acetylation of proteins in cell as well as in the mitochondria. Mitochondrial protein acetylation regulates enzymatic activity of some very important proteins such as TFAM, PGC-1α and p53 [66].

Pankaj et al. has also reported the decreased activity of SIRT-3, a mitochondrial sirtuin in heart during diabetic cardiomyopathy. Restoring the SIRT-3 activity in diabetic hearts using SIRT-3 activator, resveratrol, leads to reversal of cardiomyopathy [67, 68]. Now a day, SIRT-3 activator is considered as an attractive target to treat mitochondrial dysfunction in cardiomyopathy which can also be extended to other cardiac diseases.

25.5.4 Inflammatory Cytokines

Inflammatory cytokines, TNF alpha, interleukin 6 and interleukin 1 beta affects the function of mitochondria and cellular energy production. Wendy et al. reported that, treatment of adipocytes with these cytokines in vitro leads to mitochondrial dysfunction and increases ROS production [69].

Inflammatory cytokines are critical part of immune responses and plays an important role in more than one system in cell. Directly neutralizing these cytokines could be detrimental for the cell. Hence, the receptors, which controls the secretion of these cytokines becomes promising target for mitochondrial dysfunction. Recently we have shown that, inhibition of toll like receptor 4 (TLR4) in cardiac hypertrophy model in rats could be novel therapeutic approach to attenuate mitochondrial dysfunction and cardiac hypertrophy [43]. Other TLRs (i. e. TLR9) are also being explored to establish as novel target for the attenuation of mitochondrial dysfunction in heart failure [70].

25.5.5 Defective Mitochondrial Membrane

Defects in the mitochondrial membrane can lead to increased proton leak from mitochondria and decreased energy production efficiency. Apart from proton leak, defective mitochondrial membrane may also leak cytochrome C in the cytoplasm, which in turn may induce cellular apoptosis in cardiomyocytes.

Replacement of defective mitochondrial membrane components with externally administered phospholipid can rescue mitochondria from dysfunction in this condition.

25.5.6 Mitochondrial DNA Mutations

Mitochondria contain a separate small circular DNA in its matrix. It does not have any specific compartment to keep its DNA safe. Hence mitochondrial DNA is always exposed to all metabolites and ROS that generated in mitochondria during oxidative phosphorylation. Due to high exposure to ROS, mitochondrial DNA is highly prone to mutations. Many a time, even though mutation may occur in DNA of few mitochondria, the phenotype doesn't occur due to the presence of healthier DNA copies in rest mitochondria [71]. However, if the specific mutation is present in all copies of mtDNA, phenotype becomes apparent and mitochondria may become dysfunctional depending upon the gene effected by mutation.

There are many reports of mitochondrial mutations leading to mitochondrial dysfunction [70, 72]. The best available treatment in this case is symptomatic and mainly depends upon combination supply of nutrients.

25.6 Mitochondrial Dysfunction and Oxidative Stress: Crucial Factors in Cardiac Hypertrophy and Heart Failure

The myocardium is the most mitochondria-dense and oxygen-consuming tissue of the body. These features make heart a site vulnerable to development of mitochondrial dysfunction and oxidative stress. Hypertrophy in heart is considered as a compensatory response to increased workload, which when unmitigated, becomes pathological and transcends to heart failure. This transition is associated with extensive myocardial remodelling involving increased protein turnover, myocyte hypertrophy, fibrosis and apoptosis. Mitochondrial dysfunction is considered to be an important determinant of the transition from hypertrophy to failure, and this may be due to its role in the development of energy deficit which leads to impaired contractile activity. Energy generation in cells through mitochondrial electron transport chain is associated with 'leakage' of oxygen free radicals despite tightly linked ETC enzyme complexes. Generation of ROS from mitochondrial ETC is known to increase in failing hearts, apart from its other sources such as NADPH oxidases, uncoupled NOS and xanthine oxidases.

Mitochondrial respiratory dysfunction in cardiac hypertrophy and heart failure has been extensively reported in both rodents and humans. Du et al. (2017) showed that doxorubicin induced cardiac hypertrophy in rats involved mitochondrial dysfunction including mitochondrial permeability transition pore (mPTP) opening, loss of mitochondrial membrane potential ($\Delta\Psi m$), and respiration dysfunction. However, Sirt3 overexpression ameliorated hypertrophy by inhibiting these respiratory chain defects [73]. In another experimental study, Wust et al. (2016) showed that, right ventricular hypertrophy and failure in monocrotaline induced pulmonary artery hypertension induced in rats was accompanied by ~3.5 fold and two-fold reduction in Complex I and Complex II coupled respiration, respectively [74]. Mitochondrial remodeling was also evident in ultrastructural studies where the authors concluded that these impairments could mainly be attributed to dysfunctioning at Complex I. Apart from mitochondrial respiratory chain defects, other factors such as impaired calcium handling, metabolic remodeling and structural dynamics of mitochondria have also been cited as determinants of heart failure. In a mouse model of heart failure, Santulli et al. (2015) found that diastolic Ca^{2+} leak from sarcoplasmic reticulum causes mitochondrial calcium overload and dysfunction [75]. Wai et al. (2015) highlighted the importance of balanced mitochondrial fission and fusion in mice hearts, where processing of long isoform of optic atrophy protein 1 (L-OPA1) by OMA1 and mitochondrial fragmentation led to dilated cardiomyopathy and heart failure [76]. Lately, it has also been recognised that heart failure is a state of metabolic impairment of the myocardium, an independent risk factor that is also linked to mitochondria. In the metabolomic profiling study, Hunter et al. (2016) identified novel circulatory markers of dysregulated fatty acid oxidation which were differentially elevated in cases of heart failures with preserved as well as reduced ejection fraction [77].

Oxidative stress, independent of mitochondrial dysfunction, resulting from overproduction of free radicals or exhausted endogenous antioxidants or both, is frequently reported in models of cardiac dysfunction such as diabetic cardiomyopathy, dilated cardiomyopathy and myocardial infarction, both in experimental as well as clinical studies. All these conditions involve cardiac hypertrophy, ventricular remodeling and progression to failure, where oxidative injury plays a key role. Cardiac hypertrophy in streptozotocin-treated type 1 diabetic rats was found to be coupled with oxidative stress measured as increased lipid peroxidation and proteins oxidation, and activation of c-Jun Nuclear Kinase-1 [78]. We reported similar observations in a model of type 2 diabetes, where 8 week-long fructose feeding to rats resulted in cardiac hypertrophy and myocardial oxidative stress through increased NOX activity and ROS content [65]. Low superoxide dismutase (SOD) activity and high malondialdehyde levels have been reported in serum of rats with isoproterenol-induced cardiac hypertrophy [79]. These findings were linked with expression of pro-inflammatory and fibrotic markers such as collagen 1 and 3, TNF alpha, IL6, IFN-γ, phospho-IκBα, NF-κB p65, JAK2 and STAT3. Increase in membrane lipid peroxidation and decrease in expression of antioxidant genes like NRF2, HO-1 and NQO-1 has been reported in abdominal aortic constriction model (AAC) of cardiac hypertrophy in rats [80]. These changes were seen to be associated with pressure overload-induced inflammation and fibrosis in rat hearts. Oxidative injury is known to stimulate activation of matrix metalloproteinases, a class of proteolytic enzymes involved in tissue remodeling which is seen to be activated in failing hearts. Apart from its link with structural impairments resulting from inflammation, fibrosis and remodeling, oxidative stress also affects cardiac function. Yokoe et al. (2017), showed that oxidative stress upregulates protein von Hippel-Lindau expression to induce degradation of phospholamban – a key modulator of cardiac contractility – in mouse hearts with dilated cardiomyopathy [81]. In adult db/db mice, improvement of cardiac fractional shortening by mTOR inhibition was reported to be linked with attenuation of oxidative stress [82].

From all studies described above, strong link between mitochondrial dysfunction and oxidative stress is evident in the context of cardiac hypertrophy and heart failure. Disruption of the components of mitochondrial respiratory and metabolic reactions results in oxidative stress, while oxidative stress may itself compromise mitochondrial structure and function. Benderdour et al. (2004) reported that cardiac hypertrophy in coarctated rats is associated with decreased mitochondrial isocitrate dehydrogenase activity, protein, and gene expression, possibly through an adduct formation with 4-hydroxynonenal (4-HNE) [83]. Mice with knockout of mitochondrial isocitrate dehydrogenase (idh2$^{-/-}$), a regulator of mitochondrial redox balance, display cardiac hypertrophy, apoptosis, mitochondrial dysfunction and heart failure [84]. Caldas et al. (2015) showed that diazoxide (a mitoKATP channel opener) reduced oxidative stress in isoproterenol-treated mice [85]. All these studies prove that, oxidative stress and mitochondrial dysfunction plays a crucial role in the pathogenesis of cardiac hypertrophy and failure. Inhibition of these aetiological factors could be very beneficial to attenuate cardiac hypertrophy and failure.

25.7 Inflammation and Apoptosis: Two Major Mechanisms Linked to Oxidative Stress and Mitochondrial Dysfunction

Mitochondrial dysfunction and oxidative stress may activate multiple pathways in cardiomyocyte, which all together leads the cell towards senescence. As stated above, mitochondrial dysfunction itself can lead to increased oxidative stress, which leads to a decrease in nicotinamide adenine dinucleotide (NAD) concentration in the cell. NAD is very important cofactor for sirtuin activity. Decreased NAD concentration leads to decreased sirtuin activity in the cell and mitochondria. Sirtuins, which regulates the acetylation of many important proteins in the cell including p53 and tubulin, are very important for cell survival. Decreased sirtuin activity in cell, follows increased acetylation of p53, which then translocate to nucleus to initiate an apoptosis cascade through caspase pathway [86].

Mitochondrial dysfunction, may render mitochondrial membrane leaky, which then can lead to perturbed mitochondrial membrane potential and release of cytochrome C in the cytoplasm. Cytochrome C can bind to its other partners in the cytoplasm and increases the active IL-1 beta concentration through the activation of NLRP3 –caspase 1 pathway. IL-1 beta then can lead the cell towards inflammation and apoptosis.

During cellular damage and mitochondrial dysfunction, mitochondrial DNA may be leaked into cytoplasm. Mitochondrial DNA is a potent activator of TLR pathway. Through the activation of TLR pathway, it increases the production of proinflammatory cytokines including IL-6 and TNF alpha by activating transcription

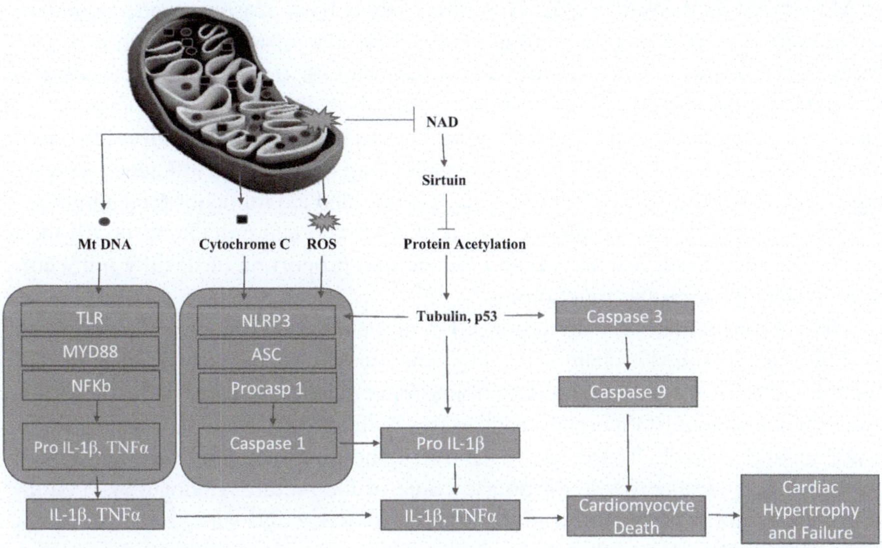

Fig. 25.3 Mechanism of cardiac hypertrophy and failure by mitochondrial dysfunction

factor, nuclear factor kappa B (NFkB). These inflammatory cytokines may then lead to apoptosis in cardiomyocyte through caspase activation.

Altogether, the ultimate fate of cell moves it towards senescence after mitochondrial dysfunction. Cardiomyocytes are terminally differentiated cells and hence, apoptosis of these cells is crucial factor in development of cardiac hypertrophy and its transition towards failure. It has been reported that, cardiac hypertrophy to failure progression is majorly dominated by inflammation and apoptosis of cardiomyocytes (Fig. 25.3) [87].

25.8 Treatment Strategy to Attenuate Mitochondrial Dysfunction and Oxidative Stress in Cardiac Hypertrophy and Heart Failure

Mitochondrial damage is crucial in the pathogenesis of heart failure. Mitochondrial dysfunction plays a central role to induce mitochondrial damage, which in turn impairs the contraction of heart. A reduced energy supply from mitochondria due to age related oxidative stress leads to a diminished contractile function of cardiomyocytes. Therefore, it becomes necessary to control and restore the mitochondrial health in heart failure [88].

Treating mitochondrial dysfunction is very difficult task as compared to targeting membrane receptors as the drug which is meant for targeting mitochondria has to cross biological cell membrane as well as mitochondrial membrane. Due to this condition, only highly lipophilic drugs are used for this condition.

Mitochondrial dysfunction develops due to three basic reasons. Inadequate number of mitochondria being the leading cause followed by insufficient supply of mitochondrial substrates or a dysfunction of the electron transport chain leading to insufficient ATP supply. Mitochondrial equilibrium in the cell is maintained by three basic mechanisms, (1) Fusion of two dysfunctional mitochondria leading to mixing up of their undamaged components leading to increased efficiency of fused mitochondria (2) mitochondrial division (fission), and (3) mitochondrial autophagy (mitophagy) leading to complete removal of dysfunctional mitochondria. All these events play important role in maintaining enough proportion of healthy mitochondria in the cell for proper functioning.

Most of the cellular ATP are produced in mitochondria. The proton gradient created by electron transfer chain is used to generate ATP in the mitochondria by ATP synthase. One of the important effect of improper electron transfer in the mitochondria is generation of ROS, highly reactive free radicals, which can be detrimental for mitochondria and cell. Therefore, to neutralize these ROS, endogenous antioxidant enzymes such as superoxide dismutase come in the picture. Endogenous cellular antioxidants are mitochondrial glutathione peroxidase and superoxide dismutase among others. Apart from this, dietary antioxidants provide an additional protection against ROS. These natural antioxidants can be exploited to counter the excessive ROS produced during mitochondrial dysfunction.

Understanding the molecular mechanism is very important to attenuate the mitochondrial dysfunction. We have discussed these molecular mechanisms in upper section in detail. From detailed literature review, we have proposed the following treatment strategies that could be effective for the mitochondrial dysfunction (Fig. 25.4). Grossly we can divide the treatment strategies into two main headings as antioxidants therapy and agents that improve ETC efficiency.

Many supplements and vitamins are being used in the treatment of mitochondrial dysfunction and diseases. Among all of the nutrients like vitamins, cofactors, enzymes, enzyme inhibitors and herbs which are being used in this ailment, only few are truly effective. Here we enlisted few of the promising supplements for mitochondrial dysfunction. After complete analyses, we concluded that, combination of nutrient supplements, vitamins and co factors could be a better treatment option rather than mono therapy.

25.8.1 Alpha-Lipoic Acid

Alpha- lipoic acid is an antioxidant, ion chelator and anti-inflammatory agent. It is an important co-factor for tricarboxylic acid cycle (TCA) enzyme called alpha ketoglutaric acid dehydrogenase. It is one of the important component of oxidative phosphorylation in mitochondria. Alpha lipoic acid is used clinically as a supplement in diabetes mellitus. It has several beneficial effects in diabetes such as in neuropathy and inflammation. Most of these effects are attributed to modulation of gene regulation of glucose uptake and metabolism as well as its antioxidant property.

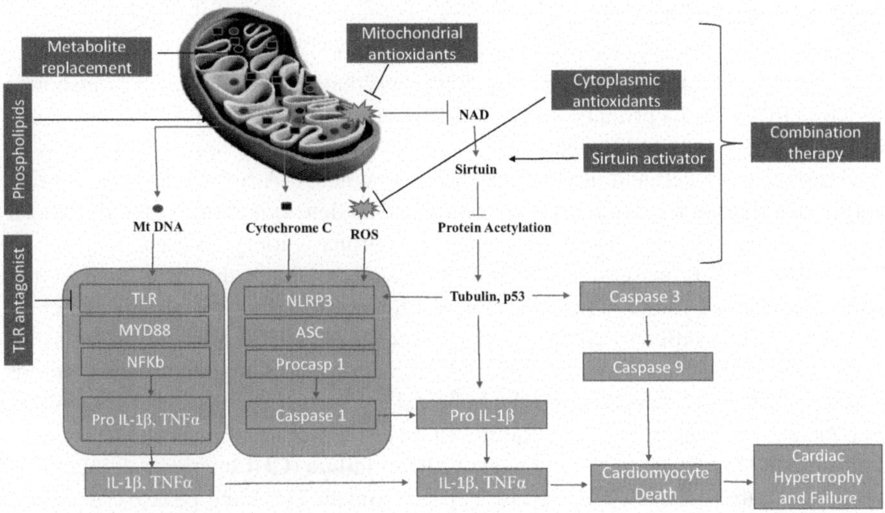

Fig. 25.4 Novel targets to reduce mitochondrial dysfunction in cardiac hypertrophy and heart failure

Accumulation of sphingolipids, particularly ceramide in mitochondria due to aging or any chronic disease condition could be fatal for cardiomyocyte. Ceramide accumulation in the cardiomyocyte is known to retard electron transport chain in the mitochondria. Alpha lipoic acid was found to decrease the ceramide level in vascular endothelium of cardiac muscle by inhibiting sphingomyelinase activity, leading to increased levels of mitochondrial glutathione and electron transport chain function. It has been found that, alpha lipoic acid is beneficial in diabetic complications. Clinical trial showed that neuropathic complications were improved significantly after alpha lipoic acid treatment.

Antioxidant property of alpha lipoic acid gives many beneficial effects in the cell, including decreased ROS and reduced activation nuclear factor kappa B (NFkB), which in turn reduces many cytokines and inflammatory gene expression. Being a transition metal chelator, alpha lipoic acid can remove excess amount of many metals including, copper, iron and lead, which are involved in many chronic diseases like, renal failure and Parkinson's disease.

In one of the recent clinical trial it has been found that alpha lipoic acid can improves cardiac dysfunction and prevents from development of diabetic cardiomyopathy [89]. The use of alpha lipoic acid in cardiac hypertrophy and heart failure has not been studied in clinical trials, but researchers believe that the molecule has great potential for mitigating the cardiac complications and mitochondrial dysfunction. Energy efficiency of mitochondria can be improved up to 30% using treatment strategies based upon increasing glucose oxidation and decreasing fatty acid metabolism [88]. Many studies have confirmed that alpha lipoic acid can attenuate the progression of cardiac remodelling and improve cardiac function by improving mitochondrial function and energy balance [88].

25.8.2 L-Carnitine

L-carnitine (3-hydroxy-4-N-trimethylaminobutyrate) is a fatty acid transporter found in the cells. Its primary function is to transport fatty acids into the mitochondrial matrix for beta oxidation. It also plays a crucial role in removal of excess of acyl groups from cell and modulating the coenzyme Co-A homeostasis balance. L carnitine is essential for the cell functioning and its deficiency can lead to decreased mitochondrial function, insulin resistance and coronary artery diseases.

L carnitine may increase the overall fatty acid uptake in the mitochondria and reduce the carbohydrate dependency of cell for energy production. This switch in the substrate utilization by cardiomyocytes could be beneficial during cardiac insult. L carnitine is being used in many disorders including cardiomyopathy, renal failure and sepsis. A small placebo controlled clinical trial was performed with L carnitine in patients with congestive heart failure. In this trial propionyl L carnitine administration found to be beneficial in congestive heart failure (CHF) patients. It increased peak heart rate (mean 12%), exercise capacity (mean 21%) and peak oxygen consumption (mean 45%) after L carnitine administration [90].

L carnitine can increase the aging impaired rate of mitochondrial oxidative phosphorylation. It has been found that, administration of acetyl L carnitine in old rats increases the fatty acid metabolism. It also restored decreased GSH level and improved mitochondrial complex IV activity in skeletal muscles.

It has been found that, administration of L carnitine in ischemic cardiomyopathy patients improved ejection fraction (mean 4.5%) as compared to placebo group [91]. In a meta-analysis of randomised controlled clinical trials, it was found that L carnitine treatment in congestive heart failure patients is effective. In participants with muscle weakness, fatigue and impaired mobility, treatment with L carnitine significantly improved physical fatigue, fatigue severity and mental fatigue. It also improved physical capacity and cognitive activity in these people. Few other trials also indicate that, administration of L carnitine can have a beneficial effect on mental as well as physical health (Table 25.1).

25.8.3 Coenzyme Q10

Coenzyme Q is a crucial cofactor in electron transport chain, where it improves the function of electron transport along with other mitochondrial complexes. It is also known as an antioxidant in reduced form. It plays an important role in regulation of gene expression of certain genes involved in metabolism and transport.

Several clinical trials were carried out to find the effect of coenzyme Q administration in physical exercise, hypertension and heart failure. Most of the studies showed moderate improvement in exercise capacity after coenzyme Q administration [109, 110]. However, nine randomised clinical trials that carried out to examine the effect of coenzyme Q administration in heart failure patients showed no significant improvement in ejection fraction or mortality [109, 111]. Another randomised double blind clinical trial was performed by Rosenfeldt et al. for a duration of

Table 25.1 Experimental evidences for treatment of mitochondrial dysfunction

Sr. No.	Treatment	Subjects	Outcome	References
1	Mitochondria-targeted antioxidants	Cardiac ischemia reperfusion disease model in sheep, Guinea pigs and rabbit	Reduced the infract size	[92–94]
2	Alpha lipoic acid	Patients with diabetic cardiomyopathy	Improves cardiac dysfunction	[89]
3	L-carnitine	Patients with congestive heart failure	It increased peak heart rate (12%), exercise capacity (21%) and peak oxygen consumption (45%)	[95]
4	Coenzyme Q10	1. Hypertension and heart failure	Improvement in exercise capacity	[96]
		2. Heart failure patients	Decreased mortality and improvements in symptoms	[97]
5	Nicotinamide adenine dinucleotide	Mice model of dilated cardiomyopathy	Preserves the heart from failing	[98]
6	Membrane phospholipids	Fibromyalgia and chronic fatigue syndrome	Improved mitochondrial function	[99–101]
7	Combination of oral supplement	1. Intractable fatigue and mitochondrial dysfunction	Behavioural, cognitive, sensory and improved muscle strength	[102, 103]
		2. Patients with congested heart failure	Significant decrease in left ventricular mass (17.1%) and increase in left ventricular ejection fraction (5.3%)	[104]
8	Targeting post translational modifications	1. Patients with heart failure	Improving metabolic and skeletal muscle function (results awaited)	[105]
		2. Patients with heart failure	Potential anti-remodelling agent (results awaited)	[106]
9	Aqueous extract of garlic	Rat model of cardiac hypertrophy	Attenuate the cardiac hypertrophy and restored mitochondrial health	[107, 108]
10	TLR4 inhibitor	Isoproterenol induced cardiac hypertrophy in rat	Attenuate mitochondrial dysfunction and cardiac hypertrophy	[43]

3 months to see the effect of coenzyme Q administration in heart failure. The study showed significant improvements in symptoms and mean exercise times [110].

Apart from cardiovascular diseases, coenzyme Q has also been used in neurological disorders. It has been shown that, coenzyme Q supplementation in Alzheimer's disease in rats significantly delayed brain atrophy and beta amyloid plaque formation. In a randomised, placebo controlled clinical trial, it has been showed that coenzyme Q treatment along with other supplements significantly reduces oxidative stress markers.

25.8.4 Nicotinamide Adenine Dinucleotide

Nicotinamide adenine dinucleotide (NAD) is a substrate for many enzymes and plays a crucial role in many redox reactions in the cell. Presence of NADH in the cell at appropriate concentration is essential, deficiency of which may lead to pellagra, characterised by dermatitis, dementia, diarrhoea and death. NADH in mitochondria delivers electrons from hydrolysed substrates to electron transport chain. In a reduced form, NADH acts as an antioxidant. Administration of NADH orally is not viable, since it gets degraded in the stomach. Therefore, it is generally administrated in the form of niacin, nicotinamide or nicotinic acid.

NAD plays a vital role in regulating the activity of sirtuins, a deacetylators in cell. Proper functioning of sirtuins requires NAD as a cofactor. Sirtuins deacetylase many important proteins in the cell including p53, MnSOD and NFkB, which are involved in many cellular functions ranging from cellular metabolism, cell survival and cell death. It has been reported that exogenous administration of NAD can block development of pathological hypertrophy in rats [112]. Diguet et al. reported that, administration of nicotinamide riboside preserves cardiac function in mice with dilated cardiomyopathy [98]. Sirtuin 3, which is located in the mitochondria, regulates the function and activity of mitochondrial complexes. Therefore, function and activity of sirtuin 3 is very much crucial in energy homeostasis in the mitochondria. NAD, maintain this energy homeostasis indirectly by regulating the function of sirtuin 3 in mitochondria.

Apart from cardiac diseases, NAD is also known to play an important role in the neurological disorders. Recently, it has been shown that, stable NADH administration could improve cognitive function in Alzheimer's diseases [113]. In another clinical trial, alzheimer patients were administered stable NADH. The study found significant improvement in visual construction and verbal fluency as compared to placebo group. However, this study failed to give any evidence of improvements in attention or memory [114].

25.8.5 Replacement of Membrane Phospholipids

Membrane phospholipids plays an important role in cellular homeostasis and signalling. Increased reactive oxygen species/reactive nitrogen species in cytoplasm may oxidise membrane phospholipids. Oxidation of phospholipids in cell and cellular organelles renders the membrane dysfunctional. Functionally damaged phospholipid membrane in mitochondria may fail to perform basic functions and hinder the signalling, and therefore, becomes leaky. Thus, it becomes necessary to replace damaged portions of membrane to avoid any further cellular damage and increase energy production by decreasing proton leak from mitochondria.

The dietary replacement of mitochondrial membrane phospholipid using an exogenous molecules has been proven very efficacious in improving mitochondrial function [99, 115, 116]. Mitochondrial function has been shown to be improved after oral administration of membrane phospholipids. This improvement is associated with decreased fatigue in fibromyalgia and chronic fatigue syndrome [101].

25.8.6 Combination Oral Supplement

Combination of antioxidant, phospholipids and mitochondrial cofactors could be a very efficient treatment to treat mitochondrial dysfunction. A formulation named ATP Fuel, a combination of membrane phospholipids, coenzyme Q, microencapsulated NADH, L carnitine, alpha ketoglutaric acid and other nutrients has been tried in a 2 month clinical trial to treat an intractable fatigue and mitochondrial dysfunction [102, 103]. Different parameters like behavioural, cognitive and sensory, and treatment effectiveness were assessed in the study. Investigator found that, there was a significant improvement in all the parameters after 2 months of treatment with ATP Fuel as compared to control group.

In another double blind, placebo controlled, randomised clinical trial conducted by Witte et al. to find the effect of micronutrient supplementation in congested heart failure patients for a period of 9 months. At the end of 9 months, there was a significant decrease in left ventricular mass (17.1%) and increase in left ventricular ejection fraction (5.3%) as compared to placebo controlled group. Whereas, six-minute walk test and inflammatory marker expression remained unchanged in the intervention group as compared to placebo group [104].

25.8.7 Mitochondria-Targeted Antioxidants

Mitochondria are the substantial source of ROS in heart. Therefore, targeting ROS directly in mitochondria, that is at the site of origin, could be a better treatment strategy to avoid further escalation of the damage in cardiomyocytes. ROS from mitochondria is known to play crucial role in the pathogenesis of cardiac hypertrophy and heart failure through mitochondrial dysfunction.

A novel class of antioxidants which can target ROS more specifically in mitochondria could be of great use in these conditions. Among the other options, mitochondrial targeted antioxidant peptides have been proven to be efficacious in ischemic reperfusion (I/R) injury [117]. In cardiac I/R disease model in sheep, guinea pigs and rabbit, SS-31 analogues, a mitochondrial antioxidant, reduced the infract size as compared to control [92].

25.8.8 Post Translational Modifications

The enzyme activity of many mitochondrial proteins is regulated by their acetylation status [118]. Acetylation of proteins at lysine residue in mitochondria is generally a negative regulator of mitochondrial enzymes. Mitochondria produces a large amount of acetyl Co A, which in turn renders mitochondrial proteins highly susceptible to acetylation.

A group of histone deacetylases named Sirtuins, which performs the function of deacetylation of histone and non-histone proteins. SIRT-3, SIRT-4 and SIRT-5 are known to present in mitochondria. Among them, SIRT-3 have robust deacetylating activity, which regulates the acetylation status of almost 80–90% of proteins in mitochondria [68]. SIRT-3 is an antiaging gene and known to be involved in multiple cellular signalling events including regulation of oxidative stress, mitochondrial biogenesis, metabolic activity, apoptosis and cardiac hypertrophy [119–121]. Decreased SIRT-3 activity in heart may lead to cardiac complications due to hyperacetylation of proteins in mitochondria [122]. In addition, SIRT-3 plays an important role in regulating whole body energy homeostasis by regulating the metabolic pathways in all organs [123].

Recently, we found decreased SIRT-3 activity and increased acetylation of many mitochondrial proteins including Mn-SOD and TFAM in diabetic heart [44]. This was followed by decreased activity of mitochondrial enzymes and mitochondrial dysfunction. This correlates with decreased myocardial SOD activity as observed in diabetic group. Mitochondrial dysfunction was attenuated after oral administration garlic homogenate in these rats. We found that, garlic homogenate increased the SIRT-3 activity in the heart of these animals which lead to restoration of mitochondrial enzyme activity [44].

25.8.9 Toll like Receptor (TLR) Inhibitors

The role of innate immunity as critical component of "adaptive cardiac biology" is being recently recognized [124]. The innate immunity pathway seems to play an important role in heart failure and pressure-overload-mediated cardiac decompensation. Damage associated molecular patterns (DAMPs) released from injured cardiac cells, which in turn act as a danger signal in the innate system [125]. Recently it has shown that, necrotic heart tissue may release DAMPs, which are potent enough to initiate the chronic inflammation in the myocardium [126]. Pathogen recognition receptors (PRRs), part of innate immunity, are critical in the identification of danger signals released from damaged cardiomyocytes [127]. Mitochondrial DNA released from cardiac cells during cardiac damage may act as a strong DAMP in heart and may initiate an inflammatory cascade and mitochondrial dysfunction. Inflammatory cytokine released by activation of PRRs in heart, activates NF-kB pathway which in turn increases expression of many cytokine including IL-1β, IL-6 and TNF-α [128], which are known to cause mitochondrial dysfunction. Recently, many studies have shown that TLR4 plays an important role in the cardiac adaptation during decompensated state of the system.

Recently, we investigated mitochondrial health in cardiac hypertrophy and found that TLR4 inhibition in isoproterenol induced cardiac hypertrophy can attenuate mitochondrial dysfunction [43]. To confirm the role of TLR4 on mitochondrial function in cardiac hypertrophy, we have analysed the protein expression of ETC complexes in heart of these animals. Protein expression of Complex-I, III, and V was significantly decreased in hypertrophy heart as compared to control. TLR4 inhibition in hypertrophy animals retained the mitochondrial complexes protein expression to normal as compared to disease group animals. Our data also suggest that mitochondrial enzyme activity is the direct outcome of TLR4 modulation. The activity of mitochondrial enzymes in hearts was found to be significantly decreased in hypertrophy animals. However, TLR4 inhibition in hypertrophy animals preserved the activity of these enzymes [43].

25.9 Conclusion

Mitochondria are cellular powerhouse and provide constant supply of ATP for heart. Constant exposure of mitochondria to high workload and oxidative stress makes it very susceptible to dysfunction and mutations. Being a largest source of ROS in the cell, mitochondria need a large pool of antioxidants to neutralize all the ROS that has been generated during ETC electron transfer. Maintaining mitochondrial antioxidant status alone may not be enough to avoid the malfunctioning of ETC chain. Apart from ROS, dietary deficiency of some important nutrients may also cause mitochondrial dysfunction. Heart being the highest energy consuming organ in the body, even mild dysfunction mitochondria may affect cardiac function. During the pathological states like cardiac hypertrophy and failure, when, heart is already facing energy deficit, mitochondrial dysfunction is a critical component of disease progression.

Among many attempts that have been done to target mitochondrial dysfunction in the treatment of cardiac hypertrophy and failure, few of them found to be successful. Monotherapy of mitochondrial dysfunction is difficult because of severity and complexity of the disease. Therefore, we found that, combination therapy with nutritional agents, antioxidants and phospholipid membrane components could be the better strategy. However, targeting mitochondrial dysfunction in chronic diseases like cardiac hypertrophy and failure where mortality is very high, extending the life expectancy even by few years could be a big achievement. Targeting mitochondrial dysfunction in these disease condition is still in its infancy, which warrant's more study in this area. More clinical trials considering mitochondrial targeted therapy in future may give more information regarding effectiveness of these therapy in cardiac hypertrophy and heart failure.

Acknowledgement PBK is thankful to Indian Council of Medical Research (ICMR) for awarding senior research fellowship (SRF). HLN is thankful to Council for Scientific and Industrial Research (CSIR) for awarding senior research fellowship (SRF).

Conflict of Interest The authors declare that they have no competing interests.

References

1. Kühlbrandt W (2015) Structure and function of mitochondrial membrane protein complexes. BMC Biol 13(1):89
2. Hill S, Sataranatarajan K, Van Remmen H (2018) Role of signaling molecules in mitochondrial stress response. Front Genet 9:225
3. Pagidipati NJ, Gaziano TA (2013) Estimating deaths from cardiovascular disease: a review of global methodologies of mortality measurement. Circulation 127(6):749–756
4. Lorell BH, Carabello BA (2000) Left ventricular hypertrophy: pathogenesis, detection, and prognosis. Circulation 102(4):470–479
5. McMullen JR, Jennings GL (2007) Differences between pathological and physiological cardiac hypertrophy: novel therapeutic strategies to treat heart failure. Clin Exp Pharmacol Physiol 34(4):255–262
6. Hernández-Reséndiz S, Buelna-Chontal M, Correa F, Zazueta C (2012) Oxidative stress and mitochondrial dysfunction in cardiovascular diseases. Oxidative Stress and Diseases 8:157–188. Available from: https://www.intechopen.com/books/oxidative-stress-and-diseases/oxidative-stress-and-mitochondrial-dysfunction-in-cardiovascular-diseases
7. Piquereau J, Caffin F, Novotova M, Lemaire C, Veksler V, Garnier A, Ventura-Clapier R, Joubert F (2013) Mitochondrial dynamics in the adult cardiomyocytes: which roles for a highly specialized cell? Front Physiol 4:102
8. Wang K, Xu Y, Sun Q, Long J, Liu J, Ding J (2018) Mitochondria regulate cardiac contraction through ATP-dependent and independent mechanisms. Free Radic Res 52:1–10
9. Hajnóczky G, Csordás G, Das S, Garcia-Perez C, Saotome M, Roy SS, Yi M (2006) Mitochondrial calcium signalling and cell death: approaches for assessing the role of mitochondrial Ca2+ uptake in apoptosis. Cell Calcium 40(5–6):553–560
10. Koppers AJ, De Iuliis GN, Finnie JM, McLaughlin EA, Aitken RJ (2008) Significance of mitochondrial reactive oxygen species in the generation of oxidative stress in spermatozoa. J Clin Endocrinol Metab 93(8):3199–3207
11. Sharma P, Jha AB, Dubey RS, Pessarakli M (2012) Reactive oxygen species, oxidative damage, and antioxidative defense mechanism in plants under stressful conditions. J Bot 2012:194–209
12. Nita M, Grzybowski A (2016) The role of the reactive oxygen species and oxidative stress in the pathomechanism of the age-related ocular diseases and other pathologies of the anterior and posterior eye segments in adults. Oxidative Med Cell Longev 2016:3164734
13. Moris D, Spartalis M, Tzatzaki E, Spartalis E, Karachaliou G-S, Triantafyllis AS, Karaolanis GI, Tsilimigras DI, Theocharis S (2017) The role of reactive oxygen species in myocardial redox signaling and regulation. Ann Transl Med 5(16):324
14. Santos CX, Anilkumar N, Zhang M, Brewer AC, Shah AM (2011) Redox signaling in cardiac myocytes. Free Radic Biol Med 50(7):777–793
15. Dai D-F, Johnson SC, Villarin JJ, Chin MT, Nieves-Cintrón M, Chen T, Marcinek DJ, Dorn GW, Kang YJ, Prolla TA (2011) Mitochondrial oxidative stress mediates angiotensin II–induced cardiac hypertrophy and Gαq overexpression–induced heart failure. Circ Res 108(7):837–846
16. Rosca MG, Tandler B, Hoppel CL (2013) Mitochondria in cardiac hypertrophy and heart failure. J Mol Cell Cardiol 55:31–41

17. Phaniendra A, Jestadi DB, Periyasamy L (2015) Free radicals: properties, sources, targets, and their implication in various diseases. Indian J Clin Biochem 30(1):11–26
18. Santulli G (2017) Mitochondrial dynamics in cardiovascular medicine, vol 982. Springer, Cham
19. Kohlhaas M, Nickel AG, Maack C (2017) Mitochondrial energetics and calcium coupling in the heart. J Physiol 595(12):3753–3763
20. Gunga H-C, von Ahlefeld VW, Coriolano H-JA, Werner A, Hoffmann U (2016) Cardiovascular system, red blood cells, and oxygen transport in microgravity. Springer, Cham
21. Katz AM (2010) Physiology of the heart. Lippincott, Williams, and Wilkins, Philadelphia, pp 427–461
22. Brand T (2003) Heart development: molecular insights into cardiac specification and early morphogenesis. Dev Biol 258(1):1–19
23. Dorn GW II (2013) Mitochondrial dynamics in heart disease. Biochim Biophys Acta 1833(1):233–241
24. Friehs I, Barillas R, Vasilyev NV, Roy N, McGowan FX, del Nido PJ (2006) Vascular endothelial growth factor prevents apoptosis and preserves contractile function in hypertrophied infant heart. Circulation 114(1_suppl):I-290–I-295
25. Shiflett A, Johnson PJ (2010) Mitochondrion-related organelles in parasitic eukaryotes. Annu Rev Microbiol 64:409
26. Skulachev VP (1999) Mitochondrial physiology and pathology; concepts of programmed death of organelles, cells and organisms. Mol Asp Med 20(3):139–184
27. Brown DA, Perry JB, Allen ME, Sabbah HN, Stauffer BL, Shaikh SR, Cleland JG, Colucci WS, Butler J, Voors AA (2017) Expert consensus document: mitochondrial function as a therapeutic target in heart failure. Nat Rev Cardiol 14(4):238
28. Aon MA, Tocchetti CG, Bhatt N, Paolocci N, Cortassa S (2015) Protective mechanisms of mitochondria and heart function in diabetes. Antioxid Redox Signal 22(17):1563–1586
29. Doenst T, Nguyen TD, Abel ED (2013) Cardiac metabolism in heart failure: implications beyond ATP production. Circ Res 113(6):709–724
30. Towbin JA, Lipshultz SE (1999) Genetics of neonatal cardiomyopathy. Curr Opin Cardiol 14(3):250
31. Nicolson GL (2014) Mitochondrial dysfunction and chronic disease: treatment with natural supplements. Integ Med 13(4):35
32. Zorov DB, Juhaszova M, Sollott SJ (2014) Mitochondrial reactive oxygen species (ROS) and ROS-induced ROS release. Physiol Rev 94(3):909–950
33. Bayeva M, Gheorghiade M, Ardehali H (2013) Mitochondria as a therapeutic target in heart failure. J Am Coll Cardiol 61(6):599–610
34. Violi F, Cangemi R (2005) Antioxidants and cardiovascular disease. Curr Opin Investig Drugs 6(9):895–900
35. Bhatti JS, Bhatti GK, Reddy PH (2017) Mitochondrial dysfunction and oxidative stress in metabolic disorders—a step towards mitochondria based therapeutic strategies. Biochim Biophys Acta (BBA) – Mol Basis Dis 1863(5):1066–1077
36. Sheeran FL, Pepe S (2006) Energy deficiency in the failing heart: linking increased reactive oxygen species and disruption of oxidative phosphorylation rate. Biochim Biophys Acta Bioenergetics 1757(5–6):543–552
37. Hauck AK, Bernlohr DA (2016) Oxidative stress and lipotoxicity. J Lipid Res. https://doi.org/10.1194/jlr.R066597
38. Kang J, Pervaiz S (2012) Mitochondria: redox metabolism and dysfunction. Biochem Res Int 2012:1–14
39. Kasahara T, Kato T (2017) What can mitochondrial DNA analysis tell us about mood disorders? Biol Psychiatry 9(83):731–738
40. Seddon M, Looi YH, Shah AM (2007) Oxidative stress and redox signalling in cardiac hypertrophy and heart failure. Heart 93(8):903–907
41. Sag CM, Santos CX, Shah AM (2014) Redox regulation of cardiac hypertrophy. J Mol Cell Cardiol 73:103–111

42. Kowaltowski AJ, de Souza-Pinto NC, Castilho RF, Vercesi AE (2009) Mitochondria and reactive oxygen species. Free Radic Biol Med 47(4):333–343
43. Katare PB, Bagul PK, Dinda AK, Banerjee SK (2017) Toll-like receptor 4 inhibition improves oxidative stress and mitochondrial health in isoproterenol-induced cardiac hypertrophy in rats. Front Immunol 8:719
44. Sultana MR, Bagul PK, Katare PB, Mohammed SA, Padiya R, Banerjee SK (2016) Garlic activates SIRT-3 to prevent cardiac oxidative stress and mitochondrial dysfunction in diabetes. Life Sci 164:42–51
45. Ide T, Tsutsui H, Kinugawa S, Utsumi H, Kang D, Hattori N, Uchida K, Arimura K-i, Egashira K, Takeshita A (1999) Mitochondrial electron transport complex I is a potential source of oxygen free radicals in the failing myocardium. Circ Res 85(4):357–363
46. Conrad M, Jakupoglu C, Moreno SG, Lippl S, Banjac A, Schneider M, Beck H, Hatzopoulos AK, Just U, Sinowatz F (2004) Essential role for mitochondrial thioredoxin reductase in hematopoiesis, heart development, and heart function. Mol Cell Biol 24(21):9414–9423
47. Stanley BA, Sivakumaran V, Shi S, MacDonald I, Lloyd D, Watson WH, Aon MA, Paolocci N (2011) Thioredoxin reductase 2 is essential for keeping low levels of H_2O_2 emission from isolated heart mitochondria. J Biol Chem. https://doi.org/10.1074/jbc.M111.284612
48. Dai D-F, Santana LF, Vermulst M, Tomazela DM, Emond MJ, MacCoss MJ, Gollahon K, Martin GM, Loeb LA, Ladiges WC (2009) Overexpression of catalase targeted to mitochondria attenuates murine cardiac aging. Circulation 119(21):2789–2797
49. Brandes RP, Weissmann N, Schröder K (2010) NADPH oxidases in cardiovascular disease. Free Radic Biol Med 49(5):687–706
50. Brown DI, Griendling KK (2009) Nox proteins in signal transduction. Free Radic Biol Med 47(9):1239–1253
51. Santos CX, Nabeebaccus AA, Shah AM, Camargo LL, Filho SV, Lopes LR (2014) Endoplasmic reticulum stress and Nox-mediated reactive oxygen species signaling in the peripheral vasculature: potential role in hypertension. Antioxid Redox Signal 20(1):121–134
52. Zhao Y, McLaughlin D, Robinson E, Harvey AP, Hookham M, Shah AM, McDermott BJ, Grieve DJ (2010) Nox2 NADPH oxidase promotes pathologic cardiac remodeling associated with Doxorubicin chemotherapy. Cancer Res 2664:2010
53. Kuroda J, Ago T, Matsushima S, Zhai P, Schneider MD, Sadoshima J (2010) NADPH oxidase 4 (Nox4) is a major source of oxidative stress in the failing heart. Proc Natl Acad Sci 107(35):15565–15570
54. Chin K-T, Kang G, Qu J, Gardner LB, Coetzee WA, Zito E, Fishman GI, Ron D (2011) The sarcoplasmic reticulum luminal thiol oxidase ERO1 regulates cardiomyocyte excitation-coupled calcium release and response to hemodynamic load. FASEB J 25(8):2583–2591
55. Kaludercic N, Carpi A, Nagayama T, Sivakumaran V, Zhu G, Lai EW, Bedja D, De Mario A, Chen K, Gabrielson KL (2014) Monoamine oxidase B prompts mitochondrial and cardiac dysfunction in pressure overloaded hearts. Antioxid Redox Signal 20(2):267–280
56. Kaludercic N, Carpi A, Menabò R, Di Lisa F, Paolocci N (2011) Monoamine oxidases (MAO) in the pathogenesis of heart failure and ischemia/reperfusion injury. Biochim Biophys Acta (BBA) – Mol Cell Res 1813(7):1323–1332
57. Landmesser U, Dikalov S, Price SR, McCann L, Fukai T, Holland SM, Mitch WE, Harrison DG (2003) Oxidation of tetrahydrobiopterin leads to uncoupling of endothelial cell nitric oxide synthase in hypertension. J Clin Invest 111(8):1201–1209
58. Lu D, Ma Y, Zhang W, Bao D, Dong W, Lian H, Huang L, Zhang L (2012) Knockdown of cytochrome P450 2E1 inhibits oxidative stress and apoptosis in the cTnTR141W dilated cardiomyopathy transgenic mice. Hypertension 60(1):81–89
59. Chung HY, Baek BS, Song SH, Kim MS, Im Huh J, Shim KH, Kim KW, Lee KH (1997) Xanthine dehydrogenase/xanthine oxidase and oxidative stress. AGE 20(3):127–140
60. Nishino T, Okamoto K, Eger BT, Pai EF, Nishino T (2008) Mammalian xanthine oxidoreductase–mechanism of transition from xanthine dehydrogenase to xanthine oxidase. FEBS J 275(13):3278–3289

61. Minhas KM, Saraiva RM, Schuleri KH, Lehrke S, Zheng M, Saliaris AP, Berry CE, Vandegaer KM, Li D, Hare JM (2006) Xanthine oxidoreductase inhibition causes reverse remodeling in rats with dilated cardiomyopathy. Circ Res 98(2):271–279
62. Halliwell B, Gutteridge JM (2015) Free radicals in biology and medicine. Oxford University Press, New York
63. Tsutsui H, Kinugawa S, Matsushima S (2008) Mitochondrial oxidative stress and dysfunction in myocardial remodelling. Cardiovasc Res 81(3):449–456
64. Tsutsui H, Kinugawa S, Matsushima S (2011) Oxidative stress and heart failure. Am J Phys Heart Circ Phys 301(6):H2181–H2190
65. Bagul PK, Deepthi N, Sultana R, Banerjee SK (2015) Resveratrol ameliorates cardiac oxidative stress in diabetes through deacetylation of NFkB-p65 and histone 3. J Nutr Biochem 26(11):1298–1307
66. Parodi-Rullán RM, Chapa-Dubocq XR, Javadov S (2018) Acetylation of mitochondrial proteins in the heart: the role of SIRT3. Front Physiol 9:1094
67. Bagul PK, Middela H, Matapally S, Padiya R, Bastia T, Madhusudana K, Reddy BR, Chakravarty S, Banerjee SK (2012) Attenuation of insulin resistance, metabolic syndrome and hepatic oxidative stress by resveratrol in fructose-fed rats. Pharmacol Res 66(3):260–268
68. K Bagul P, K Banerjee S (2013) Insulin resistance, oxidative stress and cardiovascular complications: role of sirtuins. Curr Pharm Des 19(32):5663–5677
69. Hahn WS, Kuzmicic J, Burrill JS, Donoghue MA, Foncea R, Jensen MD, Lavandero S, Arriaga EA, Bernlohr DA (2014) Proinflammatory cytokines differentially regulate adipocyte mitochondrial metabolism, oxidative stress, and dynamics. Am J Physiol Endocrinol Metab 306(9):E1033–E1045
70. Oka T, Hikoso S, Yamaguchi O, Taneike M, Takeda T, Tamai T, Oyabu J, Murakawa T, Nakayama H, Nishida K (2012) Mitochondrial DNA that escapes from autophagy causes inflammation and heart failure. Nature 485(7397):251
71. Taylor RW, Turnbull DM (2005) Mitochondrial DNA mutations in human disease. Nat Rev Genet 6(5):389
72. Lee SR, Kim N, Noh Y, Xu Z, Ko KS, Rhee BD, Han J (2016) Mitochondrial DNA, mitochondrial dysfunction, and cardiac manifestations. Front Biosci 21:1410–1426
73. Du Q, Zhu B, Zhai Q, Yu B (2017) Sirt3 attenuates doxorubicin-induced cardiac hypertrophy and mitochondrial dysfunction via suppression of Bnip3. Am J Transl Res 9(7):3360
74. Wüst RC, de Vries HJ, Wintjes LT, Rodenburg RJ, Niessen HW, Stienen GJ (2016) Mitochondrial complex I dysfunction and altered NAD (P) H kinetics in rat myocardium in cardiac right ventricular hypertrophy and failure. Cardiovasc Res 111(4):362–372
75. Santulli G, Xie W, Reiken SR, Marks AR (2015) Mitochondrial calcium overload is a key determinant in heart failure. Proc Natl Acad Sci 112(36):11389–11394
76. Wai T, García-Prieto J, Baker MJ, Merkwirth C, Benit P, Rustin P, Rupérez FJ, Barbas C, Ibañez B, Langer T (2015) Imbalanced OPA1 processing and mitochondrial fragmentation cause heart failure in mice. Science 350(6265):aad0116
77. Hunter WG, Kelly JP, McGarrah RW III, Khouri MG, Craig D, Haynes C, Ilkayeva O, Stevens RD, Bain JR, Muehlbauer MJ (2016) Metabolomic profiling identifies novel circulating biomarkers of mitochondrial dysfunction differentially elevated in heart failure with preserved versus reduced ejection fraction: evidence for shared metabolic impairments in clinical heart failure. J Am Heart Assoc 5(8):e003190
78. Adebiyi AO, Adebiyi OO, Owira PM (2016) Naringin mitigates cardiac hypertrophy by reducing oxidative stress and inactivating c-Jun nuclear kinase-1 protein in type I diabetes. J Cardiovasc Pharmacol 67(2):136–144
79. Zhao L, Wu D, Sang M, Xu Y, Liu Z, Wu Q (2017) Stachydrine ameliorates isoproterenol-induced cardiac hypertrophy and fibrosis by suppressing inflammation and oxidative stress through inhibiting NF-κB and JAK/STAT signaling pathways in rats. Int Immunopharmacol 48:102–109
80. Zhou S, Sun W, Zhang Z, Zheng Y (2014) The role of Nrf2-mediated pathway in cardiac remodeling and heart failure. Oxidative Med Cell Longev 2014:260429

81. Yokoe S, Asahi M (2017) Phospholamban is downregulated by pVHL-mediated degradation through oxidative stress in failing heart. Int J Mol Sci 18(11):2232
82. Das A, Durrant D, Koka S, Salloum FN, Xi L, Kukreja RC (2013) mTOR inhibition with rapamycin improves cardiac function in type 2 diabetic mice: potential role of attenuated oxidative stress and altered contractile protein expression. J Biol Chem. https://doi.org/10.1074/jbc.M113.521062
83. Benderdour M, Charron G, Comte B, Ayoub R, Beaudry D, Foisy S, Des Rosiers C (2017) Decreased cardiac mitochondrial NADP+-isocitrate dehydrogenase activity and expression: a marker of oxidative stress in hypertrophy development. Am J Phys Heart Circ Phys 287(5):H2122–H2131
84. Ku HJ, Ahn Y, Lee JH, Park KM, Park J-W (2015) IDH2 deficiency promotes mitochondrial dysfunction and cardiac hypertrophy in mice. Free Radic Biol Med 80:84–92
85. Caldas FRL, Leite IMR, Filgueiras ABT, de Figueiredo Júnior IL, de Sousa TAGM, Martins PR, Kowaltowski AJ, Facundo HTF (2015) Mitochondrial ATP-sensitive potassium channel opening inhibits isoproterenol-induced cardiac hypertrophy by preventing oxidative damage. J Cardiovasc Pharmacol 65(4):393–397
86. Hori YS, Kuno A, Hosoda R, Horio Y (2013) Regulation of FOXOs and p53 by SIRT1 modulators under oxidative stress. PLoS One 8(9):e73875
87. Hou J, Kang YJ (2012) Regression of pathological cardiac hypertrophy: signaling pathways and therapeutic targets. Pharmacol Ther 135(3):337–354
88. Skibska B, Goraca A (2015) The protective effect of lipoic acid on selected cardiovascular diseases caused by age-related oxidative stress. Oxidative Med Cell Longev 2015:313021
89. Hegazy SK, Tolba OA, Mostafa TM, Eid MA, El-Afify DR (2013) Alpha-lipoic acid improves subclinical left ventricular dysfunction in asymptomatic patients with type 1 diabetes. The review of diabetic studies. Rev Diabet Stud 10(1):58
90. Anand I, Francis G, Maseri A, Milazzotto F, Pepine C, Yusuf S, Beaufils P, Ferrari R, Poole-Wilson P, Remme W (1999) Study on propionyl-L-carnitine in chronic heart failure. Eur Heart J 20(1):70–76
91. Gürlek A, Tutar E, Akçil E, Dinçer İ, Erol Ç, Kocatürk PA, Oral D (2000) The effects of L-carnitine treatment on left ventricular function and erythrocyte superoxide dismutase activity in patients with ischemic cardiomyopathy. Eur J Heart Fail 2(2):189–193
92. Kloner RA, Hale SL, Dai W, Gorman RC, Shuto T, Koomalsingh KJ, Gorman JH III, Sloan RC, Frasier CR, Watson CA (2012) Reduction of ischemia/reperfusion injury with bendavia, a mitochondria-targeting cytoprotective peptide. J Am Heart Assoc 1(3):e001644
93. Dai W, Cheung E, Alleman RJ, Perry JB, Allen ME, Brown DA, Kloner RA (2016) Cardioprotective effects of mitochondria-targeted peptide SBT-20 in two different models of rat ischemia/reperfusion. Cardiovasc Drugs Ther 30(6):559–566
94. Kloner RA (2017) Mitochondrial protective agents for ischemia/reperfusion injury. Am Heart Assoc 10(9):e005805
95. Song X, Qu H, Yang Z, Rong J, Cai W, Zhou H (2017) Efficacy and safety of L-carnitine treatment for chronic heart failure: a meta-analysis of randomized controlled trials. Biomed Res Int 2017. https://doi.org/10.1155/2017/6274854
96. Belardinelli R, Muçaj A, Lacalaprice F, Solenghi M, Seddaiu G, Principi F, Tiano L, Littarru GP (2006) Coenzyme Q10 and exercise training in chronic heart failure. Eur Heart J 27(22):2675–2681
97. Mortensen SA, Rosenfeldt F, Kumar A, Dolliner P, Filipiak KJ, Pella D, Alehagen U, Steurer G, Littarru GP, investigators Q-Ss (2014) The effect of coenzyme Q10 on morbidity and mortality in chronic heart failure: results from Q-SYMBIO: a randomized double-blind trial. JACC Heart Fail 2(6):641–649
98. Diguet N, Trammell SA, Tannous C, Deloux R, Piquereau J, Mougenot N, Gouge A, Gressette M, Manoury B, Blanc J (2018) Nicotinamide riboside preserves cardiac function in a mouse model of dilated cardiomyopathy. Circulation 137(21):2256–2273

99. Agadjanyan M, Vasilevko V, Ghochikyan A, Berns P, Kesslak P, Settineri RA, Nicolson GL (2003) Nutritional supplement (NT Factor™) restores mitochondrial function and reduces moderately severe fatigue in aged subjects. Syndrome 11(3):23–36
100. Nicolson G (2003) Lipid replacement as an adjunct to therapy for chronic fatigue, anti-aging and restoration of mitochondrial function. J Am Nutraceut Assoc 6:22–26
101. Ellithorpe RR, Settineri R, Nicolson GL (2003) Pilot study: reduction of fatigue by use of a dietary supplement containing glycophospholipids. J Am Nutraceutical Assoc 6(1):23–28
102. Nicolson GL, Settineri R, Ellithorpe R (2012) Glycophospholipid formulation with NADH and CoQ10 significantly reduces intractable fatigue in Western blot-positive 'chronic lyme disease' patients: preliminary report. Funct Foods Health Dis 2(3):35–47
103. Nicolson GL, Settineri R, Ellithorpe R (2012) Lipid replacement therapy with a glycophospholipid formulation with NADH and CoQ10 significantly reduces fatigue in intractable chronic fatiguing illnesses and chronic Lyme disease patients. Int J Clin Med 3(03):163
104. Witte KK, Nikitin NP, Parker AC, von Haehling S, Volk H-D, Anker SD, Clark AL, Cleland JG (2005) The effect of micronutrient supplementation on quality-of-life and left ventricular function in elderly patients with chronic heart failure. Eur Heart J 26(21):2238–2244
105. https://clinicaltrials.gov/ct2/show/NCT03525379 (2018) Evaluating the clinical efficacy of resveratrol in improving metabolic and skeletal muscle function in patients with heart failure (REV-HF). clinicaltrialsgov
106. https://clinicaltrials.gov/ct2/show/NCT01914081 (2018) Resveratrol: a potential anti-remodeling agent in heart failure, From Bench to Bedside. clinicaltrialsgov
107. Khatua TN, Padiya R, Karnewar S, Kuncha M, Agawane SB, Kotamraju S, Banerjee SK (2012) Garlic provides protection to mice heart against isoproterenol-induced oxidative damage: role of nitric oxide. Nitric Oxide 27(1):9–17
108. Khatua TN, Adela R, Banerjee SK (2013) Garlic and cardioprotection: insights into the molecular mechanisms. Can J Physiol Pharmacol 91(6):448–458
109. Lei L, Liu Y (2017) Efficacy of coenzyme Q10 in patients with cardiac failure: a meta-analysis of clinical trials. BMC Cardiovasc Disord 17(1):196
110. Rosenfeldt F, Hilton D, Pepe S, Krum H (2003) Systematic review of effect of coenzyme Q10 in physical exercise, hypertension and heart failure. Biofactors 18(1–4):91–100
111. Fotino AD, Thompson-Paul AM, Bazzano LA (2012) Effect of coenzyme Q10 supplementation on heart failure: a meta-analysis. Am J Clin Nutr 97(2):268–275
112. Pillai VB, Sundaresan NR, Kim G, Gupta M, Rajamohan SB, Pillai JB, Samant S, Ravindra P, Isbatan A, Gupta MP (2010) Exogenous NAD blocks cardiac hypertrophic response via activation of the SIRT3-LKB1-AMP-activated kinase pathway. J Biol Chem 285(5):3133–3144
113. Birkmayer J (1996) Coenzyme nicotinamide adenine dinucleotide: new therapeutic approach for improving dementia of the Alzheimer type. Ann Clin Lab Sci 26(1):1–9
114. Demarin V, Podobnik S, Storga-Tomic D, Kay G (2004) Treatment of Alzheimer's disease with stabilized oral nicotinamide adenine dinucleotide: a randomized, double-blind study. Drugs Exp Clin Res 30(1):27–33
115. Nicolson GL (2007) Metabolic syndrome and mitochondrial function: molecular replacement and antioxidant supplements to prevent membrane peroxidation and restore mitochondrial function. J Cell Biochem 100(6):1352–1369
116. Nicolson GL, Settineri R (2011) Lipid replacement therapy: a functional food approach with new formulations for reducing cellular oxidative damage, cancer-associated fatigue and the adverse effects of cancer therapy. Funct Foods Health Dis 1(4):135–160
117. Dai D-F, Chen T, Szeto H, Nieves-Cintrón M, Kutyavin V, Santana LF, Rabinovitch PS (2011) Mitochondrial targeted antioxidant peptide ameliorates hypertensive cardiomyopathy. J Am Coll Cardiol 58(1):73–82
118. Anderson KA, Hirschey MD (2012) Mitochondrial protein acetylation regulates metabolism. Essays Biochem 52:23–35
119. Hirschey M, Shimazu T, Huang J-Y, Schwer B (2011) Verdin E SIRT3 regulates mitochondrial protein acetylation and intermediary metabolism. In: Cold Spring Harbor symposia on quantitative biology. Cold Spring Harbor Laboratory Press, New York, p a010850

120. Matsushima S, Sadoshima J (2015) The role of sirtuins in cardiac disease. Am J Phys Heart Circ Phys 309(9):H1375–H1389
121. Koentges C, Pfeil K, Schnick T, Wiese S, Dahlbock R, Cimolai MC, Meyer-Steenbuck M, Cenkerova K, Hoffmann MM, Jaeger C (2015) SIRT3 deficiency impairs mitochondrial and contractile function in the heart. Basic Res Cardiol 110(4):36
122. Hirschey MD, Shimazu T, Jing E, Grueter CA, Collins AM, Aouizerat B, Stančáková A, Goetzman E, Lam MM, Schwer B (2011) SIRT3 deficiency and mitochondrial protein hyperacetylation accelerate the development of the metabolic syndrome. Mol Cell 44(2):177–190
123. Dittenhafer-Reed KE, Richards AL, Fan J, Smallegan MJ, Siahpirani AF, Kemmerer ZA, Prolla TA, Roy S, Coon JJ, Denu JM (2015) SIRT3 mediates multi-tissue coupling for metabolic fuel switching. Cell Metab 21(4):637–646
124. Smith RM, Lecour S, Sack MN (2002) Innate immunity and cardiac preconditioning: a putative intrinsic cardioprotective program. Cardiovasc Res 55(3):474–482
125. De Haan J, Smeets M, Pasterkamp G, Arslan F (2013) Danger signals in the initiation of the inflammatory response after myocardial infarction. Mediat Inflamm 2013:206039
126. Zhang W, Lavine KJ, Epelman S, Evans SA, Weinheimer CJ, Barger PM, Mann DL (2015) Necrotic myocardial cells release damage-associated molecular patterns that provoke fibroblast activation in vitro and trigger myocardial inflammation and fibrosis in vivo. J Am Heart Assoc 4(6):e001993
127. Suresh R, Mosser DM (2013) Pattern recognition receptors in innate immunity, host defense, and immunopathology. Adv Physiol Educ 37(4):284–291
128. Srikrishna G, Freeze HH (2009) Endogenous damage-associated molecular pattern molecules at the crossroads of inflammation and cancer. Neoplasia 11(7):615–628

Oxidative Stress in Hypertension and Cardiovascular-Renal Remodeling: Focus on the Renin-Angiotensin-Aldosterone System

26

Giuseppe Maiolino, Verdiana Ravarotto, and Lorenzo A. Calò

26.1 Introduction

The last report of the World Health Organization highlights that ischaemic heart disease and stroke, which account for 15.2 millions of death in 2016 combined, are the two leading causes of mortality worldwide. Both diseases are strictly related to atherosclerosis and hypertension where oxidative stress and the renin-angiotensin-aldosterone system (RAAS) play a crucial role.

The RAAS regulates various physiological components of the cardiovascular-renal systems but it is also critical in engendering the pathophysiological response into the same systems *via* hypertension, progression of atherosclerosis, vascular remodeling. The strict interplay between the RAAS components is fine tuned by hormonal stimuli and the feedback of baroreceptors in the kidneys, carotid arteries, and the heart. The baroreceptor reflex senses the blood pressure decrease inducing an increase of the sympathetic tone with ensuing vasoconstriction. Furthermore, the reduction of the renal perfusion pressure stimulates the renin production by juxtaglomerular cells in the afferent arterioles of the kidneys, which activates the conversion of the liver generated angiotensinogen into angiotensin (Ang) I. The latter is then converted through the hydrolysis of the angiotensin-converting enzyme (ACE) into the octapeptide Ang II that is the crucial mediator of multiple cardiovascular and renal effects. In fact, Ang II exerts its actions *via* its main receptors, i.e. the Ang II receptor 1 (AT1R) and 2 (AT2R) and stimulates various intracellular signaling pathways leading to vasoconstriction/vasodilation, hypertrophy, insulin resistance/sensitivity, and vascular remodeling. Moreover, the activation of the renin-angiotensin system, along with serum potassium, is the main stimulus to the secretion from the adrenal cortex *zona glomerulosa* of the steroid aldosterone, the last player of the RAAS, which promotes the transcription of proteins inducing sodium

G. Maiolino · V. Ravarotto · L. A. Calò (✉)
Department of Medicine, Nephrology, University of Padova, Padova, Italy
e-mail: renzcalo@unipd.it

S. Chakraborti et al. (eds.), *Oxidative Stress in Heart Diseases*,
https://doi.org/10.1007/978-981-13-8273-4_26

reabsorption and potassium excretion from the kidneys, leading to blood pressure increase.

Reactive oxygen species (ROS), a product generated by multiple enzymatic reactions, are considered one of the crucial contributor to a multiplicity of diseases. In pathological conditions, the exaggerated formation of ROS that commonly participate to the reduction-oxidation (redox) reactions, overwhelms the recovery capacity of the cells, inducing oxidative stress and cellular damage [1]. Therefore, it is not surprising their involvement in the detrimental effects of the RAAS in the cardiovascular and renal pathophysiology, where they contribute *via* oxidative damage of lipids, proteins and deoxyribonucleic acids to myocardial hypertrophy and fibrosis, atherosclerosis and vascular remodeling, and chronic kidney disease [2].

26.2 Oxidative Stress Sources and the Renin-Angiotensin-Aldosterone System

Oxidative stress is due to an imbalance between pro-oxidant and antioxidant factors, with the formers prevailing due to the loss of the intracellular redox homeostasis and leading to damage of proteins, membrane lipids, and oxidation of nucleic acids [3].

Reactive oxygen species are produced during the incomplete reduction of molecular oxygen. Superoxide ($O_2^{\bullet-}$) is the primary ROS deriving from enzymes such as nicotinamide adenine dinucleotide phosphate (NAD(P)H) oxidase, xantine oxidase (XO), uncoupled nitric oxide synthase (NOS), P-450 monooxygenase, lipooxygenase (LOX), cyclooxygenase (COX), and from the respiratory chain in the mitochondria.

NAD(P)H oxidases (Nox) entail a family of enzymes catalyzing the electrons transfer from NAD(P)H to flavin and heme moieties to molecular oxygen, with five isoforms located in cardiovascular and renal tissues: Nox1, Nox2, Nox3, Nox4, and Nox5 [4]. These enzymes comprise a complex of transmembrane and cytoplasmic proteins, which are assembled together to create the catalytic subunit. Nox2 ($gp91^{phox}$), the prototypical NAD(P)H oxidase, is found in phagocytes and throughout the cardiovascular system and requires five subunits to create the active cytochrome b558: $p47^{phox}$, $p67^{phox}$, $p40^{phox}$, $p22^{phox}$, and the small G protein Rac 1/2 [5]. Nox2 induces ROS generation only upon an adequate stimulation whereas non-phagocytic Nox's are constitutively active and produce $O_2^{\bullet-}$ triggering intracellular signaling pathways, influencing transcription factors and molecules involved in cell growth, inflammation, and contraction [6, 7].

Superoxide *per se,* has a very short half-life and might spontaneously or catalytically dismutate to hydrogen peroxide (H_2O_2), a ROS with higher stability, which can cross the biological membranes and act as a signaling molecule along with $O_2^{\bullet-}$. The latter promotes also the formation of the highly reactive and toxic compound peroxynitrite ($ONOO^-$), while hydroxyl radical ($OH^\bullet$) is formed by catalytic decomposition of H_2O_2 [8].

Xanthine oxidase represent a crucial enzyme in purine degradation, catalyzing the sequential hydroxylation of hypoxanthine to yield xanthine and uric acid. It entails two forms, e.g. the dehydrogenase and, under a pro-inflammatory status, the oxidase form, which provides electrons to molecular oxygen, generating $O_2^{\bullet-}$ and H_2O_2.

Nitric oxide synthase comprises multiple enzymes catalyzing the nitric oxide (NO) production exploiting the cofactor tetrahydrobiopterin to reduce and incorporate molecular oxygen into L-arginine. However, in conditions of depleted tetrahydrobiopterin, NOS reduces molecular oxygen rather than L-arginine, producing $O_2^{\bullet-}$ instead of NO.

Oxidative stress and RAAS are strictly intertwined as suggested by pathological conditions characterized by RAAS upregulation, e.g. atherosclerosis and hypertension, where Ang II activates Nox with ensuing ROS increase [9, 10]. AT1R activates at least two profibrotic cell signaling pathways that involve also NAD(P)H oxidase stimulation, e.g. the extracellular signal-regulated kinase (ERK) 1/2, and the RhoA/Rho kinase. *In vitro* studies demonstrated the effect of Nox2 activation in the Ang II-mediated signaling of the serine-threonine kinase Akt, which is crucial for cardiomyocyte hypertrophy development [11]. The inactivation of Nox2 in fact, abolished both Ang II-induced $O_2^{\bullet-}$ generation and cardiomyocyte hypertrophy by reducing Akt activity. *In vivo* studies by Li and colleagues on guinea pigs with left ventricular hypertrophy (LVH), supported the impact of overexpression of NAD(P)H oxidase subunits and the generated ROS in the induction of hypertrophy. Alongside the increased free radicals levels, they observed significant activation of ERK 1/2, ERK 5, and other kinases involved in inflammatory signal and fibrosis pathways such as c-Jun NH_2-terminal kinase 1/2, and p38 mitogen-activated protein kinase [12]. Conversely, Grieve et al., in Nox2 knockout mice exposed to pressure overload, observed a reduction of interstitial fibrosis and cardiac contractile dysfunction, compared to wild-type littermates [13]. Furthermore, in a mineralocorticoid-dependent hypertension model Johar et al. demonstrated that aldosterone signaling *via* the mineralocorticoid receptor contributes to the Ang II cardiac profibrotic effects via Nox2 activation [14]. Nox2 activation, although clearly involved in this process, is not mandatory for the development of LVH thus, other mechanisms triggered by Ang II are implicated in cellular functions reprogramming [15, 16]. Finally, in a model of RAAS pathway overstimulation, Wistar-Kyoto (WKY) rats infused with Ang II showed endothelial dysfunction and hypertrophy of vascular smooth muscle cells and cardiomyocytes through RhoA/Rho kinase increase of oxidative stress [17].

Although, these animal models are a useful tool to understand the impact of Ang II signaling and oxidative stress in the development of hypertension and cardiovascular diseases [14, 18, 19], they do not exhibit all the features of human hypertension, making it difficult to extend the results to humans. Two renal tubular disorders, i.e. Bartter's and Gitelman's syndromes (BS/GS) [20], characterized by normal or low blood pressure and absence of endothelial dysfunction and cardiovascular remodeling [21] despite the activation of the RAAS, might shed new light on this field. Extensive studies on BS/GS have shown biochemical abnormalities of Ang II short- and long-term signaling providing useful insights on the pathophysiology of vascular tone control and hypertension, as well as its complications such as atherosclerosis and cardiovascular remodeling. Moreover, BS/GS patients, despite the over activation of the RAAS, possess reduced oxidative stress and ROS production, and increased antioxidant defenses.

Conversely, in essential hypertensive, chronic kidney disease (CKD), and end stage renal diseases (ESRD) patients, oxidative stress is increased and triggers multiple inflammatory pathways [22, 23]. These diseases have been useful to investigate the physiological response to hypertension in terms of oxidative stress and antioxidant defenses [24, 25].

Studies performed on oxidative stress and hypertension are therefore of considerable relevance because of the possibility that therapies against ROS generation or increasing nitric oxide availability and antioxidants might be useful in preventing vascular injury and renal dysfunction and hypertensive end-organ damage.

26.3 Oxidative Stress and Hypertension

Compelling evidences demonstrate that ROS are a frequent feature of hypertension, although, it is still under scrutiny their role in its pathogenesis, in that it is controversial if ROS are determined by or cause high blood pressure. Different animal models have improved our understanding of hypertension pathogenesis. They have been useful to investigate potential pharmacological targets and drugs and have been manipulated to create the factors hypothesized to influence the onset of hypertension, such as excessive salt intake, overactivation of the RAAS, and genetic predisposition [26, 27].

One of the first demonstration of the role played by ROS in hypertension derived by studies in spontaneously hypertensive rats (SHR), that have doubled vascular $O_2^{\bullet-}$ production compared to WKY rats. In these animal models treatment with the superoxide dismutase mimetic tempol has been shown to normalize ROS levels, reduce renal sympathetic nerve activity in both SHR and WKY, decrease the mean arterial blood pressure in both groups of rats by increasing the plasma total antioxidant capacity overall and preventing the progression of hypertension [28, 29].

The relationship between hypertension, RAAS, and oxidative stress has been suggested even earlier in a study by Laursen et al. [30]. The authors exploited a rat model of hypertension through Ang II or norepinephrine infusion. Despite a similar increase in blood pressure Ang II was associated to increased vascular ROS, whereas norepinephrine was not. Moreover, superoxide dismutase treatment significantly reduced blood pressure in Ang II-infused rats, with no effect on norepinephrine-infused rats. Based on these results the authors concluded that Ang II-mediated hypertension is at least in part mediated by $O_2^{\bullet-}$, probably through degradation of NO. In SHR and stroke-prone SHR (SPSHR) the expression of thioredoxin, a redox regulating protein, was reduced compared to WKY rats, and in an *in vitro* experiment Ang II treatment of peripheral blood mononuclear cells decreased thioredoxin more in SHR and SPSHR compared with WKY [31].

Further support to the interplay between RAAS and oxidative stress in hypertension was provided by treatment of SPSHR with an AT1R blocker (ARB), i.e. irbesartan, which reduced blood pressure similarly to amlodipine and hydrochlorothiazide/hydralazine, but was the most efficacious drug at reducing oxidative stress as demonstrated by a decrease of $O_2^{\bullet -}$ and of p22^{phox} production [32]. In SHR and SPSHR experimental evidences demonstrated that hypertension is associated to NAD(P)H oxidases activity through Ang II signaling and to reduced NO bioavailability or dysfunctional endothelial NOS (eNOS), supported by the evidence that the treatment with antioxidants and ARBs decreases superoxide production and hypertension progression [33]. The contribution of each component of the RAAS to the development of hypertension and oxidative stress have been supported also by studies carried out in SHR rats where treatment with the renin inhibitor aliskiren [34] and the ACE inhibitor enalapril [35] showed ROS reduction.

In hypertensive humans clinical studies demonstrated an increase of ROS and a decrease of the antioxidant defense mechanisms *in vivo* and in vascular smooth muscle cells of arteries *in vitro* [36]. Systolic and diastolic blood pressure increased with raising ROS and decreasing plasma antioxidant status [37, 38]. Moreover, ROS contribute to hypertension-induced target organ damage through increase of pro-inflammatory gene expression and modulation of cell proliferation and death pathways [39]. However, the role of oxidative stress in hypertension might be limited to advanced stages and trivial in early stages of the disease, as suggested by the finding of similar levels of urinary ROS in never-treated mild-to-moderate hypertensives compared to healthy controls [40].

Among the multiple sources relevant to the pathogenesis of human hypertension a crucial role is played by the production of ROS under Ang II stimulation [2, 41]. In fact, in the long-term AT1R activation favors the onset of hypertension complications such as changes in cardiovascular structure and induction of atherosclerosis, *via* increasing oxidative stress through NAD(P)H oxidases stimulation [25]. In human vascular smooth muscle cells isolated from peripheral arteries of essential hypertensive patients, Ang II stimulation increases ROS levels and reduces antioxidant defenses [42]. Moreover, in multiple clinical studies the pharmacological inhibition of the RAAS system through ACE inhibitors or ARBs together with the blood pressure lowering effects was able not only to reduce inflammatory markers [24, 43] and oxidative stress [44], but also to increase the antioxidant defenses [24, 45].

Supported by these evidences the therapeutic use of antioxidant therapy supplementation in hypertensive patients, mainly by means of vitamins C and E, has been advocated to decrease oxidative stress and lower blood pressure. In a randomized clinical trial completed in 110 patients with grade 1 essential hypertension assigned to either vitamins C and E or placebo for 8 weeks, the treatment significantly decreased blood pressure and increased antioxidant capacity [46]. These data have been corroborated by a meta-analysis including 29 randomized clinical trials on short term vitamin C supplementation that demonstrated a significant reduction of systolic and diastolic blood pressure (−3.8 mmHg and − 1.4 mmHg, respectively) [47]. However, the results of all large clinical trials testing the hypothesis of a positive effect of antioxidant therapy on cardiovascular disease demonstrated no effect whatsoever on cardiovascular endpoints or blood pressure decrease [33]. These results might be explained by the advanced cardiovascular disease of the patients enrolled, the antioxidant vitamin selected for the treatment, and its suboptimal dose [33].

26.4 Oxidative Stress and Cardiovascular Remodeling

26.4.1 Atherosclerosis

Atherosclerosis is a chronic inflammatory disease recognizing as an early pathogenic step endothelial dysfunction, characterized by endothelial cells changes induced by "irritative" stimuli (e.g., cigarette smoking, dyslipidemia, diabetes, and hypertension) [48]. The endothelial damage induces the expression of adhesion and chemotactic molecules, mediating the leucocyte migration into the arterial wall, and increases the permeability to macromolecules. This favors the low-density lipoprotein (LDL) particles entrance into the arterial subintimal extracellular matrix where they are trapped and oxidized by resident vascular cells [49]. The oxidized LDLs (oxLDL) exert a proatherogenic effect [50] stimulating the endothelial cells to produce monocyte chemotactic protein-1, macrophage and granulocyte colony-stimulating factors that recruit monocytes and promote their conversion into macrophages. The latters promote the LDLs complete oxidation [51] allowing their recognition by macrophages scavenger receptors, which internalize them and transform into foam cells, the hallmark of the atherosclerosis [52].

The RAAS plays a crucial role in the pathogenesis of atherosclerosis affecting the endothelium and promoting inflammatory reactions, thrombosis, and oxidant injury [53]. The Ang II mediated injury is at least partially mediated by ROS produced by NAD(P)H oxidase activation causing endothelial dysfunction through oxidative damage of membrane lipids, which promotes inflammatory cytokine production such as tumor necrosis factor-α. This effect as well as arterial wall production of endothelial cells adhesion molecules, monocyte chemoattractant protein 1, and macrophage stimulating factors are increased *via* NF-κB activation. Moreover, an overproduction of $O_2^{\bullet-}$ by NAD(P)H oxidase reacts with NO to form the highly reactive intermediate $ONOO^-$, which is responsible of subsequent NOS uncoupling, tyrosin nitration of prostacyclin synthase (PGI_2S), and vasoconstriction [54].

Endothelial progenitor cells (EPCs), a bone marrow or cord blood derived cell population, are deemed to be involved in vascular injury repair *via* production of a cell patch at sites of vessel injury [55]. Hypertension is characterized by a decrease of the EPCs number and impairment of their function, constituting an additional risk factor for cardiovascular events [24]. The detrimental role of the RAAS in the vascular system might be related also to the Ang II induction of EPCs accelerated senescence, as opposed to the vasorelaxant calcitonin gene-related peptide (CGRP) that prevents their ageing [56, 57]. In clinical studies carried out in hypertensive patients ARBs, in particular olmesartan medoxomil, which has indirectly been shown to possess vasoprotective, anti-inflammatory, and anti-atherosclerotic properties in the EUTOPIA, VIOS, MORE, and OLIVUS clinical trials, is able to significantly increase the antioxidant heme oxygenase (HO)-1 alongside the CGRP and to raise the number of circulating EPCs, thereby favoring vascular repair and preventing atherosclerosis [24].

26.4.2 Heart Failure

Heart failure is a pathological condition characterized by myocardial functional and structural alterations, sodium retention, endothelial dysfunction, activation of the RAAS and the sympathetic nervous system [58]. The increased oxidative stress in heart failure has been demonstrated in many studies both in animal models and in humans, proving that the sources of ROS in the heart and the vasculature are numerous, e.g. XO, Nox2 and Nox4, mitochondrial electron transport chain activity, and uncoupled NOS [58, 59].

The detrimental effects of ROS in heart failure are attributable to derangements of redox-regulated ion handling proteins, which affects the excitation-contraction coupling leading to contractile and relaxation dysfunction. A crucial regulator of proteins involved in excitation-contraction coupling is the Ca^{2+}/calmodulin-dependent kinase II (CaMKII), whose expression is increased in heart failure and activated by ROS [60, 61]. This leads to an increased sodium influx via voltage-gated channels, leading to intracellular accumulation with ensuing prolongation of the action potential, and altered calcium handling by the sarcoplasmic reticulum, followed by its intracellular increase and subsequent reduced contractility and arrhythmias [61]. Furthermore, oxidative stress is involved in ventricular remodeling, mediated by the activation of CAMKII and mitogen-activated protein kinases, and in cardiomyocyte apoptosis and necrosis. Finally, heart failure is characterized by endothelial dysfunction partially secondary to NOS uncoupling with a shift from NO to $O_2^{\bullet-}$ production, as demonstrated by the beneficial effects of eNOS blunting in this setting [62]. The decrease of myocardial NO leads to accumulation of cytosolic calcium affecting, as reported above, myocardial relaxation and contractility [63].

The beneficial effects of the RAAS blockers on reduced ejection fraction heart failure have been demonstrated more than three decades ago [64] and this is probably related, at least partially, to the reduced oxidative stress obtained by this therapy. In fact, in a mouse model of heart failure Ang II induced cardiomyocyte increase of mitochondrial ROS with ensuing cardiac hypertrophy, fibrosis, and diastolic dysfunction, which was blunted by the antioxidant effect of mitochondrial catalase overexpression [65]. The positive effect of the RAAS inhibition is attributable also to the inhibition of the eNOS activity. In fact, in a post myocardial infarction animal model of heart failure the ACE inhibitor trandolapril and the ARB irbesartan were able to reduce the impaired endothelium-dependent relaxation normalizing the aortic eNOS expression, at least partially *via* reduction of $O_2^{\bullet-}$, which was determined only by the ARB treatment [66].

Taken together, these observations further highlight the impact of oxidative stress promoted by Ang II signaling in the onset of all the molecular events that induce cardiovascular remodeling and heart failure.

26.5 Oxidative Stress and Kidney Disease

Kidney diseases are associated with both traditional risk factors, e.g. diabetes, hypertension, obesity, smoke, alcohol, and non-traditional cardiovascular disease risk factors such as endothelial dysfunction, inflammation, and oxidative stress, which arise very early in renal disorders. In these stages oxidative stress related-products are also biomarkers of disease progression and, although renal replacement treatments and kidney transplantation increase patients survival, cardiovascular risk factors have a strong impact on patients life expectancy [67, 68].

Several studies have shown that many components of the immune system have a critical role in the progression of CKD, as suggested by elevated levels of circulating interleukin(IL)-6, IL-8, and C-reactive protein (CRP) [69]. In fact, in a toll-like receptors (TLRs) knockout model, mice are protected against acute kidney injury and renal dysfunction, whereas antigen presenting dendritic cells, which are crucial for the activation of T cells and their mediated glomerular inflammation, are over stimulated in animal models of glomerulonephritis [70–72]. The interplay between immune system and RAAS activation is supported by evidences in mice lacking AT1R on immune cells, where Ang II induced proliferation in splenocytes and the specific blockade of AT1R reversed this effect [73]. Moreover, blockade of AT1R and ACE inhibitor treatment in SPSHR significantly decreased the thickening and degeneration of glomeruli and tubules, decreasing inflammatory cells infiltration of the glomeruli [74].

In CKD RAAS activation contributes to oxidative damage through increased ROS production and impairment of antioxidant defenses. In fact, Sprague-Dawley rats with CKD are characterized not only by lipid peroxidation, NF-kB activation, mononuclear cell infiltration, upregulation of chemoattractant protein-1, Noxs, COX, and 12-lipoxygenase upregulation, but also by glutathione depletion and nuclear factor erythroid 2-related factor-2 (Nrf2) reduced activity. The latter regulates the induction of genes encoding for antioxidant enzymes and related proteins, such as superoxide dismutase, catalase, HO-1 and others, demonstrating that antioxidant system decreased function in the early stages of CKD is critical for its progression and worsens with the disease severity [75, 76].

Renal replacement therapies such as dialysis, affects the endothelium integrity through production of toxic and profibrotic molecules. Thus, many studies have been performed on hemodialysis, CKD, and ESRD patients in order to evaluate the impact of renal failure and the kidney replacement therapy on oxidative stress and its complications in cardiovascular-and renal remodeling.

In patients on chronic dialysis the use of vitamin E-coated dialyzer reduced mRNA levels of $p22^{phox}$ and hydroperoxyde, increasing the total antioxidant power [77, 78]. Haemodiafiltration with online regeneration of ultrafiltrate, which is reported to be effective at reducing cytokines such as IL-1, lowers oxidative stress as demonstrated by reduced $p22^{phox}$, atherothrombogenic plasminogen activator inhibitor (PAI)-1 expression, and oxLDLs, promoting the increase of antioxidant defenses like HO-1 [77–80]. The treatment of oxidative stress reduces the proteins involved in fibrosis and cardiovascular remodeling as it has been shown in hemodialysis patients supplemented with green tea [81]. In fact, the antioxidant treatment reduced ERK1/2 activation, a kinase eliciting hypertrophic responses via phosphorylation of nuclear targets (c-myc, c-jun, and ATF-2) and leading to transcriptional reprogramming of genes associated with hypertrophy. Most of hemodialysis patients display LVH, which is correlated with the ROS markers oxLDL and $p22^{phox}$, but the antioxidant therapy is able to reduce significantly the cardiac mass alongside the oxLDL and to blunt macrophage mobility, preventing their accumulation, and myocardial fibrosis [81, 82].

Another pathway deeply involved in the fibrotic responses both in kidney and cardiovascular disease *via* induction of oxidative stress is the Rho A/Rho kinase pathway, which also mediates the upregulation of ROS through induction of NADPH oxidases. In dialysis and CKD patients Rho A/Rho kinase is over activated, particularly in patients who had already LVH compared to those with normal left ventricular mass, in agreement with data observed in hypertensive patients [23, 83]. The incubation of circulating leukocytes from CKD and dialysis patients with fasudil, a Rho A/Rho kinase inhibitor, reduced dose dependently its activity, suggesting that the latter could be a very useful target for the prevention of cardiovascular-renal remodeling [23].

26.6 Bartter's and Gitelman's Syndromes

The physiological signaling of the RAAS is critical for the control and regulation of vascular tone and blood pressure. In particular, as above reported, Ang II signaling through AT1R is pivotal for the induction of a wide spectrum of intracellular pathways entangled in oxidative stress responses, profibrotic alterations, insulin resistance, and cardiovascular-renal remodeling [84]. Two rare genetic diseases, characterized by renal electrolytic derangement, display activation of the RAAS, but yet exhibit reduced peripheral vascular resistance, normal or low blood pressure, and absence of cardiovascular-renal remodeling. Bartter's and Gitelman's (BS/GS) syndromes provide a mirror image of hypertension because of a blunted signaling of Ang II despite its high plasma levels and normal Ang II receptors [85]. Both BS/GS are characterized by hypokalemia, metabolic alkalosis, intravascular volume depletion due to renal salt wasting, with some key features that allow differential diagnosis. BS shows electrolyte abnormalities similar to those induced by the treatment with furosemide or other drugs that inhibit the $Na^+K^+2Cl^-$ cotransporter in the thick ascending limb of Henle's loop, and can be classified into 5 types based on the cotransporter or channel affected by loss-of-function mutation. These patients are characterized by normomagnesemia and hyper-normocalciuria and typically arise in infancy or childhood. GS mutations affect the

$Na^{+}Cl^{-}$ cotransporter in the distal convoluted tubule of the nephron, therefore, their abnormalities are similar to those induced by the treatment with thiazides. GS is characterized by the concomitant presence of hypomagnesemia and hypocalciuria and a later onset of the symptoms in childhood or adulthood. Notwithstanding some different clinical features, both syndromes are a human model of blunted Ang II signaling as demonstrated by the reduced oxidative stress, the increased antioxidant defenses, and the lack of cardiovascular remodeling, suggesting that Ang II signaling is blocked or interrupted at the post-receptor level or very close to the central switch control of Ang II signaling [86].

In hypertension the binding of Ang II with its heterotrimeric G-protein-coupled receptors, Gq and Gi proteins, promotes the increase of free intracellular Ca^{2+}, mediates the activation of phospholipase C (PLC), which stimulates the protein kinase C (PKC) and the Rho A/Rho kinase pathway. BS/GS have decreased gene and protein expression of the α subunit of Gq protein and blunted downstream intracellular events that promote Ca^{2+} release and PKC activation [85, 87, 88]. The downregulation of the Gq protein signaling can be explained by the observation that, contrarily to what has been described in hypertension, the regulators of G protein signaling (RGS)-2 is increased in BS/GS and is crucial for the vasodilatatory activity of NO [89, 90].

The pathways involved in the long-term signaling of Ang II that, mainly *via* oxidative stress, lead to cardiovascular-renal remodeling, hypertension, atherosclerosis, heart and kidney failure, have been extensively studied in BS/GS. These patients are characterized by decreased $p22^{phox}$ gene and protein expression, that implies lower production of ROS, particularly of $O_2^{\bullet-}$, and by increased gene expression of the antioxidant HO-1 [86]. Reduced oxidative stress in BS/GS is demonstrated by the decreased susceptibility of LDLs to oxidation [91], as proved by reduced levels of oxLDLs, coupled with increased NO production and decreased PAI-1 [92]. Finally, among the profibrotic effects promoted by Ang II signaling, the cytokine transforming growth factor-β that induces cell differentiation/proliferation and fibrosis is significantly reduced in BS/GS [86].

A crucial player of cardiovascular-renal pathophysiology is the balance between the Rho A/Rho kinase pathway and the NO system that are involved in processes such as induction/decrease of oxidative stress, regulation of PAI-1, control of neointimal formation, inhibition/activation of phosphoinositol 3 kinase (PI3K)/Akt, increase/decrease of eNOS activity, and regulation of glucose transport and metabolism [84]. Experimental evidences have shown that in BS/GS Rho A/Rho kinase activity is blunted and associated with increased NO and eNOS expression [92, 93]. This contention is supported by the finding that p63RhoGEF gene and protein expression is reduced along with the phosphorylation of myosin phosphatase target protein-1 (MYPT-1), a marker of Rho A/Rho kinase activation, which is conversely increased in hypertensive subjects [22].

The low oxidative stress in BS/GS is not associated to a reduced anti-inflammatory capacity, as demonstrated by unchanged levels of CRP and other inflammatory mediators such as vascular cell adhesion molecule (VCAM), intracellular cell adhesion molecule (ICAM), IL-6, and NF-kB [94, 95].

Taken together, the studies in BS/GS have shed light on key regulatory elements of Ang II signaling crucial in the vascular tone regulation and cardiovascular-renal remodeling. Thus, the importance of the Rho A/Rho kinase pathway and its relationship with the NO system is determinant not only for the induction of hypertension and insulin resistance, but also for the long term effects of Ang II signaling as seen above.

26.7 Conclusions

The over activation of the RAAS is the switch for many intracellular pathways that induce hypertension and in the long term cardiovascular-renal injury through increased oxidative stress.

Based on the relevance of oxidative stress on the onset of endothelial and vascular dysfunction, several experimental evidences demonstrated that its decrease might be the best strategy to prevent complications related to hypertension and cardiovascular diseases. A wide range of prospective cohort studies have shown that the differences in cardiovascular morbidity and mortality among diverse populations are at least partially attributable to differences in the antioxidant intake from foods and beverages [54]. The risk factors for cardiovascular and renal diseases are numerous, such as unhealthy lifestyle and diet habits, but at a molecular level there is a strict connection between inflammation, blood pressure, and endothelial dysfunction. This awareness should be kept in mind when considering pharmacological interventions for the treatment of hypertension and related cardiovascular-renal diseases, which can be integrated with antioxidant supplementation.

References

1. Montezano AC, Dulak-Lis M, Tsiropoulou S et al (2015) Oxidative stress and human hypertension: vascular mechanisms, biomarkers, and novel therapies. Can J Cardiol 31:631–641. https://doi.org/10.1016/J.CJCA.2015.02.008
2. Maiolino G, Azzolini M, Paolo Rossi G et al (2015) Bartter/Gitelman syndromes as a model to study systemic oxidative stress in humans. Free Radic Biol Med 88:51–58. https://doi.org/10.1016/J.FREERADBIOMED.2015.02.037
3. Sies H (1985) Oxidative stress. Academic, London
4. Lassègue B, Clempus RE (2003) Vascular NAD(P)H oxidases: specific features, expression, and regulation. Am J Physiol Integr Comp Physiol 285:R277–R297. https://doi.org/10.1152/ajpregu.00758.2002
5. Bokoch GM, Zhao T (2006) Regulation of the phagocyte NADPH oxidase by Rac GTPase. Antioxid Redox Signal 8:1533–1548. https://doi.org/10.1089/ars.2006.8.1533
6. Cosentino F, Francia P, Camici GG et al (2008) Final common molecular pathways of aging and cardiovascular disease: role of the p66Shc protein. Arterioscler Thromb Vasc Biol 28:622–628. https://doi.org/10.1161/ATVBAHA.107.156059
7. Cruzado MC, Risler NR, Miatello RM et al (2005) Vascular smooth muscle cell NAD(P)H oxidase activity during the development of hypertension: effect of angiotensin II and role of insulinlike growth factor-1 receptor transactivation. Am J Hypertens 18:81–87. https://doi.org/10.1016/J.AMJHYPER.2004.09.001

8. Nistala R, Whaley-Connell A, Sowers JR (2008) Redox control of renal function and hypertension. Antioxid Redox Signal 10:2047–2089. https://doi.org/10.1089/ars.2008.2034
9. Montezano AC, Nguyen Dinh Cat A, Rios FJ, Touyz RM (2014) Angiotensin II and vascular injury. Curr Hypertens Rep 16:431. https://doi.org/10.1007/s11906-014-0431-2
10. Nguyen Dinh Cat A, Touyz RM (2011) Cell signaling of angiotensin II on vascular tone: novel mechanisms. Curr Hypertens Rep 13:122–128. https://doi.org/10.1007/s11906-011-0187-x
11. Hingtgen SD, Tian X, Yang J et al (2006) Nox2-containing NADPH oxidase and Akt activation play a key role in angiotensin II-induced cardiomyocyte hypertrophy. Physiol Genomics 26:180–191. https://doi.org/10.1152/physiolgenomics.00029.2005
12. Li J-M, Gall NP, Grieve DJ et al (2002) Activation of NADPH oxidase during progression of cardiac hypertrophy to failure. Hypertension 40:477–484. https://doi.org/10.1161/01.HYP.0000032031.30374.32
13. Grieve DJ, Byrne JA, Siva A et al (2006) Involvement of the nicotinamide adenosine dinucleotide phosphate oxidase isoform Nox2 in cardiac contractile dysfunction occurring in response to pressure overload. J Am Coll Cardiol 47:817–826. https://doi.org/10.1016/J.JACC.2005.09.051
14. Johar S, Cave AC, Narayanapanicker A et al (2006) Aldosterone mediates angiotensin II-induced interstitial cardiac fibrosis via a Nox2-containing NADPH oxidase. FASEB J 20:1546–1548. https://doi.org/10.1096/fj.05-4642fje
15. Byrne JA, Grieve DJ, Bendall JK et al (2003) Contrasting roles of NADPH oxidase isoforms in pressure-overload versus angiotensin II–induced cardiac hypertrophy. Circ Res 93:802–805. https://doi.org/10.1161/01.RES.0000099504.30207.F5
16. Maytin M, Siwik DA, Ito M et al (2004) Pressure overload-induced myocardial hypertrophy in mice does not require gp91phox. Circulation 109:1168–1171. https://doi.org/10.1161/01.CIR.0000117229.60628.2F
17. Higashi M, Shimokawa H, Hattori T et al (2003) Long-term inhibition of Rho-kinase suppresses angiotensin II-induced cardiovascular hypertrophy in rats in vivo: effect on endothelial NAD(P)H oxidase system. Circ Res 93:767–775. https://doi.org/10.1161/01.RES.0000096650.91688.28
18. Doerries C, Grote K, Hilfiker-Kleiner D et al (2007) Critical role of the NAD(P)H oxidase subunit p47phox for left ventricular remodeling/dysfunction and survival after myocardial infarction. Circ Res 100:894–903. https://doi.org/10.1161/01.RES.0000261657.76299.ff
19. Satoh M, Ogita H, Takeshita K et al (2006) Requirement of Rac1 in the development of cardiac hypertrophy. Proc Natl Acad Sci U S A 103:7432–7437. https://doi.org/10.1073/pnas.0510444103
20. Calò LA, Pessina AC, Semplicini A (2005) Angiotensin II signalling in Bartter's and Gitelman's syndromes: a negative human model of hypertension. High Blood Press Cardiovasc Prev 12:17–26. https://doi.org/10.2165/00151642-200512010-00004
21. Calò LA (2006) Vascular tone control in humans: insights from studies in Bartter's/Gitelman's syndromes. Kidney Int 69:963–966. https://doi.org/10.1038/sj.ki.5000253
22. Calò LA, Davis PA, Pagnin E et al (2014b) Increased level of p63RhoGEF and RhoA/Rho kinase activity in hypertensive patients. J Hypertens 32:331–338. https://doi.org/10.1097/HJH.0000000000000075
23. Calò LA, Vertolli U, Pagnin E et al (2016) Increased Rho kinase activity in mononuclear cells of dialysis and stage 3-4 chronic kidney disease patients with left ventricular hypertrophy: cardiovascular risk implications. Life Sci 148:80–85. https://doi.org/10.1016/j.lfs.2016.02.019
24. Calò LA, Dal Maso L, Pagnin E et al (2014a) Effect of olmesartan medoxomil on number and survival of circulating endothelial progenitor cells and calcitonin gene related peptide in hypertensive patients. J Hypertens 32:193–199. https://doi.org/10.1097/HJH.0b013e32836522c3
25. Ravarotto V, Pagnin E, Maiolino G et al (2015b) The blocking of angiotensin II type 1 receptor and RhoA/Rho kinase activity in hypertensive patients: effect of olmesartan medoxomil and implication with cardiovascular-renal remodeling. JRAAS – J Renin-Angiotensin-Aldosterone Syst 16:1245–1250. https://doi.org/10.1177/1470320315594324

26. Badyal DK, Lata H, Dadhich AP (2003) Animal models of hypertension and effect of drugs. Indian J Pharmacol 35:349–362. http://medind.nic.in/ibi/t03/i6/ibit03i6p349.pdf
27. Leong X-F, Ng C-Y, Jaarin K (2015) Animal models in cardiovascular research: hypertension and atherosclerosis. Biomed Res Int 2015:1–11. https://doi.org/10.1155/2015/528757
28. Park JB, Touyz RM, Chen X, Schiffrin EL (2002) Chronic treatment with a superoxide dismutase mimetic prevents vascular remodeling and progression of hypertension in salt-loaded stroke-prone spontaneously hypertensive rats. Am J Hypertens 15:78–84
29. Shokoji T, Nishiyama A, Fujisawa Y et al (2003) Renal sympathetic nerve responses to tempol in spontaneously hypertensive rats. Hypertension (Dallas, Tex 1979) 41:266–273
30. Laursen JB, Rajagopalan S, Galis Z et al (1997) Role of superoxide in angiotensin II-induced but not catecholamine- induced hypertension. Circulation 95:588–593. https://doi.org/10.1161/01.CIR.95.3.588
31. Tanito M, Nakamura H, Kwon Y-W et al (2004) Enhanced oxidative stress and impaired thioredoxin expression in spontaneously hypertensive rats. Antioxid Redox Signal 6:89–97. https://doi.org/10.1089/152308604771978381
32. Brosnan MJ, Hamilton CA, Graham D et al (2002) Irbesartan lowers superoxide levels and increases nitric oxide bioavailability in blood vessels from spontaneously hypertensive stroke-prone rats. J Hypertens 20:281–286
33. Touyz RM (2004) Reactive oxygen species, vascular oxidative stress, and redox signaling in hypertension: what is the clinical significance? Hypertension 44:248–252. https://doi.org/10.1161/01.HYP.0000138070.47616.9d
34. Zhang W, Han Y, Meng G et al (2014) Direct renin inhibition with aliskiren protects against myocardial ischemia/reperfusion injury by activating nitric oxide synthase signaling in spontaneously hypertensive rats. J Am Heart Assoc 3:e000606. https://doi.org/10.1161/JAHA.113.000606
35. Chandran G, Sirajudeen KNS, Yusoff NSN et al (2014) Effect of the antihypertensive drug enalapril on oxidative stress markers and antioxidant enzymes in kidney of spontaneously hypertensive rat. Oxidative Med Cell Longev 2014:608512. https://doi.org/10.1155/2014/608512
36. Ahmad A, Singhal U, Hossain MM et al (2013) The role of the endogenous antioxidant enzymes and malondialdehyde in essential hypertension. J Clin Diagn Res 7:987–990
37. Rodrigo R, Bächler JP, Araya J et al (2007) Relationship between (Na + K)-ATPase activity, lipid peroxidation and fatty acid profile in erythrocytes of hypertensive and normotensive subjects. Mol Cell Biochem 303:73–81
38. Russo C, Olivieri O, Girelli D et al (1998) Anti-oxidant status and lipid peroxidation in patients with essential hypertension. J Hypertens 16:1267–1271
39. Togliatto G, Lombardo G, Brizzi MF (2017) The future challenge of reactive oxygen species (ROS) in hypertension: from bench to bed side. Int J Mol Sci 18:1988
40. Cracowski JL, Baguet JP, Ormezzano O et al (2003) Lipid peroxidation is not increased in patients with untreated mild-to-moderate hypertension. Hypertension 41:286–288
41. Harrison DG, Gongora MC (2009) Oxidative stress and hypertension. Med Clin North Am 93:621–635. https://doi.org/10.1016/j.mcna.2009.02.015
42. Touyz RM, Schiffrin EL (2001) Increased generation of superoxide by angiotensin II in smooth muscle cells from resistance arteries of hypertensive patients: role of phospholipase D-dependent NAD(P)H oxidase-sensitive pathways. J Hypertens 19:1245–1254. Lippincott Williams & Wilkins
43. Fliser D, Buchholz K, Haller H, EUropean Trial on Olmesartan and Pravastatin in Inflammation and Atherosclerosis (EUTOPIA) Investigators (2004) Antiinflammatory effects of angiotensin II subtype 1 receptor blockade in hypertensive patients with microinflammation. Circulation 110:1103–1107. https://doi.org/10.1161/01.CIR.0000140265.21608.8E
44. Ahmad KA, Yuan Yuan D, Nawaz W et al (2017) Antioxidant therapy for management of oxidative stress induced hypertension. Free Radic Res 51:428–438. https://doi.org/10.1080/10715762.2017.1322205
45. Hornig B, Landmesser U, Kohler C et al (2001) Comparative effect of ACE inhibition and angiotensin II type 1 receptor antagonism on bioavailability of nitric oxide in patients with

coronary artery disease: role of superoxide dismutase. Circulation 103:799–805. https://doi.org/10.1017/9781316274385.007
46. Rodrigo R, Prat H, Passalacqua W et al (2008) Decrease in oxidative stress through supplementation of vitamins C and E is associated with a reduction in blood pressure in patients with essential hypertension. Clin Sci 114:625–634. https://doi.org/10.1042/CS20070343
47. Juraschek SP, Guallar E, Appel LJ et al (2012) Effects of vitamin C supplementation on blood pressure: a meta-analysis of randomized controlled trials. Am J Clin Nutr 95:1079–1088. https://doi.org/10.3945/ajcn.111.027995
48. Libby P, Ridker PM, Hansson GK (2011) Progress and challenges in translating the biology of atherosclerosis. Nature 473:317–325
49. Navab M, Berliner JA, Watson AD et al (1996) The Yin and Yang of oxidation in the development of the fatty streak: a review based on the 1994 George Lyman Duff memorial lecture. Arterioscler Thromb Vasc Biol 16:831–842
50. Choi SH, Harkewicz R, Lee JH et al (2009) Lipoprotein accumulation in macrophages via toll-like receptor-4-dependent fluid phase uptake. Circ Res 104:1355–1363. https://doi.org/10.1161/CIRCRESAHA.108.192880
51. Parhami F, Fang ZT, Fogelman AM et al (1993) Minimally modified low density lipoprotein-induced inflammatory responses in endothelial cells are mediated by cyclic adenosine monophosphate. J Clin Invest 92:471–478. https://doi.org/10.1172/JCI116590
52. Henriksen T, Mahoney EM, Steinberg D (1981) Enhanced macrophage degradation of low density lipoprotein previously incubated with cultured endothelial cells: recognition by receptors for acetylated low density lipoproteins. Proc Natl Acad Sci U S A 78:6499–6503. https://doi.org/10.1073/pnas.78.10.6499
53. Husain K (2015) Inflammation, oxidative stress and renin angiotensin system in atherosclerosis. World J Biol Chem 6:209–217. https://doi.org/10.4331/wjbc.v6.i3.209
54. Münzel T, Gori T, Bruno RM, Taddei S (2010) Is oxidative stress a therapeutic target in cardiovascular disease? Eur Heart J 31:2741–2748. https://doi.org/10.1093/eurheartj/ehq396
55. Hill JM, Zalos G, Halcox JPJ et al (2003) Circulating endothelial progenitor cells, vascular function, and cardiovascular risk. N Engl J Med 348:593–600. https://doi.org/10.1056/NEJMoa022287
56. Imanishi T, Hano T, Nishio I (2005) Angiotensin II accelerates endothelial progenitor cell senescence through induction of oxidative stress. J Hypertens 23:97–104
57. Zhou Z, Peng J, Wang C-J et al (2010) Accelerated senescence of endothelial progenitor cells in hypertension is related to the reduction of calcitonin gene-related peptide. J Hypertens 28:931–939. https://doi.org/10.1097/HJH.0b013e3283399326
58. Münzel T, Camici GG, Maack C et al (2017) Impact of oxidative stress on the heart and vasculature. J Am Coll Cardiol 70:212–229. https://doi.org/10.1016/j.jacc.2017.05.035
59. Ungvári Z, Gupte SA, Recchia FA et al (2005) Role of oxidative-nitrosative stress and downstream pathways in various forms of cardiomyopathy and heart failure. Curr Vasc Pharmacol 3:221–229
60. Hudmon A, Schulman H (2002) Structure-function of the multifunctional Ca2+/calmodulin-dependent protein kinase II. Biochem J 364:593–611. https://doi.org/10.1042/BJ20020228
61. Wagner S, Rokita AG, Anderson ME, Maier LS (2013) Redox regulation of sodium and calcium handling. Antioxid Redox Signal 18:1063–1077. https://doi.org/10.1089/ars.2012.4818
62. Mollnau H, Oelze M, August M et al (2005) Mechanisms of increased vascular superoxide production in an experimental model of idiopathic dilated cardiomyopathy. Arterioscler Thromb Vasc Biol 25:2554–2559. https://doi.org/10.1161/01.ATV.0000190673.41925.9B
63. Pacher P, Beckman JS, Liaudet L (2007) Nitric oxide and peroxynitrite in health and disease. Physiol Rev 87:315–424. https://doi.org/10.1152/physrev.00029.2006
64. CONSENSUS Trial Study Group (1987) Effects of Enalapril on mortality in severe congestive heart failure. N Engl J Med 316:1429–1435. https://doi.org/10.1056/NEJM198706043162301
65. Dai DF, Johnson SC, Villarin JJ et al (2011) Mitochondrial oxidative stress mediates angiotensin II-induced cardiac hypertrophy and Gαq overexpression-induced heart failure. Circ Res 108:837–846. https://doi.org/10.1161/CIRCRESAHA.110.232306

66. Schäfer A, Fraccarollo D, Tas P et al (2004) Endothelial dysfunction in congestive heart failure: ACE inhibition vs. angiotensin II antagonism. Eur J Heart Fail 6:151–159. https://doi.org/10.1016/j.ejheart.2003.10.009
67. Park S-H, Stenvinkel P, Lindholm B (2012) Cardiovascular biomarkers in chronic kidney disease. J Ren Nutr 22:120–127. https://doi.org/10.1053/j.jrn.2011.10.021
68. Sun J, Axelsson J, Machowska A et al (2016) Biomarkers of cardiovascular disease and mortality risk in patients with advanced CKD. Clin J Am Soc Nephrol 11:1163–1172. https://doi.org/10.2215/CJN.10441015
69. Imig JD, Ryan MJ (2013) Immune and inflammatory role in renal disease. Compr Physiol 3:957–976. https://doi.org/10.1002/cphy.c120028
70. Robertson J, Wu J, Arends J et al (2005) Activation of glomerular basement membrane-specific B cells in the renal draining lymph node after T cell-mediated glomerular injury. J Am Soc Nephrol 16:3256–3263. https://doi.org/10.1681/ASN.2005040421
71. Yasuda H, Leelahavanichkul A, Tsunoda S et al (2008) Chloroquine and inhibition of Toll-like receptor 9 protect from sepsis-induced acute kidney injury. Am J Physiol Renal Physiol 294:F1050–F1058. https://doi.org/10.1152/ajprenal.00461.2007
72. Zhang B, Ramesh G, Uematsu S et al (2008) TLR4 signaling mediates inflammation and tissue injury in nephrotoxicity. J Am Soc Nephrol 19:923–932. https://doi.org/10.1681/ASN.2007090982
73. Nataraj C, Oliverio MI, Mannon RB et al (1999) Angiotensin II regulates cellular immune responses through a calcineurin-dependent pathway. J Clin Invest 104:1693–1701. https://doi.org/10.1172/JCI7451
74. Matsumoto K, Morishita R, Moriguchi A et al (1999) Prevention of renal damage by angiotensin II blockade, accompanied by increased renal hepatocyte growth factor in experimental hypertensive rats. Hypertension (Dallas, Tex 1979) 34:279–284
75. Kim HJ, Vaziri ND (2010) Contribution of impaired Nrf2-Keap1 pathway to oxidative stress and inflammation in chronic renal failure. Am J Physiol Renal Physiol 298:F662–F671. https://doi.org/10.1152/ajprenal.00421.2009
76. Libetta C, Sepe V, Esposito P et al (2011) Oxidative stress and inflammation: implications in uremia and hemodialysis. Clin Biochem 44:1189–1198. https://doi.org/10.1016/j.clinbiochem.2011.06.988
77. Calò LA, Naso A, Pagnin E et al (2004a) Vitamin E-coated dialyzers reduce oxidative stress related proteins and markers in hemodialysis–a molecular biological approach. Clin Nephrol 62:355–361
78. Calò LA, Naso A, D'Angelo A et al (2011) Molecular biology-based assessment of vitamin E-coated dialyzer effects on oxidative stress, inflammation, and vascular remodeling. Artif Organs 35:33. https://doi.org/10.1111/j.1525-1594.2010.01125.x
79. Calo LA, Naso A, Carraro G et al (2007) Effect of haemodiafiltration with online regeneration of ultrafiltrate on oxidative stress in dialysis patients. Nephrol Dial Transplant 22:1413–1419. https://doi.org/10.1093/ndt/gfl783
80. Calò LA, Naso A, Davis PA et al (2010) Hemodiafiltration with online regeneration of ultrafiltrate: effect on heme-oxygenase-1 and inducible subunit of nitric oxide synthase and implication for oxidative stress and inflammation. Artif Organs 35:183–187. https://doi.org/10.1111/j.1525-1594.2010.01045.x
81. Calo LA, Vertolli U, Davis PA et al (2014) Molecular biology based assessment of green tea effects on oxidative stress and cardiac remodelling in dialysis patients. Clin Nutr 33:437–442. https://doi.org/10.1016/j.clnu.2013.06.010
82. Aoki J, Ikari Y, Nakajima H et al (2005) Clinical and pathologic characteristics of dilated cardiomyopathy in hemodialysis patients. Kidney Int 67:333–340. https://doi.org/10.1111/j.1523-1755.2005.00086.x
83. Gabrielli L, Winter JL, Godoy I et al (2014) Increased Rho-kinase activity in hypertensive patients with left ventricular hypertrophy. Am J Hypertens 27:838–845. https://doi.org/10.1093/ajh/hpt234

84. Ravarotto V, Pagnin E, Fragasso A et al (2015a) Angiotensin II and cardiovascular-renal remodelling in hypertension: insights from a human model opposite to hypertension. High Blood Press Cardiovasc Prev 22:215–223. https://doi.org/10.1007/s40292-015-0082-7
85. Calò LA, Davis PA, Rossi GP (2014c) Understanding the mechanisms of angiotensin II signaling involved in hypertension and its long-term sequelae. J Hypertens 32:2109–2119. https://doi.org/10.1097/HJH.0000000000000321
86. Calò LA, Pagnin E, Davis PA et al (2003) Oxidative stress-related factors in Bartter's and Gitelman's syndromes: relevance for angiotensin II signalling. Nephrol Dial Transplant 18:1518–1525
87. Calò L, Ceolotto G, Milani M et al (2001) Abnormalities of Gq-mediated cell signaling in Bartter and Gitelman syndromes1∗1See editorial by Warnock, p. 1197. Kidney Int 60:882–889. https://doi.org/10.1046/j.1523-1755.2001.060003882.x
88. Calò L, Davis PA, Semplicini A (2002) Reduced content of α subunit of Gq protein content in monocytes of Bartter and Gitelman syndromes: relationship with vascular hyporeactivity. Kidney Int 61:353–354. https://doi.org/10.1046/j.1523-1755.2002.00128.x
89. Calò LA, Pagnin E, Davis PA et al (2004b) Increased expression of regulator of G protein signaling-2 (RGS-2) in Bartter's/Gitelman's syndrome. A role in the control of vascular tone and implication for hypertension. J Clin Endocrinol Metab 89:4153–4157. https://doi.org/10.1210/jc.2004-0498
90. Semplicini A, Lenzini L, Sartori M et al (2006) Reduced expression of regulator of G-protein signaling 2 (RGS2) in hypertensive patients increases calcium mobilization and ERK1/2 phosphorylation induced by angiotensin II. J Hypertens 24:1115–1124. https://doi.org/10.1097/01.hjh.0000226202.80689.8f
91. Calò L, Sartore G, Bassi A et al (1998) Reduced susceptibility to oxidation of low-density lipoprotein in patients with overproduction of nitric oxide (Bartter's and Gitelman's syndrome). J Hypertens 16:1001–1008
92. Pagnin E, Davis PA, Sartori M et al (2004) Rho kinase and PAI-1 in Bartter's/Gitelman's syndromes: relationship to angiotensin II signaling. J Hypertens 22:1963–1969
93. Calò LA, Pessina AC (2007) RhoA/Rho-kinase pathway: much more than just a modulation of vascular tone. Evidence from studies in humans. J Hypertens 25:259–264. https://doi.org/10.1097/HJH.0b013e328010d4d2
94. Caló LA, Davis PA, Pagnin E et al (2008) Linking inflammation and hypertension in humans: studies in Bartter's/Gitelman's syndrome patients. J Hum Hypertens 22:223–225. https://doi.org/10.1038/sj.jhh.1002309
95. Davis PA, Mussap M, Pagnin E et al (2006) Early markers of inflammation in a high angiotensin II state – results of studies in Bartter's/Gitelman's syndromes. Nephrol Dial Transplant 21:1697–1701. https://doi.org/10.1093/ndt/gfl112

The manufacturer's authorised representative in the EU is Springer Nature Customer Service Centre GmbH, Europaplatz 3, 69115 Heidelberg, Germany. If you have any concerns regarding our products, please contact ProductSafety@springernature.com

Printed and bound by CPI Group (UK) Ltd, Croydon, CR0 4YY
30/09/2026
02225249-0005